Computational Studies of RNA and DNA

CHALLENGES AND ADVANCES IN COMPUTATIONAL CHEMISTRY AND PHYSICS

Volume 2

Series Editor:

JERZY LESZCZYNSKI
Department of Chemistry, Jackson State University, U.S.A.

The titles published in this series are listed at the end of this volume

Computational Studies of RNA and DNA

Edited by

J. Šponer
Institute of Biophysics, Academy of Sciences of the Czech Republic

and

F. Lankaš
Institute for Mathematics B, EPFL (Swiss Federal Institute of Technology), Lausanne, Switzerland

 Springer

A C.I.P. Catalogue record for this book is available from the Library of Congress.

ISBN-10 1-4020-4794-0 (HB)
ISBN-13 978-1-4020-4794-7 (HB)
ISBN-10 1-4020-4851-3 (e-book)
ISBN-13 978-1-4020-4851-3 (e-book)

Published by Springer,
P.O. Box 17, 3300 AA Dordrecht, The Netherlands.

www.springer.com

Printed on acid-free paper

CONTENTS

PREFACE

Nucleic acids have been subject of intense experimental research for many years. Due to unprecedented development of computer hardware and simulation algorithms, we have witnessed an expansion of computational studies of nucleic acids in the past decade that provides a powerful complement to experimental studies. While meaningful computations on biomolecules were very limited a decade ago, today's state-of-the-art computations can substantially extend our knowledge about nucleic acids. The very first realistic atomic resolution simulations on a short B-DNA double helix with explicit inclusion of solvent and ions were published in 1995, with trajectories extended to one nanosecond. In modern studies, the first nanosecond of simulation is mostly considered a part of the equilibration protocol and leading contributions in the field now report several hundreds of nanoseconds of simulation within a single study. Around 1995, the very first base pairing and base stacking ab initio quantum chemical calculations with inclusion of electron correlation were considered as a major breakthrough. With small basis sets these calculations achieved a medium level of accuracy and required large super-computers like the liquid-nitrogen cooled vectorized Cray Y-MP supercomputer. Today, such calculations can be, with appropriate software, executed within a few hours on a laptop computer. Contemporary QM calculations are carried out with extrapolations to the complete basis set and achieve a close to ultimate accuracy.

In this book we cover these exciting developments. We have selected a wide range of leading researchers in the field and asked them to kindly write a contribution for this book. Only very few authors were unable to accept the invitation while there are many others that we could not invite as the size of the book is limited (we ended up more than 200 pages above the originally contracted size).

While contacting the authors, we gave them complete freedom to select the scope of their contribution and express their own view on the field. We assumed that, considering the quality of the invited researchers, we should minimize our interventions into the writing process, except of keeping eyes on the deadlines. Thus, we intentionally did not try to remove eventual overlaps among different contributions. We believe that it is profitable to the reader if key information is treated several times from different perspectives by various leading researchers in the field. This is almost certainly a more justified approach than to try to cover each and every corner and tiny sub-area of computational chemistry related to nucleic acids. Not all published computational studies bring biologically relevant informa-tion and not always the methods are applied with proper consideration of their limitations.

It is useful to keep in mind that computers always give us some numbers and structures. Computational chemistry and computational molecular biology can be successful only if researchers pay maximum attention to the quality of their work, applicability and limitations of the methods, and biological or biochemical relevance of the computations.

Although this book covers the state of the art research in the area of nucleic acid computations, it is not primarily directed to specialists in the field. This book provides a useful guide to modern computa-tions for students as well as for experimental researchers interested in computations or computational results. Many chapters in this book bring not only the recent results obtained by leading groups in the field, but contain also substantial introductory information written at the undergraduate level.

Thus, the very first chapter of the book contains basic introduction to nucleic acid structure, crystallography and crystal databases. Only after this introduction one starts to speak about com-putations, and the book logically continues with atomistic modelling directly complementing the atomic-resolution structural experiments. The following chapters introduce main molecular mechanical program packages, principles of explicit solvent simu-lations, provide in-depth insights into continuum solvent approaches, useful guides in trajectory analysis, and many recent results related to DNA simulations. Then, additional chapters introduce principles of RNA molecular organization and a wide range of RNA simulation applications including studies on the ribosome.

After the MD part of the book, we reduce the size of the system but increase the accuracy, and the reader will find several chapters devoted mainly to quantum chemical calculations on base stacking, pairing, cation interactions, excited states and NMR para-meters. Also the fascinating area of electron transfer through DNA is reviewed. Finally, we make a sharp turn and the book finishes with a series of chapters devoted to mesoscopic modelling, covering systems like DNA molecules many hundreds or thousands of basepairs long, or the whole chromatin fibre. These systems cannot at present be treated at atomistic level, and in most cases it would even not be desirable. What is needed instead are models on mesoscopic scales comprising only variables that are most relevant for the problem under study. Such models, however, contain parameters typically describing mechanical properties of the components and interactions between them, and those have to be supplied to make the model complete. Such parameters are conventionally sequence-independent and taken from experiment, but modern simulations offer yet another possibility, namely, to extract the parameters and their sequence dependence from more detailed computations at shorter length scales. Examples in the book include the use of atomistic simulations for deducing sequence-dependent deformability of nucleic acids or structural and energetic characteristics of DNA overstretching.

It is our great pleasure to thank all the contributing authors for devoting their time and energy to such a demanding work. We are grateful to the editors at Springer for excellent cooperation, to our co-workers for kind help, and to our families and friends for patience and support.

Brno and Lausanne, December 2005

Jiří Šponer, Filip Lankaš

Chapter 1

BASICS OF NUCLEIC ACID STRUCTURE
Concepts, Tools, and Archives

Bohdan Schneider[1] and Helen M. Berman[2]

1) Center for Biomolecules and Complex Molecular Systems, Institute of Organic Chemistry and Biochemistry, Academy of Sciences of the Czech Republic, Flemingovo n. 2, CZ-16610 Prague, Czech Republic; 2) Rutgers University, Department of Chemistry and Chemical Biology, Piscataway, NJ-08854-8087, U.S.A.

Abstract: Basic structural features of nucleic acids are reviewed. Based mostly on crystal structures, the following are described: nucleic acid building blocks, double helical DNA (B-, A-, Z-) and RNA forms (A-), and other repetitive nucleic acid forms such as tetraplexes and folded RNA forms, as well as selected folded nucleic acids, tRNA, and ribozymes. Also discussed are complexes of nucleic acids with small molecules and proteins. Briefly shown are web resources dedicated to nucleic acid structures, mainly the Nucleic Acid Database.

Key words: DNA; RNA; nucleic acid; structure; 3D; X-ray crystallography; NMR; NDB; Nucleic Acid Database; PDB; Protein Data Bank

1. BACKGROUND

In 1944, Avery, MacLeod, and McCarty discovered that a nucleic acid, namely DNA, was the transforming agent responsible for heredity[1], a theory which was later confirmed in 1952 by Hershey and Chase[2]. Investigating this nucleic acid structurally, X-ray diffraction studies of fibrous DNA demonstrated the helical nature of DNA[3,4]. The subsequent discovery of the double helix structure of DNA by Watson and Crick[5] then laid the foundation for the new science of molecular biology, in which it became possible to understand replication, transcription, and translation in molecular terms. Following Watson and Crick's landmark discovery, structural studies using a variety of methods including X-ray, NMR, and electron microscopy have provided a

1

J. Šponer and F. Lankaš (eds.), Computational Studies of RNA and DNA, 1–44.
© 2006 *Springer.*

visual gallery of nucleic acids as well as insight into their function in the cell. Additionally, computer modeling simulations have shed even more details on the structures of DNA, RNA, and their complexes.

There are several longer reviews of nucleic acid structure and interactions including Saenger[6], Neidle[7] Calladine and Drew[8], and the Oxford Handbook on Nucleic Acid Structure edited by Neidle[9]. Branden and Tooze[10] contains important information about structures of protein-DNA complexes.

In this chapter we outline the principles of nucleic acid structure, present an overview of the current state of knowledge on nucleic acid structure determined using X-ray crystallography, and provide a concise summary of online resources for nucleic acid structures.

2.   THE BUILDING BLOCKS

2.1   Chemical Composition

Polymeric nucleic acids are built from monomeric units called nucleotides. A nucleotide consists of three chemically and structurally distinct components: a base, a sugar, and a phosphate group (Figure 1-1).

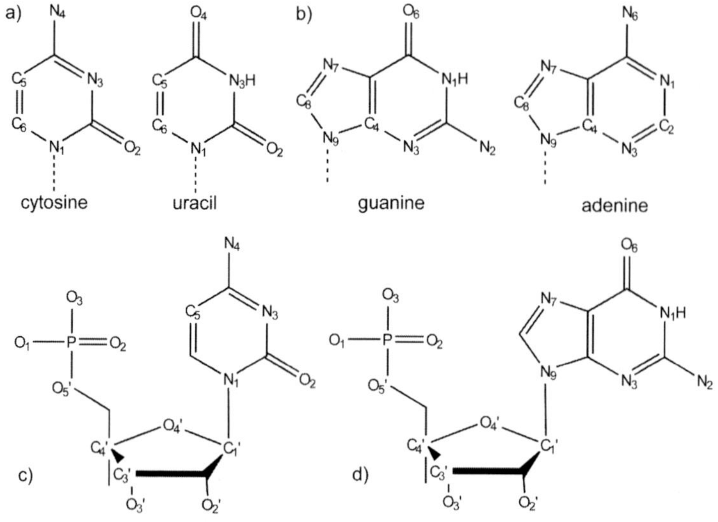

Figure 1-1. Chemical composition and nomenclature of nucleic acid components. a) Pyrimidines. Uracil occurs in RNA, DNA base thymine has a methyl group attached to C5. b) Purines. c) A pyrimidine nucleotide, cytidine-5'-phosphate. d) A purine nucleotide, guanosine-5'-phosphate.

The phosphate group gives nucleic acids their acidic properties, as they are fully ionized at the physiological pH. There are two natural sugars in nucleic acids. Both are cyclic pentoses — β-D-deoxyribose in DNA and the closely related β-D-ribose in RNA. The bases are nitrogenous aromatic rings with lipophilic flat faces, and feature several hydrogen bond donors and acceptors along their edges.

2.2   Structure

In natural nucleic acids, the nucleotide building block is first composed of a nucleoside subunit, formed when a base and sugar are linked by a C-N glycosidic bond in the β stereochemistry. These nucleosides are then linked through phosphate groups that are attached to the 3' carbon of one nucleotide and the 5' carbon of the other, so that the full repeating unit in a nucleic acid is a 3',5'-nucleotide.

Chemical analyses of cellular genetic material showed that there are four types of nucleotides, which only differ by the attached nitrogenous base. Two nucleotides, adenosine-5'-phosphate (A) and guanosine-5'-phosphate (G), contain fused-ring purines. Cytosine-5'-phosphate (C) and thymine-5'-phosphate (T) are single-ring pyrimidines. Thymine is replaced by its de-methylated form, uracil, in RNA. Purines are labeled as R, pyrimidines as Y.

Nucleic acid and oligonucleotide sequences use single-letter codes for the five unit nucleotides, A, T, G, C and U, with intervening phosphate groups designated as "p". Polynucleotide chains are numbered from the 5' end; for example, ApGpCpT has a 5' terminal adenosine nucleoside, a free hydroxyl at its 5' position, and the 3' end thymine has a free 3' terminal hydroxyl group. Chain direction may be emphasized with 5' and 3' labels:

$^{5'}$ApGpCpT

$^{3'}$TpCpGpA

The shorthand nomenclature of this duplex is AGCT, where the phosphate linkage and second strand codes are omitted. An "r" or a "d" prefix denotes whether the oligonucleotide contains ribose or deoxyribose sugars, respectively.

Early on, chemical analysis of DNA from genetic material showed that the molar ratio of guanine/cytosine and adenine/thymine are close to one[11]. This information was critical for the determination of the DNA double helical structure, as Watson and Crick recognized that the two pairs of bases could hydrogen bond and form an antiparallel double helix.

The bases have exocyclic oxygens in keto form, and exocyclic nitrogens in amino form. Experimental structural information suggests that only these tautomeric forms are significantly found in the condensed phase, at "room temperature" around 290 K and physiological pH around 7. Outliers may exist, as a recent neutron diffraction study of a B-DNA decamer d(CCATTAATGG) (PDB code 1WQZ[12]) exhibited hydrogens in the positions corresponding to the keto-form tautomers.

2.2.1   Nucleotide Components

The three chemical components of nucleotides, the phosphate group, the (deoxy)ribose sugar, and the bases, have distinct structural features.
- **(deoxy)ribose**

The intrinsically non-planar (deoxy)ribose ring has a limited conformational flexibility[13,14]. The mathematical description of the pentose conformational space is complex[15], but can be reduced to two parameters: pseudorotation phase angle P, and puckering amplitude (the maximum degree of pucker) v_M. The value of P indicates the type of pucker, since P is defined in terms of the five cyclic ribose torsion angles v_0 - v_4:

$$tgP = \left[(v_4 + v_1) - (v_3 + v_0)\right]/2\ v_2(\sin 36° + \sin 72°)$$

The pseudorotation phase angle can take any value between 0° and 360°. If v_2 is larger than 180°, then 180° is added to the value of P. The puckering amplitude, v_M, is defined by

$$v_M = v_2/(\cos P)$$

Values of v_M indicate the amplitude of puckering of the ring; for example, experimental values from high resolution crystal structures of nucleosides and mononucleotides[16] are in the range 25°-45°. Chiral and substitution constraints of the (deoxy)ribose ring limit v_M into a narrow unimodal range, so that the sugar pucker can be described by the pseudorotation phase angle P. Phase angle P correlates quite tightly with the backbone torsion angle δ, C5'-C4'-C3'-O3', so that sugar pucker is effectively connected to the backbone conformation.

Several distinct deoxyribose ring pucker geometries have been observed experimentally. It is most common for two atoms to deviate from the plane of the remaining ring atoms, but rings with only one puckered atom have been observed. The direction of atomic displacement from the plane is important. If the major displacement is on the same side as the base and C4'-C5' bond, then the ring pucker involved is termed *endo*. If it is on the opposite side, it is called *exo*.

The most commonly observed puckers in crystal structures of isolated nucleosides and nucleotides are close to either C2'-*endo* or C3'-*endo* types (Figure 1-2). The C2'-*endo* family of puckers have P values in the range 140° to 185°. The C3'-*endo* domain has P values in the range –10° to +40°. In high resolution structures, these pure envelope forms are rarely observed. The puckers are best described in terms of twist conformations, where two ring atoms deviate from the plane of the remaining three. When the major out-of-plane deviation is on the *endo* side, a minor deviation lies in the opposite (*exo*) side. Describing twist (deoxy)ribose conformations uses the convention that the major out-of-plane deviation is followed by the minor one, for example C2'-*endo*, C3'-*exo*.

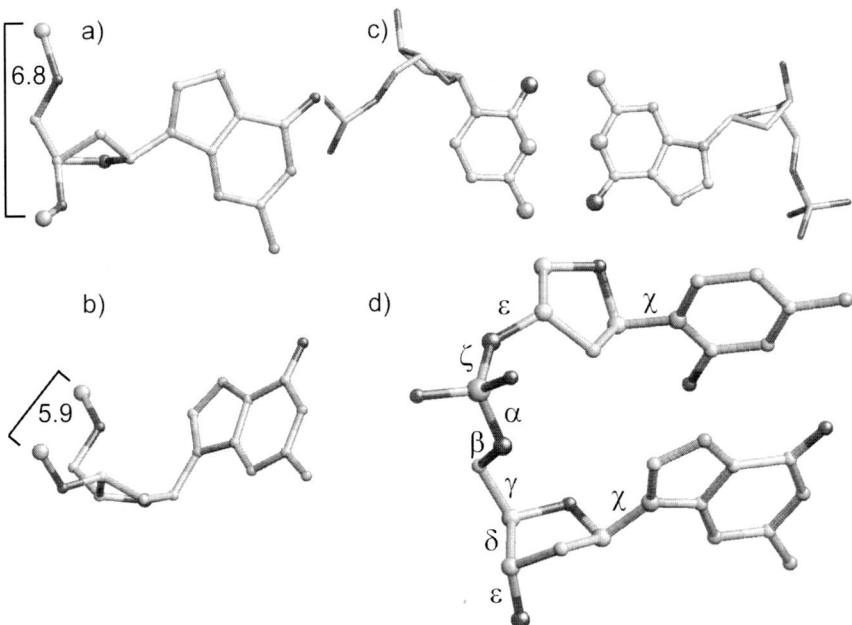

Figure 1-2. Structural features of nucleic acid components. a) C2'-*endo* sugar pucker, typical for B-DNA. Atom C4' is concealed by C3' and atom C2' points in the direction of C5'. A typical distance between two consecutive P atoms is about 6.0 Å. b) C3'-*endo* sugar pucker, typical for A-form nucleic acids (shown for deoxyribose, as in a). In this case, atom C3' is "above" the plane of the remaining four sugar ring atoms. The P-P distance is shorter than in the case of the C2'-*endo* pucker, about 5.9 Å. c) A G•C Watson-Crick base pair; hydrogen bonded atoms are highlighted as small spheres. d) A dinucleotide fragment with stacked cytosine and guanine. The backbone torsions are labeled: α = O3'-P-O5'-C5, β = P-O5'-C5'-C4', γ = O5'-C5'-C4'-C3', δ = C5'-C4'-C3'-O3', ε = C4'-C3'-O3'-P, ζ = C3'-O3'-P-O5'. Torsions around the glycosidic bonds are defined as χ = O4'-C1'-N9-C4 for purines and χ = O4'-C1'-N1-C2 for pyrimidines.

Each major sugar pucker, C2'-*endo* and C3'-*endo*, leads to very different relative positions of phosphate groups at the C5' and O3' ends of the sugar ring (Figure 1-2a, b), with consequences for the overall architecture of the resulting helical conformation.

- **Bases**

Nucleic acid bases are planar aromatic rings. A measurement of infrared spectral data of isolated bases[17] as well as analysis of nucleotide crystal structures[18] have revealed a limited non-planarity of exocyclic groups, mainly the amino groups of adenine, cytosine, and guanine. These modest deformations from the base planarity justify the common consideration of bases as "planar" molecules.

The nucleoside subunit of nucleotides is formed when a base connects to a (deoxy)ribose ring by a glycosidic bond between the endocyclic base nitrogen and the sugar's C1' atom. In natural nucleic acids, the glycosidic bond has the β-stereochemistry, causing the base to point in the same orientation relative to the sugar ring's C4'-C5' exocyclic bond, while the C3'-O3' bond points in the opposite direction (Figure 1-2a, b).

Rotation of the base relative to the sugar around the glycosidic bond is described by the torsion angle χ, O4'-C1'-N9-C4 in purines and O4'-C1'-N1-C2 in pyrimidines. χ has two populated regions, *syn* and *anti*. The majority of nucleotides adopt the *anti* orientation, with χ between 180° and 300°, while the minority is in the *syn* orientation, with χ between 30° and 90°.

- **Phosphate**

The phosphate group has the arrangement of a deformed tetrahedron[19,20]. Both O-P-O-C linkages show a preference for torsional arrangement at +60° and +300° due to the so-called gauche effect[21,22]. Further steric restrictions by the attached nucleoside lead to the prevalence of conformations around both O3'-P and P-O5' bonds to be near 300° (Figure 1-3). Although these are the most common values for these phosphodiester angles, DNA and especially single-stranded RNA can assume different values.

- **Nucleotide**

The conformational space of a nucleotide is defined by six torsion angles: α, β, γ, δ, ε, ζ (for definition, see Figure 1-2) and the glycosidic torsion angle χ.

The variability of nucleotide conformations can be assessed by an analysis of available crystal structures of nucleic acids. For a representative group of about 140 conformationally diverse RNA molecules, the distribution of torsions α and γ is trimodal, δ, χ, and ε are bimodal, and β has a wide unimodal distribution with two small buried subpopulations. The most complex behavior is exhibited by torsion ζ around the phosphodiester bond O3'-P (Figure 1-3). However, the steric requirements of the phosphodiester linkage reduce a potential number of nucleic acid conformations. When ζ values are plotted against torsion α at the other phosphodiester bond, P-O5', its seemingly featureless histogram shows that only a handful of ζ/α combinations are experimentally observed. These two torsion angles, in combination with γ, δ, and χ, produce typical "imprints" that can be used to characterize local DNA[23] as well as RNA[24,25] conformations.

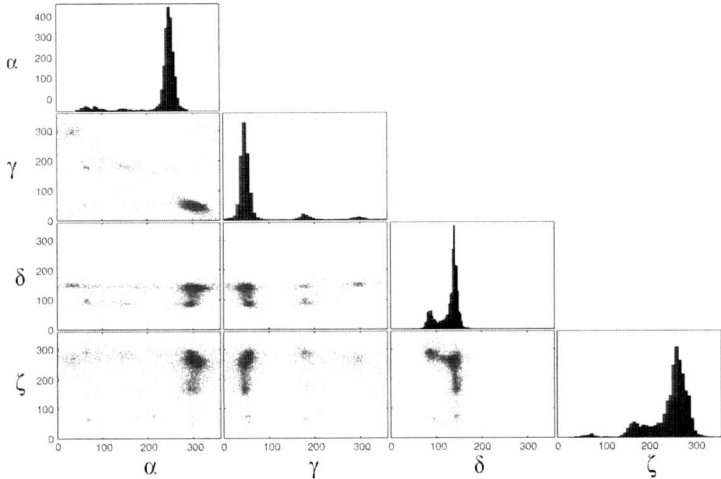

Figure 1-3. One- and two-dimensional distributions of the backbone torsion angles α, γ, δ, and ζ in about eight thousand nucleotides from 450 DNA crystal structures (Svozil, personal communication, 2005). α, γ, and ζ have trimodal distributions with one dominating peak, δ has a bimodal distribution corresponding to C2'-*endo* ($\delta \sim 140°$) and C3'-*endo* ($\delta \sim 80°$) sugar puckers.

Reliable high resolution crystal structures of nucleotides and nucleosides from structures in the Cambridge Structural Database[26] and the Nucleic Acid Database[27] have been used to determine the average values of the valence distances and angles of all nucleic acid building blocks[16,28,29]. These values have been incorporated[30] in the programs used for molecular dynamics simulations and in computer packages for both crystallographic and NMR structural analyses, X-Plor[31], AMBER[32], and CHARMM[33].

2.2.2   Base Pairing

Nitrogenous bases can form pairs via hydrogen bonds between their polar groups. There are many possible hydrogen bonded base pair arrangements[34], but among all, the "canonical" or Watson-Crick pairs are of fundamental importance. Watson-Crick pairing between G and C and between A and T/U is the physical carrier of genetic information, and demonstrates perhaps the most obvious link between molecular structure and function in biology. Watson-Crick pairing prevails in both DNA and RNA. The large functional and structural diversity of RNA molecules is also reflected by the larger variability of their base pair patterns. However, even extremely structurally diverse ribosomal RNA has about 60% of all bases in Watson-Crick pairs.

The unique role of the Watson-Crick pairing in nucleic acids is a consequence of several factors:

i) High stabilization energies via their hydrogen bonds.

ii) Isostericity. Both pairs, G•C and A•T, have similar dimensions along their long axes. For example, C1'-C1' distances across the base pair are typically 10.1-10.9 Å.

iii) Minimal steric hindrance between the optimal attachment of a base to the backbone, glycosidic torsion, and the optimal conformation of the sugar phosphate backbone, as observed in the right-handed double helical conformations.

Some non-Watson-Crick base pairs have comparable hydrogen-bonding interactions[35], but they are either not isosteric or impose energy constraints on the backbone.

Base pairs are connected by two or three relatively weak and flexible hydrogen bonds, allowing their geometries to deviate quite significantly from the optimal near-planar arrangement. Base pair arrangements are described by a set of parameters defined by the Cambridge Accord[36]. These parameters describe geometries of an isolated pair: propeller twist, buckle, inclination, and the X and Y displacement; and of two pairs: rise, helical twist, roll, tilt, and slide. Analogous definitions can be applied to other non-standard base pairings in a duplex, including purine-purine and pyrimidine-pyrimidine pairs, but these geometric descriptors are typically used for description of Watson-Crick pairs.

Propeller twist (ω) between bases is the dihedral angle between the normal vectors to the bases, when viewed along the long axis of the base pair. The angle has a negative sign under normal circumstances.

Buckle (κ) is the dihedral angle between bases, along their short axis, after propeller twist has been set to 0°. The sign of buckle is defined as positive if the distortion is convex in the direction 5'→3' of the first strand. The difference between the buckle at a given step and that of the preceding one, is termed cup.

Inclination (η) is the angle between the long axis of a base pair and a plane perpendicular to the helix axis. This angle is defined as positive for right-handed rotation about a vector from the helix axis towards the major groove.

X and Y displacements (**dx, dy**) define translations of a base pair within its mean plane in terms of the distance of the midpoint of the base pair's long axis from the helix axis. Positive X displacement is towards the major groove, positive Y displacement is towards the first nucleic acid strand of the duplex.

Helical twist (Ω) is the angle between successive base pairs, measured as the change in orientation of the C1'-C1' vectors going from one base pair to the next, projected down the helix axis.

Roll (ρ) is the dihedral angle for rotation of one base pair with respect to its neighbor, about the long axis of the base pair. A positive roll angle opens up a base pair step towards the minor groove.

Tilt (τ) is the corresponding dihedral angle along the short axis of the base pair.

Slide (**Dy**) is the relative displacement of one base pair compared to another, in the direction of the first nucleic acid strand measured between the midpoints of each base pair's long axis.

These definitions led to different values for base morphology parameters when calculated by different programs[37-47]. This meant that it was virtually impossible to compare any two structures by using the numbers in the published literature, and that any analysis needed to recalculate base pair morphology. Inconsistent definitions have been revised by the Tsukuba Workshop on Nucleic Acid Structure and Interactions[48] so that a single reference frame would be used to calculate base morphology parameters and generate consistent values of base pair parameters when calculated by different programs. Now all the programs are amended so that they produce very similar values for these parameters.

3. NUCLEIC ACID CONFORMATIONS

In the following paragraphs, we describe some well-characterized confor–mational types of non-complexed nucleic acids. The discussion is based on selected structures, most solved by X-ray crystallography. Discussion of a particular structure is, besides citation, accompanied by the PDB and NDB codes to facilitate searches of these structures on the web. Figure 1-4 depicts most typical DNA and RNA structural classes.

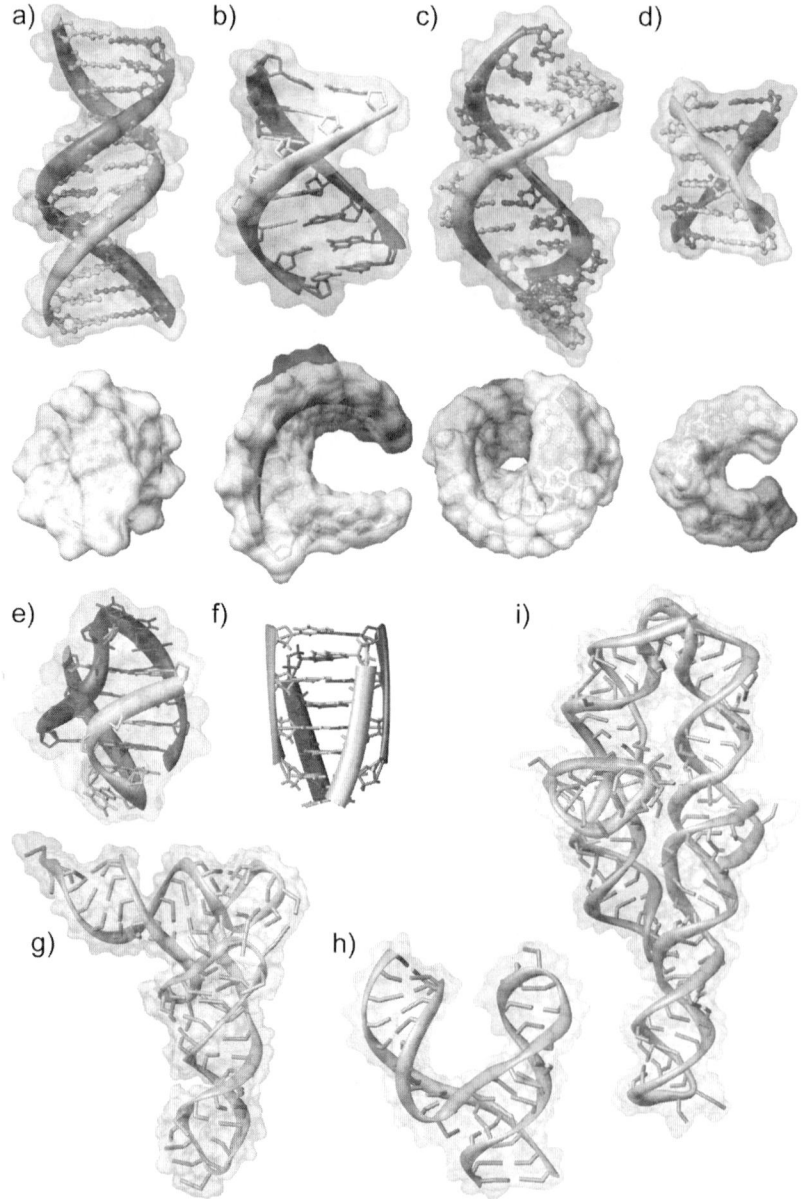

Figure 1-4. Well-characterized conformational types of nucleic acids. a) B-DNA dodecamer (1BNA/BDL001[49]). b) A-DNA octamer (2ANA/ADH006[50]). c) A-RNA dodecamer with non-Watson-Crick base pairs in the double helix (255D/ARL037[51]). d) Z-DNA hexamer (1DCG/ZDF002[52]). e) DNA guanine tetraplex. Four guanine tetramers are flanked by two dT4 loops (1JPQ/UD0013[53]). f) i-motif or cytosine tetraplex (190D/UDD024[54]). g) Phenylalanine transfer RNA (6TNA/TRNA04[55]). h) Hammerhead ribozyme. The shorter strand is DNA (1HMH/UHX026[56]). i) group I intron ribozyme (1GID/URX053[57]).

3.1   Double Helical Forms

Right-handed, antiparallel double helices are the most prevalent forms of both DNA and RNA[58]. The helical arrangement allows nucleotides to adopt optimal or near-optimal conformations of the sugar-phosphate backbone, as well as of the bases to form hydrogen bonding pairs, mostly of the Watson-Crick type with their hydrophobic faces shielded from the solvent by mutually attractive stacking interactions. The distance across the double helix, e.g. measured by a distance between C1' atoms of the paired bases, is very similar in both G•C and A•T/U Watson-Crick pairs. This allows the backbone to run smoothly regardless of the sequence.

The charged phosphate groups are fully exposed to the solvent at the edge of the backbone. The edge divides the double helix into two grooves. Groove walls are mostly hydrophilic, lined by CH and CH_2 groups of (deoxy)ribose. Groove bottoms are faced by edges of flat nitrogenous bases with exposed hydrogen bond donors (NH2 of G and C), but mostly acceptors (carbonyl O of C, T/U, immino N of G and A). The grooves are non-equivalent: the groove with exposed base atoms N3(R) and O2(Y) is called the minor groove; the major groove has exposed purine atoms C8, N7 and O6/N6, and pyrimidine C6, C5, N4/O4 (Figure 1-4a-d). The different chemical composition of the grooves is reflected by differences in their shape and dimension.

The average geometries for well-defined double helical forms, A-, B-, Z-, are known from fiber diffraction[59], and also from comparative studies based on single crystal structures[23].

X-ray diffraction studies of single crystals of oligonucleotides demonstrate that double helical forms are inherently flexible. Large deformations of the double helical form are observed especially in complexes between nucleic acids and proteins. DNA remains in a double helical form with its predominantly Watson-Crick base pairing intact, and the sometime extreme bend of the helix is absorbed by deformations in the backbone. Comparative studies have uncovered a few trends for sequence-dependent structural flexibility[60]. Smooth bending typical for the B-DNA double helix has not been observed in RNA. RNA can accommodate long stretches of non-Watson-Crick base pairs or bulged-out bases into its double helix without losing its overall shape. Changes of helical direction in RNA are accomplished by clusters of nucleotides in non-helical conformations. RNA nucleotides have a tendency to form interactions with distant parts of the molecule in a way similar to amino acids in proteins, in other words, allowing RNA to fold. Therefore, DNA structure is characterized as local, whereas RNA is global.

3.1.1   Right-Handed Double Helices, B- and A-Forms

The DNA form most stable under physiological conditions is called B-DNA (Figure 1-4a). It is formed by the right-handed double helix with parallel Watson-Crick A•T and G•C base pairs almost exactly perpendicular to the helical axis. The helix diameter is approximately 20 Å. One full helix turn is completed every 10 base pairs so that the average helical twist is 36°. Because base pairs are perpendicular to the helical axis, the pitch between two stacked base pairs, 3.4 Å, is the same as the helical rise. The deoxyribose pucker is predominantly in the C2'-*endo* region, and the glycosidic torsion angle χ is in the "high *anti*" region, which is typical for the C2'-*endo* pucker, between 240° and 270°.

The minor and major grooves have similar depths, while the minor groove is narrower than the major. In general, the major groove is richer in base substituents, O6, N6 of purines and N4, O4 of pyrimidines. This, together with the steric differences between the two grooves, has important consequences for interaction with other molecules.

Over 250 single crystal structures with B-DNA conformations have been determined since the first such structure was published (1BNA/BDL001[49]), reviewed by Berman[61]. The average values of the base morphology parameters are close to the canonical values derived from fiber studies[59]. However, the individual B-DNA structures are capable of large deformations from the ideal B-DNA geometry, and vary in their local geometry as well as in their overall shape.

As reviewed by Dickerson[62], many double helical DNA structures are bent by as much as 15° and have variable groove widths[63]. The bending may be a consequence of nucleotide sequence, crystal packing, and solvent effects. The bend can be mathematically described by the base pair parameters twist and roll, as attempts to relate these base morphology parameters to the sequence of the bases have shown trends. For example, twist and roll appear to be correlated with regressions that depend on the nature of the bases in the step, as purine-pyrimidine, purine-purine, and pyrimidine-purine steps show distinctive differences[64,65]. Additionally, A-T base pairs in AA steps have high propeller twist and bifurcated hydrogen bonds[66], and stretches of A in sequences appear to be straight[67]. On the other hand, certain steps such as TA and CA show a very large variability due to either crystal packing, sequence context, or both (1D60/BDJB43[68]). To summarize, despite large local variability, sequence dependence is moderate, and although clear-cut rules do not exist, certain sequences do indeed show structural propensities. A typical example is the so-called AA track, opposed to the stiffer CG regions.

Interestingly, base mispairings cause only local perturbations to nucleic acid structure, with the overall conformation remaining the same as the corresponding parent structure. Thus, destabilization seems more thermodynamic than structural.

The B-DNA backbone exists in two main conformations known as BI and BII[49,69]. On average, BI is closer to the fiber-derived B-DNA conformations[59]. BI is more common than BII, forming about 75% of all nucleotides in double helical B-DNA crystal structures. BI can combine with both BI and BII conformations in the neighboring nucleotide backbone, and shows no sequence dependence. In contrast, BII-BII dinucleotides are observed rarely, and BII occurs mostly in purine nucleotides. BII differs from BI mainly in torsions β, ε, and ζ[23], large differences in ε and ζ between BI and BII almost cancel each other. Both B-conformations prefer C2'-*endo* sugar pucker but a BI nucleotide can also adopt C3'-*endo* deoxyribose.

The canonical A-form double helix (Figure 1-4b) is very similar in both DNA and RNA. The values of the backbone torsion angles in both forms are virtually identical[25]. Figure 1-4b shows an A-DNA octamer d(GGGGCCCC)$_2$ (2ANA/ADH006[50]). The A-form has C3'-*endo* sugar puckers, which brings consecutive phosphate groups on the nucleotide chains closer together –5.9 Å compared to the 7.0 Å in B-form DNA – and alters the glycosidic angle from high *anti* (~270°) typical to B-DNA to *anti* (~200°). Base pairs are displaced nearly 5 Å from the helix axis and are tilted relative to the helical axis so that the helical rise is reduced to 2.6 Å, compared to 3.4 Å for B-DNA. The combination of base pair tilt, with respect to the helix axis, and base pair displacement from the axis, results in a wider helix than in the B-form, an 11 base pair helical repeat, and different groove characteristics. The major groove is now deep and narrow, and the minor is wide and shallow.

Another difference between the A- and B- forms is that A-helices, though capable of a small degree of bending (up to ca 15°), do not undergo the large-scale bending seen often in B-DNA, for example in the A-tract of B-DNA or in DNA wrapped around the histone core particle (1KX3/PD0285[70]).

The A-form nucleic acid conformation has significant biological roles as the only double helical RNA form as well as a likely recognition motif in some protein-DNA complexes. DNA in the TATA binding protein-DNA complex (1YTB/PDT012[71], PDT009[72]) has the conformation of an A-B chimera[73]. The apparent deformability of the base geometry seen in A-DNA oligonucleotide crystal structures may prove to be an advantage in forming protein-DNA complexes.

3.1.2   Double Helices with Non-Watson-Crick Base Pairs and Bulges

The A-RNA double helix can accommodate relatively long stretches of non-Watson-Crick base pairs without significant deformation of the overall double helical arrangement, but with large local conformational changes of the backbone and sugar pucker. This ability to accommodate non-Watson-Crick base pairs, bulged-out single bases, and short internal loops is a prerequisite for the ability of RNA molecules to form molecules with complex three-dimensional folds, as seen in tRNA or ribozymes. The occurrence of different types of base pairs, including base-sugar edge pairs, has been studied by Leontis and Westhof[74]. For instance, the G•U base pair is an important element of large RNA structures due to its stability[75]. It is known as the "wobble" pair, since a G in the first position of a codon can accept either a C or a U in the third anticodon (wobble) position.

Inclusion of non-Watson-Crick base pairs into RNA oligonucleotide sequences has been a subject of many detailed structural studies. These studies aimed at understanding the influence of "mismatched" base pairs on the local structure and base stacking. For instance, in crystal the dodecamer sequence r(GGACUUCGGUCC) forms a base-paired duplex (255D/ARL037[51]), with U•C and G•U base pairs (Figure 1-4c). Its double helix has a significantly wider major groove than in canonical A-RNA, possibly on account of water molecules that are strongly associated with the mismatched base pairs. Similarly, four non-Watson-Crick base pairs in the center of the dodecamer r(GGCCGAAAGGCC) (283D/URL051[76]) form an internal loop with sheared G•A and A•A base pairs. As a result, the structure is strongly distorted from canonical A-RNA.

3.1.3   Left-Handed Double Helix

Until 1980, all DNA models based on fiber diffraction data were modeled as right-handed double helices with smooth-running backbone. Nonetheless, the first oligonucleotide single-crystal structure solved[77] was, quite unexpectedly, a left-handed duplex with an irregular zig-zag pattern of the backbone (Figure 1-4d, 1DCG/ZDF002[52]). This DNA form is now called Z-DNA, and its architecture has several remarkable features. The repeating unit is a dinucleotide rather than a nucleotide as in A- and B-type conformations, since Z-DNA strongly prefers alternating pyrimidine-purine, commonly CG, sequences. The Watson-Crick pairing is achieved by a combination of several structural features unusual in other DNA duplexes: the second base (purine) takes the rare *syn* orientation, several backbone torsions (mainly α at the first and γ at the second nucleotide) acquire rare conformations (~180°), and sugar puckers alternate between C2'-*endo* at the first and C3'-*endo* at the second nucleotide.

Besides being left-handed, the Z-DNA double helix is quite distinct from both B- and A-form helices. The dinucleotide repeating unit and the purine *syn* orientation push the major groove base edges away from the helical axis so that they form one concave surface with the backbone phosphates. The minor groove, in contrast, forms a deep and narrow crevice suitable for docking of small molecules and ions.

Z-DNA is known to occur in two closely related conformational subclasses, the dominant ZI and the less common ZII[78]. ZI differs from ZII in all backbone torsion angles, and the two conformations are best distinguished by values of torsion ζ at the purine step. In ZI, ζ is near 300°; in ZII it is close to 50°[23].

Z-DNA is thermodynamically less stable than B- and A-forms, as it exists only in high salt conditions to compensate for short phosphate-phosphate distances. Its actual presence *in vivo* has not been observed, but indications exist that Z-DNA may play a role in the regulation of transcription and translation[79].

3.1.4   Role of Solvent

Solvation is directly related to the conformation and properties of nucleic acids. Water is an integral part of the nucleic acid structure[80]. Additionally, ionic metal and other cations such as polyamines influence the most stable DNA conformation, and may be necessary for RNA to fold.

It has been estimated that about two layers of water molecules are tightly bound to the DNA surface[81]. The number of water molecules with restricted mobility is estimated to be between 5 and 12 per nucleotide; release of this partially ordered water is a source of entropic or hydration force[82,83].

Many crystallographic studies stress the importance of structural features of the solvation shell around nucleic acids. A well organized network of water molecules has been observed in high resolution complex dCG/proflavine[84,85], and hydration of double helical DNA shows prominent features. "The spine of hydration" in the minor groove of B-DNA[86] is an important part of the B-DNA structure. In A-DNA, a network of edge-linked pentagons has been observed[87]. Single water molecules from these networks connect phosphate groups, an arrangement that is not possible in B-DNA since each phosphate is hydrated individually. This hydration pattern inspired the concept of the "economy of hydration"[88]. Observed as the humidity is lowered, B to A transition can be driven by rearrangement of the backbone and consequent reduction extension of phosphate hydration[89]. In A-RNA, hydration of the deep minor group is connected via the RNA-specific hydroxyl O2'H to hydration of the phosphate groups[90].

A series of systematic studies of hydration of DNA double helices has revealed that water molecules are well organized around both bases[91,92] and phosphates[93], and that the observed hydration patterns depend on the DNA conformational type as well as base identity. These "hydration sites" represent sites of preferred polar binding to DNA. For instance, water and the polar groups of drugs and amino acids bind to similar sites in the B-DNA minor groove[94]. Calculated hydration sites can also be used to predict polar binding of amino acids to DNA in protein/DNA complexes[95].

A methodologically similar study has shown that metal cations such as Na^+, Mg^{++}, Ca^{++} or Zn^{++} also have preferred binding sites around phosphate groups, and that these sites are characteristic for each metal[96].

3.2   Other Nucleic Acid Conformations

• **Quadruplex**

The quadruplex is thought to be an important structural element at the ends of telomeres, sequences ending eukaryotic chromosomes and essential for their replication and stability. Some telomeres are G-rich and may contain sequences with repeating G-tetrads. For example, it has been shown by both NMR (156D[97]) and crystallography (1JPQ/UD0013[53]) that oligonucleotides containing repeat d(TTTTGGGG) form DNA quadruplexes with guanine bases hydrogen bonded into quartets, with alternating *syn* and *anti* glycosyl conformations (Figure 1-4e). Two strands form a quadruplex with an antiparallel diagonal arrangement; thymines form loops. Each guanine quartet is stabilized by alkali metals organized in a linear channel, with square antiprismatic coordination to eight guanine O6 oxygen atoms. Both sodium and potassium quadruplex forms are known.

• **i-motif or cytosine tetraplex**

Crystal structures containing tetrads of cytosines (190D/UDD024[54]) are four-stranded, similarly to structures with guanine tetrads (Figure 1-4f). Two opposite strands run in parallel and have their cytosines paired, the other two parallel strands run in the opposite direction with their cytosine pairs intercalated between pairs of the first two strands. The molecule has a slight right-handed twist. Each duplex is stabilized by hemi-protonated C-C+ base pairing between its parallel strands, and a string of water molecules bridge the cytosine N4 atoms to phosphate O atoms.

• **Four-way junction**

The four-way junction, or Holliday junction, is a four-way branched nucleic acid topology that occurs as an intermediate during genetic recombination of chromosomes, including site-specific recombination by integrases. Topologically and structurally different four-way junctions have been crystallized. Specific DNA sequences can form stable four-way junctions with two short inter-helical links of antiparallel alignment between two DNA helices (1L6B/UD0019[98], 1M6G/UD0021[99]) Several crystal structures of DNA Holliday junctions complexed with their specific recombinases are also known. For instance, structures of Cre recombinase bound to a nicked Holliday junction (2CRX/PD0103[100]) reveal a nearly planar, twofold-symmetric DNA four-way intermediate.

- **Triple helix**
  Although an extensive triple helix example has not been observed in either DNA or RNA crystal structures, several models have been deduced from fiber diffraction measurements[59]. There are also several structures determined using NMR methods.

- **Adenine zipper**
  Certain sequences of single-stranded DNA are known to be important during replication and have been proposed to have defined 3D structure. The structure of one such sequence is known, nonamer d(GCGAAAGCT) (376D/UDIB70[101]). This structure reveals an unexpected zipper-like motif in the middle of a standard B-DNA duplex. The zipper is formed by four central adenines, intercalated and stacked on top of each other without any interstrand Watson-Crick base pairing; they are flanked by two non-Watson-Crick G.A pairs. The arrangement closes the minor groove and exposes the intercalated and unpaired adenines to the solvent and DNA-binding proteins.

3.3 Nucleic Acids with Modified Components

Modified nucleic acids are important for their potential medicinal use as drugs, and also for research on the molecular origins of life. Evolutionary ancient molecules of tRNA contain some naturally modified bases, such as pseudouridine and N2-dimethylguanine, as well as the radically chemically different wybutosine. Some base modifications known to cause cancer have been studied. For instance, 8-oxo-guanosine forms a stable G(*anti*)-A(*syn*) base pair stabilized by water molecules (153D/BDLB53[102]), but the overall, double helix remains B-DNA. Thymines can be linked into intrastrand covalent dimers by ultraviolet light (1SM5/UD0053[103]); severe chemical modification of the neighboring thymine bases is accommodated by their rotation around the glycosidic bond, while the rest of the backbone remains close to B-type.

Studies with a modified sugar in some or all nucleotides are relatively common. Thioribose replaces ribose, or a fluoro-substituent replaces a ribose hydroxyl group. For instance, 2'-fluoroarabino-furanosyl can be incorporated into a double helix without deformation of the B-DNA conformation

(1DPN/BD0030[104]), while chemically larger modification of ribose to bicyclo-arabinose (1EI4/BD0038[105]) leads to sugar conformations intermediate between those typical for B- and A-type double helices. Nevertheless, the structural change is quite localized, leaving the double helix in overall B conformation. Studies with completely different sugars, e.g. hexoses, are important for research on the origins of life (481D/HD0001[106]), with the resulting double helix different from both A- and B-types.

Phosphate groups have been modified, e.g. to sulphonates and phosphonates, in order to enhance resistance of the resulting molecules to degradation. The crystal structure of the chiral phosphorothioate shows a B-like double helix, with the backbone alterations between the BI and BII conformations (1D97/BDFP24[107]). A single 3'-methylene phosphonate insertion in the backbone of a DNA octamer moderately modified its original A-DNA conformation by opening the major groove (1D26/ADHP36[108]). Phosphoramidate DNA dodecamer duplex (363D/ADLS105[109]) has a double helical form similar to the A-DNA double helix.

A more dramatic modification in nucleotide composition is represented by so called "peptide nucleic acids". These molecules have standard aromatic bases linked by chiral or achiral peptide links. The bases form Watson-Crick base pairs, and two strands can arrange into a narrow double helix with small helical twist (1PNP/UPNA56[110]).

A complex between peptide nucleic acid and DNA demonstrates a potential use of peptide nucleic acids. The structure (1PNN/PNA001[111]) is a triple helix formed by one DNA and two PNA strands, and demonstrates the possibility of the self-assembly of a nucleic acid double helix with xenobiotic and biologically stable oligomers of peptide nucleic acid.

3.4   Folded Single-Stranded Nucleic Acids

Folded forms are mostly acquired by RNA. The biological roles of ribonucleic acids are greatly variable. Understanding and manipulating RNA function can have a great role in medicine, since RNA molecules are involved in essential biological processes, such as splicing of pre-mRNA and the synthesis of proteins in ribosomes. In principle, such tasks can be achieved by synthesis of "small molecule" drugs and antibiotics, or less conventionally, by designing specific ribozymes and antisense nucleic acid mimicking drugs.

RNA molecules are structurally as diverse as their functions. They have complex 3D architectures, often consisting of domains and, in analogy to proteins, must be properly folded to be functional. The A-form double helix is ubiquitous, but molecules of large folded RNAs contain a significant proportion of "unusual" conformations with many non-Watson-Crick base pairs, base triples, three- and four-way junctions, and loops of variable length. Extensive single-stranded regions serve as hinges between double helices or as folding anchors by forming base pairs with distant parts of the molecule. An open issue in RNA structural biology remains how to find and define structural motifs other than the double helix.

A handful of dinucleotide[25] and ribose-to-ribose[24] structural motifs have been characterized, but relatively few oligonucleotide-size motifs have been identified. One important motif is represented by loop insertions into minor grooves of A-type helices, now known as the A-minor motif[112] and previously observed in several ribozyme structures[57,113]. The motif involves adenines interacting in the minor groove. These adenines are often highly conserved and presumably essential for preserving an RNA molecule's functional architecture.

The kink-turn motif[114], an elbow-like helix-loop-helix motif, is another well-characterized motif. Recently, a related structural and possibly functional motif, "reversed kink-turn", has been characterized[115] in the group I intron (1U6B/PR0133[116]). Its sequence resembles the more common elbow-like "regular" kink-turn, but it bends in the opposite direction, toward the major groove of the flanking helices. The existence of these sequentially and topologically similar, but structurally distinct, motifs suggests the importance of flexibility for RNA function.

3.4.1   Transfer RNA

Independently determined over thirty years ago by two groups, the three dimensional structure of phenylalanine tRNA[117,118], was the only known structure of natural RNA until the determination of the first ribozyme structures. Both tRNA structures show a relatively short RNA oligonucleotide folded into a complex shape with two perpendicular arms. The prevailing structural feature of this evolutionary old molecule is still an A-RNA double helix (Figure 1-4g, 6TNA/TRNA04[55]), but several buldged bases and short loops form additional interactions between sequentially distant parts of the molecule that stabilize the fold. The longer arm contains the anticodon triplet; the shorter arm is formed by the helix of the acceptor stem with the stacked T-arm helix. Such helix-helix stacking has turned out to be a universal way of building other complex RNAs. The topologically complicated elbow region contains many non-Watson-Crick base pairs and base triplets.

Subsequently solved crystal structures of tRNAs have proven that the L-shape observed for the phenylalanine tRNA is invariant.

3.4.2   Ribozymes

The ability of RNA molecules to catalyze chemical reactions was first observed in the organism *Tetrahymena*[119], in the so-called self-splicing group I introns. They, along with most other RNA enzymes, ribozymes, catalyze splitting or formation of phosphodiester bonds in the polynucleotide chain, and usually operate on themselves. In addition, specific RNA molecules can be designed to catalyze other reactions, making the idea of an RNA world[120,121] conceivable.

The first ribozyme to be solved by X-ray crystallography was a relatively small "hammerhead" ribozyme (Figure 1-4h, 1HMH/UHX026[56]). The catalytic RNA strand is observed in complex with the second "substrate" strand. In this specific structure, the substrate is DNA in order to inhibit the reaction. The overall fold of the hammerhead ribozyme is simple, with three A-RNA stems encompassing its catalytic core. However, the structural details, especially of the catalytic site, show an intricate web of hydrogen bonding and stacking base interactions, as well as unusual backbone conformations.

The group I intron of *Tetrahymena* consists of two domains of approximately the same size, P4-P6 and P3-P9, and the crystal structure of the former is known (1GID/URX053[57]). The 160-nucleotide single-stranded RNA comprises two helical regions that pack side by side (Figure 1-4i). A bulge formed mostly by invariant adenosines links two parallel helical sides. The bulge is responsible for folding of the whole P4-P6 domain, and four of its adenosine residues are involved in contacts with the minor groove side of the C•G pairs of the P4 domain. This structural motif, called A-minor[112] or ribose zipper[57] and characteristic by inclusion of adenine bases into the deep and narrow minor groove of A-RNA at a C•G step, seems to be one of a few universal motifs by which large RNA molecules are built.

The tight packing of the two helical regions and resulting close phosphate-phosphate distances are mediated by hydrated magnesium cations. Metal-phosphate coordination is another general feature of folded RNA molecules, and divalent cations are especially essential for packing of RNA into its functional form. Divalent metals are also required for both folding and catalysis of the yeast Ai5γ group II self-splicing intron, whose catalytically essential domains (5 and 6) have been determined crystallographically (1KXK/UR0019[122]).

Ribonuclease P is a ribozyme conserved in all kingdoms of life, as it is responsible for maturation of tRNA. Crystal structures of two forms of its RNA component, the 154-nucleotide specificity domain A and B, are known (1NBS/UR0027[123], 1U9S/UR0040[124]). Despite the large similarity in sequence and 2D structure of these molecules, their 3D folds show significant differences, stressing the importance of detailed structural studies.

It has been shown that ribozymes can catalyze reactions other than phosphodiester transfer reactions. For instance, the crystal structure of a ribozyme catalyzing enantioselective carbon-carbon bond formation by the Diels-Alder reaction (1YKV/UR0054[125]). This ribozyme adopts a lambda-shaped pseudoknot fold dictated by extensive stacking and hydrogen bonding. The catalytic pocket is highly hydrophobic, and its shape is complementary to the reaction product.

4. NUCLEIC ACIDS IN COMPLEXES

Formation of all well defined conformational classes of nucleic acids as double, triple, and quadruple helices, as well as folded RNA as tRNA or ribozymes, is a consequence of the self-recognition of nucleotides. Most nucleic acid self-recognition interactions are driven by "direct readout" of one nucleotide sequence by another. Direct readout is best conceptualized as interactions via hydrogen bonds of nucleic acid bases. In this sense, direct readout is recognition of digital sequence.

Indirect readout is the analogous readout of features not directly connected to nucleotide sequence. All structural characteristics used for indirect readout are ultimately sequence dependent, but consist of features like groove dimensions and backbone conformations that are much less sequence dependent than the hydrogen bonding potential of base edges.

The following general structural features are important in recognition of nucleic acids by other molecules:

i) Phosphate-phosphate distances are different in B- and A-forms and, more subtly, between BI and BII conformations.

ii) Minor groove width tends to be narrower in A/T rich sequences.

iii) Electron density. C-G pairs are electron rich compared to T-A, and have a higher electric dipole. Pyrimidine-purine dinucleotide steps tend to have smaller stacking attraction than purine-pyrimidine steps.

iv) Stereochemical repulsion. Similar hydrogen bonding potential of G-C and A-T in the minor groove is dramatically altered by the presence of the guanine NH_2 group. Similarly, the thymine methyl group protrudes into the major groove and restricts the stereochemistry of an approaching group.

Figure 1-5 shows a few selected structures of nucleic acids complexed with drugs. These structures represent some typical structural features observed in known structures.

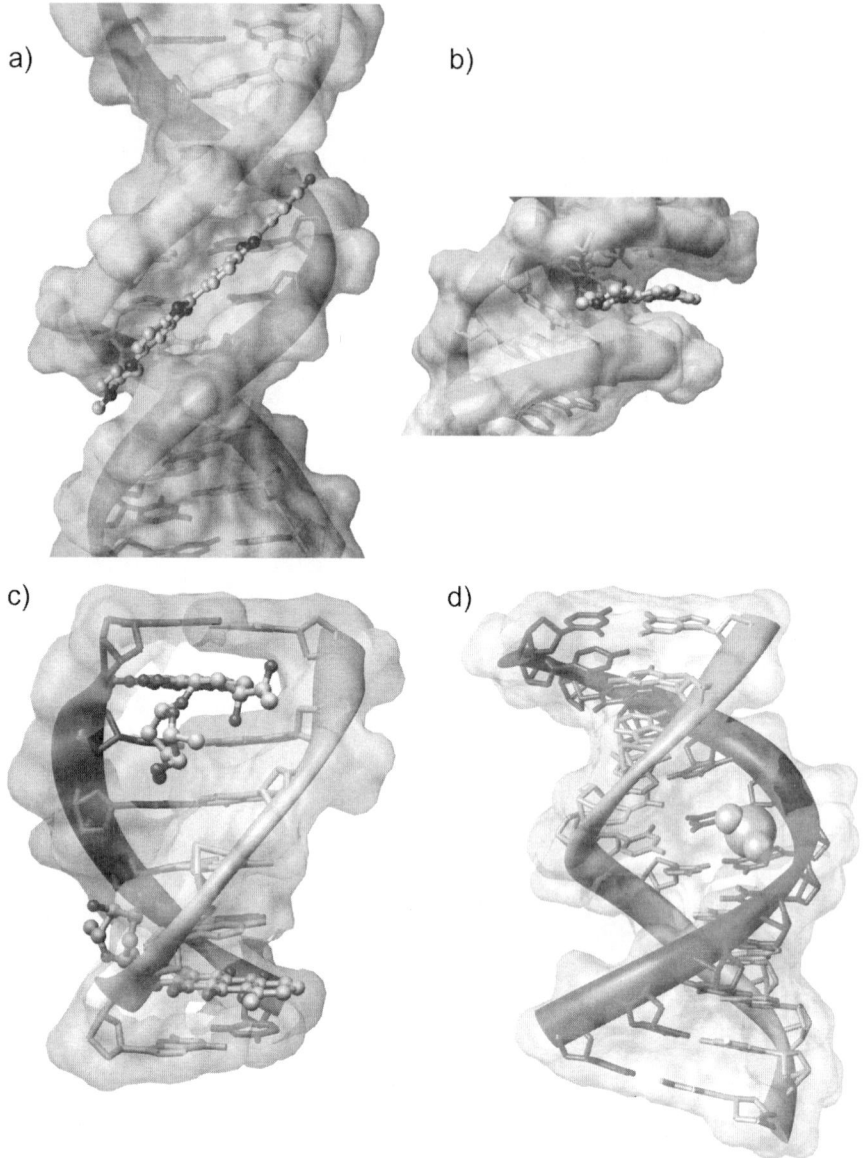

Figure 1-5. Drug molecules interacting with double helical DNA. a) Binding to the minor groove of B-DNA (1D43/GDL0101[26]). b) A view into the minor groove of 1D43/GDL010 showing how a flat drug molecule fits into the minor groove. c) Intercalation between bases of B-DNA octamer (1DA0/DDF0181[27]). d) Cis platinum bound to two guanine bases, shown as large spheres. Two Pt-N7(G) bonds bend DNA double helix (1AIO/DDLB731[28]).

4.1   Recognition of Nucleic Acids by Small Molecules

Many small molecules are able to form complexes with nucleic acids. Some, such as polyamines, are naturally present in the cell and participate in normal DNA and/or RNA function. More or less specific binding of other, xenobiotic molecules can interfere with nucleic acid functions and serve as drugs or poisons.

4.1.1   Drugs Interacting with Double Helical DNA

Interactions of small molecules ("drugs") with double helical nucleic acids can conveniently be divided into several groups according to the type of interaction between the nucleic acid and drug molecules: intercalation, minor groove binding, covalent DNA modification, and metal binding.

Intercalation is an interaction between the π-electron system of nucleic acid bases and a flat, aromatic drug molecule. The intercalation mode was first observed in a co-crystal between UpA and ethidium bromide (DRBB12[129]). Now, many more related structures have been determined, thus shedding light on the effects of drug binding on nucleic acid structure. An intercalating drug squeezes between two stacked nucleic acid bases and pushes them away from each other, thereby increasing the rise from about 3.4 to 7 Å. This local deformation is easily accommodated by the backbone, but effectively unwinds the double helix.

Simple intercalators, such as proflavine, show a limited selectivity for the CG step and have limited clinical use. Selectivity of intercalators is increased by attached side chains. Such side chains are often quite complex peptide oligomers, as in actinomycin (173D/DDH048[130], 1I3W/DD0039[131]), or sugar derivatives, as in nogalamycin (1D21/DDFA14[132]) or daunomycin (Figure 1-5c, 1DA0/DDF018[127]). The side chains reside in the helical grooves, mostly in the minor groove, where they interact with DNA bases by van der Waals contacts and hydrogen bonds. The side chain of an intercalator bound in the major groove has been observed for an acridine derivative, DACA (452D/DD0012[133]).

Minor groove-binding drug molecules usually are shaped complementary to the minor groove of B-DNA. Often a crescent shape, they are unbranched molecules of variable length. Almost a hundred such structures have been solved since the first complex between netropsin and the Dickerson-Drew dodecamer was solved (6BNA/GDLB05[134]). Groove binders show strong sequence specificity, due to multiple hydrogen bonding and hydrophobic interactions with the groove walls and base edges[135]. A typical example of a groove binder is a prototype drug, Hoechst 33258, which is a benzimidazole derivative. This flat, linear molecule fits well into the minor groove (Figure

1-5a, 1D43/GDL010[126]), but takes on several different binding geometries depending on nucleotide sequence and details of the molecule. Distamycin, a minor groove binder with similar characteristics to Hoechst 33258, lines up along the minor groove walls as expected (2DND/GDL003[136]). Quite surprisingly, it also binds in tandem, as in the structure 159D/GDHB25[137], where it binds in a widened minor groove of a B-DNA octamer.

Most known minor groove-binding drugs preferably recognize A/T sequences. However, design of sequence-specific drugs requires recognition of G/C pairs, as well[138]. A current approach to the design of such drugs is the incorporation of peptide or amide links into a chain of aromatic heterocycles, as in the structure of a six base pair B-DNA complexed with imidazole-pyrrole polyamide (407D/BDD002[139]). An analogical approach designs doubly charged lexitropsins, where the netropsin =C- groups are replaced by = N- ones (1LEX/GDL037[140]). Design of sequence-specific minor groove binders has been extensively reviewed by Dervan[141,142].

One pharmaceutically important drug is an anti-tumor agent cis platinum, cis-dichlorodiammino platinum(II). Platinum, like other heavy metals, binds preferably to purines, and most specifically to guanine N7. The stereochemistry of cis platinum allows it to form covalent bonds with two guanine bases in the major groove. Both intra- (Figure 1-5d, 1AIO/ DDLB73[128], 1LU5/DD0050[143]) and across-the-strand (DDJ075/1A2E[144]) complexes between cis-Pt and DNA oligomers are known. In both cases, the double strand is strongly bent as a result of two Pt-N7 bonds.

Direct covalent adducts between carcinogenic drugs and DNA are important causes of mutagenic lesions and cancer. One of a few known structures is that of anthramycine bound to a decamer sequence (274D/ GDJB29[145]).

4.1.2   Drugs Interacting with Folded Nucleic Acids

Functionally variable RNA molecules are an important target for drugs. Bacterial ribosomes are a classic target for clinically important antibiotics. With the determination of ribosomal crystal structures, structural details about their interactions with different drugs have become abundant[146], thus shedding light on structural details of translational machinery as well as giving hints to the discovery of new antibiotics.

The ribosome can be targeted at different sub-sites by chemically diverse molecules. For example, amino-glycoside antibiotics binding at the decoding site of the 16S rRNA (A-site) in the small ribosomal subunit compromises the fidelity of protein synthesis of the targeted bacteria. Structures of complexes between the small subunit and clinically important drugs as tetracyclines (1HNW/RR0020[147]) or paromomycin (1FJG/RR0016[148]) are known. Quite significantly, key structural features of drug/RNA interactions in these large ribosome–drug complexes are preserved at the oligonucleotide level in both solution (1PBR[149]) and crystal structures (1J7T/DR0006[150]), opening a way to use these smaller systems for drug design.

Some of the clinically most important antibiotics, e.g. macrolides, steptogramins, chloramphenicol, and oxazolidones, block the peptidyl-transferase catalytic site and/or peptide exit tunnel in the 23S rRNA of the large ribosomal subunit[151]. Extensive structural studies of ribosome inhibition by these drugs (1K8A/RR0043[152], 1NWY/RR0069[153], 1YHQ/RR0112[154]) have revealed structural details about their binding inside the peptide exit tunnel of the 23S rRNA.

4.2   Recognition Between Nucleic Acids and Proteins

Gene replication, expression, and translation are controlled by complicated biological machinery in which regulatory and enzymatic proteins must recognize unique, but short, nucleotide sequences. Toward this aim, nucleic acids and proteins have developed several different strategies in order to recognize each other. Many different recognition motifs are known today in both proteins and nucleic acids, which supports the early theory by Matthews summarized by his maxim "no code for recognition"[155]. Matthews' theory posits that protein/nucleic acid recognition is not digital, as in the self-recognition of DNA and RNA double helical forms.

Structural studies at atomic resolution reflect the chemical composition and architecture of nucleic acid and protein recognition[156]. The overall negative charge of nucleic acids implies the importance of interactions of polar and positively charged amino acids, such as arginines and lysines. Positively charged amino acids preferably select phosphate groups for their negative charge, but they may also approach either helical groove and interact with polar edges of bases. On the other hand, non-polar amino acids may interact with less polar groove walls. Interactions with base edges are usually considered more specific, and sometimes are referred to as "direct readout", while amino acid interactions with the phosphodiester backbone, mostly with the phosphate groups, is called "indirect readout". A protein surface interacting with nucleic acid is more hydrophilic than a protein surface interacting with another protein. All polar recognition may be water-mediated.

4.2.1   Protein Interaction with the Double Helical Forms of DNA

Character of binding is different in specifically bound transcription factors and enzymes. The binding interface is tighter but smaller in the former class, while enzymes encompass DNA with a large interaction area but lower specificity[157]. Enzymes also often interact with DNA via oligopeptide loops.

An average double helical DNA deviates from the canonical B-type conformation in protein/DNA complexes more than naked B-DNA structures[158]. There are three basic types of deformations of double helical geometry[42]:

i) They concentrate into local severe disruptions, "kinks", which add up to form a bent DNA structure as observed in CAP/DNA complexes (1CGP/PDT049[159]).

ii) Overall bending cumulates over several base steps, with relatively small local distortions of the regular geometry.

iii) Multiple local kinks do not add up, and the overall path of the helical axis is almost linear as in the native B-DNA (EcoRI endonucleases, e.g. 1ERI/ PDE001[160]).

In all these examples, DNA is deformed by opening the minor or major groove, thus bending the helix away from the widened groove. When the helix bends toward the major groove, DNA can bend away from the interacting protein or wrap around it. When bending towards the minor groove, DNA is only known to bend towards the protein.

An extensive bend of the DNA double helix by "nonspecific" binding of proteins is best exemplified by complexes between the histone core particle and DNA (1AOI/PD0001[161]). A high resolution complex (1KX3/PD0285[70]) revealed many water molecules at the protein/DNA interface. In fact, water-mediated hydrogen bonds form roughly half of all protein/DNA contacts. These contacts accommodate intrinsic DNA conformational variation, thus limiting the sequence dependency of nucleosome positioning while enhancing mobility.

Topologically and structurally complicated constructs called four-way junctions or Holliday junctions are important in recombination, or crossing over, where blocks of homologous genes are exchanged. Structures of four-way junctions made of naked DNA (1L6B/UD0019[98], 1M6G/UD0021[99]) as well as in protein-DNA complexes (2CRX/PD0103[100], 1BDX/PD0098[162]) both show the ability of double helical DNA to locally unravel its regular structure.

These different ways of double helical DNA deformation show that the double helix is an extremely flexible molecule capable of different ways of induced fit conformations with interacting molecules.

4.2.2 Protein Motifs Recognizing DNA

Protein motifs recognizing double helical DNA are morphologically variable and formed by all types of protein secondary structures, such as helices, sheets, and loops. Functionally and structurally well characterized motifs are mainly helix-turn-helix, zinc fingers, zipper motifs, and β-strand recognizing motifs[10,163,164].

- **Helix-turn-helix[169] and helix-loop-helix[170] motifs**

  Both are common binding motifs of proteins regulating transcription or replication. Their structure was first observed in the prokaryotic cro-repressor protein/DNA complex[171]. The motifs are formed by two, usually short α-helices at an angle of about 120°, and are linked by a short peptide turn. The second helix recognizes the major groove of B-DNA by forming both polar and hydrophobic contacts, mostly to the base edges but also to the backbone atoms, mainly phosphate groups. Loops or other protein structures at the N- or C-periphery of the motif can make additional contacts to the minor groove, thus enhancing the specificity of interaction (1TUP/PDR027[172]). Some transcription factors bend DNA significantly, for instance catabolite gene protein (CAP) bends DNA in two sharp 40° kinks. Figure 1-6a shows the tight fit of the recognition helix of protein SAP-1 into the B-DNA major groove in structure 1BC8/PD0020[165].

  Prokaryotic transcription factors usually bind to palindromic DNA sequences as dimers enhancing the motif's binding specificity; eukaryotic regulation factors, like yeast homeodomains MATα1 and MATα2, bind as heterodimers.

- **Zinc-fingers recognizing DNA[173]**

  Zinc fingers are eukaryotic binding motifs formed by an α-helix in combination with anti-parallel β-sheet or loop(s). Structural integrity of the motif is achieved by specific histidine and cysteine amino acid residues tetrahedrally complexed to a zinc cation. In the best characterized system, the "finger" region, makes the most specific as well as non-specific contacts with the DNA major groove, is about twelve amino acids long, and is located between the second β-strand and N-terminal part of the α-helix. Contacts are made to both base edges as well as phosphate groups.

  Zinc fingers can bind in tandems or oligomers. Three zinc-fingers interacting in succession with atoms in the DNA major groove are shown in Figure 1-6b (1A1H/PDT057[166]). Note how the binding α-helix is positioned differently in the major groove compared to the recognition α-helix of the helix-turn-helix motif. Zinc fingers can bind to DNA as dimers, in a way similar to the helix-turn-helix motif, as in the dimeric GAL4 transcription factor (1D66/PDT003[174]).

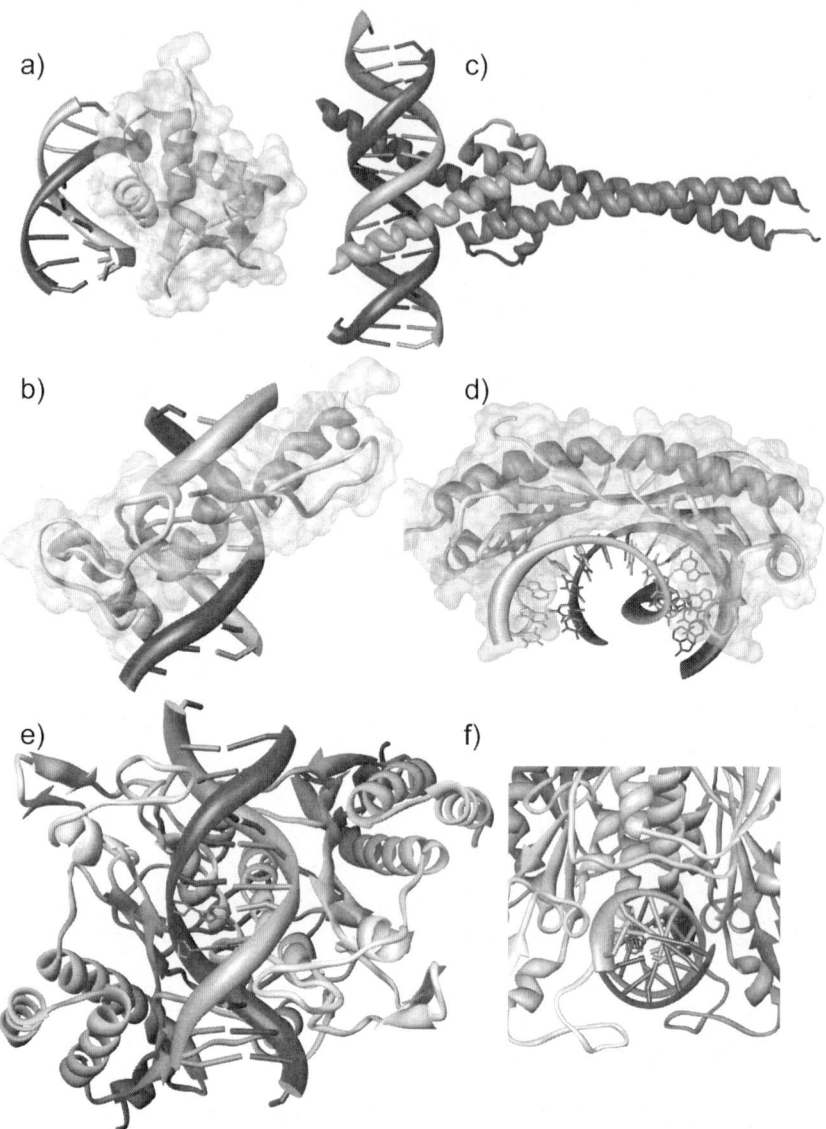

Figure 1-6. Examples of complexes between proteins and double helical DNA. a) Helix-turn-helix motif of transcription factor SAP-1 bound to the DNA major groove (1BC8/PD0020[165]). b) Three zinc-fingers interacting with the major groove of B-DNA. Zinc atoms are displayed as spheres (1A1H/PDT057[166]). c) α-helices of the leucine-zipper motif bound to the major groove (1NKP/PD0386[167]). d) An extensive β-sheet of transcription initiation factor TFIID-1 opens up DNA minor groove (1QNA/PD0160[168]). e) Restriction endonuclease makes contacts to an extensive surface of DNA, mainly in the major groove (1ERI/PDE001[160]). f) The same structure as in e) shown in a perpendicular view stresses large contact surface between the enzyme and DNA.

- **Leucine zipper motif** [175]

Labeled b/zip, the leucine zipper is a basic dimeric motif found in many eukaryotes. Two long α-helices join to form a dimer of coiled coils by intertwinning their leucine side chains. Dimerization of either identical or different α-helices alters the motif's DNA-binding specificity.

The structure of a leucine zipper/DNA complex was first observed in GCN4 mammalian transcription factor (1YSA/PDT002[176]), and other sequentially related factors C/EBP, Fos, and Jun. As in the helix-turn-helix motif, the key element of the recognition process is its α-helix interacting with the DNA major groove.

The leucine zipper motif can combine with the helix-loop-helix motif in some eukaryotic transcription factors, as in the factor Myc, which is closely related to Fos and Jun. This motif, labeled b/HLH/zip, is a four-helix-bundle usually formed by a heterodimer. Figure 1-6c shows two α-helices of the proto-oncogenic Myc-Max/DNA complex (1NKP/PD0386[167]) bound to the DNA major groove half-turn (about five base pairs).

- **β-strands**

β-strands can recognize either groove of a double helical DNA. TATA box binding proteins, TBPs, an important class of transcription factors binding to promoter DNA sequences, always recognize the minor groove as dimers[177]. The recognizant β-sheet of a TBP is usually quite extended, for instance the ten-fold sheet in 1YTB/PDT012[71]. The saddle-like β-sheet system can fit only into a severely deformed, widely open minor groove. DNA bases are partially untwisted, and the whole double helix bends away from the protein. This bend is concentrated in two kinks. Protein/DNA interactions are spread over an extensive area, and many contacts are hydrophobic, stressed by semi-intercalation of two aromatic phenylalanine rings between two pairs of bases. Figure 1-6d shows interaction between an extensive β-sheet of transcription initiation factor TFIID-1 with the widely open minor groove (1QNA/PD0160[168]).

Less extensive β-sheets recognize the DNA major groove, such as in met repressor-operator (1CMA/PDR008[178]) or in arc repressor-operator complex (1PAR/PDR012[179]). In these structures, the DNA is not severely distorted and retains its B-form.

4.2.3 Interaction of Proteins with RNA

RNA/protein interactions are almost as variable as RNA's biological roles, of which protein/RNA recognition is an abundant and essential activity.

While most protein/DNA complexes contain canonical double helical DNA, complexes with RNA exhibit myriad RNA structural features: double helices, loops, bulges, single-stranded hinges, etc. Protein/RNA structures also show other important differences from their protein/DNA counterparts[180]. Positively charged and hydrogen bonding amino acids still form a large proportion of contacts to RNA nucleotides, but many more contacts are made by aromatic amino acids to unstacked RNA bases and, overall, a larger proportion of contacts are non-polar van der Waals interactions, especially to single-stranded RNA regions with unstacked bases.

Proteins bind to RNA in two main modes[181]: into the A-type major groove and to single-stranded RNA motifs. The groove binders bind to the often-deformed deep and narrow major groove via interaction with a protein's α-helix, β-sheet, or loop. The double helix in these complexes can be quite short, even only a half-turn, so that the normally quite narrow major groove is open to incoming protein residues. The regular double helix is also often interrupted by non-Watson-Crick base pairs that may induce modulation essential for protein recognition.

In the second category of complexes, a protein β-sheet contacts single-stranded portions of RNA. Unstacked RNA bases are often bound in hydrophobic protein pockets, typically stacked on an aromatic side chain. In a complex between the RNA-binding domain of the U1A spliceosomal protein and an RNA hairpin (1URN/PRV002[182]), the β-sheets form multiple hydrogen bonding interactions with the RNA loop, which by itself is unstructured. Significantly, the loop's three bases are stacked on conserved amino acids of U1A.

4.2.4   Ribosome Structures

Ribosomes are complex RNA/protein complexes responsible for protein synthesis in all cells. A ribosome consists of two subunits: the small subunit contains about 20 proteins and a single ribosomal RNA (rRNA) molecule of around 1500 nucleotides in length; the large one consists of about fifty proteins, a 120-nucleotide and a ~3000-nucleotide rRNA. In prokaryotes, the small subunit is referred to as 30S by its sedimentation coefficient, the large subunit 50S, and the whole ribosome 70S; its molecular weight is about 2.5 MDaltons. The primary function of the small subunit is to read out the messenger RNA, recognize appropriate tRNA, and position them. The large subunit executes the peptide bond formation.

In the last few years, several groups have successfully determined the structures of ribosomal subunits, the 30S[183,184], 50S[185], as well as the whole ribosome[186]. The overall shape of both subunits is determined by their rRNA molecules, while the proteins serve to stabilize the tertiary RNA fold and to support and enhance the RNA control and catalytic functions. The majority of ribosomal proteins consists of a relatively small globular region and a long flexible arm typically protruding into the rRNA, thus helping to stabilize the RNA fold. In both subunits, the rRNA mediates the essential functions of both subunits, not the ribosomal proteins. 16S ribosomal RNA of the 30S subunit is organized into four well-defined and relatively independent domains implying considerable flexibility between them. This flexibility is needed to ensure the movement of messenger and tRNAs[148]. In contrast, 23S rRNA of the 50S subunit[185], although formed of six domains on the basis of its secondary structure, is remarkably globular and conformationally rigid.

All structural features observed in other nucleic acids are observed in the rRNA molecules. By the sheer size of rRNA, we can observe structural motifs at different scales within one structure, such as large domains made of hundreds of nucleotides, inter-helical stacks, loops, and bulges of different sizes, non-Watson-Crick base pairs, base triplets, conformations of individual nucleotides, and even atom-atom interactions as in different types of base pairs and metal-nucleotide interactions.

5. WEB RESOURCES FOR NUCLEIC ACID STRUCTURES

The information of primary importance for study of nucleic acid molecular structures, namely their three-dimensional coordinates, is in three principal archives. Structures of mononucleotides and their components, as well as of modified fragments of nucleic acids, are available in the Cambridge Structural Database[187], an excellent licensed stand-alone database system. All "macromolecular" nucleic acid structures, dinucleotides and larger, determined by experimental techniques, namely by X-ray crystallography, NMR techniques, and electron microscopy and diffraction, are available in the Protein Data Bank (PDB[188]) and in the Nucleic Acid Database (NDB[27]). Experimental data related to structures determined by the NMR techniques are deposited in the BioMagResBank, BMRB[189] (http://www.bmrb.wisc.edu/). Many other useful resources available on the world wide web are linked from the principal sites (Table 1-1).

Table 1-1. Web resources related to nucleic acid structures.

Resource	web address http://	Characterization
PDB	www.rcsb.org/	primary archive of macromolecular structures, searchable databases, web services, links
NDB	ndbserver.rutgers.edu/	database of NA structures, web services, links
SCOR	scor.lbl.gov/	hierarchical database of RNA structural motifs
RNA Society	www.rnasociety.org/	links, web services

5.1   The Primary Archives of Experimental Molecular Structures – PDB and NDB

The PDB and NDB are the primary archives containing experimental nucleic acid structures, and are both maintained by the Research Collabor-atory for Structural Bioinformatics, RCSB (http://www.rcsb.org/). The RCSB has developed generalized software for processing, archiving, querying, and distributing structural data for nucleic acid-containing structures.

Both RCSB maintained databases adopted the Macromolecular Crystallographic Information File (mmCIF[190]) as their data standard. This format has several advantages from the point of view of building a database:

1) The definitions for the data items are based on a comprehensive dictionary of crystallographic terminology and molecular structure description.

2) mmCIF is self-defining.

3) The syntax enforces relationships between data items.

This latter feature is important because it allows for rigorous checking of the data.

5.1.1   Data Deposition

Structures are deposited to the PDB and NDB databases using a web-based interface, AutoDep Input Tool (ADIT[191]; http://pdb.rutgers.edu/adit/). ADIT operates on top of the mmCIF dictionary and can be easily customized for different use. For instance, the RCSB team has developed deposition interfaces specific for different experimental techniques, namely X-ray crystallography, solution NMR, and electron microscopy.

5.1.2   Validation of Nucleic Acid Structures

A validation procedure specifically designed for nucleic acids, NUCheck[192], has been developed within the NDB project. The procedure integrates checks specific to nucleic acids with those developed for proteins. A structure is checked for features common to any molecular structure. Close intra- and inter-molecular atom-atom contacts are calculated, valence geo-metry is compared to the dictionary values, nucleotide torsion angles are calculated and plotted against their distributions in well-characterized double helical forms, and all chiral centers in (deoxy)riboses and configurations of prochiral oxygens at phosphates are checked.

This validation procedure can be run as a part of the deposition process or used as a standalone checking protocol.

5.2 Content of the NDB

Structures available in the NDB (http://ndbserver.rutgers.edu/) include RNA and DNA oligonucleotides with two or more bases either alone or complexed with ligands, natural nucleic acids such as tRNA, protein-nucleic acid complexes, and chemically synthesized nucleic acid derivatives such as peptide nucleic acids. The NDB currently contains about 2800 structures, which in total comprises over half a million nucleotides.

5.2.1 Database Searches and Reports in the NDB

The core of the NDB project is a relational database in which all of the primary and derived data items are organized into tables, allowing all tables to be searchable. For example, tables with primary information detail literature references related to the structures, the names and sequences of nucleic acids and proteins, and items related to NMR or crystallographic experimental conditions. Secondary information relates mostly to derived chemical and structural data, such as base pair parameters, torsion angles, or the presence of specific structural motifs like four-way junctions or a quadruple helix.

A Web interface was designed to make the query capabilities of the NDB as widely accessible as possible. Interaction with the database is a two-step process. In the first step, the user defines the criteria for selecting structures by combining different database items. Once the structures that meet the constraint criteria have been selected, various reports may be written. Two styles of the interface have been developed. First, a search with a compact interface allowing the user to set conditions on the most important properties of the structures, such as author names, the type of experiment, sequences, or the presence of structural motifs. A more comprehensive "Integrated Search" gives access to most database tables via easily navigated pull down menus. From either search, a variety of reports may be created.

5.2.2 NDB Atlas Pages

Browsing through the Atlas pages (http://ndbserver.rutgers.edu/atlas/) gives a quick yet comprehensive overview of the NDB content. They summarize the most important information about the structure: Names and sequences of the present biomolecules, authors and reference, crystallographic, and experimental, a molecular view of the biological unit and a crystal packing picture, and a link to the coordinate file.

The Atlas pages are conveniently categorized into groups of related structures that can be navigated visually by thumbnail figures. For instance, RNA structures categorized as ribozymes give an instant distinction between their different structural types, such as hammerhead ribozyme and larger intron-type ribozymes. Atlas pages can also be generated after the NDB is searched as a specific type of user-generated report.

5.2.3   NDB Archives

The NDB stores and distributes archival files of coordinates and experimental diffraction and NMR data:

    http://ndbserver.rutgers.edu/download_data/
    ftp://ndbserver.rutgers.edu

It also makes available tools and programs for calculation and visualization of nucleic acid structural characteristics like base pairing, standard geometries, and valence parameters for nucleotides, and building blocks for solvation of double helical DNA forms. Most tools and programs developed as a part of the NDB or RCSB PDB programs are available for download, including the standalone validation and pre-deposition software suite.

5.3   Structural Classification of RNA (SCOR)

SCOR is a database classifying RNA structural motifs[193]. It provides a survey of 3D motifs, interactions, and their relationships extracted from the NMR and crystal structures deposited in the PDB and NDB. Classification is performed for different types of internal loops, hairpins, as well as RNA motifs reported in the literature as kink-turn motifs, A-minor, or GNRA loops. The classified elements are described and represented by cartoons.

The classified elements are organized into a database with the "directed acyclic graph architecture", allowing multiple parent classes of a motif. The database can be either browsed or searched by PDB and NDB ID, oligonucleotide sequence or keywords.

ACKNOWLEDGEMENTS

Support from the Ministry of Education of the Czech Republic via a grant LC512 is greatly acknowledged. This work was supported in part by a grant from the National Science Foundation for the Nucleic Acid Database Project. The careful review of this paper by David Micallef is very much appreciated.

REFERENCES

1. Avery, O.T., MacLeod, C.M. & McCarthy, M. (1944). Studies on the chemical nature of the substance inducing transformation of pneumococcal types: induction of transformation by a desoxyribonucleic acid fraction isolated from Pneumococcus Type III. *J.Exp.Med.* **79**, 137-158.
2. Hershey, A. & Chase, M. (1952). *Cold Spring Harbor Symp. Quant. Biol.* **16**, 445-456.
3. Wilkins, M.H.F., Stokes, A.R. & Wilson, H.R. (1953). Molecular structure of deoxypentose nucleic acids. *Nature* **171**, 738-740.
4. Franklin, R.E. & Gosling, R.G. (1953). Molecular configuration in sodium thymonucleate. *Nature* **171**, 740-741.
5. Watson, J.D. & Crick, F H.C. (1953). A structure for deoxyribose nucleic acid. *Nature* **171**, 737-738.
6. Saenger, W. (1984). *Principles of nucleic acid structure.* Springer Advanced Texts in Chemistry (Cantor, C.R., Ed.), Springer-Verlag.
7. Neidle, S. (2002). *Nucleic Acid Structure and Recognition*, Oxford University, Oxford.
8. Calladine, C.R. & Drew, H.R. (1997). *Understanding DNA. The Molecule & How It Works*, second edition, Academic, London.
9. Neidle, S., Ed. (1999). Oxford handbook of nucleic acid structure. Oxford University, Oxford.
10. Branden, C. & Tooze, J. (1999). *Introduction to Protein Structure*, second edition, Garland, New York.
11. Zamenhof, S., Brawermann, G. & Chargaff, E. (1952). On the Desoxypentose Nucleic Acids from Several Microorganisms. *Biochim.Biophys.Acta* **9**, 402-405.
12. Arai, S., Chatake, T., Ohhara, T., Kurihara, K., Tanaka, I., Suzuki, N., Fujimoto, Z., Mizuno, H. & Niimura, N. (2005). Complicated water orientations in the minor groove of the B-DNA dodecamer d(CCATTAATGG)2 observed by neutron diffraction measurement. *Nucleic Acid Res.* **33**, 3017-3024.
13. Altona, C. & Sundaralingam, M. (1972). Conformational analysis of the sugar ring in nucleosides and nucleotides. A new description using the concept of pseudorotation. *J.Am.Chem.Soc.* **94**, 8205-8212.
14. Murray-Rust, P. & Motherwell, S. (1978). Computer retrieval and analysis of molecular geometry. III. Geometry of the β-1`-aminofuranoside fragment. *Acta Cryst.* **B34**, 2534-2546.
15. Dunitz, J.D. (1972). Approximate relationships between conformational parameters in 5- and 6-membered rings. *Tetrahedron* **28**, 5459-5467.
16. Gelbin, A., Schneider, B., Clowney, L., Hsieh, S.-H., Olson, W.K. & Berman, H.M. (1996). Geometric parameters in nucleic acids: sugar and phosphate constituents. *J.Am.Chem.Soc.* **118**, 519-528.
17. Dong, F. & Miller, R.E. (2002). Vibrational transition moment angles in isolated biomolecules: A structural tool. *Science* **298**, 1227-1230.
18. Luisi, B.F., Orozco, M., Šponer, J., Luque, F.J. & Shakked, Z. (1998). On the potential role of the amino nitrogen atom as a hydrogen acceptor in macromolecules. *J.Mol.Biol.* **279**, 1123-1136.
19. Baur, W.H. (1974). The Geometry of Polyhedral Distortions. Predictive Relationships for the Phosphate Group. *Acta Cryst.* **B30**, 1195-1215.
20. Schneider, B., Kabelac, M. & Hobza, P. (1996). Geometry of the Phosphate Group and Its Interactions with Metal Cations in Crystals and ab initio calculations. *J.Am.Chem.Soc.* **118**, 12207-12217.

21. Lemieux, R.U. (1971). Effects of unshared pairs of electrons and their solvation on conformational equilibria. *Pure Appl.Chem.* **25**.

22. Wolfe, S. (1972). The gauche effect. Some stereochemical consequences of adjacent electron pairs and polar bonds. *Acc.Chem.Res.* **5**, 102-111.

23. Schneider, B., Neidle, S. & Berman, H.M. (1997). Conformations of the sugar-phosphate backbone in helical DNA crystal structures. *Biopolymers* **42**, 113-124.

24. Murray, L.J., Arendall, W.B., 3rd, Richardson, D.C. & Richardson, J.S. (2003). RNA backbone is rotameric. *Proc.Natl.Acad.Sci.USA* **100**, 13904-13909.

25. Schneider, B., Moravek, Z. & Berman, H.M. (2004). RNA conformational classes. *Nucleic Acids Res.* **32**, 1666-1677.

26. Allen, F.H., Bellard, S., Brice, M.D., Cartright, B.A., Doubleday, A., Higgs, H., Hummelink, T., Hummelink-Peters, B.G., Kennard, O., Motherwell, W.D.S., Rodgers, J.R. & Watson, D.G. (1979). The Cambridge Crystallographic Data Centre: Computer-based search, retrieval, analysis and display of information. *Acta Cryst.* **B35**, 2331-2339.

27. Berman, H.M., Olson, W.K., Beveridge, D.L., Westbrook, J., Gelbin, A., Demeny, T., Hsieh, S.-H., Srinivasan, A.R. & Schneider, B. (1992). The Nucleic Acid Database - A Comprehensive Relational Database of Three-Dimensional Structures of Nucleic Acids. *Biophys.J.* **63**, 751-759.

28. Taylor, R. & Kennard, O. (1982). The molecular structures of nucleosides and nucleotides part I. *J.Mol.Struct.* **78**, 1-28.

29. Clowney, L., Jain, S.C., Srinivasan, A.R., Westbrook, J., Olson, W.K. & Berman, H.M. (1996). Geometric Parameters In Nucleic Acids: Nitrogenous Bases. *J.Am.Chem.Soc.* **118**, 509-518.

30. Parkinson, G., Vojtechovsky, J., Clowney, L., Brunger, A.T. & Berman, H.M. (1996). New parameters for the refinement of nucleic acid containing structures. *Acta Cryst.* **D52**, 57-64.

31. Brünger, A.T. (1992). X-PLOR, version 3.1, a system for X-ray crystallography and NMR 3.1 edit. Yale University Press, New Haven, CT.

32. Weiner, P. & Kollman, P. (1981). AMBER. *J.Comput.Chem.* **2**, 287-303.

33. Brooks, B.R., Bruccoleri, R.E., Olafson, B.D., States, D.J., Swaminathan, S. & Karplus, M. (1983). CHARMM: a program for macromolecular energy, minimization, and dynamics calculations. *J.Comput.Chem.* **4**, 187-217.

34. Leontis, N.B. & Westhof, E. (2001). Geometric nomenclature and classification of RNA base pairs. *RNA* **7**, 499-512.

35. Šponer, J., Jurecka, P. & Hobza, P. (2004). Accurate interaction energies of hydrogen-bonded nucleic acid base pairs. *J.Am.Chem.Soc.* **126**, 10142-10151.

36. Dickerson, R.E., Bansal, M., Calladine, C.R., Diekmann, S., Hunter, W.N., Kennard, O., von Kitzing, E., Lavery, R., Nelson, H.C.M., Olson, W., Saenger, W., Shakked, Z., Sklenar, H., Soumpasis, D.M., Tung, C.-S., Wang, A.H.-J. & Zhurkin, V.B. (1989). Definitions and nomenclature of nucleic acid structure parameters. *EMBO J.* **8**, 1-4.

37. Lu, X.-J., El Hassan, M.A. & Hunter, C.A. (1997). Structure and conformation of helical nucleic acids: analysis program (SCHNAaP). *J.Mol.Biol.* **273**, 668-680.

38. El Hassan, M.A. & Calladine, C.R. (1995). The assessment of the geometry of dinucleotide steps in double-helical DNA: a new local calculation scheme with an appendix. *J.Mol.Biol.* **251**, 648-664.

39. Gorin, A.A., Zhurkin, V.B. & Olson, W.K. (1995). B-DNA twisting correlates with base-pair morphology. *J.Mol.Biol.* **247**, 34-48.

40. Kosikov, K.M., Gorin, A.A., Zhurkin, V.B. & Olson, W.K. (1999). DNA stretching and compression: Simulations of double helical structures. *J.Mol.Biol.* **289**, 1301-1326.

41. Lavery, R. & Sklenar, H. (1989). Defining the structure of irregular nucleic acids: conventions and principles. *J.Biomol.Struct.Dyn.* **6**, 655-667.
42. Dickerson, R.E. (1998). DNA bending: the prevalence of kinkiness and the virtues of normality. *Nucleic Acids Res.* **26**, 1906-1926.
43. Tung, C.-S., Soumpasis, D.M. & Hummer, G. (1994). An extension of the rigorous base-unit oriented description of nucleic acid structures. *J.Biomol.Struct.Dyn.* **11**, 1327-1344.
44. Soumpasis, D.M. & Tung, C.S. (1988). A rigorous base pair oriented description of DNA structures. *J.Biomol.Struct.Dyn.* **6**, 397-420.
45. Bansal, M., Bhattacharyya, D. & Ravi, B. (1995). NUPARM and NUCGEN: software for analysis of sequence dependent nucleic acid structures. *CABIOS* **11**, 281-287.
46. Bhattacharyya, A. & Lilley, D.M. (1989). The contrasting structures of mismatched DNA sequences containing looped-out bases (bulges) and multiple mismatches (bubbles). *Nucleic Acids Res.* **17**, 6821-6840.
47. Babcock, M.S., Pednault, E.P.D. & Olson, W.K. (1994). Nucleic acid structure analysis. Mathematics for local cartesian and helical structure parameters that are truly comparable between structures. *J.Mol.Biol.* **237**, 125-156.
48. Olson, W.K., Bansal, M., Burley, S.K., Dickerson, R.E., Gerstein, M., Harvey, S.C., Heinemann, U., Lu, X.-J., Neidle, S., Shakked, Z., Sklenar, H., Suzuki, M., Tung, C.-S., Westhof, E., Wolberger, C. & Berman, H.M. (2001). A standard reference frame for the description of nucleic acid base-pair geometry. *J.Mol.Biol.* **313**, 229-237.
49. Drew, H.R., Wing, R.M., Takano, T., Broka, C., Tanaka, S., Itakura, K. & Dickerson, R.E. (1981). Structure of a B-DNA dodecamer: conformation and dynamics. *Proc.Natl.Acad.Sci.USA* **78**, 2179-2183.
50. McCall, M., Brown, T. & Kennard, O. (1985). The crystal structure of d(G-G-G-G-C-C-C-C). A model for poly(dG)•poly(dC). *J.Mol.Biol.* **183**, 385-396.
51. Holbrook, S.R., Cheong, C., Tinoco Jr., I. & Kim, S.-H. (1991). Crystal structure of an RNA double helix incorporating a track of non-Watson-Crick base pairs. *Nature* **353**, 579-581.
52. Gessner, R.V., Frederick, C.A., Quigley, G.J., Rich, A. & Wang, A.H.-J. (1989). The molecular structure of the left-handed Z-DNA double helix at 1.0-Å atomic resolution. *J.Biol.Chem.* **264**, 7921-7935.
53. Haider, S., Parkinson, G.N. & Neidle, S. (2002). Crystal structure of the potassium form of an Oxytricha nova G-quadruplex. *J.Mol.Biol.* **320**, 189-200.
54. Chen, L., Cai, L., Zhang, X. & Rich, A. (1994). Crystal Structure of a Four-Stranded Intercalated DNA: d(C4). *Biochemistry* **33**, 13540-13546.
55. Sussman, J.L., Holbrook, S.R., Warrant, R.W. & Kim, S.-H. (1978). Crystal structure of yeast phenylalanine transfer RNA. I. Crystallographic refinement. *J.Mol.Biol.* **123**, 607-630.
56. Pley, H.W., Flaherty, K.M. & McKay, D.B. (1994). Three-dimensional structure of a hammerhead ribozyme. *Nature* **372**, 68-74.
57. Cate, J.H., Gooding, A.R., Podell, E., Zhou, K.H., Golden, B.L., Kundrot, C.E., Cech, T.R. & Doudna, J.A. (1996). Crystal structure of a group I ribozyme domain: Principles of RNA packing. *Science* **273**, 1678-1685.
58. Dickerson, R.E., Drew, H.R., Conner, B.N., Wing, R.M., Fratini, A.V. & Kopka, M.L. (1982). The Anatomy of A-, B-, and Z-DNA. *Science* **216**, 475-485.
59. Chandrasekaran, R. & Arnott, S. (1989). *Landolt-Börnstein Numerical Data and Functional Relationships in Science and Technology, Group VII/1b, Nucleic Acids* (Saenger, W., Ed.), Springer-Verlag, Berlin.

60. Olson, W.K., Gorin, A.A., Lu, X.-J., Hock, L.M. & Zhurkin, V.B. (1998). DNA sequence-dependent deformability deduced from protein-DNA crystal complexes. *Proc.Natl.Acad.Sci.USA* **95**, 11163-11168.
61. Berman, H.M. (1997). Crystal studies of B-DNA: the answers and the questions. *Biopolymers* **44**, 23-44.
62. Dickerson, R.E., Goodsell, D. & Kopka, M.L. (1996). MPD and DNA bending in crystals and in solution. *J.Mol.Biol.* **256**, 108-125.
63. Heinemann, U., Alings, C. & Hahn, M. (1994). Crystallographic studies of DNA helix structure. *Biophys.Chem.* **50**, 157-167.
64. Gorin, A. & Zhurkin, V.B. (1995). DNA structure and flexibility in DNA-protein complexes. A comparative study based on x-ray data. *Mathematical Approaches to DNA*.
65. Quintana, J.R., Grzeskowiak, K., Yanagi, K. & Dickerson, R.E. (1992). Structure of a B-DNA decamer with a central T-A step: C-G-A-T-T-A-A-T-C-G. *J.Mol.Biol.* **225**, 379-395.
66. Yanagi, K., Prive, G.G. & Dickerson, R.E. (1991). Analysis of local helix geometry in three B-DNA decamers and eight dodecamers. *J.Mol.Biol.* **217**, 201-214.
67. Young, M.A., Ravishanker, G., Beveridge, D.L. & Berman, H.M. (1995). Analysis of local helix bending in crystal structures of DNA oligonucleotides and DNA-Protein complexes. *Biophys.J.* **68**, 2454-2468.
68. Lipanov, A., Kopka, M.L., Kaczor-Grzeskowiak, M., Quintana, J. & Dickerson, R.E. (1993). The Structure of the B-DNA Decamer C-C-A-A-C-I-T-T-G-G in Two Different Space Groups: Conformational Flexibility of B-DNA. *Biochemistry* **32**, 1373-1380.
69. Privé, G.G., Yanagi, K. & Dickerson, R.E. (1992). Structure of the B-DNA decamer C-C-A-A-C-G-T-T-G-G and comparison with isomorphous decamers C-C-A-A-G-A-T-T-G-G and C-C-A-G-G-C-G-T-G-G. *J.Mol.Biol.* **217**, 177-199.
70. Davey, C.A., Sargent, D.F., Luger, K., Maeder, A.W. & Richmond, T.J. (2002). Solvent mediated interactions in the structure of the nucleosome core particle at 1.9 a resolution. *J.Mol.Biol.* **319**, 1097-1113.
71. Kim, Y., Geiger, J.H., Hahn, S. & Sigler, P.B. (1993). Crystal structure of a yeast TBP/TATA-box complex. *Nature* **365**, 512-520.
72. Kim, J.L., Nikolov, D.B. & Burley, S.K. (1993). Co-crystal structure of TBP recognizing the minor groove of a TATA element. *Nature* **365**, 520-527.
73. Guzikevich-Guerstein, G. & Shakked, Z. (1996). A novel form of the DNA double helix imposed on the TATA-box by the TATA-binding protein. *Nat.Struct.Biol.* **3**, 32-37.
74. Leontis, N., Stombaugh, J. & Westhof, E. (2002). The non-Watson-Crick base pairs and their associated isostericity matrices. *Nucleic Acids Res.* **30**, 3497-3531.
75. Varani, G. & McClain, W.H. (2000). The G x U wobble base pair. A fundamental building block of RNA structure crucial to RNA function in diverse biological systems. *EMBO Rep.* **1**, 18-23.
76. Baeyens, K.J., DeBondt, H.L., Pardi, A. & Holbrook, S.R. (1996). A curved RNA helix incorporating an internal loop with G center dot A and A center dot A non-Watson-Crick base pairing. *Proc.Natl.Acad.Sci.USA* **93**, 12851-12855.
77. Wang, A.H.-J., Quigley, G.J., Kolpak, F.J., Crawford, J.L., van Boom, J.H., van der Marel, G.A. & Rich, A. (1979). Molecular structure of a left-handed double helical DNA fragment at atomic resolution. *Nature* **282**, 680-686.
78. Wang, A.H.-J., Quigley, G.J., Kolpak, F.J., van der Marel, G.A., van Boom, J.H. & Rich, A. (1981). Left-handed double helical DNA: variations in the backbone conformation. *Science* **211**, 171-176.

79. Schwartz, T., Behlke, J., Lowenhaupt, K., Heinemann, U. & Rich, A. (2001). Structure of the DLM-1-Z-DNA complex reveals a conserved family of Z-DNA-binding proteins. *Nat.Struct.Biol* **8**, 761-765.

80. Westhof, E. (1988). Water: An integral part of nucleic acid structure. *Ann.Rev.Biophys.Chem.* **17**, 125-144.

81. Chalikian, T.V. & Breslauer, K.J. (1998). Thermodynamic analysis of biomolecules: a volumetric approach. *Curr.Opin.Struct.Biol.* **8**, 657-664.

82. Leikin, S., Parsegian, V.A. & Rau, D.C. (1993). Hydration forces. *Annu.Rev.Phys.Chem.* **44**, 369-395.

83. Rau, D.C. & Parsegian, V.A. (1992). Direct measurement of the intermolecular forces between counterion-condensed DNA double helices. *Biophys.J.* **61**, 246-259.

84. Neidle, S., Achari, A., Taylor, G.L., Berman, H.M., Carrell, H.L., Glusker, J.P. & Stallings, W. (1977). The structure of a proflavine: dinucleoside phosphate complex - a model for nucleic acid drug interaction. *Nature* **269**, 304-307.

85. Schneider, B., Ginell, S.L. & Berman, H.M. (1992). Low temperature structures of dCpG-proflavine: conformational and hydration effects. *Biophys.J.* **63**, 1572-1578.

86. Drew, H.R. & Dickerson, R.E. (1981). Structure of a B-DNA Dodecamer III. Geometry of Hydration. *J.Mol.Biol.* **151**, 535-556.

87. Kennard, O., Cruse, W.B.T., Nachman, J., Prange, T., Shakked, Z. & Rabinovich, D. (1986). Ordered water structure in an A-DNA octamer at 1.7A resolution. *J.Biomol.Struct.Dyn.* **3**, 623-647.

88. Saenger, W., Hunter, W.N. & Kennard, O. (1986). DNA conformation is determined by economics in the hydration of phosphate groups. *Nature* **324**, 385-388.

89. Eisenstein, M. & Shakked, Z. (1995). Hydration patterns and intermolecular interactions in A-DNA crystal structures. Implications for DNA recognition. *J.Mol.Biol.* **248**, 662-678.

90. Egli, M., Portmann, S. & Usman, N. (1996). RNA hydration: a detailed look. *Biochemistry* **35**, 8489-8494.

91. Schneider, B., Cohen, D.M., Schleifer, L., Srinivasan, A.R., Olson, W.K. & Berman, H.M. (1993). A systematic method for studying the spatial distribution of water molecules around nucleic acid bases. *Biophys.J.* **65**, 2291-2303.

92. Schneider, B. & Berman, H.M. (1995). Hydration of the DNA bases is local. *Biophys.J.* **69**, 2661-2669.

93. Schneider, B., Patel, K. & Berman, H.M. (1998). Hydration of the phosphate group in double helical DNA. *Biophys.J.* **75**, 2422-2434.

94. Moravek, Z., Neidle, S. & Schneider, B. (2002). Protein and drug interactions in the minor groove of DNA. *Nucleic Acids Res.* **30**, 1182-1191.

95. Woda, J., Schneider, B., Patel, K., Mistry, K. & Berman, H.M. (1998). An analysis of the relationship between hydration and protein-DNA interactions. *Biophys.J.* **75**, 2170-2177.

96. Schneider, B. & Kabelac, M. (1998). Stereochemistry of binding of metal cations and water to a phosphate group. *J.Am.Chem.Soc.* **120**, 161-165.

97. Schultze, P., Smith, F.W. & Feigon, J. (1994). Refined solution structure of the dimeric quadruplex formed from the *Oxytricha* telomeric oligonucleotide d(GGGGTTTTGGGG). *Structure* **2**, 221-233.

98. Vargason, J.M. & Ho, P.S. (2002). The effect of cytosine methylation on the structure and geometry of the Holliday junction - The structure of d(CCGGTACm(5)CGG) at 1.5 A resolution. *J.Biol.Chem.* **277**, 21041-21049.

99. Thorpe, J.H., Gale, B.C., Teixeira, S.C. & Cardin, C.J. (2003). Conformational and hydration effects of site-selective sodium, calcium and strontium ion binding to the DNA Holliday junction structure d(TCGGTACCGA)(4). *J.Mol.Biol.* **327**, 97-109.

100 Gopaul, D.N., Guo, F. & Van Duyne, G.D. (1998). Structure of the Holliday junction intermediate in Cre-loxP site-specific recombination. *EMBO J.* **17**, 4175-4187.

101. Shepard, W., Cruse, W.B.T., Fourme, R., de la Fortelle, E. & Prangé, T. (1998). A zipper-like duplex in DNA: the crystal structure of d(GCGAAAGCT) at 2.1A resolution. *Structure* **6**, 849-861.

102. Ginell, S.L., Vojtechovsky, J., Gaffney, B., Jones, R. & Berman, H.M. (1994). The crystal structure of a mispaired dodecamer d(CGAGAATTC(O^6Me)GCG)$_2$, containing a carcinogen O^6-methylguanine. *Biochemistry* **33**, 3487-3493.

103. Park, H., Zhang, K., Ren, Y., Nadji, S., Sinha, N., Taylor, J.-S. & Kang, C. (2002). Crystal Structure of a DNA Decamer Containing a Thymine-dimer. *Proc.Natl.Acad.Sci.USA* **99**, 15965-15970.

104. Egli, M., Tereshko, V., Teplova, M., Minasov, G., Joachimiak, A., Sanishvili, R., Weeks, C.M., Miller, R., Maier, M.A., An, H., Cook, P.D. & Manoharan, M. (1998). X-ray crystallographic analysis of the hydration of A-and B-Form DNA at atomic resolution. *Biopolymers* **48**, 234-252.

105. Minasov, G., Teplova, M., Nielsen, P., Wengel, J. & Egli, M. (2000). Structural basis of cleavage by RNase H of hybrids of arabinonucleic acids and RNA. *Biochemistry* **39**, 3525-3532.

106. Declercq, R., Van Aerschot, A., Read, R.J., Herdewijn, P. & Van Meervelt, L. (2002). Crystal structure of double helical hexitol nucleic acids. *J.Am.Chem.Soc.* **124**, 928-933.

107. Cruse, W.B.T., Salisbury, S.A., Brown, T., Cosstick, R., Eckstein, F. & Kennard, O. (1986). Chiral Phosphorothioate Analogues of B-DNA. The Crystal Structure of Rp-d(Gp(S)CpGp(S)CpGp(S)C). *J.Mol.Biol.* **192**, 891-905.

108. Heinemann, U., Rudolph, L.-N., Alings, C., Morr, M., Heikens, W., Frank, R. & Bloecker, H. (1991). Effect of a single 3'-methylene phosphonate linkage on the conformation of an A-DNA octamer double helix. *Nucleic Acids Res.* **19**, 427-433.

109. Tereshko, V., Gryaznov, S. & Egli, M. (1998). Consequences of replacing the DNA 3'-oxygen by an amino group: high-resolution crystal structure of a fully modified N3'->P5' phosphoramidate DNA dodecamer duplex. *J.Am.Chem.Soc.* **120**, 269-283.

110. Rasmussen, H. & Sandholm, J. (1997). Crystal structure of a peptide nucleic acid (PNA) duplex at 1.7 Å resolution. *Nat.Struct.Biol.* **4**, 98-101.

111. Betts, L., Josey, J.A., Veal, J.M. & Jordan, S.R. (1995). A Nucleic-Acid Triple-Helix Formed by a Peptide Nucleic-Acid DNA Complex. *Science* **270**, 1838-1841.

112. Nissen, P., Ippolito, J.A., Ban, N., Moore, P.B. & Steitz, T.A. (2001). RNA tertiary interactions in the large ribosomal subunit: the A-minor motif. *Proc.Natl.Acad.Sci.USA* **98**, 4899-4903.

113. Battle, D.J. & Doudna, J.A. (2002). Specificity of RNA-RNA helix recognition. *Proc.Natl.Acad.Sci.USA* **99**, 11676-11681.

114. Klein, D.J., Schmeing, T.M., Moore, P.B. & Steitz, T.A.. (2001). The Kink-Turn: A new RNA secondary structure motif. *EMBO J.* **20**, 4214-4221.

115. Strobel, S.A., Adams, P.L., Stahley, M.R. & Wang, J. (2004). RNA kink turns to the left and to the right. *RNA* **10**, 1852-1854.

116. Adams, P.L., Stahley, M.R., Kosek, A.B., Wang, J. & Strobel, S.A. (2004). Crystal structure of a self-splicing group I intron with both exons. *Nature* **430**, 45-50.

117. Robertus, J.D., Ladner, J.E., Finch, J.T., Rhodes, D., Brown, R.S., Clark, B.F.C. & Klug, A. (1974). Structure of yeast phenylalanine tRNA at 3 A resolution. *Nature* **250**, 546-551.

118. Kim, S.-H., Suddath, F.L., Quigley, G.J., McPherson, A., Sussman, J.L., Wang, A.H.-J., Seeman, N.C. & Rich, A. (1974). Three-dimensional tertiary structure of yeast phenylalanine transfer RNA. *Science* **185**, 435-440.

119. Cech, T., Zaug, A. & Grabowski, P. (1981). In vitro splicing of the ribosomal RNA precursor of Tetrahymena: involvement of a guanosine nucleotide in the excision of the intervening sequence. *Cell* **27**, 487-496.

120. Gilbert, W. (1986). Origin of Life - the RNA World. *Nature* **319**, 618-618.

121. Gesteland, R.F., Cech, T.R. & Atkins, J.F. (1999). *The RNA World*, second edition. Cold Spring Harbor monograph series, 37, Cold Spring Harbor Laboratory Press, Cold Spring Harbor.

122. Zhang, L. & Doudna, J.A. (2002). Structural insights into group II Intron catalysis and branch-site selection. *Science* **295**, 2084-2088.

123. Krasilnikov, A.S., Yang, X., Pan, T. & Mondragon, A. (2003). Crystal structure of the specificity domain of ribonuclease P. *Nature* **421**, 760-764.

124. Krasilnikov, A.S., Xiao, Y., Pan, T. & Mondragon, A. (2004). Basis for structural diversity in homologous RNAs. *Science* **306**, 104-107.

125. Serganov, A., Keiper, S., Malinina, L., Tereshko, V., Skripkin, E., Hobartner, C., Polonskaia, A., Phan, A.T., Wombacher, R., Micura, R., Dauter, Z., Jaschke, A. & Patel, D.J. (2005). Structural basis for Diels-Alder ribozyme-catalyzed carbon-carbon bond formation. *Nat.Struct.Mol.Biol.* **12**, 218-224.

126. Quintana, J.R., Lipanov, A.A. & Dickerson, R.E. (1991). Low-temperature crystallographic analysis of the binding of the Hoechst-33258 to the double-helical DNA dodecamer C-G-C-G-A-A-T-T-C-G-C-G. *Biochemistry* **30**, 10294-10306.

127. Moore, M. H., Hunter, W.N., Langlois d'Estaintot, B. & Kennard, O. (1989). DNA-Drug interactions. The crystal structure of d(CGATCG) complexed with daunomycin. *J.Mol.Biol.* **206**, 693-705.

128. Takahara, P.M., Rosenzweig, A.C., Frederick, C.A. & Lippard, S.J. (1995). Crystal Structure of Double-Stranded DNA Containing the Major Adduct of the Anticancer Drug Cisplatin. *Nature* **377**, 649-652.

129. Jain, S.C., Tsai, C.-C. & Sobell, H.M. (1977). Visualization of drug-nucleic acid interactions at atomic resolution. II. Structure of an ethidium/dinucleoside monophosphate crystalline complex, ethidium: 5-iodocytidylyl(3'-5')guanosine. *J.Mol.Biol.* **114**, 317-331.

130. Kamitori, S. & Takusagawa, F. (1994). Multiple binding modes of anticancer drug actinomycin D: x-ray, molecular modeling, and spectroscopic studies of d(GAAGCTTC)2-actinomycine D complexes and its host DNA. *J.Am.Chem.Soc.* **116**, 4154-4165.

131. Robinson, H., Gao, Y.G., Yang, X., Sanishvili, R., Joachimiak, A. & Wang, A.H.-J. (2001). Crystallographic analysis of a novel complex of actinomycin D bound to the DNA decamer CGATCGATCG. *Biochemistry* **40**, 5587-5592.

132. Gao, Y.-G., Liaw, Y.-C., Robinson, H. & Wang, A.H.-J. (1990). Binding of the antitumor drug nogalamycin and its derivatives to DNA: structural comparison. *Biochemistry* **29**, 10307-10316.

133. Todd, A.K., Adams, A., Thorpe, J.H., Denny, W.A., Wakelin, L.P. & Cardin, C.J. (1999). Major groove binding and 'DNA-induced' fit in the intercalation of a derivative of the mixed topoisomerase I/II poison N-(2-(dimethylamino)ethyl)acridine-4-carboxamide (DACA) into DNA: X-ray structure complexed to d(CG(5-BrU)ACG)2 at 1.3-A resolution. *J.Med.Chem.* **42**, 536-540.

134. Kopka, M.L., Yoon, C., Goodsell, D., Pjura, P. & Dickerson, R.E. (1985). Binding of an antitumor drug to DNA. Netropsin and C-G-C-G-A-A-T-T-BrC-G-C-G. *J.Mol.Biol.* **183**, 553-563.

135. Tabernero, L., Bella, J. & Alemán, C. (1996). Hydrogen bond geometry in DNA-minor groove binding drug complexes. *Nucleic Acids Res.* **24**, 3458-3466.

136. Coll, M., Frederick, C.A., Wang, A.H.-J. & Rich, A. (1987). A Bifurcated Hydrogen-Bonded Conformation in the d(A•T) Base Pairs of the DNA Dodecamer

d(CGCGAAATTTGCG) and Its Complex With Distamycin. *Proc.Natl.Acad.Sci.USA* **84**, 8385-8389.

137. Chen, X., Ramakrishnan, B., Rao, S.T. & Sundaralingam, M. (1994). Side by side binding of two distamycin a drugs in the minor groove of an alternating B-DNA duplex. *Nat.Struct.Biol.* **1**, 169-175.

138. Dervan, P.B. & Buerli, R.W. (1999). Sequence-specific DNA recognition by polyamides. *Curr.Opin.Chem.Biol.* **3**, 688-693.

139. Kielkopf, C.L., White, S., Szewczyk, J.W., Turner, J.M., Baird, E.E., Dervan, P.B. & Rees, D.C. (1998). A structural basis for recognition of A.T and T.A base pairs in the minor groove of B-DNA. *Science* **282**, 111-115.

140. Goodsell, D.S., Ng, H.L., Kopka, M.L., Lown, J.W. & Dickerson, R.E. (1995). Structure of a dicationic monoimidazole lexitropsin bound to DNA. *Biochemistry* **34**, 16654-16661.

141. Dervan, P.B. & Edelson, B.S. (2003). Recognition of the DNA minor groove by pyrrole-imidazole polyamides. *Curr.Opin.Struct.Biol.* **13**, 284-299.

142. Dervan, P.B. (2001). Molecular recognition of DNA by small molecules. *Bioorg.Med.Chem.* **9**, 2215-2235.

143. Silverman, A.P., Bu, W.M., Cohen, S.M. & Lippard, S.J. (2002). 2.4-A crystal structure of the asymmetric platinum complex {Pt(ammine)(cyclohexylamine)}(2+) bound to a dodecamer DNA complex. *J.Biol.Chem.* **277**, 49743-49749.

144. Coste, F., Malinge, J.M., Serre, L., Shepard, W., Roth, M., Leng, M. & Zelwer, C. (1999). Crystal structure of a double-stranded DNA containing a cisplatin interstrand cross-link at 1.63 A resolution: Hydration at the platinated site. *Nucleic Acids Res.* **27**, 1837-1846.

145. Kopka, M.L., Goodsell, D.S., Baikalov, I., Grzeskowiak, K., Cascio, D. & Dickerson, R.E. (1994). Crystal structure of a covalent DNA-drug adduct: anthramycin bound to C-C-A-A-C-G-T-T-G-G and a molecular explanation of specificity. *Biochemistry* **33**, 13593-13610.

146. Hermann, T. (2005). Drugs targeting the ribosome. *Curr.Opin.Struct.Biol.* **15**, 355-366.

147. Brodersen, D.E., Clemons Jr., W.M., Carter, A.P., Morgan-Warren, R., Wimberly, B.T. & Ramakrishnan, V. (2000). The structural basis for the action of the antibiotics Tetracycline, Pactamycin, and Hygromycin B on the 30S ribosomal subunit. *Cell* **103**, 1143-1154.

148. Carter, A.P., Clemons, W.M.J., Brodersen, D.E., Wimberly, B.T., Morgan-Warren, R. & Ramakrishnan, V. (2000). Functional insights from the structure of the 30S ribosomal subunit and its interactions with antibiotics. *Nature* **407**, 340-348.

149. Fourmy, D., Recht, M.I., Blanchard, S.C. & Puglisi, J.D. (1996). Structure of the a Site of Escherichia Coli 16S Ribosomal RNA Complexed with an Aminoglycoside Antibiotic. *Science* **274**, 1367-1371.

150. Vicens, Q. & Westhof, E. (2001). Crystal structure of paromomycin docked into the eubacterial ribosomal decoding A site. *Structure* **9**, 647-658.

151. Harms, J.M., Bartels, H., Schlunzen, F. & Yonath, A. (2003). Antibiotics acting on the translational machinery. *J.Cell.Sci.* **116**, 1391-1393.

152. Hansen, J.L., Ippolito, J.A., Ban, N., Nissen, P., Moore, P.B. & Steitz, T.A. (2002). The structures of four macrolide antibiotics bound to the large ribosomal subunit. *Mol.Cell.* **10**, 117-128.

153. Schlunzen, F., Harms, J.M., Franceschi, F., Hansen, H.A., Bartels, H., Zarivach, R. & Yonath, A. (2003). Structural basis for the antibiotic activity of ketolides and azalides. *Structure* **11**, 329-338.

154. Tu, D., Blaha, G., Moore, P.B. & Steitz, T.A. (2005). Structures of MLS$_B$K antibiotics bound to mutated large ribosomal subunits provide a structural explanation for resistance. *Cell* **121**, 257-270.

155. Matthews, B.W. (1988). No code for recognition. *Nature* **335**, 294-295.

156. Luscombe, N.M., Laskowski, R.A. & Thornton, J.M. (2001). Amino acid-base interactions: a three-dimensional analysis of protein-DNA interactions at an atomic level. *Nucleic Acids Res.* **29**, 2860-2874.

157. Nadassy, K., Wodak, S.J. & Janin, J. (1999). Structural features of protein-nucleic acid recognition sites. *Biochemistry* **38**, 1999-2017.

158. Jones, S., van Heyningen, P., Berman, H.M. & Thornton, J.M. (1999). Protein-DNA interactions: A structural analysis. *J. Mol. Biol.* **287**, 877-896.

159. Schultz, S.C., Shields, G.C. & Steitz, T.A. (1991). Crystal structure of a CAP-DNA complex: the DNA is bent by 90°. *Science* **253**, 1001-1007.

160. Kim, Y., Grable, J.C., Love, R., Greene, P.J. & Rosenberg, J.M. (1990). Refinement of Eco RI endonuclease crystal structure: A revised protein chain tracing. *Science* **249**, 1307-1309.

161. Luger, K., Mader, A.W., Richmond, R.K., Sargent, D.F. & Richmond, T.J. (1997). Crystal structure of the nucleosome core particle at 2.8 A resolution. *Nature* **389**, 251-260.

162. Hargreaves, D., Rice, D.W., Sedelnikova, S.E., Artymiuk, P.J., Lloyd, R.G. & Rafferty, J.B. (1998). Crystal structure of E.coli RuvA with bound DNA Holliday junction at 6 A resolution. *Nat.Struct.Biol.* **5**, 441-446.

163. Harrison, S.C. (1991). A structural taxonomy of DNA-binding domains. *Nature* **353**, 715-719.

164. Travers, A.A. (1992). DNA conformation and configuration in protein-DNA complexes. *Curr.Opin.Struct.Biol.* **2**, 71-77.

165. Mo, Y., Vaessen, B., Johnston, K. & Marmorstein, R. (1998). Structures of SAP-1 bound to DNA targets from the E74 and c-fos promoters: insights into DNA sequence discrimination by Ets proteins. *Mol.Cell* **2**, 201-212.

166. Elrod-Erickson, M., Benson, T.E. & Pabo, C.O. (1998). High-resolution structures of variant Zif268-DNA complexes: implications for understanding zinc finger-DNA recognition. *Structure* **6**, 451-464.

167. Nair, S.K. & Burley, S.K. (2003). X-ray structures of Myc-Max and Mad-Max recognizing DNA. Molecular bases of regulation by proto-oncogenic transcription factors. *Cell* **112**, 193-205.

168. Patikoglou, G.A., Kim, J.L., Sun, L., Yang, S.H., Kodadek, T. & Burley, S.K. (1999). TATA element recognition by the TATA box-binding protein has been conserved throughout evolution. *Genes Dev.* **13**, 3217-3230.

169. Brennan, R.G. (1992). DNA Recognition by the Helix-Turn-Helix Motif. *Curr.Opin.Struct.Biol.* **2**, 100-108.

170. Littlewood, T.D. & Evan, G.I. (1995). Transcription factors 2: helix-loop-helix. *Protein Profile* **2**, 621-702.

171. Anderson, W.F., Ohlendorf, D.H., Takeda, Y. & Matthews, B.W. (1981). Structure of the cro repressor from bacteriophage λ and its interaction with DNA. *Nature* **290**, 754-758.

172. Cho, Y., Gorina, S., Jeffrey, P.D. & Pavletich, N.P. (1994). Crystal structure of a p53 tumor suppressor-DNA complex: understanding tumorigenic mutations. *Science* **265**, 346-355.

173. Schmiedeskamp, M. & Klevit, R.E. (1994). Zinc finger diversity. *Curr.Opin.Struct.Biol.* **4**, 28-35.

174. Marmorstein, R., Carey, M., Ptashne, M. & Harrison, S.C. (1992). DNA recognition by GAL4: structure of a protein-DNA complex. *Nature* **356**, 408-414.

175. Pathak, D. & Sigler, P.B. (1992). Updating structure-function relationships in the bZip family of transcription factors. *Curr.Opin.Struct.Biol.* **2**, 116-123.

176. Ellenberger, T.E., Brandl, C.J., Struhl, K. & Harrison, S.C. (1992). The GCN4 basic region leucine zipper binds DNA as a dimer of uninterrupted α helices - crystal structure of the protein-DNA complex. *Cell* **71**, 1223-1237.

177. Phillips, S.E.V. (1994). The β-ribbon DNA recognition motif. *Ann.Rev.Biophys. Biomol.Struct.* **23**, 671-701.

178. Somers, W.S. & Phillips, S.E.V. (1992). Crystal structure of the met repressor-operator complex at 2.8 Å resolution reveals DNA recognition by β-strands. *Nature* **359**, 387-393.

179. Raumann, B.E., Rould, M.A., Pabo, C.O. & Sauer, R.T. (1994). DNA recognition by β-sheets in the arc repressor operator crystal structure. *Nature* **367**, 754-757.

180. Jones, S., Daley, D.T.A., Luscombe, N.M., Berman, H.M. & Thornton, J.M. (2001). Protein-RNA interactions: A structural analysis. *Nucleic Acids Res.* **29**, 934-954.

181. Draper, D. (1999). Themes in RNA-protein recognition. *J.Mol.Biol.* **293**, 255-270.

182. Oubridge, C., Ito, H., Evans, P.R., Teo, C.H. & Nagai, K. (1994). Crystal structure at 1.92 Å resolution of the RNA-binding domain of the U1A spliceosomal protein complexed with an RNA hairpin. *Nature* **372**, 432-438.

183. Schluenzen, F., Tocilj, A., Zarivach, R., Harms, J., Gluehmann, M., Janell, D., Bashan, A., Bartels, H., Agmon, I., Franceschi, F. & Yonath, A. (2000). Structure of functionally activated small ribosomal subunit at 3.3 A resolution. *Cell* **102**, 615-623.

184. Wimberly, B.T., Brodersen, D.E., Clemons Jr., W.M., Morgan-Warren, R., Carter, A.P., Vonrhein, C., Hartsch, T. & Ramakrishnan, V. (2000). Structure of the 30S ribosomal subunit. *Nature* **407**, 327-339.

185. Ban, N., Nissen, P., Hansen, J., Moore, P.B. & Steitz, T.A. (2000). The complete atomic structure of the large ribosomal subunit at 2.4 Å resolution. *Science* **289**, 905-920.

186. Yusupova, G.Z., Yusupov, M.M., Cate, J.H.D. & Noller, H.F. (2001). The path of messenger RNA through the ribosome. *Cell* **106**, 233-241.

187. Allen, F.H., Davies, J.E., Galloy, J.J., Johnson, O., Kennard, O., Macrae, C.F., Mitchell, E.M., Mitchell, G.F., Smith, J.M. & Watson, D.G. (1991). The development of versions 3 and 4 of the Cambridge Structural Database System. *J.Chem.Inf.Comp.Sci.* **31**, 187-204.

188. Berman, H.M., Westbrook, J., Feng, Z., Gilliland, G., Bhat, T.N., Weissig, H., Shindyalov, I.N. & Bourne, P.E. (2000). The Protein Data Bank. *Nucleic Acids Res.* **28**, 235-242.

189. Doreleijers, J.F., Mading, S., Maziuk, D., Sojourner, K., Yin, L., Zhu, J., Markley, J.L. & Ulrich, E.L. (2003). BioMagResBank database with sets of experimental NMR constraints corresponding to the structures of over 1400 biomolecules deposited in the Protein Data Bank. *J.Biomol.NMR* **26**, 139-46.

190. Bourne, P., Berman, H.M., Watenpaugh, K., Westbrook, J.D. & Fitzgerald, P.M.D. (1997). The macromolecular Crystallographic Information File (mmCIF). *Meth.Enzymol.* 277, 571-590.

191. Rutgers University, New Brunswick, NJ. (1998). AutoDep Input Tool. Westbrook, J. NDB-406.

192. Rutgers University, New Brunswick, NJ. (1998). NUCheck. Feng, Z., Westbrook, J. & Berman, H.M. NDB-407.

193. Klosterman, P.S., Tamura, M., S.R., H. & Brenner, S.E. (2002). SCOR: a structural classification of RNA database. Nucleic Acids Res. 30, 392-394.

Chapter 2

USING AMBER TO SIMULATE DNA AND RNA

Thomas E. Cheatham, III[1] and David A. Case[2]

[1]*Departments of Medicinal Chemistry and of Pharmaceutics and Pharmaceutical Chemistry and of Bioengineering, 2000 East, 30 South, Skaggs Hall 201, University of Utah, Salt Lake City, UT 84112, USA, tec3@utah.edu;* [2]*Department of Molecular Biology, TPC15, The Scripps Research Institute, 10550 N. Torrey Pines Rd., La Jolla, CA 92037, USA, case@scripps.edu*

Abstract: The use of molecular dynamics and free energy simulation model nucleic acid structure, dynamics, and interactions are discussed from the authors AMBER-centric viewpoint.

Key words: AMBER; molecular dynamics; DNA; RNA; biomolecular simulation; MM-PBSA; force fields; Ewald

1. AMBER APPLIED TO NUCLEIC ACIDS

AMBER, "Assisted Model Building with Energy Refinement", is a suite of programs that has evolved over the past three decades to enable the application of molecular dynamics (and free energy) simulation methods to bio-molecular systems including nucleic acids, proteins, and more recently carbohydrates. Starting with the first simulations of DNA in explicit water[1] and progressing to recent large-scale explicit solvent simulations of components of the ribosome[2, 3] [and also see Chapter 12], AMBER or its associated force fields have arguably been applied in the majority of published simulations involving nucleic acids to date. This relates to the performance of the AMBER-related force fields[4-6] (and their free availability) and the relatively early adoption of fast and efficient particle mesh Ewald methods[7, 8] for the proper treatment of long-range electrostatic interactions. These advances allowed stable simulation of nucleic acids in explicit solvent[9, 10] including an accurate representation of the surroundings

45

J. Šponer and F. Lankaš (eds.), Computational Studies of RNA and DNA, 45–71.
© 2006 *Springer.*

and sequence dependent structure and dynamics[11-18]. More recently — and in addition to the extensive and emerging sets of simulation in explicit solvent—there has been resurgence in the application of generalized Born implicit solvent methods[19-21] in nucleic acid simulation[22]. This has been facilitated by the development of better methods and force fields in both CHARMM[23] and AMBER[24].

In this chapter we highlight the use of molecular dynamics simulation applied to nucleic acids from our limited author- and AMBER-centric viewpoint. For more thorough background related to AMBER development and application, see related review articles[24, 25]. For a complete review of MD methods applied to nucleic acids and their success, a number of comprehensive reviews have been published[26-31]. For a more complete review of nucleic acid force fields and their performance, see our work[32], or work by MacKerell[33], Karplus[34], Langley[35], van Gunsteren[36], or also the ABC consortium[18]. In addition, a set of guides to nucleic acid simulation have been published that may help introduce the field and methods[37]. To aid in learning and using AMBER, a number of tutorials (which include applications to nucleic acids) are distributed with the program and available on the WWW (see http://amber.scripps.edu).

It is useful to note that AMBER has not been developed or designed as a program that can do everything or that aims to include *all* molecular dynamics methods. Instead it is focused into a series of specialized programs that allow set-up and limited model building (LEaP, antechamber), accurate and efficient molecular dynamics and free energy simulation (sander, PMEMD) and analysis (ptraj, carnal, mm-pbsa). A brief introduction to each of these general capabilities (in the context of nucleic acid simulation) is provided below. AMBER is also not a force field, although a number of force fields developed in the developer's and other research labs are distributed with the suite of programs.

1.1   Setting up AMBER MD Simulations

LEaP is a freely available (X windows or text based) program that facilitates setting up the parameter/topology and coordinate input files for subsequent MD studies. Only very limited model building is provided and the normal starting point is the reading of specific force field specifications (for a variety of available force fields) and coordinates (in the form of a PDB file). Since AMBER 7.0, the antechamber program (also freely available) has been distributed. This program, in addition to facilitating file conversion from Gaussian and other file formats into those readable by LEaP, includes access to the GAFF (general AMBER force field) force field[38] that is intended to provide a means to specify decent molecular mechanical parameters for drug-like molecules or new residues. Details into its power and applicability are described elsewhere[39, 40]; the point of discussing the program here is to point out that now it is significantly easier to model DNA modifications or nucleic acid-ligand interactions than previously since the difficult part of obtaining parameters for a new residue is greatly simplified. Both of these programs do not provide significant model building capability and therefore users typically rely on experimental structures as starting points or must generate their own reliable model structures. To generate arbitrary nucleic acid models, NAB (nucleic acid builder) is a general purpose molecular manipulation programming language that has been developed[41, 42]. It is intended for the building of both helical and non-helical models with up to 100s of nucleotides using a combination of rigid body transformations and distance geometry; the examples supplied with the program show how to build a DNA mini-circle such as is discussed later in this chapter. The program can also perform molecular mechanics including generalized Born implicit solvent methods. Other common programs used for building nucleic acid models include 3DNA[43], ERNA-3D[44, 45], MANIP[46], MC-SYM[47], and NAMOT[48].

1.2 AMBER Dynamics

The main molecular dynamics engine is the `sander` program which is focused primarily in two directions: (1) accurate and parallel efficient simulation of explicitly solvated periodically replicated unit cells and (2) fast/efficient simulation of non-periodic systems in implicit solvent. Development is driven largely by research and the needs of the associated developers; yet, despite being primarily a research code, the programs are widely utilized throughout the world. Specialized methods that are highlighted in `sander` include a general facility for including distance, angle, torsion, NOESY volume, chemical shift, and residual dipolar coupling restraints, an evolving efficient QM/MM implementation[49, 50], a specially optimized parallel version of the PME code (PMEMD by Robert Duke), a new long-range electrostatics methods developed by Wu called isotropic periodic (IPS) boundary conditions[51], an efficient energy-mixing based thermodynamic integration facility, and various enhanced sampling methods. The enhanced and/or path sampling methods include:

- Umbrella sampling using the NMR-based restraints in `sander`.
- LES: locally enhanced sampling[52, 53] is a method which lowers barriers to conformational exchange by generating copies of sub-regions of a molecule that are seen by the rest of the molecule as an average.

- Replica-exchange: Two facilities, either internally or externally via the MMTSB toolkit[54], are provided which allow simultaneous running of separate calculations with exchange of information (such as temperature).
- Targeted-MD: Restraints have been added that use the RMSd (best fit root-mean-squared deviation) as a reaction coordinate.

As shown consistently in the literature, an accurate description of the aqueous environment is essential for realistic bio-molecular simulations. However, this easily becomes very expensive computationally. For example, an adequate representation of the solvation of a medium-size protein typically requires thousands of discrete water molecules to be placed around it. While some of the cost is ameliorated by using fast efficient methods for treating the long-ranged electrostatics (such as PME[8, 55] or IPS) and specially optimized code like PMEMD, an alternative replaces the discrete water molecules by "virtual water"– an infinite continuum medium with the (some of the) dielectric and hydrophobic properties of water. Implicit solvent methodology comes at a price of number of approximations whose effects are often hard to estimate. Some familiar descriptors of molecular interaction, such as solute–solvent hydrogen bonds, are no longer explicitly present in the model; instead, they come in implicitly, in the mean-field way *via* a linear dielectric response, and contribute to the overall solvation energy. However, despite the fact that the methodology represents an approximation at a fundamental level, it has in many cases been successful in calculating various macromolecular properties[56-58]. Here we outline the implicit solvent methodologies implemented in AMBER.

Within the framework of the continuum model, a numerically exact way to compute the electrostatic potential $\phi(r)$ produced by molecular charge distribution $\rho_m(r)$, is based on the Poisson-Boltzmann (PB) approach in which the following equation (or its equivalent) must be solved; for simplicity we give its linearized form:

$$\nabla\varepsilon(r)\nabla\phi(r) = -4\pi\rho_m(r) + \kappa^2\varepsilon(r)\phi(r). \qquad (1)$$

Here, $\varepsilon(r)$ represents the position-dependent dielectric constant which equals that of bulk water far away from the molecule, and is expected to decrease fairly rapidly across the solute/solvent boundary. The electrostatic screening effects of (monovalent) salt enter via the second term on the right-hand side of Eq. 1, where the Debye-Hückel screening parameter $\kappa \approx 0.1\text{Å}^{-1}$ at physiological conditions. Once the potential $\phi(r)$ is computed, the electrostatic part of the solvation free energy is given by:

$$\Delta G_{el} = \frac{1}{2}\sum_i q_i \left[\varphi(r_i) - \varphi(r_i)_{vac}\right] \qquad (2)$$

where q_i are the partial atomic charges at positions r_i that make up the molecular charge density and $\phi(r_i)|_{vac}$ is the electrostatic potential computed for the same charge distribution in the absence of the dielectric boundary, *e.g.* in vacuum. Full accounts of this theory are available elsewhere[59, 60].

The analytic generalized Born (GB) method is an approximate way to compute ΔG_{el}. The methodology has become popular, especially in MD applications, due to its relative simplicity and computational efficiency, compared to the more standard numerical solution of the PB equation[57, 61]. Within the GB models currently available in AMBER and NAB, each atom in a molecule is represented as a sphere of radius ρ_i with a charge q_i at its center; the interior of the atom is assumed to be filled uniformly with material of dielectric constant of 1. The molecule is surrounded by a solvent of a high dielectric ε_w (78.5 for water at 300 K). The GB model approximates ΔG_{el} by an analytical formula[19, 57],

$$\Delta G_{el} = -\frac{1}{2} \sum_{ij} \frac{q_i q_j}{f^{GB}(r_{ij}, R_i, R_j)} \left(1 - \frac{e^{-\kappa f^{GB}}}{\varepsilon_w}\right) \qquad (3)$$

where r_{ij} is the distance between atoms i and j, R_i is the so-called *effective Born radii* of atom i, and f^{GB} is a certain smooth function of its arguments. The electrostatic screening effects of (monovalent) salt are incorporated [62] into Eq. 3 via the Debye-Hückel screening parameter $\kappa[Å^{-1}] \approx 0.316 \sqrt{[salt][mol/L]}$.

A common choice[19] of f^{GB} is:

$$f^{GB} = \left[r_{ij}^2 + R_i R_j \exp(-r_{ij}^2 / 4 R_i R_j)\right]^{1/2} \qquad (4)$$

although other expressions have been tried[63-66]. The effective Born radius of an atom reflects the degree of its burial inside the molecule: for each charge, the "perfect" effective radius R_i satisfies:

$$\Delta G_i^{PB} = -\frac{1}{2}\left(1 - \frac{1}{\varepsilon_w}\right)\frac{q_i^2}{R_i} \qquad (5)$$

where ΔG_i^{PB} is the solvation energy (computed from a numerical solution to Eq. 1) for a single charge q_i in the dielectric environment of the full system. The effective radii depend on the molecule's conformation, and these need to be re-computed every time the conformation changes.

The efficiency of computing the effective radii is therefore a critical issue, and various approximations are normally made to accelerate the calculations. In particular, the so-called *Coulomb field approximation*, is often used, which approximates the electric displacement around an atom by the Coulomb field $D_i^0(r) \equiv q_i r/r^3$. Within this assumption, the following expression for R_i can be derived[57, 67]:

$$R_i^{-1} = \frac{1}{\rho_i} - \frac{1}{4\pi} \int\limits_{solute} \theta(|r| - \rho_i) \frac{1}{r^4} d^3 r \qquad (6)$$

where the integral is over the solute volume, excluding a sphere of radius ρ_i around atom i. Over the years, a number of slightly different intrinsic radii ρ_i have been proposed. A good set is expected to be transferable or to perform reasonably well in different types of problems. One example is the Bondi radii set originally proposed in the context of geometrical analysis of macromolecular structures, but later found to be useful in continuum electrostatics models as well[68].

For a realistic molecule, computing the integral in Eq. 6 is anything but trivial, so approximations are often made to obtain a closed-form analytical expression. The AMBER and NAB programs adopt the pairwise approach of Hawkins, Cramer and Truhlar[20, 69, 70], where the integral in Eq. 6 is approximated by a sum of terms over all other atoms ($j \neq i$) in the molecule:

$$R_i^{-1} = \frac{1}{\rho_i} + \sum_{j \neq i} f(d_{ij}, \rho_i, \rho_j) \qquad (7)$$

Here d_{ij} is the distance between atoms i and j, and detailed expressions for the functions $f()$ are given elsewhere[70, 71] These terms are exact for two overlapping spherical regions, and are approximate for points in space that are inside more than two atomic spheres. Although the scheme is approximate, it is completely analytical, so that forces can be computed as the derivatives of the energy, allowing standard applications of minimization and molecular dynamics to be carried out[21, 22].

1.3   AMBER Analysis

Another area of focus is to facilitate the analysis of molecular dynamics trajectories; this includes the ability to process, manipulate and analyze MD trajectory and coordinate files (ptraj) and also to extract (free) energetics (mm-pbsa). ptraj is a program that provides a set of "actions" that are performed sequentially on each snapshot of a set of trajectories, the format of which are auto-detected. Currently ptraj supports AMBER formats for trajectory and restart files, PDB, CHARMM binary trajectories of either endian (byte) order, and a simple binary format. The "actions" that can be performed include best RMSd fitting, measurement of distances, angles, ring puckers, vectors, and atomic positional fluctuations, construction of radial distribution functions, calculation of mean-squared displacements (diffusion), 2D RMSd maps, and output of coordinate averaged structures. Trajectory manipulations supported include general periodic imaging, stripping atoms, translation, scaling, running averages, and saving the closest solvent molecules. Data that is accumulated during the trajectory processing (such as time series of scalars or vectors or matrices) is analyzed after the trajectories are processed sequentially with a series of "analyze" commands. More specialized features of the program include the hydration analysis (grid), hydrogen bond analysis (hbond), matrix facility, and clustering (cluster), each of which we describe in more detail below. If one is familiar with the C language, by following the detailed comments in the code it is relatively straightforward to add new actions that act on coordinate snapshots, or to add new analysis commands that work on the accumulated stack of scalars or vectors. Additionally, adding support for different coordinate file formats (for input or output) is also possible (albeit slightly less straightforward). The ease of extensibility of this code has led a number of different groups to extend the code.

Hbond: This facility provides a means to keep track of the time series, lifetime and occupancy of specific pair or triple distances and angles. To do this for all possible pair interactions—as a function of time—quickly becomes intractable (due to memory demands) so the set of interactions to consider is limited as specified by the user using the donor and acceptor commands (based on atom/residue name matching or general atom selection). Interactions between *electron pair* donors (single atoms) and *electron pair* acceptors (two atoms) are monitored (in contrast to *hydrogen* acceptors and donors). The following examples will keep track of all pair interactions between atoms named N3 in residues named GUA and the bond/angle formed to atoms N4 and H42 in residues named CYT.

```
donor GUA N3
acceptor CYT N4 H42
```

Multiple donor and acceptor commands may be specified. If the same atom name is specified twice in an acceptor interaction, the angle is ignored (i.e. only the distance is calculated); this is useful for monitoring pair interactions of single atoms to donors such as the interaction of ions (where one might specify "acceptor CIO Na+ Na+" to select atoms named Na+ in residues named CIO). In addition to specific pair interactions, a concept of general interactions has also been implemented to monitor solvent interactions. In this case, rather than keeping track of every separate solvent molecule and its interaction (which quickly becomes memory intensive particularly if solvent-solvent interactions are monitored), the program only keeps track of a list (equal to the value of `solventneighbors` specified) of solvent molecules interacting with each donor or acceptor group. By default, six solvent interactions are stored, where solvent is defined arbitrarily by the user as a set of molecules (such as residues named WAT or residues named Na+). An example usage of the hbond command (post definition of the solvent and specification of the acceptor and donor lists) follows:

```
hbond series hb out hbond_wat.out \
   solventdonor WAT O \
   solventacceptor WAT O H1 \
   solventacceptor WAT O H2 \
   time 1.0 angle 120.0 distance 3.5
```

The keywords above turn on the time series (`series`) and name it hb, output a summary file called `hbond_wat.out`, setup the list of `solventdonor` and `solventacceptors`, specify that there is 1 ps between frames, set the angle cutoff such that the donor – acceptorH – acceptor angle is greater than 120.0 (although the angle stored internally and output is 180 minus this value) and that the donor – acceptor distance is less than 3.5 Å. A truncated version of the output (omitting standard deviations and more details) looks somewhat like the following:

```
DONOR     ACCEPTOR      %occ   dist   angle   lifetime
|  :5@N1  |:40@H3  :40@N3 |  99.93 2.955 15.50 1281.4 3159
|  :40@O4 |  :5@H61  :5@N6 |  98.26 2.975 17.90   60.3 517
|  :39@O4 |  :5@H61  :5@N6 |   1.86 3.296 50.73    1.1 5
|  :5@O4' |  solv accptr  |  36.64 3.016 33.69    1.8 35
|  :5@N7  |  solv accptr  | 132.69 2.958 26.33    3.0 67
|  :5@N3  |  solv accptr  |  98.88 2.916 24.69    7.4 129
|solv dnr |  :5@H62  :5@N6 |  93.75 3.085 29.61    3.0 48
```

The data shown is for some of the interactions with residue 5 of the DNA duplex with sequence d($CGA_4T_4CGA_4T_4CG$) from a ~27 ns MD simulation (in explicit solvent within a truncated octahedron unit cell applying the particle mesh Ewald method and standard MD simulation protocols [32]). Listed are the atoms involved in the hydrogen bond (or a designation that specific that the interaction is with generic solvent), the percent occupation, the average distance and angle (away from linear or 180.0°), the lifetime in picoseconds and the maximum number of continuous frames for which the interaction was present. For interaction with water molecules, the occupancy can be greater than 100% signifying that multiple waters are interacting. Planned extensions to the hbond facility include storing the distribution of distances (to avoid strict cutoffs on the hydrogen bond distance) and enhancements that will facilitate dumping out and analysis of the individual time series. The hbond facility provides a means to track specific pair or triple interactions over the course of a simulation. The output shown above is augmented by an ascii representation of the time series of the interaction (where "darker" characters imply greater occupancy, i.e. " " for 0-5%, "." for 5-20%, "-" for 20-40%, "x" for 60-80%, "*" for 80-95% and "@" for 95-100% occupancy over equal spaced time intervals from the trajectory). In the first case, full occupancy is seen over the trajectory (representing one of the AT base pair hydrogen bonds which also transiently forms with the preceding thymine residue, and partially occupied water interactions.

```
|   :40@O4 |  :5@H61  :5@N6 |@@@@@@@@@@@@@@@@@@@@@@@|
|   :39@O4 |  :5@H61  :5@N6 |                       |
|   :5@O4' |  solv accptr   |---o--o---------o-o-o|
|solv dnr  |  :5@H62  :5@N6 |******@****@**@***@*|
```

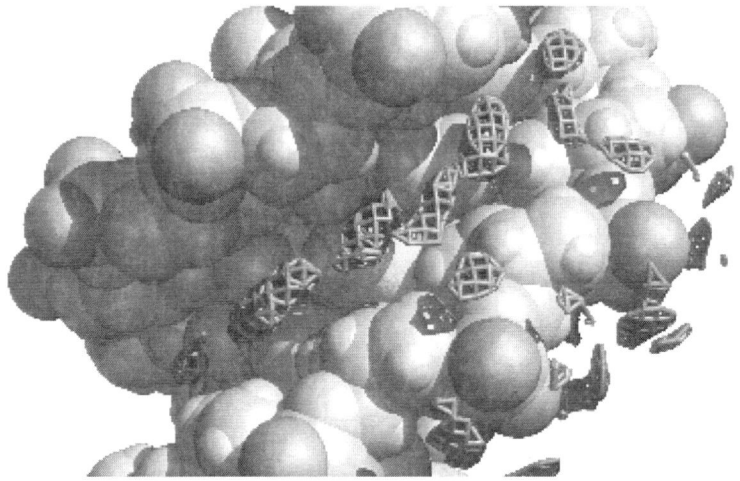

Figure 2-1. From ~27 ns of MD simulation of d(CGA₄T₄CGA₄T₄CG)₂ the water oxygen density at ~3x (light gray) and ~ 4x bulk water (darker gray) is displayed (as a patchwork grid) on top of the straight coordinate averaged structure from 10-15ns with a view into the minor groove. The phosphate oxygens are highlighted (as a different shade) as is adenine residue 5 atom N3; the 100 x 100 x 100, 0.5 Å spacing grid, was centered on residues 4-6. In addition to the spine of hydration in the minor groove, hydration of the phosphates is evident.

Grid: The hydrogen bond analysis gives an indication of pair interactions but does not provide a visualization of how, for example, water interacts with a particular group. To look at solvation or ion density or effectively any atomic density around a particular group of atoms, a very simple grid procedure was developed that constructs an arbitrarily sized cubic grid and counts the number of atoms (of a given type chosen by the user) within each grid element. The resulting grid can be output in X-PLOR density format for subsequent visualization by Chimera (see Figure 2-1), MIDAS, or VMD. In practice, prior to constructing the grid the region of interest needs to be centered, the trajectory fit to a common reference frame (usually the region interest), and the atoms imaged (as the current grid procedure does not image on the fly and therefore will put holes of density if atoms are in image cells). This method provides a simple and qualitative means to view regions of high density around the molecule of interest (such as in the grooves of DNA). To provide more quantitative estimates (less biased by motion of the region upon which the grid is centered and the chosen reference frame), radial distribution functions can be calculated (with the `radial` command).

Matrix/vector: With AMBER version 8.0, the vector facility of ptraj was greatly expanded and a new matrix facility developed by Holger Gohlke[25]. In addition, the analyze section of the code was greatly enhanced. It is now possible to follow the time correlation of arbitrarily defined vectors (or including vectors perpendicular to a least-squares fit plane) and to build distance, covariance, and related matrices which can subsequently be processed to estimate entropies, and to analyze and project estimated modes of vibration[25].

Cluster: A general purpose clustering library, implementing nine differrent clustering algorithms (hierarchical, single linkage, centroid linkage, complete linkage, K-means, centripetal, COBWEB, Bayesian, and self-organizing maps), has been integrated into the `ptraj` program and allows clustering of the trajectory snapshots based on pairwise RMSd or distance matrix comparisons. New trajectories (and/or average and/or representative structures) are output for each trajectory. To make clustering of large trajectories tractable, sieving has been implemented in a 2-pass approach such that every n^{th} frame is initially clustered then the initially skipped frames are added to the cluster that is most representative. Users can specify how many clusters are desired or set-up a pairwise distance metric cutoff value (representing the distance between clusters) to dynamically choose how many clusters should be formed. More detailed discussion of the performance is available elsewhere[72].

MM-PBSA: In addition to the general analysis provided by `ptraj`, MM-PBSA is a method developed to post-processes MD trajectory data to extract approximate free energies. The name—an abbreviation for molecular mechanics with Poisson-Boltzmann and surface area terms—came from collaboration between the authors (the Case and Kollman groups) and a test of the idea that performing simple energy analysis over a series of snapshots obtained from explicit solvent simulations and averaging of the results (to dampen the noise) might provide insight into the energetics[73]. Rather than including the dominant and largely fluctuating water-water and water-solvent molecular mechanical energies, explicit water is stripped and replaced by an implicit model (PB or GB). This simple method provides a quick and easy means to understand binding free energies, the relative stability between two different conformations of the same molecule, or the effect of mutation (alanine scanning)[74]. This analysis is equivalent to the ES/IS method of Hermans[75, 76], similar to the LIE approaches of Aqvist[77, 78], and a direct extension (that involves averaging the results over an ensemble of configurations or coordinate snapshots) of well-known methods for estimating free energy[79].

$$G = < E_{solute} - TS + G_{solvation} > \qquad (8)$$

As summarized in the equation above, the free energy is estimated as the average over a series of snapshots/configurations from the MD trajectory of the molecular mechanical energy of the solute (E_{solute}), an estimate of the entropy of the solute (TS) either from minimization and normal mode analysis or quasi-harmonic estimates from the covariance matrix, and the solvation free energy as calculated from an implicit solvent method such as Poisson-Boltzmann or generalized Born methods ($G_{solvation}$, potentially with the addition of a hydrophobic surface area term). All of these energy terms can be calculated within current versions of AMBER and this is done via a series of Perl language scripts supplied with the program. In application to nucleic acids, the methods have proven useful for estimating free energy differences between A- and B- DNA in water or water/ethanol solution[73], different sequences[80], different loop geometries[53, 81], and ligand interaction[82, 83], among other applications. The accuracy of the method is limited by the approximations. In general, this includes omitting all of the explicit solvent and ions. If a specific ion or water is important to mediate stability or a particular interaction, such molecules cannot be ignored as has been shown in studies of DNA minor groove binders[83] and also quadruplex DNA formation[84]. However, these molecules can be included as part of the solute. An additional limitation is the approximation of the entropy which can be noisy and difficult to converge in quasiharmonic approaches. This is most significant when attempting to calculate absolute entropies of binding and estimating the rotational and translation entropy loss upon binding. As discussed previously[84], estimates from both theory and experiment of the rotational and translational entropy loss upon binding of a drug-like molecule span a large range (from 3.0-30.0 kcal/mol). The standard harmonic approach to estimating the vibrational entropy (with the program nmode in AMBER) currently requires minimization in vacuo and this tends to significantly distort highly charged structures like nucleic acids. A promising advance is the inclusion of second derivatives for the generalized Born model as is currently supported in NAB.

2.   MOLECULAR DYNAMICS OF NUCLEIC ACIDS

As discussed in a recent review[30], this past decade may aptly be called the 10 ns era of nucleic acid MD simulation. An extensive set of simulation results for a wide variety of nucleic acid systems, on the 1-10 ns, have been published. We highlight some of these results and some unpublished work in what follows focusing on benchmarking the performance, reliability and accuracy of nucleic acid MD simulation.

After the initial successes in the 1994-1995 timeframe, applying Ewald methods to simulate nucleic acids in solution[9, 10, 85], a wide variety of simulations were initiated to assess the performance, complement experiment, and learn something new about sequence specific nucleic acid structure, dynamics, bending, backbone modification and unusual structures.

The initial studies investigated standard DNA and RNA duplexes and triplexes, including modified backbones, benchmarking the performance of the methods and investigating the results on different sequences, including various structures determined experimentally and poly-adenine tracts[12, 14, 17, 85-93]. These initial studies were able to show the differential structure and dynamics of DNA-RNA hybrids which adopted a mixed A/B-DNA structure compared to the flexible B-DNA and rigid A-RNA of the pure duplexes[12]. These studies also clearly demonstrated specific and differential ion association and hydration in the grooves depending on sequence and structure[12, 90]. A significant next step was to test whether the empirical force fields and the DNA in explicit solvent were sensitive to changes in the environment or surroundings. The obvious test was to investigate the A-B DNA transition which is caused by changes in relative humidity or water activity. Although some of the earlier force fields had shown unexpected spontaneous B-DNA to A-DNA transitions in explicit water[94], the Cornell et al. parm94.dat force field distributed with AMBER showed rapid A-DNA to B-DNA transitions on a nanosecond time scale[11] (and even faster transitions in implicit solvent[22]). Stabilizing the expected geometry was not sufficient to show the generality of the force field. To do this, it was necessary to demonstrate that A-DNA could be stabilized under conditions expected to stabilize A-DNA (such as in mixed water/ethanol, high salt or with tightly associated polyvalent ions) and ideally spontaneous B-DNA to A-DNA transitions could be observed. We were able to demonstrate stabilization of A-DNA in mixed water/ethanol[16, 95] and also spontaneous B-DNA to A-DNA transitions when hexaamminecobalt(III) ions were bound[13]. Although we never succeeded with the Cornell et al. force field to see spontaneous B-DNA to A-DNA transitions under high monovalent salt conditions or in mixed water/ethanol (which is not unreasonable given the expected 10 μs or greater time scales for conformational transition[96, 97]), we were encouraged by the ability of the simple empirical force fields (lacking any explicit polarization) to represent these subtle effects of the surroundings. In an attempt to overcome these limitations, Langley iteratively optimized (based on extensive MD simulation) the BMS nucleic acid force field to facilitate rapid A-B transitions in small DNA duplex structures under the appropriate conditions [35]. Although this force field does appear to facilitate the A-B transitions, overall the structures sampled appear more rigid and crystal like than the softer geometries sampled by the Cornell et al. force field[98] and the time scales for the B to A transition may be too rapid. Work on better understanding the A-B equilibrium continues to date with various force fields[99, 100] and minimal hydration models[101, 102].

A next step towards validating the methods was the study of unusual nucleic acid structures such as DNA quadruplexes (as discussed in more detail in the Chapter by Spackova and co-workers), i-DNA and zipper structures and investigation of sequence dependent structure (bending, twisting) and dynamics. In general, the methods have proven incredibly robust and well able to handle virtually any type of nucleic acid structure in MD simulation. Emerging limitations, such as the convergence to incorrect loop geometries in G-DNA tetraplex structures[103], relate to force field limitations and the limited sampling afforded by computational limits which only allow routine simulation in the 10-50 ns time scales.

In addition to correct representation of structure, the MD simulations have also allowed more detailed understanding of sequence and structure specific dynamics. In our early simulations of DNA-RNA hybrids, we showed that A-form geometries are relatively rigid compared to B-form duplexes, while mixed A/B geometries as seen in the hybrids have a flexibility between pure A- and B-form duplexes[12]. A more detailed analysis of sequence specific dynamics, demonstrated trends consistent with experiment, excepting that the Cornell et al. force field is slightly more flexible than expected[98, 104]. We have also demonstrated that the structural fluctuations also effect the electronic structure using DFT calculations to analyze snapshots extracted from MD simulation[105, 106]. The sequence specific structure and dynamics have also been investigated in a large scale collaborative effort (the ABC consortium) of a number of different research groups; two papers have resulted that highlight the performance of the methods[18, 107].

Although water and ions are integral to nucleic acid structure, and most of the studies discussed so far in this chapter included explicit ions and counterions, there has been considerable progress in the development of implicit solvent models (as mentioned) and they have proven worthwhile in nucleic acid simulation. The first molecular dynamics simulations carried out using the AMBER generalized Born (GB) codes were on duplex DNA[22], and our early experiences with this solvation model have been reviewed[21]. In general, the basic goal of a continuum solvent model, namely that the results should closely mimic those using explicit solvent models, are achieved, at least for stable/regular helical structures of DNA and RNA. As an additional example, we discuss here both unpublished GB and explicit solvent simulations of the B-DNA duplex d(GTGACTGACTGACTG)-d(CAGTCA GTCAGTCAC). The extended (up to 80 ns) explicit solvent simulations were carried out as a part of an extensive study of sequence-dependence effects in DNA duplexes, whose results are reported elsewhere [18, 107]. Since the reported results (on the 15 ns time scale) have already been extensively characterized, comparisons to 200+ ns simulations with the GB model can focus on differences between the implicit and explicit solvent models. The GB calculations described here used the *igb = 2* model in AMBER (originally applied to proteins)[108], setting a salt concentration to 0.1 M, and using a nonbonded cutoff of 20 A. Langevin dynamics were used, with a target temperature of 300K and a friction constant of 5 ps^{-1}. This friction constant is less than that expected for water, but is one that is appropriate for rapid exploration of configuration space[109]. Hence, one should not expect the time-dependence of the GB results to have physical significance, but the configurations explored should reflect those preferred by our force field.

As we have found in our earlier DNA simulations, the basic double-helical character remains nearly the same in explicit and implicit solvent simulations, at least for 1-10 nanoseconds of simulation, and that over the course of longer simulations the all-atom profiles of the RMSd versus time are similar. During the MD, the structures quickly move to a metastable set of structures on the range of 6-7 Å (measured over all 15 base pairs) from the starting structure. Figure 2-2 shows the time-dependence of the all-atom mass-weighted RMSd difference between the starting structure (which is an idealized B-form helix) and snapshots taken at ps time intervals during the simulation over the first 80 ns (with the entire GB simulation out to over 200 ns shown on the inset).

A comparison of the average all-atom RMSd over the 20-40 ns interval to the starting structure is 5.6+/–0.8 Å for the explicit water simulation and 6.7+/–0.7 Å for the GB simulation. Although these RMSd differences may appear alarmingly large, the main differences between the sampled configurations and the starting structures are primarily related to the slightly smaller twist at each basepair step in the simulation compared to an idealized form, a slight tendency to roll into the major groove (which shifts the RMSd closer towards an A-form like helix despite maintaining B-form sugar puckers and helicoidal parameters), some terminal group fraying in the GB simulations, and population of unexpected α, γ backbone sub-states; these preferences for lower twist (which arise from the use of the Cornell et al., AMBER 94 force field)[4] have been discussed in detail elsewhere[11, 32, 110, 111].

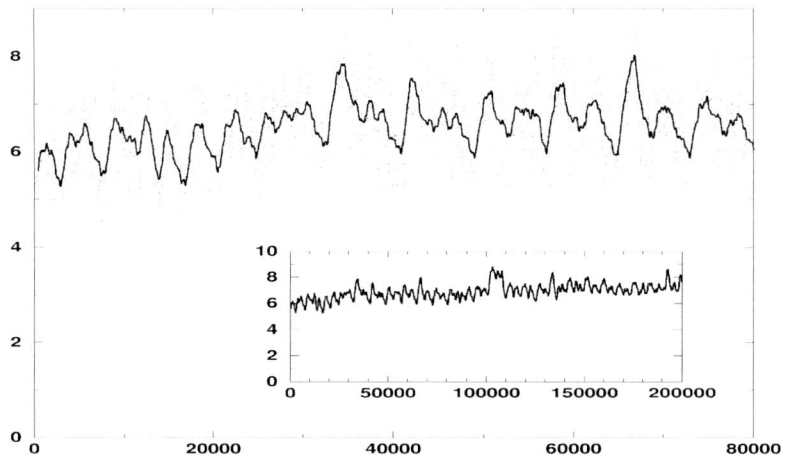

Figure 2-2. All-atom mass-weighted RMSd (Å) vs. time (ps) of the MD snapshots at ps intervals compared to the idealized canonical B-DNA starting model. The explicit water simulation is shown in gray and the GB simulation in black. In both plots, a running average has been applied over 1000 ps to smooth the data for easier visualization with the raw data at 10 ps intervals in the outer plot shown as dots.

If instead the RMSd is focused on the central 5 base pairs, the average all-atom RMSd over the first 40 ns for the explicit solvent simulation is 2.6+/–0.4 Å, and 2.9+/–0.4 Å for the GB simulations.

Additionally, the average structures for the two simulations (GB and explicit solvent) are rather close (1.78 Å, all-atom, for the 9-10 ns straight coordinate averaged structures) until later time frames in the GB simulation where (sometimes reversible) fraying of the helix ends is observed. This is shown in Figure 2-3 with snapshots taken after 13, 14 and 15 ns of the GB MD simulation. All of the structures in the first 13 ns resemble the one at the left, with all Watson-Crick hydrogen bonds intact and a stable helix.

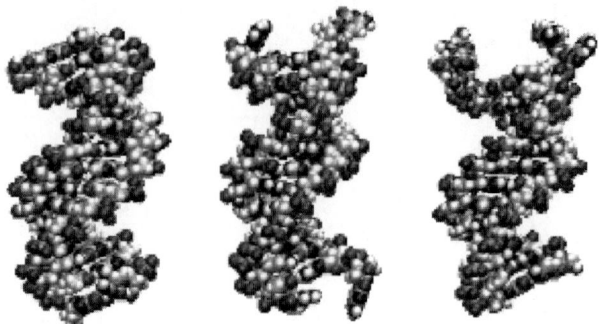

Figure 2-3. Snapshots of the simulated DNA duplex structure from the GB simulations at 13, 14 and 15 nanoseconds.

Starting at about 10 ns in this simulation, some fraying of the end groups becomes apparent, as shown in the middle part of the figure above. This is not unexpected, although the correct extent of end-fraying in 15-mer DNA duplexes is not known. In some cases, as at the bottom of the right-hand view in the figure, the Watson-Crick bonds re-form, but in other cases the end fraying is irreversible on the time scales investigated here and as the simulation is extended beyond 100 ns, end-group fraying is seen at both ends of the helix with the frayed bases tucking back into the grooves. Complicating the discussion is the existence of transitions about the backbone α and γ torsion angles, which undergo a "crankshaft" motion that appears to be irreversible on the time scales sampled here in both the implicit and explicit solvent simulations.

Because of these problems with the long-term behavior of the helix, and since the initial ABC simulations of this sequence were analyzed in detail on the 15 ns time scale, we have chosen here to use the middle portion of the helix for the first 15 ns to analyze internal fluctuations. One set of parameters of great interest to the interpretation of NMR data are order parameters for C-H bonds, which are computed from time-correlation functions[112]:

$$S^2 = \lim_{\tau \to \infty} < P_2(\cos \theta(\tau)) > \tag{9}$$

Here $\theta(\tau)$ is the angle between the direction of the C-H vector at time 0 and its direction at time τ, and P_2 is a second-order Legendre polynomial. The brackets indicate an average over all members of an equilibrium ensemble, and the time delay is long enough to include all internal motions; overall rotational tumbling is assumed to be an independent motion, and has been removed from the analysis. These order parameters are convenient measures of the extent of internal fluctuations that can be determined from heteronuclear NMR measurements, and also be readily determined from molecular dynamics simulations. They thus form a common ground for comparisons of MD simulations to NMR data[113].

Figure 2-4 shows some characteristic behavior for these sorts of time correlation functions, using the C8-H8 vector of two adenine bases for examples. At zero time, the angle θ is zero, and the correlation function must go to unity.

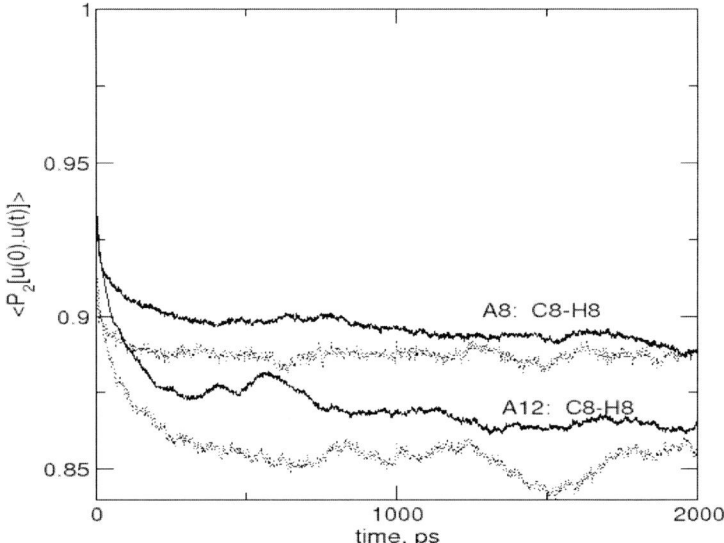

Figure 2-4. Time correlation functions for the C8-H8 inter-nuclear vector in adenine residue 8 (A8), and the corresponding vector in adenine residue 12 (A12). Solid lines show results from the explicit solvent simulations, dotted lines are from the GB results.

At very short times, not visible in this picture, the angle changes due to fast (sub-picosecond) vibrational motion; after this, there may be additional decays of the correlation function arising from larger scale conformational changes. In most cases, as seen here, there is an approximate plateau value that is usually achieved in the first 1-2 nanoseconds of motion. This plateau gives the value of S^2. The examples in Figure 2-4 are representative ones: values of the plateau region are clearly evident in an approximate fashion, but noise in the simulations limits the precision to which they can be determined. Nevertheless, it is clear that the values reached are smaller for A12 than for A8, and that explicit and implicit results are roughly the same. As expected, the rate at which these plateau values are reached is clearly longer for the explicit solvent simulation, although even in the generalized Born calculation it can take 0.5 ns for the fluctuations to build up.

More of an overall view of the amplitudes of these motions are shown in Figures 2-5 and 2-6, which plot the order parameters for bases and sugars for the explicit and implicit solvent models. The extent of base motion is quite similar in the two simulations, with high order parameters (less motion) in explicit solvent being very strongly predictive of a high order parameter in the GB simulation. The situation is a little more complex for the C1'-H1' internuclear vector in the sugars. The general trend in the two simulations is about the same, with more motion (as expected) near the ends of each strand, and less motion in the middle. But here the GB simulation is more smooth as a function of position within the helix, whereas the explicit solvent results show more structure, with purine sugars tending to be more mobile than the pyrimidines. Further studies will be required to see if this is a general trend, and to identify what differences in the solvent models might lead to this sort of behavior.

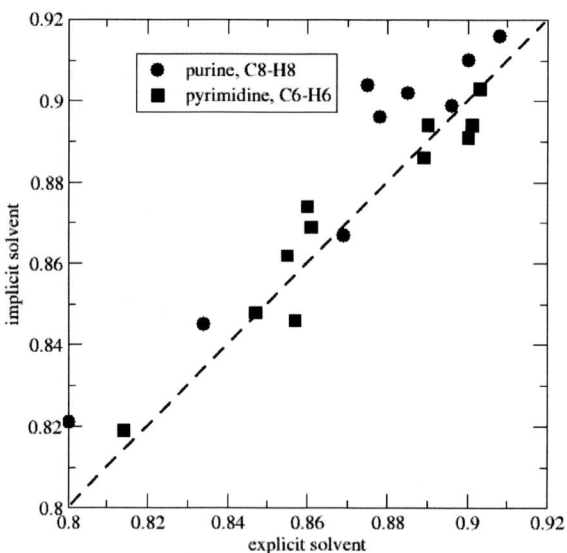

NMR order parameters

Figure 2-5. Comparison of order parameters computed for C-H vectors purine and pyrimidine bases, comparing results for the explicit and implicit solvent simulations.

Overall, generalized Born solvation models are still quite new, and development continues that have them be a more faithful description of explicit solvent results. The examples shown here illustrate that the overall behavior for nucleic acids is acceptable (so that these force fields are attractive models for X-ray or NMR refinement studies, for example[114, 115]) but that one must still treat details of the simulations as suspect.

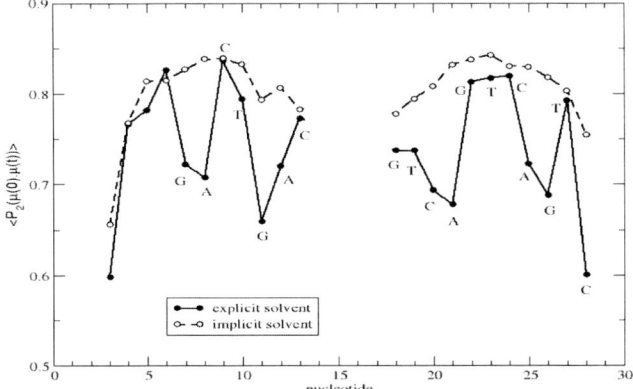

Figure 2-6. Order parameters for the C1'-H1' vector in sugar residues, comparing explicit solvent (solid lines) with generalized Born results (dashed lines).

3. CHALLENGES FOR THE FUTURE

A focusing and mystifying question that readily arises when studying nucleic acids is: *what is the correct answer?* This is a broad question that relates not only to the accuracy and reliability of the simulation results, but also to experiment and our incomplete understanding of nucleic acid structure, dynamics and interactions. Experiment is sometimes inconclusive and potentially influenced by artifacts (due to packing, the surroundings, end-effects, etc); moreover, the structures of nucleic acids are profoundly sensitive to their environment. This sensitivity provides challenge for the methods. Despite this caveat, in the 1-10 ns time scale, simulation has given tremendous insight into ligand interaction, unusual nucleic acid structures, backbone modification and general sequence specific structure and dynamics. However, as we push the simulation methods to longer size and time scales, key limitations become more readily apparent, including the need to obtain more data from experiment to better verify the simulation methods. Key questions whose answers will aid in the optimization of the nucleic acid force fields include:

- What is the rate of sugar repuckering?
- What is the rate (time scale) and nature of the A-B transition for a small DNA duplex in solution?
- What is the rate of B_I to B_{II} or α/γ crankshaft backbone transitions?
- What sub-states of the sugar ring and backbone should be populated and to what extent?
- Can additive models properly represent ionic effects (ionic strength, identify) and specific ion/solvent interaction?
- Do ions cause DNA to bend and how long do ions and solvent associate with nucleic acids?

The lack of complete understanding makes nucleic acids a fun and challenging system for study and justifies the application of theoretical methods in an attempt to better understand sequence dependent structure, dynamics and recognition. To date we have had clear successes. Examples include A–>B transitions, drug-nucleic acid interaction, and accurate representation of sequence specific dynamics and structural differences. On the other hand, the force fields all do not agree regarding repuckering rates or base pair opening time scales/energetics, or backbone sub-state populations. Moreover, as we push to longer time and size scales, hidden problems may emerge and sampling limitations may hinder understanding. An example is the unexpectedly high population of α,γ:g +,t states in DNA duplex simulation. These issues provide impetus to improve the simulation methods and force fields applied to nucleic acids.

ACKNOWLEDGEMENTS

We wish to thank all the AMBER developers over the years for their efforts and especially acknowledge the vision and dynamic of Peter Kollman. TEC would like to point out that he never would have continued on a life-long path of nucleic acid studies if it were not for the first few days with the PME method and A-B transitions that so excited Peter such that his enthusiasm did inspire and still infects me! An additional note of thanks goes to Tom Darden (NIEHS) for his tremendous efforts in getting particle mesh Ewald methods into AMBER and the ABC consortium of DNA simulators for working together to better understand applications of molecular dynamics methods to nucleic acids. TEC would like to acknowledge support from NSF CHE-0326027 and generous grants of computer time from the LRAC (large resources allocations committee) MCA01S027 at the Pittsburgh Super-computing Center, the National Center for Supercomputing Applications at the University of Illinois, and the San Diego Supercomputing Center, and also significant resources on the Arches metacluster at the Center for High Performance Computing at the University of Utah (NIH-1S10RR017214-01). DAC was partially supported by NIH grant GM-57513.

REFERENCES

1. Seibel, G.L., U.C. Singh, and P.A. Kollman (1985). A molecular dynamics simulation of double-helical B-DNA including counterions and water. *Proc. Nat. Acad. Sci.* **82**, 6537-6540.

2. Sanbonmatsu, K.Y. and S. Joseph (2003). Understanding discrimination by the ribosome: Stability testing and goove measurement of codon-anticodon pairs. *J. Mol. Biol.* **328**, 33-47.

3. Tung, C.S. and K.Y. Sanbonmatsu (2004). Atomic model of the Thermus thermophilus 70S ribosome developed in silico. *Biophys J.* **87**, 2714-2722.

4. Cornell, W.D., *et al.* (1995). A second generation force field for the simulation of proteins, nucleic acids, and organic molecules. *J. Amer. Chem. Soc.* **117**, 5179-5197.

5. Cheatham, T.E., III, P. Cieplak, and P.A. Kollman (1999). A modified version of the Cornell *et al.* force field with improved sugar pucker phases and helical repeat. *J. Biomol. Struct. Dyn.* **16**, 845-862.

6. Wang, J., P. Cieplak, and P.A. Kollman (2000). How well does a restrained electrostatic potential (RESP) model perform in calculating conformational energies of organic and biological molecules? *J. Comp. Chem.* **21**, 1049-1074.

7. Darden, T.A., D.M. York, and L.G. Pedersen (1993). Particle mesh Ewald - An N log(N) method for Ewald sums in large systems. *J. Chem. Phys.* **98**, 10089-10092.

8. Essmann, U., L. Perera, M.L. Berkowitz, T. Darden, H. Lee, and L.G. Pedersen (1995). A Smooth Particle Mesh Ewald Method. *J. Chem. Phys.* **103**, 8577-8593.

9. Cheatham, T.E., III, J.L. Miller, T. Fox, T.A. Darden, and P.A. Kollman (1995). Molecular dynamics simulations on solvated biomolecular systems - The particle mesh Ewald method leads to stable trajectories of DNA, RNA and proteins. *J. Amer. Chem. Soc.* **117**, 4193-4194.

10. York, D.M., W. Yang, H. Lee, T.A. Darden, and L. Pedersen (1995). Toward accurate modeling of DNA: the importance of long-range electrostatics. *J. Amer. Chem. Soc.* **117**, 5001-5002.

11. Cheatham, T.E., III and P.A. Kollman (1996). Observation of the A-DNA to B-DNA transition during unrestrained molecular dynamics in aqueous solution. *J. Mol. Biol.* **259**, 434-444.

12. Cheatham, T.E., III. and P.A. Kollman (1997). Molecular dynamics simulations highlight the structural differences in DNA:DNA, RNA:RNA and DNA:RNA hybrid duplexes. *J. Amer. Chem. Soc.* **119**, 4805-4825.

13. Cheatham, T.E., III and P.A. Kollman (1997). Insight into the stabilization of A-DNA by specific ion association: Spontaneous B-DNA to A-DNA transitions observed in molecular dynamics simulations of d[ACCCGCGGGT]2 in the presence of hexaammine cobalt(III). *Structure* **5**, 1297-1311.

14. Young, M.A., G. Ravishanker, and D.L. Beveridge (1997). A 5-nanosecond molecular dynamics trajectory for B-DNA: Analysis of structure, motions and solvation. *Biophys. J.* **73**, 2313-2336.

15. Cheatham, T.E., III and P.A. Kollman (1998). *Molecular dynamics simulation of nucleic acids in solution: How sensitive are the results to small perturbations in the force field and environment*, in *Structure, motion, interactions and expression of biological macromolecules*, M. Sarma and R. Sarma, Editors, Adenine Press: Schenectady, NY. p. 99-116.

16. Sprous, D., M.A. Young, and D.L. Beveridge (1998). Molecular dynamics studies of the conformational preferences of a DNA double helix in water and an ethanol/water mixture: Theoretical considerations of the A->B transition. *J. Phys. Chem. B.* **102**, 4658-4667.

17. Young, M.A. and D.L. Beveridge (1998). Molecular dynamics simulations of an oligonucleotide duplex with adenine-tracts phased by a full helix turn. *J. Mol. Biol.* **281**, 675-687.

18. Beveridge, D.L., *et al.* (2004). Molecular dynamics simulations of the 136 unique tetranucleotide sequences of DNA oligonucleotides. I. Research design and results on d(CpG) steps. *Biophys. J.* **87**, 3799-3813.

19. Still, W.C., A. Tempczyk, R.C. Hawley, and T. Hendrickson (1990). Semi analytical treatment of solvation for molecular mechanics and dynamics. *J. Amer. Chem. Soc.* **112**, 6127-6128.

20. Hawkins, G.D., C.J. Cramer, and D.G. Truhlar (1995). Pairwise solute descreening of solute charges from a dielectric medium. *Chem. Phys. Lett.* **246**, 122-129.

21. Tsui, V. and D.A. Case (2001). Theory and applications of the generalized Born solvation model in macromolecular simulations. *Biopol.* **56**, 275-291.

22. Tsui, V. and D.A. Case (2000). Molecular dynamics simulations of nucleic acids with a generalized Born solvation model. *J. Amer. Chem. Soc.* **122**, 2489-2498.

23. Brooks, B.R., R.E. Bruccoleri, B.D. Olafson, D. States, J,S. Swaminathan, and M. Karplus (1983). CHARMm: A program for macromolecular energy, minimization, and dynamics calculations. *J. Comp. Chem.* **4**, 187-217.

24. Pearlman, D.A., D.A. Case, J.W. Caldwell, W.S. Ross, T.E. Cheatham, S. Debolt, D. Ferguson, G. Seibel, and P. Kollman (1995). AMBER, a package of computer programs for applying molecular mechanics, normal mode analysis, molecular dynamics and free energy calculations to simulate the structure and energetic properties of molecules. *Comp. Phys. Comm.* **91**, 1-41.

25. Case, D.A., *et al.* (2005). The AMBER biomolecular simulation programs. *J. Comp. Chem.* **26**, 1668-1688.

26. Beveridge, D.L. and K.J. McConnell (2000). Nucleic acids: theory and computer simulation, Y2K. *Cur. Op. Struct. Biol.* **10**, 182-196.

27. Cheatham, T.E., III and P.A. Kollman (2000). Molecular dynamics simulation of nucleic acids. *Ann. Rev. Phys. Chem.* **51**, 435-471.

28. Orozco, M., A. Perez, A. Noy, and F. Javier Luque (2003). Theoretical methods for the simulation of nucleic acids. *Chem. Soc. Rev.* **32**, 350-364.

29. Cheatham, T.E., III (2004). Simulation and modeling of nucleic acid structure, dynamics and interactions. *Cur. Opin. Struct. Biol.* **14**, 360-367.

30. Cheatham, T.E., III (2005). Molecular modeling and atomistic simulation of nucleic acids. *Ann. Reports Comp. Chem.* **1**, 75-89.

31. Norberg, J. and L. Nilsson (2002). Molecular dynamics applied to nucleic acids. *Acc. Chem. Res.* **35**, 465-472.

32. Cheatham, T.E., III and M.A. Young (2001). Molecular dynamics simulations of nucleic acids: Successes, limitations and promise. *Biopolymers* **56**, 232-256.

33. MacKerell, A.D., Jr. (2004). Empirical force fields for biological macromolecules: Overview and issues. *J. Comp. Chem.* **25**, 1584-1604.

34. Reddy, S.Y., F. Leclerc, and M. Karplus (2003). DNA polymorphism: A comparison of force fields for nucleic acids. *Biophys. J.* **84**, 1421-1449.

35. Langley, D.R. (1998). Molecular dynamics simulations of environment and sequence dependent DNA conformation: The development of the BMS nucleic acid force field and comparison with experimental results. *J. Biomol. Struct. Dyn.* **16**, 487-509.

36. Soares, T.A., P.H. Hunenberger, M.A. Kastenholz, V. Krautler, T. Lenz, R.D. Lins, C. Oostenbrink, and W.F. van Gunsteren (2005). An improved nucleic acid parameter set for the GROMOS force field. *J. Comp. Chem.* **26**, 727-737.

37. Cheatham, T.E., III, B.R. Brooks, and P.A. Kollman (2000). *Molecular modeling of nucleic acid structure*, in *Current Protocols in Nucleic Acid Chemistry*, S.L. Beaucage, et al., Editors, Wiley: New York. p. 7.5.1-7.5.12.

38. Wang, J., R.M. Wolf, J.W. Caldwell, P.A. Kollman, and D.A. Case (2004). Development and testing of a general amber force field. *J. Comp. Chem.* **25**, 1157-1174.

39. Jakalian, A., B.L. Bush, D.B. Jack, and C.I. Bayly (2000). Fast, efficient generation of high quality atomic charges. AM1-BCC model: I. method. *J. Comp. Chem.* **21**, 132-146.

40. Jakalian, A., D.B. Jack, and C.I. Bayly (2002). Fast, efficient generation of high-quality atomic charges. AM1-BCC model: II. Parameterization and validation. *J Comput Chem* **23**, 1623-41.

41. Macke, T.J. and D.A. Case, *NAB (Nucleic Acid Builder)*. 1996, The Scripps Research Institute: La Jolla, CA.

42. Macke, T. and D.A. Case (1998). *Modeling unusual nucleic acid structures*, in *Molecular modeling of nucleic acids*, N.B. Leontis and J. Santa Lucia, Editors, ACS: Washington. p. 379-393.

43. Lu, X.J. and W.K. Olson (2003). 3DNA: a software package for the analysis, rebuilding and visualization of three-dimensional nucleic acid structures. *Nucleic Acids Res* **31**, 5108-21.

44. Mueller, F. and R. Brimacombe (1997). A new model for the three-dimensional folding of Escherichia coli 16 S ribosomal RNA. I. Fitting the RNA to a 3D electron microscopic map at 20 A. *J. Mol. Biol.* **271**, 524-44.

45. Burks, J., C. Zwieb, F. Muller, I. Wower, and J. Wower (2005). Comparative 3-D Modeling of tmRNA. *BMC Mol. Biol.* **6**, 14.

46. Massire, C. and E. Westhof (1998). MANIP: an interactive tool for modelling RNA. *J. Mol. Graph. Model.* **16**, 197-205, 255-7.

47. Major, F., D. Gautheret, and R. Cedergren (1993). Reproducing the three-dimensional structure of a tRNA molecule from structural constraints. *Proc. Natl. Acad. Sci.* **90**, 9408-12.

48. Tung, C.S. and E.S. Carter, 2nd (1994). Nucleic acid modeling tool (NAMOT): an interactive graphic tool for modeling nucleic acid structures. *Comput. Appl. Biosci.* **10**, 427-33.

49. Diaz, N., D. Suarez, T.L. Sordo, and K.M. Merz, Jr. (2001). A theoretical study of the aminolysis reaction of lysine 199 of human serum albumin with benzylpenicillin: consequences for immunochemistry of penicillins. *J. Am. Chem. Soc.* **123**, 7574-83.

50. Park, H., E.N. Brothers, and K.M. Merz, Jr. (2005). Hybrid QM/MM and DFT investigations of the catalytic mechanism and inhibition of the dinuclear zinc metallo-b-lactamase CcrA from Bacteroides fragilis. *J. Am. Chem. Soc.* **127**, 4232-41.

51. Wu, X. and B.R. Brooks (2005). Isotropic periodic sum: a method for the calculation of long-range interactions. *J. Chem. Phys.* **122**, 44107.

52. Roitberg, A. and R. Elber (1991). Modeling side chains in peptides and proteins: Application of the locally enhanced sampling and the simulated annealing methods to find minimum energy conformations. *J. Chem. Phys.* **95**, 9277-9286.

53. Simmerling, C., J.L. Miller, and P.A. Kollman (1998). Combined locally enhanced sampling and particle mesh Ewald as a strategy to locate the experimental structure of a non-helical nucleic acid. *J. Amer. Chem. Soc.* **120**, 7149-7155.

54. Feig, M., J. Karanicolas, and C.L. Brooks, 3rd (2004). MMTSB Tool Set: enhanced sampling and multiscale modeling methods for applications in structural biology. *J. Mol. Graph Model* **22**, 377-95.

55. Sagui, C. and T.A. Darden (1999). Molecular dynamics simulations of biomolecules: Long-range electrostatic effects. *Ann. Rev. Biophys. Biomol. Struct.* **28**, 155-179.

56. Cramer, C.J. and D.G. Truhlar (1999). Implicit Solvation Models: Equilibria, Structure, Spectra, and Dynamics. *Chem Rev* **99**, 2161-2200.

57. Bashford, D. and D.A. Case (2000). Generalized Born models of molecular solvation effects. *Ann. Rev. Phys. Chem.* **51**, 129-152.

58. Feig, M. and C.L.I. Brooks (2004). Recent advances in the development and application of implicit solvent models in biomolecule simulations. *Curr. Opin. Struct. Biol.* **14**, 217-224.

59. Simonson, T. (2003). Electrostatics and dynamics of proteins. *Rep. Prog. Phys.* **66**, 737-787.

60. Baker, N.A. (2004). Poisson-Boltzmann methods for biomolecular electrostatics. *Meth. Enzym.* **383**, 94-118.

61. Feig, M., W. Im, and C.L. Brooks, 3rd (2004). Implicit solvation based on generalized Born theory in different dielectric environments. *J. Chem. Phys.* **120**, 903-11.

62. Srinivasan, J., M.W. Trevathan, P. Beroza, and D.A. Case (1999). Application of a pairwise generalized Born model to proteins and nucleic acids: inclusion of salt effects. *Theor. Chem. Acc.* **101**, 426-434.

63. Jayaram, B., Y. Liu, and D.L. Beveridge (1998). A modification of the generalized Born theory for improved estimates of solvation energies and pK shifts. *J. Chem. Phys.* **109**, 1495-1471.

64. Onufriev, A., D.A. Case, and D. Bashford (2002). Effective Born radii in the generalized Born approximation: The importance of being perfect. *J. Comp. Chem.* **23**, 1297-1304.

65. Grycuk, T. (2003). Deficiency of the Coulomb-field approximation in the generalized Born model: An improved formula for Born radii evaluation. *J. Comp. Chem.* **119**, 4817-4826.

66. Sigalov, G., P. Scheffel, and A. Onufriev (2005). Incorporating variable dielectric environments into the generalized Born model. *J. Chem. Phys.* **122**, 094511.

67. Onufriev, A., D. Bashford, and D.A. Case (2000). Modification of the generalized Born model suitable for macromolecules. *J. Phys. Chem. B.* **104**, 3712-3720.

68. Bondi, A. (1964). van der Waals volumes and radii. *J. Phys. Chem.* **68**, 441-451.

69. Hawkins, G.D., C.J. Cramer, and D.G. Truhlar (1997). Parameterized model for aqueous free energies of solvation using geometry-dependent atomic surface tensions with implicit electrostatics. *J. Phys. Chem. B.* **101**, 7147-7157.

70. Hawkins, G.D., C.J. Cramer, and D.G. Truhlar (1996). Parameterized models of aqueous free energies of solvation based on pairwise descreening of solute atomic charges from a dielectric medium. *J. Phys. Chem.* **100**, 19824-19839.

71. Schaefer, M. and C. Froemmel (1990). A precise analytical method for calculating the electrostatic energy of macromolecules in aqueous solution. *J Mol Biol* **216**, 1045-66.

72. Tanner, S.W., N. Thompson, and T.E. Cheatham, III (2005). Cluster analysis of molecular dynamics trajectories: Insight into nucleic acid structure, dynamics and interactions. *[submitted]*.

73. Srinivasan, J., T.E. Cheatham, III, P. Cieplak, P.A. Kollman, and D.A. Case (1998). Continuum solvent studies of the stability of DNA, RNA and phosphoramidate helices. *J. Amer. Chem. Soc.* **120**, 9401-9409.

74. Kollman, P.A., *et al.* (2000). Calculating structures and free energies of complex molecules: Combining molecular mechanics and continuum models. *Acc. Chem. Res.* **33**, 889-897.

75. Vorobjev, Y.N., J.C. Almagro, and J. Hermans (1998). Discrimination between native and intentionally misfolded conformations of proteins: ES/IS, a new method for calculating conformational free energy that uses both dynamics simulations with an explicit solvent and an implicit solvent continuum model. *Proteins.* **32**, 399-413.

76. Vorobjev, Y.N. and J. Hermans (1999). ES/IS: Estimation of conformational free energy by combining dynamics simulations with explicit solvent with an implicit solvent continuum model. *Biophys. Chem.* **78**, 195-205.

77. Froloff, N., A. Windemuth, and B. Honig (1997). On the calculation of binding free energies using continuum methods: Application to MHC class I protein-peptide interactions. *Prot. Sci.* **6**, 1293-1301.

78. Hansson, T., J. Marelius, and J. Aqvist (1998). Ligand binding affinity prediction by linear interaction energy methods. *J. Comp. Aided Mol. Des.* **12**, 27-35.

79. Misra, V.K., K.A. Sharp, R.A. Friedman, and B. Honig (1994). Salt effects on ligand-DNA binding. Minor groove binding antibiotics. *J. Mol. Biol.* **238**, 245-263.

80. Cheatham, T.E., III, J. Srinivasan, D.A. Case, and P.A. Kollman (1998). Molecular dynamics and continuum solvent studies of the stability of polyG-polyC and polyA-polyT DNA duplexes in solution. *J. Biomol. Struct. Dyn.* **16**, 265-280.

81. Srinivasan, J., J.L. Miller, P.A. Kollman, and D.A. Case (1998). Continuum solvent studies of the stanbility of RNA hairpin loops and helices. *J. Biomol. Struct. Dyn.* **16**, 671-682.

82. Gouda, H., I.D. Kuntz, D.A. Case, and P.A. Kollman (2003). Free energy calculations for theophylline binding to an RNA aptamer: Comparison of MM-PBSA and thermodynamic integration methods. *Biopol.* **68**, 16-34.

83. Spackova, N., T.E. Cheatham, III, F. Ryjacek, F. Lankaš, L. van Meervelt, P. Hobza, and J. Šponer (2003). Molecular dynamics simulations and thermodynamic analysis of DNA-drug complexes. Minor groove binding between 4',6-diamidino-2-phenylindole (DAPI) and DNA duplexes in solution. *J. Amer. Chem. Soc.* **125**, 1759-1769.

84. Stefl, R., T.E. Cheatham, III, N. Spackova, E. Fadrna, I. Berger, J. Koca, and J. Šponer (2003). Formation pathways of a guanine-quadruplex DNA revealed by molecular dynamics and thermodynamical analysis of the substates. *Biophys. J.* **85**, 1787-1804.

85. Cheng, Y.-K. and B.M. Pettitt (1995). Solvent effects on model d(CG-G)₇ and d(TA-T)₇ DNA triplex helices. *Biopol.* **35**, 457-473.

86. Young, M.A., J. Srinivasan, I. Goljer, S. Kumar, D.L. Beveridge, and P.H. Bolton (1995). Structure determination and analysis of local bending in an A-tract DNA duplex: Comparison of results from crystallography, nuclear magnetic resonance, and molecular dynamics simulation on d(CGCAAAAATGCG). *Methods in Enzymology* **261**, 121-144.

87. Norberg, J. and L. Nilsson (1996). Constant pressure molecular dynamics simulations of the dodecamers: d(GCGCGCGCGCGC)2 and r(GCGCGCGCGCGC)2. *J. Chem. Phys.* **104**, 6052-6057.

88. Shields, G.C., C.A. Laughton, and M. Orozco (1997). Molecular dynamics simulations of the d(TAT) triplex helix. *J. Amer. Chem. Soc.* **119**, 7463-7469.

89. Flatters, D., M. Young, D.L. Beveridge, and R. Lavery (1997). Conformational properties of the TATA-box binding sequence of DNA. *J. Biomol. Struct. Dyn.* **14**, 757-765.

90. Young, M.A., B. Jayaram, and D.L. Beveridge (1997). Intrusion of counterions into the spine of hydration in the minor groove of B-DNA: Fractional occupancy of electronegative pockets. *J. Amer. Chem. Soc.* **119**, 59-69.

91. Spector, T.I., T.E. Cheatham, III, and P.A. Kollman (1997). Unrestrained molecular dynamics of photodamage DNA in aqueous solution. *J. Amer. Chem. Soc.* **119**, 7095-7104.

92. Luo, J. and T.C. Bruice (1998). Nanosecond molecular dynamics of hybrid triplex and duplex of polycation deoxyribonucleic guanidine strands with a complementary DNA strand. *J. Amer. Chem. Soc.* **120**, 1115-1123.

93. Shields, G.C., C.A. Laughton, and M. Orozco (1998). Molecular dynamics simulation of a PNA-DNA-PNA triple helix in aqueous solution. *J. Amer. Chem. Soc.* **120**, 5895-5904.

94. Yang, L.Q. and B.M. Pettitt (1996). B to A transition of DNA on the nanosecond time scale. *J. Phys. Chem.* **100**, 2564-2566.

95. Cheatham, T.E., III, M.F. Crowley, T. Fox, and P.A. Kollman (1997). A molecular level picture of the stabilization of A-DNA in mixed ethanol-water solutions. *Proc. Natl. Acad. Sci.* **94**, 9626-9630.

96. Jose, D. and D. Porschke (2005). The Dynamics of the B-A transition of natural DNA double helices. *J. Am. Chem. Soc.* **127**, 16120-16128.

97. Jose, D. and D. Porschke (2004). Dynamics of the B-A transition of DNA double helices. *Nuc. Acids Res.* **32**, 2251-8.

98. Lankaš, F., J. Šponer, J. Langowski, and T.E. Cheatham, III (2003). DNA base-pair step deformability inferred from molecular dynamics simulation. *Biophys. J.* **85**, 2872-2883.

99. Pastor, N. (2005). The B- to A-DNA transition and the reorganization of solvent at the DNA surface. *Biophys. J.* **88**, 3262-75.

100. Banavali, N.K. and B. Roux (2005). Free energy landscape of A-DNA to B-DNA conversion in aqueous solution. *J. Am. Chem. Soc.* **127**, 6866-76.

101. Mazur, A.K. (2002). DNA dynamics in a water drop without counterions. *J. Am. Chem. Soc* **124**, 14707-15.

102. Mazur, A.K. (2003). Titration *in silico* of reversible B<->A transitions in DNA. *J. Amer. Chem. Soc.* **125**, 7849-7859.

103. Fadrna, E., N. Spackova, R. Stefl, J. Koca, T.E. Cheatham, III, and J. Šponer (2004). Molecular dynamics simulations of guanine quadruplex loops: Advances and force field limitations. *Biophys. J.* **87**, 227-242.

104. Lankaš, F., J. Šponer, J. Langowski, and T.E.I. Cheatham (2004). DNA deformability at the base pair level. *J. Amer. Chem. Soc.* **126**, 4124-4125.

105. Lewis, J.P., J. Pikus, T.E. Cheatham, III, E.B. Starikov, H. Wang, J. Tomfohr, and O.F. Sankey (2002). A comparison of electronic states in periodic and aperiodic poly(dA)-poly(dT) DNA. *Phys. Stat. Sol. (b)* **233**, 90-100.

106. Lewis, J.P., T.E. Cheatham, III, H. Wang, E.B. Starikov, and O.F. Sankey (2003). Dynamically amorphous character of electronic states in poly(dA)-poly(dT) DNA. *J. Phys. Chem. B* **107**, 2581-2587.

107. Dixit, S.B., *et al.* (2005). Molecular dynamics simulations of the 136 unique tetranucleotide sequences of DNA oligonucleotides. II. Sequence context effects on the dynamical structures of the 10 unique dinucleotide steps. *Biophys J*, [in press].

108. Onufriev, A., D. Bashford, and D.A. Case (2004). Exploring protein native states and large-scale conformational changes with a modified generalized born model. *Proteins* **55**, 383-94.

109. Loncharich, R.J., B.R. Brooks, and R.W. Pastor (1992). Langevin dynamics of peptides: the frictional dependence of isomerization rates of N-acetylalanyl-N'-methylamide. *Biopolymers* **32**, 523-35.

110. Feig, M. and B.M. Pettitt (1997). Experiment vs force fields: DNA conformation from molecular dynamics simulations. *J. Phys. Chem.* **101**, 7361-7363.

111. Real, A.N. and R.J. Greenall (2000). B->A->B transitions in a molecular dynamics trajectory of low salt DNA solution. *J. Mol. Model.* **6**, 654-658.

112. Lipari, G. and A. Szabo (1982). Model-free approach to the interpretation of nuclear magnetic resonance relaxation in macromolecules. 1. Theory and range of validity. *J. Am. Chem. Soc.* **104**, 4546-4559.

113. Case, D.A. (2002). Molecular dynamics and NMR spin relaxation in proteins. *Acc. Chem. Res.* **35**, 325-31.

114. Xia, B., V. Tsui, D.A. Case, H.J. Dyson, and P.E. Wright (2002). Comparison of protein solution structures refined by molecular dynamics simulation in vacuum, with a generalized Born model, and with explicit water. *J. Biomol. NMR* **22**, 317-31.

115. Moulinier, L., D.A. Case, and T. Simonson (2003). Reintroducing electrostatics into protein X-ray structure refinement: bulk solvent treated as a dielectric continuum. *Acta Crystallogr D. Biol Crystallogr* **59**, 2094-103.

Chapter 3

THEORETICAL STUDIES OF NUCLEIC ACIDS AND NUCLEIC ACID-PROTEIN COMPLEXES USING CHARMM

Alexander D. MacKerell, Jr.[1] and Lennart Nilsson[2]

1) Department of Pharmaceutical Sciences, School of Pharmacy, University of Maryland, Baltimore, MD 21201, USA, 2) Department of Biosciences at NOVUM, Karolinska Institutet, S-141 57 Huddinge, Sweden

Address correspondence to: Alexander D. MacKerell, Jr., 20 Penn Street, Baltimore, MD 21201, Tel. 706 410-7442; Fax. 410 706-5017; E-mail: alex@outerbanks.umaryland.edu

Abstract: Empirical force field calculations of nucleic acids have become a standard method for studying the relationship of structure and dynamics to the activity of these biologically essential molecules. One of the major tools used to perform such calculations is the program CHARMM along with the CHARMM all-atom force fields designed for biomolecules, including nucleic acids. In this chapter the utility of CHARMM and the associated empirical force field as a modeling tool to study nucleic acids is presented. In addition, an overview of studies performed using these utilities is presented.

Key words: DNA, RNA, CHARMM, empirical force field, proteins

1. INTRODUCTION

Understanding the relationship of the structure and dynamics of nucleic acids with their wide variety of biological functions represents a major thrust in the field of biophysics.[1] Experimentally, information on the structures of nucleic acids, equilibria between different conformations, and the transitions rates between those conformers can be measured and related to biological

J. Šponer and F. Lankaš (eds.), Computational Studies of RNA and DNA, 73–94.

activity. Such studies can target conformational transitions of nucleic acids alone as well as those induced upon their binding to proteins. Supplementing the data from experimental methods are theoretical studies. Theoretical methods applied to nucleic acids include those based on empirical force fields,[2] which are often applied in the context of molecular dynamics (MD) simulations from which atomic details and free energies associated with conformational changes may be obtained. In addition, quantum mechanical (QM) methods are useful to understand the relationship of structure to energy for small molecules representative of nucleic acids (e.g. nucleosides and nucleotides) as well as for probing chemical reaction mechanisms, often in the context of quantum mechanical/molecular mechanical (QM/MM) approaches.

An obvious requirement for applying theoretical methods to nucleic acids is the proper tools to perform the calculations. These tools can be put into two classes: 1) the software packages required to perform the modeling study and 2) the level of theory or theoretical model being applied to the system. With respect to software packages, a number of programs are available for studying nucleic acids. In the context of empirical force field calculations are the widely used AMBER[3] and CHARMM[4] packages as well as other programs including TINKER,[5] GROMOS,[6] GROMACS[7] and DISCOVER.[8] Software packages for QM studies of nucleic acids include the widely used GAUSSIAN package,[9] GAMESS,[10] NWCHEM,[11] QCHEM,[12] and JAGUAR,[13] among others. Concerning the level of theory, empirical force fields models most commonly used are the CHARMM[14-17] and AMBER[18,19] force fields, followed by the force field from Bristol-Myers Squibb (BMS)[20] and, recently, a force field associated with the program GROMOS, [21] although the latter has not seen significant use as of the writing of this chapter. QM studies of nucleic acid related compounds are typically limited to 6-31G* and similar size basis sets, with recent increases in computing power allowing for calculations, including energy optimizations, that apply explicit treatment of electron correlation via either DFT or MP2 methods.[22] Finally, QM/MM methods typically employ either the AMBER or CHARMM force fields in combination with semiempirical QM methods, though examples exist where *ab initio* approaches have been used.[23]

Clearly, there are a variety of computational tools and approaches available to study nucleic acids. In this chapter we will focus on the use and application of the program CHARMM[4,24] for calculations involving nucleic acids. Included will be an overview of the tools available in CHARMM for nucleic acid calculations as well as a discussion of the CHARMM force field for nucleic acids. This will be followed by an overview of computational studies of nucleic acids using CHARMM, including nucleic acid-protein interactions. The reader is referred to the remainder of this book as well as a recent review of empirical force fields[2] for additional information on the different force fields in use.

2. CHARMM AS A TOOL FOR MODELING STUDIES OF NUCLEIC ACIDS

CHARMM (Chemistry at Harvard for Macromolecular Modeling) is a general package for modeling, minimization and simulation studies of molecules, most often being applied to biological macromolecules. Typically, CHARMM based calculations will be initiated from experimental structures, often obtained from the nucleic acid databank (NDB)[25] or protein databank (PDB)[26]. Alternatively, structures can be built in CHARMM using the internal coordinate (IC) utilities. For example, the CHARMM27 nucleic acid force field will allow for a single DNA strand to be built; however, the program is not designed for creation of a wide variety of nucleic acid structures, though other packages are available to perform this function.[27,28] Once the Cartesian coordinates for a structure are obtained, CHARMM may be used for energy minimization and MD simulation studies, similar to the majority of modeling packages. Such calculations may be performed in the presence of a wide variety of constraints and restraints, including restraints that allow for systematic sampling of conformational space from which free energy profiles may be obtained via potential of mean force (PMF) calculations (see below). Alternatively, free energy changes associated with chemical perturbations (e.g. conversion between base types) may be obtained via a variety of free energy perturbation methods[29] in the program. These include the PERT, BLOCK and TSM modules. An important aspect of the MD capabilities in CHARMM is the availability of a variety of integrators. These include methods that give formally correct ensembles that allow for direct comparison with experimentally derived thermodynamic properties.[30] A publication on CHARMM, including an extensive overview of new utilities in the program, is expected to be published in 2006 (M. Karplus and coworkers, personal communication).

Scripts are typically used to drive CHARMM to create a DNA duplex (or any other oligonucleotide) and then perform energy minimization, dynamics or other functions, as required. CHARMM initially requires reading of the topology and parameter files that contain information on the nucleic acid atom types, chemical connectivity, atomic charges and empirical force field parameters. This is followed by input of the sequence of one strand of the DNA, which may also be read from a pdb or ndb file, followed by generation of the strand (i.e. creation of the representation of the molecule in the program). This procedure is then repeated for the second strand following which, in the case of DNA, the 2' hydroxyls are replaced by hydrogens to create the deoxy state. Generation of the molecule is then followed by reading the Cartesian coordinates for the molecule, often obtained from the Protein or Nucleic Acid Databanks. If necessary, coordinates for hydrogens or other missing atoms can be added to the molecule and, as is generally done for nucleic acid simulations, the system immersed in a box or sphere of water by applying periodic or stochastic boundary conditions. The system is then subjected to energy minimization to relax bad contacts in the DNA or RNA and between the oligonucleotide and the added solvent. Once minimized, a molecular dynamics simulation may then be performed allowing a detailed investigation of the structural dynamics of the oligonucleotide as a function of time. It should be noted that conformational changes in oligonucleotides may also be performed via normal mode analysis[31] or with one of the many implicit solvent molecules available to avoid the computational expense of simulating explicit waters.[32] However, it should be noted that implicit solvent models have not been rigorously tested in MD simulations with polyanionic molecules such as oligonucleotides, although their use in combined calculations where free energies are evaluated for nucleic acids on structures obtained from explicit solvent MD simulations seems reasonable.[33-35]

Importantly, CHARMM contains a variety of utilities to manipulate the system (e.g., imposing restraints) and to analyze different static and time-dependent properties of the system (e.g., torsion angles and other geometric characteristics, RMSD comparisons, radial distribution functions, hydrogen bonding patterns, interaction energies, correlation functions). Several stand alone programs for analysis, including PTRAJ (Cheatham, T.E., III, personal communication) and a version of FREEHELIX[36] modified to read CHARMM format trajectory files, and for visualization (VMD[37]) is also available. Scripts for solvation of oligonucleotides, generation of protein-nucleic acid complexes and for different types of analysis may be obtained from the CHARMM forum at www.charmm.org. In addition, questions concerning the use of the CHARMM may be submitted to the forum and, importantly, the forum may be searched for answers to previously discussed questions.

While the CHARMM force fields will not be discussed in detail in this chapter, a recent addition to the energy function is worth noting. This addition involves a two-dimensional dihedral surface energy correction map (CMAP) that allows for nearly exact reproduction of any target energy surface, typically a potential energy surface obtained for QM calculations.[38-40] This addition has been shown to significantly improve the treatment of the peptide backbone in the CHARMM protein force field (i.e. the ϕ, ψ map or Ramachandran diagram) allowing for better reproduction of experimental data with respect to the ϕ, ψ dihedral angles as well as order parameters from NMR experiments. This implementation is anticipated to greatly improve the accuracy of protein-nucleic acid simulations. Furthermore, the technology may be applied to nucleic acids to, for example, accurately treat the α and ζ dihedrals in the phosphodiester backbone.

3. CHARMM NUCLEIC ACID FORCE FIELDS

Force fields for nucleic acids included with the CHARMM package have undergone several rounds of evolution. The earliest model was based on an ad hoc collection of parameters from proteins and small molecules. Its use was limited to gas phase normal mode calculations.[41] Subsequently, a more comprehensive parameter development effort was undertaken, leading to the development of what may be considered the first official CHARMM nucleic acid force field.[17] This version, termed EF2, included both extended atom and all-atom models, used an explicit hydrogen bonding term and was able to be used in vacuum calculations[42,43] together with "hydrated counterions"[44], or with experimental constraints. A notable application of this force field was its use in a seminal study on the use of NOEs from NMR in combination with MD simulations for structure determination.[45] The versatility of the CHARMM program is indicated by the fact that although this application had not been foreseen in the program design, restraints could be incorporated in the simulation and a structure in agreement with the experimental data was obtained without any additional programming. A free energy perturbation study in explicit water comparing binding of the drug netropsin in the minor groove of B-form d(AATT)$_2$ and an analogue of d(GGCC)$_2$ with this force field predicted a quite reasonable preference of 3.7 kcal/mol in favor of the netropsin:d(AATT)$_2$ complex[46]. As computational resources continued to grow, interest for a nucleic acid force field that was designed for the explicit treatment of the aqueous environment developed, leading to the CHARMM22 all-atom nucleic acid force field.[14] This force field was used in a variety of studies[47-51]; however, the model had an intrinsic bias towards A form structures in high water activity where the B form should be stable.[52-54] This motivated the next generation CHARMM27 all-atom force field,[15,16,55] which is the current additive force field in CHARMM. Optimization of this force field included results from simulations of oligonucleotides in both crystals and solution, allowing for condensed phase effects to be included in the optimization process. This force field has been subjected to a variety of tests and has been shown to yield the B form in solutions of high water activity[56] and to model the difference between the A and B forms of DNA as a function of water activity.

Future nucleic acid force fields in CHARMM are anticipated to explicitly include the treatment of electronic polarizability. Efforts in our laboratory are focusing on the use of a classical Drude Oscillator[57-60]. These efforts have lead to development of water models, termed SWM4-DP [61] and SWM4-NDP[62] and models for ethanol and alkanes.[63] Notable, the Drude based polarizable model has been used to successfully simulate DNA in solution including counterions using a preliminary force field.[64] These successes are currently being extended to produce a highly optimized polarizable force field for nucleic acids as well as proteins, lipids and sugars.

4.   COMPUTATIONAL STUDIES OF SMALL NUCLEIC ACIDS AND RELATED COMPOUNDS

Central to the success of empirical force field methods in applications on nucleic acids is the proper calibration of those force fields. Towards this end, theoretical studies of nucleosides, mono-, di- and tri-nucleotides and other related model compounds have been undertaken. These calculations are typically designed to reproduce a variety of target data, including results from both experimental work and QM studies. Concerning the later, MacKerell and coworkers, among others have performed a number of studies. These studies have targeted the conformational energies associated with puckering of the furanose ring, the glycosyl linkage (i.e. the χ dihedral) and of the many rotatable bonds in the phosphodiester backbone (i.e. α, β, γ, δ, ε ζ). Notable in these studies was the analysis of the conformational energies of the rotatable bonds in nucleotides as a function of sugar pucker, allowing for the relationship of the conformational properties of the backbone to the canonical A, B and Z forms of DNA to be understood. In general, it was observed that the intrinsic conformational flexibility of the rotatable bonds in the phosphodiester backbone as well as the glycosyl linkage is such that the low energy regions identified in model compounds are occupied in the canonical forms of DNA, although exceptions do exist. A particularly interesting observation from these studies was the inherent bias of the cytosine base to energetically favor the A form of DNA. This observation helps explain, in part, the tendency of GC base pairs to favor the A form of DNA and emphasizes the role that the intrinsic conformational properties of nucleic acids have on the structural properties of nucleic acids.

A number of studies on small molecules have used potential of mean force approaches. A hybrid QM/MM investigation using umbrella sampling of bond cleavage in a phosphodiester model compound (dianionic oxyphosphorane) clearly indicated a lowering of the barrier when solvent was added, and in particular in the presence of Mg^{2+} ion [65]. Potential of mean force calculations using the CHARMM22 force field for the stacked/unstacked equilibrium for all 32 ribo- and deoxyribose dinucleotides were in good agreement with experimental data concerning the influence of temperature, solvent and dinucleotide composition on the stability of the stacked state.[66-69] Although interactions with the backbone did contribute to the relative stabilities, the difference between ribose and deoxyribose in the backbone was rather minor.

In a recent study[70] of all 8 ribo- and deoxyribonucleosides, using long simulations (50 ns in explicit water), equilibrium distributions of the accessible conformations as well as structural interconversion rates were found to be in good agreement with available experimental data. Differences in sugar conformational preferences as compared to *ab initio* gas phase results can be understood when the stabilizing influence of water molecules is taken into account. Comparisons with the AMBER force field, using CHARMM with the same simulation protocol, gave rather similar results, although the AMBER force field seems to underestimate the barrier for rotation around the glycosidic bond at $\chi = 360°$. Overall the CHARMM27 force field is well balanced, with a particularly good representation of the ribose. Replica exchange molecular dynamics (REMD), with a Generalized Born implicit solvent model, and standard MD simulations in explicit solvent on the modified nucleoside 2-thiouridine reproduce the experimentally observed increased rigidity of the ribose moiety as well as the preference for the north conformation.[71]

Base flipping has been studied[72] at two sites in a dsRNA system, the gluR-B pre-mRNA, in which an ADAR (adenosine deaminase acting on RNA) enzyme specifically deaminates one, but not the other site. Umbrella sampling along a reaction coordinate defined by a pseudo-dihedral relating the centers of mass of the neighboring base-pair, the sugar and the base of the flipped-out base,[73] was used to obtain the free energy for the flipping out process. The barrier for a full flip was similar for the two sites, but in the preferentially deaminated site a local minimum was found ca 50° into the minor groove, stabilized by hydrogen bonds that cannot form at the other site.[72] Consistent with this observation a significant fraction of partial flips into the minor groove were also observed in free MD simulations of the same system.

5.   SIMULATIONS STUDIES
OF OLIGONUCLEOTIDES

A large number of simulations of DNA duplexes as well as noncanonical structures have been performed using the CHARMM22 and CHARMM27 force fields. These studies have yielded a variety of insights into the properties of oligonucleotides that are not readily accessible to experimental studies. Highlights of those observations follow.

Studies on canonical forms of DNA and RNA have used both CHARMM22 and CHARMM27. As mentioned above, CHARMM22 has a propensity to sample A form conformations,[52-54,74] while yielding stable A form RNA structures.[51] It has been applied to model RNA hairpin structure,[75-77] and the interactions between the tRNA[Phe] anticodon stem-loop and both cognate and near-cognate codon triplets in solution.[78]

CHARMM27 yields the expected B form in high-water activity conditions as well as maintains the A form in low water activity due to either 4 M Na^{+20} or ethanol[16] and yields stable A form structures of duplex RNA[79,80]. CHARMM27 has also been used in stability studies of RNA hairpins with base modifications (inosine, 2-aminopurine),[81,82] One limitation that has been pointed out with CHARMM27 for canonical DNA simulations is a tendency to overestimate the width of the minor groove,[56] although controversy concerning the impact of ions on the minor groove width in experimentally determined structures leaves this a somewhat open question.[83,84] In addition, a tendency of the force field to overpopulate the B_I subconformation of canonical DNA has been noted (N. Foloppe & L. Nilsson, unpublished data; [85]).

The A-B transition in DNA has been studied recently using CHARMM by two groups. Pastor[86] reports results from an 11 ns simulation with the CHARMM22 force field, in which the transition from B-form to A-form occurs spontaneously in a 24-mer. Although the number of DNA-water hydrogen bonds remains constant during the transition, there is a 40% increase in the number of DNA-water-DNA bridges.

Conformational properties of DNA molecules attached to a solid support may influence the binding characteristics when used in hybridization assays. This issue was investigated by Wong and Pettitt[87] in a simulation of a 12-mer duplex DNA attached to epoxide-coated silica. Although the DNA conformation remained near the B-form, there were changes in helicoidal parameters of the DNA-end that was bound to the surface. The whole DNA molecule exhibited large-scale tilting (>55°) on a ns timescale. At the end of the 40 ns simulation the linker between the surface and the DNA collapsed on to the surface.

An important quality of the CHARMM force fields are their ability to yield stable structures using both Ewald based methods, such as particle mesh Ewald,[88] and atom truncation methods for treating the electrostatic interactions.[16,89] Use of atom truncation methods requires an adequate nonbond truncation distance (12 Å or more) as well as use of the appropriate smoothing functions to gradually bring the interaction energies and forces to zero at the truncation distance.[90] The ability to perform stable MD simulations using atom truncation allows for the use of stochastic boundary simulations methods[91,92] with the nucleic acid force field.

A limited number of simulation studies have performed direct comparisons with experiment with respect to dynamic properties. Calculations with CHARMM22 by Norberg and Nilsson showed satisfactory agreement with data from NMR relaxation measurements on d(TCGCG)$_2$ and d(CGCGCG)$_2$.[93] In an earlier study, reasonable agreement was also observed for the GpU dimer with NMR relaxation data using both the extended atom EF2 and CHARMM22 all-atom force fields.[94]

A number of simulation studies using CHARMM have focused on the solvent environment of oligonucleotides; CHARMM27 has been shown to yield overall hydration numbers of DNA and RNA in agreement with various experiments.[16] Example of solvation studies include the impact of Mg^{+2} concentration on solvent properties showing that increased Mg^{+2} doesn't decrease the hydration number of the DNA, but does decrease the mobility of the waters hydrating the oligonucleotide,[47] an effect that may have a role in the influence of water activity on DNA conformation. Pettitt and coworkers have performed several studies looking at DNA hydration. These include the role of solvent, as a function of sodium concentration, on the stabilization of A versus B DNA[95] and the identification of novel spine of hydration in Triplex DNA.[96,97] In another study, the properties of water and cations around DNA was studied in the presence of polyamines and Na^+ counterions; however, that study used the SPC water model rather than the TIP3P model[98] that is recommended for use with the CHARMM force fields.

CHARMM force fields have also been used to study a variety of noncanonical DNA structures in addition to the noncanonical RNA simulation study mentioned above. An interesting study applied potential of mean force (PMF)(e.g., umbrella sampling) calculations with CHARMM22 to investigate the force acting on DNA as a function of extension of the DNA duplex in an effort to understand the force-extension profiles obtained from atomic force microscopy experiments.[49] Results from the extension PMF calculations showed the extension to occur from shorter than the A form through the B form out to approximately twice the length of the B form without significant barriers. The "barrierless" transition from the A to B forms has subsequently been reproduced using the CHARMM27 force field on a DNA hexamer.[85] This study showed an energy difference of 2.8 kcal/mol between the A and B forms with a fairly shallow minimum around the B form. PMF calculations have also been used to investigate the free energy associated with the flipping of a base out of DNA.[73] Results were in near quantitative agreement with NMR imino proton exchange experiments and a novel mechanism for impacting the equilibria between the closed and open states via interactions of the flipping base with the minor or major grooves was identified. The same methodology has been used for a number of studies of base flipping in the presence of a protein (see below) and, recently, was used to quantitatively evaluate the CHARMM27, AMBER4.1[19] and BMS[20] force fields with respect to the base opening equilibrium.[99] Those studies showed CHARMM to be in the best agreement with experiment closely followed by AMBER with BMS in significant disagreement. Base flipping PMF calculations also suggest that NMR imino proton exchange experimental are primarily monitoring opening of purine bases rather than opening of both bases (U.D. Priyakumar and A.D. MacKerell, Jr., submitted). CHARMM27 has also been used to study an adenine bulge in DNA; good agreement with NMR and X-ray structural data was observed.[100] In another study the CHARMM force field was used to investigate the impact of charge on DNA bending in combination with linear-scaling QM methods.[101] Finally, models of a DNA triplex have been developed using CHARMM22 and shown to be stable on the nanosecond time scale.[102] The success of these studies point towards the utility of the CHARMM force fields for studying noncanonical oligonucleotide structures.

Chemical modifications of DNA and RNA are often used to probe the relation of structure and dynamics to function in these molecules. In addition, living organisms contain a number of DNA repairs proteins that deal with the various chemical alterations that DNA can undergo under physiological conditions that can lead to mutations. Accordingly, CHARMM has been used for a number of simulation studies of modified oligonucleotides. In the majority of the following studies, the CHARMM force fields were systematically extended to treat the modifications, using prescribed methods to yield parameters consistent with the remainder of the force field.[103,104] Such consistency is important to insure that the results from the models are representative of the experimental regimen.

Barsky and coworkers looked at the impact of abasic sites on the structure and dynamics of DNA, suggesting that increased flexibility of the region of the protein in the vicinity of the DNA facilitates binding to abasic repair proteins.[50] Modeling studies using CHARMM have been performed on modifications of the sugar in DNA based on carbocyclic sugar analogs that can be used to enforce the sugar to assume either the north or south sugar pucker.[105] In the initial study, MD simulations on an abasic site with furanose, a south constrained and a north constrained carbocyclic sugar analog were used to develop a model where the enhanced binding of the south constrained analog to the methyltransferase from *Hha*I (M. *Hha*I) was due to the ability of that sugar conformation to spontaneously assume bound conformations in the absence of the protein.[106] This model was subsequently refined in a collaborative study involving MD and X-ray crystallography showing that M. *Hha*I binds to the south sugar in a partially flipped state that the MD simulations indicated was a transition state for base flipping.[107] A number of simulation studies have looked at modification of the phosphodiester backbone. Examples include peptide nucleic acids (PNA) and phosphoramidate DNA. PNA studies involved both homo- and heteroduplexes of DNA, showed good agreement with NMR and X-ray data, with the results indicating that the structures of the DNA analogs are dominated by base pairing and stacking interactions.[108] Additional PNA calculations looked at single stranded oligonucleotides revealing, interestingly, the PNA single strands have less flexibility than either RNA or DNA, despite the lack of the sugar moiety, as well as having increased base stacking.[109] Simulations of phosphoramidate DNA, in conjunction with quantum mechanical calculations on representative model compounds,[110] allowed for a better understanding of the differential properties of N3'- versus N5'-phosphoramidate DNA, including both intrinsic energetic and hydration contributions.[111] Other studies using chemically modified DNA including studies on duplexes containing the base inosine instead of adenosine, shedding light on the properties of TATA box DNA,[48] and studies on DNA tethered to a surface have been undertaken.[112] Such studies, along with the various investigations on non-canonical structures discussed above, emphasize the utility of the CHARMM force fields in studying a wide range of chemically – modified oligonucleotides beyond the canonical forms associated with DNA and RNA.

6.   SIMULATION STUDIES OF PROTEIN NUCLEIC ACID COMPLEXES

Structural, dynamic and thermodynamic aspects of protein-nucleic acid complexes have been investigated by MD simulations in explicit solvent for a number of systems, including enzymes (RNase T1[113-116] and EcoR1: DNA[117,118]), transcription factors (hormone receptors[119-122]; HMG protein[123,124]; homeodomains[125]; the p53 DNA-binding domain[126]), and the U1A ribonucleoprotein RNA-binding domain in complex with RNA[77]. The studies of RNase T1 involved mononucleotides represent some of the earliest work on protein nucleic acid complexes. Results from the study yielded insights into specific protein-nucleic acids interactions important for nucleotide binding, local conformational changes that may contribute to nucleotide dissociation as well as the interpretation of experimental time-resolved fluourescence data on the nature of the motion of the single tryptophan in the protein.[113,114,127] A subsequent study on that protein combined with free energy perturbation analysis represented one of the first efforts in which the relative binding of inhibitors to a protein as a function of pH was calculated.[116]

The remainder of the proteins listed in the preceding paragraph bind in the major or minor grooves of canonical B-form DNA duplexes, introduce a kink or a bend in the duplex, as well as bind an RNA hairpin structure. From the simulations it has been possible to identify factors that contribute to the stability of the complexes, as well as for the proteins or DNA/RNA molecules. In several cases, where the individual molecules have a different conformation or are unstructured, complex formation stabilizes a well-ordered conformation, often with a significant number of water molecules in the interface. Key roles for individual amino acid residues in determining the specificity of the interaction could also be identified, sometimes modulated by other nearby residues.[125] The Zn^{2+} ion in the L2 loop of the p53 DNA-binding domain was found to be directly involved in DNA binding through a phosphate group, and also to help keep the L2 loop structured,[126] which may be essential to prevent protein aggregation. By using different pressures to shift the binding equilibrium, the role of direct and water-mediated interactions in the complex of the *Bam*HI restriction endonuclease with its cognate DNA sequence were studied by MD simulations using the CHARMM27 force field.[128] A very large system (>200 000 atoms) consisting of the *lac* repressor (LacI) and two DNA operator segments connected via a loop of DNA modeled as an elastic ribbon in explicit water has recently been studied in a multiscale simulation using the CHARMM27 force field.[129] Strain induced by the DNA loop was efficiently accommodated by *lac* head group rotations, which were deemed to be essential for LacI function.

Several studies have been published on base flipping in the presence of M. *Hha*I, again using PMF calculations.[130] These studies revealed that the protein facilitates the base flipping process by effectively lowering the barrier to flipping via, in part, destabilization of the WC ground state and exclusion of the flipping base from solvent.[131,132] Sequence specific recognition of the DNA by M.*Hha*I has been shown to be associated with the ability of the protein to lower the energy barrier to flipping in the cognate sequence.[133] In addition, the base flipping PMF calculations indicate the pathway of flipping to be via the major groove of the DNA. In the collaborative study involving X-ray crystallography and MD simulations discussed above it was observed that binding of DNA to M. *Hha*I with an abasic south carbocyclic at the site of the flipping cytosine has the sugar moiety located in the major groove approximately half-way between the normal canonical and fully flipped conformation, offering the first experimental verification of the flipping pathway predicted by the PMF calculations.[107]

More recently, a simulation study was undertaken on the Runt domain protein in a ternary complex with its binding partner CBFβ and DNA as well as all possible binary complexes and monomers.[35] Results from these simulations combined with the Generalized-Born model of aqueous solvation were used to calculate the free energies of binding along with interpret mutational data on the system. Notable was the agreement with experimental data on the cooperativity of binding in the trimer, emphasizing the capability of doing free energy component analysis[34] with CHARMM using either GB or Poisson-Boltzmann continuum solvent approaches.[134]

While not a protein-nucleic acid complex, it is worth noting a study of DNA-lipid interactions by Klein and coworkers.[135] This study, motivated by the field of gene therapy in which it is necessary for oligonucleotides to cross lipid membranes, showed thr combined DNA-lipid model to yield stable structures. This, somewhat unique combination further emphasizes the utility of the CHARMM force fields to study heterogeneous macro-molecular systems.

7.   SUMMARY

As is evident from the above overview, the CHARMM nucleic acid force fields have been successfully applied to a variety of oligonucleotides, both alone and in the presence of proteins. The ability of the force field to treat, with acceptable accuracy, canonical and non-canonical forms of DNA and RNA, including transitions between various forms, chemically modified oligonucleotides and DNA and RNA-protein complexes speaks to the adaptability of the force field. The adaptability is due to a consistent force field development approach allowing compatibility with other, non nucleic acid portions of the force field as well as extensions of the force field and to the inclusion of both small molecule and duplex DNA and RNA structural and energetic information in the force field optimization process.[2,136] The success of the many studies of interactions of proteins or lipids with nucleic acids validates the consistent force field development approach used in the CHARMM force fields.

Future developments of the CHARMM force fields are anticipated to occur in the additive model as well as in extensions to polarizable force fields. Ongoing studies with the additive force field are looking at the minor groove width and equilibrium between the B_I and B_{II} conformations of duplex DNA (A.D. MacKerell, Jr. and L. Nilsson, Work in Progress). In the area of polarizable force fields a simulation of DNA has already been performed for 5 ns [64](I. Vorobyov and A.D. MacKerell, Jr. Personal communication). Expectedly, the model distorted significantly from a B form structure due to the direct application of the majority of the parameters from the additive force field to the polarizable model. However, the ability to successfully simulate DNA in solution using a polarizable model, combined with ongoing efforts towards the optimization of the parameters for a polarizable nucleic acid force field indicate that a successful non-additive force field will be available in the foreseeable future. Such advances strongly indicate that empirical force field based calculations of nucleic acids will continue to improve in accuracy, making the approach applicable to an increase number of systems.

ACKNOWLEDGEMENTS

Financial support is acknowledged by ADM from the NIH (GM 51501), by LN from the Swedish Research Council, and we thank DOD ACS Major Shared Resource Computing and PSC Pittsburgh Supercomputing Center for their generous CPU allocations.

REFERENCES

1. Becker, O. M., MacKerell, A. D., Jr., Roux, B. & Watanabe, M., Eds. (2001). Computational Biochemistry and Biophysics. New York: Marcel-Dekker, Inc.
2. MacKerell, A. D., Jr. (2004). Empirical Force Fields for Biological Macromolecules: Overview and Issues. *J. Comput. Chem.* **25**, 1584-1604.
3. Pearlman, D. A., Case, D. A., Caldwell, J. W., Ross, W. S., Cheatham, T. E., Debolt, S., Ferguson, D., Seibel, G. & Kollman, P. (1995). AMBER. *Comp. Phys. Commun.* **91**, 1-41.
4. Brooks, B. R., Bruccoleri, R. E., Olafson, B. D., States, D. J., Swaminathan, S. & Karplus, M. (1983). CHARMM: A Program for Macromolecular Energy, Minimization, and Dynamics Calculations. *J. Comput. Chem.* **4**, 187-217.
5. Ponder, J. (2002). Tinker: Software Tools for Molecular Design. Washington University School of Medicine, St. Louis.

6. Van Gunsteren, W. F. (1987). GROMOS. Groningen Molecular Simulation Program Package. University of Groningen, Groningen.
7. Lindahl, E., Hess, B. & Van der Spoel, D. (2001). GROMACS 3.0: A package for molecular simulation and trajectory analysis. *J. Mol. Mod.* **7**, 306-317.
8. Ewig, C. S., Berry, R., Dinur, U., Hill, J.-R., Hwang, M.-J., Li, H., Liang, C., Maple, J., Peng, Z., Stockfisch, T. P., Thacher, T. S., Yan, L., Ni, X. & Hagler, A. T. (2001). Derivation of Class II Force Fields. VIII. Derivation of a General Quantum Mechanical Force Field for Organic Compounds. *J. Comput. Chem.* **22**, 1782-1800.
9. Frisch, M. J., Trucks, G. W., Schlegel, H. B., Scuseria, G. E., Robb, M. A., Cheeseman, J. R., Zakrzewski, V. G., Montgomery, J. A., Jr., Stratmann, R. E., Burant, J. C., Dapprich, S., Millam, J. M., Daniels, A. D., Kudin, K. N., Strain, M. C., Farkas, O., Tomasi, J., Barone, V., Cossi, M., Cammi, R., Mennucci, B., Pomelli, C., Adamo, C., Clifford, S., Ochterski, J., Petersson, G. A., Ayala, P. Y., Cui, Q., Morokuma, K., Malick, D. K., Rabuck, A. D., Raghavachari, K., Foresman, J. B., Cioslowski, J., Ortiz, J. V., Baboul, A. G., Stefanov, B. B., Liu, G., Liashenko, A., Piskorz, P., Komaromi, I., Gomperts, R., Martin, R. L., Fox, D. J., Keith, T., Al-Laham, M. A., Peng, C. Y., Nanayakkara, A., Gonzalez, C., Challacombe, M., Gill, P. M. W., Johnson, B., Chen, W., Wong, M. W., Andres, J. L., Gonzalez, C., Head-Gordon, M., Replogle, E. S. & Pople, J. A. (1998). Gaussian 98. Gaussian, Inc., Pittsburgh, PA.
10. Schmidt, M. W., Baldridge, K. K., Boatz, J. A., Elbert, S. T., Gordan, M. S., Jensen, J. H., Koseki, S., Matsunaga, N., Nguyen, K. A., Su, S., Windus, T. L., Dupuis, M. & Montgomery, J. A. (1993). General Atomic and Molecular Electronic Structure System. *J. Comput. Chem.* **14**, 1347-1363.
11. Harrison, R. J., Nichols, J. A., Straatsma, T. P., Dupuis, M., Bylaska, E. J., Fann, G. I., Windus, T. L., Apra, E., Anchell, J., Bernholdt, D., Borowski, P., Clark, T., Clerc, D., Dachsel, H., Jong, B. D., Deegan, M., Dyall, K., Elwood, D., Fruchtl, H., Glendenning, E., Gutowski, M., Hess, A., Jaffe, J., Johnson, B., Ju, J., Kendall, R., Kobayashi, R., R. Kutteh, Z. L., Littlefield, R., X. Long, B. M., Nieplocha, J., Niu, S., Rosing, M., Sandrone, G., Stave, M., Taylor, H., Thomas, G., Lenthe, J. V., Wolinski, K., Wong, A. & Zhang, Z. (2000). NWChem, A Computational Chemistry Package for Parallel Computers, Version 4.0 4.0 edit. Pacific Northwest National Laboratory, Richland, WA, 99352-0999, USA.
12. Kong, J., White, C. A., Krylov, A. I., Sherrill, C. D., Adamson, R. D., Furlani, T. R., Lee, M. S., Lee, A. M., Gwaltney, S. R., Adams, T. R., Ochsenfeld, C., Gilbert, A. T. B., Kedziora, G. S., Rassolov, V. A., Maurice, D. R., Nair, N., Shao, Y., Besley, N. A., Maslen, P. E., Dombroski, J. P., Daschel, H., Zhang, W., Korambath, P. P., Baker, J., Byrd, E. F. C., Voorhis, T. V., Oumi, M., Hirata, S., Hsu, C.-P., Ishikawa, N., Florian, J., Warshel, A., Johnson, B. G., Gill, P. M. W., Head-Gordon, M., & Pople, J. A. (2000). Q-Chem 2.0: A high-performance ab initio electronic structure program. *J. Comput. Chem.* **21**, 1532-1548.
13. Jaguar. (1991-2000). 4.1 edit. Schrödinger, Inc., Portland, OR.
14. MacKerell, A. D., Jr., Wiórkiewicz-Kuczera, J. & Karplus, M. (1995). An all-atom empirical energy function for the simulation of nucleic acids. *J. Am. Chem. Soc.* **117**, 11946-11975.
15. Foloppe, N. & MacKerell, A. D., Jr. (2000). All-atom empirical force field for nucleic acids: 1) Parameter optimization based on small molecule and condensed phase macromolecular target data. *J. Comput. Chem.* **21**, 86-104.
16. MacKerell, A. D., Jr. & Banavali, N. K. (2000). All-atom empirical force field for nucleic acids: 2) Application to solution MD simulations of DNA. *J. Comput. Chem.* **21**, 105-120.

17. Nilsson, L. & Karplus, M. (1986). Empirical Energy Functions for Energy Minimization and Dynamics of Nucleic Acids. *J. Comput. Chem.* **7**, 591-616.

18. Cheatham, T. E., III, Cieplak, P. & Kollman, P. A. (1999). A modified version of the Cornell et al. force field with improved sugar pucker phases and helical repeat. *J. Biomol. Struct. Dyn.* **16**, 845-861.

19. Cornell, W. D., Cieplak, P., Bayly, C. I., Gould, I. R., Merz, K. M., Ferguson, D. M., Spellmeyer, D. C., Fox, T., Caldwell, J. W. & Kollman, P. A. (1995). A Second Generation Force Field for the Simulation of Proteins, Nucleic Acids, and Organic Molecules. *J. Am. Chem. Soc.* **117**, 5179-5197.

20. Langley, D. R. (1998). Molecular dynamics simulations of environment and sequence dependent DNA conformation: The development of the BMS nucleic acid force field and comparison with experimental results. *J. Biomol. Struct. Dyn.* **16**, 487-509.

21. Soares, T. A., Hunenberger, P. H., Kastenholz, M. A., Kraeutler, V., Lenz, T., Lins, R., Oostenbrink, C. & Van Gunsteren, W. (2005). An improved nucleic acid parameter set for the GROMOS force field. *J. Comput. Chem.* **26**, 725-737.

22. Foloppe, N., Hartmann, B., Nilsson, L. & MacKerell, A. D., Jr. (2002). Intrinsic Conformational Energetics Associated with the Glycosyl Torsion in DNA: a Quantum Mechanical Study. *Biophys. J.* **82**, 1554-1569.

23. Garcia-Viloca, M., Gao, J., Karplus, M. & Truhlar, D. G. (2004). How Enzymes Work: Analysis by Modern Rate Theory and Computer Simulations. *Science* **303**, 186-195.

24. MacKerell, A. D., Jr., Brooks, B., Brooks, C. L., III, Nilsson, L., Roux, B., Won, Y. & Karplus, M. (1998). CHARMM: The Energy Function and Its Paramerization with an Overview of the Program. In *Encyclopedia of Computational Chemistry* (Schreiner, P. R., ed.), Vol. 1, pp. 271-277. John Wiley & Sons, Chichester.

25. Berman, H. M., Olson, W. K., Beveridge, D. L., Westbrook, J., Gelbin, A., Demeny, T., Hsieh, S.-H., Srinivasan, A. R. & Schneider, B. (1992). The Nucleic Acid Database: A comprehensive relational database of three-dimensional structures of nucleic acids. *Biophys. J.* **63**, 751-759.

26. Berman, H. M., Battistuz, T., Bhat, T. N., Bluhm, W. F., Bourne, P. E., Burkhardt, K., Feng, Z., Gilliland, G. L., Iype, L., Jain, S., Fagan, P., Marvin, J., Padilla, D., Ravichandran, V., Schneider, B., Thanki, N., Weissig, H., Westbrook, J. D. & Zardecki, C. (2002). The protein data bank. *Acta Crystallogr D Biol Crystallogr* **58**, 899-907.

27. Macke, T. J. & Case, D. A. (1998). Modeling Unusual Nucleic Acid Structures. In *Molecular Modeling of Nucleic Acids* (J. SantaLucia, J., ed.), Vol. 682, pp. 379-392. American Chemical Society, Washington, DC.

28. Lu, X. J. & Olson, W. K. (2003). 3DNA: a software package for the analysis, rebuilding and visualization of three-dimensional nucleic acid structures. *Nucleic Acids Res* **31**, 5108-5121.

29. Beveridge, D. L. & DiCapua, F. M. (1989). Free Energy via Molecular Simulations: Applications to Chemical and Biomolecular Systems. *Annu.Rev.Biophys.Biophys.Chem.* **18**, 431-492.

30. Tuckerman, M. E. & Martyna, G. J. (2000). Understanding Modern Molecular Dynamics: Techniques and Applications. *J. Phys. Chem. B* **104**, 159-178.

31. Ma, J. (2005). Usefulness and limitations of normal mode analysis in modeling dynamics of biomolecular complexes. *Structure (Camb)* **13**, 373-380.

32. Feig, M. & Brooks, C. L., III. (2004). Recent advances in the development and application of implicit solvent models in biomolecular simulations. *Curr. Opin. Struct. Biol.* **14**, 217-224.

33. Srinivasan, J., Cheatham, I., T.E., Ceiplak, P., Kollman, P. A. & Case, D. A. (1998). Continuum Solvent Studies of the Stability of DNA, RNA, and Phosphoramidate-DNA Helicies. *J. Amer. Chem. Soc.* **120**, 9401-9409.

34. Kollman, P. A., Massova, I., Reyes, C., Kuhn, B., Huo, S., Chong, L., Lee, M., Lee, T., Duan, Y., Wang, W., Donini, O., Cieplak, P., Srinivasan, J., Case, D. A. & Cheatham, T. E., III. (2000). Calculating Structures and Free Energies of Complex Molecules: Combining Molecular Mechanics and Continuum Models. *Acc. Chem. Res.* **33**, 889-897.

35. Habtemariam, B., Anisimov, V. M. & MacKerell, A. D., Jr. (2005). Cooperative Binding of DNA and CBFβ to the Runt Domain of the CBFα Studied via MD simulations. *Nucleic Acid Res.* **33**, 4212-4222.

36. Dickerson, R. E. (1998). DNA bending: the prevalence of kinkiness and the virtures of normality. *Nucleic Acids Res.* **26**, 1906-1926.

37. Humphrey, W., Dalke, A. & Schulten, K. (1996). VMD: Visual molecular dynamics. *J. Mol. Graph.* **14**, 33-38.

38. Feig, M., MacKerell, A. D., Jr. & Brooks, C. L., III. (2002). Force field influence on the observation of π-helical protein structures in molecular dynamics simulations. *J. Phys. Chem. B* **107**, 2831-2836.

39. MacKerell, A. D., Jr., Feig, M. & Brooks, C. L., III. (2004). Accurate treatment of protein backbone conformational energetics in empirical force fields. *J Am Chem Soc* **126**, 698-699.

40. MacKerell, A. D., Jr., Feig, M. & Brooks, C. L., III. (2004). Extending the treatment of backbone energetics in protein force fields: limitations of gas-phase quantum mechanics in reproducing protein conformational distributions in molecular dynamics simulations. *J. Comp. Chem.* **25**, 1400-1415.

41. Tidor, B., Irikura, K. K., Brooks, B. R. & Karplus, M. (1983). Dynamics of DNA Oligomers. *J. Biomol. Struct. Dyn.* **1**, 231-252.

42. Nordlund, T. M., Andersson, S., Nilsson, L., Rigler, R., Graeslund, A. & McLaughlin, L. W. (1989). Structure and Dynamics of a Fluorescent DNA Oligomer Containing the Ecori Recognition Sequence: Fluorescence, Molecular Dynamics, and NMR Studies. *Biochemistry.* **28**, 9095-103.

43. Nilsson, L., Åhgren-Stålhandske, A., Sjögren, A. S., Hahne, S. & Sjöberg, B.-M. (1990). Three-Dimensional Model and Molecular Dynamics Simulation of the Active Site of the Self-Splicing Intervening Sequence of the Bacteriophage T4 nrdb Messenger RNA. *Biochemistry.* **29**, 10317-22.

44. Singh, U. C., Weiner, S. J. & Kollman, P. (1985). Molecular Dynamics Simulations of d(C-G-C-G-A)⎮middle dot⎮ d(T-C-G-C-G) with and without "Hydrated" Counterions. *Proc. Natl. Acad. Sci. USA* **82**, 755-759.

45. Nilsson, L., Clore, G. M., Gronenborn, A. M., Brünger, A. T. & Karplus, M. (1986). Structure Refinement of Oligonucleotides by Molecular Dynamics with Nuclear Overhauser Effect Interproton Distance Restraints: Application to 5' d(C-G-T-A-C-G)$_2$. *J. Mol. Biol.* **188**, 455-475.

46. Härd, T. & Nilsson, L. (1992). Free-Energy Calculations Predict Sequence Specificity in DNA-Drug Complexes. *Nucleosides & Nucleotides* **11**, 167-173.

47. MacKerell, A. D., Jr. (1997). Influence of Magnesium Ions on Duplex DNA Structural, Dynamic, and Solvation Properties. *J Phys Chem B* **101**, 646-650.

48. Pastor, N., MacKerell, A. D., Jr. & Weinstein, H. (1999). TIT for TAT: The Properties of Inosine and Adenosine in TATA Box DNA. *J. Biomol. Struct. & Design* **16**, 787-810.

49. MacKerell, A. D., Jr. & Lee, G. U. (1999). Structure, Force and Energy of Double-Stranded DNA Oligonucleotides Under Tensile Loads. *Eur. J. Biophys.* **28**, 415-426.

50. Barsky, D., Foloppe, N., Ahmadia, S., Wilson, D. M., III & MacKerell, A. D., Jr. (2000). New Insights into the Structure of Abasic DNA from Molecular Dynamics Simulations. *Nucleic Acids Res.* **28**, 2613-2626.

51. Norberg, J. & Nilsson, L. (1996). Constant Pressure Molecular Dynamics Simulations of the Dodecamers: *d*(GCGCGCGCGCGC)$_2$ and *r*(GCGCGCGCGCGC)$_2$. *J. Chem. Phys.* **104**, 6052-6057.

52. Yang, L. & Pettitt, B. M. (1996). B to A Transition of DNA on the Nanosecond Time Scale. *J. Phys. Chem.* **100**, 2550-2566.

53. Feig, M. & Pettitt, B. M. (1998). Structural Equilibrium of DNA represented with Different Force Fields. *Biophys. J.* **75**, 134-149.

54. MacKerell, J., A. D. (1998). Observations on the A versus B Equilibrium in Molecular Dynamics Simulations of Duplex DNA and RNA. In *Molecular Modeling of Nucleic Acids* (SantaLucia, J., J., ed.), Vol. 682, pp. 304-311. American Chemical Society, Washington, DC.

55. MacKerell, A. D., Jr., Banavali, N. B. & Foloppe, N. (2001). Development and Current Status of the CHARMM Force Field for Nucleic Acids. *Biopolymers* **56**, 257-265.

56. Cheatham, T. E., III & Young, M. A. (2001). Molecular Dynamics Simulation of Nucleic Acids: Successes, Limitations, and Promise. *Biopolymers* **56**, 232-256.

57. Dick, B. G., Jr. & Overhauser, A. W. (1958). Theory of the Dielectric Constants of Alkali Halide Crystals. *Phys. Rev.* **112**, 90-103.

58. Cochran, W. (1959). Lattice Dynamics of Alkali Halides. *Phil. Mag.* **4**, 1082-1086.

59. Rick, S. W. & Stuart, S. J. (2002). Potentials and Algorithms for Incorporating Polarizability in Computer Simulations. *Rev. Comp. Chem.* **18**, 89-146.

60. Lamoureux, G. & Roux, B. (2003). Modelling Induced Polarizability with Drude Oscillators: Theory and Molecular Dynamics Simulation Algorithm. *J. Chem. Phys.* **119**, 5185-5197.

61. Lamoureux, G., MacKerell, A. D., Jr. & Roux, B. (2003). A simple polarizable model of water based on classical Drude oscillators. *J. Chem. Phys.* **119**, 5185-5197.

62. Lamoureux, G., Harder, E., Vorobyov, I. V., Roux, B. & MacKerell, A. D., Jr. (2005). A polarizable model of water for molecular dynamics simulations of biomolecules. *Chem. Phys. Lett.* **In press**.

63. Vorobyov, I. V., Anisimov, V. M. & MacKerell, A. D., Jr. (2005). Polarizable Empirical Force Field for Alkanes Based on the Classical Drude Oscillator Model. *J. Phys. Chem. B* **109**, 18988-18999.

64. Anisimov, V. M., Lamoureux, G., Vorobyov, I. V., Huang, N., Roux, B. & MacKerell, A. D., Jr. (2005). Determination of Electrostatic Parameters for a Polarizable Force Field Based on the Classical Drude Oscillator. *J. Chem. Theory. Comp.* **1**, 153-168.

65. Lahiri, A. & Nilsson, L. (1997). Properties of dianionic oxyphosphorane intermediates from hybrid QM/MM simulation: implications for ribozyme reactions. *J. Mol. Struct.-Theochem* **419**, 51-55.

66. Norberg, J. & Nilsson, L. (1995). Potential of Mean Force Calculations of the Stacking Unstacking Process in Single-Stranded Deoxyribodinucleoside Monophosphates. *Biophys. J.* **69**, 2277-2285.

67. Norberg, J. & Nilsson, L. (1995). Stacking Free Energy Profiles For All 16 Natural Ribodinucleoside Monophosphates in Aqueous Solution. *J. Am. Chem. Soc.* **117**, 10832-10840.

68. Norberg, J. & Nilsson, L. (1995). Temperature Dependence of the Stacking Propensity of Adenylyl-3' 5'-Adenosine. *J. Phys. Chem.* **99**, 13056-13058.

69. Norberg, J. & Nilsson, L. (1998). Solvent Influence on Base Stacking. *Biophys J.* **74**, 394-402.

70. Foloppe, N. & Nilsson, L. (2005). Towards a full characterization of nucleic acid components in aqueous solution: simulations of nucleosides. *J. Phys. Chem. B.* **109**, 9119-9131.

71. Lahiri, A., Sarzynska, J., Nilsson, L. & Kulinski, T. (2005). Molecular Dynamics Simulation of the Preferred Conformations of 2-Thiouridine in Aqueous Solution. *Theor. Chem. Acc., submitted.*

72. Hart, K., Nyström, B., Öhman, M. & Nilsson, L. (2005). Molecular dynamics simulations and free energy calculations of base flipping in dsRNA. *RNA* **11**, 609-618.

73. Banavali, N. K. & MacKerell, A. D., Jr. (2002). Free Energy and Structural Pathways of Base Flipping in a DNA GCGC containing sequence. *J. Mol. Biol.* **319**, 141-160.

74. Feig, M. & Pettitt, B. M. (1997). Experiment vs Force Fields: DNA Conformation from Molecular Dynamics Simulations. *J. Phys. Chem. B* **101**, 7361-7363.

75. Sarzynska, J., Kulinski, T. & Nilsson, L. (2000). Conformational Dynamics of a 5S rRNA Hairpin Domain Containing Loop D and a Single Nucleotide Bulge. *Biophys. J.* **79**, 1213-1227.

76. Nina, M. & Simonson, T. (2002). Molecular Dynamics of the tRNAAla Acceptor Stem: Comparison between Continuum Reaction Field and Particle-Mesh Ewald Electrostatic Treatments. *J. Phys. Chem. B* **106**, 3696-3705.

77. Tang, Y. & Nilsson, L. (1999). Molecular dynamics simulations of the complex between human U1A protein and hairpin II of U1 small nuclear RNA and of free RNA in solution. *Biophys. J.* **77**, 1284-305.

78. Lahiri, A. & Nilsson, L. (2000). Molecular Dynamics of the Anticodon Domain of Yeast tRNA[Phe]: Codon-anticodon Interaction. *Biophys. J.* **79**, 2276-2289.

79. Pan, Y. & MacKerell, A. D., Jr. (2003). Altered Structural Fluctuations in Duplex RNA versus DNA: A conformational switch involving base pair opening. *Nucleic Acid Res.* **31**, 7131-7140.

80. Pan, Y., Priyakumar, D. & MacKerell, A. D., Jr. (2005). Conformational Determinants of Tandem GU Mismatches in RNA: Insights from molecular dynamics simulations and quantum mechanical calculations. *Biochemistry* **44**, 1433-1443.

81. Sarzynska, J. & Kulinski, T. (2005). Dynamics and Stability of GCAA Tetraloops with 2-Aminopurine and Purine Substitutions. *J. Biomol. Struct. Dyn.* **22**, 425-439.

82. Sarzynska, J., Nilsson, L. & Kulinski, T. (2003). Effects of Base Substitutions in an RNA Hairpin from Molecular Dynamics and Free Energy Simulations. *Biophys. J.* **85**, 3445-3459.

83. McFail-Isom, L., Sines, C. C. & Williams, L. D. (1999). DNA structure: cations in charge? *Curr. Opin. Struct. Biol.* **9**, 298-304.

84. Hamelberg, D., Williams, L. D. & Wilson, W. D. (2001). Influence of the dynamic positions of cations on the structure of the DNA minor groove: sequence-dependent effects. *J. Am. Chem. Soc.* **123**, 7745-7755.

85. Banavali, N. K. & Roux, B. (2005). Free energy landscape of A-DNA to B-DNA conversion in aqueous solution. *J. Am. Chem. Soc.* **127**, 6866-6876.

86. Pastor, N. (2005). The B- to A-DNA Transition and the Reorganization of Solvent at the DNA Surface. *Biophys. J.* **88**, 3262-3275.

87. Wong, K.-Y. & Pettitt, B. M. (2004). Orientation of DNA on a surface from simulation. *Biopolymers* **73**, 570-578.

88. Darden, T. A., York, D. & Pedersen, L. G. (1993). Particle mesh Ewald: An Nlog(N) method for Ewald sums in large systems. *J. Chem. Phys.* **98**, 10089-10092.

89. Norberg, J. & Nilsson, L. (2000). On the Truncation of Long-Range Electrostatic Interactions in DNA. *Biophys. J.* **79**, 1537-1553.

90. Steinbach, P. J. & Brooks, B. R. (1994). New Spherical-Cutoff Methods of Long-Range Forces in Macromolecular Simulations. *J. Comput. Chem.* **15**, 667-683.

91. Brooks, C. L., III & Karplus, M. (1983). Deformable Stochastic Boundaries in Molecular Dynamics. *J. Chem. Phys.* **79**, 6312-6325.

92. Beglov, D. & Roux, B. (1994). Finite Representation of an Infinite Bulk System: Solvent Boundary Potential for Computer Simulations. *J. Chem. Phys.* **100**, 9050-9063.

93. Norberg, J. & Nilsson, L. (1996). Internal mobility of the oligonucleotide duplexes d(TCGCG)$_2$ and d(CGCGCG)$_2$ in aqeous solution from molecular dynamics simulations. *J. Biomol. NMR* **7**, 305-314.

94. Norberg, J. & Nilsson, L. (1995). NMR Relaxation Times, Dynamics, and Hydration of a Nucleic Acid Fragment from Molecular Dynamics Simulations. *J. Phys. Chem.* **99**, 14876-14884.

95. Feig, M. & Pettitt, B. M. (1999). Sodium and chlorine ions as part of the DNA solvation shell. *Biophys. J.* **77**, 1769-1781.

96. Mohan, V., Smith, P. E. & Pettitt, B. M. (1993). Evidence for a New Spine of Hydration: Solvation of DNA Triple Helices. *J. Am. Chem. Soc.* **115**, 9297-9298.

97. Mohan, V., Smith, P. E. & Pettitt, B. M. (1993). Molecular dynamics simulation of ions and water around triplex DNA. *J. Phys. Chem.* **97**, 12984-12990.

98. Jorgensen, W. L., Chandrasekhar, J., Madura, J. D., Impey, R. W. & Klein, M. L. (1983). Comparison of Simple Potential Functions for Simulating Liquid Water. *J. Chem. Phys.* **79**, 926-935.

99. Priyakumar, U. D. & MacKerell, A. D., Jr. (2005). Base Flipping in a GCGC Containing DNA Dodecamer: A Comparative Study of the Performance of the Nucleic Acid Force Fields, CHARMM, AMBER and BMS. *J. Chem. Theory Comput.* **In Press**.

100. Feig, M., Zacharias, M. & Pettitt, B. M. (2001). Conformations of an Adenine Bulge in a DNA Octamer and Its Influence on DNA Structure from Molecular Dynamics Simulations. *Biophys. J.* **81**, 353-370.

101. Range, K., Mayaan, E. & Maher, L. J., III. (2005). The contribution of phosphate-phosphate repulsions to the free energy of DNA bending. *Nucleic Acid Res.* **33**, 1257-1268.

102. Weerasinghe, S., Smith, P. E., Mohan, V., Cheng, Y.-K. & Pettitt, B. M. (1995). Nanosecond dynamics and structure of a model DNA triple helix in salt water solution. *J. Amer. Chem. Soc.* **117**, 2147-2158.

103. MacKerell, A. D., Jr. (2001). Atomistic Models and Force Fields. In *Computational Biochemistry and Biophysics* (Watanabe, M., ed.), pp. 7-38. Marcel Dekker, Inc., New York.

104. MacKerell, A. D., Jr. (2005). http://www.pharmacy.umaryland.edu/faculty/amackere/param/force_field_dev.htm.

105. Marquez, V. E., Wang, P., Nicklaus, M. C., Maier, M., Manoharan, M., Christman, J. K., Banavali, N. K. & MacKerell, A. D., Jr. (2001). Inhibition of (cytosine C5)-methyltransferase by oligonucleotides containing flexible (cyclopentane) and conformationally constrained (bicyclo[3.1.0]hexane) abasic sites. *Nucleos. Nucleot. & Nucl. Acids* **20**, 451-459.

106. Wang, P., Nicklaus, M. C., Marquez, V. E., Brank, A. S., Christman, J., Banavali, N. K. & MacKerell, A. D., Jr. (2000). Use of Oligodeoxyribonucleotides with Conformationally Constrained Abasic Sugar Targets to Probe the Mechanism of Base Flipping by HhaI DNA (Cytosine C5)-Methyltransferase. *J. Amer. Chem. Soc.* **122**, 12422-12434.

107. Horton, J. R., Ratner, G., Banavali, N. K., Huang, N., Choi, Y., Maier, M. A., Marquez, V. E., MacKerell, A. D., Jr. & Cheng, X. (2004). Caught in the act: visualization of an

intermediate in the DNA base-flipping pathway induced by HhaI methyltransferase. *Nucleic Acids Res.* **32**, 3877-3886.

108. Sen, S. & Nilsson, L. (1998). Molecular Dynamics of Duplex Systems Involving PNA: Structural and Dynamical Consequences of the Nucleic Acid Backbone. *J. Amer. Chem. Soc.* **120**, 619-631.

109. Sen, S. & Nilsson, L. (2001). MD Simulations of Homomorphous PNA, DNA and RNA Single Strands: Characterization and Comparison of Conformations and Dynamics. *J. Am. Chem. Soc.* **123**, 7414-7422.

110. Banavali, N. K. & MacKerell, A. D., Jr. (2001). Reevaluation of Stereoelectric Contributions to the Conformational Properties of the Phosphodiester and N3'-Phosphoramidate Moieties of Nucleic Acids. *J. Am. Chem. Soc.* **128**, 6747-6755.

111. Banavali, N. K. & MacKerell, A. D., Jr. (2001). Reexamination of the Intrinsic, Dynamic, and Hydration Properties of Phosphoramidate DNA. *Nucl. Acids. Res.* **29**, 3219-3230.

112. Wong, K.-Y. & Pettitt, B. M. (2001). A study of DNA tethered to a surface by an all-atom molecular dynamics simulation. *Theor. Chem. Acc.* **106**, 233-235.

113. MacKerell, A. D., Nilsson, L., Rigler, R. & Saenger, W. (1988). Molecular Dynamics Simulations of Ribonuclease T1: Analysis of the Effect of Solvent on the Structure, Fluctuations, and Active Site of the Free Enzyme. *Biochemistry* **27**, 4547-4556.

114. MacKerell, A. D., Jr., Rigler, R., Nilsson, L., Heinemann, U. & Saenger, W. (1988). Molecular dynamics simulations of ribonuclease T1: Effect of solvent on the interaction with 2'GMP. *European Biophysics Journal* **16**, 287-297.

115. MacKerell, A. D., Jr., Rigler, R., Hahn, U. & Saenger, W. (1991). Thermodynamic analysis of the equilibrium , association and dissociation of 2'GMP and 3'GMP with ribonuclease T1 at pH 5.3. *Biochim. Biophy. Acta* **1073**, 357-365.

116. MacKerell, J., A. D., Sommer, M. S. & Karplus, M. (1995). pH Dependence of Binding Reactions from Free Energy Simulations and Macroscopic Continuum Electrostatic Calculations: Application to 2'GMP/3'GMP Binding to Ribonuclease T1. *J. Mol. Biol.* **247**, 774-807.

117. Sen, S. & Nilsson, L. (1999). Structure, interaction, dynamics and solvent effects on the DNA-EcoRI complex in aqueous solution from molecular dynamics simulation. *Biophys. J.* **77**, 1782-1800.

118. Sen, S. & Nilsson, L. (1999). Free energy calculations and molecular dynamics simulations of wild- type and variants of the DNA-EcoRI complex. *Biophys. J.* **77**, 1801-10.

119. Eriksson, M. A. & Nilsson, L. (1999). Structural and dynamic differences of the estrogen receptor DNA-binding domain, binding as a dimer and as a monomer to DNA: molecular dynamics simulation studies [published erratum appears in Eur Biophys J 1999;28(4):356]. *Eur Biophys. J.* **28**, 102-11.

120. Eriksson, M. A. L. & Nilsson, L. (1995). Structure, Thermodynamics and Cooperativity of the Glucocorticoid Receptor DNA-Binding Domain in Complex With Different Response Elements - Molecular Dynamics Simulation and Free Energy Perturbation Studies. *J. Mol. Biol.* **253**, 453-472.

121. Eriksson, M. A. L., Hard, T. & Nilsson, L. (1995). Molecular Dynamics Simulations of the Glucocorticoid Receptor DNA-Binding Domain in Complex With DNA and Free in Solution. *Biophys. J.* **68**, 402-426.

122. Eriksson, M. A. & Nilsson, L. (1998). Structural and dynamic effects of point mutations in the recognition helix of the glucocorticoid receptor DNA-binding domain. *Protein Eng.* **11**, 589-600.

123. Tang, Y. & Nilsson, L. (1999). Effect of G40R Mutation on the Binding of Human SRY Protein to DNA: A Molecular Dynamics View. *Proteins* **35**, 101-113.

124. Tang, Y. & Nilsson, L. (1998). Interaction of human SRY protein with DNA: A molecular dynamics study. *Proteins* **31**, 417-433.

125. Duan, J. & Nilsson, L. (2002). The role of residue 50 and hydration water molecules in homeodomain DNA recognition. *Eur. Biophys. J.* **31**, 306-316.

126. Duan, J. & Nilsson, L. (2005). The Effect of Zn Ion on p53 DNA Recognition and Stability of the DNA Binding Domain. *submitted.*

127. MacKerell, J., A. D., Rigler, R., Nilsson, L., Hahn, U. & Saenger, W. (1987). Protein Dynamics: A time-resolved fluorescence, energetic and molecular dynamics study of ribonuclease T1. *Biophysical Chemistry* **26**, 247-261.

128. Lynch, T. W., Kosztin, D., McLean, M. A., Schulten, K. & Sligar, S. G. (2002). Dissecting the Molecular Origins of Specific Protein-Nucleic Acid Recognition: Hydrostatic Pressure and Molecular Dynamics. *Biophys. J.* **82**, 93-98.

129. Villa, E., Balaeff, A. & Schulten, K. (2005). Structural dynamics of the lac repressor-DNA complex revealed by a multiscale simulation. *Proc. Natl. Acad. Sci. USA* **102**, 6783-6788.

130. Priyakumar, U. D. & MacKerell, A. D., Jr. (2006). Computational Approaches for Investigating Base Flipping in Oligonucleotides. *Chem. Rev.* **106**, 489-505.

131. Huang, N. & MacKerell, A. D., Jr. (2004). Atomistic view of base flipping in DNA. *Phil Trans Royal Soc, London, A* **362**, 1439-1460.

132. Huang, N., Banavali, N. K. & MacKerell, A. D., Jr. (2003). Protein-facilitated base flipping in DNA by cytosine-5-methyltransferase. *Proc. Natl. Acad. Sci. U S A* **100**, 68-73.

133. Huang, N. & MacKerell, A. D., Jr. (2005). Specificity in protein-DNA interactions: energetic recognition by the (cytosine-C5)-methyltransferase from HhaI. *J. Mol. Biol.* **345**, 265-274.

134. Banavali, N. K. & Roux, B. (2002). Atomic Radii for Continuum Electrostatic Calculations on Nucleic Acids. *J. Phys. Chem. B* **106**, 11026-11035.

135. Bandyopadhyay, S., Tarek, M. & Klein, M. L. (1999). Molecular Dynamics Study of a Lipid-DNA Complex. *J. Phys. Chem. B* **103**, 10075-10080.

136. MacKerell, A. D., Jr. & Nilsson, L. (2001). Nucleic Acid Simulations. In *Computational Biochemistry and Biophysics* (Watanabe, M., ed.), pp. 441-464. Marcel Dekker, Inc., New York.

Chapter 4

CONTINUUM SOLVENT MODELS TO STUDY THE STRUCTURE AND DYNAMICS OF NUCLEIC ACIDS AND COMPLEXES WITH LIGANDS

Martin Zacharias

School of Engineering and Sciences, International University Bremen, Campus Ring 6, D-28759 Bremen

Abstract: The aqueous environment has an important influence on the structure and function of nucleic acids. The explicit inclusion of many solvent molecules and ions during simulation studies on nucleic acids can lead to prohibitively expensive computational demands and limits the maximum simulation time. Many applications such as systematic conformational searches and ligand-receptor docking approaches used for example in drug design efforts require computationally rapid implicit treatment of solvation. Ideally, such implicit continuum solvation models should still be accurate enough to be useful for a realistic evaluation of generated structures or docked complexes. An overview is given on recent developments concerning continuum solvent modeling of nucleic acids and complexes with ligands. This includes applications of Poisson-Boltzmann based as well as Generalized Born type continuum solvent models. A number of studies indicate that conformational search procedures based on continuum solvent models could be useful for structure prediction of nucleic acid motifs and during ligand-nucleic acid docking studies. Shortcomings of continuum solvent modeling approaches as well as possible future improvements are also discussed.

Key words: Molecular modeling, nucleic acid structure, non-helical structures, molecular dynamics, conformational analysis, implicit solvation, nucleic acid drug design

J. Šponer and F. Lankaš (eds.), Computational Studies of RNA and DNA, 95–119.
© 2006 *Springer.*

1.   INTRODUCTION

Nucleic acids play a key role in many biological processes ranging from storage of genetic information and replication for DNA to transcription regulation, mRNA splicing and protein synthesis in case of RNA. Most cellular DNA adopts primarily a double-stranded (B-form) helical structure. The interaction with proteins is mostly mediated through major groove or minor groove recognition. Especially in eukaryotes the accessibility of DNA is also controlled by the condensation with nuclear proteins (histones) to compact structures. In contrast, many cellular RNA molecules form complex three-dimensional folds that consist only partially of double stranded (A-form) base-paired regions. The double-stranded regions are frequently interrupted by non-helical motifs such as extra-unmatched nucleotides, mismatched base pairs, bulge structures or hairpin loop structures that cap the end of helices[1-5]. These structural motifs are often of functional importance as protein, small ligand or ion binding sites or to mediate tertiary interactions in folded RNAs.

In recent years, structural knowledge of RNA and RNA in complex with proteins, small organic ligands and ions has increased dramatically since many new structures have been solved by X-ray crystallography or Nuclear Magnetic Resonance (NMR) spectroscopy[6-12]. The rapid increase of structural information has also helped to collect and better understand the types of interactions in folded nucleic acid structures and at nucleic acid – ligand interfaces[3-6]. Since RNA and DNA are of functional importance in many cellular structures nucleic acids are also increasingly recognized as possible drug targets[13-17]. However, the structural information on folded nucleic acid structures as well as complexes with ligands alone does not allow to directly investigate the driving forces and energetic contributions to structure formation and association. In addition, experimental high-resolution structure determination methods have only a limited time resolution and allow only limited insights into the dynamics of nucleic acids and the process of structure formation and association.

Computer simulation studies that are based on a molecular mechanics force field are increasingly being used to investigate nucleic acids[18-21]. These approaches can complement experimental studies in cases in which the timescales or the molecular properties of interest are difficult to access experimentally. Simulation studies include molecular dynamics methods to follow the dynamics of biomolecules at high spatial and time resolution as well as structure optimization approaches and conformational search methods. Provided that the force field to describe the molecular interactions is sufficiently accurate, these methods could allow a more direct study of the driving forces of association and structure formation. Molecular dynamics simulation studies typically include the nucleic acid under study at atomic resolution. During the simulation Newton's equations of motion are solved numerical in small time steps of ~1fs (10^{-15} s) in a periodic box including a large number of surrounding water molecules and ions. The large number of atoms and the short time step used to accurately follow the atomic motions limits current molecular dynamics simulation methods to the nanosecond time regime. From a computational point of view the explicit inclusion of many solvent molecules and ions around a biomolecule can lead to prohibitively expensive simulations and limits the maximum simulation time. In addition, the equilibration of the solvent around a solute requires long simulation times which in turn limits the precision of calculated structural and thermodynamic averages and the chance to observe biologically relevant conformational changes of the molecule. One possibility to reduce the computational demand is to remove explicit solvent molecules and ions and to modify the force field approach to include the effect of the solvent and ions around a biomolecule implicitly. Many applications such as systematic conformational searches and ligand-receptor docking approaches used, for example, in drug design efforts require computationally rapid implicit treatment of solvation. Ideally, such continuum solvation models should still be accurate enough to be useful for a realistic evaluation of generated structures or docked complexes. The smaller number of particles can significantly accelerate the simulation and allows improved sampling of possible conformational states during molecular dynamics simulations. It is hoped that such methods help to improve structure prediction, ligand design and model building of nucleic acids. In the following an overview and some recent progress on implicit solvent modeling will be given with a focus on conformational search applications and modeling of nucleic acids and complexes with ligands.

Also, structure prediction methods based on systematic conformational searches as well as ligand-nucleic acid docking approaches could strongly benefit from an implicit treatment of solvation effects. These methods are difficult to apply in the presence of explicit water molecules.

2. COMPUTER SIMULATION OF NUCLEIC ACIDS USING CONTINUUM SOLVENT MODELS

2.1 Molecular Mechanics Force Fields to Study Biomolecular Structure and Dynamics

The application of quantum mechanical approaches is currently limited by the large size and complexity of nucleic acids and other biomolecules. Computer simulation methods based on model energy functions that use the position of whole atoms as variables (instead of electrons and nuclei in case of quantum mechanics) are now frequently used to investigate nucleic acids[17-21]. In most biomolecular simulation studies the interatomic interactions are described using a classical molecular mechanics force field:

$$V(\mathbf{r}^N) = \sum_{i=1}^{Nbonds} \frac{1}{2} k_i (b_i - b_{io})^2 +$$

$$\sum_{i=1}^{Nangles} \frac{1}{2} k_i (\theta_i - \theta_{io})^2 +$$

$$\sum_{i=1}^{Ndihedrals} \sum_n K_n (1 + \cos(n\tau_i + \delta_n)) +$$

$$\sum_{i>j}^{Npairs} \left(\frac{A_{ij}}{r_{ij}^{12}} - \frac{B_{ij}}{r_{ij}^{6}} + \frac{q_i q_j}{r_{ij}} \right) \qquad (1)$$

The potential energy of a molecule as a function of the atom positions ($\mathbf{r}^N$) includes bonded (first three summations in Eq. (1)) and non-bonded contributions, last term in Eq. (1). Usually quadratic penalty terms with appropriate force constants (k_b and k_θ) respectively, are used to control all bond length (b_i) and bond angle (θ_i) deviations from reference values (b_{io} and θ_{io}). A linear combination of periodic functions is employed to control dihedral torsion angles (τ_i). Additional non-bonded terms describe van der Waals and Coulomb interactions (as a double sum over all non-bonded pairs of atoms). The form of the energy function allows a rapid evaluation of the potential energy of a molecule and the calculation of gradients necessary for energy optimization and molecular dynamics simulations based on a solution of the classical equations of motion. Simulation studies on biomolecules require the inclusion of surrounding aqueous solvent and ions. However, the inclusion of a sufficiently large number of explicit water molecules signify-cantly increases the computational demand. The possibility to implicitly account for solvent effects in molecular mechanics calculations is of great interest because it would allow longer simulation times and a better convergence of calculated thermodynamic averages. One simple modification of the Coulomb term of the force field is the use of a distance dependent dielectric function to mimic electrostatic screening effects. Although frequently applied also in the field of nucleic acid modeling a distance-dependent dielectric does not account for desolvation effects due to burying polar or charged groups in the interior of a biomolecule.

2.2   Continuum Solvent Modeling

One possibility to approximately include solvation effects in molecular mechanics calculations are hydration shell models or models based on the solvent accessible surface area of a molecule[22-25]. Such models typically involve the calculation of the exposed solvent accessible surface of each atom. The contribution of each atom to the solvation of the molecule depends on the type of atom and in most approaches linearly on the amount of exposed surface area of the atom[23]. Solvent accessible surface area (SASA) models have been frequently used to estimate the influence of solvation on protein stability and to evaluate the effect of mutations[23]. Such models have also been used successfully in simulation studies of peptide and small protein folding[24,25]. One principle short coming is the fact that polar and charged chemical groups or atoms once completely buried in the interior of a biomolecule do not influence the hydration properties of the molecule in the framework of SASA models. It is, however, well known that the position of a charged group inside a biomolecule relative to the surface can significantly affect its hydration properties and stability. Nucleic acid molecules, in particular, are highly charged giving rise to long range electrostatic effects that require a more accurate treatment of solvation beyond the possibilities of "pure" SASA methods.

2.2.1 Poisson-Boltzmann Dielectric Continuum Solvent Modeling

An alternative approach based on macroscopic solvation concepts describes the protein interior as a medium with low dielectric permittivity embedded in a high dielectric continuum representing the aqueous solution. The dielectric solute solvent boundary and presence of charges on the solute molecule cause a polarisation of the high dielectric continuum that stabilize the charges of the molecule. The electrostatic potential ($\Phi(r)$) as a function of position (r) including the solvent effect as a reaction field can be obtained from a solution of Poisson's equation (2) for the charge distribution of the molecule ($\rho(r)$) and the position dependent dielectric constant ($\varepsilon(r)$)[26-30].

$$-\nabla \left(\varepsilon(r) \nabla \phi(r) \right) = 4\pi \rho(r) \qquad (2)$$

The mean effect of a salt atmosphere can be included by solving the Poisson-Boltzmann (PB) equation (3).

$$-\nabla \left(\varepsilon(r) \nabla \phi(r) \right) = 4\pi \rho(r) + \sum_i q_i c_i \exp(q_i \phi(r)/k_B T) \qquad (3)$$

Here, it is assumed that the energy of an ion is just its charge times the potential. Counter and co-ions (with charges q_i and bulk concentration c_i) then redistribute according to the Boltzmann weight of the electrostatic potential (times ion charge) around the biomolecule.

Taking only the linear term of the Taylor expansion of the ion distribution term yields the linearized PB equation which is frequently used in calculations on biomolecules[27]. An analytic solution of the Poisson or Poisson-Boltzmann equation is only possible for simple regular geometries of the dielectric boundaries and charge distributions[27]. In case of the irregular form of biomolecules the PB equation needs to be solved numerically. The most common approach is the application of finite-difference (FD) methods where the molecule is embedded in a regular grid[26-33]. In the FDPB method the charges on the atoms are distributed onto neighbouring grid points (using a linear interpolation scheme) and each grid point is assigned a value for the dielectric constant (often smoothed at the boundary between molecule and solution). A variety of numerical methods has been applied to efficiently solve the FDPB equation[26,31-33]. Other approaches employ finite-element and boundary-element techniques to represent the molecule surface and to numerically solve the Poisson equation[34-36]. The solution of the Poisson and PB equation allows to calculate the reaction field due to the charges of the biomolecule and the shape of the dielectric boundary. The reaction field of the solvent stabilizes the charge distribution.

However, solvation of nonpolar groups (hydrophobic solvation) which is unfavorable is not yet included. In continuum solvent modeling nonpolar contributions can be calculated as an additional term typically as an interfacial free energy proportional to the solvent accessible surface area (SASA) of the molecule,

$$\Delta G^{solv}_{nonpolar} = \gamma SASA + b \qquad\qquad (4)$$

where γ is a surface area proportionality constant (termed surface tension parameter) and b is a constant. Both parameters can be obtained from a fit of experimental hydration free energies of alkanes vs. SASA[29,37-39]. The total energy of a molecule structure in aqueous solution is given by its vacuum energy calculated using a classical molecular mechanics force field, Eq. (1), supplemented with an electrostatic solvation contribution calculated using the Poisson (2) or PB equation (3) and including a SASA-dependent nonpolar solvation contribution (4) (illustrated in Figure 4-1).

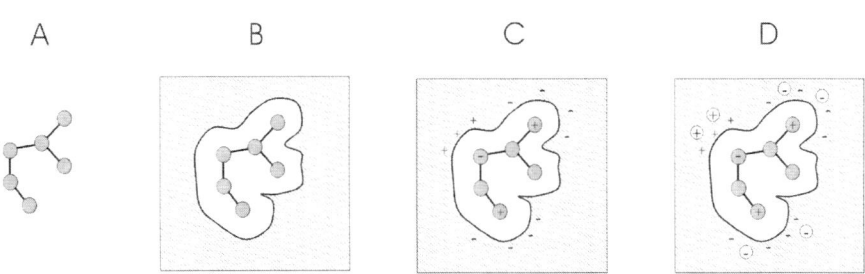

Figure 4-1. Illustration of the contributions to the total free energy of a molecule using a continuum solvent model. (A) The vacuum energy of a molecule includes bonded and non-bonded contribution from a molecular mechanics force field. (B) Transfer to an aqueous environment adds a contribution due to cavity formation in water and solute-solvent dispersion interactions assumed to be proportional to the exposed surface area. (C) Solution of the Poisson equation for the molecule embedded in a high-dielectric continuum adds a reaction field contribution. (D) The mean redistribution of ions can be obtained from a solution of the Poisson-Boltzmann equation (see text for details).

Such models have been extensively tested on small polar and nonpolar organic solute molecules[39,40]. With an appropriate choice of parameters for atomic radii a very reasonable correlation between calculated and experimental hydration free energies can been obtained[39,40]. This can be achieved at a fraction of the computational cost for simulations with explicit solvent molecules. Many parameterizations for different force fields and partial charge assignments on biomolecules have been developed.

The Poisson-Boltzmann continuum model has been frequently applied to estimate the stability of folded proteins, to study the binding of ligands to proteins and to calculate salt contributions to the stability and association of biomolecules[30,39]. Employing the finite-difference form of the nonlinear Poisson-Boltzmann equation good agreement between calculated and experimental salt dependencies of DNA-protein and DNA-ligand interactions has been obtained[41-43]. This was also found for salt dependencies of RNA-ligand complexes[44,45].

Although for most nucleic acid modeling problems changes in ionization states are not expected it should be mentioned that one of the most important application of Poisson-Boltzmann continuum solvent models is the calculation of protonation states of ionizable groups in biomolecules. A number of studies have shown that with a relatively large interior effective dielectric constant (~10-20) acceptable correlation with experiment can be reached for a number of protein examples[46,47]. It is possible to obtain valuable hints on the pK's of certain amino acids in folded proteins for which the experimental determination is difficult or currently impossible. During the calculation of the stability of various ionization states the biomolecule is rigid that is any conformational relaxation due to the charge changes is not possible. The large value of the effective dielectric constant might be required because it approximately accounts for the conformational relaxation of the biomolecule upon changing ionization states. Recently, a number of approaches have been developed to allow changes in protonation states during MD simulations (reviewed in[48]). These techniques allow to directly investigate the coupling of conformational changes and associated electrostatic potential changes to protonation changes of ionizable groups. Such methods may also find applications in the nucleic acid simulation field in cases where the ionization states of nucleo-bases can vary.

Many molecular mechanics calculations require not only energies of a given biomolecular structure but also gradients (forces) with respect to the atom positions. Unfortunately, it turns out to be difficult to extract sufficiently accurate forces from finite-difference solutions of the Poisson-Boltzmann equation[49]. One principle limitation is here the correct extrapolation of forces at solute-solvent boundaries on neighbouring atoms that limits the applicability in molecular dynamics calculations and energy minimization[49]. A possibility to avoid the requirement to calculate forces is to use Monte Carlo type conformational search methods that only require the calculation of energies[50]. However, alternative methods to approximately calculate electrostatic solvation contributions have been developed (e.g., the Generalized Born method, see below). These methods allow to rapidly calculate gradients of the solvation contribution and to use them in molecular mechanics applications.

2.2.2   Improved Treatment of Nonpolar Solvation

The assumption of a linear relationship between nonpolar solvation and solvent accessible surface area is largely based on the experimental observation that the free energy of hydration of n-alkanes increases linearly with alkane length. However, already in case of cyclic alkanes the model based on parameters derived from n-alkanes significantly overestimates hydration free energies of cyclic alkanes. Gallicio et al.[51] have found that this can be due to more favorable solute-solvent dispersion interactions of cyclic alkanes compared to linear alkanes (of similar surface area). Zacharias[52] has developed a rapid continuum modeling approach to calculate non-polar solvation contributions based on a separation of hydrophobic contributions due to water reordering around a molecule and solute-solvent dispersion interactions. The hydrophobic part is assumed to be proportional to the surface area whereas the solute-solvent dispersion interactions are rapidly calculated using a surface integral method[52]. This approach showed excellent agreement of calculated free energies of hydration and experiment for linear, branched and cyclic alkanes. In addition, it also compared much better to explicit solvent simulations on the contribution of nonpolar solvation to the conformational transitions. Levy et al.[53] found that solute solvent dispersion interactions per surface area can largely differ between different proteins and different conformations of proteins. An efficient method also based on a separation of surface area based cavity contribution and favourable solute-water dispersion interactions has been developed termed AGBNP indicating improved performance to model nonpolar solvation[54].

2.2.3   Generalized Born Continuum Solvent Modeling

In recent years a variety of approximate continuum solvent methods has been developed that allow both a reasonably accurate calculation of reaction field contributions and the calculation of solvation forces. One of the most successful methods is the Generalised Born (GB) approach originally developed and applied to organic molecules by Still et al.[55]. The method has been modified and extended by Hawkins et al.[56,57] and Schäfer & Karplus[58] and to include salt effects by Srinivasan et al.[59]. Excellent overviews on many more recent developments on GB models can be found in[60,61]. In this method the electrostatic energy (ΔG_{el}) of a molecule in solution is approximated by,

$$\Delta G_{el} = -\frac{1}{2}\left(1-\frac{1}{\varepsilon_w}\right)\sum_{i,j}\frac{q_i\,q_j}{f^{GB}(r_{ij},R_i,R_j)} + \sum_{i,j,i>j}\frac{q_i\,q_j}{r_{ij}} \tag{5}$$

where r_{ij} is the distance between charges q_i and q_j, respectively, ε_w is the solvent dielectric constants, and R_i is the effective solvation (Born) radius of atom i (respectively j). The first part of Eq. (5) represents the electrostatic solvation contribution to the molecule in aqueous solution. The generalized Born function (6), is a function that interpolates between effective Born radii short and long distances. The second sum in the equation corresponds to the Coulomb interactions between charges in the molecule. The first sum can be split into two parts: a sum of self-energy contributions (ΔG_i^{self}) to solvation (i = j),

$$f^{GB}(r_{ij},R_i,R_j) = \sqrt{r_{ij}^2 + R_i\,R_j\,e^{(-r_{ij}^2/4R_i R_j)}} \tag{6}$$

$$\Delta G_i^{self} = -\left(1-\frac{1}{\varepsilon_w}\right)\frac{q_i^2}{2R_i} \tag{7}$$

and interaction contributions (ΔG_{ij}int) to solvation:

$$\Delta G_{ij}^{int} = -\left(1-\frac{1}{\varepsilon_w}\right)\frac{q_i q_j}{\sqrt{r_{ij}^2 + R_i R_j \, e^{\left(-r_{ij}^2/4R_i R_j\right)}}} \tag{8}$$

In the hypothetical case of no interactions between the charges the total electrostatic solvation free energy would be given by the sum of self-energy contributions which are given as a sum of the familiar Born ion solvation free energies (Eq. 7). The self energy of the charge (ΔG_iself) corresponds to the electrostatic energy of the charge at its location in the molecule in solution vs. its energy in the gas phase. The effective solvation radii for each atom are (ideally) chosen such that the Born energy of a charge with effective Born radius R_i is equal to the self-energy of the isolated charge in the solute cavity. The effective Born radii, R_i, can be thought of an average distance of the selected atom from the solvent or from the solvent accessible surface of the molecule. Numerical as well as approximate analytical approaches to calculate effective Born radii of atoms from the distribution of atoms in the molecule have been developed. Among the most popular is the analytical method by Hawkins et al.[56,57], to calculate effective Born radii approximately from interatomic distances. This method also allows straight forward calculation of gradients of effective Born radii with respect to atomic coordinates.

The electrostatic solvation free energy of a charge distribution is not simply the sum of the contributions of each individual charge (a dipole generates a much smaller reaction field than two isolated charges). This is accounted for by the interaction contribution to solvation (ΔG_{ij}int, Eq. 8) given by the f[GB] interpolation function. Both the effective calculation of Born radii as well as the f[GB] interpolation function are main approximations of the GB model. Onufriev et al.[62] have shown that the original f[GB] function by Still et al.[55] is in fact a quite reasonable approximation if one uses "exact" effective Born radii that can be obtained from a finite-difference solution of the Poisson equation. Improvements of the GB model are therefore expected from an improved calculation of effective Born radii. Such improvements have for example been suggested by Onufriev et al.[62, 63] and others (e.g. Galliccchio & Levy[54]). A detailed comparison of the various GB and PB models has been performed by Feig et al.[64]. For more details on the GB model and a more comprehensive overview on recent developments the reader is referred to several excellent recent reviews[60,61].

The GB electrostatic solvation can be calculated relatively rapidly and a main advantage compared to finite-difference and boundary-element solutions of the Poisson and PB equations is the possibility to calculate forces due to the electrostatic solvation on each atom. It is therefore possible to apply this method in molecular mechanics calculations such as energy minimization, conformational searches and molecular dynamics simulations. The GB model in combination with a surface area dependent nonpolar solvation term is often refereed to as GB/SA model. However, the approximations inherent to continuum solvent models in general and the generalised Born method in particular needs to extensively evaluated and tested for various possible applications.

2.3 Molecular Dynamic Simulation on Nucleic Acids Using Continuum Models

Implicit solvation models based either only on the solvent accessible surface area or on the GB/SA model have been used quite extensively in molecular dynamics simulations of proteins and peptides. Simulations including explicit solvent molecules are time consuming due to the larger number of particles compared to implicit solvent simulations on the one hand but also due to the viscosity of the solvent that slows down conformational transitions during the simulation. Typically dynamics simulations using implicit solvent models employ a Langevin dynamics algorithm that adds a stochastic component to the force (calculated from the derivative of the energy function) to mimic collisions with solvent molecules and a damping term to represent the viscosity of the solvent. By adjusting the implicit viscosity parameter it is possible to "speed up" the "diffusive" motion and accelerate conformational transitions. Implicit solvent models have been used successfully for folding simulations of peptides and small proteins[25,65,66] reaching simulation times in the range of μs[25]. The number of continuum solvent applications on dynamics simulations of nucleic acids is significantly smaller. Tsui and Case[67] compared the dynamics of double-stranded (ds)DNA using the GB model with simulations in the presence of explicit solvent and counter ions. On the timescale of several nanoseconds quite similar dynamical behaviour and similar average structures were observed. However, Giudice and Lavery[20] reported that on other DNA sequences the agreement is less satisfactory. Williams & Hall[68,69] have used MD simulations combined with the GB approach to study the energetics and dynamics of UUCG RNA hairpin tetraloops. Pande and coworkers[70,71] studied folding events of the GNRA tetraloop using massively parallel simulations approaches. However, in a recent simulation study using a similar approach but in explicit solvent by the same group[72] a different structure formation mechanism has been proposed as in the study using a GB model. Cheng et al.[73] have combined the LES-technique (locally enhanced multi-copy sampling) with the GB approach and applied it successfully to study the formation of an UUCG tetraloop (starting from an incorrect structure) in good agreement with experiment. Realistic MD simulations of nucleic acid structures using continuum solvent models are highly desirable to improve the sampling of conformational sub-states and to investigate larger scale conformational transitions than currently possible using explicit solvent methods.

2.3.1   Poisson-Boltzmann/Generalized Born Calculations on Conformational Snapshots

Continuum solvent methods allow to calculate an equilibrium solvent response (reaction field) for a given structure and charge distribution of a molecule. Explicit solvent simulations on the other hand show large fluctuations of the solvent reaction field even in case of a static solute molecule. This is due to the superposition of the rapidly fluctuating electrostatic fields of many water molecules in permanent motion during the simulation. Hence, long simulation times are required to calculate an equilibrated interaction between solute and the many surrounding explicit solvent molecules. However, explicit solvent simulations provide a principally more accurate description of the biomolecule and its environment. It is assumed that these simulations result in more realistic molecule conformations than simulations that employ a continuum solvent description. In the MM-PBSA (Molecular Mechanics Poisson Boltzmann Surface Area) method one tries to combine the advantages of explicit solvent simulations to sample more realistic structures with the advantage of continuum models to provide a more rapidly converging and less fluctuating solvation contribution, reviewed in[19]. In MM-PBSA calculations water molecules are removed from the trajectory generated during an explicit solvent MD simulation. The sampled solute conformations are reevaluated using a GB (MM-GBSA) or FDPB (MM-PBSA) continuum solvent model. In combination with methods to estimate the conformational entropy of the biomolecule under study it is possible to approximately calculate a free energy of the biomolecule. This allows to compare free energies of different conformational states, e.g. ensembles that represent folded and unfolded conformational states, or to study ligand binding to biomolecules. The method has found broad applications on many types of biomolecules including applications to nucleic acids and complexes with proteins or organic ligands[19,74-80]. However, in MM/PBSA calculations for example on ligand binding to a biomolecule the (free) energy of binding corresponds typically to a small difference of large quantities for the bound and unbound states of the biomolecule and ligand. This requires careful analysis of the convergence of the calculated MM/PBSA (free) energies and can involve long simulation times to achieve converged results. The neglect of the discrete nature of water molecules can also limit the accuracy of the approach especially in case of nucleic acids. Recent improvements involve the inclusion of a few explicit waters to account for the discrete nature of water close to the region of interest[81].

2.4 DNA and RNA as Drug Targets

The exploration of new target structures for anti-bacterial and anti-viral agents is of importance due to the increasing emergence of bacterial and viral strains that are resistant to current drug treatments. The regular double-stranded structure of RNA shows only little sequence dependent structural variation and is difficult to target specifically by small ligands. In case of dsDNA, however, impressive progress has been achieved in recent years by using polyamide derivatives that bind to the minor groove of DNA in a sequence dependent manner[82]. With such ligands a sequence dependent reading of the DNA minor groove and interference with protein-DNA binding appears to be possible.

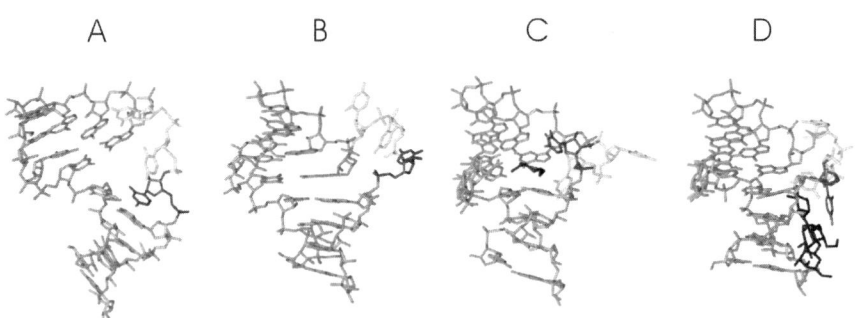

Figure 4-2. Structural variability of the HIV-TAR bulge region.(A) NMR structure of the free TAR bulge[86], (pdb1anr);(B) X-ray structure of the free TAR bulge in the presence of Ca2+[85] (pdb397d); (C) TAR-NMR structure in the presence of argininamid[86] (pdb1arj); (D) solutions structure with bound Neomycin B[87] (pdb1qd3). For clarity in each case only the TAR bulge region (sequence UCU) and four flanking base pairs are shown (5'-C_{19}AGAUCUGAGC/5'-G_{36}CUCUCUG; numbering according to HIV sequence). The U_{23} bulge base and bound ligands are in dark grey and other bulge bases are in light grey.

Folded RNAs contain often non-helical structural motifs such as bulges, internal loops or mismatches that interrupt double stranded regions with unique cavities and surface structures that may allow specific recognition by proteins or small organic molecules. The presence of such unique interaction surfaces for possible specific ligand binding makes it an interesting target for drug design[14-17]. In addition to experimental drug screening approaches, the availability of structural information on many RNA containing molecules may also allow structure-based computational nucleic acid drug design[83]. However, the rational computer-aided design or identification of putative ligands that specifically bind to nucleic acid structural motifs faces a number of difficulties that need to be appropriately tackled by any computational design method. In general a computational docking or design approach needs to be rapid enough to allow systematic screening of many putative ligands, ligand conformations and ligand binding sites on the nucleic acid receptor. On the other hand ligand-nucleic acid interactions and the effect of the solvent environment need to be described sufficiently accurate to allow realistic predictions of possible ligands, binding modes and conformations. This highly demanding computational task can only be tackled using sufficiently accurate continuum solvent models. However, ligand design in case of nucleic acid motifs faces the additional problem that often non-helical motifs in nucleic acids are flexible and can adopt a variety of conformations. This is exemplified for the human immunodeficiency virus (HIV) trans-activation response (TAR) bulge motif[84]. This motif located near the 5'-end of the viral RNA is recognized by the viral Tat protein and host proteins to activate transcription and is essential for effecient viral replication. A number of studies have focused on the design of ligands that bind to HIV-TAR and inhibit its function[84]. The HIV-TAR bulge is an example for a motif with a high degree of conformational flexibility. Several experimental studies of the TAR bulge element either in free form[85,86] or in complex with argininamid[86], neomycin B[87] or synthetic drug molecules[88] indicate a variety of possible motif structures. The bulge structure ranges from fully looped out bulge bases seen in the crystal structure of the free TAR bulge to partially stacked bulge bases in the NMR solution structure (illustrated in Figure 4-2). In the presence of peptides derived from the TAR binding domain of the Tat protein or upon addition of milli-molar amounts of argininamid the TAR element undergoes a conformational change and adopts conformations with one of the bulge bases in a position such that it may form a base triple with an adjacent stem base pair (Figure 4-2c). The apparent high flexibility of the TAR bulge structure or small energetic penalty to adopt a variety of structures indicates a high capacity for conformational adaptation to bind a molecule with high affinity. This has, however, also an effect on the variety of molecules that are bound with high affinity. In case of the TAR element chemically and structurally very different ligand molecules can bind in micromolar and submicromolar range[86-88]. Experimental as well as theoretical studies on other non-helical motifs indicate that a high conformational flexibility is not unique to the TAR bulge. For example conformational search studies and molecular dynamics simulations on bulges in nucleic acids[89,90] indicate a significant conformational freedom of the bulge bases compared to base-paired regions. As a consequence of this observation docking methods to predict possible binding sites and binding geometries on nucleic acid motifs need to allow for conformational flexibility of the RNA motif structure. In addition, the conformational flexibility of some motifs can also affect the specifity of ligand binding such that a high affinity ligand designed for one motif may also bind to other motifs that have a similar capacity for conformational adaptation.

2.5 Application of Continuum Solvent Models to Nucleic Acid Motif Structure Prediction

Double-stranded regions in biological RNA molecules are frequently interrupted by non-helical structural motifs such as internal loops, bulges and hairpin loops[3]. These motifs are of functional importance as recognition sites for proteins and other ligands and can also mediate tertiary interactions in large RNA-protein complexes such as the ribosome. Many nucleic acid structural motifs have been studied using NMR spectroscopy or X-ray crystallography[5-12]. However, there is still a demand for structure prediction of unknown motifs or sequence variants of already structurally characterised motifs. Also, as described in the previous paragraph flexible motifs may undergo conformational changes upon ligand-binding. Computational docking of ligands to structural motifs may require a coupled conformational search to identify possible motif conformations compatible with ligand binding. The accessible time scale of current molecular dynamics simulations limits its applicability for structure prediction of nucleic acid motifs. Replica-exchange molecular dynamics[70] and local enhanced sampling (LES) methods[73] could be useful in the future to tackle structure prediction of nucleic acid motifs. In the LES method part of the nucleic acid is represented by several conformational copies. This method has recently been combined with the GB continuum model[73] and applied to study the structure of a UUCG tetraloop starting from an incorrect loop structure. A structure close to experiment was obtained after a relatively short LES-MD simulation time. We have developed rapid conformational search approaches for generating sterically possible conformations for a given structural motif surrounded by regular helical RNA or DNA structures[89,91]. These methods are based on combining nucleic acid substructures to build sterically possible confor-mations that can be embedded into or close (in case of hairpins) a regular helix and serve as start structures for energy minimization. For the search a modified version of the program Jumna (Junction minimization of nucleic acids)[92] developed by Lavery and coworkers has been used. This program uses a combination of helical and internal coordinates to describe and to energy minimize nucleic acid structures. It is especially useful for splitting nucleic acid structures into rigid and flexible parts, a systematic search is only performed for the mobile part of the structure.

Extensive conformational searches have for example been performed on three experimentally known trinucleotide hairpin loops (AGC, AAA and GCA) closed by a four base pair stem[91]. A GB/SA (and FDPB/SA) solvation model has been employed during energy minimization and conformational searches. The three loop sequences with known structures[93-95] served as test systems to investigate the usefulness of FDPB and GB based methods for structure prediction. For all three loop sequences conformations close to experiment were found as lowest energy structures among several thousand alternative energy minima[91]. The inclusion of electrostatic solvation contributions was found to be important for a realistic conformer ranking. Simpler electrostatic models such as using a constant or distance-dependent dielectric constant failed to select a structure close to experiment as lowest energy conformer. An efficient stochastic search and a single dihedral angle flip scanning method has recently been suggested based on a GB continuum model[96]. It was successfully applied to predict structures of a DNA-triloop and the UUCG RNA tetraloop in good agreement with experiment starting from various random start structures.

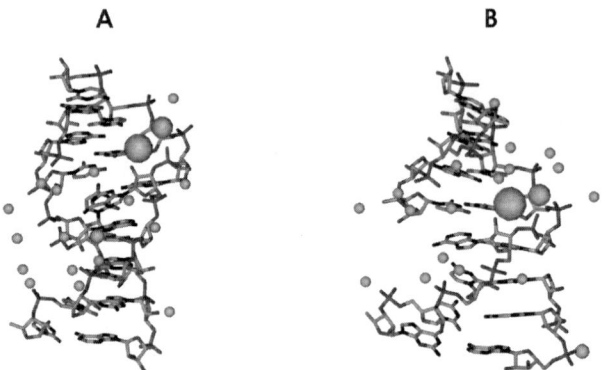

Figure 4-3. Lowest energy K⁺ ion binding sites (spheres) obtained for two AA-platform motifs from the *Tetrahymena* ribozyme P4-P6 domain[104] (pdb1gid) using a systematic search employing a continuum solvent model and allowing for conformational relaxation of the RNA[100]: (A) AA-platform motif corresponding to nucleotides 221-229/245-252 of the P4-P6 domain; (B) AA-platform formed by nucleotides 215-223 (first strand) and 103-105/250-253 (second strand). In each case the largest sphere corresponds to the most favorable calculated binding sites and agreed with experiment to within a Rmsd of 1.5 Å[100].

Such searches employing several thousand energy minimizations in internal coordinates can be performed within 1-2 days of computer time. It needs to be stressed that because of computational limitations such systematic conformational searches cannot be performed using explicit solvent methods.

2.6 Docking of Ligands to Nucleic Acids

Continuum solvent methods are frequently employed in ligand-receptor docking simulations including binding of ligand and ions to nucleic acids[97-102]. Rapid prediction of possible ion binding sites on nucleic acids could be quite useful for the setup of MD simulations. Using NMR spectroscopy or low resolution X-ray crystallography it is often difficult to determine the identity and exact localisation of ions bound to nucleic acids. Hermann and Westhof[103] have successfully used Brownian dynamics simulations and a pre-calculated electrostatic potential to identify possible ion binding sites in RNA. Binding of monovalent cations to two adenine-adenine platform structures from the *Tetrahymena* group I intron ribozyme have been studied based on GB and FDPB continuum solvent approaches[100]. The adenine-adenine platform RNA motif forms an experimentally characterised monovalent ion binding site important for ribozyme folding and function[104]. Systematic energy minimization docking simulations starting from several hundred initial placements of potassium ions at the surface of platform containing RNA fragments identified a binding geometry in close agreement with experiment as the lowest energy binding site[100] (Figure 4-3). This study included both structural relaxation of the RNA during the search and desolvation contributions. Qualitative agreement between calculated and experimentally available data on ion binding selectivity was obtained. The inclusion of solvation effects turned out to be important to obtain low energy structures and ion binding placements in agreement with experiment. The calculations indicated that differences in solvation of the isolated ions contribute to the calculated ion binding preference. However, Coulomb attraction and van der Waals interactions due to ion size differences and RNA conformational adaptation also influenced the calculated ion binding affinity. The calculated alkali ion binding selectivity for both platforms followed a typical Eisenman selectivity series (IV) of the order: $K^+ > Na^+ > Rb^+ > Cs^+ > Li^+$ which changed to a selectivity series V in case of using rigid RNA. It indicates that in this case binding specificity is largely determined by a cavity-dependent balance between desolvation and ion-RNA-Coloumb attraction[100]. The calculations also predicted in agreement with experiment that divalent ions are unlikely to bind at the AA platform ion-binding site due to a large desolvation penalty.

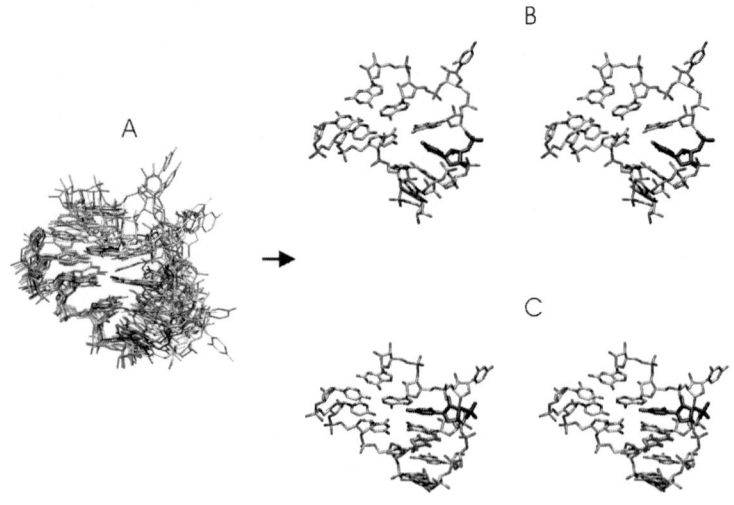

Figure 4-4. Conformational search on the BIV-TAR bulge motif[105]: (A) Superposition of a subset of low energy structures obtained from a conformational search employing a GB continuum solvent model implemented in the Jumna program[91,92]; (B) Lowest energy structure (stereo view) obtained in the absence of ligand showing the uridine bulge base (in bold) in a stacked conformation in agreement with the experimental observation. (C) Second lowest energy structure (in stereo) obtained from a conformational search in the presence of an arginine side chain (in bold and dark grey) located close to the Guanine base of the central G:C base pair. For clarity only the central BIV-TAR motif including to uridine bulges separated by a single G:C base pair and two flanking base pairs are shown.

As indicated in paragraph 2.5 and Figure 4-2 ligand binding can induce significant conformational changes in the RNA target structure. Methods to predict possible ligand binding sites on RNA molecules need to appropriately account for such induced fit conformational changes during docking of ligands to nucleic acid structures. It is possible to combine conformational search approaches employing a GB solvation approach with ligand docking studies. As an example the application on the binding of an arginine side chain to the bovine immuno deficiency virus (BIV)-TAR bulge is illustrated in Figure 4-4. The BIV-TAR bulge motif has a similar function as described in the section 2.5 for the HIV-TAR element and consists of two single uridine bulges separated by a single G:C base pair[105,106]. Similar to the HIV-TAR bulge the BIV-TAR bulge can undergo significant conformational changes in the presence of Arginin-containing peptides derived from the BIV-Tat protein. In the ligand free state the first uridine adopts an intra-helical stacked structure[106] (the second uridine was found to adopt a looped out state in both free and ligand bound structures[106]). Ligand binding induces a conformational transition of the first uridine bulge from a stacked structure to a base triple conformation (in the major groove) with the next-nearest A:U base pair along the stem[106] (Figure 4-4). A conformational search on the first uridine bulge motif and neighboring nucleotides (as described in section 2.6) resulted in ~350 distinct energy minima. During the search a GB continuum model was employed. Among the minima a stacked conformation as shown in Figure 4-4b was found as the conformation of lowest energy. This agrees with the experimental finding of a stacked conformation in the absence of Arginin-rich BIV-Tat-derived peptides or a single Argininamid ligand[106]. The observed minima were reminimized in the presence of an Arginin side chain. During a first minimization the Arginin side chain was weakly restrained to stay close to the guanine base (major groove side) of the G:C base pair between the two uridine bulges. This guanine base is known to contact a key Arginin side chain in the Tat-peptide-TAR complex[106]. In a second energy minimization the system was free to move to the closest energy minimum. In this search a base triple conformation (Figure 4-4c) was selected as the second best energy minimum. The base triple form showed a relatively high ranking already in the absence of the ligand (rank 6). The base triple form showed good agreement with the experimental structure in the presence of a Tat-peptide-bound BIV-TAR complex (Rmsd of the nucleotides contacting a critical Arginin side chain <1.5 Å). The stacked structure found to be the preferred conformation in the absence of the ligand was not among the ten top-ranked structures of the conformational search in the presence of the ligand. The result demonstrates that such rapid conformational search (~6 h computer time on a PC-workstation) could be useful to study putative ligand binding sites in nucleic acid structural motifs representing flexibility as an ensemble of target structures. It is of course possible to limit the reminimization in the presence of ligands to a small subset of low-energy structures after the initial search. Also, the generated minima can be used for docking several different ligands or for evaluating various putative binding regions.

2.7   Conclusions and Outlook

Macroscopic continuum solvation concepts in particular the Poisson-Boltzmann approach have an enormous influence on the current theoretical understanding of biomolecular structure formation and association processes. The application of continuum solvent models to investigate nucleic acids range from the analysis of solvation contributions and salt effects on structural stability to the study of ligand binding and structure prediction. A number of studies indicate that conformational search procedures based on continuum solvent models could be useful for structure prediction of nucleic acid motifs and during ligand-nucleic acid docking studies. At the expense of a less accurate solvent representation compared to explicit solvent methods, continuum solvent approaches also hold the potential for much longer simulation times in MD simulations and broader sampling of possible conformational states. However, improvements of current methods, e.g. the explicit inclusion of a number of critical water molecules, are desirable. Further testing and careful comparison with simulations using explicit solvent are necessary to evaluate the accuracy of current continuum models for dynamics simulation applications.

ACKNOWLEDGEMENTS

Research work presented has been performed in part by C. Burkardt, F. Pineda and G. Villescas-Diaz in my group. I also want to thank A. Barthel, D. Roccatano and H. Sklenar for many helpful discussions. The work was funded in part by the Deutsche Forschungsgemeinschaft (grant ZA 153/3-1).

REFERENCES

1. Schroeder, R., Barta, A. & Semrad, K. (2004). Strategies for RNA folding and assembly. *Nat. Rev. Mol. Cell Biol.* **5**, 908-919.
2. Ferre D'Amare, A. R. & Doudna, J. A. (1999). RNA folds: insights from recent crystal structures, *Annu Rev. Biophys. Biomol. Struct.* **28**, 57-73.
3. Zacharias, M. (2000). Simulation of the structure and dynamics of nonhelical RNA motifs. *Curr. Opin. Struct. Biol.* **10**, 307-311.
4. Hall, K. B. (2002). RNA-protein interactions. *Curr. Opin. Struct. Biol.*, **12**, 283-288.
5. Leontis, N. B. & Westhof, E. (2003). Analysis of RNA motifs. *Curr. Opin. Struct. Biol.*, **13**, 300-308.
6. Perez-Canadillas, J. M., & Varani, G. (2001). Recent advances in RNA-protein recognition. *Curr Opin Struct Biol.* **11**, 53-58.
7. Zidek, L., Stefl, R. & Sklenar, V. (2001). NMR methodology for the study of nucleic acids.*Curr Opin Struct Biol.* **11**, 275-281.
8. Moore, P. B. & Steitz, T. A. (2002). The involvement of RNA in ribosome function. *Nature* **418**, 229-235.
9. Furtig, B., Richter, C., Wöhnert, J. & Schwalbe, H. (2003). NMR spectroscopy of RNA. Chembiochem. **4**, 936-962.
10. Egli, M. (2004). Nucleic acid crystallography: current progress. *Curr. Opin. Chem. Biol.* **8**, 580-591.
11. Ke, A., Doudna, J. A. (2004). Crystallization of RNA and RNA-protein complexes. *Methods.* **34**, 408-414.
12. Wu, H., Finger, L. D., Feigon, J. (2005). Structure determination of protein/RNA complexes by NMR. *Methods Enzymol.* **394**, 525-545.
13. Afshar, M., Prescott, C. D. & Varani, G. (1998). RNA as drug target: chemical, modeling, and evolutionary tools *Curr. Opin. Biotech.* **9**, 66-73.

14. DeJong, E. S., Luy, B. & Marino, J. P. (2002) RNA and RNA-protein complexes as targets for therapeutic intervention. *Curr. Top. Med. Chem.* **2**, 289-302.

15. Schroeder, R., Waldsich, C. & Wank, H. (2000). Modulation of RNA function by aminoglycoside antibiotics. *EMBO J.* **19**, 1-9.

16. Vicens, Q. & Westhof, E. (2003). RNA as a drug target: the case of aminoglycosides. *Chembiochem.* **4**, 1018-1023.

17. Zacharias, M. (2003). Perspectives of Drug design that targets RNA. *Curr. Med. Chem-AIA* **2**, 161-172.

18. Cheatham, T. E. & Kollman, P. A. (2000). Molecular dynamics simulation of nucleic acids. *Annu. Rev. Phys. Chem.* **51**, 435-471.

19. Kollman, P. A., Massova, I., Reyes, C., Kuhn, B., Huo, S., Chong, L., Lee, M., Duan, Y., Wang,W., Donini, O., Cieplak, P., Srinivasan, J., Case, D. A. & Cheatham 3rd. T. E. (2000). Calculating structures and free energies of complex molecules: combining molecular mechanics and continuum models. *Acc. Chem. Res.* **33**, 889–897.

20. Guidice, E. & Lavery, R. (2002). Simulation of nucleic acids and their complexes. *Acc. Chem. Res.* **35**, 350-357.

21. Cheatham, T. E. (2004). Simulation and modeling of nucleic acid structure, dynamics and interactions. *Curr. Opin. Struct. Biol.* **14**, 360-367.

22. Eisenberg, D. & McLachlan, A. D. (1986). Solvation energy in protein folding and binding. *Nature* **319**, 199-203.

23. Juffer, A. H., Eisenhaber, F., Hubbard, S. J., Walter, D. & Argos, P. (1995). Comparison of atomic solvation parametric sets: applicability and limitations in protein folding and binding. *Prot. Sci.* **4**, 2499-2509.

24. Ferrara, P., Apostolakis, J. & Caflisch, A. (2002). Evaluation of a fast implicit solvent model for molecular dynamics simulations. *Proteins* **46**, 24-33.

25. Paci, E., Cavalli, A., Vendruscolo, M. & Caflisch, A. (2003) Analysis of the distributed computing approach applied to the folding of a small beta peptide. *Proc. Natl. Acad. Sci U S A.* **100**, 8217-8222.

26. Warwicker, J. & Watson, H. C. (1982). Calculation of the electric potential in the active site cleft due to alpha-helix dipoles. *J. Mol. Biol.* **157**, 671-679.

27. Davis, M. E. & McCammon, J. A. (1990). Electrostatics in biomolecular structure and dynamics. *Chem. Rev.* **90**, 509-521.

28. Sharp, K. A. & Honig, B. (1990). Calculating total electrostatic energies with the nonlinear Poisson-Boltzmann equation. *J. Phys. Chem.* **94**, 7684-7692.

29. Honig, B., Sharp K. A. & Yang, A-S. (1993). Macroscopic models of aqueous solutions: biological and chemical implications. *J. Phys. Chem.* **97**, 1101-1109.

30. Gilson, M. K. (1995). Theory of electrostatic interactions in macromolecules. *Curr. Opin. Struc. Biol.* **5**, 216-223.

31. Madura, J. D., Davis, M. E., Wade, R., Luty, B. A., Ilin, A., Anosiewicz, A., Gilson, M. K., Bagheri, B., Ridgway-Scot, R. & McCammon, J. A.(1995). Electrostatics and diffusion of molecules in solution: simulations with University of Houston Brownian dynamics program. *Comput. Phys.Commun.* **91**, 57-95.

32. Nicholls, A., Sharp, K. A. & Honig, B. (1991). Protein folding and association: insights from the interfacial and thermodynamic properties of hydrocarbons. *Proteins.* **11**, 281-296.

33. Holst, M., Kozack, R. E., Saied, F. & Subramaniam, S. (1994). Treatment of electrostatic effects in proteins: multigrid-based Newton iterative method for solution of the full nonlinear Poisson-Boltzmann equation. *Proteins* **18**, 231-245.

34. Zauhar, R. J. & Morgan, R. S. (1985). A new method for computing the macromolecular electric potential. *J. Mol. Biol.* **86**, 815-820.

35. Bharadwaj, R., Windemuth, A., Sridharan, S., Honig, B. & Nicholls, A. (1995). The fast multipole boundary element method for molecular electrostatics: An optimal approach for large systems. *J. Comput. Chem.* **16**, 898-913.

36. Lu, B., Zhang, D. & McCammon, J. A. (2005). Computation of electrostatic forces between solvated molecules determined by the Poisson-Boltzmann equation using a boundary element method. *J. Chem. Phys.* **122**, 21410-21412.

37. Sitkoff, Do., Sharp, K. A. & Honig, B. (1994). Correlating salvation free energies and surface tensions of hydrocarbon solutions. *Biophys Chem.* 51, 397-403.

38. Schmidt, A. B. & Fine, R. M. (1996). Size effects in nonpolar salvation: lessons from two simple models. *Biophys. Chem.* **57**, 219-224.

39. Sitkoff, D., Sharp, K. A. & Honig, B. (1994). Accurate calculation of hydration free energies using macroscopic solvent models. *J. Phys. Chem.* **98**, 1978-1988.

40. Jayaram, B., Sprous, D. & Beveridge, D. L. (1998). Solvation free energies of biomacromolecules: parameters for a modified generalized Born model consistent with the Amber force field. *J. Phys. Chem.* **102**, 9571-9576.

41. Zacharias, M., Luty, B. A., Davis, M. E., McCammon, J. A. (1992). Poisson-Boltzmann analysis of the lambda-repressor-operator interaction. *Biophys. J.* **63**, 1280-1285.

42. Misra, V. K., Sharp, K. A., Friedman, R. A. & Honig, B. (1994). Salt effects on ligand-DNA binding. Minor groove binding antibiotics. *J. Mol. Biol.* **238**, 245-263.

43. Sharp K. A., Friedman R. A., Misra V., Hecht J. & Honig B. (1995). Salt effects on polyelectrolyte-ligand binding: comparison of Poisson-Boltzmann, and limiting law/counterion binding models. *Biopolymers* **36**, 245-262.

44. Misra, V. & Draper, D. E. (2000). Mg(2+) binding to tRNA revisited: the nonlinear Poisson-Boltzmann model. *J. Mol. Biol.* **299**, 813-825.

45. Garcia-Garcia, C. & Draper, D. E. (2003). Electrostatic interactions in a peptide--RNA complex. *J. Mol. Biol.* **331**, 75-88.

46. Demchuk, E. & Wade R. C. (1996). Improving the continuum dielectric approach to calculating pKs of ionizable groups in proteins. *J. Phys. Chem.* **100**, 17373-17387.

47. Antosiewicz, J, McCammon, J. A. & Gilson, M. K. (1996). The determinants of pKas in proteins. *Biochemistry* **35**, 7819-7833.

48. Morgan, J. & Case, D. A. (2005). Biomolecular simulations at constant pH. *Curr. Opin. Struc.t Biol.* **15**, 157-163.

49. Gilson, M. K., Davis, M. E. Luty, B. A. & McCammon, J. A. (1993). Computation of electrostatic forces on solvated molecules using the Poisson-Boltzmann equation. *J. Phys. Chem.* **97**, 3591-3600.

50. Zacharias, M., Luty, B. A., Davis, M. E., & McCammon, J. A. (1994). Combined conformational search and finite-difference Poisson-Boltzmann approach for flexible docking. *J. Mol. Biol.* **238**, 455-465.

51. Gallicchio, E., Kubo, M. M. & Levy, R. M. (2000) Enthalpy-entropy and cavity decomposition of alkane hydration free energies: Numerical results and implications for theories of hydrophobic solvation. *J. Phys. Chem. B* **104**, 6271-6285.

52. Zacharias, M. (2003). Continuum solvent modeling of non-polar solvation: Improvement by separating surface area dependent cavity and dispersion contributions. *J. Phys. Chem. A*, **107**, 3000-3004.

53. Levy, R. M., Zhang, L. Y., Gallicchio, E., Felts, A. K. (2003). On the nonpolar hydration free energy of proteins: surface area and continuum solvent models for the solute-solvent interaction energy. *J. Am. Chem. Soc.* **25**, 9523-9527.

54. Gallicchio, E., M. & Levy, R. M. (2004). AGBNP: An analytical implicit solvent model suitable for molecular dynamics simulations and high-resolution modeling. *J. Comput. Chem.* **25**, 479-499.

55. Still, W. C., Tempczyk, A., Hawley, R. C. & Hendrikson, T. (1990). Semianalytical treatment of solvation for molecular mechanics and dynamics. *J. Am. Chem. Soc.* **112**, 6127-6129.

56. Hawkins G. D., Cramer C. J., Truhlar D. G. (1995). Pairwise solute descreening of solute charges from a dielectric continuum. *Chem. Phys. Lett.* **246**, 122-129.

57. Hawkins, G. D., Cramer C. J., Truhlar D. G. (1996). Parameterized models of aqueous free energies of solvation based on pairwise descreening of solute atomic charges from a dielectric medium. *J. Phys. Chem.* **100**, 19824-19839.

58. Schaefer, M. & Karplus, M. (1996). A comprehensive analytical treatment of continuum electrostatics. *J. Phys. Chem.* **100**, 1578-1599.

59. Srinivasan, J., Trevathan, M. W., Beroza, P. & Case, D. A. (1999). Application of a pairwise generalized Born model to proteins and nucleic acids: inclusion of salt effects. *Theor. Chem. Acc.* **101**, 426-434.

60. Bashford, D. & Case, D. A., (2000). Generalized Born models of macromolecular solvation effects. *Annu. Rev. Phys. Chem.* **51**, 129-152.

61. Tsui, V. & Case, D. A. (2001). Theory and application of the generalized Born solvation model in macromolecular simulations, *Biopolymers* **56**, 275-291.

62. Onufriev, A., Case, D. A. & Bashford, D. (2002). Effective Born Radii in the Generalized Born Approximation: The importance of being perfect, *J. Comput. Biol.* **23**, 1297-1304.

63. Onufriev A., Bashford, D. & Case, D. A. (2000). Modification of the Generalized Born model suitable for macromolecules. *J. Phys. Chem. B* **104**, 3712-3720.

64. Feig, M., Onufriev, A., Lee, S. M., Im, W., Case, D. A., Brooks III, C. L. (2003). Performance comparison of generalized Born and Poisson Methods in the calculation of electrostatic solvation energies for protein structures. *J. Comput. Chem.* **25**, 265-284.

65. Ferrara, P. & Caflisch, A. (2000). Folding of a three-stranded antiparallel beta-sheet peptide. Proc. Natl. Acad. Sci. USA **97**, 10780-10785.

66. Nymeyer, H., Gnanakaran, S. & Garcia, A. E. (2003). Atomic simulations of protein folding, using the replica exchange algorithm. *Methods Enzymol.* **383**, 119-149.

67. Tsui, V. & Case, D. A. (2000). Molecular dynamics simulations of nucleic acids with a generalized Born solvation model. *J. Am. Chem. Soc.* **122**, 2489-2498.

68. Williams D. J. & Hall K. B. (1999). Unrestrained stochastic dynamics simulations of the UUCG tetraloop using an implicit solvation model. *Biophys. J.* **76**, 3192-3205.

69. Williams D. J. & Hall K. B. (2000). Experimental and theoretical studies of the effects of deoxyribose substitutions on the stability of the UUCG tetraloop. *J. Mol. Biol.* **297**, 251-265.

70. Sorin, E. J., Min Rhee, Y., Nakatani, B. J. & Pande, V. S. (2003). Insights into nucleic acids conformational dynamics from massively parallel stochastic simulations, *Biophys. J.* **85**, 790-803.

71. Sorin, E. J., Engelhardt, M. A., Herzschlag, D. & Pande, V. S., (2002). RNA simulations: probing hairpin unfolding and the dynamics of a GNRA tetraloop, *J. Mol. Biol.* **317**, 493-506.

72. .Sorin, E. J., Rhee, Y. M. & Pande, V. S. (2005). Does water play a structural role in the folding of small nucleic acids? *Biophys J.* **88**, 2516-2524.

73. Cheng, X., Hornak, V. & Simmerling, C. (2004). Improved conformational sampling through an efficient combination of mean-field simulation approaches. *J. Phys. Chem. B* **108**, 426-437.

74. Srinivasan, J., Miller, J., Kollman, P. A. &. Case, D. A. (1998). Continuum solvent studies of the stability of RNA hairpin loops and helices. *J. Biomol. Struct. Dyn.* **16**, 671-682.

75. Gouda, H., Kuntz, I. D., Case, D. A. & Kollman, P.A. (2003). Free energy calculations for theophylline binding to an RNA aptamer: comparison of MM-PBSA and thermodynamic integration methods. *Biopolymers* **68**, 16-34.

76. Tsui, V. & Case, D. A. (2001). Calculation of absolute free energies of binding RNA and metal ions using molecular dynamics and continuum electrostatics, *J. Phys. Chem. B* **105**, 11314-11325.

77. Villescas, G. & Zacharias, M. (2003). Sequence context dependence of tandem guanine: adenine mismatch conformations in RNA: A continuum solvent analysis. *Biophys. J.* **85**, 1311-1321.

78. Reyes, C. M., Nifosi, R., Frankel, A. D. & Kollman, P. A. (2001). Molecular dynamics and binding specificity analysis of the bovine immunodeficiency virus BIV Tat-TAR complex. *Biophys J.* **80**, 2833-2842.

79. Fogolari, F. & Tosatto, S. C. (2005). Application of MM/PBSA colony free energy to loop decoy discrimination: toward correlation between energy and root mean square deviation.*Protein Sci.* **14**, 889-901.

80. Kuhn, B., Gerber, P., Schulz-Gasch, T. & Stahl, M. (2005). Validation and use of the MM-PBSA approach for drug discovery. *J Med Chem.* **48**, 4040-4048.

81. Spackova, N., Cheatham, T. E. III, Ryjacek, F., Lankaš, F., van Meervelt, L., Hobza, P. & Šponer, J. (2003). Molecular dynamics simulations and thermodynamic analysis of DNA-drug complexes. Minor groove binding between 4',6-diamidino-2-phenylindole (DAPI) and DNA duplexes in solution. *J. Am. Chem. Soc.* **125**, 1759-1769.

82. Dervan, P. B. & Burli, R. W. (1999). Sequence-specific DNA recognition by polyamides. *Curr Opin. Chem. Biol.* **3**, 688-693.

83. Froeyen, M. & Herdewijn, P. (2002). RNA as a target for drug design, the example of Tat-TAR interaction. *Curr. Top. Med. Chem.* **2**, 1123-1145.

84. Gait, M. J. & Karn, J. (1993). RNA recognition by the human immunodeficiency virus Tat and Rev proteins.*Trends Biochem. Sci.* **18**, 255-259.

85. Ippolito, J. A. & Steitz, T. A. (1998). A 1.3-A resolution crystal structure of the HIV-1 trans-activation response region RNA stem reveals a metal ion-dependent bulge conformation. *Proc. Natl. Acad. Sci. USA*, **95**, 9819-9824.

86. Aboul-ela, F., Karn, J. & Varani, G. (1995). The structure of the human immunodeficiency virus type-1 TAR RNA reveals principles of RNA recognition by Tat protein. *J. Mol. Biol.* **253**, 313-332.

87. Faber, C., Sticht, H., Schweimer, K. & Roesch, P. J. (2000). Structural rearrangements of HIV-1 Tat-responsive RNA upon binding of neomycin B. *Biol. Chem.* **275**, 20660-20667.

88. Du, Z., Lind, K. E. & James, T. L. (2002). Structure of TAR RNA complexed with a Tat-TAR interaction nanomolar inhibitor that was identified by computational screening. *Chem. Biol.* **9**, 707-712.

89. Zacharias M, & Sklenar, H. (1999). Conformational analysis of single-base bulges in A-form DNA and RNA using a hierarchical approach and energetic evaluation with a continuum solvent model. *J. Mol. Biol.* **289**, 261-275.

90. Feig, M., Zacharias, M. & Pettitt, M. E. (2001). Conformation of an adenine bulge in a DNA octamer and its influence on DNA structure from molecular dynamics simulations. *Biophys. J.* **81**, 352-370.

91. Zacharias, M. (2001). Conformational analysis of DNA-trinucleotide-hairpin loop structures using a continuum solvent model. *Biophys. J.* **80**, 2350-2363.

92. Lavery, R., Zakrzewska, K. & Sklenar. H. (1995) JUMNA (Junction minimization of nucleic acids). *Comput . Phys. Com.* **91**, 135-158.

93. Zhu, L., Chou, S. H. & Reid, B. R. (1995). Structure of a single-cytidine hairpin loop formed by the DNA triplet GCA. *Nature Struct. Biol.* **2**, 1012-1017.

94. Chou, S.-H., Zhu, L., Gao, Z., Cheng, J.-W. & Reid, B. R. (1996). Hairpin loops consisting of single adenine residues closed by sheared A:A an G:G pairs formed by DNA triplets AAA and GAG: solution structures of the d(GTACAAAGTAC) hairpin. *J. Mol. Biol.* **264**, 981-1001.

95. Chou, S.-H., Tseng, Y.-Y. & Chu, B.-Y. (1999). Stable formation of a pyrimidine-rich loop hairpin in a cruciform promoter. *J. Mol. Biol.* **292**, 309-320.

96. Villescas, G. & Zacharias, M. (2004). Efficient search approaches on energy minima for structure prediction of nucleic acid motifs. *J. Biomol. Struct. Dyn.* **22**, 355-364.

97. Zakrzewska K., Madami A. & Lavery R. (1996). Poisson-Boltzmann calculations for nucleic acids and nucleic acid complexes. *Chem. Phys.* **204**, 263-269.

98. Jayaram B., McConnell, K., Dixit, S, B., Beveridge D. L. (1999). Free energy analysis of protein-DNA binding: the EcoRI endonuclease-DNA complex. *J. Comput. Phys.* **151**, 333-357.

99. Rohs R., Sklenar H., Lavery R. & Röder, B. (2000). Methylene blue binding to DNA with alternating GC base sequence: a modeling Study. *J. Am. Chem. Soc.* **122**, 2860-2866.

100. Burkhardt, C., & Zacharias, M. (2001). Modeling ion binding to AA platform motifs in RNA: a continuum solvent study including conformational adaptation. *Nucleic Acids Res.* **29**, 3910-3918.

101. Pineda , F. & Zacharias, M. (2002). DAPI binding to the DNA minor groove: A continuum solvent analysis. *J. Mol. Recog.* **15**, 209-220.

102. Filikov, A. V., Mohan, V., Vickers, T. A., Griffey, R. H., Cook, P. D., Abagyan, R. A. & James, T. L. (2000). Identification of ligands for RNA targets via structure-based virtual screening: HIV-1 TAR. *J Comput, Aided Mol, Des.* **14**, 593-610.

103. Hermann, T. & Westhof, E. (1998). Exploration of metal ion binding sites in RNA folds by Brownian-dynamics simulations. *Structure* **6**, 1303-1314.

104. Basu, S., Rambo, R. P., Strauss-Soukup, J., Cate, J. H., Ferre-D' Amare, A., Strobel, S. A., & Doudna, J. A. (1998). A specific monovalent metal ion integral to the AA platform of the RNA tetraloop receptor. *Nature Struct. Biol.* **5**, 927-992.

105. Chen, L. & Frankel, A. D. (1994). An RNA-binding peptide from bovine immunodeficiency virus Tat protein recognizes an unusual RNA structure. *Biochem.* **33**, 2708-2715.

106. Puglisi, J. D., Chen, L., Blanchard, S. & Frankel, A. D. (1995). Solution structure of a bovine immunodeficiency virus Tat-TAR peptide-RNA complex. *Science* **270**, 1200-1203.

Chapter 5

DATA MINING OF MOLECULAR DYNAMIC TRAJECTORIES OF NUCLEIC ACIDS

Modesto Orozco[1], Agnes Noy[1], Tim Meyer[1], Manuel Rueda[1], Carles Ferrer[1], Antonio Valencia[1], Alberto Pérez[1], Oliver Carrillo[1], Juan Fernandez-Recio[1], Xavier de la Cruz[1] and F. Javier Luque[2]

[1] *Institut de Recerca Biomèdica Barcelona. Parc Científic de Barcelona. Josep Samitier 1-5. Barcelona E-08028. Spain & Departament de Bioquímica i Biologia Molecular. Facultat de Química. Universitat de Barcelona. Martí i Franquès 1. Barcelona E-08028. Spain; [2]Departament de Farmàcia. Unitat Fisicoquímica. Universitat de Barcelona. Avgda Diagonal 643. Barcelona E-08028. Spain.*

Abstract: Analysis, storage and transfer of molecular dynamic trajectories are becoming the bottleneck of computer simulations. In this paper we discuss different approaches for data mining and data processing of huge trajectory files generated from molecular dynamic simulations of nucleic acids.

Keywords: DNA models, molecular dynamics, conformational sampling, data mining, essential dynamics, flexibility, thermodynamics

1. INTRODUCTION

The discovery[1] of the structure of DNA by Watson and Crick around five decades ago opened the development of modern biochemistry and molecular biology. Since then, hundreds of nucleic acid (NA) structures have been determined experimentally, which has contributed to increase dramatically our knowledge on the structural details of NAs in different environments. This information reveals that NAs, specially DNA, are very flexible and can adapt their structure to changes in the environment. The intrinsic flexibility and dynamical properties are then crucial aspects to understand the functional roles of these molecules[2].

J. Šponer and F. Lankaš (eds.), Computational Studies of RNA and DNA, 121–145.
© 2006 *Springer.*

High resolution structural data, particularly X-ray crystallography and NMR spectroscopy, have sustained the most important advances in the knowledge of the structural features of NAs. However, those techniques are less suitable to shed light into the flexibility of NAs. In fact, the experimental set-up in X-ray and NMR experiments mainly pursues to restrict the accessible conformational space, masking then the intrinsic flexibility of the macromolecule. Theoretical methods, particularly molecular dynamics (MD), appear then to be ideal to complement experimental methods and gain insight into the structure and flexibility of NAs.

The first simulations of a NA were reported in 1983 and involved short trajectories (less than 100 ps) of a DNA duplex by Levitt's[3] and Karplus's[4] groups, and of transfer RNA by Harvey and McCammon's groups[5, 6]. These simulations were reported 7 years later than those performed for proteins, which illustrates the intrinsic problems associated to the simulation of highly charged, flexible systems such as NAs. In fact, the history of MD simulations of NAs can be divided into two periods, whose frontier is roughly located in the middle nineties. The first period was characterized by the problems in obtaining stable structures in simulations that attempted to mimic physiological conditions. Since the middle nineties, the development of new force-fields and specially the implementation of efficient methods to treat long-range electrostatic effects opened a new era where MD simulations are capable to explore reliable conformational states of NAs in a time scale (10-100 ns) which approaches the range of biological importance[7].

Here we will briefly review current MD simulation techniques, focusing our attention on the methods that can be used to extract all the information contained in MD trajectories. We plan to show the problems in handling trajectory files that can approach the Terabyte disk space, and we will present data mining techniques techniques which can help to extract human-readable information from these large data files.

2.   THE CLASSICAL APPROACH

Classical approaches assume that molecular interactions can be represented by a series of simple classical equations (i.e., the force-field) relating the nuclear structure of the system with its energy. The level of accuracy in the representation of NAs and in the force-field leads to a variety of simulation methods, which can be classified in three categories[8-10]: i) ideal-elastic (macroscopic), ii) mesoscopic (intermediate), and iii) microscopic.

Macroscopic elastic models assume that the NA is a flexible ideal rod described using principles of macroscopic mechanics[8]. This very simple model assumes an oversimplified structure, which ignores for example differences in stiffness of different sequences. However, they are very efficient from a computational point of view and provide a coarse-grained picture of the flexibility of very large pieces of DNA or even the entire ribosomal RNAs[11].

Mesoscopic models[8,9,12,13] divide the polynucleotide in small rod elements, which are considered to be rigid entities connected by flexible links that can be deformed by external or internal forces. The deformation energy is computed using elastic potentials, which are typically defined to distinguish between different type of deformations and even different types of junctions. An example of elastic potential is shown in Eqs. (1)-(3)[12], where the subscript "0" denotes the equilibrium values for bending angle (ß), twisting angle (τ) and stretching distance (l) of a given NA fragment, and the constants B, C and S define the rigidity in front of those deformations. All these parameters can be derived by fitting the model to experimental data, or to MD simulations.

$$E_{bending} = 0.5(B/l_0)(\beta - \beta_0)^2 \tag{1}$$

$$E_{twisting} = 0.5(C/l_0)(\tau - \tau_0)^2 \tag{2}$$

$$E_{stretching} = 0.5(S/l_0)(l - l_0)^2 \tag{3}$$

Mesoscopic methods have provided very valuable information on the structure and flexibility of NAs. However, many functional aspects depend on atomic details of the NA structure, which justifies the development of microscopic simulation models. As described elsewhere[2], microscopic methods can be classified based on: i) *the level of representation of the NA (i.e., the degrees of freedom considered)*, ii) *the force-field used*, and iii) *the simulation engine*.

2.1   Level of Representation

NAs can be described at i) *atomic resolution* or using ii) *collective variables*. In the former the atoms are free to move subject to the forces dictated by the force-field. They typically work in Cartesian coordinates and are intended to explore the fine structural details of relatively short NAs. In most cases the solvent (water and counterions) is explicitly treated at the same level of accuracy. When collective variables are used, they take advantage of the previous knowledge on the flexibility of NAs, which allows to eliminate degrees of freedom that contribute little to the global flexibility of the system, like stretching, bending or out-of-the-plane deformation of nucleobases. Some collective variable schemes[14-17] use the helical space to reduce the complexity of the simulation problem. This is a very powerful approach for the study of canonical duplexes, but caution is required in the analysis of non-regular structures.

2.2   The Force Field

The force-field is an analytical expression which connects the structure of the system with its potential energy. Typically, the potential energy is evaluated using an expression similar to that shown in Eq. (4), where E_{str} and E_{bnd} stands for tstretching and bending energies (Eqs. 5-6), E_{tor} stands for the torsional energy (Eq. 7), E_{nb} accounts for the non-bonded interaction energies, and E_{other} includes any other type of interactions considered in the calculations, such as improper torsions (in united-atom force-fields) experimental or symmetry restrains, external potentials, etc.

$$E = E_{str} + E_{bnd} + E_{tor} + E_{nb} + E_{other} \tag{4}$$

$$E_{str} = \sum_{bonds} K_{str}(l - l_o)^2 \tag{5}$$

$$E_{bnd} = \sum_{angless} K_{ang}(\Theta - \Theta_o)^2 \tag{6}$$

where K_{str} and K_{bnd} stand for stretching and bending constants, and l_0 and Θ_0 are equilibrium bond lengths and angles.

$$E_{tor} = \sum_{tor} \sum_{n=1}^{3} \frac{V_n}{2}(1 + \cos n\Phi - \gamma) \tag{7}$$

where n is the periodicity of the Fourier term, Φ is the torsional angle, γ is the phase angle and V_n is the torsional barrier.

Non-bonded interactions generally include a van der Waals term (E_{vw}), which accounts for dispersion-repulsion interactions between atoms, and an electrostatic term (E_{ele}), which is used to represent charge-charge interactions. The van der Waals term is typically represented by a Lennard-Jones formalism (Eq. 8), while the electrostatic term is represented using a Coulomb potential (Eq. 9). Electrostatics is essential in the representation of NAs and most of the success of a force-field depends on the quality of its charge representation. The newest force-fields, in their non-polarized versions, use charges derived from HF/6-31G(d) wavefunctions. The popularity of this medium-level QM calculation obeys to its well known tendency to overestimate polarity, thus mimicking the polarizing effect of water.

$$E_{vw} = \sum_{i,j} \left(\frac{A_{ij}}{R_{ij}^{12}} - \frac{C_{ij}}{R_{ij}^{6}} \right) \tag{8}$$

where the sum extends for all non-bonded pairs.

$$E_{ele} = \sum_{m,n} \frac{Q_m Q_n}{\varepsilon(R_{mn}) R_{mn}} \tag{9}$$

where the sum extends for all non-bonded pairs, and ε is the dielectric constant.

Traditionally, force-fields like AMBER [18,19] or CHARMM [20,21] are used in conjunction with explicit representations of the solvent (TIP3P[22] and SPC[23] are the two most popular models of water), avoiding then the need to use effective dielectric constants. In the last years, however, the need to improve the computational efficiency of microscopic methods has stimulated the implementation of continuum models[24]. The most popular of these methods is the Generalized-Born/solvent accessible surface (GB/SA) approach, originally developed by Clark and Still[25], and now available in slightly different implementations (see[24] for discussion). The GB/SA approach assumes that the solvation free energy is determined as a combination of steric and electrostatic effects (Eq. 10). Steric contributions (including cavitation and dispersion terms) are typically represented by means of an empirical linear relationship with the SAS using either universal or atom-specific surface tension parameters (ξ in Eq. 11). The electrostatic term is computed using an empirical generalization of the Born's equation (Eqs. 12-14).

$$\Delta G_{sol} = \Delta G_{ster} + \Delta G_{ele} \tag{10}$$

$$\Delta G_{ster} = \sum_{k} \xi_k SAS_k \tag{11}$$

$$\Delta G_{ele} = -\frac{1}{2}\left(1 - \frac{1}{\varepsilon_\infty}\right) \sum_{m \ n} \frac{Q_i Q_j}{\Gamma_{GB}} \tag{12}$$

where the empirical GB screening function is computed as:

$$\Gamma_{GB} = \sqrt{R_{ij}^2 + \alpha_i \alpha_j \exp(-D_{ij})} \qquad (13)$$

with

$$D_{ij} = \frac{R_{ij}^2}{d\alpha_i \alpha_j} \qquad (14)$$

The scaling constant d in Eq. (14) is typically set to 4^{24} and the Born's radii (α) determining the average distance from a given atom (i or j) to the solvent are typically derived using analytical approaches like that suggested by Hawkins et al.[26].

GB/SA methods have become very popular when coupled to MD simulation protocols owing to the good balance between computational efficiency and accuracy. However, in our experience caution is necessary since the neglect of explicit water molecules in the microscopic simulation of highly charged systems like NAs can lead to some problems in long MD trajectories. Caution and common sense is necessary in the use of MD GB/SA simulation of NAs.

2.3   Simulation Methods

The analysis of the flexibility properties of NAs requires the determination of a representative ensemble of conformations under controlled conditions, which can be obtained using i) Monte Carlo (MC) and ii) MD techniques. The first approach is usually applied in the Metropolis formalism, which generates a Markov's chain of configurations, where a new configuration, {X}, randomly generated from a previous one, {X$_0$}, is accepted or rejected based only on the relative energy of configurations {X} and {X$_0$}. The method works in internal coordinates, and the degrees of freedom sampled in the simulation are directly controlled by the user, which makes MC methods well suited for the collective variables approach[10,15,16]. Methods for the simulation of NAs at the atomic level have been only recently developed[27], and their use is still very limited.

MD is clearly the most powerful theoretical method for the atomic exploration of NAs. The method relies on the numerical integration of Newton's equations of motion, with forces being computed analytically every 0.5-2 fs. Current MD protocols include algorithms to fix temperature and pressure, allowing the simulation of NAs under conditions close to the physiological ones. It is also possible to incorporate other environmental effects in the form of external fields, restrains, or constraints. The method works in the Cartesian space, but variants using internal coordinates have also been suggested[28]. As noted above, continuum models like the GB/SA can be easily incorporated into the MD algorithms and some authors have shown promising results in collective-variable MD simulations of micro-solvated DNAs[29]. Current *state-of-the-art* MD simulations of NAs use the Particle Mesh Ewald approach (PME;[30]), which allows to capture long-range electrostatic effects.

The time scale of MD simulations is continuously increasing owing to i) better accessibility to most powerful and less expensive computers and ii) parallelization of the MD codes. Few groups are currently reaching simulation times of 0.1 μs for 10-20 mer duplexes treated at the microscopic level with explicit solvent. The size of the simulated system also increases continuously, and very large systems including micro-cycles and nucleosomes are now accessible to calculation[31]. The increase in the length of the trajectories and in the size of the simulated system will approach more the MD protocols to real experimental problems, but we should be prepared to detect new problems in these simulations, which were not so evident in short trajectories of small oligonucleotides.

2.4 Analysis of MD Trajectories

Multi-ns simulations of macromolecules generate very large trajectories, whose analysis is difficult, since the information of interest is hidden in many Gigabytes (up to Terabytes) of noise. Our guess is that in a close future the limiting step for most MD simulations will be not the derivation of the trajectory, but its analysis. We will describe here current data mining techniques for the analysis of large MD trajectories of NAs.

2.4.1 MD-Averaged Information

The MD sampling is mainly used to obtain an MD-average structure, which is expected to represent the global conformational space of the NA. The typical procedure for the derivation of the MD-averaged structure relies on three steps: i) root-mean square (RMS) fitting of the snapshots to define a common reference system, ii) computation of the average Cartesian coordinates, iii) restricted molecular mechanics optimization to relax unrealistic positions of very mobile groups. However, the derivation of such MD-averaged conformation is not always trivial, since it might yield a structure which does not correspond to any physically accessible conformation. The first difficulty arises when trajectories sample different states (Figure 5-1), where the averaging procedure will provide a "transition state-like" structure. Clearly, one needs a previous knowledge of the configurational space to detect clusters of structures that can define reference states. This information can be obtained from two-dimensional cross-RMSd plots, which shows the RMSd deviation of each snapshot with respect to the remaining ones (Figure 5-2). Clustering techniques then permits to determine a restricted number of different states, and the configurational space can be as a combination of those MD-averaged reference states (Eq. 15).

$$\{X_i\}_t \approx \sum_k P_k^t X_k^{av}$$ (15)

where $\{X_i\}_t$ stands for the global configurational space sampled along the trajectory for time t, X_k^{av} stands for the MD-averaged conformation obtained from snapshots pertaining to cluster k, and P_k^t is the time-population of cluster k.

A second problem in the standard averaging procedure stems from the use of Cartesian coordinates, which increases the probability of obtaining chemically unreliable conformations after step 2 (see above), which might not be fully corrected in the restricted optimization. This problem can be alleviated by performing the averaging in a set of distinguished internal (for example selected dihedrals) or helical coordinates, but for very flexible systems there is not guarantee that these approaches yield to better results than simple Cartesian averaging.

Once the average structure(s) is (are) obtained, a large number of analysis can be performed[2,7] with standard routines implemented in most simulation packages, including intra- or intermolecular distances, backbone dihedrals, H-bonds, solvent accessible surface, and radii of gyration. For standard NAs, helical analysis is especially useful, since it allows a dramatic reduction in the degrees of freedom of the system. Many excellent programs such as Curves[32], NewHelix[33] or 3DNA[34] are available to perform such an analysis

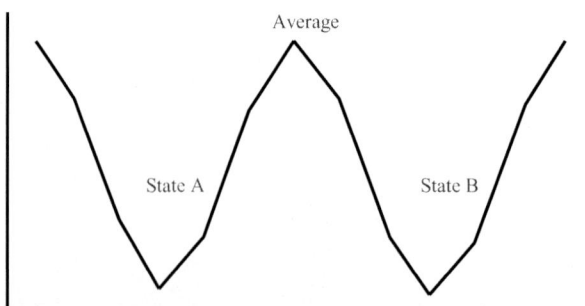

Figure 5-1. Schematic representation of a system that can be found in two interchangeable conformational spaces (A and B).

DNA 2D RMS

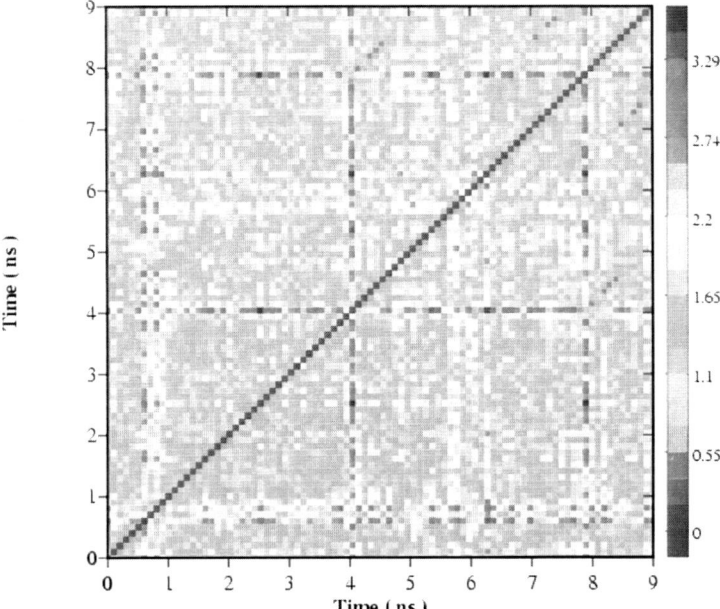

Figure 5-2. Two dimensional RMSd map for snapshots collected during 10 ns of an unrestrained MD simulation of Dickerson's dodecamer.

in a very efficient way. The average structure can be also used to analyse the non-bonded reactive properties of NAs using GRID-based approaches[35], which determine the interaction energy between the NA and different probes placed in thousands of points around the molecule.

2.4.2   Dynamic Information

Dynamical information can be obtained from the analysis of the snapshots collected along the trajectory. Concepts of crucial importance like flexibility, accessible conformational space, entropy or stiffness are implicitly included in the trajectory and need to be extracted to obtain a complete description of the system.

2.4.2.1   Determination of Optimum Samplings from MD Simulations
A first question that arises in the analysis of any MD simulation is the required length of the trajectory. A "rule of thumb" is that the length of the trajectory should be at least 10 times larger than the time scale of the process that is being studied. In practice, this means that only processes happening in

the ns time scale can be studied with current computational resources and that in principle longer trajectories are expected to yield always to results of greater quality. rules to be used to select the snapshots, that is, how many snapshots suffice to represent a trajectory, are less clearly defined. For historical reasons snapshots are generally collected each 1 ps, but there is not any indication that this spacing is a good compromise between: i) the accuracy in the representation of relevant structures and ii) the reduction in the size of the files obtained.

In principle, for a simulation of finite time (t_{TOT}) the best statistic is obtained when the snapshots are collected equally spaced. Then, assuming that we are not interested in very fast movements nor in solvent rearrangements, the optimum collection spacing should be determined by the number of points needed to represent relevant movements of the NA. Accordingly, taking around 10 snapshots for each degree of freedom, the total number of snapshots for a system with N atoms should be 10 x (3N-6) and the correct spacing should be given by Eq. (16). However, since atomic displacements are limited by covalent linkages and assuming that there are only 7 (torsional) degrees of freedom for each nucleotide (14 for each base pair), for a duplex of length M only 140 x M snapshots should be required. Since each base pair has on average 65.5 atoms, we can conclude (from a purely statistical point of view) that the number of snapshots needed to represent the trajectory is around 2 times the number of atoms in the system (Eq. 17). For a 10 ns trajectory of a dodecamer, this implies that data should be collected at least every 6 ps, yielding to ~1600 snapshots, and a conservative protocol will be to store data every 2-5 ps. However, as noted in Figure 5-3 and Tables 5-1-5-4, even more sparse collection (up to 1 structure each 10 ps for a 10 ns trajectory of duplex DNA) can provide a good representation of the geometrical properties and essential dynamics of normal DNAs (see below).

Table 5-1. Eigenvectors (in $Å^2$) associated to the first 10 essential movements of a 12-mer DNA duplex from a 10 ns trajectory.

	1ps	2ps	5ps	10ps
1	185.47	185.94	183.17	179.26
2	125.07	125.05	125.42	127.65
3	71.96	71.96	71.93	73.90
4	50.43	50.43	50.01	50.30
5	40.42	40.29	40.92	40.69
6	32.46	32.64	32.11	32.01
7	26.39	26.48	26.19	27.41
8	18.17	18.24	18.38	18.43
9	16.33	16.30	16.74	17.10
10	13.89	13.98	14.00	14.30

Table 5-2. Root mean square deviations (in Å) of a MD trajectory (snapshots collected every 1-10 ps) of a 12-mer DNA with respect to MD-averaged DNA and RNA structures. Standard deviations (also in Å) are displayed in parenthesis.

Rmsd	1ps	2ps	5ps	10ps
DNA avg	1.168(0.29)	1.165(0.29)	1.164(0.29)	1.167(0.28)
RNA avg	4.375(0.48)	4.361(0.48)	4.357(0.48)	4.358(0.47)

Table 5-3. Average nucleobase-nucleobase interaction energies (Hydrogen bond and stacking) for a MD trajectory (snapshots collected every 1-10 ps) of a 12-mer DNA computed from ensembles collected every 1,2,5 or 10 ps. The standard deviations (also in kcal/mol) in the averages are displayed in parenthesis.

Energy	1ps	2ps	5ps	10ps
Hbond	−204.63(4.70)	−204.58(4.74)	−204.69(4.54)	−204.68(4.64)
Stacking	−136.67(4.77)	−136.7(4.79)	−136.74(4.71)	−136.79(4.82)

Table 5-4. Selected average helical parameters for a 12-mer DNA in B-conformation derived from 10 ns MD trajectories with snapshots collected at different time-spacings. Displacement values are in Å and angles are in degrees. Standard deviations (in the same units) are displayed in parenthesis.

	1ps	2ps	5ps	10ps
Tilt	−0.015(4.485)	0.004(4.481)	0.002(4.487)	−0.02(4.501)
Roll	1.99(6.30)	2.04(6.30)	2.04(6.31)	2.03(6.29)
Twist	33.13(8.84)	33.08(8.93)	33.08(8.92)	33.06(8.94)
Slide	−0.32(0.61)	−0.32(0.61)	−0.32(0.60)	−0.32(0.61)
Shift	0.001(0.718)	0.000(0.711)	0.002(0.713)	0.004(0.718)
Rise	3.33(0.36)	3.33(0.36)	3.33(0.36)	3.33(0.36)
Delta	129.2(18.0)	129.3(17.9)	129.3(17.9)	129.3(18.0)
Chi	247.99(17.88)	248.12(17.93)	248.08(17.90)	247.99(17.98)

$$\Delta t_{col} = \frac{t_{TOT}}{10 \times (3N - 6)} \qquad (16)$$

$$\Delta t_{col} = \frac{t_{TOT}}{2N} \qquad (17)$$

The use of Eq. (17) permits to reduce the size of the database where trajectory is stored and helps in the mining for useful information. However, caution is necessary, since the time step for data collection should be always shorter that the time scale of the movement of interest. For instance, the use of sparse samplings should be avoided if the solvent structure around NAs is of interest, since a collection interval of 5 ps can be too large to capture relaxation movements of water molecules around the solute.

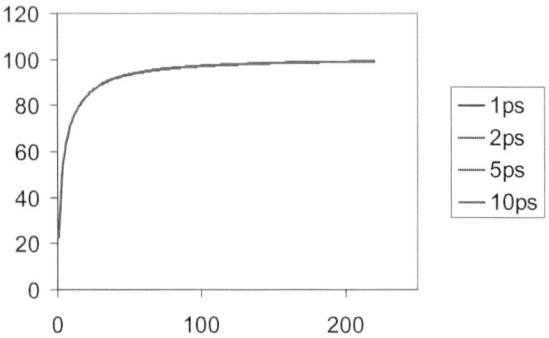

Figure 5-3. Percentage of cumulative variance when an increasing number of eigenvectors are considered for a 10 ns MD of Dickerson's dodecamer. Eigenvectors are obtained by diagonalization of covariance matrices created collecting data every 1, 2, 5 and 10 ps.

2.4.2.2   Dynamics Analysis of Solvent

Once a set of selected snapshots is stored, it can be analyze to determine many dynamic properties of both solute (NA) and solvent (typically water). The analysis of the solvent environment is typically focused on the determination of preferential solvation maps, which signals the regions where the NA has an unusually large concentration of water molecules (Figure 5-4). Solvation maps are determined by centering all the collected snapshots in a common structure, and defining a regular grid around the different snapshots, which permits to compute the population of water molecules associated to each grid point along the dynamics. The final apparent density is computed as shown in Eq. (18), where $N^{i,j,k}$ stands for the population of water in the grid cube element i,j,k of volume $Vol^{i,j,k}$, and α is a conversion factor to change from units of molecule/Å^3 to more usual density measures (like g/cm^3).

$$\rho_{ap}^{i,j,k} = \alpha \frac{N_w^{i,j,k}}{Vol^{i,j,k}} \qquad (18)$$

Solvation maps do not contain time information and do not discriminate between it regions where a water molecule is tightly bound during long periods of time from those regions that are well hydrated, but subject to a fast interchange between water molecules. Such time-dependent information can be gained from the analysis of water residence times for highly hydrated sites around polar groups of the NA.

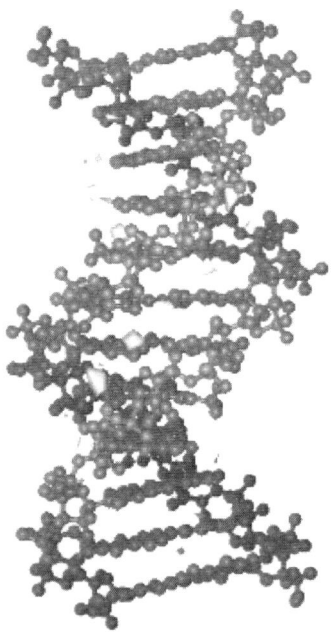

Figure 5-4. Regions of preferential solvation obtained from 10 ns MD simulation of Dickerson's dodecamer. Solvation contours are projected on the MD-averaged conformation.

2.4.2.3   Dynamic Analysis of the NA

The MD trajectory allows us to compute both time-dependent and time-independent dynamical information of the NA. Important conformational transitions are unlikely to happen in ~10 ns simulation periods, and it is even more difficult to properly sample those events. However, transitions such as base breathing as in difluorotoluene-containing NAs[36], or the 2'-deoxyribose repuckering in DNA·RNA hybrids ([37]; Figure 5-5) have been detected. In these cases, transition times and frequencies can be determined, but in the general (see bottom panel in Figure 5-5) such a time-dependent information is not accessible with current MD simulation protocols.

When the starting conformation of the NA is not far from the equilibrium, time-independent dynamical information can be obtained from relatively short (5-10 ns) trajectories. A particularly relevant aspects is the flexibility of NAs. A first index to capture the flexibility is the B-factor (Eq. 19), which provides is a rough atomic-based estimate, but with the advantage that can be compared with experimental measures. A more global index is the average deviation (Eq. 20) defined in terms of the average temporal

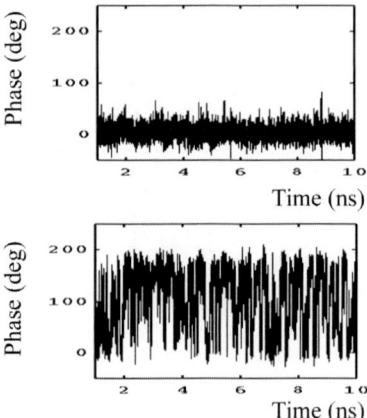

Figure 5-5. Variation of the phase angle along the time for TOP one 2'deoxyribose and BOTTOM one ribose in a 10 ns MD of a DNA·RNA hybrid with Dickerson's sequence.

position of each snapshot in the Cartesian or mass-weigthed coordinate schemes ($\langle x_l \rangle_t$). Manipulation of this equation shows that the average deviation is equal to the variance of the trajectory and the deviation between the average and the reference structures (Eq. 21). The index provides then a compact measure of how a trajectory deviates from a reference conformation, and how much it fluctuates around its average.

$$B = (8\pi^2 / 3) < \Delta r^2 > \tag{19}$$

where Δr is the atomic fluctuation with respect to the average position.

$$\left\langle RMSd^2 \right\rangle_t = \frac{1}{S}\sum_{k=1}^{S}\left[\frac{1}{N}\sum_{l=1}^{3N}(x_{kl} - x_l)^2 \right]$$

$$= \frac{1}{N}\sum_{l=1}^{3N}\left[\frac{1}{S}\sum_{k=1}^{S}(x_{kl} - \langle x_l \rangle_t - x_l + \langle x_l \rangle_t)^2 \right] \tag{20}$$

$$\left\langle RMSd^2 \right\rangle_t = RMSd_{ref}^2 + \frac{1}{N}\sum_{l=1}^{3N}\frac{1}{S}\sum_{k=1}^{S}(x_{kl} - \langle x_l \rangle_t)^2$$

$$= RMSd_{ref}^2 + \frac{1}{N}\sum_{l=1}^{3N}var_l \tag{21}$$

where x_{kl} stands for the l-coordinate in structure k, x_l is the value of the l-coordinate in the reference structure and S is the total number of snapshots representing the sampling during simulation time t.

2.4.2.4   Essential Dynamics

It is probably the most powerful approach to describe NA flexibility, since it determines the natural motions of a structure, that is, those motions that explain most of the variance detected during the trajectory. Technically, this implies the diagonalization of the covariance matrix leading to a set of 3N (N = number of atoms in the system) eigenvalues and eigenvectors. The eigenvalue (λ_i) represents the percentage of variance explained by the corresponding eigenvector (v_i) and can be easily manipulated to derive frequencies (obtained when the covariance matrix is determined from mass-weighted Cartesian coordinates). Furthermore, eigenvalues can be translated into harmonic deformation constant using Eq. (22), which permits to reduce the flexibility of the DNA to a set of harmonic potentials applied on essential deformation modes.

$$K_i = k_B T / \lambda_i \qquad\qquad (22)$$

where the index i stands for an essential movement, k_B is Boltzmann factor, T is the absolute temperature and λ is the eigenvalue associated to deformation i in distance^2 units.

The eigenvalues represent the amount of displacement along a mode accessible at a given temperature. Thus, multiplying the eigenvalues associated to a set of m selected eigenvectors, we can estimate the accessible configurational volume ([38]; Eq. 23). Unfortunately, the direct use of Eq. (23) is ambiguous since the value depends on the number of selected eigenvalues (m). Furthermore, if all eigenvalues are considered, i.e. m = 3N, the volume is always equal to zero since most of the eigenvectors have very small eigenvalues. Accordingly, it is often better to define the accessible conformational space in terms of dimensionality and generalized volume. Dimensionality is a measure of the number of eigenvectors needed to describe all the important flexibility space[38]. It is given by the number of eigenvectors that need to be introduced in Eq. (23) to maximize the volume. It roughly corresponds to the first eigenvector along which the expected displacement (at temperature T) is less than 1 Å (for variance measures in Å^2). The generalized volume (Eq. 24) is an *ad hoc* modification of Eq. (23) which introduces suitable limit conditions and guarantees that results will be not corrupted by very small eigenvalues. Accordingly, it can be computed considering all the eigenvectors, removing the ambiguity in the choice of m.

$$vol = \prod_{i=1}^{m} \lambda_i \qquad\qquad (23)$$

$$vol_{gen} = \prod_{i=1}^{m} (1 + \lambda_i)^{1/2} \tag{24}$$

The eigenvectors obtained after diagonalization of the Cartesian covariance matrix are very informative, since they are represent the nature of the essential motions. The eigenvectors of one trajectory can be compared with those of another trajectory, which permits to define quantitative measures of similarity between the essential motions of two independent trajectories using Hess metrics as shown in Eq. $(24)^{2, 7, 38, 39}$, which gives an absolute similarity measure (note that γ_{AB} should range from 0 for orthogonal trajectories to 1 for trajectories with identical essential movements), and Eq. 25, which is a relative similarity index[2, 7, 38].

$$\gamma_{AB} = \frac{1}{n} \sum_{j=1}^{n} \sum_{i=1}^{n} (v_i^A \bullet v_j^B)^2 \tag{24}$$

where n is the minimum number of eigenvectors needed to define the "important" eigenvector spaces $\{\mu_i\}_0^A$ and $\{\mu_i\}_0^B$ that explain a given variance threshold (80-90%) and v_i^A stands for the i unitary eigenvector of trajectory X (i.e $v_i^X = \mu_i^X \big/ |\mu_i^X|$).

$$\Gamma_{AB} = 2 \frac{\gamma_{AB}}{(\gamma_{AA} + \gamma_{BB})} \tag{25}$$

where the self-similarity indexes γ_{AA} and γ_{BB} are obtained by comparing first and second halves of the trajectories.

Eqs. (24) and (25) are powerful measures of similarity in the deformation patterns of NAs [2,7,38,40,43], but they present some obvious limitations due to the fact that they assume that all eigenvectors in the "important space" contribute equally to the flexibility, i.e. the method is insensitive to any permutation of eigenvectors. Very recently[38], we developed a slightly more evolved metrics (Eqs. 26 and 27) where each mode is weighted according to its associated Boltzman's factor for a harmonic energy given by Eq. (22) and a common displacement in all modes. This displacement (Δx) is defined following this procedure: i) determine for both trajectories the number of eigenvectors (n') that explain a given level of the variance, and ii) take Δx as the minimum displacement that makes the contribution of the n' + 1 mode irrelevant (less than 1% weight) in the calculation of the similarity index. This procedure guarantees that all eigenvectors in the important space contribute to the similarity according to their importance in describing the global flexibility of the molecule.

$$\xi_{AB} = \frac{2\sum\limits_{i=1}^{i=z}\sum\limits_{j=1}^{j=z}\left(v_i^A \circ v_j^B\right)\left[\dfrac{\exp\left\{-\dfrac{(\Delta x)^2}{\lambda_i^A}-\dfrac{(\Delta x)^2}{\lambda_j^B}\right\}}{\sum\limits_{i=1}^{i=z}\exp\left\{-\dfrac{(\Delta x)^2}{\lambda_i^A}\right\}\sum\limits_{j=1}^{j=z}\exp\left\{-\dfrac{(\Delta x)^2}{\lambda_j^B}\right\}}\right]^2}{\sum\limits_{i=1}^{i=z}\left(\dfrac{\exp\left\{-2\dfrac{(\Delta x)^2}{\lambda_i^A}\right\}}{\left(\sum\limits_{i=1}^{i=z}\exp\left\{-\dfrac{(\Delta x)^2}{\lambda_i^A}\right\}\right)^2}\right)^2 + \sum\limits_{j=1}^{j=z}\left(\dfrac{\exp\left\{-2\dfrac{(\Delta x)^2}{\lambda_j^B}\right\}}{\left(\sum\limits_{j=1}^{j=z}\exp\left\{-\dfrac{(\Delta x)^2}{\lambda_j^B}\right\}\right)^2}\right)^2} \tag{26}$$

where the sum can be extended to all $(z = m)$ or the important eigenvectors $(z=n)$ giving similar results.

$$\delta_{AB} = 2\frac{\xi_{AB}}{(\xi_{AA}^T + \xi_{BB}^T)} \tag{27}$$

Essential dynamics is also powerful to compress the trajectory itself, a procedure necessary for efficient MD data storage, manipulation and transfer. The strategy[44] is based on the following steps: i) determination of the essential movements, ii) selection of a small set of eigenvectors able to represent a given level of variance (typically 95-99%), iii) projection of each snapshot in this new coordinate system. The original trajectory can be recovered by a simple back-projection of the snapshots to the Cartesian coordinate space. This procedure transform the trajectory $t \times \Delta t \times 3N$ trajectory file into three small files: a) the $m \times 3N$ eigenvectors files, b) the $3N$ file containing the average structure and c) a $t \times \Delta t \times m$ file containing the m-projections of each snapshot. For normal NAs a very small number of eigenvectors (around 100 for a duplex dodecamer) are able to explain 95-99%, which implies a dramatic reduction in the size of the stored trajectory with no significant lost of information[44]. Furthermore, the user can always select the level of accuracy by modifying the number of important eigenvectors (note that when $m = 3N$ the transformation is exact, so that original and back-projected trajectories are absolutely equivalent).

Finally, we cannot forget the possibility to use essential deformation movements to perform MC or MD simulations on these essential modes. Within this approach the deformation energy associated to each displacement (ΔX_i) along the essential movement space is computed from Eq. (28). The implementation of the technique is especially simple in the

MC framework, since Metropolis test can be easily adapted to obtain an ensemble of movements in the essential space. The technique can provide exhaustive sampling of the configurational space, which might facilitate finding of uncommon configurations. However, this approach is not expected to yield conformations very different to those sampled during the atomistic trajectory from which essential movements were determined.

$$E = \sum_{i=1}^{m} \left(k_B T / \lambda_i \right) \Delta X_i^2 \tag{28}$$

Essential dynamics is then one of the most powerful methods for the analysis of MD trajectories. However, we must emphasize that caution and common sense are needed in the analysis, since the technique is very sensible to the length of the trajectory and can neglect important local distortions in favor of general, but perhaps less biologically relevant movements. Systematic studies performed in our group dividing large trajectories of DNA in smaller non-overlapped subtrajectories found similarity indexes around 0.7-0.9 when two equal portions of the same trajectory are compared, indicating a non-negligible dependence of the results on the initial conditions and numerical errors in the simulation. Additional sources of error exist in the definition of a common reference system for the trajectories, and on the elimination of translational and rotational degrees of freedom of the molecule.

2.4.2.5   Stiffness Analysis

Canonical NAs can be described in general terms using a reduced helical coordinate system[32-34]. These coordinates (see above) are close to chemical intuition and allow a more compact structural description of NAs. Olson[9,45] and later Lankaš[46,47] have developed approaches to derive stiffness constants associated to deformation in helical space. MD simulations can be used to provide equilibrium samplings from which covariance matrices in helical space can be determined. The inversion of these matrices provides the corresponding harmonic stiffness matrices for helical deformations of regular NAs (Eq. 29). Note that the determination of elastic deformation energies from stiffness matrices is straightforward (Eq. 30), which allows to use a simple potential to explore long fragments of regular NAs.

$$K = kTC^{-1} \tag{29}$$

where C is the stiffness matrix with elements K_{ij}.

$$E_{def} = \sum_i \frac{K_{ii}}{2}(X_i - X_{i_o})^2 + \sum_{i \neq j} \frac{K_{ij}}{2}(X_i - X_{i_o})(X_j - X_{j_o}) \qquad (30)$$

where K_{ij} are force-constant and X_i and X_{i0} are helical coordinates and their equilibrium values.

Stiffness analysis methods have provided valuable information on the relative flexibility of DNA, RNA and DNA·RNA hybrids[37, 48, 49]. Lankaš and coworkers[47] have derived sequence-specific stiffness constants, would allow mesoscopic simulations of large fragments of biologically important DNAs. Very promising results were obtained[47], but there are still questions regarding the minimum sequence size that can be used to define transferable stiffness variables, the length of the trajectories necessary to collect reliable data and the accuracy of current force-fields to provide this very fine sequence-specific information. These topics constitute one of the major research goals of the Ascona B-DNA consortium (ABC)[50].

Helical stiffness analysis is close to chemical intuition and facilitates translation of data extracted from MD simulations to the experimental world. However, caution is again necessary to avoid an over-representation of the results: i) the method relies on an helical description of NAs, which might not be valid for anomalous structures, ii) the values might depend on the size of the database (when helical analysis is performed on databases of crystal structures), or on the length of the trajectory used to build the covariance matrix, iii) for severe distortions the harmonic approach implicit in the model must be inaccurate, and iv) real deformations might not follow pathways purely defined by helical parameters.

2.4.2.6 Stability Calculations

MD-derived samplings can be used to evaluate the relative stability of different conformations of NAs. The calculation is done by assuming that the free energy of the system can be divided into intramolecular and solvation contributions (Eq. 31). The intramolecular contribution is determined as the addition of enthalpic and entropic terms (Eq. 32). Neglecting volume effects the intramolecular enthalpy is just the average intramolecular energy (Eq. 32). Intramolecular entropy is more difficult to evaluate, but can be reasonably estimated from quasi-harmonic models[51, 52], which diagonalize the mass-weighted covariance matrix obtained from MD simulations (Eqs. 33 and 34) and transform them into entropies by assuming quantum[51] or hybrid quantum-classical[52] oscillator models. Both approaches provide similar results and share common limitations: i) the harmonic treatment of vibrations is strictly valid only for structures fluctuating around a single minimum, ii) entropy estimates are dependent on the length of the simulation (the longer the trajectory, the greater the entropy). If the trajectory behaves in the harmonic regime, an empirical relationship can be found between the length of the trajectory and the computed entropy ([2,7,53]; Eq. 34), which permits to estimate the entropy of the structure at infinite simulation time.

$$G = G_{int\,ra} + G_{solv} \tag{31}$$

$$G_{int\,ra} = H_{int\,ra} + TS_{int\,ra} \approx \left\langle E_{int\,ra} \right\rangle + TS_{int\,ra} \tag{32}$$

$$S = k \sum_i \frac{\alpha_i}{e^{\alpha_i} - 1} - \ln(1 - e^{-\alpha_i}) \tag{33}$$

$$S = 0.5k \sum_i \ln\left(1 + \frac{e^2}{\alpha^2}\right) \tag{34}$$

where $\alpha_i = \hbar\omega_i / kT$, ω being the eigenvalues (in frequency units) obtained by diagonalization of the mass-weigthed covariance matrix, and the sum extends to all the non-trivial vibrations of the system.

$$S(t) = S_\infty - \frac{a}{t^b} \tag{35}$$

where a and b are fitted parameters obtained by fitting partial entropy estimates at different simulation times (S(t)) and S_∞ is the entropy at infinite time.

The calculation of the solvation free energy is typically performed using continuum models applied to the ensemble of structures collected along the trajectory (Eq. 36). Traditionally, both GB/SA and PB/SA methods have been used (see discussion in refs.[2,7,24]). In our experience both approaches provide similar qualitative results, though more consistent results are in general obtained from PB/SA calculations performed with a small grid-spacing. An alternative approach exploits the linear response theory approximation with surface correction[54,55]. The advantage of the method is that it does not rely on electrostatic continuum models, which require the assigment of dielectric constants and solute/solvent boundaries, since the solvation free energy (for each configuration) is determined from explicit solute-solvent interactions, setting the solvation calculation into the same physical grounds than the MD trajectory (Eq. 37).

$$G_{solv} = \langle G_{solv}(X) \rangle \tag{36}$$

where X stands for a conformation detected during the MD trajectory

$$G_{solv}(X) = \sum_{i=1}^{N} \xi_i \, SAS_i + \sum_{i=1}^{N} \sum_{j=1}^{M} \frac{Q_i Q_j}{R_{ij}} \tag{37}$$

where ξ is the tension factor, SAS is the solvent accessible surface assigned to solute atoms, M is the number of solvent atoms, Q stands for the atomic charge and R_{ij} is the interatomic distance.

To our knowledge methods based on Eqs. (31) and (32) were first suggested by Kollman's group ([56]and references therein) and are very popular[56-60] due to their ability to describe the different contributions to the stability of NAs or their complexes. However, the reader should be warned on the large numerical uncertainties implicit to the calculations of the different contributions. In fact, in our experience the method is too noisy unless very similar systems are considered or robust statistical analysis are performed. For example, for non inter-changeable regular structures of polymers of DNA with common-repetitive sequences, our group uses a strategy based on the parallel calculation of trajectories of oligonucleotides of different sizes placed in the conformations of interest. The total free energies (for each conformation) follow a perfect linear relationship with the length of the oligonucleotide, and relative stabilities can then be discussed from the slope and intercept of the regression lines[42,43,57,58], reducing dramatically the statistical noise in the calculation.

3. CHALLENGES FOR THE FUTURE

The extension of MD simulations to larger systems and larger simulation times cannot be stopped. Systems with around 10^5 atoms will be soon analyzed in routine 100 ns simulation periods. Many groups have already even longer (around 1 µs) unpublished simulations for 10-12 mer duplexes. In last decade we have witnessed a 10^2–fold increase in the length of the trajectory considered as "state-of-the-art" in the field. With this progression ms MD simulations will be reachable in less than 10-15 years. As CPU limitations will not be a real bottle-neck, new challenges can be envisaged for MD simulations of NAs.

3.1 Force-Field Refinement

There is a consensus that force-field for NAs are very good, which justify their longevity. However, longer simulations are going to illustrate their current limitations, and this will force refinements using higher level quantum mechanical data as reference, and incorporating more accurate experimental data on the validation. Two issues appear as an important challenge in the field: i) derivation of new additive force-field accepted by the majority of the simulation community, and ii) generalization of new force-fields including multi-centered charges and polarization corrections.

3.2   Reactivity in Nucleic Acids

A large effort has been spent in enzymatic reactivity, but little has been done with respect to reactivity in NAs, which is a common process with large biological impact. The integration of QM/MM protocols in NAs (mostly RNA) appears as a challenge for the next years.

3.3   Complexes with Proteins

Many of the functions of NAs depend on their interactions with proteins. Many of the proteins that interacting with NAs induce dramatic structural distortions, and how flexibility and dynamical properties of NAs facilitate binding or are altered upon binding to proteins are critical aspects to fully understand properties such as binding affinity and sequence preference. Of special interest are proteins with the ability to pump DNA or to untwist it and all the machinery responsible for replication, transcription, splicing and translation. The increase in the computational power and the number of experimentally determined NA-protein complexes, the MD study of these complexes will be possible very soon.

3.4   Unusual and Non-Regular Structures

We and others have found that MD simulations are able to represent uncommon conformations of DNA like parallel duplexes, triplexes, quadruplexes or even hybrids of DNA with other polymers. In most cases, the relative stability of different conformers was well reproduced. A future question is whether or not MD protocols might represent all types of unusual structures of NAs. Of specially relevance is the ability to predict correct folds for physiological RNAs, and to reproduce major conformational transitions induced by environmental changes.

3.5   Transfer to Mesoscopic Levels

Irrespective of the increase in computer resources, very large DNA or RNA strands will be not accessible to atomistic MD simulations. However, MD data obtained for small oligonucleotides might be incorporated into mesoscopic algorithms to create a continuum of simulations. We can even consider new methods where parts of the system are described at the microscopic level, while the rest is analyzed by coarse-grained mesoscopic models. Such integration between micro and mesoscopic approaches appears to be one of the most exciting challenges for the next years.

REFERENCES

1. Watson, J. D. & Crick, F. H. (1953). Molecular structure of nucleic acids; a structure for deoxyribose nucleic acid. *Nature* **171**, 737-738.
2. Orozco, M., Pérez, A., Noy, A., & Luque, F. J. (2003). Theoretical methods for the simulation of nucleic acids. *Chem. Soc. Rev.* **32**, 350-364.
3. Levitt, M. (1983). Computer simulation of DNA double-helix dynamics. *Cold. Spring. Harb. Symp. Quant. Biol.* **47**, 251-262.
4. Tidor, B., Irikura, K. K., Brooks, B. R. & Karplus, M. (1983). Dynamics of DNA oligomers. *J. Biomol. Struct. Dyn.* **1**, 231-252.
5. Prabhakaran, M., Harvey, S. C., Mao, B. & McCammon, J. A. (1983). Molecular dynamics of phenylalanine transfer RNA. *J. Biomol. Struct. Dyn.* **1**, 357-369.
6. Harvey, S. C., Prabhakaran, M., Mao, B. & McCammon, J. A. (1984). Phenylalanine transfer RNA: molecular dynamics simulation. *Science.* **223**, 1189-1191.
7. Schleyer, P. R. (1998). *Encyclopedia of computational chemistry*, John Wiley, Chichester; New York.
8. Olson, W. K. (1996). Simulating DNA at low resolution. *Curr. Opin. Struct. Biol.* **6**, 242-256.
9. Olson, W. K. & Zhurkin, V. B. (2000). Modeling DNA deformations. *Curr. Opin. Struct. Biol.* **10**, 286-297.
10. Lafontaine, I. & Lavery, R. (1999). Collective variable modelling of nucleic acids. *Curr. Opin. Struct. Biol.* **9**, 170-176.
11. Malhotra, A., Tan, R. K. & Harvey, S. C. (1994). Modeling large RNAs and ribonucleo-protein particles using molecular mechanics techniques. *Biophys. J.* **66**, 1777-1795.
12. Bruant, N., Flatters, D., Lavery, R. & Genest, D. (1999). From atomic to mesoscopic descriptions of the internal dynamics of DNA. *Biophys. J.* **77**, 2366-2376.
13. Matsumoto, A. & Olson, W. K. (2002). Sequence-dependent motions of DNA: a normal mode analysis at the base-pair level. *Biophys. J.* **83**, 22-41.
14. Lavery, R., Zakrzewska, K. & Sklenar, H. (1995). JUMNA (Junction minimisation of nucleic acids). *Comput. Phys. Commun.* **91**, 135-158.
15. Harvey, S. C., Wang, C., Teletchea, S. & Lavery, R. (2003). Motifs in nucleic acids: molecular mechanics restraints for base pairing and base stacking. *J. Comput. Chem.* **24**, 1-9.
16. Zhurkin, V. B., Ulyanov, N. B., Gorin, A. A. & Jernigan, R. L. (1991). Static and statistical bending of DNA evaluated by Monte Carlo simulations. *Proc. Natl. Acad. Sci. USA* **88**, 7046-7050.
17. Kosikov, K. M., Gorin, A. A., Zhurkin, V. B. & Olson, W. K. (1999). DNA stretching and compression: large-scale simulations of double helical structures. *J. Mol. Biol.* **289**, 1301-1326.
18. Cornell, W. D., Cieplak, P., Bayly, C. I., Gould, I. R., Merz Jr., K. M., Ferguson, D. M., Spellmeyer, D. C., Fox, T., Caldwell, J. W. & Kollman, P. A. (1995). A Second Generation Force Field for the Simulation of Proteins, Nucleic Acids, and Organic Molecules. *J. Am. Chem. Soc.* **117**, 5179-5197.
19. Cheatham 3rd, T. E., Cieplak, P. & Kollman, P. A. (1999). A modified version of the Cornell et al. force field with improved sugar pucker phases and helical repeat. *J. Biomol. Struct. Dyn.* **16**, 845-862.

20. Foloppe, N. & Mackerell Jr, A. D. (2000). All-Atom Empirical Force Field for Nucleic Acids: I Parameter Optimization Based on Small Molecule and Condensed Phase Macromolecular Target Data. *J. Comput. Chem.* **21**, 86-104.
21. Mackerell Jr, A. D. & Banavali, N. K. (2000). All-Atom Empirical Force Field for Nucleic Acids: II Application to Molecular Dynamics Simulations of DNA and RNA in Solution. *J. Comput. Chem.* **21**, 105-120.
22. Jorgensen, W. L. C., Madura, J. D., Impey, R. W. & Klein, M. L. (1983). Comparison of simple potential functions for simulating liquid water. *J. Chem. Phys.* **79**, 926-935.
23. Pullman, B. (1981). *Intermolecular forces: proceedings of the Fourteenth Jerusalem Symposium on Quantum Chemistry and Biochemistry held in Jerusalem, Israel, April 13-16, 1981.* The Jerusalem symposia on quantum chemistry and biochemistry ; v. 14, D. Reidel ;
    Sold and distributed in the U.S.A. and Canada by Kluwer Boston, Dordrecht, Holland; Boston Hingham, MA.
24. Orozco, M. & Luque, F. J. (2000). Theoretical Methods for the Description of the Solvent Effect in Biomolecular Systems. *Chem. Rev.* **100**, 4187-4226.
25. Still, W. C. T., Hawley, R. C. & Hendrickson, T. (1990). Semianalytical treatment of solvation for molecular mechanics and dynamics. *J. Am. Chem. Soc.* **112**, 6127-6129.
26. Hawkins, G. D. C. & Truhlar, D. G. (1995). Pairwise solute descreening of solute charges from a dielectric medium. *Chem. Phys. Lett.* **246**, 122-129.
27. Ulmschneider, J. P. & Jorgensen, W. L. (2004). Polypeptide folding using Monte Carlo sampling, concerted rotation, and continuum solvation. *J. Am. Chem. Soc.* **126**, 1849-1857.
28. Mazur, A. K. (1998). Accurate DNA Dynamics without Accurate Long-Range Electrostatics. *J. Am. Chem. Soc.* **120**, 10928-10937.
29. Mazur, A. K. (2002). DNA dynamics in a water drop without counterions. *J. Am. Chem. Soc.* **124**, 14707-14715.
30. Darden, T. A. & Pedersen, L. G. (1993). Molecular modeling: an experimental tool. *Environ. Health Perspect.* **101**, 410-412.
31. Bishop, T. C. (2005). Molecular dynamics simulations of a nucleosome and free DNA. *J. Biomol. Struct. Dyn.* **22**, 673-686.
32. Lavery, R. & Sklenar, H. (1989). Defining the structure of irregular nucleic acids: conventions and principles. *J. Biomol. Struct. Dyn.* **6**, 655-667.
33. Dickerson, R. E. (1992). NewHelix Program. University of California at Los Angeles.
34. Lu, X. J. & Olson, W. K. (2003). 3DNA: a software package for the analysis, rebuilding and visualization of three-dimensional nucleic acid structures. *Nucl. Acids Res.* **31**, 5108-5121.
35. Gelpi, J. L., Kalko, S. G., Barril, X., Cirera, J., de La Cruz, X., Luque, F. J. & Orozco, M. (2001). Classical molecular interaction potentials: improved setup procedure in molecular dynamics simulations of proteins. *Proteins* **45**, 428-437.
36. Cubero, E., Sherer, E. C., Luque, F. J., Orozco, M. & Laughton, C. A. (1999). Observation of Spontaneous Base Pair Breathing Events in the Molecular Dynamics Simulation of a Difluorotoluene-Containing DNA Oligonucleotide. *J. Am. Chem. Soc.* **121**, 8653-8654
37. Noy, A., Pérez, A., Marquez, M., Luque, F. J. & Orozco, M. (2005). Structure, recognition properties, and flexibility of the DNA.RNA hybrid. *J. Am. Chem. Soc.* **127**, 4910-4920.
38. Pérez, A., Blas, J. R., Rueda, M., Lopez-Bes, J. M., de la Cruz, X. & Orozco, M. (2005). Exploring the Essential Dynamics of B-DNA. *J. Chem. Theor. Comput.* **1**, 790-800.
39. Hess, B. (2000). Similarities between principal components of protein dynamics and random diffusion. *Phys. Rev. E* **62**, 8438-8448.
40. Rueda, M., Luque F. J. & Orozco, M. (2005). Nature of minor-groove binders-DNA complexes in the gas phase. *J. Am. Chem. Soc.* **127**, 11690-11698.
41. Rueda, M., Kalko, S. G., Luque, F. J. & Orozco, M. (2003). The structure and dynamics of DNA in the gas phase. *J. Am. Chem. Soc.* **125**, 8007-8014.

42. Cubero, E., Abrescia, N. G., Subirana, J. A., Luque, F. J. & Orozco, M. (2003). Theoretical study of a new DNA structure: the antiparallel Hoogsteen duplex. *J. Am. Chem. Soc.* **125**, 14603-14612.

43. Cubero, E. L., Luque, F. J. & Orozco, M. (2005). In Press. *Biophys J.*

44. Noy, A., Meyer, T., Rueda, M., Ferrer, C., Valencia, A., Pérez, A., de la Cruz, X., Lopez-Bes, J. M., Luque, F. J. & Orozco, M. (2005). In Press. *J. Biomol. Struct. Dynam.*

45. Olson, W. K., Gorin, A. A., Lu, X. J., Hock, L. M. & Zhurkin, V. B. (1998). DNA sequence-dependent deformability deduced from protein-DNA crystal complexes. *Proc Natl Acad Sci USA* **95**, 11163-11168.

46. Lankaš, F., Šponer, J., Hobza, P. & Langowski, J. (2000). Sequence-dependent elastic properties of DNA. *J. Mol. Biol.* **299**, 695-709.

47. Lankaš, F., Šponer, J., Langowski, J. & Cheatham 3rd, T. E. (2003). DNA basepair step deformability inferred from molecular dynamics simulations. *Biophys. J.* **85**, 2872-2883.

48. Noy, A., Pérez, A., Lankaš, F., Javier Luque, F. & Orozco, M. (2004). Relative flexibility of DNA and RNA: a molecular dynamics study. *J. Mol. Biol.* **343**, 627-638.

49. Pérez, A., Noy, A., Lankaš, F., Luque, F. J. & Orozco, M. (2004). The relative flexibility of B-DNA and A-RNA duplexes: database analysis. *Nucl. Acids Res.* **32**, 6144-6151.

50. Beveridge, D. L., Barreiro, G., Byun, K. S., Case, D. A., Cheatham 3rd, T. E., Dixit, S. B., Giudice, E., Lankaš, F., Lavery, R., Maddocks, J. H., Osman, R., Seibert, E., Sklenar, H., Stoll, G., Thayer, K. M., Varnai, P. & Young, M. A. (2004). Molecular dynamics simulations of the 136 unique tetranucleotide sequences of DNA oligonucleotides. I. Research design and results on d(CpG) steps. *Biophys. J.* **87**, 3799-3813.

51. Andricioaei, I. & Karplus, M. (2001). On the calculation of entropy from covariance matrices of the atomic fluctuations. *J. Chem. Phys.* **115**, 6289-6292

52. Schlitter, J. (1993). Estimation of absolute and relative entropies of macromolecules using the covariance matrix *Chem. Phys. Lett.* **215**, 617-621.

53. Harris, S. A., Gavathiotis, E., Searle, M. S., Orozco, M. & Laughton, C. A. (2001). Cooperativity in drug-DNA recognition: a molecular dynamics study. *J. Am. Chem. Soc.* **123**, 12658-12663.

54. Morreale, A., de la Cruz, X., Meyer, T., Gelpi, J. L., Luque, F. J. & Orozco, M. (2005). Partition of protein solvation into group contributions from molecular dynamics simulations. *Proteins* **58**, 101-109.

55. Morreale, A., de la Cruz, X., Meyer, T., Gelpi, J. L., Luque, F. J. & Orozco, M. (2004). Linear response theory: an alternative to PB and GB methods for the analysis of molecular dynamics trajectories? *Proteins* **57**, 458-467.

56. Wang, W., Donini, O., Reyes, C. M. & Kollman, P. A. (2001). Biomolecular simulations: recent developments in force fields, simulations of enzyme catalysis, protein-ligand, protein-protein, and protein-nucleic acid noncovalent interactions. *Annu. Rev. Biophys. Biomol. Struct.* **30**, 211-243.

57. Cubero, E., Luque, F. J. & Orozco, M. (2001). Theoretical studies of d(A:T)-based parallel-stranded DNA duplexes. *J. Am. Chem. Soc.* **123**, 12018-12025.

58. Cubero, E., Avino, A., de la Torre, B. G., Frieden, M., Eritja, R., Luque, F. J., Gonzalez, C. & Orozco, M. (2002). Hoogsteen-based parallel-stranded duplexes of DNA. Effect of 8-amino-purine derivatives. *J. Am. Chem. Soc.* **124**, 3133-3142.

59. Sherer, E. C., Harris, S. A., Soliva, R., Orozco, M. & Laughton, C. A. (1999). Molecular Dynamics Studies of DNA A-Tract Structure and Flexibility. *J. Am. Chem. Soc.* **121**, 5981-5991.

60. Spackova, N., Cheatham 3rd, T. E., Ryjacek, F., Lankaš, F., Van Meervelt, L., Hobza, P. & Šponer, J. (2003). Molecular dynamics simulations and thermodynamics analysis of DNA-drug complexes. Minor groove binding between 4',6-diamidino-2-phenylindole and DNA duplexes in solution. *J. Am. Chem. Soc.* **125**, 1759-1769.

Chapter 6

ENHANCED SAMPLING METHODS FOR ATOMISTIC SIMULATION OF NUCLEIC ACIDS

Catherine Kelso[%o‡] and Carlos Simmerling[†‡*]
‡Center for Structural Biology †Department of Chemistry

Stony Brook University, Stony Brook, 11794-3400

%oWard Melville High School, East Setauket, NY 11733

Email: carlos.simmerling@stonybrook.edu

Abstract: This review summarizes recent work in atomic-detail simulations of nucleic acids, with emphasis on methods to improve sampling of conformational ensembles, such as Locally Enhanced Sampling and Replica Exchange molecular dynamics. Studies of several model oligonucleotide systems are described in detail to compare these different approaches.

Key words: Locally Enhanced Sampling, Replica Exchange Method, Replica Exchange Molecular Dynamics, Parallel Tempering, Free Energy Landscapes

1. INTRODUCTION

Computer simulations have assisted in the development of structural biology, enhancing the results attained by experimental methods; in simulating biological molecules, molecular dynamics (MD) simulations have moved from re-creating experimental structures to predicting secondary and folded structures and displaying the pathways by which they occur, discussed throughout this volume, though both experimental and theoretical methods still face many limitations[1]. When simulating transitions and pathways, structures are frequently trapped in local energy minima where the thermal energy of the system is insufficient to traverse the energy barriers during an affordable amount of simulation time. These energy minima exist at a higher energy level than the native state or conformation determined

147

J. Šponer and F. Lankaš (eds.), Computational Studies of RNA and DNA, 147–167.
© 2006 Springer.

experimentally, but regular MD simulations are often incapable of freeing the structures from local minima. For this reason, Locally Enhanced Sampling (LES) and the multiple variations of the Replica Exchange Method (REM) have been developed to make computer simulations more efficient in reducing barriers between local minima and native states.

In order to illustrate their strengths and weaknesses, this chapter will compare results from recently developed methods as applied to standard duplexes, single-stranded RNA, and modified DNA. These provide examples of simulations that require enhanced sampling techniques and problems that are currently being addressed using the various methods.

Accurate treatment of solvation effects is critical to simulation of nucleic acids. The most commonly used approach involves explicit inclusion of solvent molecules in the simulation, using water models such as TIP3P[2] or SPC[3]. These simulations are typically performed using periodic boundary conditions where the system simulated corresponds to a single element in an infinite lattice made of up shapes such as a cube or a truncated octahedron. Long-range electrostatic interactions in these periodic systems can be efficiently calculated using methods such as Particle Mesh Ewald (PME)[4]. Inclusion of these long-range electrostatic interactions was shown to result in improved stability of DNA duplexes as compared to simulations using a cutoff on long-range interactions[5]. As described in other chapters, explicit solvent models can provide an accurate representation of solvation effects; however the calculations are computationally demanding due to the large increase in the number of atoms for which forces need to be calculated.

Standard molecular dynamics (MD) simulations (on the nanosecond timescale) with explicit solvent are able to model only some important conformational changes. Cheatham et al. reported that MD simulations in explicit solvent were able to model the A-DNA→B-DNA transition under conditions that stabilize the B-DNA form in solution[6]. The same force field was able to simulate the reverse as well, the B-DNA→A-DNA transition in conditions favorable to A-DNA. Similar calculations were performed in an attempt to model the transition between A-RNA and B-RNA. Although the B-RNA form has not been experimentally characterized, this structure remained stable during the entire 10ns simulation, as did an A-RNA duplex. This is most likely due to the higher energy barriers arising from the extra hydrogen bonding possible with the O2' hydroxyl group present in RNA.

Other difficulties in modeling transitions in RNA during standard MD simulation have been reported. One system that has served as a model for several sampling methods is a 12 residue UNCG tetraloop with U6 in the variable position. Two alternate structures have been published, with an early structure[7] subsequently being replaced by a more accurate model[8] that differed in the details of the U5:G8 base pairing in the tetraloop, shown in

Figure 6-1. Several simulations have been carried out with various models in an attempt to model the spontaneous transition from the incorrect conformation (I) to that with the correct hydrogen bond pattern (C). Success would indicate that the methods could form a useful component of structure refinement protocols. However, standard MD simulations using explicit solvent and PME were not able to simulate the I→C transition at 300K within 2.5ns[9]. With the removal of the 2'hydroxyl hydrogens of the loop residues, however, the I→C transition occurred spontaneously on this timescale.

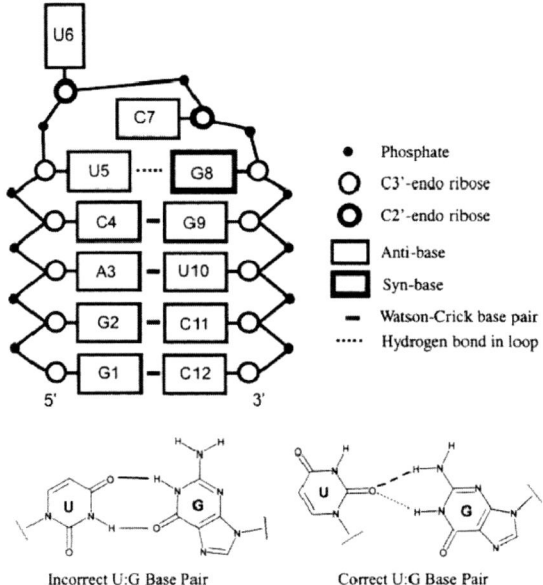

Figure 6-1. A schematic diagram of the topology of the RNA tetraloop model system. The solid lines on the lower images represent incorrect hydrogen bonds while the dashed lines are correct ones. (See text for details)

These examples (and others described in more detail in other chapters of this volume) demonstrate that MD simulations of nucleic acid systems in explicit solvent can provide accurate models, but they are also highly susceptible to becoming kinetically trapped in local minima during simulations on affordable timescales. Improved sampling techniques are required if we desire simulations to be able to reliably locate low-energy conformations that are separated from initial conformations by high energy barriers. This is particularly true for RNA, where the O2' hydroxyl group can inhibit rapid conformational change and also results in a wider variety of possible conformations that may need to be sampled.

2.   METHODS FOR IMPROVING SAMPLING EFFICIENCY

2.1   Continuum Solvation with Low Viscosity

Continuum solvent models can improve sampling through several effects. First, simulations with continuum solvents involve fewer atoms and therefore significantly less computational effort is spent evaluating forces and integrating equations of motion. This simplification means that longer simulations can be obtained for the same computational expense, thus permitting modeling of events that occur on timescales that are not accessible to simulations in explicit solvent. Since they represent average solvent properties and thus do not explicitly include friction arising from solute diffusion through the solvent, continuum models can also accelerate conformational changes for motions that are affected by solvent viscosity. If desired, friction effects can be included (at least approximately) through the use of Langevin dynamics.

Among the various continuum models that have been developed, the Generalized Born (GB) model[10;11;12] has become widely used for biomolecular simulation. The popularity of GB is likely due to the relative simplicity of the model, as well as its computational efficiency that allows its use at each step of a molecular dynamics simulation. Several variants of the GB model have been developed[13;14;15] and most current biomolecular simulation packages have implemented at least one GB model.

In simulations reported by Tsui and Case[16], the GB model was used to simulate A-form duplex DNA, converging to the B-form over 20 times faster than was required with explicit water. GB simulations of the RNA UUCG tetraloop described above converted from I to C in 1.2ns using GB[17] even though this transition did not occur during 2.5ns MD in explicit water. These studies demonstrate that the transition from an incorrect conformation to the one that has been experimentally determined occurs more rapidly with GB.

Recent simulations of DNA duplexes containing the common DNA lesion 8-oxoguanine (8o x oG) focused on the local conformation of 8o x oG either paired with cytosine or in a mismatch with adenine. This purine:purine mismatch can be accommodated by a Hoogsteen pair with a *syn* glycosidic bond rotamer for 8o x oG[18]. Crystal structures[19] indicate that the 8o x oG adopts an *anti* conformation when the same duplex forms a complex with the MutY glycosylase that selectively acts on 8o x oG:A pairs.

In order to investigate the conformational preferences of these base pairs containing 8o x oG, GB simulations were performed for 13-mer duplexes containing G:C, 8o x oG:C or 8o x oG:A as the central base pair. Initial structures were built in the *anti* conformation and an *anti→syn* transition for 8oxoG in duplexes containing 8o x oG:A was reproducibly simulated within 10ns. In contrast, the duplex with 8o x oG:A was stable in both *anti* and *syn* conformations when explicit solvent was used, with no interconversion seen.

One advantage to the improved sampling and reduced cost of GB is that more events can be simulated with available computational resources. In the case of the 8o x oG:A duplexes, the ability to perform multiple simulations revealed that two *anti→syn* transition pathways were sampled, as shown in Figure 6-2. The expected external pathway involved spontaneous flipping (extrusion) of the 8o x oG base into the major groove, rotation around the glycosidic bond and subsequent reinsertion of the base back into the duplex in the *syn* conformation. This was the major pathway, comprising 85% of the *anti→syn* events.

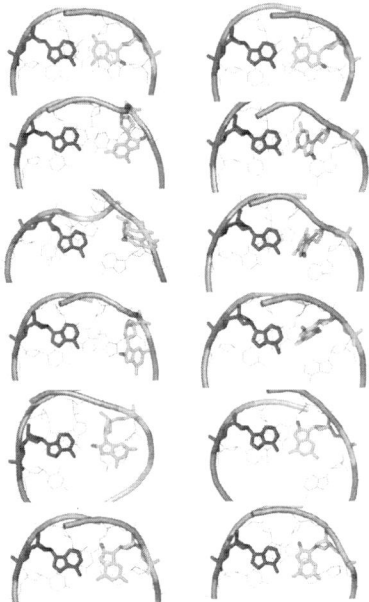

Figure 6-2. Snapshots down the helical axis at time points during the *anti/syn* transition for the two transition pathways. The left column shows the external rotation while the right column shows the internal pathway.

An alternate internal pathway, with reduced occurrence, was less local in nature and involved the A:T pair flanking 8oxoG. Spontaneous but transient opening of this A:T pair occurred prior to change in the 8o x oG glycosidic angle (likely induced by backbone strain and major groove widening due to the purine:purine mismatch). Opening of the flanking pair created sufficient space within the duplex for 8o x oG to rotate 180° without extrusion. 8o xo G subsequently formed Hoogsteen hydrogen bonds with the partner adenine and the flanking A:T pair was then re-established. In both cases, the result was a stable B-form duplex with 8o x oG:A in a *syn/anti* conformation.

The reduced cost of the GB simulations also enabled a more detailed investigation of the influence of the flanking pair sequence on the transition pathway Four alternate flanking pairs were simulated (with either G:C or A:T on the 5' or 3' side of 8o x oG). The percent of structures that made the *anti→syn* transition on the 10 ns timescale varied with the flanking base pair sequence. We made the surprising observation that an increased stability of the flanking base pairs (more G:C content) corresponded to an <u>increase</u> in the probability of the sequence making the *anti→syn* transition. As expected, however, more stable flanking pairs were correlated with increasing reliance on the pathways involving extrusion of 8oxoG during the transition. These simulations demonstrate the ability of the generalized Born solvent model to permit a more statistically detailed analysis of transitions though they also appear to be more sensitive to careful equilibration than simulations done in explicit solvent. GB methods are approximate and consequently have drawbacks[12] arising from the lack of finite sized solvent molecules and the ad hoc nature of the GB equations as opposed to Poisson-Boltzmann continuum descriptions and other models of similar rigor.

2.2   Locally Enhanced Sampling

Locally Enhanced Sampling (LES)[20], which has been described in detail in the past[20;21;22;23], has also been successfully used to improve sampling in nucleic acid simulations. This method is of particular interest for systems in which a small portion is of interest, such as several bases in a DNA or RNA sequence. For simulations in which enhanced sampling is needed for the entire structure the Replica Exchange Method, which will be discussed in detail later in this chapter, would likely be more suitable.

Briefly, LES is a mean-field approach that separates the system into two primary components (Figure 6-3): a region of interest for which enhanced sampling is desired, and the remainder (which may be a stable region where large changes are not expected to occur). The atoms in the region of interest are replicated, with each of the copies interacting with the remainder of the system but not with the other copies. The non-LES region feels the average force of the copies. Thus a single simulation of the LES system will provide multiple trajectories for the region of interest (one for each copy). These trajectories are obtained at a reduced cost as compared to repeating the simulation, since the larger non-LES region is simulated only once.

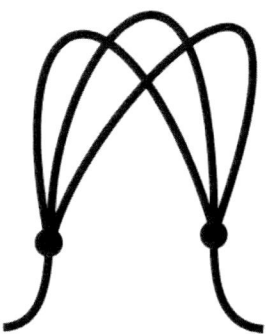

Figure 6-3. A diagram of a three copy LES simulation of the UUCG tetraloop with a single stem region.

It has been shown that the global energy minimum for the LES system is the same as that for the original (non-LES) system. This occurs when every copy in a LES system occupies identical coordinates with the conformation corresponding to the global minimum of the original system[21]. Thus LES is a useful tool for structure optimization, since no mapping procedures are needed to interpret the LES results.

An additional benefit of LES is that the barriers to conformational transitions are reduced as compared to the original system. The reduction is significant, approximately $1/N$ for N copies[24], arising from the mean-field nature of LES. Since the energy of the LES system when all copies are together is the same as the original system in the same conformation, the energy/force felt by each copy is reduced and each can be moved individually with a lower barrier compared to moving all copies together.

LES was applied to the I→C transition in the UUCG RNA hairpin described above, with neutralizing Na^+ counterions, ~2300 TIP3P water molecules and the ff94 all atom force field[25]. All atoms were coupled to a thermostat to maintain a constant temperature of 300K.

As described above, standard MD simulations were able to undergo the transition from I→C only when the 2'hydroxyl hydrogen atoms had been removed[9], implying their involvement in the delayed transition. For this reason, the initial application of LES made 5 copies only of these hydrogens (based on previous work that showed 5 LES copies was a reasonable choice)[26]. Although these hydroxyl hydrogen atoms were able to move apart and undergo frequent rotational transitions, the reduction in the key barriers was insufficient as there was no resulting change in the loop conformation within 2 ns of MD.

The next simulation used five LES copies of the entire UUCG loop to see if the larger LES region would improve the likelihood of a transition occurring. Each copy was attached to the stem and started with the same incorrect initial coordinates. At 175 ps the RMSD as compared to the correct NMR structure went to ~0.7Å, showing the conversion from the incorrect conformation to the correct loop structure (Figure 6-4). The simulation was repeated with different initial coordinates for the LES copies and a successful transition to C was observed at ~200 ps. The third LES simulation was initiated with all five copies in structure C and there was no significant change during 750ps. In comparison to the inability of the hydrogen LES copies to make the transition, the larger LES region greatly improved the sampling of the system. Applying LES to the loop region was also more efficient than using standard MD with GB solvation (which took ~1ns to undergo the transition, as described above).

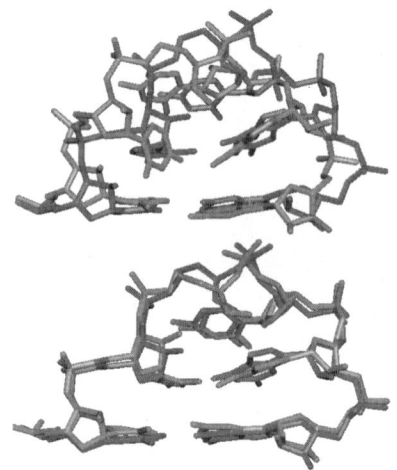

Figure 6-4. Comparison of the loop in the correct NMR structure (light) and the average LES structure (dark) from GB simulations using three LES copies of the UUCG loop region. Only the loop and the first base pair of the stem are shown (residues 4-9 except the mobile U6 base). The upper image shows the initial incorrect structure. In addition to the base pair hydrogen bond differences shown in Figure 6-4, there is severe buckling of the U5:G8 base pair, as well as other differences in the backbone on the 5' end of the loop. The lower image shows the same comparison after LES optimization.

In order to test the sensitivity of the number of LES copies a system used, two copies were implemented in place of the previous five. The transition occurred, though on a longer time scale (~600 ps) than the system with five copies. A larger LES region was used to investigate the sensitivity to LES region size, this time replicating the central six nucleotides (CUUCGG), including the CG base pair at the top of the stem in addition to the residues that had previously been copied. The I→C transition occurred within 100 ps and the resulting structure differed from that obtained with the smaller LES region by 0.7Å.

Overall, these experiments established that even as few as two LES copies greatly improved the sampling of a specific region. As long as the UUCG loop was copied LES appears not to be very sensitive to the size of the LES region or the number of LES copies, although there were some differences in the time required for the transition.

In addition, the multiple trajectories provided by each LES simulation indicated the presence of alternate transition pathways, as was evident in the 8o x oG simulations described above. In that case, however, multiple simulations were required to sample the alternate paths since LES was not used. The LES copies directly demonstrated two potential routes for the I→C transition. The first route contained a semi-stable intermediate where the U5:G8 base pair was broken with a flipped out G8 base. The second route went more directly from the incorrect to the correct conformation.

In another application of LES to nucleic acid simulation in explicit solvent, LES simulations of a 13 base pair, non-conventional DNA strand with a pyrene substitution were performed in an attempt to provide a structural interpretation for the lack of NOEs observed for the adenine on the 5' flanking side of the pyrene[27]. This base was disordered in the family of structures refined using the NMR data. Simulations in explicit solvent starting from NMR models lacking the A8:T19 pair spontaneously formed this pair on the nanosecond timescale. These pairs formed in both *anti* and *syn* conformations for A8, with Watson-Crick and Hoogsteen geometries, respectively. MM-PBSA energy analysis suggested similar stability (0.8 kcal/mol in preference of the *anti* conformation), in contrast to a >2 kcal/mol preference for *anti* using the same analysis on an A:T pair for a control duplex lacking pyrene. However, no spontaneous transitions between the two were observed in these standard MD simulations.

Ten LES copies of the A8:T19 base pair flanking the pyrene were employed, using one of the NMR models in which base pairing for these residues was not present. The same LES approach was applied to the A:T pair in the control sequence. With LES, spontaneous transitions between the alternate base pair geometries were observed, and both forms were sampled in each run (Figure 6-5). The resulting net population for *syn* was 30% in the pyrene system (<1% in the control), providing an estimated free energy difference of 0.5 kcal/mol in favor of the *anti* conformation in good agreement with the MM-PBSA energy analysis of the standard simulations that were kinetically trapped. The ability of the LES simulation to sample both base pair geometries in a single run is reassuring, particularly when the relative populations are consistent with energy analysis of standard simulations.

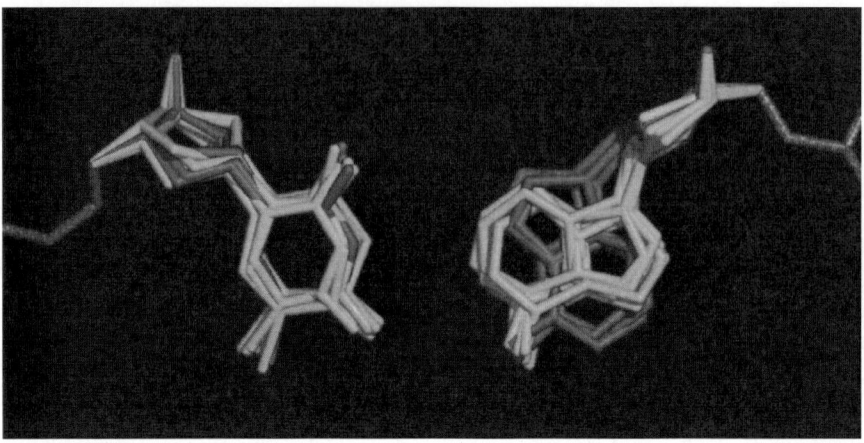

Figure 6-5. A snapshot from LES MD simulation of the A:T pair flanking a pyrene substitution. Multiple conformations are sampled, consistent with energy analysis and experimental data. 10 copies of the A:T pair were employed.

These simulations all exploited the ability of LES to improve sampling in systems with explicit inclusion of solvent molecules[26]. However, one limitation of LES with explicit solvent relates to the solvent environment surrounding each of the LES copies. When the LES copies at a specific time do not occupy the same coordinates, explicit solvent molecules are excluded from the volume occupied by any of the copies. This arises because the solvent molecules are not copied, and thus they feel the average force from any of the LES copies and cannot overlap with one copy in order to directly interact with another. As a result, individual copies may experience perturbations to their solvation shells based on the conformation of other copies. In short, although LES lowers the energy barriers permitting more sampling of structures, the distribution of solvent molecules around copies that do not occupy the same coordinates are not the same as for the corresponding atoms in the non-LES systems. This is not a problem for the global free energy minimum or any other local minimum in which the copies all occupy the same coordinates and thus the water interacts with each of them in an identical fashion. It does, however, introduce an indirect coupling between the copies that tends to inhibit independent motion.

Due to this potential drawback to LES, we investigated alternate solvation methods that would provide a description of the solvation of individual copies that more closely corresponded to the environment experienced by non-LES system sampling those conformations. This led to investigation of a combination of LES with continuum solvent models. As described above, LES increases sampling through a reduction in barrier height[21], creating a smoother potential energy surface and allowing structures to escape local energy minima more frequently. Continuum solvents such as GB also accelerate conformational change. Cheng et al. comined these methods and obtained synergistic benefits (GB + LES)[28]. With GB it became possible for the solvation of each copy to exactly correspond to the environment that would be felt by this copy in the absence of the other copies (i.e. for the standard system in the same conformation). Thus each of the copies was independently solvated and moved with significantly increased independence as compared to LES with explicit solvent.

The RNA UUCG stem-loop system was chosen as a model to test the GB + LES combination, due to its use for previously published sampling tests for both LES and GB alone. Simulations varied in the number of replicas of the sampled region as well as the size of the region in order to classify the specificity of this method. At 300K all base pairs were lost for the C and I initial structures within 400 ps. This may have been a response to weakening of the Watson-Crick hydrogen bonds because of the scaling of partial charges and Lennard-Jones well depth parameters in LES. In the explicit solvent LES simulations, these fluctuations were likely damped by solvent viscosity. For this reason, the remaining simulations were performed at varying temperatures lower than 300K. Results demonstrated that the combination of GB and LES has the same ability to reproduce the correct conformational change, but with greater computational efficiency than obtained with either GB or LES alone. However, the sensitivity of the simulations to temperature[22; 23] indicates that the choice of an appropriate temperature for LES simulations is a challenge when applying this efficient mean-field method to new systems.

2.3 Replica Exchange Molecular Dynamics

Replica Exchange is another algorithm used to increase sampling and assist in releasing trapped structures from local minima. In parallel tempering and replica exchange molecular dynamics (REMD)[29;30], several noninteracting copies (replicas) are independently and simultaneously simulated at different temperatures. These temperatures span a range from the temperature of interest (such as 280K or 300K) up to a temperature at which the system can rapidly traverse potential energy barriers to conformational transitions (such as 600K). Unlike LES copies in which only a piece of the simulation is copied, in regular REMD each replica is an individual copy of the entire system. At intervals during the otherwise standard simulations, conformations of the system being sampled at different temperatures are exchanged based on a Metropolis-type equation that considers the probability of sampling each conformation at the alternate temperature. In this way, REMD is less hampered by the local minima problem, since the low-temperature simulations (replicas) have the potential to escape kinetic traps by jumping to minima that are being efficiently sampled by the higher-temperature replicas. Moreover, the high-energy regions of conformational basins sampled by the high-temperature replicas can be relaxed in a way similar to temperature annealing. The transition probability is constructed such that canonical ensemble properties are maintained during each simulation, providing potentially useful information about conformational probabilities as a function of temperature. Replica exchange of many nontrivial systems has been demonstrated as a powerful tool to optimize structure, sample equilibrium thermodynamic quantities, and estimate pathways that could be sampled in nonequilibrium simulations (such as peptide folding or secondary structure formation). However, few applications of REMD to nucleic acids have been reported[31;32;33].

One drawback to REMD in its application to larger systems is that when the system size increases the number of replicas needed to effectively span a given temperature range also grows[29;34], proportionally with the square root of the number of degrees of freedom. This increase, along with the inherent growth in computational requirements for performing MD for each replica of the larger system, can rapidly make application of REMD prohibitive. To address this difficulty, PREMD and LREMD were developed as similar strategies to reduce the number of replicas needed for simulation.

Partial Replica Exchange Molecular Dynamics (PREMD) is designed for application to systems in which only a portion needs enhanced sampling, while the remainder is expected to be weakly coupled to conformational changes in the region of interest. This tool would be appropriate for sampling several base pairs in a DNA duplex, an RNA loop region, and other small regions of systems that are too large for practical application of standard REMD. Similar to REMD, the entire system is replicated. The difference lies in the separation of the system into a "focus" region that spans a range of temperatures, and a "remainder" region that is maintained at the same temperature for all replicas (Figure 6-6). During the simulation, periodic exchanges are performed based on an exchange probability that is calculated using only the energy terms involving the focus region and its interaction with the remainder. The essential approximation is that the remainder region has a similar conformation for all replicas and that energy differences in this region can be neglected in the exchange probability. This reduces the perceived system size for the exchange calculation, and the number of replicas required scales with the size of the focus region rather than the entire system. If the focus region in the system is much smaller than the remainder, then the number of replicas in a PREMD simulation would be significantly less than that required for standard REMD.

An additional advantage to PREMD can arise from maintaining the remainder regions of the replicas at lower temperatures. In many systems, it is desirable to <u>avoid</u> extensive conformational change in certain regions. For example, when sampling alternate RNA loop conformations one may specifically not want to change the conformation of the non-loop portion. PREMD allows this option without the need to restrain the conformation of the atoms outside the region of interest, which could inhibit even a weak coupling.

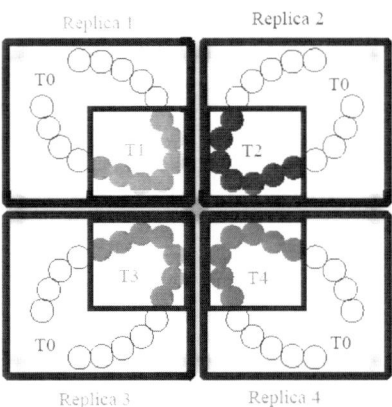

Figure 6-6. A schematic diagram of the Partial Replica Exchange Method. Only atoms in the focus region (inner squares) are coupled to different temperature baths over a range of T, while the remainder is coupled to a single bath at T0. Sampling of the focus region is enhanced.

The efficiency of REMD and PREMD were demonstrated using the RNA UUCG stem-loop system. The UUCG loop residues were defined as the focus region for PREMD. As discussed previously, the PREMD setup requires fewer replicas to cover the same temperature range as conventional REMD. Whereas eight replicas were required to span from 266 K to 405 K in regular REMD, only five replicas were used for the PREMD system. Each replica was individually coupled to thermostats at 266, 295, 327, 363, and 403 K, for an expected acceptance probability of 10%. The remainder (the stem region) was maintained at 300 K for all replicas.

A free energy landscape (Figure 6-7) was constructed using two-dimensional histograms with the RMSD of the loop region (as compared to the correct structure) and the U5 glycosidic angle as reaction coordinates. The resulting surface has three distinct transitions pathways, consistent with the multiple pathways sampled by the LES MD simulations of the RNA hairpin. Figure 6-7 also visualizes the three pathways on the energy landscape by adding white spheres to show the location of the structure at various times throughout MD simulations that spontaneously sampled the I→C transition. This mapping of pathways sampled during non-equilibrium trajectories onto landscapes from equilibrium REMD data can provide useful insights into the behavior of the system.

The first two pathways are similar in that they sample only structures that are close to the incorrect and correct basins. The transitions involve direct crossing of the barrier between two minima, but detailed analysis shows that the two pathways cross at different locations. The first path involves slight flipping out of the U5 base, rotation of its glycosidic bond, and conversion to the correct structure. The second path only involves minor reorganization in the loop region with no flipping. On the basis of the PREMD energy landscape, the estimated free energy barrier for the first path involving significant rotation of glycosidic bond is ~1.7 kcal/mol at 327 K, while that for the second path is ~2.2 kcal/mol. The barrier separating the basins at lower temperatures is not populated enough to permit reliable estimation of the free energy. In such cases, other techniques (such as umbrella sampling) would be better suited for determination of energy barriers.

The third pathway is noticeably different from the other two early on in the simulation. This pathway samples a much broader region of the energy landscape before adopting the correct conformation. It is interesting to note that upper basin in the figure was not sampled in any of the I→C conversions, and thus appears to represent an unproductive off-pathway basin for this particular transition.

Local Replica Exchange Molecular Dynamics (LREMD) combines the "exchange part of the Hamiltonian" aspect of PREMD with the LES approach. Rather than simulating the remainder portion of the system multiple times, as described in PREMD, the replicas of the focus region are created using LES, with the remainder region corresponding to the non-LES atoms. The LES-replicated region performs the same functions as the PREMD focus region. Thus the entire set of REMD replicas is performed in a single MD simulation. While PREMD included a remainder region for each replica, permitting weak coupling, LREMD assumes that this coupling is weak enough that a single representation of the remainder region can be used for all replicas of the focus region. Since the LES focus region is the same size as would be used for PREMD, the number of LES copies needed is also fewer than would be used for standard REMD.

LREMD has several advantages over PREMD. In an LREMD simulation the non-LES region, which is replicated in a PREMD simulation, is only calculated once, saving a large fraction of the calculations required by PREMD. By this method a full set of replica exchange results can be obtained at nearly the same cost as a single, standard MD simulation. A positive affect of the smoother energy surface is the need for a smaller temperature range, and consequently fewer replicas than REMD or PREMD.

LREMD is based on the LES method; consequently it uses approximations beyond those inherent in PREMD. A significant assumption used by both REMD variants is that the conformation of the majority of the system is relatively insensitive to the smaller region of interest to which replica exchange is applied. This approximation is taken further with LREMD as compared to PREMD. LES also results in a smoothing of the energy surface, as described earlier in this chapter. This smoothing makes it impossible to derive quantitative thermodynamic properties from the free energy landscape of an LREMD simulation. Despite this disadvantage, LREMD is still able to provide qualitative information, and since the global energy minimum is exact using LES, LREMD can be a highly efficient optimization tool that draws on all of the advantages of LES and REMD yet requires only a single MD simulation.

LREMD was applied to the UUCG RNA loop system. LES was used to create five copies of the UUCG loop, each coupled to a different thermostat. Temperatures were 80, 88, 99, 108 and 120 K, optimized to result in 10% exchange acceptance. The non-LES region was maintained at 100 K. These temperatures are much lower than those of the PREMD simulations due to the sensitivity of LES to temperature as described earlier in this chapter.

Since the LREMD method is an extension of LES and designed to ameliorate its temperature limitation, the results of these two simulations must be compared to validate the benefits as compared to standard LES. Standard LES simulations were run at 80K and 120K, starting from all copies in the I conformation. Each simulation employed five LES copies of the UUCG loop. At 80K, by 4 ns two of the five copies were still in the incorrect conformation, while the others had successfully switched to C. At 120 K for all LES copies, the stem-loop system was unstable and resulted in a fully extended RNA strand with no base pairing.

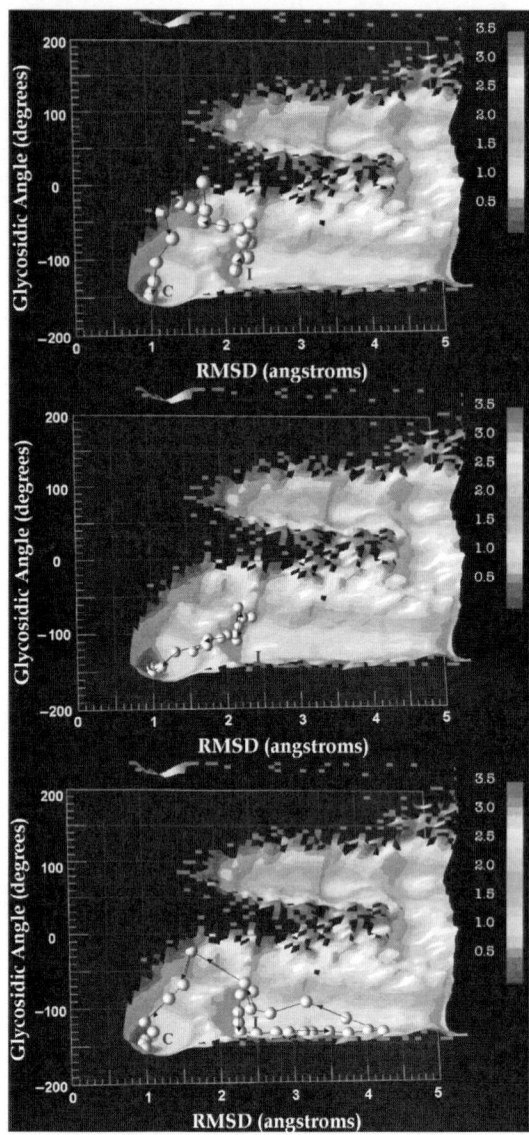

Figure 6-7. Free energy landscape from PREMD simulations at 327 K. The white spheres are the location of structures at different times during non-REMD simulation that spontaneously sampled the I→C structural transition. The top is the base-flipping pathway; middle, minor reorganization pathway; bottom, partial unfolding than refolding pathway.

LREMD was run using the same temperature range of 80-120K. In this case the stem region remained intact, and all copies converted from I→C much more rapidly than in the 80K LES simulation. We made the interesting observation that the LREMD copies did not remain in C, instead exhibiting multiple transitions between the correct and other structures during the entire simulation, providing a temperature-dependent sampling of non-native loop conformations.

In order to validate the extent to which these REMD variants were able to properly model the RNA hairpin behavior, standard REMD was performed. The REMD setup consisted of eight replicas at 266, 282, 300, 318, 338, 359, 381, and 405 K. Similar to the others, the temperatures were chosen to provide 10% exchange acceptance. In order to maintain the structure of the stem region, distance restraints were required for the Watson-Crick hydrogen bonds between the first three base pairs in the stem region. These restraints were unnecessary in the previously discussed PREMD and LREMD simulations as these variants permitted coupling of this region to a thermostat at a lower temperature. Although PREMD and LREMD both have the complexity of requiring the user to choose a focus region of interest, selecting atoms to restrain in standard REMD simulations involves a similar requirement to define a subset of the system for which enhanced sampling is desired.

To compare PREMD and LREMD to standard REMD, free energy landscapes were generated from 2-dimensional histograms of the ensembles generated using each method (Figure 6-8). Different rows are from different temperatures. Three temperatures are shown: the closest to 300K (to match previous MD studies[9;17;24;28], the middle temperature of the range, and the highest. For LREMD, the lowest, middle and highest temperatures are shown. All of the surfaces show a very similar pattern. The higher energy regions of the landscapes are qualitatively similar between all of the methods. Only one deep free energy minimum is observed at low temperature in all REMD methods, which corresponds to the NMR-based correct structure (recall that all simulations were initiated with the incorrect structure). This suggests that PREMD and LREMD can be successfully used to locate the native conformation with less computational effort than needed for standard REMD.

At the middle temperatures, both REMD and PREMD surfaces show two major minima. The correct structure is highly dominant while the incorrect structure is only a meta-stable structure corresponding to a much shallower minimum on the free energy surface. This is consistent with the spontaneous I→C conversions sampled during standard MD as described earlier in this chapter. As temperature increases, more disordered structures are sampled, with broad basins corresponding to flexible unfolded structures. Differences in the landscapes may arise from the PREMD approximation or the restraints required for normal REMD.

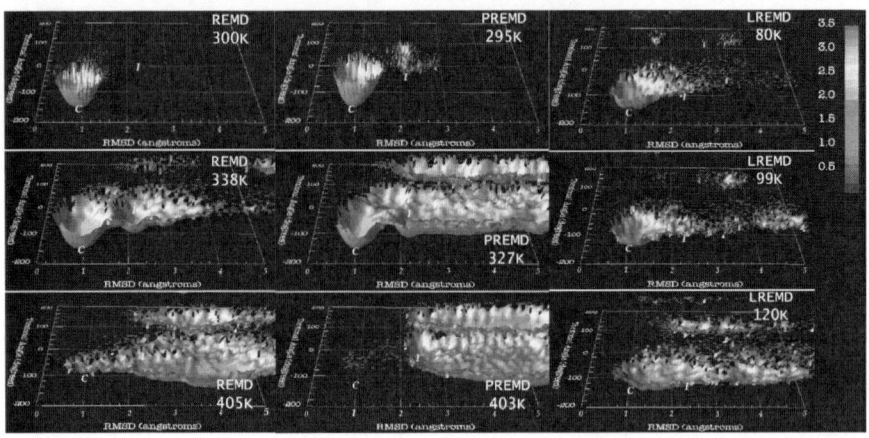

Figure 6-8. Free energy landscapes using the loop region rmsd (as compared to the correct NMR structure, X) and the U5 glycosidic angle (Y). The left column is standard REMD, the middle PREMD and the right LREMD. All three simulations began in the incorrect structure, with the basin on the free energy landscapes labeled I, and ended in the global minimum corresponding to the correct NMR structure, noted C. The surface of the landscape on the z-axis, which corresponds to a variation in color, shows the difference in free energy at different points.

The LREMD results differ the most from the others. In particular, the LREMD surfaces are notably smoother as compared to those of the PREMD and REMD landscapes, consistent with the barrier reduction properties of LES described earlier in this chapter. The temperature range chosen for the LREMD simulations also appears to not reach as high an effective temperature as the ~400 K shown for PREMD and REMD, as indicated by the presence of a significant population in the native basin at the highest temperature (120 K) in the LREMD simulations. Although choice of a temperature is simplified with the range allowed by LREMD as compared to choosing a single temperature for LES, direct mapping of this range to "real" temperatures remains challenging.

3.   SUMMARY

The simulation methods discussed in this chapter have extended the applicability of all-atom simulation to structure optimization for nucleic acids systems. Each is available in Amber and was shown to be applicable to nucleic acids. These methods provide significant advantages over standard molecular dynamics simulations, yet each has potential disadvantages that must be considered during method choice, system preparation and interpretation of the results. With the continued development of enhanced sampling simulation methods and computational resources, the range of problems that can be addressed by these types of simulation will also continue to grow.

ACKNOWLEDGEMENT

The work was supported by National Institutes of Health GM6167803 and NPACI MCA02N028. C.S. is a Research Corporation Cottrell Scholar.

REFERENCES

1. Liu, H., Spielmann, H. P., Ulyanov, N. B., Wemmer, D. E. & James, T. L. (1995). Interproton distance bounds from 2D NOE intensities: Effect of experimental noise and peak integration errors. *J Biomol Nmr* 6, 390-402.
2. Jorgensen, W. L., Chandrasekhar, J., Madura, J. & Klein, M. L. (1983). Comparison of Simple Potential Functions for Simulating Liquid Water. *J Chem Phys* 79, 926-935.
3. Berendsen, H. J. C., Postma, J. P. M., Vangunsteren, W. F. & Hermans, J., Eds. (1981). Interation models for water in relation to protein hydration. Intermolecular Forces. Edited by B, P. Dordrecht, Reidel.
4. Darden, T., York, D. & Pedersen, L. (1993). Particle Mesh Ewald - an N.Log(N) Method for Ewald Sums in Large Systems. *J Chem Phys* 98, 10089-10092.
5. Cheatham, T. E., Miller, J. L., Fox, T., Darden, T. A. & Kollman, P. A. (1995). Molecular-Dynamics Simulations on Solvated Biomolecular Systems - the Particle Mesh Ewald Method Leads to Stable Trajectories of DNA, Rna, and Proteins. *J Am Chem Soc* 117, 4193-4194.
6. Cheatham, T. E. & Kollman, P. A. (1996). Observation of the A-DNA to B-DNA transition during unrestrained molecular dynamics in aqueous solution. *J Mol Biol* 259, 434-444.
7. Varani, G., Cheong, G. & Tinoco, I. J. (1991). Structure of an Unusually Stable RNA Hairpin. *Biochemistry-Us* 30, 3280-3289.
8. Allain, F. H. T. & Varani, G. (1995). Structure of the P1 Helix from Group-I Self-Splicing Introns. *J Mol Biol* 250, 333-353.
9. Miller, J. L. & Kollman, P. A. (1997). Theoretical studies of an exceptionally stable RNA tetraloop: Observation of convergence from an incorrect NMR structure to the correct one using unrestrained molecular dynamics. *J Mol Biol* 270, 436-450.
10. Still, W. C., Tempczyk, A., Hawley, R. C. & Hendrickson, T. (1990). Semianalytical Treatment of Solvation for Molecular Mechanics and Dynamics. *J Am Chem Soc* 112, 6127-6129.
11. Cramer, C. J. & Truhlar, D. G. (1999). Implicit solvation models: Equilibria, structure, spectra, and dynamics. *Chem Rev* 99, 2161-2200.
12. Simonson, T. (2001). Macromolecular electrostatics: continuum models and their growing pains. *Curr Opin Struc Biol* 11, 243-252.

13. Onufriev, A., Case, D. A. & Bashford, D. (2002). Effective Born radii in the generalized Born approximation: The importance of being perfect. *J Comput Chem* 23, 1297-1304.

14. Feig, M., Onufriev, A., Lee, M. S., Im, W., Case, D. A. & Brooks, C. L. (2004). Performance comparison of generalized born and Poisson methods in the calculation of electrostatic solvation energies for protein structures. *J Comput Chem* 25, 265-284.

15. Zhu, J., Alexov, E. & Honig, B. (2005). Comparative study of generalized Born models: Born radii and peptide folding. *J Phys Chem B* 109, 3008-3022.

16. Tsui, V. & Case, D. A. (2000). Molecular dynamics simulations of nucleic acids with a generalized born solvation model. *J Am Chem Soc* 122, 2489-2498.

17. Williams, D. J. & Hall, K. B. (1999). Unrestrained stochastic dynamics simulations of the UUCG tetraloop using an implicit solvation model. *Biophys J* 76, 3192-3205.

18. Kouchakdjian, M., Bodepudi, V., Shibutani, S., Eisenberg, M., Johnson, F., Grollman, A. & Patel, D. (1991). NMR structural studies of the ionizing-radiation adduct 7-hydro-8-oxodeoxyguanosine (8-oxo-7H-DG) opposite deoxyadenosine in a DNA duplex-8-oxo-7H-DG(syn).DA(anti) alignment at lesion site. *Biochemistry-Us* 30, 1403-1412.19. Lipscomb, L. A., Peek, M. E., Morningstar, M. L., Verghis, S. M., Miller, E. M., Rich, A., Essigmann, J. M. & Williams, L. D. (1995). X-Ray Structure of a DNA Decamer Containing 7,8-Dihydro-8-Oxoguanine. *P Natl Acad Sci USA* 92, 719-723.

20. Elber, R. & Karplus, M. (1990). Enhanced Sampling in Molecular-Dynamics - Use of the Time-Dependent Hartree Approximation for a Simulation of Carbon-Monoxide Diffusion through Myoglobin. *J Am Chem Soc* 112, 9161-9175.

21. Roitberg, A. & Elber, R. (1991). Modeling Side-Chains in Peptides and Proteins - Application of the Locally Enhanced Sampling and the Simulated Annealing Methods to Find Minimum Energy Conformations. *J Chem Phys* 95, 9277-9287.

22. Ulitsky, A. & Elber, R. (1993). The Thermal-Equilibrium Aspects of the Time-Dependent Hartree and the Locally Enhanced Sampling Approximations - Formal Properties, a Correction, and Computational Examples for Rare-Gas Clusters. *J Chem Phys* 98, 3380-3388.

23. Straub, J. E. & Karplus, M. (1991). Energy Equipartitioning in the Classical Time-Dependent Hartree Approximation. *J Chem Phys* 94, 6737-6739.

24. Simmerling, C., Miller, J. L. & Kollman, P. A. (1998). Combined locally enhanced sampling and Particle Mesh Ewald as a strategy to locate the experimental structure of a nonhelical nucleic acid. *J Am Chem Soc* 120, 7149-7155.

25. Cornell, W. D., Cieplak, P., Bayly, C. I., Gould, I. R., Merz, K. M., Ferguson, D. M., Spellmeyer, D. C., Fox, T., Caldwell, J. W. & Kollman, P. A. (1995). A 2nd Generation Force-Field for the Simulation of Proteins, Nucleic-Acids, and Organic-Molecules. *J Am Chem Soc* 117, 5179-5197.

26. Simmerling, C. & Elber, R. (1994). Hydrophobic Collapse in a Cyclic Hexapeptide - Computer-Simulations of Chdlfc and Caaaac in Water. *J Am Chem Soc* 116, 2534-2547.

27. Cui, G. L. & Simmerling, C. (2002). Conformational heterogeneity observed in simulations of a pyrene-substituted DNA. *J Am Chem Soc* 124, 12154-12164.

28. Cheng, X. L., Hornak, V. & Simmerling, C. (2004). Improved conformational sampling through an efficient combination of mean-field simulation approaches. *J Phys Chem B* 108, 426-437.

29. Hansmann, U. H. E. (1997). Parallel tempering algorithm for conformational studies of biological molecules. *Chem Phys Lett* 281, 140-150.

30. Sugita, Y. & Okamoto, Y. (1999). Replica-exchange molecular dynamics method for protein folding. *Chem Phys Lett* 314, 141-151.

31. Sorin, E. J., Rhee, Y. M., Nakatani, B. J. & Pande, V. S. (2003). Insights into nucleic acid conformational dynamics from massively parallel stochastic simulations. *Biophys J* 85, 790-803.

32. Cheng, X. L., Cui, G. L., Hornak, V. & Sinnnerling, C. (2005). Modified replica exchange simulation methods for local structure refinement. *J Phys Chem B* 109, 8220-8230.

33. Cheng, X. L., Kelso, C., Hornak, V., de los Santos, C., Grollman, A. P. & Simmerling, C. (2005). Dynamic behavior of DNA base pairs containing 8-oxoguanine. *J Am Chem Soc* 127, 13906-13918.

34. Fukunishi, H., Watanabe, O. & Takada, S. (2002). On the Hamiltonian replica exchange method for efficient sampling of biomolecular systems: Application to protein structure prediction. *J Chem Phys* 116, 9058-9067.

Chapter 7

MODELING DNA DEFORMATION

Péter Várnai[1,2] and Richard Lavery[1]

[1]*Laboratoire de Biochimie Théorique, CNRS UPR 9080, Institut de Biologie Physico-Chimique, 13 rue Pierre et Marie Curie, Paris 75005, France;* [2]*University of Cambridge, Department of Chemistry, Lensfield Road, Cambridge CB2 1EW, United Kingdom*

Abstract: Deformations of DNA contribute to its essential biological function. In our laboratory, we have been studying both local and global deformations of DNA and their relationship to base sequence by molecular modeling and simulation techniques. In the current chapter, we first give an overview of the various approaches used in our laboratory to build DNA models and to control DNA deformations. Notably, we discuss the JUMNA program that uses internal and helicoidal variables, and also umbrella sampling free energy simulations used to follow DNA deformations. In the second part, we summarize the results these techniques enabled us to obtain, starting from the large scale deformations, such as stretching, twisting and bending, down to the more local changes involving base opening and flipping and backbone conformations. A separate section deals with the sequence specific recognition of DNA by proteins and the role of DNA deformation in the process. We hope to show the reader that theoretical studies can play a significant role in obtaining a better understanding of this fascinating biopolymer.

Key words: DNA deformation; recognition, base flipping; single molecule manipulation; internal coordinates; JUMNA; AMBER; umbrella sampling; free energy; MM-PBSA

1.   DNA DEFORMATION AND ITS BIOLOGICAL INTEREST

At first sight, DNA seems to be a relatively simple biopolymer. While it is a heteropolymer, it is composed of only four different nucleotides, a small number compared to the 20 amino acids which constitute the polypeptide chain of proteins. This simplicity led early researchers to initially reject DNA as the potential carrier of genetic information. While the beautiful

169

J. Šponer and F. Lankaš (eds.), Computational Studies of RNA and DNA, 169–210.
© 2006 *Springer.*

double helical structure proposed by Watson and Crick[1,2] and the subsequent discovery of the triplet genetic code,[3,4] explained how DNA could stock enormous amounts of information, it again suggested that structurally there was not much to study. At the core of the double helical structure was the observation that the spiral phosphodiester backbones could accommodate any Watson-Crick base pair sequence without deformation.

The first step to refining this viewpoint comes from realizing that DNA must be packed quite densely to fit into a cell. This is easily illustrated in the case of human cells which contain around 1 m of DNA (corresponding to 4×10^9 base pairs) in a nucleus within a diameter of only a few microns. Within sperm heads, the packing density is even higher. A partial explanation of how this is achieved comes from modeling DNA as a flexible rod, which naturally forms a random coil to increase its conformational entropy. But this factor alone only is not enough to account for the packing that occurs within the nucleus. As we now know, the remainder is due to protein-induced superhelical compaction leading to the complex and hierarchical structure of chromatin.

A second type of deformation was detected early in the study of DNA and concerns its overall helical form. Fiber diffraction studies already showed that the double helical structure could be modified as a function of its solvent and counterion environment. The A and B forms of the double helix first named by Rosalind Franklin[5] are now structurally well-characterized and they have been joined by many other conformational families which go even further in tampering with DNA structure, by modifying its helical chirality, changing its number of strands, its base pairing and its relative strand orientations. In recent years, structural studies have been joined by single molecule manipulation experiments which offer us a new way to directly probe the mechanical properties of DNA.[6] These experiments have again showed that DNA is more complex than initially expected and that, when pulled or twisted, it can undergo transitions to new and unexpected conformations.

At this stage, it still seems possible to treat DNA as a regular polymer with structural and mechanical properties only defined at a macroscopic level. However, as the first single crystal structure of DNA showed,[7] the base sequence does in fact modulate the structure of the canonical double helix. Base pairs are not necessarily flat and perpendicular to the helical axis, and are not necessarily arranged in a helically regular way around this axis. The axis itself can moreover be influenced by the base sequence leading, in certain cases, to significant, intrinsic curvature. On a finer level, certain base pair steps have been identified as capable of enhancing helical flexibility[8-10] and this capacity is reflected by variations in the mechanical properties of DNA fragments having regular repeating sequences.[11-13] DNA can also be deformed at the level of the phosphodiester backbones. In part, backbone deformations reflect the local inhomogeneity brought about by specific base sequences, but it has now been recognized that the backbone in itself can adopt a variety of conformational states which represent another degree of classification within each of the major structural families of DNA.

Local deformations arise when interfaces are formed between different DNA conformations, such as B-A or B-Z junctions or, at a more subtle level, the B-B' junctions connected with A-tract curvature.[14-16] Very recent work on minicircle formation[17] for fragments much shorter than the standard persistence length of DNA also seems to confirm the proposal made 30 years ago by Crick and Klug[18] that, like macroscopic thin rods, DNA can yield under stress leading to local kinks (although some questions regarding these experiments remain to be answered[19]). Other local deformations are caused by base pair disruption. Base pair opening or, in its more extreme variant, base pair flipping is of major biological importance, notably for understanding how mutated or mispaired bases are detected and repaired or how DNA becomes selectively methylated.

In summary, it appears that DNA deformation is present, and biologically important, at almost every scale, from a single nucleotide up to fragments containing many thousands of base pairs. It is thus perhaps surprising that we are still lacking a real understanding of deformation, and especially of its links to base sequence, more than 50 years after the discovery of the double helix. This is partly due to experimental difficulties. The very fact that DNA resembles a long thin rod makes it susceptible to externally induced deformations, one of which is the packing environment created within DNA crystals. Thus, while X-ray structures have been of major importance in identifying structural families and in observing local, sequence related deformations, they must be interpreted with care when the overall form of the molecule is considered.[20,21] Similarly, until the recent development of residual dipolar coupling methods,[22,23] NMR studies of DNA have also been hindered by a lack of information on the longer-range interatomic distances which are vital for describing overall deformations such as axial bending. A second problem is connected to the quantity of information necessary to understand sequence effects. Until now most models of sequence-induced changes in structure have assumed that only nearest neighbor interactions need to be taken into account. This implies that structural heterogeneities can be broken down into a series of base pair step properties. Although models at this level have been quite successful in explaining sequence-induced curvature,[24-26] a quick look at the available high-resolution structures is enough to show that given base pair steps do not have unique, well-defined conformations. This suggests that it is at least necessary to move to models based on next-nearest neighbor interactions, implying that heterogeneities could be explained in terms of overlapping tetranucleotide steps. Unfortunately, while there are only

10 unique dinucleotides, there are 136 unique tetranucleotides, and present structural databases are still far from presenting a balanced statistical sample of this number of sequence elements. Moreover, if we turn from structural to mechanical or dynamic heterogeneity the situation is still worse and there is not even enough data to build meaningful dinucleotide models.

Given this situation, theoretical studies in general, and molecular modeling and simulation in particular, have a major role to play in obtaining a better understanding of this fascinating biopolymer. Naturally, since studying DNA poses experimental problems, it is not surprising that it also poses theoretical problems, amongst which are its sensitivity to the solvent-counterion environment and the wide variety of time scales covered by its conformational fluctuations. In addition, the combinatorial problems posed by studying a sufficient number of sequence elements in order to explain sequence-dependent properties apply equally well to modeling as to experiment. Despite these difficulties considerable progress has been made in recent years and is reflected in the many chapters of this book.

Since our work with DNA has been particularly centered on studying various aspects of its deformability, the current chapter concentrates on this problem. We will notably argue that the very particular structure of the DNA double helix means that modeling techniques can take advantage of this structure in a way that is not so obvious for more complex conformations, such as those of globular proteins. We therefore present a summary of the theoretical methods we have used for studying DNA deformations at the local and global level, before moving on to a discussion of the results these techniques have enabled us to obtain. In discussing the many deformations which are of interest for DNA, we have chosen to start with the largest and most global changes and move down through the range of deformations ending with the smallest, most local, and most subtle cases.

2.   MODELING STRATEGIES

2.1   Building Models of DNA

In contrast with developments in polymer physics, most studies of biological macromolecules have been carried out using atomic-scale models combined with the Cartesian degrees of freedom. This implies that $3N-6$ variables are required to represent the internal conformation of an N-atom system (the 6 remaining degrees of freedom describing its overall position with respect to a fixed spatial reference). There are multiple reasons for this choice. First, the biological systems studied to-date, although large (often containing thousands, or tens of thousands, of atoms), are still within the computational scope of all-atom representations. Second, biological polymers are heteropolymers which adopt specific conformations and form specific interactions that are both strongly dependent on their detailed atomic structure. Third, given an atomic force field containing terms based on the chemical (bond lengths, bond angles, torsions, ...) and spatial (electrostatics, van der Waals interactions, ...) structure of a biopolymer, it is easy to formulate both the energy of the system and the forces acting on each of its atoms in terms of Cartesian coordinates. Building on these calculations, it is also relatively easy to develop molecular mechanics and dynamics algorithms designed to locate the optimal conformations of a system or to study its time evolution.

If the fine atomic details are not relevant, other models, much closer to those used in polymer physics, become interesting. This has been the case for DNA, where the behavior of long fragments (or large rings) can be modeled without reference to the underlying base sequence. Such modeling includes elastic rods, worm-like chains, and bead models, each of which can be parameterized on the basis of experimental measurements of the elastic properties of the DNA double helix. In the case of single molecule DNA stretching experiments (which will be described later), the worm-like chain model has been shown to be capable of accurately reproducing the elastic behavior of DNA at low force regimes (up to several pN). Similarly, elastic rod models have been successfully used to describe the conformations and the dynamics of DNA plasmids containing thousands to tens of thousands of base pairs and bead models, where one bead already represents many turns of the double helix, have been used to describe DNA structuring within the nucleus of eukaryotic cells.[27] It should however be stressed that since these models depend on a limited number of macroscopic parameters (such as the bending and torsional elasticity of DNA), they will not describe local or global conformational transitions which dramatically change these properties. This is the case, for example, when DNA stretching at higher forces leads to a sudden overall restructuring of the double helix[28,29] or when the formation of very small rings apparently leads to local breakdowns in structure.[17] It should also be noted that in some cases, even where DNA remains in a canonical conformation, it would be useful to refine such models to take into account fine structural changes induced by the base sequence. A good example of this is the influence of a locally curved segment on the overall conformation of a supercoiled plasmid, where it is easy to see that the intrinsically curved segment will prefer to lie at a position of matching curvature within the supercoil, typically at the extremity of a loop.[30]

In between these two extreme modeling approaches, there is also a middle ground which attempts to simplify the representation of the double helix, while maintaining an atomic representation and an all-atom force field. This approach, which was pioneered in the groups of Ivanov[31] and of Miller,[32] involves replacing Cartesian coordinates with a set of so-called internal coordinates. Basic internal coordinates, simply represent the "chemical" degrees of freedom of the molecular system: bond lengths, bond angles and torsions. In simple linear systems, the number of such variables necessary to exactly specify an internal conformation equals the number of Cartesian degrees of freedom (3N−6). The problem becomes a little more complicated in the case of branched molecules or rings. For the five atoms of a sugar ring, there are 3 x 5−6 = 9 variables, as for the corresponding linear system. However, if we count the bond lengths, bond angles and torsions, we arrive at 15 variables. The explanation for this apparent disagreement is that ring closure leads to dependency between the internal variables. In fact, only 9 variables are necessary to position all the atoms: 4 bond lengths, 3 bond angles and 2 torsions. This is the same situation as for the linear system. The remaining variables: 1 bond length, 2 bond angles and 3 torsions are dependent variables, whose value depends on the 9 true degrees of freedom.

Given this added complexity, why bother with internal coordinates? There are in fact two good reasons. First, using "chemical" variables corresponds to a partitioning which has an energetic sense. A glance at a table of typical force constants shows that torsions are "soft" degrees of freedom, bond angles are intermediate and bond lengths are stiffer. In contrast, atomic Cartesian coordinates are all "stiff". Moving any atom along any Cartesian axis will generally stretch one or more chemical bonds and deform one or more bond angles. A 3N−6 dimensional energy hypersurface in Cartesian space will therefore be made up of numerous steep valleys. The same surface in internal coordinates will have a number of much shallower valleys, corresponding to the "soft" degrees of freedom, along which the system can easily be displaced. This leads to the second advantage of "chemical" variables; their natural partitioning makes it easy to eliminate subsets of variables, and, notably, those corresponding to stiff degrees of freedom. Bond lengths would obviously be the first candidate for elimination. If we return to the example of a 5-membered ring, fixing all bond lengths eliminates 5 degrees of freedom (actually, 4 variables, plus 1 constraint - the dependent bond length) leading to a 4 variable model. In the linear system, eliminating bond lengths, leaves 5 variables. If we went further and eliminated bond angles, the linear system would have only 2 degrees of torsional freedom left (corresponding to the situation with typical molecular models).

While bond lengths can generally be fixed with little loss of accuracy, bond angles are more problematic since early studies showed that dynamical behavior was strongly influenced by coupling between bond angles and torsions. Fixing bond angles is also a problem in small rings. Returning to our 5-membered ring, fixing both bond lengths and bond angles is impossible since this leads to 9−5−5 = −1 variables, implying that all these constraints cannot be satisfied simultaneously. In large molecules it is however often possible to reduce the number of independent valence angles as will be described shortly. One other possibility of simplification with internal coordinates involves conjugated rings (such as the aromatic bases of DNA). Such moieties are naturally stiff and it is consequently possible to eliminate all their internal variables and treat them as rigid bodies. Putting together, these various simplifications, it is generally possible to arrive at a model with ten times fewer degrees of freedom than the corresponding Cartesian representation. Combined with a smoother energy hypersurface, this leads to a model well-adapted to both energy minimization,[33,34] Monte Carlo (MC) simulations[35] and molecular dynamics[36] (MD).

Naturally, the advantages of internal coordinates do not come without some drawbacks. First, internal variable changes imply the movement of a set of "chemically connected" atoms. This means that the chemical topology of the molecule has to be analyzed and converted into a convenient form. It also means that energy derivatives (or, with a sign change, forces) will be needed with respect to the internal variables. In fact, this task is not so difficult, if one begins by calculating the forces acting on each atom, as in a Cartesian variable approach. If we know which atoms move in response to the change in an internal variable and how they move (e.g. translation along a stretched bond or rotation around the central bond of a torsion), it is easy to assemble the forces on these atoms to generate the total force (or torque) acting on the variable in question.[37] A second, and more fundamental, problem with internal variables involves their long range effect within polymeric systems. In polymers, changing an internal variable can result in moving many hundreds of chemically connected atoms over a large distance in space. This, in turn, often leads to atomic collisions, creating steep valleys within the energy hypersurface, and damaging one of the main advantages of internal variable representations.

In the case of DNA, we have proposed a solution to this problem by combining internal variables with helicoidal variables which better describe overall deformations of DNA. This is the basis of the JUMNA program.[33] JUMNA (JUnction Minimization of Nucleic Acids), builds a model of DNA by placing each nucleotide with respect to a common axis system using three translational (x- and y-displacements, rise) and three rotational variables (inclination, tip, twist). The conformation of each nucleotide is described by a set of internal variables comprising all single bond torsions and a limited set of valence angles (within the sugar rings and along the main phosphodiester backbone). The bases are treated as rigid bodies. Junctions between successive nucleotides (and sugar rings) are closed using harmonic bond length restraints. This hybrid approach has the advantage of introducing still "softer" degrees of freedom into the model (corresponding to DNA deformations such as twisting and stretching), while also limiting the range of the internal variables, which now act directly only on the atoms of a single nucleotide. This approach has been very successful in modeling both local and global conformational transitions in DNA, as will be described later.

2.2   Controlling Deformations

DNA deformations cover a wide variety of size and time scales and many cannot be expected to occur spontaneously during typical molecular simulations. Until very recently most simulations of DNA in explicit solvent lasted only a few nanoseconds. It is now possible to push a few simulations out to a hundred ns or more, but this is still exceptional (see, for example, ref[38,39]). The current limit of "spontaneous" is therefore around 10 ns. In fact, quite a lot goes on below this limit. Most helical parameters and backbone parameters reach equilibrium values on this time scale and the same is true for water distributions.[40] There are nevertheless gradations within this set of movements with, for example, sugar puckers changing much faster (≈ 1 ps) than a B_I-B_{II} backbone flip (≈ 500 ps) or axis bending (≈ 1 ns). Even large conformation changes can occur on the ns timescale, if they are associated with low free energy barriers, a good example being the B-A allomorphic transition.[41-43]

Naturally, processes associated with higher activation barriers occur on longer timescales. This is the case for $\alpha\gamma$ backbone flips,[44] where the passage from g^-g^+ to g^+t or tt has an activation energy of several kcal mol^{-1} and consequently only equilibrates well beyond the 10 ns timescale. This is a problem for current simulations which cannot easily sample such transitions.[40,45] Still higher free energies mean even slower processes, as in the case of base pair opening, where the activation barrier at room temperature is roughly 20 kcal mol^{-1} and the characteristic times are of the order of ms.[46] The B-Z transition which involves base pair rotation is even slower and can take minutes or hours depending on the nature of the DNA fragments studied.[41,47] For all these processes, it is necessary to enhance the probability of a specific transition in some way. There are a number of ways of achieving this. If a simple restraint can be devised to induce the transition it is possible to use a biasing potential to force sampling of intermediate points along the pathway. This has been used in a number of applications involving DNA, and, within our group, to study base pair opening,[48-50] flipping[51] and backbone transitions[44] which are discussed in detail in a later section.

Given this range of deformations, it is often necessary to force the system to reach appropriate conformations. In the case of Cartesian coordinates, all such forcing implies imposed restraints. With internal or hybrid coordinate representations, there are more choices. First, the use of internal coordinate generally implies that certain stiff variables have been eliminated as mathematical constraints of the system. Selected bond angles or torsions can also be specifically constrained by simply removing them from the set of defined variables rather than by introducing supplementary restraints. It is also possible to impose symmetry constraints by simply using a single variable in the place of a set of symmetry-related variables. For DNA this can lead to very significant reductions in the number of variables necessary to model the system. In the simplest case of mononucleotide symmetry, where the internal conformation of every nucleotide, and its relative position with respect to the helical axis system, is identical, it is possible to model a turn of DNA with roughly 20 variables, roughly a 100 times less than in the corresponding Cartesian representation.

JUMNA offers the possibility of pushing symmetry constraints one step further, by replacing helical symmetry with superhelical symmetry. In this case, the axis of the molecule itself becomes a helix which winds around a linear superhelical axis with a defined radius and pitch. Beyond grouping together sets of variables which are related by the superhelical symmetry, this also requires a redefinition of the rise and twist variables.[52] The real interest of superhelical symmetry is that is makes it possible to study the conformational and energetic impact of DNA curvature in a perfectly controlled way. It should be noted that, in contrast to typical helical symmetries, superhelical symmetry links the variables describing nucleotides separated by a complete turn of the double helix. This means that superhelical symmetry does not imply smooth DNA curvature and can involve kinks as long as they occur in phase with the symmetry of the superhelix. It is also worth noting that the overall superhelical conformation is defined not only by the radius and pitch of the superhelix, but also by the rotational state of DNA around its superhelical pathway. We have termed this variable "rotational register". Since it can also be made a constraint of the system it is easy to study its energetics for various base sequences and this turns out to be a useful way of characterizing intrinsic curvature.[52,53]

Although constraints play an important role with internal coordinate representations, imposing chosen conformational characteristics essentially "for free", restraints also have their uses. JUMNA proposes a wide variety of restraints which have been developed to study specific DNA deformations. Amongst these we will cite two examples. The first involves restraining the total value of a group of variables. This can be useful, for example, in controlling the overall deformation of a DNA fragment (e.g. stretching or overtwisting) with a single parameter, without imposing mononucleotide helical symmetry. In this case, a quadratic restraint acts only on the sum of the variables, leaving the individual variables to vary with respect to one another (under the influence of their position within the structure or under the influence of the base sequence). Similarly, restraining differences of variables can be useful, for example, to fix base pair properties such as buckle (defined as the difference of the two base inclinations). The second example involves base opening, which is the subject of one of the following application sections. Base opening is more difficult to treat since it does not correspond to a clearly defined conformational change. In general terms, it requires breaking the Watson-Crick base pairs and moving one base out of helical stack into either the major or minor groove of the double helix. Our first attempt at achieving this involved rotating a chosen nucleotide around a space-fixed axis perpendicular to the base pair, passing through the centre of the sugar ring.[54] Although the resulting restraint (the axial rotation angle) indeed opened the base pair, it did not allow the sugar ring to change its distance from the helical axis of the molecule. This led to overestimated steric hindrance with the paired base when opening was attempted into the minor groove.[54,55] We consequently reformulated the restraint to apply to the angle between the glycosidic bond of the opening base and the virtual bond between the C1' atoms of the base pair.[56] This angle was additionally projected into a plane perpendicular to the local helical axis to ensure that changes in the restrained angle corresponded to base opening and not to buckling of the base pair. Another approach was adopted by the group of MacKerell who restrained a virtual torsion angle formed by the centers of mass of the opening base, the associated sugar ring, the sugar ring of the 3'-nucleotide and the 3'-base pair.[57] Our modified restraint and that proposed by MacKerell give similar results suggesting that they have both captured the nature of the opening process.

It should be noted that some of the restraints discussed above in the context of JUMNA can equally well be applied to modeling or simulation using Cartesian coordinates. This is the case when the restraints involve parameters which can easily be obtained from the coordinates (distances, angles, centers of mass, ...), but it is not the case when helicoidal parameters enter directly into the formulation of the restraint.

2.3   Enthalpy and Free Energy

Most biological processes take place as a result of a complex interplay between enthalpy (ΔH) and entropy (ΔS) of the system, including the molecule of interest and its environment, at room temperature. It has been noted that the entropy of a process often changes in such a way that it compensates changes in enthalpy. The phenomenon has been observed for a given process with varying temperatures but also for homologous processes at constant temperature; biological examples include protein[58,59] and RNA[60] folding, protein-DNA interactions[61] and DNA opening[62]. As a result, enthalpy-entropy compensation leads to processes with a smaller net free energy change (ΔG). It should be remarked that if the measured ΔG varies in a rather narrow range, the large fluctuations in ΔH obviously enforce compensating changes in ΔS.[63]

In some processes, involving nucleic acid deformations, enthalpy clearly dominates the overall free energy changes. In these cases, molecular modeling using JUMNA provides an efficient way to test alternative molecular mechanisms for the transition. For example, we were able to show that the deformations of the TATA-box binding site induced by its specific protein can originate from local DNA stretching and unwinding events that result in kinks within the target sequence.[64] These kinks occurred at the sites where the protein intercalates specific side chains leading to the hypothesis that such interactions occur at intrinsically weak points in the sequence. In addition, it was shown that DNA bending was energetically facilitated by neutralization of the phosphate charge on the minor groove face.

The dynamical aspects of a process become clearly important if one needs to quantitatively assess thermodynamic and kinetic parameters. While obtaining reliable enthalpies require a well-parameterized force field, evaluation of the systems's entropy is largely dependent on proper sampling of the configuration space even far from the low energy states. Since higher energy states are not normally sampled in conventional molecular dynamics simulations, the use of special techniques is required. An additional difficulty stems from the fact that sampling becomes particularly challenging for systems with the explicit inclusion of a great number of solvent molecules and counterions. Note that fluctuations of the total energy of the system are much larger than free energy changes in computer simulations and hence the uncertainties involved in free energy calculations are orders of magnitude smaller than those of the corresponding enthalpies.

There exist a number of well-established methods to evaluate the free energy changes in a system using molecular simulations.[65,66] We shall focus here on the use of a biasing potential to sample rare conformations associated with higher energy. This method accelerates a chosen slow movement and assumes that all other degrees of freedom equilibrate within the available simulation time. In practice, in umbrella sampling simulations,[67] a reaction coordinate must be defined to describe the progress and dynamics of the molecular motion of interest. The system is then restrained around a given value of this coordinate by a harmonic "umbrella" potential to increase conformational sampling. Simulation of the whole process consists of consecutive overlapping "windows" of restrained MD between the initial and final states along this coordinate. The final structure obtained in a given window can be used as the starting point in the subsequent one to keep the system as close as possible to the equilibrium pathway.

To obtain the free energy (or the potential of mean force) along the pathway, the magnitude of the restrained parameter is recorded during the simulation of each window. The distributions from the separate windows can be combined and the bias introduced by the umbrella potential removed using an iterative procedure known as the constant temperature weighted-histogram analysis method (WHAM).[68,69] This procedure finds the optimal constants to combine the full set of data in multiple dimensions. The protocol can also be used to map the free energy on surfaces defined by coordinates not directly restrained in the simulation.[43,50] As an example, in order to map free energy from restrained simulations along ξ_1 onto unrestrained degrees of freedom (ξ_2, ξ_3), we first generate a guess for the unbiased probability histogram $P(\xi_1, \xi_2, \xi_3)$ from the complete biased probability histogram, $P_i^*(\xi_1, \xi_2, \xi_3)$ as follows:

$$P(\xi_1, \xi_2, \xi_3) = \frac{\Sigma P_i^*(\xi_1, \xi_2, \xi_3)}{\Sigma n_i \exp([F_i - V_i(\xi_1)]/k_B T)} \qquad (1)$$

where n_i is the number of data points in window i, $V_i(\xi_1)$ is the biasing potential and the constants F_i are defined by

$$F_i = -k_B T \ln \int P(\xi_1, \xi_2, \xi_3) \exp[-V_i(\xi_1)/k_B T] d\xi_1 \, d\xi_2 \, d\xi_3 \qquad (2)$$

All sums run over the total number of windows sampled. Since Eq. (2) contains the unknown $P(\xi_1, \xi_2, \xi_3)$, the optimal unbiased probability distribution can be obtained through iterations of Eqs. (1) and (2). Once self-consistency is achieved, we can integrate out coordinate ξ_1 using

$$P(\xi_2, \xi_3) = \int P(\xi_1, \xi_2, \xi_3) d\xi_1 \qquad (3)$$

And finally, the free energy W can be obtained as a function of (ξ_2, ξ_3) from

$$W(\xi_2, \xi_3) = -k_B T \ln P(\xi_2, \xi_3) + C \qquad (4)$$

The method described above not only results in the free energy difference between two end conformations, but also gives insight into the structural and energetic changes throughout the transformation. There is, however, no guarantee that a reaction coordinate defined *a priori* for a complex molecular motion provides the lowest free energy pathway between the end points. A more elaborate method uses transition-path sampling to obtain reactive trajectories that can help to identify suitable coordinates.[70] Nevertheless, description of complex molecular processes often requires multiple or combined reaction coordinates that can subsequently be optimized by a variational procedure.[71] Free energy surfaces projected onto these coordinates then allow the determination of accurate rate constants, offering a calculated quantity that can be directly compared with experiment. Obtaining the transition state in a complex biological system is far from trivial, and consequently a number of new methodologies are currently being developed for this purpose.[72-74] In a recent study of base opening in a terminal base pair,[75] a number of reaction coordinates were tested, including those explicitly involving solvent molecules. Although no single coordinate was found to describe the complete process, two molecular pathways were readily identified.

In cases, where only an estimate for the free energy difference between the end states is required, simpler methods are available using ensembles of structures generated by all-atom MD simulations. In many cases, such as ligand binding to macromolecules where electrostatic effects often dominate, the free energy difference can be calculated by the linear response approximation.[76] This method evaluates the reorganization energy by averaging the energy difference between the two states in the initial and final configurations.[77] The entropic contribution associated with the transformation is often evaluated in a separate calculation. In order to improve convergence of the results, the fluctuations of the large number of solvent molecules can be replaced by the average polarization of the solvent, represented by dipoles or continuum description.

In a similar fashion, the recently introduced molecular mechanics Poisson-Boltzmann surface area (MM-PBSA) model calculates the average energies from ensembles of structures for different states of interest.[78] Although a rigorous approach to calculate binding free energy would require separate simulations for the components and for the complex, more consistent results were obtained when structures were analyzed from a single trajectory.[79] The method partitions the total free energy into a sum of enthalpic and entropic contributions.[80] For the enthalpic term, the averaged gas phase molecular mechanics energy of the solute is used, while the electrostatic solvation free energy is calculated using the Poisson-Boltzmann equations. The non-polar part of the solvation free energy is calculated from the solvent accessible surface area of the solute. In addition, to obtain the total free energy, the changes in conformational entropy are evaluated separately by a harmonic, or other more elaborate, analysis of the motions of the solute.[81,82] This method was applied to nucleic acids and reproduced experimentally observed trends for the conformational preferences of DNA and RNA helices.[83,84] Since then the methodology has been applied to a variety of systems to study the relative stabilities of RNA tetraloop structures,[85] rationalize the origin of binding specificity in protein-RNA,[86] protein-DNA,[87] metal ion-RNA[88] and drug-DNA complexes.[89] Despite the surprising success of this simple method, problems arose mainly due to the structured and polarized ions and solvent molecules in the vicinity of the solute. For this reason, these molecules were later considered as an integral part of the structure and included explicitly in the molecular mechanics energy calculations. A challenging example is the G-DNA quadruplex structure that involves intrinsically bound monovalent cations where the MM-PBSA methodology correctly predicted the stability of the stem formation, but favored an incorrect loop conformation due to force field limitations.[90] Unfortunately, few comparisons of the MM-PBSA method with other, more rigorous methods exist today. In one case, the thermodynamics of ligand binding to an RNA aptamer was calculated by MM-PBSA and the conventional thermodynamic integration (TI) technique in explicit ionic and solvent environment.[91] The comparison showed that MM-PBSA ranks non-polar ligand binding affinities qualitatively well, however, as expected, the results from TI are quantitatively accurate. The discrepancy between the two methods was mainly attributed to the deficiencies in the description of the first solvation shell in the continuum representation of water.

3.   PRACTICAL APPLICATIONS - FROM THE MACROSCOPIC TO THE MICROSCOPIC

3.1   Large Scale Helical Deformations

The study of very large deformations of the DNA double helix was revolutionized by the introduction of single molecule experiments.[6,92,93] DNA was an ideal target for single molecule manipulation since it could easily be obtained in defined lengths with a defined base sequence, short pieces being directly synthesized and longer pieces being extracted from biological sources, notably bacteriophages. In addition, biochemistry provided an arsenal of techniques for chemically modifying DNA in order to facilitate its manipulation. Although early experiments studied DNA stretched out on a surface, techniques were rapidly developed for studying DNA in solution. For micron-scale DNA, the ends of a DNA fragment containing several tens of thousands of base pairs (e.g. λ-bacteriophage has a DNA genome roughly 15 μm in length, containing approximately 48,000 base pairs) were attached to a solid surface at one end and to a microbead at the other. Connections were made using specific antigen-antibody pairs, different pairs defining which end of the molecule would become attached to which support. The antigen is fixed to the DNA by first creating a single-stranded hanging-end with an exonuclease. A complementary oligonucleotide, chemically modified to carry many copies of the antigen, is then added. This oligonucleotide pairs with the hanging-end and the backbone nick is repaired with a ligase. Microbeads and the glass surface are then prepared by binding the appropriate antibodies. When mixed, the bead-DNA-surface system will autoassemble. Fluid flow can be used to remove unassembled components. Viewing the system through an optical microscope enables DNA-attached beads to be located. The beads can then be manipulated using a micropipette or laser tweezers, enabling forces to be applied to the DNA molecule (which is naturally invisible through the microscope). It is remarked that similar experiments replace the fixed surface with a second bead and then trap each bead independently in laser tweezers.

The first stretching experiments on DNA led to a surprising result. At low forces (several pN), the molecule behaved as expected, being straightened out, against entropic resistance, to its full contour length and then undergoing elastic deformation. However, at a force of roughly 70 pN, the molecule suddenly stretched to 1.7 times its normal contour length without any significant increase in force. After this event, the molecule could only be stretched further by sharply increasing the applied force, which, if it went too far, led to rupture of the system.[28] To better understand what happened to the DNA double helix at 70 pN, we modeled DNA stretching using JUMNA. This involved restraining the distance between the ends of the molecule and successively increasing this distance, carrying out an energy minimization at each step. Since the phosphodiester strand of DNA has two chemically distinct ends, the 3'-end and the 5'-end, and canonical DNA has two antiparallel strands, there are actually several ways to stretch the molecule in a computer experiment. One can pull DNA by acting on the distance between its two 3'-ends or between its two 5'-ends, or between both ends of a single strand, or between both ends of both strands simultaneously.

All these pulling techniques were tried. The results had one common feature; DNA could be stretched by at least 70% with little damage to the base pairs. When the energy curves for stretching were derived with respect to the length of the molecule they led to a force curve closely related to that obtained experimentally, although the force plateau occurred at higher forces in the computer simulation (around 200 pN).[28,94] However, the nature of the deformation depended on the way the molecule was pulled and, to some extent, on its base sequence. Two major types of deformation were seen: a ribbon form, where the double helix had almost completely unwound, and a narrow fiber, where the base pairs were strongly inclined, becoming almost aligned with the helical axis of the molecule. 3'-3' stretching always led to the ribbon form, while 5'-5' stretching always led to the fiber form. Other stretching techniques could go to either ribbon or fiber structures depending on the sequence. These results explained how DNA could undergo such a large extension. Basically, although the separation between successive base pairs along the axis of DNA is roughly 3.4 Å, the length of the intervening phosphodiester backbone is roughly 7 Å. The twist of the double helix absorbs this difference by helically wrapping the backbone around the outside of the molecule. Upon stretching this length "reservoir" can be accessed either by unwinding the helix or by inclining the base pairs and thus reducing the diameter of the helix. Stretched DNA now carries the name of S-DNA. It has been integrated into all subsequent physical models of DNA micro-manipulation,[6] although whether it actually resembles the ribbon or fiber forms remains unclear. It is probable that both 3'- and 5'-ends of DNA are attached at both extremities of the molecule. Single strand nicks can occur along its length, but this can be avoided by having a ligase present in solution. Under these circumstances, it has been shown that the force plateau is sensitive to the twist.[95] In contrast, x-ray diffraction experiments on stretched fibers have shown a strongly reduced diameter.[96] It is also remarked that other groups propose that stretched DNA is unpaired and should be thought of as two loosely associated single strands.[97-99] Computer simulations continue in this area and ongoing molecular dynamics simulations with explicit solvent and counterions (M. Zacharias, private communication) should help to resolve this question.

A new way of micromanipulating DNA resulted from the use of paramagnetic beads which could be both pulled and rotated using a magnetic field.[100,101] This enabled DNA to be stretched and twisted in a single experiment and has led to a wealth of information on both isolated DNA and protein-DNA complexes.[102,103] Over- or undertwisting DNA is clearly of biological interest since DNA is generally supercoiled in living organisms: negatively in most species, but positively in certain extremophiles.[104,105] The first experiments carried out on isolated DNA again led to surprising results.

If DNA was only weakly stretched, negative or positive winding led to length contraction (followed by observing the vertical position of the attached bead through a microscope). This corresponds to the expected formation of plectonemes (in the same way that a twisted telephone cable coils up), and reflects the fact that DNA can writhe in space more easily than it can change its intrinsic twist. If the stretching force was increased, the underwound DNA could be prevented from contracting, showing that the molecule had underwound, leading to unpaired strands. The surprise was that, at still higher forces, the molecule could again be prevented from contracting but, this time, after strong positive winding. Since overwinding cannot lead to strand separation, it appeared that a new conformation was being created. Calculations showed that the twist per base pair in this state could be increased from its usual value of around 34.5° to a value close to 160°. This was very difficult to envisage, since increasing the twist in normal DNA much above 80° leads to stretching the phosphodiester backbone and compressing the base pairs together. Modeling with JUMNA again led to an explanation. Beyond roughly 80° of twist, the Watson-Crick base pairs of the double helix were broken and the bases moved to the outside of the helix, allowing the backbones to move close to the helical axis. This enables the backbone to become much more strongly twisted, the new limiting value being around 160° per nucleotide, close to the value observed experimentally. This very unusual form of DNA is now termed P-DNA.[106] The "P" was chosen to recall the unsuccessful model proposed for DNA in 1952 by Linus Pauling,[107] not long before the publication of Watson and Crick's famous double helix.[1] The Pauling model was actually a triple helix, but, like our overwound model, it had the phosphodiester backbones in the middle and the bases on the outside. Pauling's mistake was to assume that the phosphate groups would be neutral and could hydrogen bond together (as in clays for example). The fact that phosphate groups are actually anionic in aqueous solution would normally make Pauling DNA unstable, but external forces can overcome this repulsion as shown in the micromanipulation experiment. It is worth adding that there exists strong evidence that a similar DNA conformation can exist within a specific class of bacteriophages, such as pf1,[108,109] where protein packaging leads the single-stranded circular genome to twist into a form very close to that of P-DNA. Naturally, since this DNA is single-stranded, it does not have the possibility of extensive pairing and double helix formation, but the phosphate-phosphate repulsion still exists and must consequently be counterbalanced by interactions with the coat protein of the bacteriophage.

On a smaller scale than the stretching or twisting necessary to create S- or P-DNA, we come to bending. Bending is nevertheless a vital form of DNA deformation since it plays a significant role in DNA packaging within the cell. In eukaryotes, histone proteins twist DNA into a tight left-handed supercoil with a radius of 42 Å and a pitch of 26 Å with 147 base pairs being wound into roughly 1.7 superhelical turns around each histone core.[41,110,111] These nucleosome particles are then packaged themselves in successively more complex structures, enabling ≈1 m of DNA to finally fit into a nucleus only a few microns in diameter. Although less sophisticated, prokaryotes contain related proteins able to package their genomes. Many other proteins which interact with DNA can cause strong local bending, which in turn can act as a genetic control mechanism by bringing distant parts of the double helix into contact. Lastly, as already mentioned, some DNA sequences can lead to intrinsic curvature. This is notably the case of so-called "A-tracts", short A- or T- repeats, separated by other sequences so that they are in-phase with the helical turns of DNA.[16]

The conformational and energetic changes associated with DNA bending, and notably the mechanism of intrinsic curvature, have been the subject of many modeling studies.[16] Molecular dynamics studies have shown that it is possible to reproduce the bending associated with A-tracts and, thanks to explicit solvent and counterion representations, have also reproduced the experimental observation that the degree of curvature depends on the type of counterion present (being smaller with Na^+ than with K^+).[112] These simulations also support the idea that curvature results mainly from junction formation between canonical B-DNA and a modified helical form adopted by the A-tracts and termed B'-DNA (characterized by strong propeller twisting and by a narrow minor groove). Both the MD studies[112,113] and earlier Monte Carlo[114] simulations have stressed that apparent curvature may be more the result of anisotropic bending fluctuations than static curvature. This seems to be the case for A-tracts whose thermally induced bending fluctuations are not higher than those of control sequences, but, by favoring bending in the direction of the minor groove of the A-tracts, are notably less isotropic. Note that, as mentioned above, although the net bending direction is towards the A-tract minor groove, this is largely the result of the combined effects of the junctions bracketing the A-tract.

The superhelical symmetry options in JUMNA enabled us to look at DNA bending in another way. Using these symmetry constraints it was possible to make a systematic study of the transition from straight to curved DNA. We carried out such modeling for uniform sequences such as poly(dA), poly(dG), poly(dAT) or poly(dGC) and also for phased A-tracts (A_6CGCG, A_4T_4CG, ...). The A_4T_4CG sequence has a particular interest since Hagerman has shown that while it is strongly curved, the related T_4A_4CG sequence is straight. As shown in Figure 7-1a, this difference was correctly reproduced by our modeling. It is worth noting that although both sequences can be made to adopt a wide variety of radii of curvature, one can

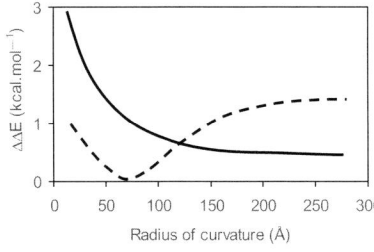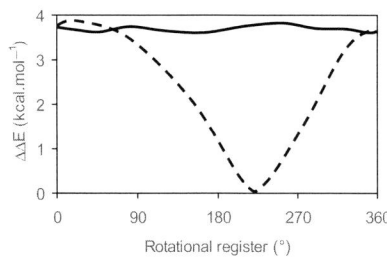

Figure 7-1. Energy as a function of radius of curvature (a) and rotational register (b) for DNA polymers with the sequences A_4T_4CG (dashed line) and T_4A_4CG (solid line).

always detect the intrinsically curved sequence by studying its rotational register, that is to say rotating the DNA around its own axis.[52] As Figure 7-1b shows, the curved DNA is anisotropic and resists being twisted away from its preferred direction of bending, while the naturally straight DNA, is isotropic with respect to twist. A detailed analysis of the bent conformations of the A_4T_4 and T_4A_4 sequences supports the idea that most of the bending comes from the junctions at the ends of the AT segments and also at the ApT and TpA step. In the A_4T_4 sequence these local deformations work together to create overall curvature, whereas in the T_4A_4 sequence that compensate one another, leading to an axis with no overall deformation.

3.2 Deformations of the Base Pairs

In the canonical B-DNA structure the bases are held in the interior of the double helix and hence protected from modification by chemical agents. In certain biological processes, however, bases must become accessible to the environment for proper functioning. Disrupting the base pair is for example a necessary step for both replication and transcription in which local "melting" of DNA (separation of strands) takes place involving several consecutive base pairs. More specific base pair disruption occurs in DNA repair and methylation processes where a single base is either excised or chemically transformed. Since the bases are held quite strongly within the double helix both by Watson-Crick hydrogen bonding and by base stacking, breaking base pairs was expected to require significant deformations including severe stretching or kinking. The actual conformational rearrangements were first observed in 1994 when the crystal structure of *HhaI* DNA methyltansferase ternary complex was solved.[115] Surprisingly, the DNA was found to be only slightly distorted, but with the target cytidine completely rotated out of the double helix into the binding pocket of the enzyme.

Base flipping was suggested to occur via minor groove pathway in the enzyme based on structural and kinetic studies.[115,116] This was later challenged based on simulation studies[117] and a crystallographic structure of the complex containing an abasic south-constrained pseudosugar.[118] Early molecular modeling studies involved restrained minimizations in vacuo[119] or using implicit solvent models[54,120] that limited the reliability of these approaches. The principal conclusion of these studies was that opening into the major groove is possible, while into the sterically more hindered minor groove was energetically prohibitive. This general view was subsequently supported by simulation studies where spontaneous base opening in the major groove direction was observed.[75,121,122]

Studying spontaneous base pair opening ("DNA breathing") experimentally involves monitoring the exchange of labile base protons with the surrounding solvent. NMR spectroscopy can be used for such purposes, in particular, for the study of imino proton exchange in thymine and guanine as a function of temperature and catalyst concentration.[46,123] It is assumed that the exchange rate is equal to the base opening rate at infinite catalyst concentration in a process that is normally interpreted according to a two-state (open-closed) model. These studies have enabled the kinetic and thermo-dynamic characterization of base opening for a number of DNA and RNA sequences. Opening of base pairs within B-DNA normally occurs once every 1-50 ms with open state lasting only a few nanoseconds, depending on the type of base pair and sequence, but bases can remain closed up to minutes in some tRNA.

It was noted that the base pair lifetimes, as defined by the proton exchange, are on similar timescales as the enzymatic turnover in methyltransferases.[124] Consequently, it was suggested that the open base can represent a new recognition motif that the enzyme exploits during its diffusive search along the DNA sequence for site-specific binding. Hence, a passive process was envisaged in which the enzyme recognizes and traps the flipped base that is formed spontaneously in free DNA. In the case of uracyl DNA glycosylase[125] and certain glycosyltransferases[126] a passive mechanism was clearly shown. Nevertheless, there is another possibility that the enzyme actively participates in the extrusion of the target base. Experimental NMR studies are, unfortunately, unable to characterize the conformational changes involved in base opening and thus the exact nature of the transient "open" state. Although, conformations with bases trapped out of helical stack have been observed crystallographically,[127] it is not clear whether these states are directly related to spontaneous base opening in solution. We therefore used molecular dynamics simulations to obtain structural and energetic insight into the base opening/flipping process and rationalize kinetic data from NMR measurements. A biasing potential was applied in the simulations that exploits the helicoidal nature of DNA (described above) without directly influencing the sugar-phosphate backbone conformations.[56] In this way, we control the opening of the base into the major or minor groove. Although only thymines and guanines are "visible" in imino proton exchange experiments, simulation studies can show how the four different types of bases open up in canonical AT and GC pairs.[48,49] In addition, the study of wobble GT pairs, that arise as mismatches in DNA and need to be corrected by a suitable repair system, was particularly useful since the presence of two imino protons in the base pair allows us to simultaneously probe both opening bases by simulation and NMR spectroscopy.[50]

The calculated base opening process is characterized by a funnel-shaped free energy profile in which the free energy rises in a quadratic fashion around the native closed state beyond which lies a quasi-linear zone. This was a surprising result as opening into the minor groove was thought to be energetically prohibitive. It should be remarked that, in line with this result, earlier continuum solvent modeling using the same restraint showed that opening into the minor groove is feasible.[56] Bases rotate in a coupled fashion into the grooves until the intervening hydrogen bonds break, marking the limit of the quadratic zone. Since this regime mainly corresponds to the strength of hydrogen bonding within the pair, the free energy required to disrupt different base pairs vary from 2-4 kcal mol^{-1} for the wobble GT pair to 6-10 kcal mol^{-1} for canonical base pairs. Bases however remain essentially in the helical stack due to reorganization of the neighboring base pairs. Interestingly, further base opening is not like pulling a card out of a stack. When a base finally loses favorable stacking, it turns out of plane to create interactions with either base or backbone sites lining the grooves. The base rotation helps shield the relatively hydrophobic ring from aqueous solution and creates non-conventional hydrogen bonds whose exact identity depends on the nature of the surrounding DNA sequence. This open arrangement is, however, attained at a gradually increasing free energy cost. Although all four bases in the canonical base pairs can open into both the major and minor grooves, the free energy profile is only approximately symmetric for the pyrimidines, while the purines show a clear energetic preference for opening into the major groove.[49] This is also the case for bases in the wobble GT pair which preferentially opens into the major groove.[50] In the case of an A-tract DNA, molecular dynamics simulations were able to reproduce the higher free energy requirement inferred from the increased experimental opening times of these AT pairs.[128] In one case we observed a tandem opening of a C/T stack into the minor groove even though only single base opening was directly induced. The opening cytosine thus provoked the coupled opening of an adjacent 3'-thymine to maintain stacking interactions. Recent experimental data related to concerted opening of adjacent AT pairs seems to be in line with this coupled opening mechanism.[62] Another example shows that a wobble GT pair inserted in the Dickerson oligomer reduces the stability of the surrounding base pairs as compared to the original oligomer with an AT pair. This coupling suggests that the strictly local nature of base pair opening may not hold for all base sequences and could point to important mechanistic aspects of how and where in the sequence helix separation or branch migration is initiated.

Opening of the base pairs leads to the creation of a number of new water binding positions between the partially open partners with increased residency times. It is interesting to note that such potentially stabilizing bridging water molecules were seen in crystallographic studies of an open GC base pair[127] and their existence has also been deduced from NMR studies of base pair opening to explain exchange rates in the absence of added catalyst.[129] These pre-open sites have been seen in simulations to trap ions from the solution for extended periods of time. At larger opening angles, however, the surrounding water molecules reorganize around the bases and form a solvent channel through the duplex, filling the gap left in the base stack. To assess the extent of DNA deformation necessary for an effective exchange with the surrounding solvent, we calculated the surface accessibility of the imino protons along the base opening pathways. Although in all the simulated pathways the intervening hydrogen bonds eventually break, this does not always lead to significant exposure to solvent. As discussed above, imino protons may be involved in other types of hydrogen bonding interactions, typically in the minor groove, or can be shielded by the neighboring bases. It was surprising however to see that imino protons gain reasonable accessibility often at modest opening angles (~50°), shortly after they move out of stack. Nevertheless, the structural ensemble corresponding to a given opening angle is heterogeneous and the imino proton accessibility falls in a rather wide range of values. Therefore the extent of opening does not always correlate well with the degree of solvent exposure. In the case of the wobble GT pair we determined the free energy required to achieve a given imino proton accessibility. A series of accessibility values was then used to define the limit between the closed and the exchange-competent open state and to compare the calculated base pair dissociation constants (K_d) with the experimental one. It was found that ~30% imino proton accessibility (with respect to the value calculated for isolated nucleosides) already provides a good correspondence between the calculated and experimental K_d values and hence a relatively low accessibility can result in an effective proton exchange.[50]

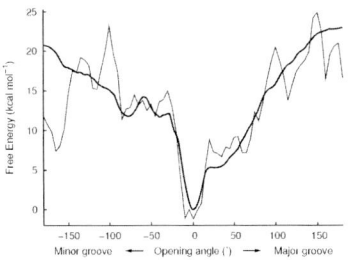 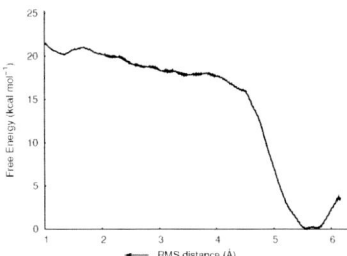

Figure 7-2. Free energy curves of base flipping (a) calculated using umbrella sampling with angular opening restraint (bold) and MM-PBSA protocol (thin). Opening angles correspond to relative angles with respect to the base position in the closed state; positive and negative values indicate movement into the major and minor groove, respectively. (b) base flipping induced by an RMSD restraint on the heavy atoms of the flipped base and its 3'/5' neighbors.

The base has to rotate even further out of stack in a flipping process as seen in the recognition sequence of *HhaI* methyltransferase. The free energy then increases to 21-23 kcal mol^{-1} for the flipped state (~180°) obtained in separate pathways through major or minor groove rotations using our angular opening restraint (Figure 7-2a).[51] Since, in the case of flipping, the conformational end-point is known from the crystal structure of the complex, we were able to compare these results with trajectories obtained by using a general biasing potential (Figure 7-2b). This procedure transforms the relaxed canonical structure to the flipped form (the "target") without reference to any specific internal coordinate of the system by successively reducing the root mean square distance of a selected set of atoms, for example, the flipping nucleotide plus the adjacent nucleotides. It was observed that the targeted MD approach can effectively induce cytosine flipping. Although this method does not control the opening directionality, multiple trajectories showed that base flipping through the major groove was predominant. It was also very encouraging to see that the free energy difference between the closed and the flipped state corresponds to the value obtained using the angular restraint, demonstrating good convergence of the simulations using entirely different driving forces.

In addition, we evaluated the free energy changes with the MM-PBSA method along the opening pathway using structures from the opening trajectories obtained with the angular opening restraint. As described earlier, this method involves the calculation of large energy contributions (±150 kcal mol^{-1}) which tend to cancel out to give reasonable free energy changes along the structural pathway. Notable cancellations include solute electrostatics with the continuum solvation electrostatics or the total electrostatic energy with the van der Waals energy. The fluctuations in the bonded energy terms contribute to the ruggedness of the free energy profile. In spite of the more approximate nature of these calculations, the total free energy changes follow surprisingly closely the free energy profile obtained with the umbrella sampling simulations, with an exception at $-160°$ where interactions between the base and the minor groove are overestimated. By calculating interaction energies between groups of atoms in the structural ensemble at a given opening angle, we found that base pairing (electrostatic interaction) is lost first along the opening pathway, followed by stacking (mainly van der Waals interaction).

We initiated independent molecular dynamics trajectories starting from the equilibrated flipped state to examine the kinetics of the base re-closing process and test whether the flipped base corresponds to a stable conformation in water. Most trajectories resulted in a closed state within 1-5 ns passing through the major groove. In some cases, however, perfect Watson-Crick base pairing could not be achieved as the flipped base rotated around its glycosyl bond to *syn* conformation before the closing event took place. We also observed in one case that the flipped base was stabilized by interactions with the sugar-phosphate backbone and could not achieve the closed state within 10 ns. Therefore, as expected from the funnel-shape free energy profile and the experimental open state lifetime in aqueous environment, base flipping takes place without an activation barrier or a stable sub-state along either pathway. Opening into the major groove is favored for small opening angles, however, the numerous stabilizing interactions in the minor groove make larger flipping into the minor groove energetically comparable. In our studies, the free energy reaches a plateau around the fully flipped ($\Delta\theta = 180 \pm 20°$) conformation at $\sim$21 kcal mol^{-1}. Importantly, on the basis of imino proton accessibility during the simulated flipping process, we have been able to show that bases become significantly exposed at less than 50° of opening. This result indicates that imino-proton exchange requires only local base pair opening with smaller free energy requirement than a full flipping process. This view is further supported by a recent experimental study of base flipping in which the melting rate of a DNA was measured in the presence of β-cyclodextrin.[130] The kinetics of this non-enzymatic flipping process was $\sim$6 orders of magnitude slower than that of base opening and hence it was concluded that the DNA conformational changes involved in the imino-proton exchange are unlikely to correspond to spontaneous base flipping. Given the high free energy cost of spontaneous flipping, the presence of a fully flipped base in solution prior to protein binding is improbable. It is possible that sequences with higher opening probability are recognized by the protein and further rotation of the open base is facilitated by interactions with the protein environment.

The flipping pathway shows the main characteristics of local opening, including the quadratic region with hydrogen bond rupture and the gradual loss of stacking interactions with neighboring base pairs. The free energy constantly rises as the base gets exposed to solvent, although much lower free energy is required with the angular restraint at intermediate positions as compared with the targeted approach. At larger opening angles important structural rearrangements take place both in the sugar-phosphate backbone around the flipping base and the stacking rearrangements of the orphan base. In the minor groove pathway, the rotated cytosine creates a planar base triplet with the adjacent 5'-G:C base pair resulting in a local minimum in the free energy profile. Similar stable triplet arrangement has already been observed in adenine bulges.[131,132] The flipping cytosine contacts both base and backbone sites up to three base pairs away on its 5'-side before finally leaving the minor groove to point to the surrounding solution. In the major groove pathway, the orphan guanine moves into the gap in the DNA helix, created by the base flipping, now filled with water, to form an unusual inter-strand stacking of three guanines in the GCGC recognition sequence. Similar rearrangements are seen in the *TaqI*-DNA complex where the orphan thymidine is shifted toward the center of the helix and occupies the space of the flipped adenine. In contrast, in the *HhaI*-DNA complex the orphan guanosine is stabilized by Gln237 and thus the DNA retains closely a B-DNA structure.

Backbone torsions in the dinucleotide junctions adjacent to the opening base often show an increased flexibility or undergo full transitions as base opening occurs. Some of these backbone conformations are not normally populated in B-DNA and are likely to be the direct consequence of the destruction of the stabilizing base stacking on either side of the opening base. Notable changes include $\alpha\gamma$ and $\varepsilon\zeta$ torsions (see below) and sugar pucker changes which were not observed in a 50 ns long unrestrained simulation of this sequence. We noted that flipping via the major groove leads to backbone conformations in striking agreement with the structural data from the protein-bound crystal complex. It seems that these changes represent the inherent flexibility of DNA rather than the effect of interactions with specific protein residues in the complex. It is also interesting to note that the DNA fragment is relatively straight in the complex crystal structure with a bending magnitude of less than 10°. Previous modeling studies, however, pointed to DNA bending upon base opening, with a marked effect on major groove opening.[56] We found in separate 5 ns long simulations that while the bending magnitude of the fully flipped state is on average similar to that in the closed conformation, for intermediate opening angles, in particular during opening into the major groove, bending significantly increases towards the major groove. In the case of intermediate minor groove opening, local untwisting was seen to help maintaining favorable stacking with the neighboring base pairs.

In a related study of cytosine flipping[57] a symmetrical, sharply rising free energy curve was obtained as the base opens into the minor or major groove until roughly 35°. Beyond this opening angle, the free energy is flat over 300° at 14 kcal mol^{-1}. It is not clear how such a free energy curve can explain the difference in rate between base opening and flipping. Nevertheless, although this study used a different opening restraint, a different force field, and a different sampling protocol, the results were qualitatively similar to those described above. Another study of cytosine flipping in a GCGG sequence reports a free energy profile where the flipped state is a stable minimum at 9 kcal mol^{-1} above the closed state with a corresponding major groove barrier of 21 kcal mol^{-1}.[133] In contrast to the previous umbrella sampling studies, this free energy profile was based on separate 2 ns simulations of the closed and the flipped state. Experimental data for imino-proton exchange kinetics[134] was used to adjust the free energy difference between the end-points of the process and the barrier to flipping was estimated based on an interpolation scheme. We have, however, shown that the thermodynamic data of imino-proton exchange does not correspond to full base flipping. Moreover, the experimental activation parameters include contributions from both the base pair motions and the exchange chemistry and thus cannot be directly compared with the calculated free energy barriers of base opening. The fitted barrier of 12 kcal mol^{-1} for the re-closing process is therefore overestimated.

3.3   Deformations of the Phosphodiester Backbone

The DNA backbone has an important role in restricting the available conformational space for stacking bases and in coupling the conformational properties of neighboring dinucleotide steps in a sequence. Crystal and solution structures of nucleic acid and their protein complexes indicate that the phosphodiester backbone can adopt a multitude of different conformations. The equilibrium between canonical and non-canonical conformations may provide flexibility in the molecule that can be exploited during the recognition process.

Conformational changes involving ε ($C_{4'}$-$C_{3'}$-$O_{3'}$-P) and ζ ($C_{3'}$-$O_{3'}$-P-$O_{5'}$) torsion angles are commonly observed in the free DNA backbone in nanosecond timescale simulations. These structurally correlated torsions define two conformational sub-states: the lower energy B_I is characterized by $\varepsilon\zeta$: tg^- with (ε–ζ) around –90° and B_{II} is characterized normally by $\varepsilon\zeta$: g^-t with (ε–ζ) around +90°. In DNA-protein complexes, however, $\varepsilon\zeta$: tt has also been seen with low (ε–ζ) values. The B_I phosphate conformation generally prevails with about 80% as measured by solution NMR spectroscopy,[135] analysis of crystallographic structures[136] and molecular dynamics simulations.[40] Certain dinucleotide steps, mostly the flexible pyrimidine-purine steps (e.g., CpA and CpG),[12] however, show a higher propensity for B_{II} conformation.[137,138] It has been suggested that base-base destacking facilitates the transition between B_I and B_{II} sub-states[139] and hence sequence-dependent stacking properties can shift B_I/B_{II} equilibrium. The role of B_I/B_{II} equilibrium in the early stage of protein-DNA recognition has been shown in several studies.[140-142]

The transition between the B_I/B_{II} conformations results in altered phosphate positions and influence groove parameters, stacking properties, and the helicoidal structure. The B_{II} state usually produces large twist and negative roll values of the corresponding step.[143-145] It is generally easier to find simultaneously two B_{II} phosphates facing one another than at consecutive junctions in the same strand.[146] This can be explained by the fact that the base pair shift parameters anticorrelate between neighboring steps and hence the negative shift corresponding to B_{II} state must be followed by a positive shift with a B_I backbone.[147] Interestingly, in the crystal structure of *HhaI* DNA methyltransferase ternary complex, discussed above, three consecutive phosphates are present in B_{II} conformations in the strand opposite to the flipped base. Here, however, the destruction of base pairing in the central pair probably relaxes the requirement for a negative shift in the B_{II} state. Facing phosphates in B_{II} sub-states have frequently been seen for certain steps, e.g. in CpA and CpG, with strong accompanying perturbation of the canonical structure.[143,148,149] This particular arrangement is characterized by positive x-disp and a very shallow major groove, negative roll and a local bend toward the minor groove. It has been shown that these sequence-dependent conformational changes in the free DNA are exploited by the NF-κB protein during specific binding.[141] Coupling between the B_I/B_{II} conformation and the adjacent and facing sugar puckers has also been observed in NMR studies and crystal structures.[150] Phosphates in B_{II} conformation appear to restrict the 5'-side sugar exclusively in the south conformation, while the 3'-side sugar can exhibit some variations away from the south.

Much less is known about the structural and energetic consequences of other transitions in the backbone, such as those which involve α ($O_{3'}$-P-$O_{5'}$-$C_{5'}$) and γ ($O_{5'}$-$C_{5'}$-$C_{4'}$-$C_{3'}$) torsions. Our analysis of the B-DNA backbone geometry in crystal structures shows almost exclusively the canonical $\alpha\gamma$ ($g^- g^+$) in B-DNA, only 2% of $g^+ g^-$ was observed, always associated with particular regions of the DNA, such as terminal base pairs or intermolecular contacts in the crystal.[44] The alternative *tt* conformation was, however, seen in A-DNA and RNA structures. In contrast to free B-DNA structures,

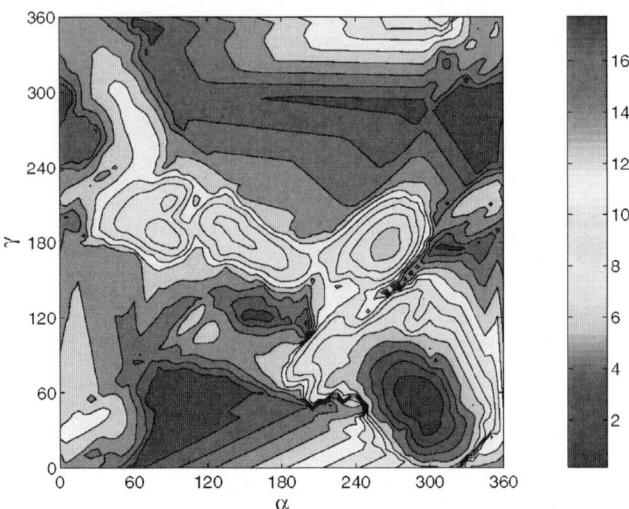

Figure 7-3. Free energy of a GpC step mapped on the $\alpha\gamma$ kcal.mol^{-1} torsional surface.

protein-bound B-DNA oligomers involve up to 10% non-canonical $\alpha\gamma$ conformations (g^+t, tt, g^+g). It is not known if these deformations are readily accessible in solution at room temperature and can guide the specific recognition of a given DNA sequence or are induced during the protein binding process.

An interesting example includes the central GpC dinucleotide step of 5'-dGTCAGCGCATGG-3', the target sequence for base-flipping *HhaI* methyltransferase. The flipping process was shown to be associated with changes in $\alpha\gamma$ values of this step both in the protein-DNA complex and during computer simulation of the flipping process without the enzyme (see above). Furthermore, our analysis of crystallographic protein–B-DNA complexes has revealed that GpC steps easily undergo $\alpha\gamma$ transitions, as 37% of these steps are found to adopt unusual $\alpha\gamma$ conformations. We have induced coupled $\alpha\gamma$ transitions in the DNA backbone to evaluate the free energy changes between the alternative $\alpha\gamma$ conformations and the effect of these transitions on the DNA structure.[44] A linear combination of the torsion angles was used to drive the system on a surface defined by the α and γ angles and preferentially sample low free energy regions. This method reduces the sampling problem to one dimension while obtaining the optimal α and γ values. As discussed above, using the WHAM technique, one can easily re-map the 1D free energy profile as a 2D surface of α and γ torsions (Figure 7-3). Five stable or metastable non-canonical $\alpha\gamma$ sub-states were found. Stable conformations include g^-t and g^+t, while tt, g^+g^+, and g^+g^-

were found to be only marginally stable. The most favorable pathway from the canonical $\alpha\gamma$ structure to any unusual form involves a counter-rotation of α and γ, via the *trans* conformation. Pathways that involve transitions of γ through the *cis* conformations were found to be heavily penalized. Curiously, the stacking energy did not show large variations along the pathways that would leave one wondering where the presumed sequence effect originates. The lowest free energy barrier to $\alpha\gamma$ flipping in the GpC step of this B-DNA oligomer sequence is above 6 kcal mol^{-1}. Thus, according to these calculations, it is unlikely that alternative $\alpha\gamma$ conformations are highly populated in a free DNA or can be exploited by the protein in the early stage of selective binding.

In a recent molecular dynamics study[40,45] of the 136 unique tetranucleotides (see below) B_I/B_{II} transitions and, surprisingly, $\alpha\gamma$ flips were observed during the 15 ns long simulations. An analysis of the B_I/B_{II} equilibrium among all the dinucleotide junctions revealed that most steps have less than 5% B_{II} conformation. A notable exception is the pyr-pur step that is characterized by 10-15% B_{II} phosphate conformation. Similarly, in the case of $\alpha\gamma$ flips, transitions from its canonical value concern only a small fraction of the base steps (<5%). However, these induce significant conformational changes in the DNA structure, notably low twist, positive slide and roll values in the adjacent 5'-step. The $\alpha\gamma$: g^+t sub-state was seen most frequently (~3%), but other arrangements including g^-t and g^+g^- were also observed. A sequence preference for these transitions was noted with the GpA step being the most likely to adopt non-canonical $\alpha\gamma$ sub-states. An important caveat is, however, that the majority of $\alpha\gamma$ transitions are not reversible on the simulation timescale, and hence their lifetimes could not be estimated. The accuracy of force fields for representing these transitions is also currently under investigation.

3.4   Deformation and Recognition

We now turn to some more subtle DNA deformations, namely those involved in protein-DNA recognition. Early models of how proteins located specific binding sites within DNA in fact took no account of DNA deformation. It was rather assumed that recognition was built up using a set of specific interactions between amino acid side chains and base sites accessible within the grooves of the double helix. If DNA deformation was necessary to form these interactions, it was assumed that the double helix was flexible enough to adapt with little energetic cost. However, as more high-resolution structures were obtained, it became clear that there were often not enough "direct" interactions to explain the number of base pairs forming the consensus protein binding sites. It was therefore necessary to consider "indirect" effects, amongst which, DNA deformation seemed to be a likely candidate. Indeed, protein binding often led to very significant DNA deformation, exemplified by a variety of structural changes such as groove widening, local untwisting and bending. It was however not easy to determine how much these deformations were likely to contribute to recognition.

In principle, modeling would be a useful tool to deconvolute interaction mechanisms, however, understanding recognition requires not just analyzing a single protein-DNA complex, but answering the question why did this protein prefer to bind to this specific base sequence? We must therefore be able to compare many base sequences for the same complex. Since estimating a binding free energy for even a single complex is already a computationally intensive task, it was necessary to search for ways of simplifying the problem. The first point to note is that the relative magnitudes of the various terms which contribute to binding free energies are not necessarily related to their importance in determining sequence dependence. As an example, given that forming a protein-DNA complex involves creating an interface with an area of several hundred $Å^2$, desolvation energies are obviously going to be very large. However, it does not follow that these energies will change very dramatically as a function of the base sequence of the fragment of DNA in question. The base sequence will however clearly have a major impact on two terms involved in binding: the difficulty of deforming the DNA fragment and protein-DNA interactions.

We consequently decided to limit our analysis of recognition mechanisms to these two terms. In order to calculate the necessary energies we used the internal/helicoidal representation in JUMNA coupled with the AMBER parm94 force field.[151] The protein-DNA complex was built starting from high-resolution crystallographic data and DNA deformation was studied by comparing the internal energy of DNA in the complex with an isolated fragment of B-DNA having the same base sequence. Despite the simplicity of this approach we were able to reproduce the order of stability changes in various protein-DNA complexes as a function of base sequence mutations.

Although protein-DNA interaction energy (ΔE_{int}) and DNA deformation energy (ΔE_{def}) can be calculated rapidly, it is still impossible to study the number of sequences necessary for understanding recognition. For a protein binding to N base pairs, there are 4^N possible base sequences. For typical specifically binding proteins, N ranges from 10-15, implying that it is necessary to study between 10^6 and 10^9 sequences in order to define specificity. In order to overcome this difficulty, we developed an original method related to multi-copy simulation algorithms. This involved modifying JUMNA to introduce so-called "lexides", nucleotides carrying not just one, but all four standard bases.[152,153] These bases are superposed in space and bound to the same sugar C1' atom. When we calculate the energy for a DNA fragment or for a protein-DNA complex using these lexides, we create a matrix which separates the energies associated with each possible base pair at each position with the DNA. Subsequently, the total energy corresponding to a given base sequence can be obtained by simply summing the appropriate matrix elements. This procedure is fast enough to allow the energies of millions of possible base sequences to be compared, without requiring any further approximations. One other useful feature of lexides is that they enable us to study DNA fragments with "neutral" sequences made up from 25% of each base pair at each position. This possibility is used in creating an optimal, sequence-neutral structure for isolated DNA fragments. It is also used for "cleaning" the protein-DNA structure of any structural details associated with the DNA sequence used for crystallization, but unessential for maintaining the protein-DNA interface.[154]

We have applied this procedure to a wide variety of protein-DNA complexes drawn from the protein data bank. In order to define a consensus sequence for each protein, we calculated our estimate of the total binding energy $\Delta E_{tot} = \Delta E_{int} + \Delta E_{def}$ for all possible base sequences. We then select those sequences falling within a chosen cutoff of the best energy (5 kcal mol^{-1} works well). We can then use this set of sequences to calculate the probability of each base at each position within the binding site, forming a weight matrix. We can also display the results graphically as a sequence logo.[154] This procedure led to good agreement with experimental results and enabled us to move on to the problem of how much DNA deformation is contributing to recognition?

In order to do this it is convenient to construct a quantitative measure of recognition. This can be done by converting the number of sequences falling within our ΔE_{tot} energy cutoff into a number of recognized base pairs. If there are S sequences within the energy cutoff, which are all considered acceptable for protein binding, this implies that the protein effectively allows a variability in the binding sequence equal to M base pairs, where $S = 4^M$. If the total length of the DNA fragment studied is N base pairs, then protein binding has recognized (and fixed) a site of length $L_{tot} = N-M = N-\log(S)/\log(4)$. How can we now find out how much protein-DNA interaction and DNA deformation contribute to defining this site length? The answer is to look at the distribution of the S binding sequences in terms of ΔE_{int} and ΔE_{def} and to count how many sequences fall below the least stable of the S sequences (S_{int} and S_{def} respectively). If this number is equivalent to M, the corresponding energy distribution resembles the ΔE_{tot} distribution and plays an important role in selecting the M binding sequences. If the number is much bigger than M, the corresponding energy distribution played only a

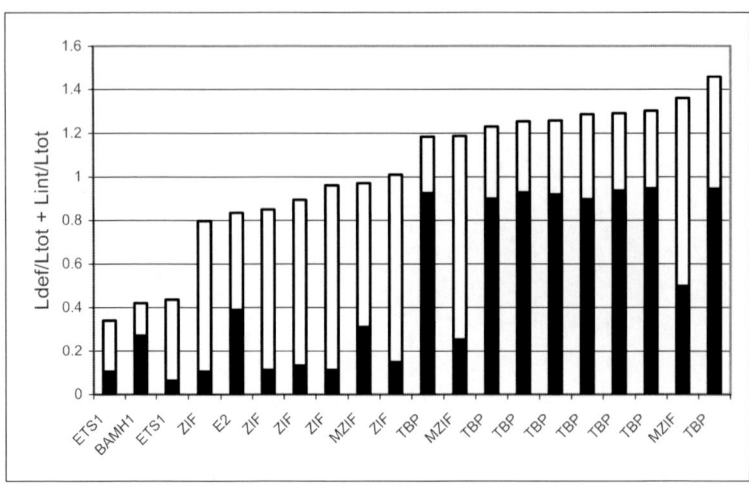

Figure 7-4. Contribution of DNA deformation (black bars) and protein-DNA interaction (white bars) to recognition in various complexes belonging to the indicated families.

minor role. By analogy with L_{tot}, we can define $L_{int} = N-log(S_{int}/4)$ and $L_{def} = N-log(S_{def}/4)$. Figure 7-4 shows the results for various complexes (in some cases for several variants of the same protein coming form different organisms) in the form of a bar chart. Note that we have normalized the length of each binding site to unity by dividing L_{int} and L_{def} by L_{tot}.

As one can see, DNA deformation plays some role in recognition for all the complexes studied. Its contribution is relatively small in complexes involving zinc-finger proteins such as Zif268, but very large in minor groove binding proteins such as TBP. It was a surprise to find that DNA deformation in recognition could play an important role even when protein binding only induced small structural perturbations (e.g. for zinc-finger proteins re-engineered to bind to AT-rich sequences (labeled MZIF in Figure 7-4).[155] It is also worth noting that one can learn something else from this analysis. The results in Figure 7-4 are ordered in terms of $L_{def}/L_{tot} + L_{int}/L_{tot}$. Whether this sum is smaller or greater than one indicates whether or not the two factors contributing to recognition are working in disaccord or in harmony (we have termed this property discordant or concordant recognition).[154]

Finally, our analysis also enables us to address the question of correlation between the sequence preferences for the successive base pair positions along a protein binding site. By looking at the S sequences selected by the energy cutoff, it is possible to ask whether the probability of finding a given dinucleotide pair XY at positions i and i + 1 is simply the product of the probabilities of finding X at position i and Y at position i + 1. It turns out that this result can be far from being satisfied, especially in cases where protein-induced DNA deformation modifies the interactions between the base pairs i and i + 1. This in turn implies that standard weight matrices do not accurately describe the binding preferences in such cases and that they need to be replaced by weight matrices based on overlapping dinucleotide probabilities, or even by more complex formulations.[156] This finding should help in improving genomic searches for protein binding sites, notably in the case of transcription factors, many of which have poorly defined consensus sequences.

3.5   Sequence Induced Fluctuations

Lastly we turn to the impact of the base sequence on the structure and dynamics of the double helix. While the initial models of DNA derived from fibre diffraction data necessarily saw the double helix as being perfectly uniform, as soon as high-resolution crystallographic data became available[7] it also became clear that specific base pairs had specific local effects on DNA structure. Unfortunately, although the number of DNA oligomer structures has grown significantly in the past 25 years, we are still far from having fully understood base sequence effects. In part, as already mentioned, this is due to a combinatorial problem. If we assume that base pair interactions can be understood on the basis of nearest neighbor interactions alone then there are only 10 possible combinations which need to by studied: GG, GA, GC, GT, AG, AA, AT, CG, CA and TA (the remaining six combinations, AC, CC, CT, TG, TC and TT, simply corresponding to the paired strand of members of the first set). Unfortunately, an analysis of the existing experimental database of DNA structures shows that although certain base pair combinations have some characteristic properties it is not possible to associate each step with clearly defined helical parameters. This implies that longer range interactions must play a role.

If we again assume that modulations in base stacking represent the main pathway from sequence to structure then the next step after a dinucleotide analysis is to pass to tetranucleotides – allowing the "context" of each central base pair step to be taken into account. While longer fragments allow a finer analysis of sequence effects, one also has to take into account their complexity. The number of unique sequences for an N-base pair fragment grows exponentially: for odd values of N there are $4^N/2$ possibilities, and for even values of N there are $(4^N + 4^{N/2})/2$ possibilities (the difference coming from the fact that for odd numbers of base pairs every sequence differs from its complement, while for even numbers of base pairs, by symmetry, N/2 sequences are autocomplementary). Thus, by moving from a dinucleotide to a tetranucleotide analysis we pass from 10 to 136 unique cases. In passing, although trinucleotide scales have been used in analyzing DNA properties[42,157-159] these are not easy to relate to stacking since in some way they must average the effects of two successive base pair steps.

Unfortunately, adopting a tetranucleotide model implies that present structural databases are no longer able to provide enough examples to make a meaningful statistical analysis of each case and theoretical approaches must be considered. While, energy minimization has already been applied at (and even beyond) the tetranucleotide level,[160-162] moving to molecular dynamics is necessary to get an idea of sequence effects within a room temperature ensemble. In an attempt to obtain this data, a group of laboratories interested in modeling nucleic acids got together in 2001 to form the "Ascona DNA Consortium" (the name having been chosen during a conference on modeling rod-like molecules and films organized in Ascona). In order to limit the number of simulations to be carried out, it was decided to pack the 136 unique tetranucleotide sequences into oligomers with regularly repeating sequences of the type: G-DABCDABCDABCD-G. In this way, 39 oligomers are enough to study the 136 tetranucleotides (each oligomer containing between one and four distinct, but overlapping tetranucleotides). These oligomers were divided between the participating groups and simulated using identical protocols. The analysis of the resulting 15 ns trajectories is the subject of two recent articles.[40,45] Although this data has led to a much finer view of sequence effects, it has also shown that 15 ns is not enough to fully sample the conformational space of the B-DNA due to long-lived backbone sub-states, notably those involving the $\alpha\gamma$ dihedrals (discussed in detail above). While this problem will require further investigation, conclusions on sequence effects beyond the dinucleotide level can already be drawn. It has thus been seen, for example, that while pyr-pur steps are effectively the most flexible, they are equally the least affected by their tetranucleotide context. In contrast, these steps have a significantly perturbed adjacent, and more rigid, pur-pur or pur-pyr steps. For detailed results see the cited ABC references.

REFERENCES

1. Watson, J. D. & Crick, F. H. (1953). Molecular structure of nucleic acids; a structure for deoxyribose nucleic acid. *Nature* **171**, 737-738.
2. Watson, J. D. & Crick, F. H. (2003). A structure for deoxyribose nucleic acid. 1953. *Nature* **421**, 397-398; discussion 396.
3. Crick, F. H., Barnett, L., Brenner, S. & Watts-Tobin, R. J. (1961). General nature of the genetic code for proteins. *Nature* **192**, 1227-1232.
4. Crick, F. H. (1962). The genetic code. *Sci. Am.* **207**, 66-74.
5. Franklin, R. E. & Gosling, R. G. (1953). Molecular configuration in sodium thymonucleate. *Nature* **171**, 740-741.

6. Lavery, R., Lebrun, A., Allemand, J.-F., Bensimon, D. & Croquette, V. (2002). Structure and mechanics of single biomolecules: experiment and simulation. *J. Phys. (Cond. Mat.)* **14**, R383-R414.

7. Wing, R., Drew, H., Takano, T., Broka, C., Tanaka, S., Itakura, K. & Dickerson, R. E. (1980). Crystal structure analysis of a complete turn of B-DNA. *Nature* **287**, 755-758.

8. Mack, D. R., Chiu, T. K. & Dickerson, R. E. (2001). Intrinsic bending and deformability at the T-A step of CCTTTAAAGG: a comparative analysis of T-A and A-T steps within A-tracts. *J. Mol. Biol.* **312**, 1037-1049.

9. Olson, W. K., Gorin, A. A., Lu, X. J., Hock, L. M. & Zhurkin, V. B. (1998). DNA sequence-dependent deformability deduced from protein-DNA crystal complexes. *Proc. Natl. Acad. Sci. USA* **95**, 11163-11168.

10. Ulyanov, N. B. & Zhurkin, V. B. (1984). Sequence-dependent anisotropic flexibility of B-DNA. A conformational study. *J. Biomol. Struct. Dyn.* **2**, 361-385.

11. Lankaš, F., Šponer, J., Hobza, P. & Langowski, J. (2000). Sequence-dependent elastic properties of DNA. *J. Mol. Biol.* **299**, 695-709.

12. Lankaš, F., Šponer, J., Langowski, J. & Cheatham, T. E., 3rd (2003). DNA basepair step deformability inferred from molecular dynamics simulations. *Biophys J.* **85**, 2872-2883.

13. Lankaš, F. (2004). DNA sequence-dependent deformability - insights from computer simulations. *Biopolymers* **73**, 327-339.

14. Tan, R. K. & Harvey, S. C. (1987). A comparison of six DNA bending models. *J. Biomol. Struct. Dyn.* **5**, 497-512.

15. Goodsell, D. S. & Dickerson, R. E. (1994). Bending and curvature calculations in B-DNA. *Nucleic Acids Res.* **22**, 5497-5503.

16. Zhurkin, V. B., Tolstorukov, M. Y., Fei, X., Colasanti, A. V. & Olson, W. K. (2005). Sequence-dependent variability of B-DNA: an update on bending and curvature. In *DNA conformation and transcription* (Ohyama, T., ed.). Landes Bioscience.Eurekah.com.

17. Cloutier, T. E. & Widom, J. (2004). Spontaneous sharp bending of double-stranded DNA. *Mol. Cell.* **14**, 355-362.

18. Crick, F. H. & Klug, A. (1975). Kinky helix. *Nature* **255**, 530-533.

19. Du, Q., Smith, C., Shiffeldrim, N., Vologodskaia, M. & Vologodskii, A. (2005). Cyclization of short DNA fragments and bending fluctuations of the double helix. *Proc. Natl. Acad. Sci. U S A* **102**, 5397-5402.

20. Dickerson, R. E., Goodsell, D. S. & Neidle, S. (1994). "...the tyranny of the lattice..." *Proc. Natl. Acad. Sci. USA* **91**, 3579-3583.

21. Dlakic, M., Park, K., Griffith, J. D., Harvey, S. C. & Harrington, R. E. (1996). The organic crystallizing agent 2-methyl-2,4-pentanediol reduces DNA curvature by means of structural changes in A-tracts. *J. Biol. Chem.* **271**, 17911-17919.

22. Bax, A., Kontaxis, G. & Tjandra, N. (2001). Dipolar couplings in macromolecular structure determination. *Methods Enzymol.* **339**, 127-174.

23. Wu, Z., Delaglio, F., Tjandra, N., Zhurkin, V. B. & Bax, A. (2003). Overall structure and sugar dynamics of a DNA dodecamer from homo- and heteronuclear dipolar couplings and ^{31}P chemical shift anisotropy. *J Biomol NMR* **26**, 297-315.

24. Calladine, C. R., Drew, H. R. & McCall, M. J. (1988). The intrinsic curvature of DNA in solution. *J. Mol. Biol.* **201**, 127-137.

25. De Santis, P., Palleschi, A., Savino, M. & Scipioni, A. (1990). Validity of the nearest-neighbor approximation in the evaluation of the electrophoretic manifestations of DNA curvature. *Biochemistry* **29**, 9269-9273.

26. Bolshoy, A., McNamara, P., Harrington, R. E. & Trifonov, E. N. (1991). Curved DNA without A-A: experimental estimation of all 16 DNA wedge angles. *Proc. Natl. Acad. Sci. USA* **88**, 2312-2316.

27. Munkel, C., Eils, R., Dietzel, S., Zink, D., Mehring, C., Wedemann, G., Cremer, T. & Langowski, J. (1999). Compartmentalization of interphase chromosomes observed in simulation and experiment. *J. Mol. Biol.* **285**, 1053-1065.

28. Cluzel, P., Lebrun, A., Heller, C., Lavery, R., Viovy, J. L., Chatenay, D. & Caron, F. (1996). DNA: an extensible molecule. *Science* **271**, 792-794.
29. Smith, S. B., Cui, Y. & Bustamante, C. (1996). Overstretching B-DNA: the elastic response of individual double-stranded and single-stranded DNA molecules. *Science* **271**, 795-799.
30. Pfannschmidt, C. & Langowski, J. (1998). Superhelix organization by DNA curvature as measured through site-specific labeling. *J. Mol. Biol.* **275**, 601-611.
31. Zhurkin, V. B., Lysov, Y. P. & Ivanov, V. I. (1979). Anisotropic flexibility of DNA and the nucleosomal structure. *Nucleic Acids Res.* **6**, 1081-1096.
32. Miller, K. J. (1979). Interactions of molecules with nucleic acids. I. An algorithm to generate nucleic acid structures with an application to the B-DNA structure and a counterclockwise helix. *Biopolymers* **18**, 959-980.
33. Lavery, R., Zakrzewska, K. & Sklenar, H. (1995). JUMNA (Junction Minimization of Nucleic-Acids). *Comput. Phys. Commun.* **91**, 135-158.
34. Lafontaine, I. & Lavery, R. (1999). Collective variable modelling of nucleic acids. *Curr. Opin. Struct. Biol.* **9**, 170-176.
35. Gabb, H. A., Prevost, C., Bertucat, G., Robert, C. H. & Lavery, R. (1997). Collective variable Monte Carlo simulation of DNA. *J. Comput. Chem.* **18**, 2001-2011.
36. Mazur, A. K. (2002). DNA dynamics in a water drop without counterions. *J. Am. Chem. Soc.* **124**, 14707-14715.
37. Lavery, R., Parker, I. & Kendrick, J. (1986). A general approach to the optimization of the conformation of ring molecules with an application to valinomycin. *J. Biomol. Struct. Dyn.* **4**, 443-462.
38. Varnai, P. & Zakrzewska, K. (2004). DNA and its counterions: a molecular dynamics study. *Nucleic Acids Res.* **32**, 4269-4280.
39. Ponomarev, S. Y., Thayer, K. M. & Beveridge, D. L. (2004). Ion motions in molecular dynamics simulations on DNA. *Proc. Natl. Acad. Sci. USA* **101**, 14771-14775.
40. Dixit, S. B., Beveridge, D. L., Case, D. A., Cheatham, T. E., 3rd Giudice, E., Lankaš, F., Lavery, R., Maddocks, J. H., Osman, R., Sklenar, H., Thayer, K. M. & Varnai, P. (2005). Molecular dynamics simulaitons of the 136 unique tetranucleotide sequences of DNA oligonucleotides. II. Sequence context effects on the dynamical structures of the 10 unique dinucleotide steps. *Biophys. J.* **89**, 3721-3740.
41. Saenger, W. (1984). *Principles of nucleic acid structure*, Springer-Verlag, New York.
42. Tolstorukov, M. Y., Ivanov, V. I., Malenkov, G. G., Jernigan, R. L. & Zhurkin, V. B. (2001). Sequence-dependent B↔A transition in DNA evaluated with dimeric and trimeric scales. *Biophys J.* **81**, 3409-3421.
43. Banavali, N. K. & Roux, B. (2005). Free energy landscape of A-DNA to B-DNA conversion in aqueous solution. *J. Am. Chem. Soc.* **127**, 6866-6876.
44. Varnai, P., Djuranovic, D., Lavery, R. & Hartmann, B. (2002). α/γ transitions in the B-DNA backbone. *Nucleic Acids Res.* **30**, 5398-5406.
45. Beveridge, D. L., Barreiro, G., Byun, K. S., Case, D. A., Cheatham, T. E., 3rd Dixit, S. B., Giudice, E., Lankaš, F., Lavery, R., Maddocks, J. H., Osman, R., Seibert, E., Sklenar, H., Stoll, G., Thayer, K. M., Varnai, P. & Young, M. A. (2004). Molecular dynamics simulations of the 136 unique tetranucleotide sequences of DNA oligonucleotides. I. Research design and results on d(CpG) steps. *Biophys J.* **87**, 3799-3813.
46. Guéron, M. & Leroy, J. L. (1995). Studies of base pair kinetics by NMR measurement of proton exchange. *Methods Enzymol.* **261**, 383-413.
47. Jovin, T. M., Soumpasis, D. M. & McIntosh, L. P. (1987). The transition between B-DNA and Z-DNA. *Annu. Rev. Phys. Chem.* **38**, 521-558.
48. Giudice, E., Varnai, P. & Lavery, R. (2001). Energetic and conformational aspects of A:T base-pair opening within the DNA double helix. *Chemphyschem* **2**, 673-677.

49. Giudice, E., Varnai, P. & Lavery, R. (2003). Base pair opening within B-DNA: free energy pathways for GC and AT pairs from umbrella sampling simulations. *Nucleic Acids Res.* **31**, 1434-1443.

50. Varnai, P., Canalia, M. & Leroy, J. L. (2004). Opening mechanism of G.T/U pairs in DNA and RNA duplexes: a combined study of imino proton exchange and molecular dynamics simulation. *J. Am. Chem. Soc.* **126**, 14659-14667.

51. Varnai, P. & Lavery, R. (2002). Base flipping in DNA: pathways and energetics studied with molecular dynamic simulations. *J. Am. Chem. Soc.* **124**, 7272-7273.

52. Sanghani, S. R., Zakrzewska, K., Harvey, S. C. & Lavery, R. (1996). Molecular modelling of $(A_4T_4NN)_n$ and $(T_4A_4NN)_n$: sequence elements responsible for curvature. *Nucleic Acids Res.* **24**, 1632-1637.

53. Sanghani, S. R., Zakrzewska, K. & Lavery, R. (1996). Modeling DNA bending induced by phosphate neutralisation. In *Biological Structure and Dynamics* (Sarma, R. H. & Sarma, M. H., eds.), Vol. 2, pp. 267-278. Adenine Press, New York.

54. Ramstein, J. & Lavery, R. (1988). Energetic coupling between DNA bending and base pair opening. *Proc. Natl. Acad. Sci. USA* **85**, 7231-7235.

55. Ramstein, J. & Lavery, R. (1990). Base pair opening pathways in B-DNA. *J. Biomol. Struct. Dyn.* **7**, 915-933.

56. Bernet, J., Zakrzewska, K. & Lavery, R. (1997). Modelling base pair opening: the role of helical twist. *J. Mol. Struct. (Theochem)* **398-399**, 473-482.

57. Banavali, N. K. & MacKerell, A. D., Jr. (2002). Free energy and structural pathways of base flipping in a DNA GCGC containing sequence. *J. Mol. Biol.* **319**, 141-160.

58. Cooper, A. (1999). Thermodynamic analysis of biomolecular interactions. *Curr. Opin. Chem. Biol.* **3**, 557-563.

59. Onuchic, J. N. & Wolynes, P. G. (2004). Theory of protein folding. *Curr. Opin. Struct. Biol.* **14**, 70-75.

60. Strazewski, P. (2002). Thermodynamic correlation analysis: hydration and perturbation sensitivity of RNA secondary structures. *J. Am. Chem. Soc.* **124**, 3546-3554.

61. Jen-Jacobson, L., Engler, L. E. & Jacobson, L. A. (2000). Structural and thermodynamic strategies for site-specific DNA binding proteins. *Structure (Camb)* **8**, 1015-1023.

62. Chen, C. & Russu, I. M. (2004). Sequence-dependence of the energetics of opening of AT basepairs in DNA. *Biophys J.* **87**, 2545-2551.

63. Sharp, K. (2001). Entropy-enthalpy compensation: fact or artifact? *Protein Sci.* **10**, 661-667.

64. Lebrun, A., Shakked, Z. & Lavery, R. (1997). Local DNA stretching mimics the distortion caused by the TATA box-binding protein. *Proc. Natl. Acad. Sci. USA* **94**, 2993-2998.

65. Beveridge, D. L. & DiCapua, F. M. (1989). Free energy via molecular simulations. Application to chemical and biomolecular systems. *Annu. Rev. Biophys. Biophys. Chem.* **18**, 431-492.

66. Rodinger, T. & Pomes, R. (2005). Enhancing the accuracy, the efficiency and the scope of free energy simulations. *Curr. Opin. Struct. Biol.* **15**, 164-170.

67. Torrie, G. M. & Valleau, J. P. (1977). Nonphysical sampling distribution in Monte Carlo free-energy estimation: Umbrella sampling. *J. Comput. Phys.* **23**, 187-199.

68. Kumar, S., Bouzida, D., Swendsen, R. H., Kollman, P. A. & Rosenberg, J. M. (1992). The weighted histogram analysis method for free energy calculations on biomolecules. I. The Method. *J. Comput. Chem.* **13**, 1011-1021.

69. Boczko, E. M. & Brooks, C. L., 3rd. (1993). Constant-temperature free energy surfaces for physical and chemical processes. *J. Phys. Chem.* **97**, 4509-4513.

70. Bolhuis, P. G., Chandler, D., Dellago, C. & Geissler, P. L. (2002). Transition path sampling: throwing ropes over rough mountain passes, in the dark. *Annu. Rev. Phys. Chem.* **53**, 291-318.

71. Best, R. B. & Hummer, G. (2005). Reaction coordinates and rates from transition paths. *Proc. Natl. Acad. Sci. USA* **102**, 6732-6737.

72. Radhakrishnan, R. & Schlick, T. (2004). Biomolecular free energy profiles by a shooting/umbrella sampling protocol, "BOLAS". *J. Chem. Phys.* **121**, 2436-2444.

73. Ma, A. & Dinner, A. R. (2005). Automatic method for identifying reaction coordinates in complex systems. *J. Phys. Chem. B.* **109**, 6769-6779.

74. Rhee, Y. M. & Pande, V. S. (2005). One-dimensional reaction coordinate and the corresponding potential of mean force from commitment probability distribution. *J. Phys. Chem. B* **109**, 6780-6786.

75. Hagan, M. F., Dinner, A. R., Chandler, D. & Chakraborty, A. K. (2003). Atomistic understanding of kinetic pathways for single base-pair binding and unbinding in DNA. *Proc. Natl. Acad. Sci. USA* **100**, 13922-13927.

76. Lee, F. S., Chu, Z. T., Bolger, M. B. & Warshel, A. (1992). Calculations of antibody-antigen interactions: microscopic and semi-microscopic evaluation of the free energies of binding of phosphorylcholine analogs to McPC603. *Protein Eng.* **5**, 215-228.

77. Sham, Y. Y., Chu, Z. T., Tao, H. & Warshel, A. (2000). Examining methods for calculations of binding free energies: LRA, LIE, PDLD-LRA, and PDLD/S-LRA calculations of ligands binding to an HIV protease. *Proteins* **39**, 393-407.

78. Kollman, P. A., Massova, I., Reyes, C., Kuhn, B., Huo, S., Chong, L., Lee, M., Lee, T., Duan, Y., Wang, W., Donini, O., Cieplak, P., Srinivasan, J., Case, D. A. & Cheatham, T. E., 3rd (2000). Calculating structures and free energies of complex molecules: combining molecular mechanics and continuum models. *Acc. Chem. Res.* **33**, 889-897.

79. Swanson, J. M., Henchman, R. H. & McCammon, J. A. (2004). Revisiting free energy calculations: a theoretical connection to MM/PBSA and direct calculation of the association free energy. *Biophys J.* **86**, 67-74.

80. Levy, R. M. & Gallicchio, E. (1998). Computer simulations with explicit solvent: recent progress in the thermodynamic decomposition of free energies and in modeling electrostatic effects. *Annu. Rev. Phys. Chem.* **49**, 531-567.

81. Andricioaei, I. & Karplus, M. (2001). On the calculation of entropy from covariance matrices of the atomic fluctuations. *J. Chem. Phys.* **115**, 6289-6292.

82. Peter, C., Oostenbrink, C., Van Dorp, A. & Van Gunsteren, W. F. (2004). Estimating entropies from molecular dynamics simulations. *J. Chem. Phys.* **120**, 2652-2661.

83. Jayaram, B., Sprous, D., Young, M. A. & Beveridge, D. L. (1998). Free energy analysis of the conformational preferences of A and B forms of DNA in solution. *J. Am. Chem. Soc.* **120**, 10629-10633.

84. Srinivasan, J., Cheatham, T. E., 3rd Cieplak, P., Kollman, P. A. & Case, D. A. (1998). Continuum solvent studies of the stability of DNA, RNA and phosphoramidate-DNA helices. *J. Am. Chem. Soc.* **120**, 9401-9409.

85. Srinivasan, J., Miller, J., Kollman, P. A. & Case, D. A. (1998). Continuum solvent studies of the stability of RNA hairpin loops and helices. *J. Biomol. Struct. Dyn.* **16**, 671-682.

86. Reyes, C. M. & Kollman, P. A. (2000). Structure and thermodynamics of RNA-protein binding: using molecular dynamics and free energy analyses to calculate the free energies of binding and conformational change. *J. Mol. Biol.* **297**, 1145-1158.

87. Jayaram, B., McConnell, K., Dixit, S. B., Das, A. & Beveridge, D. L. (2002). Free-energy component analysis of 40 protein-DNA complexes: a consensus view on the thermodynamics of binding at the molecular level. *J. Comput. Chem.* **23**, 1-14.

88. Tsui, V. & Case, D. A. (2001). Calculations of the absolute free energies of binding between RNA and metal ions using molecular dynamics simulations and continuum electrostatics. *J. Phys. Chem. B.* **105**, 11314-11325.

89. Spackova, N., Cheatham, T. E., 3rd Ryjacek, F., Lankaš, F., Van Meervelt, L., Hobza, P. & Šponer, J. (2003). Molecular dynamics simulations and thermodynamics analysis of

DNA-drug complexes. Minor groove binding between 4',6-diamidino-2-phenylindole and DNA duplexes in solution. *J. Am. Chem. Soc.* **125**, 1759-1769.

90. Fadrna, E., Spackova, N., Stefl, R., Koca, J., Cheatham, T. E., 3rd & Šponer, J. (2004). Molecular dynamics simulations of guanine quadruplex loops: advances and force field limitations. *Biophys J.* **87**, 227-242.

91. Gouda, H., Kuntz, I. D., Case, D. A. & Kollman, P. A. (2003). Free energy calculations for theophylline binding to an RNA aptamer: comparison of MM-PBSA and thermodynamic integration methods. *Biopolymers* **68**, 16-34.

92. Bustamante, C., Bryant, Z. & Smith, S. B. (2003). Ten years of tension: single-molecule DNA mechanics. *Nature* **421**, 423-427.

93. Bensimon, D. (1996). Force: a new structural control parameter? *Structure (Camb)* **4**, 885-889.

94. Lebrun, A. & Lavery, R. (1996). Modelling extreme stretching of DNA. *Nucleic Acids Res.* **24**, 2260-2267.

95. Leger, J. F., Romano, G., Sarkar, A., Robert, J., Bourdieu, L., Chatenay, D. & Marko, J. F. (1999). Structural transitions of a twisted and stretched DNA molecule. *Phys. Rev. Lett.* **83**, 1066-1069.

96. Greenall, R. J., Nave, C. & Fuller, W. (2001). X-ray diffraction from DNA fibres under tension. *J. Mol. Biol.* **305**, 669-672.

97. Rouzina, I. & Bloomfield, V. A. (2001). Force-induced melting of the DNA double helix. 1. Thermodynamic analysis. *Biophys J.* **80**, 882-893.

98. Rouzina, I. & Bloomfield, V. A. (2001). Force-induced melting of the DNA double helix. 2. Effect of solution conditions. *Biophys J.* **80**, 894-900.

99. Wenner, J. R., Williams, M. C., Rouzina, I. & Bloomfield, V. A. (2002). Salt dependence of the elasticity and overstretching transition of single DNA molecules. *Biophys J.* **82**, 3160-3169.

100. Strick, T. R., Allemand, J. F., Bensimon, D., Bensimon, A. & Croquette, V. (1996). The elasticity of a single supercoiled DNA molecule. *Science* **271**, 1835-1837.

101. Strick, T. R., Allemand, J. F., Bensimon, D. & Croquette, V. (1998). Behavior of supercoiled DNA. *Biophys J.* **74**, 2016-2028.

102. Allemand, J. F., Bensimon, D. & Croquette, V. (2003). Stretching DNA and RNA to probe their interactions with proteins. *Curr. Opin. Struct. Biol.* **13**, 266-274.

103. Saleh, O. A., Allemand, J. F., Croquette, V. & Bensimon, D. (2005). Single-molecule manipulation measurements of DNA transport proteins. *Chemphyschem* **6**, 813-818.

104. Forterre, P., Bergerat, A. & Lopez-Garcia, P. (1996). The unique DNA topology and DNA topoisomerases of hyperthermophilic archaea. *FEMS Microbiol. Rev.* **18**, 237-248.

105. Lopez-Garcia, P. & Forterre, P. (2000). DNA topology and the thermal stress response, a tale from mesophiles and hyperthermophiles. *Bioessays* **22**, 738-746.

106. Allemand, J. F., Bensimon, D., Lavery, R. & Croquette, V. (1998). Stretched and overwound DNA forms a Pauling-like structure with exposed bases. *Proc. Natl. Acad. Sci. USA* **95**, 14152-14157.

107. Pauling, L. & Corey, R. B. (1953). Structure of the nucleic acids. *Nature* **171**, 346.

108. Day, L. A., Wiseman, R. L. & Marzec, C. J. (1979). Structure models for DNA in filamentous viruses with phosphates near the center. *Nucleic Acids Res.* **7**, 1393-1403.

109. Liu, D. J. & Day, L. A. (1994). Pf1 virus structure: helical coat protein and DNA with paraxial phosphates. *Science* **265**, 671-674.

110. Davey, C. A., Sargent, D. F., Luger, K., Maeder, A. W. & Richmond, T. J. (2002). Solvent mediated interactions in the structure of the nucleosome core particle at 1.9 Å resolution. *J. Mol. Biol.* **319**, 1097-1113.

111. Richmond, T. J. & Davey, C. A. (2003). The structure of DNA in the nucleosome core. *Nature* **423**, 145-150.

112. Young, M. A. & Beveridge, D. L. (1998). Molecular dynamics simulations of an oligonucleotide duplex with adenine tracts phased by a full helix turn. *J. Mol. Biol.* **281**, 675-687.

113. Beveridge, D. L., Dixit, S. B., Barreiro, G. & Thayer, K. M. (2004). Molecular dynamics simulations of DNA curvature and flexibility: helix phasing and premelting. *Biopolymers* **73**, 380-403.

114. Olson, W. K., Marky, N. L., Jernigan, R. L. & Zhurkin, V. B. (1993). Influence of fluctuations on DNA curvature. A comparison of flexible and static wedge models of intrinsically bent DNA. *J. Mol. Biol.* **232**, 530-554.

115. Klimasauskas, S., Kumar, S., Roberts, R. J. & Cheng, X. (1994). HhaI methyltransferase flips its target base out of the DNA helix. *Cell* **76**, 357-369.

116. Daujotyte, D., Serva, S., Vilkaitis, G., Merkiene, E., Venclovas, C. & Klimasauskas, S. (2004). HhaI DNA methyltransferase uses the protruding Gln237 for active flipping of its target cytosine. *Structure (Camb)* **12**, 1047-1055.

117. Huang, N., Banavali, N. K. & MacKerell, A. D., Jr. (2003). Protein-facilitated base flipping in DNA by cytosine-5-methyltransferase. *Proc. Natl. Acad. Sci. USA* **100**, 68-73.

118. Horton, J. R., Ratner, G., Banavali, N. K., Huang, N., Choi, Y., Maier, M. A., Marquez, V. E., MacKerell, A. D., Jr. & Cheng, X. (2004). Caught in the act: visualization of an intermediate in the DNA base-flipping pathway induced by HhaI methyltransferase. *Nucleic Acids Res.* **32**, 3877-3886.

119. Keepers, J., Kollman, P. A., Weiner, P. K. & James, T. L. (1982). Molecular mechanical studies of DNA flexibility: coupled backbone torsion angles and base-pair opening. *Proc. Natl. Acad. Sci. USA* **79**, 5337-5541.

120. Chen, Y. Z., Mohan, V. & Griffey, R. H. (1998). The opening of a single base without perturbations of neighboring nucleotides: a study on crystal B-DNA duplex d(CGCGAATTCGCG)$_2$. *J. Biomol. Struct. Dyn.* **15**, 765-777.

121. Briki, F., Ramstein, J., Lavery, R. & Genest, D. (1991). Evidence for the stochastic nature of base pair opening in DNA: a Brownian dynamics simulation. *J. Am. Chem. Soc.* **113**, 2490-2493.

122. Cubero, E., Sherer, E. C., Luque, F. J., Orozco, M. & Laughton, C. A. (1999). Observation of spontaneous base pair breathing event in the molecular dynamics simulation of a difluorotoluene-containing DNA oligonucleotide. *J. Am. Chem. Soc.* **121**, 8653-8654.

123. Russu, I. M. (2004). Probing site-specific energetics in proteins and nucleic acids by hydrogen exchange and nuclear magnetic resonance spectroscopy. *Methods Enzymol.* **379**, 152-175.

124. Vilkaitis, G., Merkiene, E., Serva, S., Weinhold, E. & Klimasauskas, S. (2001). The mechanism of DNA cytosine-5 methylation. Kinetic and mutational dissection of HhaI methyltransferase. *J. Biol. Chem.* **276**, 20924-20934.

125. Cao, C., Jiang, Y. L., Stivers, J. T. & Song, F. (2004). Dynamic opening of DNA during the enzymatic search for a damaged base. *Nat. Struct. Mol. Biol.* **11**, 1230-1236.

126. Lariviere, L., Sommer, N. & Morera, S. (2005). Structural evidence of a passive base-flipping mechanism for AGT, an unusual GT-B glycosyltransferase. *J. Mol. Biol.* **352**, 139-150.

127. van Aalten, D. M. F., Erlanson, D. A., Verdine, G. L. & Joshua-Tor, L. (1999). A structural snapshot of base-pair opening in DNA. *Proc. Natl. Acad. Sci. USA* **96**, 11809-11814.

128. Giudice, E. & Lavery, R. (2003). Nucleic acid base pair dynamics: the impact of sequence and structure using free-energy calculations. *J. Am. Chem. Soc.* **125**, 4998-4999.

129. Guéron, M., Kochoyan, M. & Leroy, J. L. (1987). A single mode of DNA base-pair opening drives imino proton exchange. *Nature* **328**, 89-92.

130. Spies, M. A. & Schowen, R. L. (2002). The trapping of a spontaneously "flipped-out" base from double helical nucleic acids by host-guest complexation with beta-cyclodextrin: the intrinsic base-flipping rate constant for DNA and RNA. *J. Am. Chem. Soc.* **124**, 14049-14053.

131. Feig, M., Zacharias, M. & Pettitt, B. M. (2001). Conformations of an adenine bulge in a DNA octamer and its influence on DNA structure from molecular dynamics simulations. *Biophys J.* **81**, 352-370.

132. Zacharias, M. & Sklenar, H. (1997). Analysis of the stability of looped-out and stacked-in conformations of an adenine bulge in DNA using a continuum model for solvent and ions. *Biophys J.* **73**, 2990-3003.

133. Fuxreiter, M., Luo, N., Jedlovszky, P., Simon, I. & Osman, R. (2002). Role of base flipping in specific recognition of damaged DNA by repair enzymes. *J. Mol. Biol.* **323**, 823-834.

134. Folta-Stogniew, E. & Russu, I. M. (1994). Sequence dependence of base-pair opening in a DNA dodecamer containing the CACA/GTGT sequence motif. *Biochemistry* **33**, 11016-11024.

135. Gorenstein, D. G. (1994). Conformation and dynamics of DNA and protein-DNA complexes by ^{31}P NMR. *Chem. Rev.* **94**, 1315-1338.

136. Djuranovic, D. & Hartmann, B. (2003). Conformational characteristics and correlations in crystal structures of nucleic acid oligonucleotides: evidence for sub-states. *J. Biomol. Struct. Dyn.* **20**, 771-788.

137. Bertrand, H., Ha-Duong, T., Fermandjian, S. & Hartmann, B. (1998). Flexibility of the B-DNA backbone: effects of local and neighbouring sequences on pyrimidine-purine steps. *Nucleic Acids Res.* **26**, 1261-1267.

138. Lefebvre, A., Mauffret, O., Lescot, E., Hartmann, B. & Fermandjian, S. (1996). Solution structure of the CpG containing d(CTTCGAAG)₂ oligonucleotide: NMR data and energy calculations are compatible with a BI/BII equilibrium at CpG. *Biochemistry* **35**, 12560-12569.

139. Grzeskowiak, K., Yanagi, K., Prive, G. G. & Dickerson, R. E. (1991). The structure of B-helical C-G-A-T-C-G-A-T-C-G and comparison with C-C-A-A-C-G-T-T-G-G. The effect of base pair reversals. *J. Biol. Chem.* **266**, 8861-8883.

140. Schroeder, S. A., Roongta, V., Fu, J. M., Jones, C. R. & Gorenstein, D. G. (1989). Sequence-dependent variations in the ^{31}P NMR spectra and backbone torsional angles of wild-type and mutant Lac operator fragments. *Biochemistry* **28**, 8292-8303.

141. Tisne, C., Delepierre, M. & Hartmann, B. (1999). How NF-κB can be attracted by its cognate DNA. *J. Mol. Biol.* **293**, 139-150.

142. van Dam, L., Korolev, N. & Nordenskiold, L. (2002). Polyamine-nucleic acid interactions and the effects on structure in oriented DNA fibers. *Nucleic Acids Res.* **30**, 419-428.

143. Hartmann, B., Piazzola, D. & Lavery, R. (1993). BI-BII transitions in B-DNA. *Nucleic Acids Res.* **21**, 561-568.

144. Winger, R. H., Liedl, K. R., Pichler, A., Hallbrucker, A. & Mayer, E. (1999). Helix morphology changes in B-DNA induced by spontaneous B(I)↔B(II) substrate interconversion. *J. Biomol. Struct. Dyn.* **17**, 223-235.

145. van Dam, L. & Levitt, M. H. (2000). BII nucleotides in the B and C forms of natural-sequence polymeric DNA: a new model for the C form of DNA. *J. Mol. Biol.* **304**, 541-561.

146. Schneider, B., Neidle, S. & Berman, H. M. (1997). Conformations of the sugar-phosphate backbone in helical DNA crystal structures. *Biopolymers* **42**, 113-124.

147. Packer, M. J. & Hunter, C. A. (1998). Sequence-dependent DNA structure: the role of the sugar-phosphate backbone. *J. Mol. Biol.* **280**, 407-420.

148. Isaacs, R. J. & Spielmann, H. P. (2001). NMR evidence for mechanical coupling of phosphate B_I-B_{II} transitions with deoxyribose conformational exchange in DNA. *J. Mol. Biol.* **311**, 149-160.

149. Tisne, C., Hantz, E., Hartmann, B. & Delepierre, M. (1998). Solution structure of a non-palindromic 16 base-pair DNA related to the HIV-1 κB site: evidence for BI-BII equilibrium inducing a global dynamic curvature of the duplex. *J. Mol. Biol.* **279**, 127-142.

150. Djuranovic, D. & Hartmann, B. (2004). DNA fine structure and dynamics in crystals and in solution: the impact of BI/BII backbone conformations. *Biopolymers* **73**, 356-368.

151. Cornell, W. D., Cieplak, P., C. I., B., Gould, I. R., Merz, K. M. J., Ferguson, D. M., Spellmeyer, D. C., Fox, T., Caldwell, J. W. & Kollman, P. A. (1995). A second generation force field for the simulation of proteins, nucleic acids and organic molecules. *J. Am. Chem. Soc.* **117**, 5179-5197.

152. Lafontaine, I. & Lavery, R. (2000). ADAPT: a molecular mechanics approach for studying the structural properties of long DNA sequences. *Biopolymers* **56**, 292-310.

153. Lafontaine, I. & Lavery, R. (2000). Optimization of nucleic acid sequences. *Biophys J.* **79**, 680-685.

154. Paillard, G. & Lavery, R. (2004). Analyzing protein-DNA recognition mechanisms. *Structure (Camb)* **12**, 113-122.

155. Paillard, G., Deremble, C. & Lavery, R. (2004). Looking into DNA recognition: zinc finger binding specificity. *Nucleic Acids Res.* **32**, 6673-6682.

156. O'Flanagan R, A., Paillard, G., Lavery, R. & Sengupta, A. M. (2005). Non-additivity in protein-DNA binding. *Bioinformatics* **21**, 2254-2263.

157. Travers, A. A. (1991). DNA bending and kinking - sequence dependence and function. *Curr. Opin. Struct. Biol.* **1**, 114-122.

158. Brukner, I., Sanchez, R., Suck, D. & Pongor, S. (1995). Trinucleotide models for DNA bending propensity: comparison of models based on DNaseI digestion and nucleosome packaging data. *J. Biomol. Struct. Dyn.* **13**, 309-317.

159. Vlahovicek, K. & Pongor, S. (2000). Model.it: building three dimensional DNA models from sequence data. *Bioinformatics* **16**, 1044-1045.

160. Packer, M. J., Dauncey, M. P. & Hunter, C. A. (2000). Sequence-dependent DNA structure: dinucleotide conformational maps. *J. Mol. Biol.* **295**, 71-83.

161. Packer, M. J., Dauncey, M. P. & Hunter, C. A. (2000). Sequence-dependent DNA structure: tetranucleotide conformational maps. *J. Mol. Biol.* **295**, 85-103.

162. Gardiner, E. J., Hunter, C. A., Packer, M. J., Palmer, D. S. & Willett, P. (2003). Sequence-dependent DNA structure: a database of octamer structural parameters. *J. Mol. Biol.* **332**, 1025-1035.

Chapter 8

MOLECULAR DYNAMICS SIMULATIONS AND FREE ENERGY CALCULATIONS ON PROTEIN-NUCLEIC ACID COMPLEXES

David L. Beveridge[1], Surjit B. Dixit[1], Bethany L. Kormos[1] Anne M. Baranger[1] and B. Jayaram[2]

[1]*Department of Chemistry and Molecular Biophysics Program, Wesleyan University, Middletown, CT 06459, USA;* [2]*Department of Chemistry and Supercomputing Facility for Bioinformatics and Computational Biology, Indian Institute of Technology, Hauz Khas, New Delhi 110016, India*

Abstract: We review herein the relevant literature on molecular dynamics simulations of protein-nucleic acid complexes and discuss how the procedure can be employed to gain insight into aspects of the molecular recognition mechanism such as induced fit and the role of solvent, which are not easily forthcoming in experimental studies. We present an overview of the free energy component analysis procedure that can be used in understanding the significance of various physical forces that participate in the complexation process and provide a critical analysis of their applications.

Key words: Protein-DNA, Protein-RNA, Induced Fit, Comformational Capture, Structural Adaptation

1.   INTRODUCTION

With the advent of the fast multiprocessor computers and PC clusters, the study of protein nucleic acid complexes including explicit consideration of solvent and mobile counter and co-ions has become accessible to molecular dynamics (MD) simulations. The results reported thus far are limited to only a few systems, but provide a clear indication of the potential of MD and corresponding theoretical calculations of binding affinities and specificities

211

J. Šponer and F. Lankaš (eds.), Computational Studies of RNA and DNA, 211–234.
© 2006 *Springer.*

to contribute new knowledge. In this article we provide some general perspectives and review the literature on MD studies and related calculations of the structural and thermodynamic aspects of protein nucleic acid binding and specificity. Results from studies recently carried out in this laboratory on the complexes of catabolite activator protein (CAP) DNA complex and the U1A protein RNA complex are presented as case studies, and a recent initiative to understand the chemical forces involved in the formation of some 40 different protein DNA complexes is described.

1.1   **Protein-Nucleic Acid Complexes**

The study of protein DNA recognition was pioneered by Zubay and Doty[1] with their insight on the complementarity of the protein α-helix and the major groove of DNA, the inferences made from early crystal structures[2-4]and the first delineation of digital readout via donor and acceptor H-bond sites by Seeman et al[5]. A great deal of insight into the details of recognition is available from structural studies of protein-nucleic acid complex cocrystals based on X-ray crystallography[6]. The number of protein-nucleic acid complexes in the structural databank is rapidly expanding with the current count (as of August 2005) at about 1300 protein-nucleic acid complexes (http://www.rcsb.org)[7]. Approximately 75% of those reported are protein-DNA systems and 25% are protein-RNA complexes. Solution state NMR studies are also proving to be useful to study these complexes but are limited to smaller systems. Only about 10% of the total protein-nucleic acid complex structures in the database have been studied by NMR based methods.

Analysis of these results has yielded a typology of binding motifs[8-11] and the idea that protein DNA recognition involves both a digital code (direct readout) and an analog code (indirect readout)[12]. Useful generalities about the nature of the protein DNA complexation are emerging from broad based surveys of the crystal structure data[13;14]. Early molecular modeling studies on protein DNA complexation are due to Zakrzewska and Pullman[15] based on molecular mechanics and this laboratory using MD[16]. These studies produced a first glimpse of structural adaptation phenomena at the molecular level[17], a subject that remains today an active area of research interest[18]. The functional energetics of protein DNA complexation has been analyzed by Jen-Jacobson[19;20] who makes the general observations that measured affinities are –12 to –15 kcal/mol for specific binding and –6 to –7 kcal/mol in the case of non-specific binding, and has identified the factors involved. An issue highly relevant to the subject of this article is that the relative contributions of the various chemical forces to complex formation are not directly accessible to experimental determination, and computational modeling via MD thus has a potentially unique vantage point on contributing to understanding the nature of the complexation process. One thing is absolutely clear: studies of a protein-nucleic acid complex *per se* are not sufficient to characterize complexation. One needs knowledge of the unbound protein and DNA as well, i.e. a full description of the initial and final states of complexation to study structural adaptation and the thermodynamic origins of specificity.

The literature on molecular modeling studies of protein DNA complexes up to 1999 was reviewed by Zakrzewska and Lavery[21]. They concluded that beyond the binding motifs, the small number of amino acid residues involved, and general knowledge of the chemical forces, the problem is one of complex combinations of subtle factors, some indirect, which do not necessarily fit the simple pairwise additive models which dominated early ideas on recognition, affinity and specificity. The structural aspects of protein-DNA recognition, its relationship to specificity and application in the development of knowledge based potential for target prediction has been reviewed recently by Sarai and Kono[22]. The current evidence speaks of a special role for dynamical adaptation and the participation of metastable states preorganized for binding[23;24], the latter referred to as the conformational capture hypothesis[25]. The role of solvent effects on these phenomena is important to clarify further. Most of the MD studies on protein DNA complexes in solution to date have been directed more at specific systems rather than general principles, which of course require a larger data base of computational cases. A recent study of an RNA protein complexation[26] has yielded some provocative ideas about induced fit, conformational capture and the idea that molecular substates may code for specificity. Obtaining reliable treatments of the functional energetics of protein DNA complexation is an outstanding issue of considerable consequence. Other review articles on both protein-DNA[27] and protein-RNA complexes[23; 25] will inform future MD studies.

1.2   Molecular Dynamics

The theoretical and computational background for studies of protein nucleic acid complexes is provided in a series of review articles from this laboratory in which the topics of molecular dynamics simulations of DNA[28-30], DNA hydration[28], the ion atmosphere of DNA[31], DNA bending[29], and protein DNA interactions[32] are treated. Independent perspectives on this research area have also been collected in review articles by other groups[33-37]. MD methodology has stabilized somewhat in recent years. A revision of the *AMBER* force field, parm.94, produced by Kollman and coworkers at

UCSF[38] and developed with consideration to including solvent explicitly has proved to give accurate MD models of DNA, RNA and hybrid forms acids in solution[29]. Earlier problems with the CHARMM force field in describing nucleic acids[39;40] have been remedied[41-43]. Current perspectives on MD force field for nucleic acids is provided in two recent reviews, one by Mackerell[33] and one by Cheatham and Young[44]. The problem of long range interactions is seemingly under control with the advent of the particle mesh Ewald (PME) method[45] for boundary conditions despite some lingering concerns about long range correlations[46;47]. Systems with solvent water, counterions and coions at experimental and even physiological ionic strengths can now be treated in MD[48]. The results of MD on DNA and RNA at this level are quite encouraging, with studies at least on selected oligonucleotides supporting reasonable forms of B-form DNA in water[49], A-form DNA in ethanol-water mixtures[50], and A/B conformational transitions[50;51]. The field of MD on DNA and RNA has been the subject of several recent review articles[30;34;36;52;53].

1.3 Molecular Dynamics of Protein-Nucleic Acid Complexes

All atom molecular dynamics simulation studies of protein DNA complexes using explicit solvent description which started to appear in the early 1990's[32;54;55] were mainly evaluatory and the subsequent development in simulation methodology with regard to the treatment of long range electrostatics and improved force fields yielded the means for carrying out well behaved simulations on protein-nucleic acid complexes in a regular manner. The early MD simulations using the Ewald method is exemplified by the work of Duan et al.[56] on the DNA EcoRI and DNA-EcoRV complexes which provided an estimate of the configurational and vibrational entropy change of DNA upon complexation.

Simulations of protein-DNA complex employing description of explicit solvent molecules have provided a great deal of insight into the role of water molecules in complex formation. Reddy et al.[57] have reported on the role of water molecules as an electrostatic buffer in protein-DNA recognition based on a comprehensive analysis of 109 protein-DNA cocrystal structures and MD simulations on 35 of these complexes. Schulten and coworkers[58] have carried out MD simulation of the Estrogen receptor protein both with its consensus and non-consensus DNA sequence in water. They conclude that the marked weakening of the hydrogen bonding network at the protein-DNA interface and the inability of the protein to expel fixed water molecules from the interface may reduce the affinity of the estrogen receptor for the nonspecific DNA sequence. Recently, proximity analysis of hydration pattern around cognate and non-cognate DNA sequences binding to the BamHI protein[59] based on Grand Canonical Monte Carlo simulations predicted a sequence dependent variation in the water distribution. It has been proposed that this variation could control the entropic contribution depending on the number of water molecules released and thus constitute a hydration finger print of the DNA sequence. MD simulations of protein-DNA complex with explicit water molecules reveal that on the basis of their small size and potential hydrogen-bonding sites, water molecules act as good intermediary agents at the interface and enhance interactions. In a MD simulation of the homeodomain-DNA complex[60] it is shown that the protein-DNA interface has a large number (~30) of water molecules mediating contacts and these are shown to be mobile. On the basis of this mobility of the water molecules, it has been suggested that their presence at the interface does not result in any entropic penalty during complexation. In the work of Gutmanas and Billeter[61], multiple MD trajectories of the wild type and mutant Antp Homeodomain protein bound to specific and nonspecific DNA sites are used to confirm the water network at the protein-DNA interface observed in X-ray and NMR structures and further predict the formation of a dry hydrophobic core in the case of the mutant complex. The MD simulations of HMG Domain of LEF-1 - DNA complex presented by Drumm et al.[62] shows that a water molecules is able to temporarily replace intercalating amino acid side chain and thus maintain the bent structure of the DNA.

MD of protein-DNA complex finds application in the refinement of modeled or docked complex structure such as in the work of Balaeff et al.[63] on the chromosomal HMG domain protein complexed to DNA. Tsui et al.[64] have used MD simulation of the Transcription factor IIIA Zinc Finger protein –15 basepair DNA duplex complex in explicit water to evaluate the potential hydration sites observed in NOE data. Strong agreement between the MD and NOE data has been shown. Marco et al.[65] have employed MD of model built complex of Transcription factor Sp1 and a 14 mer DNA to study this zinc finger system where no experimentally derived structure is available. The simulations are employed to determine putative DNA binding sites for the protein. Recently, Amara et al.[66] have described the role of a dynamic and functionally important loop in the Formamidopyrimidine-DNA glycosylase (Fpg) enzyme that is not present in the cocrystal structure with DNA. MD simulations were employed in this study to describe how the presence of a damaged base alters dynamics of this system. In the work of Gorfe et al. on the Tn916 Integrase protein-DNA complex[67], MD of the free and DNA bound conformations of the protein and the complex was used to identify interactions that are poorly resolved in the NMR structures of the complex and establish a qualitative link between experimental binding data and the contribution of individual residues.

Simulations of the wild type and mutant forms of protein-DNA complex have frequently been employed to gain insight into the dynamic structure function relationships in the complex. Tang and Nilsson[68; 69] have carried out MD simulations of the Human SRY protein HMG domain complexed to DNA and the unbound forms of the protein and DNA using the Charmm 22 force field and report on how the G40R mutation in the hSRY protein could impact DNA binding. Stockner et al.[70] have carried out MD study of the Glucocorticoid receptor DNA binding domain (GR DBD) to show that two different single point mutations in the GR DBD have an allosteric effect that ultimately induces conformational change affecting DNA binding and the dimerization interface. In a series of articles, Schlick and coworkers[71;72] have reported MD simulations of DNA polymerase β with cognate and mismatched DNA sequences to decipher how the presence of mismatched DNA leads to a difference in the structural and dynamic properties of the complex with pol β and lead to the proof reading mechanism. MD simulations of various mutants of the DNA polymerase pol β shows the sensitivity of this enzyme to localized changes and the role of compensatory changes in the local environment to preserve their function.

The indirect mode of protein-DNA recognition where in the sequence directed structural properties of the DNA constitutes the important recognition factor has been analyzed using MD simulations. Recently, Hartmann and coworkers[73] have employed MD simulations to show that the structural fluctuations of the OR1 and OR2 site are altered on Phage 434 cI Repressor binding and these differences are important for the preferred binding to the OR1 site. The complex of the ETS protein with a high affinity GGAA sequence and another low affinity GGAG sequence core has been studied by Reddy et al.[74]. They note differences in the propensity of the DNA structure to occupy B_I or B_{II} substates and other conformational differences as factors governing recognition. In the case of Bovine Papillomavirus E2 Protein-DNA which has been one of the prototypical case known to be governed by the indirect mode of recognition[75], recent report of MD simulation by Djuranovic and Hartmann[76] indicate that protein binding has little impact on the intrinsic B_I/B_{II} substate and other conformational preferences of the DNA sequence. Instead, the E2 protein takes advantage of the predistorted forms of the free DNA during recognition. Weinstein and coworkers in a series of articles on MD simulations of the TATA box protein –DNA complex[77;78] explain how differences in the sequence dependent free energy cost of DNA conformational distortion of AA and TA step in the TATA box region plays an indirect role in recognition.

Induced fit effects on the formation of Even-skipped Homeodomain protein – DNA complex have been studied using MD simulations[79]. The induced fit mechanism in this system involves first conformational selection of the backbone substates followed by slower occurrence of other effects like bending of the helix axis, changes in the minor groove width and other helical parameters. MD of the human topoisomerase I-DNA complex has been used to ascertain the dynamic behavior of different regions in the enzyme[80]. Cross correlation maps are employed to describe the information flux between different regions of the complex. Broyde and coworkers[81] have employed MD simulations to show the characteristics of the active site of the Y-family DNA Polymerase – Carcinogen - DNA complex that leads to the promiscuous nature of this polymerase action.

In an interesting application of MD, the BamHI endonuclease - DNA complex has been simulated at elevated pressure and used to provide atomic level description of the hydrostatic effects in experimental high-pressure gel shift analysis studies[82]. In another interesting application, the Schulten group [83] has employed multiscale simulation involving all atom description of the lac repressor and two operator DNA segments connected by a longer DNA segment defined using an analytical elastic loop model. The simulations are used to study the influence of the strained loop on the structure and dynamics of the repressor-DNA complex. Huang and MacKerell[84] have carried out potential of mean force calculations based on MD simulations to derive an atomistic view of the base flipping mechanism in free DNA and DNA bound to the (cytosine-C5)-methyltransferase from *Hha*I. A role for the protein in facilitating the base flipping mechanism has been proposed on the basis of these simulations. MD and free energy component analysis procedure are employed to explain the cooperative interactions in the ternary complex formed between the Runt Domain DNA binding protein (RD) in complex with DNA and the core binding factor β (CBFβ)[85]. In a demanding test of the simulation method employed, Bishop[86] has attempted MD of the nucleosome complex structure and the DNA free in solution and analyzed the differences in the structure and flexibility of the DNA. While MD on short DNA sequences (within about 30 base pairs long) corresponding to the length of most regulatory sites have been well characterized, simulations of such long DNA sequences, 146 base pairs in the nucleosome structure, present new challenges.

Several projects in the area of nucleic acid protein complexes have been carried out in this laboratory. Thayer and Beveridge[87] have incorporated Boltzmann probability models of DNA sequence dependent structural information obtained from MD simulations into Hidden Markov Models and used this as a bioinformatics tool to recognize structure and sequence signal of protein binding sites in genome. As a prototypical case, the method was employed to mine the *E. coli* genome for CAP binding sites. We have addressed questions of structural adaptation in the λ repressor operator complex[88-92], U1A-RNA[26;93] and CAP-DNA[94;95], considering both bound and unbound states. Detailed free energy component analysis have been pursued on Eco RI protein DNA complex[96], λ repressor operator complex[97] and the protein DNA complex from E2 Papilloma Virus[98;99]. Our projects so far on protein RNA complexes are motivated by questions posed to us from experimental binding energy studies carried out by Baranger et al[100], and have addressed issues such as the alteration of dynamical structure in the binding region of the protein-RNA complex on mutation of key amino acid residues[93]. These studies illustrate many provocative issues about the nature of nucleic acid protein complexation which are uniquely accessible to MD simulation and related free energy calculations.

2.   CASE STUDIES

2.1   Catabolite Activator Protein (CAP)-DNA Complex

2.1.1   Structural Studies

One of the most extreme cases of DNA bending was observed in the catabolite activator protein (CAP) –DNA complex cocrystal[101], in which the DNA "makes a right turn" (Figure 8-1). Subsequent structures of the CAP DNA complex and variants thereof[102-104] have shown a range of DNA bending angles from 51° to 107°. CAP-DNA complexes such as 1CGP and 1J59 (PDB ID) are obtained in an orthorhombic unit cell and the DNA bend is about 90°, effected by large positive roll and twist at TpG steps in the conserved region. The complex structures solved with a 46 base pair DNA sequence of which 26 base pairs are resolved in a trigonal space group crystal structure (e.g. 2CGP), exhibits a smaller bend of ~60° highlighting the fact that crystal packing effects could play a role in the observed curvature of DNA. To determine how much of this is intrinsic curvature and how much is ligand induced bending requires structures of the complexed and uncomplexed forms of the DNA in solution. The structure of the uncomplexed CAP binding site has been beyond the reach of crystallography or NMR spectroscopy.

We have used all atom models of the dynamical structures of protein-DNA complexes and corresponding uncomplexed protein and DNA to compare the intrinsic and induced effects in the DNA structure and study the structural adaptation process in CAP-DNA complexes[94;95]. In the time scale of the simulations, the DNA bending angle is reasonably well stabilized in all the systems. MD simulations predict that the DNA bending on complexation with CAP is 65° to 80° in the solution state and shows significant salt concentration dependence. This is in good accord with the solution state results reported by Lutter et al.[105] and Ebright and coworkers[106] on the CAP-DNA complex. MD simulation of the 2CGP structure tends to maintain a structural form with ~55° bend. The unbound form presents a bend of about 30°. Phosphate charge neutralization studies by Maher and coworkers[107] report a 26° curvature for the unbound form of the CAP binding site. Considering the structure of the CAP bound DNA relative to a canonical B reference state with an essentially straight helix axis, the bend in protein bound form is ~40% due to intrinsic curvature in the cognate DNA and ~60% due to protein-induced bending. Another interesting question that can be addressed on the basis of the MD model pertains to the direction of the intrinsic curvature and that in the complex. The helical axis of the DNA in bound form clearly exhibits curvature originating near the two TpG kink points, similar to that observed in the original crystal structure. Interestingly, MD simulation of the unbound form of DNA too shows significant bending at the TpG positions implying preorganization of the structure for the ligand induced bending by CAP.

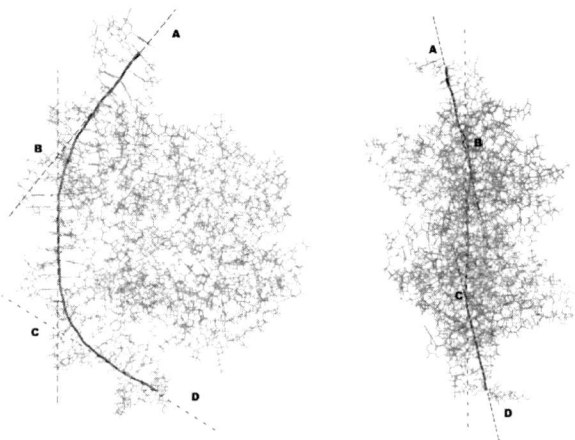

Figure 8-1. The original crystallographic structure of the CAP-DNA complex (1CGP) with the helix axis highlighted with a dark solid line. The vectors AB and CD which present the direction of the DNA at the two ends intersect at almost right angle. A ~35° out of plane bend in the helical axis is presented by the dihedral angle between the sections ABC and BCD.

The DNA axis in the co-crystal structure with CAP[101] is not planar but can be viewed as an effective dihedral angle close to 35° due to out of plane bending between the two ends of the DNA as shown in Figure 8-1. Simulations of the free DNA in solution exhibit large flopping motions of this axial dihedral angle averaging about 0° with the fluctuations ranging from +80 to −80°. On the other hand, the DNA in the complex exhibits significant localization of dihedral twist with the average about 38° in quantitative agreement with the original co-crystal structure. The observed fluctuations in the bound form are in the range of 80 to 0°, significantly reduced in comparison to the trajectory of the unbound DNA. Since the free DNA relaxes to a significantly different structural form in MD beginning with the DNA structure in the complex, this indicates that the structure of the DNA in the complex is an unstable, energetically strained form and not a stable or metastable substate. However, the dynamic range of the free DNA at its outer limit encompasses the structure of the protein bound form, indicating that the conformational capture phenomena[108] in which the protein selects out and stabilizes complex adapted conformable structures of the nucleic acid may be involved.

2.1.2   Thermodynamic Studies

The contribution of structural adaptation has both enthalpic and entropic effects to binding thermodynamics and is typically difficult if not impossible to isolate directly in an experiment. The importance of this contribution has been inferred in an analytical manner based on experimental data[109] and from theoretical estimates[56]. The entropy contribution to protein-DNA complexation originating in the changes in vibrational and configurational degrees of freedom, key components in the thermodynamics of binding, can be approximated to estimates of the "quasi-harmonic" term calculated from the MD trajectories[110]. We have carried out comparison of the catabolite activator protein (CAP) DNA complex in which the DNA sequence is highly bent[101] and the λ repressor/operator[111] in which the DNA remains relatively straight to assess the relation between structure, dynamics and thermodynamics. The balance between enthalpy and entropy leads these complexes to form with similar binding free energy changes, although the binding of CAP to cognate DNA is entropy controlled while the λ repressor/operator binding is enthalpy controlled[18]. An explanation proposed for this result is that CAP requires additional enthalpy for bending the DNA to arrive at a viable binding free energy for a regulatory process, CAP DNA binds with increased flexibility and entropy increase. However, the origin of the increased flexibility on CAP DNA complexation is not obvious, since there is typically a rigidification of structure on complex formation.

MD trajectories of the protein-DNA complexes and the corresponding unbound forms of the protein and DNA for CAP-DNA and λ repressor operator complexes form the basis of this study. This is presented in Figure 8-2 as an overlay of structures after least square fit. The dispersion of structures in the various images reflects the range of the dynamics observed in the simulations of the protein, DNA and complex in solution. Visual examination of the structures of the unbound and bound form leads to some ideas about the dynamics of the system. In both cases there appears to be a net decrease in the ordering of the DNA on binding. With respect to the protein, the CAP appears more disordered in the complex whereas the λ repressor protein appears to be more ordered in the complex. These ideas derived from inspection of the figure are verified by our calculations.

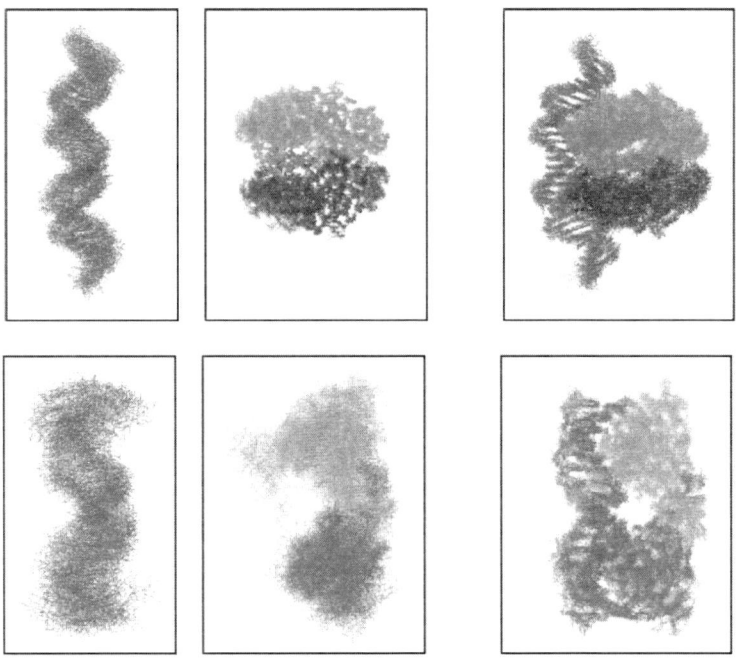

Figure 8-2. Overlay of superimposed structures from the MD simulations of the CAP-DNA (top) and λ repressor protein –OL1 operator DNA (bottom) systems. The unbound form of the DNA (left), the protein (middle) and the complex (right) are shown for both the systems.

The quasiharmonic entropy calculations allow us to quantify the extent to which these structural changes in the MD models of both CAP and λ systems translate into the quantitative thermodynamics. The quasiharmonic (QH) contribution is obtained from the determinant of the cross-correlation matrix of atomic motions for a molecule[110; 112]. The calculated differential entropy change, $\Delta\Delta S_{QH}$ (λ->CAP) = +609 J/molK, is consistent with the idea that the CAP complexation is more disordered than for λ. Simulations permit us to estimate the relative contribution of protein and DNA to the quasiharmonic entropy and investigate the nature of the difference between ΔS_{QH} for CAP and λ complexation. The $\Delta\Delta S_{QH}(D \rightarrow D^*)$ i.e. the differential entropy in DNA with regard to the CAP and λ system, and $\Delta\Delta S_{QH}(P^*+D^* \rightarrow P^*D^*)$, the effect of the correlated motions between the protein-DNA are both negative, and oppose in sign the overall entropy difference $\Delta\Delta S_{QH}(\lambda \rightarrow CAP)$. The key factor that favors is the increase in relative ordering of the protein on complexation, a negative ΔS_{QH} ($P \rightarrow P^*$) of λ exceeding that of CAP. Thus we find from the MD that CAP complexation is indeed a relative less ordering process than that of λ, but the nature of this differential entropy change lies predominantly in the contribution from protein rather than cognate DNA, and originates in the increased rigidification of the N-terminal arm region of the λ repressor on complexation. The large change in structure of the DNA induced by the protein does not translate into a major thermodynamic contribution, which issues a caveat with respect to intuiting thermodynamic behavior from observed structures.

2.2   U1A RNA Complex

2.2.1   Structural Studies

U1A binds to RNA using an RNA binding domain (RBD) (also called the ribonucleoprotein (RNP) domain or the RNA recognition motif (RRM)), one of the most common eukaryotic RNA binding motifs[113]. Extensive biochemical and structural experiments have probed RNA recognition by the N-terminal RRM of U1A, making U1A a paradigm for RNA recognition by single RRMs. The structural data available for both free and bound forms have revealed structural adaptation of one or both components upon binding. The complex of the N-terminal RRM of U1A with stem loop II (SL2) of U1 snRNA has been solved by X-ray crystallography (Figure 8-3).[114] The structure of the N-terminal RRM of U1A has been solved by NMR spectroscopy[115] and this study revealed that the α_C domain assumes a significantly different orientation in solution than that in the crystal structure of the complex. The structure of the complex shows α_C oriented perpendicular to the β sheet, whereas α_C in the NMR structure lies across the RNA binding motif. In this position α_C would obstruct the approach of the RNA.

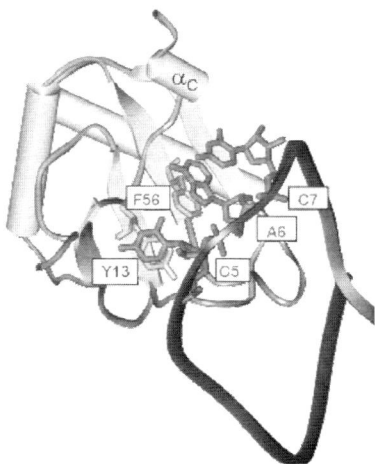

Figure 8-3. Crystal structure of the U1A-RNA complex.[114] SL2 RNA is represented as a red ribbon. Conserved aromatic amino acids, Y13 and F56 (blue sticks) stack with SL2 RNA nucleotides C5, A6 and C7 (red sticks). The C-terminal helix is in the open conformation.

Results of MD simulations from our lab[26;93] on U1A-RNA in both free and bound forms have led us to explore the question of substates coding for specificity. The U1A protein appears to exhibit substate behavior, one of which is preorganized for complex formation. Results on the free energy of U1A-RNA indicate the open form of α_C in the free U1A is a non-native conformational substate that is preorganized for RNA binding. In contrast, structural adaptation of the RNA component involves a conformation unstable in the free RNA. The RNA shows a single bound state, with the bound form unstable in the absence of protein. However, the thermal fluctuations in the structures of the unbound form in the loop region are such as to carry the bases into structures incipient of the bound form. These results support a conformational capture hypothesis for the U1A protein and an induced fit hypothesis for the SL2 RNA, and suggest aspects of protein-nucleic acid complex formation which are uniquely accessible to MD and related theoretical studies.

2.2.2   Dynamical Studies

Preliminary results[116] from our lab indicate dynamical cross correlation maps (DCCM) can be used as a means of examining covariance as well as the nature and extent of cooperative effects within the structures of bound and unbound forms of protein and RNA and gives a vantage point on explaining why long-range correlations occur. A DCCM is a graphical representation of the dynamical structure of a system obtained from the calculation of all interatomic cross correlations of atomic fluctuations from an MD simulation.[117;118] The DCCM for the U1A-RNA complex is shown in Figure 8-4. The DCCM itself indicates a striking amount of correlated motions throughout the RRM of U1A and between the protein and RNA. Additionally, the pink boxes indicate intersections of the residues and nucleotides that have been implicated in the cooperative network of interactions for binding in U1A-RNA[117;118]. Highly correlated motions are found in a significant amount of the boxes, indicating strong motional correlations between residues and nucleotides implicated in the cooperative network. This result is indicative of the potential for the use of DCCMs in mapping out as well as predicting cooperativity in protein-nucleic acid systems.

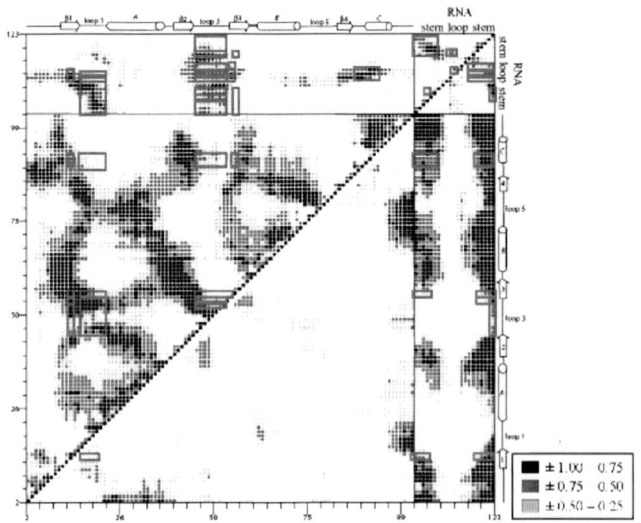

Figure 8-4. Calculated dynamical cross-correlation map of positional fluctuations of amino acid residue and nucleotide motions in the U1A-RNA complex based on a 3 ns MD trajectory. Positive correlations are collected in upper triangle of the plot, negative correlations in lower triangle. Cooperative networks of residues can be identified from certain off-diagonal through space correlations on this plot, indicated by pink boxes.

In support of the utility of DCCMs, we have compared our results to those of Crowder et al.[119] who performed a covariance analysis of 330 proteins containing RRMs and identified a network of covariant amino acid residues present within those RRMs. This result is illustrated in Figure 8-5. Figure 8-5a is the covariance result from Crowder et al.[119] and Figure 8-5b illustrates the DCCM correlations superimposed over panel a. Though not all of these covariant residues are present specifically in U1A, those that are present reveal strongly correlated motions from the U1A-RNA DCCM. The majority of the residues shown to covary exhibit motional correlations greater than 50% in the MD simulation. This result again indicates the significant amount of information contained in DCCMs and suggests the potential for making predictions based on this information. Other recent studies that illustrate the viability of this method for such applications are due to Jeruzalmi et al.[120] and Estabrook et al.[121].

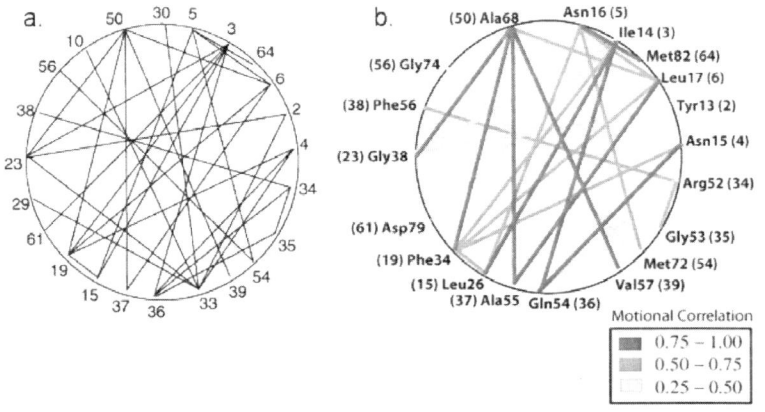

Figure 8-5. a. Covariance pairings greater than 3.5σ calculated by Crowder et al.[161] from the analysis of 330 proteins containing RRMs. b. Superposition of motional correlations from DCCM of the U1A-RNA complex onto figure from panel a. Red lines indicate the strongest motional correlations between residues and yellow indicates weaker correlations. Residues are numbered according to U1A residues, and the Crowder et al. numbering is in parentheses beside each residue.

3. FREE ENERGY CALCULATIONS

Free energy as the thermodynamic index of chemical equilibrium and spontaneous change is a matter of concern in all applications involving macromolecular stability and ligand binding. However, for complex molecular systems, it is also one of the most challenging of quantities to calculate accurately and to interpret properly at the molecular level. A formal treatment of the statistical mechanics of binding affinities has been provided by Gilson, McCammon and coworkers[122;123]. Free energy can be calculated from MD simulations using either thermodynamic integration or the perturbation method[124], but each of these involves a windowing process, a sequence of lengthy simulations just to obtain one number. The interpretation of the results in terms of chemical forces is always problematic due to the entropy decomposition problem[96;125]. In response, a method for estimating binding free energies has been introduced over the last few years. This is known by various names - the master equation approach[126], predominant state studies[127], free energy component analysis (our name for it to date), or, referring to a specific all atom implementations of this general approach, MMPBSA (Molecular Mechanics - Poisson Boltzmann - Solvent Accessibility)[128] and the variant MMGBSA in which the Generalized Born approximation[96;129;130] to the electrostatics of solvation is employed. A recent critical review is due to Aqvist et al.[131]. Specific applications to ligand binding problems have likewise been surveyed[132].

An interesting variation on the theme which has the potential of making component analysis more viable is the linear response formulation introduced by Aqvist et al.[133] and used extensively in Monte Carlo studies of ligand binding and pharmaceutical design[131; 134]. This essentially converts the difficult problem of calculating absolute binding affinities to relying on computational models for relative magnitudes of various terms. This is also not a panacea since it often requires a "free" parameter to obtain the best result. Given the complexity of the free energy problem even qualitatively correct answers may be quite useful in understanding otherwise un interpretable results and are thus useful computational tools.

The spate of structural information on protein—DNA complexes sets the stage for theoretical investigations on the molecular thermodynamics of binding aimed at identifying forces responsible for specific protein-DNA recognition. Computation of absolute binding free energies for systems of this complexity transiting from structural information is a stupendous task. Adopting some recent progresses in treating atomic level interactions in proteins and nucleic acids including solvent and salt effects, Jayaram et al.[96; 129] constructed an energy component methodology and developed a computational first atlas of the free energy contributors to binding in ~40 protein–DNA complexes representing a variety of structural motifs and functions.

The results vividly illustrate the compensatory nature of the free energy components contributing to the energetics of recognition for attaining optimal binding, and highlight unambiguously the roles played by packing, electrostatics including hydrogen bonds, ion and water release (cavitation) in protein–DNA binding. Cavitation and van der Waals contributions without exception favor complexation. The electrostatics is marginally unfavorable in a consensus view. Basic residues on the protein contribute favorably to binding despite the desolvation expense. The electrostatics arising from the acidic and neutral residues proves unfavorable to binding. An enveloping mode of binding to short stretches of DNA makes for a strong unfavorable net electrostatics but a highly favorable van der Waals and cavitation contribution. Thus, noncovalent protein–DNA association is a system-specific fine balancing act of these diverse competing forces. Zakrzewska[130] has performed a complementary analysis on the deformation energy in 71 different protein DNA complexes on the basis of internal coordinate molecular mechanics calculations and reports on the significance of backbone distortion of the DNA which contributes to about 60% of the net deformation energy.

4.   SUMMARY

MD simulations are increasingly contributing to an improved understanding of the relationship between structure, function and thermodynamics of protein-nucleic acid complexes. Subtle but significant questions about protein-nucleic acid recognition can be addressed with a combination of experimental data and computational modeling. Advances in free energy simulation methods permit detailed calculation of the binding energy in such macromolecular complexes.

ACKNOWLEDGEMENTS

DLB was supported by NIH grant GM37909 and supercomputing resource allocations from NCSA. BLK was supported by a NIH Post-Doctoral Fellowship F32 GM072345. AMB was support by NIH grant GM 56857. BJ was supported by a grant from DST.

REFERENCES

1.   Zubay, G. & Doty, P. (1959). *J. Mol. Biol* **1**, 1.
2.   Warrant, R. W. & Kim, S.-H. (1978). *Nature* **271**, 130.
3.   Mirzabekov, A. D. & Rich, A. (1979). Asymmetric lateral distribution of unshielded phosphate groups in nucleosomal DNA and its role in DNA bending. *Proc Natl Acad Sci U S A* **76**, 1118-21.
4.   Carter, C. W., Jr. & Kraut, J. (1974). A proposed model for interaction of polypeptides with RNA. *Proc Natl Acad Sci U S A* **71**, 283-7.
5.   Seeman, N. C., Rosenberg, J. M. & Rich, A. (1976). Sequence-specific recognition of double helical nucleic acids by proteins. *Proc. Natl. Acad. Sci. U. S. A* **73**, 804-8.
6.   Steitz, T. A. (1990). Structural studies of protein-nucleic acid interaction: the sources of sequence-specific binding. *Q Rev Biophys* **23**, 205-80.
7.   Berman, H. M., Battistuz, T., Bhat, T. N., Bluhm, W. F., Bourne, P. E., Burkhardt, K., Feng, Z., Gilliland, G. L., Iype, L., Jain, S., Fagan, P., Marvin, J., Padilla, D.,

Ravichandran, V., Schneider, B., Thanki, N., Weissig, H., Westbrook, J. D. & Zardecki, C. (2002). The Protein Data Bank. *Acta Crystallogr D Biol Crystallogr* **58**, 899-907.

8. Harrison, S. C. (1991). A structural taxonomy of DNA-binding domains. *Nature* **353**, 715-9.

9. Pabo, C. O. & Sauer, R. T. (1992). Transcription factors: structural families and principles of DNA recognition. *Annu Rev Biochem* **61**, 1053-95.

10. Jones, S., Daley, D. T., Luscombe, N. M., Berman, H. M. & Thornton, J. M. (2001). Protein-RNA interactions: a structural analysis. *Nucleic Acids Res* **29**, 943-54.

11. Luscombe, N. M., Austin, S. E., Berman, H. M. & Thornton, J. M. (2000). An overview of the structures of protein-DNA complexes. *Genome Biol* **1**, REVIEWS001.

12. Dickerson, R. E. (1983). The DNA Helix and How It Is Read. *Sci. Am.* **249**, 94-111.

13. Nadassy, K., Wodak, S. J. & Janin, J. (1999). Structural features of protein-nucleic acid recognition sites. *Biochemistry* **38**, 1999-2017.

14. Jones, S., van Heyningen, P., Berman, H. M. & Thornton, J. M. (1999). Protein-DNA interactions: A structural analysis. *J Mol Biol* **287**, 877-96.

15. Zakrzewska, K., Lavery, R. & Pullman, B. (1986). Theoretical studies on the interaction of proteins and nucleic acid. II. The binding of alpha-helix to B-DNA. *Biophys Chem* **25**, 201-13.

16. DiCapua, F. (1991). Molecular Dynamics and Monte Carlo Studies of Protein Stability and Protein-DNA Interactions. Ph. D. Thesis, Wesleyan University.

17. Kumar, S., Duan, Y., Kollman, P. A. & Rosenberg, J. M. (1994). Molecular dynamics simulations suggest that the Eco RI kink is an example of molecular strain. *J Biomol Struct Dyn* **12**, 487-525.

18. Jen-Jacobson, L., Engler, L. E. & Jacobson, L. A. (2000). Structural and thermodynamic strategies for site-specific DNA binding proteins. *Structure* **8**, 1015-23.

19. Jen-Jacobson, L. (1995). Structural-perturbation approaches to thermodynamics of site-specific protein-DNA interactions. *Methods Enzymol* **259**, 305-44.

20. Jen-Jacobson, L. (1997). Protein-DNA recognition complexes: conservation of structure and binding energy in the transition state. *Biopolymers* **44**, 153-80.

21. Zakrzewska, K. & Lavery, R. (1999). Modelling Protein-DNA Interactions. *Theoretical Computational Chemistry* **8**, 441-83.

22. Sarai, A. & Kono, H. (2005). Protein-DNA recognition patterns and predictions. *Annu Rev Biophys Biomol Struct* **34**, 379-98.

23. Perez-Canadillas, J. M. & Varani, G. (2001). Recent advances in RNA-protein recognition. *Curr Opin Struct Biol* **11**, 53-8.

24. Bosshard, H. R. (2001). Molecular recognition by induced fit: how fit is the concept? *News Physiol Sci* **16**, 171-3.

25. Williamson, J. R. (2000). Induced fit in RNA-protein recognition. *Nat Struct Biol* **7**, 834-7.

26. Pitici, F., Beveridge, D. L. & Baranger, A. M. (2002). Molecular dynamics simulation studies of induced fit and conformational capture in U1A-RNA binding: Do molecular substates code for specificity? *Biopolymers* **65**, 424-35.

27. Hard, T. (1999). NMR studies of protein-nucleic acid complexes: structures, solvation, dynamics and coupled protein folding. *Q Rev Biophys* **32**, 57-98.

28. Beveridge, D. L., Swaminathan, S., Ravishanker, G., Withka, J. M., Srinivasan, J., Prevost, C., Louise-May, S., Langley, D. R., DiCapua, F. M. & Bolton, P. H. (1993). Molecular Dynamics Simulations on the Hydration, Structure and Motions of DNA Oligomers. In *Water and Biological Molecules* (Westhof, E., ed.), pp. 165-225. The Macmillan Press, Ltd., London.

29. Beveridge, D. L., Young, M. A. & Sprous, D. (1998). Modeling of DNA via Molecular Dynamics Simulation: Structure, Bending, and Conformational Transitions. In *Molecular Modeling of Nucleic Acids* 682 edit. (Leontis, N. B. & Santa Lucia, J., J., eds.), Vol. 1, pp. 260-84. American Chemical Society, Washington, D.C.

30. Beveridge, D. L. & McConnell, K. J. (2000). Nucleic acids: theory and computer simulation, Y2K. *Curr Opin Struct Biol* **10**, 182-96.

31. Jayaram, B. & Beveridge, D. L. (1996). Modeling DNA in Aqueous Solution: Theoretical And Computer Simulation Studies on the Ion Atmosphere of DNA. *Ann Rev Biophys Biomol Struct* **25**, 367-94.

32. Beveridge, D. L., McConnell, K. J., Nirmala, R., Young, M. A., Vijayakumar, S. & Ravishanker, G. (1994). Molecular Dynamics Simulations of DNA and Protein-DNA Complexes Including Solvent: Recent Progress. *ACS symposium series* **568**, 381-94.

33. MacKerell, A. D., Jr., Banavali, N. & Foloppe, N. (2000). Development and current status of the CHARMM force field for nucleic acids. *Biopolymers* **56**, 257-65.

34. Giudice, E. & Lavery, R. (2002). Simulations of nucleic acids and their complexes. *Acc Chem Res* **35**, 350-7.

35. Norberg, J. & Nilsson, L. (2002). Molecular dynamics applied to nucleic acids. *Acc Chem Res* **35**, 465-72.

36. Orozco, M., Perez, A., Noy, A. & Luque, F. J. (2003). Theoretical Methods for the Simulation of Nucleic Acids. *Chem Soc Rev* **32**, 350-64.

37. Cheatham, T. E., 3rd. (2004). Simulation and modeling of nucleic acid structure, dynamics and interactions. *Curr Opin Struct Biol* **14**, 360-7.

38. Cornell, W. D., Cieplak, P., Bayly, C. I., Gould, I. R., Merz, K. M., Ferguson, D. M., Spellmeyer, D. C., Fox, T., Caldwell, J. W. & Kollman, P. A. (1995). A second generation force field for the simulation of proteins, nucleid acids and organic molecules. *J Am Chem Soc* **117**, 5179-97.

39. Mackerell Jr., A. D., Wiorkiewicz-Kuizera, T. & Karplus, M. (1995). An All-atom Empirical Energy Function for the Simulation of Nucleic Acids. *J Am Chem Soc* **117**, 11946-75.

40. Feig, M. & Pettitt, B. M. (1997). Experiment vs force fields: DNA Conformation From Molecular Dynamics Simulations. *J Phys Chem* **101**, 7361-3.

41. MacKerell, A. D., Jr. & Banavali, N. (2000). All-atom empirical force field for nucleic acids: II. Application to molecular dynamics simulations of DNA and RNA in solution. *J Comput Chem* **21**, 105-20.

42. Foloppe, N. & MacKerell, J., A. D. (2000). All-atom empirical force field for nucleic acids: I. Parameter optimization based on small molecule and condensed phase macromolecular target data. *J Comput Chem* **21**, 86-104.

43. Langley, D. R. (1998). Molecular dynamic simulations of environment and sequence dependent DNA conformations: the development of the BMS nucleic acid force field and comparison with experimental results. *J Biomol Struct Dyn* **16**, 487-509.

44. Cheatham, T. E., 3rd & Young, M. A. (2001). Molecular dynamics simulation of nucleic acids: Successes, limitations, and promise. *Biopolymers* **56**, 232-56.

45. Darden, T., Perera, L., Li, L. & Pedersen, L. (1999). New tricks for modelers from the crystallography toolkit: the particle mesh Ewald algorithm and its use in nucleic acid simulations. *Structure Fold Des* **7**, R55-60.

46. Smith, P. E. & Pettitt, B. M. (1996). Ewald Artifacts in Liquid State Molecular Dynamics Simulations. *J Chem Phys* **105**, 4289.

47. Hunenberger, P. H. & McCammon, J. A. (1999). Effect of artificial periodicity in simulations of biomolecules under Ewald boundary conditions: a continuum electrostatics study. *Biophys Chem* **78**, 69-88.

48. Young, M. A. & Beveridge, D. L. (1998). Molecular Dynamics Simulations of an Oligonucleotide Duplex with Adenine Tracts Phased by a Full Helix Turn. *J Mol Biol* **281**, 675-87.

49. Young, M. A., Ravishanker, G. & Beveridge, D. L. (1997). A 5-Nanosecond Molecular Dynamics Trajectory for B-DNA: Analysis of Structure, Motions and Solvation. *Biophys J* **73**, 2313-36.

50. Sprous, D., Young, M. A. & Beveridge, D. L. (1998). Molecular Dynamics Studies of the Conformational Preferences of a DNA Double Helix in Water and in an Ethanol/Water Mixture: Theoretical Considerations of the A/B Transition. *J Phys Chem* **102**, 4658-67.

51. Jayaram, B., Sprous, D., Young, M. A. & Beveridge, D. L. (1998). Free Energy Analysis of the Conformational Preferences of A and B forms of DNA in Solution. *J Amer Chem Soc* **120**, 10629-33.

52. Cheatham, T. E., 3rd & Kollman, P. A. (2000). Molecular dynamics simulation of nucleic acids. *Annu Rev Phys Chem* **51**, 435-71.

53. Soares, T. A., Hunenberger, P. H., Kastenholz, M. A., Krautler, V., Lenz, T., Lins, R. D., Oostenbrink, C. & van Gunsteren, W. F. (2005). An improved nucleic acid parameter set for the GROMOS force field. *J Comput Chem* **26**, 725-37.

54. Harris, L. F., Sullivan, M. R., Popken-Harris, P. D. & Hickok, D. F. (1994). Molecular dynamics simulations in solvent of the glucocorticoid receptor protein in complex with a glucocorticoid response element DNA sequence. *J Biomol Struct Dyn* **12**, 249-70.

55. Eriksson, M. A., Hard, T. & Nilsson, L. (1995). Molecular dynamics simulations of the glucocorticoid receptor DNA-binding domain in complex with DNA and free in solution. *Biophys J* **68**, 402-26.

56. Duan, Y., Wilkosz, P. & Rosenberg, J. M. (1996). Dynamic contributions to the DNA binding entropy of the EcoRI and EcoRV restriction endonucleases. *J. Mol. Biol.* **264**, 546-55.

57. Reddy, C. K., Das, A. & Jayaram, B. (2001). Do water molecules mediate protein-DNA recognition? *J Mol Biol* **314**, 619-32.

58. Kosztin, D., Bishop, T. C., Schulten, K., Beckman Institute, D. o. C. & Physics, U. o. I. a. U.-C. U. I. L. U. S. A. (1997). Binding of the estrogen receptor to DNA. The role of waters. *Biophys J* **73**, 557-70.

59. Fuxreiter, M., Mezei, M., Simon, I. & Osman, R. (2005). Interfacial water as a "hydration fingerprint" in the noncognate complex of BamHI. *Biophys J* **89**, 903-11.

60. Duan, J. & Nilsson, L. (2002). The role of residue 50 and hydration water molecules in homeodomain DNA recognition. *Eur Biophys J* **31**, 306-16.

61. Gutmanas, A. & Billeter, M. (2004). Specific DNA recognition by the Antp homeodomain: MD simulations of specific and nonspecific complexes. *Proteins* **57**, 772-82.

62. Drumm, M., Teletchea, S. & Kozelka, J. (2005). Recognition complex between the HMG domain of LEF-1 and its cognate DNA studied by molecular dynamics simulations with explicit solvation. *J Biomol Struct Dyn* **23**, 1-11.

63. Balaeff, A., Churchill, M. E. & Schulten, K. (1998). Structure prediction of a complex between the chromosomal protein HMG-D and DNA. *Proteins* **30**, 113-35.

64. Tsui, V., Radhakrishnan, I., Wright, P. E. & Case, D. A. (2000). NMR and molecular dynamics studies of the hydration of a zinc finger- DNA complex. *J Mol Biol* **302**, 1101-17.

65. Marco, E., Garcia-Nieto, R. & Gago, F. (2003). Assessment by molecular dynamics simulations of the structural determinants of DNA-binding specificity for transcription factor Sp1. *J Mol Biol* **328**, 9-32.

66. Amara, P., Serre, L., Castaing, B. & Thomas, A. (2004). Insights into the DNA repair process by the formamidopyrimidine-DNA glycosylase investigated by molecular dynamics. *Protein Sci* **13**, 2009-21.

67. Gorfe, A. A., Caflisch, A. & Jelesarov, I. (2004). The role of flexibility and hydration on the sequence-specific DNA recognition by the Tn916 integrase protein: a molecular dynamics analysis. *J Mol Recognit* **17**, 120-31.

68. Tang, Y. & Nilsson, L. (1998). Interaction of human SRY protein with DNA: a molecular dynamics study. *Proteins* **31**, 417-33.

69. Tang, Y. & Nilsson, L. (1999). Effect of G40R mutation on the binding of human SRY protein to DNA: a molecular dynamics view. *Proteins* **35**, 101-13.

70. Stockner, T., Sterk, H., Kaptein, R. & Bonvin, A. M. (2003). Molecular dynamics studies of a molecular switch in the glucocorticoid receptor. *J Mol Biol* **328**, 325-34.

71. Yang, L., Beard, W., Wilson, S., Roux, B., Broyde, S. & Schlick, T. (2002). Local deformations revealed by dynamics simulations of DNA polymerase Beta with DNA mismatches at the primer terminus. *J Mol Biol* **321**, 459-78.

72. Yang, L., Beard, W. A., Wilson, S. H., Broyde, S. & Schlick, T. (2004). Highly organized but pliant active site of DNA polymerase beta: compensatory mechanisms in mutant enzymes revealed by dynamics simulations and energy analyses. *Biophys J* **86**, 3392-408.

73. Hartmann, B., Sullivan, M. R. & Harris, L. F. (2003). Operator recognition by the phage 434 cI repressor: MD simulations of free and bound 50-bp DNA reveal important differences between the OR1 and OR2 sites. *Biopolymers* **68**, 250-64.

74. Reddy, S. Y., Obika, S. & Bruice, T. C. (2003). Conformations and dynamics of Ets-1 ETS domain-DNA complexes. *Proc Natl Acad Sci U S A* **100**, 15475-80.

75. Hegde, R. S. (2002). The papillomavirus E2 proteins: structure, function, and biology. *Annu Rev Biophys Biomol Struct* **31**, 343-60.

76. Djuranovic, D. & Hartmann, B. (2005). Molecular dynamics studies on free and bound targets of the bovine papillomavirus type I e2 protein: the protein binding effect on DNA and the recognition mechanism. *Biophys J* **89**, 2542-51.

77. Pardo, L., Campillo, M., Bosch, D., Pastor, N. & Weinstein, H. (2000). Binding mechanisms of TATA box-binding proteins: DNA kinking is stabilized by specific hydrogen bonds. *Biophys J* **78**, 1988-96.

78. Pastor, N., Weinstein, H., Jamison, E. & Brenowitz, M. (2000). A detailed interpretation of OH radical footprints in a TBP-DNA complex reveals the role of dynamics in the mechanism of sequence-specific binding. *J Mol Biol* **304**, 55-68.

79. Flader, W., Wellenzohn, B., Winger, R. H., Hallbrucker, A., Mayer, E. & Liedl, K. R. (2003). Stepwise induced fit in the pico- to nanosecond time scale governs the complexation of the even-skipped transcriptional repressor homeodomain to DNA. *Biopolymers* **68**, 139-49.

80. Chillemi, G., Fiorani, P., Benedetti, P. & Desideri, A. (2003). Protein concerted motions in the DNA-human topoisomerase I complex. *Nucleic Acids Res* **31**, 1525-35.

81. Perlow-Poehnelt, R. A., Zharkov, D. O., Grollman, A. P. & Broyde, S. (2004). Substrate discrimination by formamidopyrimidine-DNA glycosylase: distinguishing interactions within the active site. *Biochemistry* **43**, 16092-105.

82. Lynch, T. W., Kosztin, D., McLean, M. A., Schulten, K. & Sligar, S. G. (2002). Dissecting the molecular origins of specific protein-nucleic acid recognition: hydrostatic pressure and molecular dynamics. *Biophys J* **82**, 93-8.

83. Villa, E., Balaeff, A. & Schulten, K. (2005). Structural dynamics of the lac repressor-DNA complex revealed by a multiscale simulation. *Proc Natl Acad Sci U S A* **102**, 6783-8.

84. Huang, N. & MacKerell, A. D., Jr. (2004). Atomistic view of base flipping in DNA. *Philos Transact A Math Phys Eng Sci* **362**, 1439-60.

85. Habtemariam, B., Anisimov, V. M. & MacKerell, A. D., Jr. (2005). Cooperative binding of DNA and CBFbeta to the Runt domain of the CBFalpha studied via MD simulations. *Nucleic Acids Res* **33**, 4212-22.

86. Bishop, T. C. (2005). Molecular dynamics simulations of a nucleosome and free DNA. *J Biomol Struct Dyn* **22**, 673-86.

87. Thayer, K. M. & Beveridge, D. L. (2002). Hidden Markov models from molecular dynamics simulations on DNA. *Proc Natl Acad Sci U S A* **99**, 8642-7.

88. Kombo, D. C., Ravishanker, G., Rackovsky, S. & Beveridge, D. L. (1999). Computational analysis of variants of the operator binding domain of the bacteriophage lambda repressor. *Int J Quan Chem* **75**, 313-325.

89. Kombo, D. C., Young, M. A. & Beveridge, D. L. (2000). One Nanosecond Molecular Dynamics Simulation of the N-Terminal Domain of the λ–Repressor Protein. *Biopolymers* **53**, 596-605.

90. Kombo, D. C., Young, M. A. & Beveridge, D. L. (2000). Molecular dynamics simulation accurately predicts the experimentally-observed distributions of the (C, N, O) protein atoms around water molecules and sodium ions. *Proteins* **39**, 212-215.

91. Kombo, D. C., Young, M. A. & Beveridge, D. L. (2000). One nanosecond molecular dynamics simulation of the N-terminal domain of the lambda repressor protein. *Biopolymers* **53**, 596-605.

92. Kombo, D. C., McConnell, K. J., Young, M. A. & Beveridge, D. L. (2001). Molecular dynamics simulation reveals sequence-intrinsic and protein-induced geometrical features of the OL1 DNA operator. *Biopolymers* **59**, 205-25.

93. Blakaj, D. M., McConnell, K. J., Beveridge, D. L. & Baranger, A. M. (2001). Molecular dynamics and thermodynamics of protein-RNA interactions: mutation of a conserved aromatic residue modifies stacking interactions and structural adaptation in the U1A-stem loop 2 RNA complex. *J Am Chem Soc* **123**, 2548-51.

94. Dixit, S. B. & Beveridge, D. L. (2005). Axis Curvature and Ligand Induced Bending in the CAP-DNA Oligomers. *Biophys J* **88**, L04-6.

95. Dixit, S. B., Andrews, D. Q. & Beveridge, D. L. (2005). Induced fit and the entropy of structural adaptation in the complexation of CAP and lambda-repressor with cognate DNA sequences. *Biophys J* **88**, 3147-57.

96. Jayaram, B., McConnell, K., Dixit, S. B. & Beveridge, D. L. (1999). Free Energy Analysis of Protein-DNA Binding: The EcoRI Endonuclease - DNA Complex. *J. Comput. Phys.* **151**, 333-357.

97. Kombo, D. C., B., J., J., M. K. & Beveridge, D. L. (2002). Calculation of the Affinity of the lamda Repressor-Operator Complex Based on free Energy Component Analysis. *Mol Sim* **28**, 187-211.

98. Byun, K. S. & Beveridge, D. L. (2004). Molecular dynamics simulations of papilloma virus E2 DNA sequences: dynamical models for oligonucleotide structures in solution. *Biopolymers* **73**, 369-79.

99. Laura, V., Dixit, S. B. & Beveridge, D. L. (2005). Molecular dynamics studies of indirect mode of recognition in the papilloma virus E2-DNA complexes. *Manuscript in preparation*.

100. Nolan, S. J., Shiels, J. C., Tuite, J. B., Cecere, K. L. & Baranger, A. M. (1999). Recognition of an Essential Adenine at a Protein-RNA Interface: Comparison of Hydrogen Bonds and a Stacking Interaction. *J. Am. Chem. Soc.* **121**, 8951-8952.

101. Schultz, S. C., Shields, G. C. & Steitz, T. A. (1991). Crystal structure of a CAP-DNA complex: the DNA is bent by 90 degrees. *Science* **253**, 1001-7.

102. Parkinson, G., Wilson, C., Gunasekera, A., Ebright, Y. W., Ebright, R. E. & Berman, H. M. (1996). Structure of the CAP-DNA complex at 2.5 angstroms resolution: a complete picture of the protein-DNA interface. *J Mol Biol* **260**, 395-408.

103. Passner, J. M. & Steitz, T. A. (1997). The structure of a CAP-DNA complex having two cAMP molecules bound to each monomer. *Proc Natl Acad Sci U S A* **94**, 2843-7.

104. Lawson, C. L., Swigon, D., Murakami, K. S., Darst, S. A., Berman, H. M. & Ebright, R. H. (2004). Catabolite activator protein: DNA binding and transcription activation. *Curr Opin Struct Biol* **14**, 10-20.

105. Lutter, L. C., Halvorson, H. R. & Calladine, C. R. (1996). Topological measurement of protein-induced DNA bend angles. *J Mol Biol* **261**, 620-33.

106. Kapanidis, A. N., Ebright, Y. W., Ludescher, R. D., Chan, S. & Ebright, R. H. (2001). Mean DNA bend angle and distribution of DNA bend angles in the CAP-DNA complex in solution. *J Mol Biol* **312**, 453-68.

107. Hardwidge, P. R., Zimmerman, J. M. & Maher, L. J., 3rd. (2002). Charge neutralization and DNA bending by the Escherichia coli catabolite activator protein. *Nucleic Acids Res* **30**, 1879-85.

108. Leulliot, N. & Varani, G. (2001). Current topics in RNA-protein recognition: control of specificity and biological function through induced fit and conformational capture. *Biochemistry* **40**, 7947-56.

109. Spolar, R. S. & Record, M. T., Jr. (1994). Coupling of local folding to site-specific binding of proteins to DNA. *Science* **263**, 777-84.

110. Schlitter, J. (1993). Estimation of Absolute and Relative Entropies of Macromolecules Using the Covariance Matrix. *Chem. Phys. Lett.* **215**, 617-21.

111. Beamer, L. J. & Pabo, C. O. (1992). Refined 1.8 A crystal structure of the lambda repressor-operator complex. *J Mol Biol* 227, 177-96.

112. Karplus, M. & Kushick, J. N. (1981). Method for Estimating the Configurational Entropy of Macromolecules. *Macromolecules* **14**, 325-32.

113. Varani, G. & Nagai, K. (1998). RNA recognition by RNP proteins during RNA processing. *Annu Rev Biophys Biomol Struct* **27**, 407-45.

114. Oubridge, C., Ito, N., Evans, P. R., Teo, C. H. & Nagai, K. (1994). Crystal structure at 1.92 A resolution of the RNA-binding domain of the U1A spliceosomal protein complexed with an RNA hairpin. *Nature* **372**, 432-8.

115. Avis, J. M., Allain, F. H., Howe, P. W., Varani, G., Nagai, K. & Neuhaus, D. (1996). Solution structure of the N-terminal RNP domain of U1A protein: the role of C-terminal residues in structure stability and RNA binding. *J Mol Biol* **257**, 398-411.

116. Kormos, B. L., Beveridge, D. L. & Baranger, A. M. Correlated Motions, Covariance and Cooperativity in the U1A protein – Stem Loop II RNA System: Molecular Dynamics Studies. *Manuscript in preparation.*

117. Ichiye, T. & Karplus, M. (1991). Collective Motions in Proteins: A Covariance Analysis of Atomic Fluctuations in Molecular Dynamics and Normal Mode Simulations. *Proteins* **11**, 205-17.

118. Swaminathan, S., Harte, W. E. & Beveridge, D. L. (1991). Investigation of Domain Structure in Proteins via Molecular Dynamics Simulation: Application to HIV-1 Protease Dimer. *J. Am. Chem. Soc.* **113**, 2717-21.

119. Crowder, S., Holton, J. & Alber, T. (2001). Covariance Analysis of RNA Recognition Motifs Identifies Functionally Linked Amino Acids. *J. Mol. Biol.* **310**, 793-800.

120. Jeruzalmi, D., Yurieva, O., Zhao, Y., Young, M., Stewart, J., Hingorani, M., O'Donnell, M. & Kuriyan, J. (2001). Mechanism of processivity clamp opening by the delta subunit wrench of the clamp loader complex of E. coli DNA polymerase III. *Cell* **106**, 417-28.

121. Estabrook, R. A., Luo, J., Purdy, M. M., Sharma, V., Weakliem, P., Bruice, T. C. & Reich, N. O. (2005). Statistical coevolution analysis and molecular dynamics: Identification of amino acid pairs essential for catalysis. *Proc. Natl. Acad. Sci.* **102**, 994-9.

122. Gilson, M. K., Given, J. A., Bush, B. L. & McCammon, J. A. (1997). The statistical-thermodynamic basis for computation of binding affinities: a critical review. *Biophys J* **72**, 1047-69.

123. Swanson, J. M., Henchman, R. H. & McCammon, J. A. (2004). Revisiting free energy calculations: a theoretical connection to MM/PBSA and direct calculation of the association free energy. *Biophys J* **86**, 67-74.

124. Beveridge, D. L. & DiCapua, F. M. (1989). Free Energy via Molecular Simulation: Applications to Chemical and Biomolecular Systems. *Ann. Rev. Biophys. Biophys. Chem.* **18**, 431-92.

125. Smith, P. E. & van Gunsteren, W. F. (1994). When are Free Energy Components Meaningful? *J. Phys. Chem.* **98**, 13735-40.

126. Ajay & Murcko, M. A. (1995). Computational methods to predict binding free energy in ligand-receptor complexes. *J Med Chem* **38**, 4953-67.

127. Gilson, M. K., Given, J. A. & Head, M. S. (1997). A new class of models for computing receptor-ligand binding affinities. *Chem Biol* **4**, 87-92.

128. Kollman, P. A., Massova, I., Reyes, C., Kuhn, B., Huo, S., Chong, L., Lee, M., Lee, T., Duan, Y., Wang, W., Donini, O., Cieplak, P., Srinivasan, J., Case, D. A. & Cheatham, T. E., 3rd. (2000). Calculating structures and free energies of complex molecules: combining molecular mechanics and continuum models. *Acc Chem Res* **33**, 889-97.

129. Jayaram, B., McConnell, K., Dixit, S. B., Das, A. & Beveridge, D. L. (2002). Free-energy component analysis of 40 protein-DNA complexes: A consensus view on the thermodynamics of binding at the molecular level. *J Comput Chem* **23**, 1-14.

130. Zakrzewska, K. (2003). DNA deformation energetics and protein binding. *Biopolymers* **70**, 414-23.

131. Brandsdal, B. O., Osterberg, F., Almlof, M., Feierberg, I., Luzhkov, V. B. & Aqvist, J. (2003). Free energy calculations and ligand binding. *Adv Protein Chem* **66**, 123-58.

132. Simonson, T., Archontis, G. & Karplus, M. (2002). Free energy simulations come of age: protein-ligand recognition. *Acc Chem Res* **35**, 430-7.

133. Aqvist, J. (1990). Ion-Water Interaction Potentials Derived from Free Energy Perturbation Simulations. *J. Phys. Chem.* **94**, 8021-4.

134. Grootenhuis, P. D., Roe, D. C., Kollman, P. A. & Kuntz, I. D. (1994). Finding potential DNA-binding compounds by using molecular shape. *J Comput Aided Mol Des* **8**, 731-50.

Chapter 9

DNA SIMULATION BENCHMARKS AS REVEALED BY X-RAY STRUCTURES

Wilma K. Olson,[1] Andrew V. Colasanti,[1] Yun Li,[1] Wei Ge,[1] Guohui Zheng,[1] and Victor B. Zhurkin[2]

[1]*Department of Chemistry and Chemical Biology, Rutgers, the State University of New Jersey, Wright-Rieman Laboratories, 610 Taylor Road, Piscataway, New Jersey 08854-8087, USA;* [2]*National Cancer Institute, National Institutes of Health, Building 12B, Room B116, Bethesda, Maryland 20892-5677, USA*

Abstract: The inferences that can be drawn from known DNA structures provide new stimuli for improvement of nucleic acid force fields and fresh ideas for exploration of the sequence-dependent properties of DNA. The rapidly growing database of high-resolution nucleic acid crystal structures reveals long anticipated sequence-dependent variability in DNA backbone conformation. Nucleotides in specific sequence contexts exhibit decided tendencies to undergo changes of rotational state that are associated with large-scale helical transitions. In particular, the sugars attached to cytosine exhibit a clear-cut tendency to adopt A-like conformations. Overall, however, the large set of protein-bound DNA structures remains very close to the classical B form. This distinguishes the crystallographically observed DNA duplexes from computer-generated atomic-level DNA models, which are characterized by a systematic shift toward the A form. In addition, the base-pair steps found in different protein-DNA complexes span a narrower range of conformational states than those generated with state-of-the-art molecular simulations. The sequence-dependent positioning of water and the build-up of amino acid residues around the DNA bases point to mechanisms which may underlie the sequence-dependent structure and deformability of DNA in complexes with ligands.

Key words: DNA conformation; deformability; backbone conformation; base-pair parameters; sequence context; protein-DNA recognition

235

J. Šponer and F. Lankaš (eds.), Computational Studies of RNA and DNA, 235–257.
© 2006 *Springer.*

1.   INTRODUCTION

The combination of computational short cuts and increased computer power has made routine large-scale, atomic-level simulations of nucleic acids with surrounding solvent molecules and bound proteins or drugs (see other chapters in this book). To the extent that these calculations account for known sequence effects and solvent induced transitions of DNA, such efforts can provide atomic-level insights into the dynamic organization of the genome, such as which sites on DNA are more likely to bind protein and how the double helical structure responds to its environment in the course of biological processing and packaging.

The calculated properties of DNA are ultimately governed by the force fields that are applied to the constituent nucleotides and thus are subject to the limitations of these treatments. DNA is difficult to deal with in that its double helical structure is determined by both the local base-pair context and the long-range electrostatics of the sugar-phosphate backbone. Atomic models which may account satisfactorily for the folds of proteins and other molecules do not necessarily mimic the structural properties of the DNA polyanion. The sequence-dependent fine structure of the DNA double helix is much more subtle and thus more challenging to reproduce than the large-scale differences among the myriad conformational states observed in globular proteins.

Whereas atomic force fields are typically parameterized against assorted physical and quantum mechanical properties of small molecules, *e.g.*, chemical geometry, vibrational spectra, heats of sublimation, electronic charges, *etc.*,[1,2] recent improvements in nucleic acid force fields have incorporated structural information unique to DNA and RNA, including the root-mean-square atomic deviations of simulated vs. canonical A and B helices and the distributions of backbone and glycosyl torsion angles in computed vs. observed high-resolution structures.[3-5] "Correct" nucleic acid force fields should not only match these general features, but also reproduce the preferred arrangements of bases and solvent in experimentally determined structures. Small uncertainties in the internal rotations about single bonds of the chemical framework can alter the positioning of neighboring residues as well as introduce large-scale changes in global helical structure and molecular accessibility. For example, small correlated changes in the phosphodiester linkage and the glycosyl torsion angle—specifically variation in the angles ζ and χ (defined respectively by the C3′–O3′–P–O5′ sequence of backbone atoms and the O4′–C1′ –N9–C4 or O4′–C1′–N1–C1 atoms of purine or pyrimidine nucleotides) convert ordinary right-handed A DNA and B DNA into left-handed helical forms, concomitantly perturbing the displacement and twist of successive base pairs.[6-8] Other changes in ζ in combination with variation of ε (the torsion defined by the C4′–C3′–O3′–P atom sequence), *i.e.*, so-called BI→BII conformational changes,[9-11] are implicated in the overtwisting of the tenfold B-DNA helix to the ninefold C form.[12] Furthermore, potentials which permit relatively "free" pseudorotation of the sugar ring may facilitate transitions from the B to A form. Indeed, the conformational parameters of many simulated B-DNA duplexes, which are based on such force fields, lie between typical A- and B-DNA values.[13,14]

The need for reliable benchmarks increases as the technical barriers to atomic-level simulations of nucleic acid molecules are surmounted and nanosecond-length studies of several helical turns of double helical DNA in aqueous salt solution become feasible. The credibility of the computed structures depends upon the extent to which the predictions match critical data. This article presents a series of such standards collected from the growing database of high-resolution nucleic acid crystal structures. The number of well determined structures is now large enough to detect marked differences in the torsional preferences of individual nucleotides and the effect of sequence context on these properties. It is also possible to identify subtle changes in the positioning and orientation of complementary purine and pyrimidine bases in A·T vs. G·C base pairs. The conformational states reported in recent molecular simulations of B-DNA oligomer duplexes[15] can be compared against the crystallographically observed effects of sequence context on the orientation and positioning of base pairs, including the conformations of dimer steps within unique tetrameric sequences. The crystallographic data also reveal a sequence-dependent build-up of water molecules and amino acid residues around the nucleic acid bases which presumably modulates the conformation and recognition of DNA. State-of-the-art molecular simulations should aim to account for all such properties.

2.   METHODS

2.1   Database

The information reported herein is based on DNA double helical structures of 2.5 Å resolution or better available as of August 2005 in the Nucleic Acid Database (NDB).[16] Selected subsets of pure A-DNA and B-DNA structures are used in the analyses of torsion angle preferences, complementary base-pair geometry, and hydration patterns. A set of 239 protein-DNA crystal complexes with the same 2.5 Å resolution cutoff is used to study the effects of sequence context on the conformational properties of mono- and dinucleotides and to examine the chemical microenvironments around the four bases. The latter dataset includes 101 structures of double helical DNA bound to enzymes, 121 duplexes in the presence of regulatory proteins, 16 complexes with structural proteins, and one DNA associated with a multifunctional protein. The structures have been filtered to exclude over-represented complexes in order to obtain a balanced sample of spatial and functional forms. The selected DNA molecules contain 2234 G·C and 1886 A·T base pairs. Further details of the protein-DNA sample will be reported elsewhere.

2.2   Conformational Analysis

Conformational properties of double helical DNA—including (i) the backbone (α, β, γ, δ, ε, ζ) and glycosyl (χ) torsion angles, (ii) the phase angle (P) and amplitude (τ_{max}) of sugar pseudorotation, (iii) the six rigid-body parameters describing the arrangement of the Watson-Crick base pairs (Buckle, Propeller, Opening, Stretch, Shear, Stagger),[17] and (iv) the six rigid-body parameters relating sequential base-pair steps (Tilt, Roll, Twist, Shift, Slide, Rise)[17]—were obtained with the 3DNA software package.[18] The working dataset excludes "melted" residues, in which complementary base pairs are highly distorted and do not contain the requisite number or types of hydrogen bonds—*e.g.*, the vertical offset of complementary bases (Stagger) is 2.5 Å or more, the planes of bases are oriented at angles greater than 65°, the distances between hydrogen-bond donor and acceptor atoms on purine and pyrimidine bases are longer than 3.4 Å. The DNA sample also omits terminal nucleotides, *i.e.*, residues at chain ends and nicked dimer steps, chemically modified nucleotides, and nucleotides involved in non-canonical base pairs.

2.3   Deformability

The sequence-dependent deformability of base-pair steps is deduced from the scatter of computed parameters in the configurational sample. The covariance of observed data is expressed in matrix form in terms of the differences between the mean products and mean values of the six base-pair step parameters, *i.e.*, $\langle\theta_i\theta_j\rangle - \langle\theta_i\rangle\langle\theta_j\rangle$ where θ_i ($i = 1$–6) corresponds to one of the rigid-body parameters and the averages are computed over the observed states of a given dimer. The eigenvalues, λ_k^2 ($k = 1$–6) of the resulting array measure the spread of the observed data, and the product of the λ_k values provides an estimate of the conformational entropy, or the volume of conformation space accessible to the dinucleotide step, *i.e.*, $S_{XZ} = \prod \lambda_k$ where the subscript XZ refers to the chemical identity of the base-pair step.[19]

2.4   Hydration Patterns

The hydration patterns of the heterocyclic bases are derived from the positions of water molecules in 27 B-DNA structures of 2.0 Å or better resolution.[20] All bases at the ends of strands are excluded from the sample, as are those that form non-Watson-Crick base pairs or are unpaired, chemically modified, or located in the vicinity of metal ions, spermine, and other non-water molecules. The coordinates of all water molecules that lie within 3.4 Å of any of the heavy (non-hydrogen) atoms on the selected bases are expressed in a common base reference frame.[21] Complementary bases are arranged in ideal, planar A·T and G·C Watson-Crick pairs with the discrete water contact sites on the individual bases included in the composite structure. The set of observed points for each base pair is converted into a set of density ellipsoids using Fourier averaging techniques.[22] Patterns of sequence-dependent hydration are then deduced from the locations, dimensions, and orientations of the derived ellipsoids.

2.5   Protein-DNA Contacts

The pairwise contacts of heavy (non-hydrogen) atoms, one from DNA and the other from protein, within different distance ranges (cutoffs varying between 3.0 Å and 4.0 Å) have been collected from the aforementioned set of 239 protein-DNA crystal complexes. Excluded from the analysis are (i) contacts involving chemically modified and incompletely defined amino acids, bases, sugars, and phosphates and (ii) nucleotides found at the ends of helices and in non-Watson-Crick base pairs. Contacts are separated by nucleotide identity and amino acids are grouped according to standard chemical properties, *i.e.*, cationic (Arg, Lys), anionic (Asp, Glu), nonpolar (Ala, Ile, Leu, Met, Phe, Pro, Val), polar (Asn, Cys, Gln, His, Ser, Thr, Trp, Tyr).

3.   RESULTS

3.1   Torsion Angles

The chemical components of double helical B DNA adopt characteristic spatial forms with the vast majority of nucleotides sharing a common torsion angle pattern.[23,24] The predominance of the pattern is evident from the angular values seen in the 145 high-resolution B-DNA duplexes of 2.5 Å resolution or better in the current dataset. Most of the extant data, *i.e.*, 75% of the examples of each angle, fall in the following characteristic ranges: $\alpha = -60\pm18°$; $\beta = 172\pm23°$; $\gamma = 51\pm14°$; $\delta = 131\pm20°$; $\varepsilon = -175\pm27°$; $\zeta = -100\pm40°$; $\chi = -109\pm21°$; $P = 145\pm32°$; $\tau_{max} = 37\pm8°$, where the angles are defined by the standard IUPAC/IUB convention.[25] Further analysis of the torsions reveals subtle, sequence-dependent conformational excursions of individual parameters. Examples of these differences are shown in the histograms in Figure 9-1, where normalized distributions of P and ζ values are reported for the four common nucleotides. As seen from the distributions of pseudorotation phase angles P, the sugars attached to cytosine show a marked tendency to adopt the A-like, C3'-endo, or N (north), puckered form,[26] where $-90° < P < +90°$, compared to the sugars linked to adenine, guanine, and thymine (~30% of the cytosine sugars are in the N region vs. less than 4% of the other nucleotide sugars). Similarly, the phosphodiester bonds linked to the O3' atoms of guanosine have a much greater proportion of noncanonical *trans* (*t*) arrangements of ζ compared to other nucleotides, *i.e.*, $f_t = 0.61$ for G vs. 0.18-0.26 for A, C, T, where ζ lies in the range $180\pm60°$. The B-DNA dataset, however, is quite small and highly biased in terms of the types of sequences in currently available structures. There are fewer than 10 examples of 12 of the 32 unique trimers, and five other trimers are over-represented (85 CGC:GCG, 68 CGA:TCG, 48 GAA:TTC, 39 AAA: TTT, 95 AAT:TTA). The latter sequences are common to the many solved B-DNA dodecamer structures.

   The sequence-dependent differences in conformational variability are not so pronounced in corresponding images of P and ζ values in protein-bound DNA structures (data not shown). The latter dataset is much less biased than the B-DNA sample in terms of crystallized sequences, with 96 or more examples of every possible trimer. Not surprisingly, the proportion of noncanonical states in the protein-bound DNA structures is not as great as that found in currently available B-DNA structures. Nevertheless, compared to other nucleotides in the protein-DNA dataset, there is a slightly greater fraction of thymine sugars in N-puckered states (0.16 vs. 0.06-0.11) and a smaller fraction of adenine phosphodiester links with ζ in the *trans* range (0.21 vs. 0.28-0.31) compared to other nucleotides. Further analysis of the protein-bound DNA structures shows that departures from the canonical torsional values depend on sequence content.

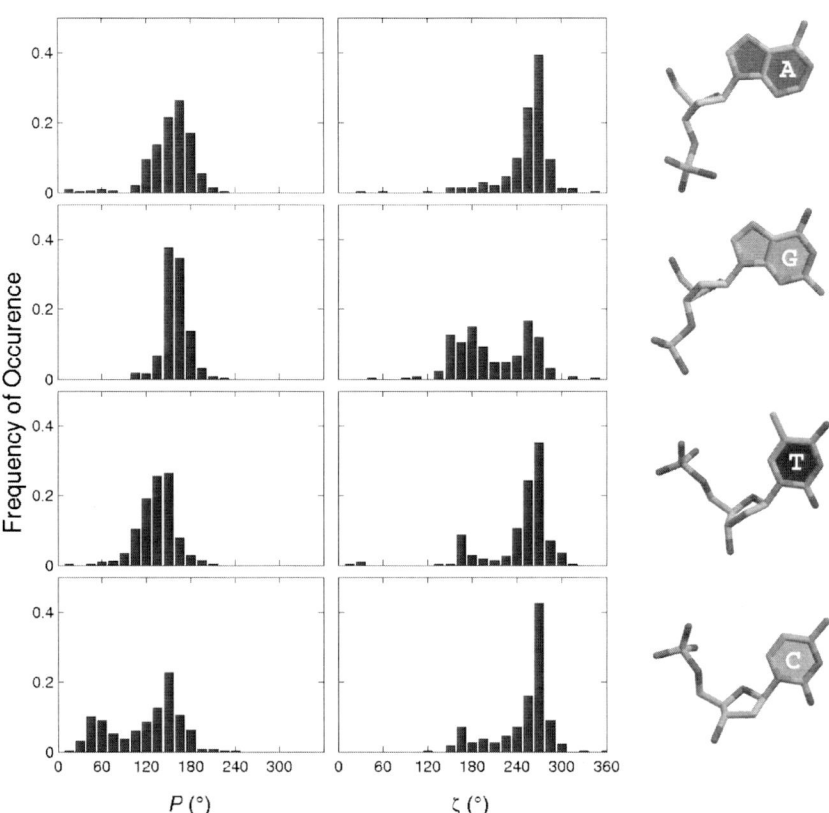

Figure 9-1. Histograms of the sequence-dependent variability of selected chemical parameters in high-resolution B-DNA structures. Values of P in the northern (N) half of the pseudorotation cycle, $-90° < P < +90°$, are indicative of nucleotide sugars in the A-like, C3′-endo puckered form.[26] Note the much larger proportion of such states in the sugars attached to cytosine compared to other nucleotides. Values of ζ in the *trans* range, $120° < \zeta < 240°$, are characteristic of the BII conformational state.[9-11] Note the greater fraction of BII phosphodiester linkages attached to guanine compared to other nucleotides. Molecular images illustrate S- and N-puckered sugars in 5′-Tmp and 5′-Cmp, respectively, and *t* and g^- states of ζ in 3′-Gmp and 3′-Amp, respectively.

The fractions of conformers with P and ζ in different conformational ranges are reported in Table 9-1 for all combinations of purine (r) and pyrimidine (y) neighbors on either side of the four nucleotides. Whereas there is no strong overall bias for N-puckered cytosine rings in the protein-bound DNA dataset, a larger than average fraction of rCy steps adopt such states ($f_N = 0.17$). On the other hand, there is a much lower occurrence of N-puckered guanine sugars in yGr and yGy sequences ($f_N = 0.03$ and 0.05, respectively). The overall propensity for N-puckered thymine sugars compared to other nucleotides reflects the relatively large numbers of yTr and rTr steps with pseudorotation angles in the range $-90° < P < +90°$. Although the population of BII-like (*trans*) ζ states is fairly low in the protein-bound DNA structures, the proportion of secondary conformers is relatively high in some trimeric contexts, *e.g.*, yTr, rGr, yCr, rCr ($f_t = 0.38$–0.44), and lower than average in others, *e.g.*, yTy, rAy, yAy ($f_t = 0.14$-0.17). The central nucleotides of three trimers—AGA, CCA and TGA—have roughly equivalent numbers of BI and BII forms of ζ ($f_t = 0.48$-0.52). The lowest fraction of *trans* ζ states occurs in GAT trimers, where $f_t = 0.09$. Notably, the trimer motifs associated with a greater proportion of *gauche* to *trans* excursions occur in DNA polymeric sequences, poly d(AG)·poly d(CT) and poly d(GGT)·poly d(ACC), long known to adopt C-type helical structures.[27]

3.2   Base-pair Parameters

3.2.1   Propeller Twisting

Complementary bases in high-resolution DNA structures are typically twisted with respect to one another like the blades of a propeller, with the C1′ atom of the sugar on the sequence strand (Strand I in Figure 9-2) shifted below and the corresponding atom on the complementary strand displaced above the mean base-pair plane, *i.e.*, negative propeller (see parameter definitions in Figure 9-2 where positive propeller is illustrated). The degree of propeller twisting depends on both base-pair and conformational context. For example, the A·T pairs are more distorted on average than the G·C pairs in B DNA (Table 9-2), with some of the most strongly propeller-twisted steps occurring in structures that contain A-tracts, *i.e.*, short stretches of 4-6 consecutive A·T base pairs. Large negative propeller twisting reorients the A·T base pair such that the N6 atom of adenine is brought into hydrogen-bonding proximity with the O4 atoms of both the complementary thymine and its predecessor, *i.e.*, 5′-neighbor, allowing for the formation of three-center, bifurcated hydrogen bonds in the major groove of the DNA. Such hydrogen-bond stabilization is also found in highly overtwisted CA·TG dimer steps.[28] By contrast, the corresponding exocyclic groups are identical in overtwisted TA dimers so that analogous inter-residue contacts would be destabilizing.[29]

Table 9-1. Effects of sequence context on the rotational isomeric state populations of selected torsional parameters in high-resolution protein-bound DNA structures.[†]

Trimer	#	f_N	f_S	f_{g+}	F_t	f_{g-}
		$-90° < P < 90°$	$90° < P < 270°$	$0° < \zeta < 120°$	$120° < \zeta < 240°$	$240° < \zeta < 360°$
rAr	428	0.13	0.87	0.03	0.30	0.67
yTy	428	0.14	0.86	0.01	**0.17**	**0.83**
rAy	412	0.08	0.92	0.00	**0.15**	**0.84**
rTy	412	0.10	0.90	0.00	0.22	0.78
yAr	403	0.13	0.87	0.00	0.26	0.74
yTr	403	**0.17**	**0.83**	0.00	**0.38**	**0.61**
yAy	504	0.12	0.88	0.00	**0.14**	**0.86**
rTr	504	**0.20**	**0.80**	0.00	0.33	0.67
rGr	345	0.09	0.91	0.00	**0.43**	**0.56**
yCy	345	0.10	0.90	0.01	0.23	0.76
rGy	357	0.08	0.92	0.00	0.25	0.75
rCy	357	**0.17**	**0.83**	0.01	0.20	0.78
vGr	314	**0.03**	**0.97**	0.01	0.34	0.65
yCr	314	0.07	0.93	0.01	**0.44**	**0.55**
yGy	357	**0.05**	**0.95**	0.00	0.21	0.79
rCr	357	0.08	0.92	0.01	**0.38**	**0.61**
xBz	6240	$0.11_{(0.05)}$	$0.89_{(0.05)}$	$0.01_{(0.01)}$	$0.28_{(0.10)}$	$0.72_{(0.10)}$

[†] Nucleotides are grouped in terms of the chemical character of immediate neighbors, i.e., purine (r) or pyrimidine (y). Sugars are divided into C3'-endo (N) and C2'-endo (S) puckered forms based on the designated value of the phase angle of pseudorotation P. The ζ phosphodiester torsions are grouped into trans (t) and gauche$^\pm$ ($g^\pm$) states corresponding to standard definitions of staggered conformational states. The number of examples # and the fractional populations f are reported for the designated trimers. Fractional values, noted in boldface, lie more than one standard deviation from the mean (xBz) values of all four nucleotides regardless of sequence context.

3.2.2   Buckling

As evident from Table 9-2, the buckling of complementary base pairs, while fixed on average near zero, shows pronounced variability, and for G·C base pairs, exhibits a notable dependence on helical conformation. The G·C base pairs tend to buckle in a positive sense in B-DNA duplexes (as depicted in Figure 9-2) and in a negative direction in A-DNA structures. The buckling of

A·T pairs follows the same trend, *i.e.*, more positive Buckle in the B form, but the data are sparse.

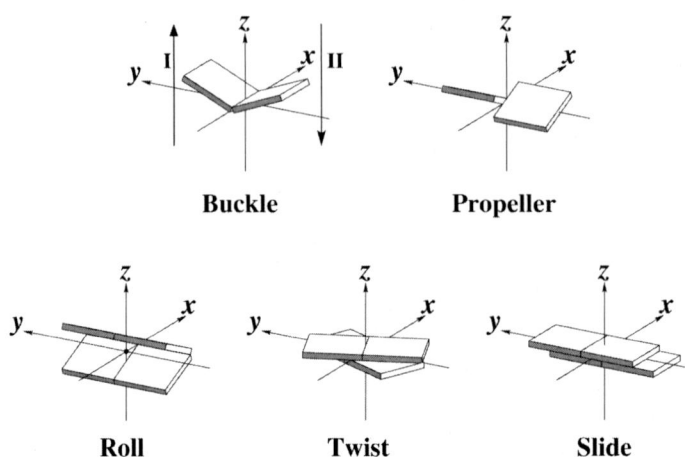

Buckle   **Propeller**

Roll   **Twist**   **Slide**

Figure 9-2. Pictorial definitions of rigid-body parameters used to describe the geometry of Watson-Crick base pairs and sequential base-pair steps.[17] Images illustrate positive values of key base-pair (Buckle, Propeller) and base-pair step parameters (Roll, Twist, Slide), as viewed from the minor groove, and the 5 -3′ directionality of the strands to which he bases are attached (arrows marked I and II). Note that all of the parameters except Buckle retain the same values if Strands I and II are exchanged (that is, if each picture is rotated 180° around the short *x*-axis of the base pair). Buckle, however, changes sign but retains its absolute value upon such a rotation. In particular, for B DNA, if the purine is in Strand I (also denoted the "sequence" strand), the average Buckle is positive (Table 9-2); if, however, the purine is placed in Strand II, the Buckle becomes negative on average.

Table 9-2. Parameters describing Watson-Crick base-pair geometry in high-resolution DNA structures.[†]

Base pair	Buckle (deg)	Propeller (deg)	Opening (deg)	Shear (Å)	Stretch (Å)	Stagger (Å)
A·T (A–DNA)	$0.3_{(6.2)}$	$-10.3_{(5.8)}$	$0.7_{(3.7)}$	$0.04_{(0.21)}$	$-0.15_{(0.07)}$	$0.03_{(0.21)}$
G·C	$-5.5_{(6.3)}$	$-11.4_{(4.8)}$	$0.4_{(2.6)}$	$-0.22_{(0.20)}$	$-0.18_{(0.10)}$	$-0.07_{(0.30)}$
A·T (B–DNA)	$1.8_{(6.5)}$	$-15.0_{(5.6)}$	$1.5_{(4.6)}$	$0.07_{(0.36)}$	$-0.19_{(0.41)}$	$0.07_{(0.28)}$
G·C	$4.9_{(7.0)}$	$-8.7_{(6.0)}$	$-0.6_{(3.3)}$	$-0.16_{(0.30)}$	$-0.17_{(0.17)}$	$0.15_{(0.26)}$

[†]See schematic illustrations in Figure 9-2 for definitions of selected parameters. Parameters are given for purine·pyrimidine (A·T and G·C) pairs. The values are identical for the corresponding pyrimidine·purine (T·A and C·G) pairs, except that Buckle and Shear are of the opposite sign. Data based on the analysis of base pairs in 143 A-DNA (29 A·T and 304 G·C) and 145 B DNA (353 A·T and 269 G·C) crystal structures. Refer to text for further details.

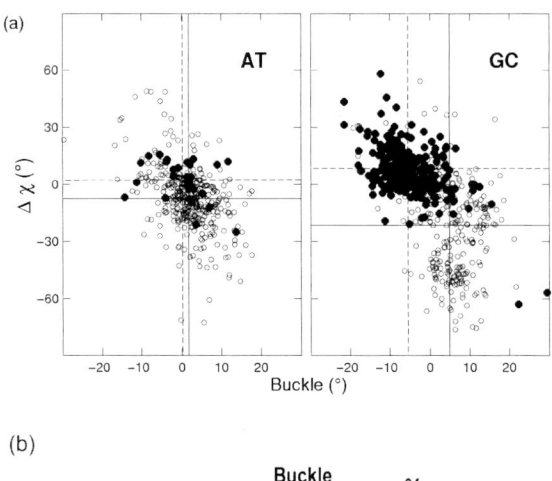

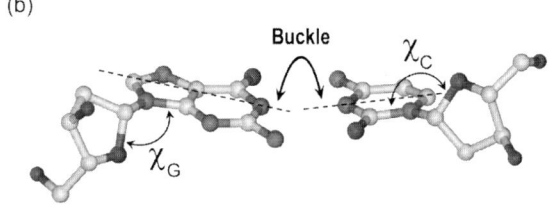

Figure 9-3. (a). Scatter plots of conformational variables distinguishing A•T and G•C Watson-Crick base pairs in A-DNA and B-DNA structures (solid and open circles, respectively). The difference, $\Delta\chi = \chi_Y - \chi_R$, in the glycosyl torsions of complementary pyrimidine (Y) and purine (R) residues underlies the observed direction of buckling. Mean values of $\Delta\chi$ and Buckle are noted by broken and solid lines for A- and B-DNA structures, respectively. (b) Minor-groove view of the G4•C7 pair from BD0023[48] illustrating the coupling of negative $\Delta\chi$ ($\chi_C < \chi_G$) and positive Buckle in B DNA (see text for details).

The sense of buckling is seemingly linked to the difference in positioning of the bases and sugars in complementary nucleotides. As follows from Figure 9-3a, there is a negative correlation between Buckle and $\Delta\chi$ the difference between the glycosyl torsion angles of complementary pyrimidines and purines. For example, the conformations in the bottom of the G·C-plate (with negative $\Delta\chi = \chi_C - \chi_G$) have almost exclusively positive Buckle. This coupling of Buckle and $\Delta\chi$ is clear from the molecular image in Figure 9-3b. As noted above, only the sugars attached to adenine, guanine, and thymine fully assume the (C2′-endo/C1′-exo) S-conformation typical of the canonical B-DNA duplex structure. By contrast, the sugars linked to cytosine show a significant population of N-type puckered states with much lower values of the pseudorotation angle P. In the particular case shown in Figure 9-3b, the guanine sugar adopts the C1′-exo conformation with $P = 140°$ while the cytosine sugar occurs in the O1′-endo form with

$P = 85°$, *i.e.*, $P_C - P_G = -55°$. Accordingly, the glycosyl torsion angles differ by a similar amount: $\chi_C - \chi_G = -50°$. The latter difference underlies the pronounced positive Buckle of G·C pairs. Indeed, imagine that we rotate the two bases in Figure 9-3b around their C–N glycosyl bonds, the cytosine clockwise and the guanine counter-clockwise. Such rotations increase χ_C and decreases χ_G and concomitantly reduce the Buckle toward zero (*i.e.*, as the two glycosyl angles become closer in value, the two bases approach coplanarity). This stereochemical consideration allows one to link the positive Buckle in the G·C pairs of B DNA to the difference in sugar conformation between G and C.

    Although many of the B-resistant nucleotides occur at points of contact in B-dodecamer structures, the unusual conformational propensity of cytosine persists at sites free of crystal packing effects in other duplexes and in high-resolution NMR structures in solution.[30] This intrinsic conformational stress in cytosine may contribute to the "A-philicity" of GC-rich DNA[31] and the curvature of DNA at the junctions between A-tracts and GC-rich sequences.[32,33]

Table 9-3. Knowledge-based estimates of sequence-dependent dimeric flexibility based on the structure of DNA in protein-DNA crystal complexes.[†]

Dimer (XZ)	# (2.5 Å)	S_{XZ} (2.5 Å)	# (2.5 Å subset)	S_{XZ} (2.5 Å subset)	# (1998)	S_{XZ} (1998)
AG	618	2.4	364	1.8	106	2.1
AA	762	1.0	421	0.7	97	2.9
GG	514	3.3	317	3.1	129	6.1
GA	643	1.9	338	1.4	117	4.5
AT	808	0.7	498	1.6	140	1.4
AC	648	0.9	418	0.8	137	2.3
GC	560	3.3	344	2.0	86	4.0
CG	512	4.9	326	3.1	88	12.1
CA	627	6.3	341	4.5	110	9.8
TA	854	7.2	574	6.6	134	6.3

[†]Analysis of flexibility based on reduced datasets which exclude outlying states of extreme bending, twisting, and/or stretching in a stepwise fashion until there are no base-pair step parameters which deviate from their averages by more than three times the root-mean-square deviation before culling. Different sets of protein-DNA complexes, with the number (#) of examples of each dimer type, are considered: 2.5 Å, 401 structures with resolution equal to or better than 2.5 Å; 2.5 Å subset, 239 structures with resolution equal to or better than 2.5 Å and filtered to exclude over-represented structure; 1998, 70 structures available in 1998 for the estimation of dimeric entropies S_{XZ}.[19]

3.3   Dimeric Structural Variability

The pyrimidine-purines are the most deformable of all base-pair steps. The conformational entropy values in Table 9-3, derived from the base-pair step parameters in various sets of protein-bound DNA structures, reveal this flexibility. The relative trends in the data persist when selected structures are omitted for reasons of redundancy or choice of resolution, and also when the most highly distorted dimers, *e.g.*, TBP-bound and other severely kinked steps, are considered (not shown). As more structural data examples have accumulated, however, the CG dimer appears to be somewhat stiffer and the TA step more deformable than originally estimated.[19] As is clear from Table 9-3, far fewer structures were available when the conformational entropies of dimer steps, S_{XZ}, were first reported, and many of the structures included in the earlier calculations were of poorer resolution than those considered here. According to the tabulated values of S_{XZ}, the pyrimidine-purine steps are 3-4 times more flexible than the purine-pyrimidine dimers and about three times more flexible than the purine-purine steps. The GC and GG dimers stand out as steps of intermediate flexibility. Contrary to the original analysis, the current data suggest that the AG step is more flexible than the GA step.

Table 9-4. Comparative deformability of CG dimers in different tetrameric contexts in protein-DNA crystal complexes vs. molecular dynamics simulations.[†]

Tetramer	#		σ(Twist) (deg)			σ(Roll) (deg)			σ(Slide) (Å)		
	R	C	R	C	MD	R	C	MD	R	C	MD
aCGa	28	29	2.2	2.3	5.7–6.9	3.4	3.4	6.6–7.5	0.37	0.38	0.8–0.9
aCGc	34	36	3.6	4.2	5.1–6.6	5.6	**10.1**	6.2–6.4	0.59	0.64	0.7–0.8
aCGg	8	9	1.9	2.5	5.4–6.1	3.2	3.3	6.5–6.9	0.22	0.33	0.7
aCGt	42	58	2.8	5.2	5.6–6.1	4.4	**15.2**	6.3–6.7	0.55	**0.88**	0.7–0.8
cCGa	8	8	3.9	3.9	6.8	6.1	6.1	6.3	0.31	0.31	0.6
cCGc	6	6	2.4	2.4	5.8	6.0	6.0	6.0	0.52	0.52	0.7
cCGg	28	42	2.0	4.1	6.7	4.0	**11.6**	5.6–6.0	0.51	**0.80**	0.6
gCGa	27	29	3.7	5.5	7.7–9.9	4.7	**14.5**	6.0–6.7	0.41	0.42	0.5–0.6
gCGc	24	26	3.9	4.4	5.9–6.4	4.8	4.9	6.1–6.3	0.42	0.41	0.8
tCGa	10	16	2.0	3.8	6.5–6.6	9.4	**15.7**	6.5–6.6	0.39	**0.99**	0.6
xCGz	326	376	3.4	4.6	–	5.3	11.4	–	0.53	0.67	–

[†]Standard deviations σ of designated step parameters in 239 protein-DNA complexes with resolution equal to or better than 2.5 Å and filtered to exclude over-represented structures: R - restricted dataset which omits "melted" dimers, residues at chain ends, nicked dimer steps, chemically modified nucleotides, nucleotides involved in non-canonical base pairs, and outlying states of extreme deformation; C - complete dataset which includes the 50 outlying, highly deformed dimer steps; MD - values reported for CG-containing tetramers from MD methods.[15] Values in boldface highlight cases where the variation of step parameters in the protein-DNA complexes exceeds that deduced from computation. Dimers grouped in terms of chemical identities of immediate neighbors (a, c, g, t).

3.4   Effects of Sequence Context on Dimeric Properties

The database of high-resolution protein-DNA complexes is large enough to study the effects of nearest neighbors on dinucleotide properties. The observed arrangements of base pairs can thus be used to check the computed behavior of dimers in different sequence contexts, such as the CG dimers reported in the context of all 10 unique tetramers.[15] The CG dimers from well resolved crystal structures span a broad range of states, including steps that are bent strongly into the major or minor groove (where Roll is respectively positive or negative) and/or displaced laterally (via ±Slide). The variation of step parameters in the protein-DNA complexes is lower than that deduced from 15-ns molecular dynamics (MD) simulations (Table 9-4),[15] except when highly deformed base-pair steps, far outside the bounds of normal B DNA are included. The base pairs in the simulated structures are also undertwisted (by ~4° on average), more positively rolled (by 2-3° on average), and displaced in the wrong sense (via negative rather than positive Slide) compared to the CG dimers in known structures (Figure 9-4).

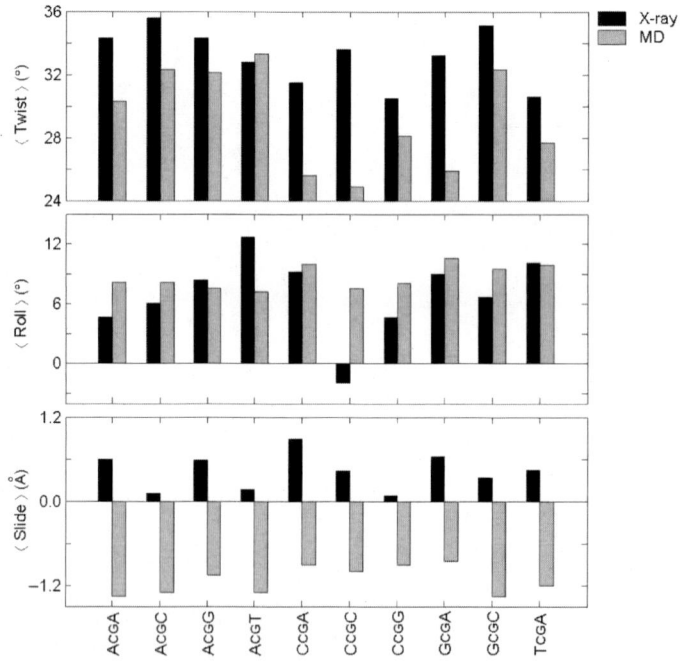

Figure 9-4. Comparative effects of sequence context on the mean values of CpG base-pair step parameters in different sequence contexts. The data from 239 protein-DNA crystal complexes (X-ray) are compared with the corresponding values obtained from MD simulations.[15] Crystallographic averages based on the "complete" dataset in Table 9-4.

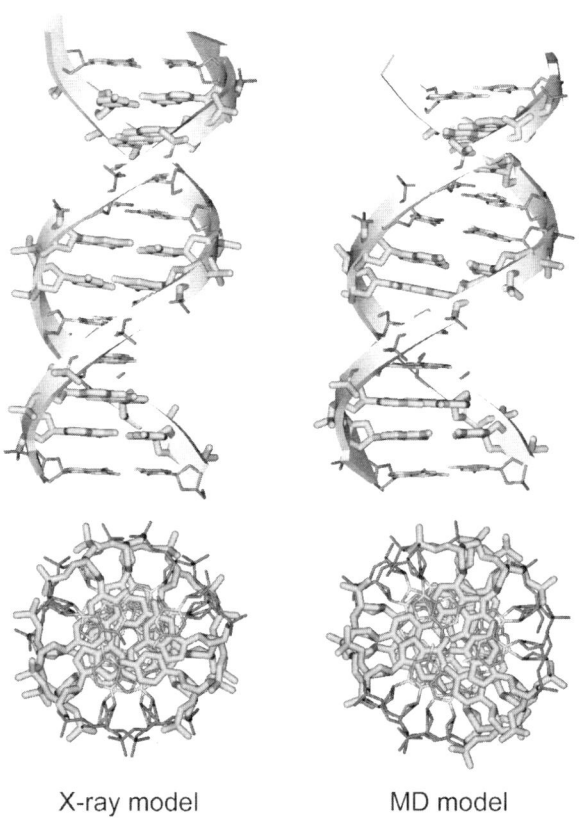

X-ray model   MD model

Figure 9-5. Comparison between the crystallographically observed (X-ray) and simulated (MD) structures of CG dimers in the context of CCGC tetramers.[15] For better visualization of the difference between the two structures, each tetramer is repeated three times, so that CCGCCCGCCCGC dodecamers are shown. The CG dimers, highlighted by thick lines, are assigned the Twist, Roll, and Slide values shown in Figure 9-4. The remaining (GC and GG:CC) dimers in the two models are assigned the average step parameters observed in protein-DNA co-crystals.[19] Values of Tilt and Shift are fixed at zero and Rise at 3.4 Å in the latter steps, but assigned the average values found in the MD simulations[15] and X-ray dataset at CG steps. The top pair of images highlights the wider minor groove and greater inclination of base pairs in the MD model compared to the X-ray model. The lower set of images illustrates the greater displacement of base-pairs in the model based on MD simulations vs. solid-state structures.

These seemingly subtle discrepancies between the two sets of base-pair step parameters (X-ray and MD) produce significant structural differences that are detectable at the level of a single helical turn of DNA (Figure 9-5). The "side" views of DNA dodecamers constructed from the repetition of "CCGC" sequences clearly show the increased minor-groove width and the greater base-pair inclination of the MD-based structure compared to the X-ray model—two global consequences of the differences in CCGC Roll and Twist values noted above. The lower views of the two structures reveal the greater displacement of the base pairs in the MD model with respect to the helical axis, compared to those in the X-ray model—a direct effect of the differences in Slide in Figure 9-4.

Based on these differences in structure, we conclude that the MD model of the CCGC tetramer is more "A-like" than the average DNA structure observed in protein-DNA complexes. At the same time, the protein-bound DNA structures are shifted toward the A form compared to "pure" B DNA.[19] Thus, the deviation of the MD-simulated structure from B DNA is even greater than that visualized in Figure 9-5. This inference is true not only for the CG dimer in particular, but also for "generic" DNA sequences as well.[14] As noted previously,[13,14] the simulated structures reflect the well known tendency of the AMBER force field[2] which was used in the calculations, to shift double helical B DNA toward intermediate B/A structural forms.

3.5   DNA Hydration

Ellipsoidal ligand-binding densities, derived from the sites of water molecules in contact with the heterocyclic bases in B-DNA structures,[20] are illustrated in Figure 9-6. The mean coordinates of hydration sites in the vicinity of the purine N3 and pyrimidine O2 proton acceptor atoms confirm the well known pseudosymmetry of Watson-Crick base pairing and the regularity of minor-groove recognition of normal duplex DNA by other molecules. Interestingly, the N3 and O2 atoms are contacted preferentially via their lower faces in most B-DNA duplexes, *i.e.*, hydrogen-bonding ligands approach individual bases from the 5′-direction, making contacts to the bases of Strand I from below and contacts to the bases in Strand II from above the common base-pair plane (see the minor-groove views in Figure 9-6). In other words, the centers of the N3 and O2 binding ellipsoids are displaced on opposite sides of the base pair, and the major (longest) axis of each ellipsoid lies roughly parallel to the base normal. Thus, the preferential approach of ligands to a purine-pyrimidine (R·Y) base pair is pincer-like, with contacts to R in the sequence strand coming from below the base-pair plane and those to Y in the complementary strand located above the base-pair plane. The offset of the minor-groove recognition sites, toward the 5′-phosphate groups of the antiparallel strands, promotes an economy of interaction between the bases and the sugar oxygens of the preceding residues on each strand, *i.e.*, the bridging of N3 or O2 and the neighboring sugar by a common water molecule. By contrast, the approach of ligands to the exposed minor-groove edges of A-DNA base pairs shows more lateral, "in-plane" character.[20] The out-of-plane component of the N3 and O2 binding centers of the four bases in A-DNA structures is closer to zero and the major axes of the binding ellipsoids span a broad range of orientations.

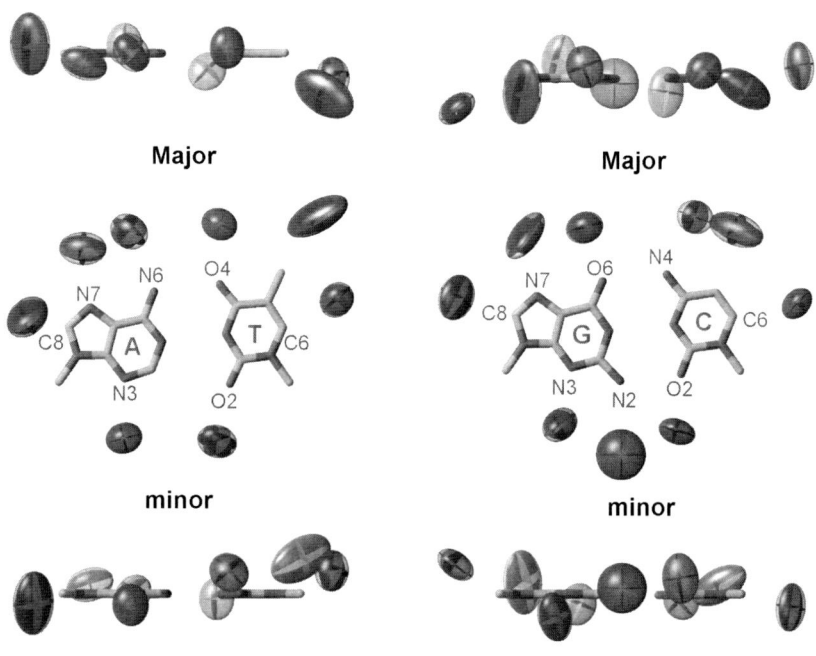

Figure 9-6. Knowledge-based ellipsoidal representations of the sites of waters bound to A•T and G•C Watson-Crick base pairs in B-DNA helical structures. Views looking toward the (top) major- and (bottom) minor-groove edges and (middle) perpendicular to the base-pair planes. Labeled atoms are discussed in the text.

As seen in Figure 9-6, the water molecules approach A·T pairs differently from G·C pairs in the B-DNA major groove. The long axes of the guanine O6 and N7 binding ellipsoids are consistently parallel to the normal of G, as opposed to the more lateral approach of waters to the corresponding N6 and N7 sites on adenine. The directionality of interactions of the pyrimidines tends to be opposite to that of the complementary purines. That is, the cytidine N4 is contacted more laterally and the thymine O4 is approached from above or below, *i.e.*, parallel to the base normal. These tendencies may be related to intrinsic chemical features of the exocyclic carbonyl and amino groups (since the approach to the G and T oxygens and the A and C nitrogens are comparable) or to well known sequence-dependent differences in DNA major groove width and accessibility, *i.e.*, the major groove of G·C helical stretches in B-DNA structures is typically wider than that of A·T stretches.[34]

The interactions of ligands with the purine C8 and pyrimidine C6 atoms on the outer edges of the Watson-Crick base pairs depend upon helical context. There are no clusters of waters near such sites in A-DNA helices.[20] The 5′-phosphorus atoms of A DNA lie roughly in the same plane as the bases attached to the same sugar,[35] leaving little space for water near the C6 or C8 atoms. The A-DNA phosphate oxygens apparently displace the C6/C8 water clusters of the B-form structure, allowing the DNA to act as its own solvent in the dehydrated A form. These intra-molecular C–H···O hydrogen bonds contribute a previously unrecognized component to the well known economy of hydration around the A-DNA phosphates.[36]

3.6   Protein Recognition

Figure 9-7 provides an overview of the close contacts between amino acid and base atoms in the current database of protein-DNA crystal complexes. The contacts to base constitute roughly a third of the intermolecular interactions in these structures. As is clear from the plotted data, the proteins contact guanine more frequently than any other base, and most of these contacts involve the atoms of cationic amino acids (Arg, Lys). The preference for cationic residues is expected from the polyanionic character of DNA and is also well known from previous surveys of the molecular contacts in smaller sets of protein-DNA structures[37-41] and from analyses of the electrostatic surfaces of proteins that bind nucleic acids.[42-45] By contrast, the contacts of anionic amino acids (Asp, Glu) with DNA, although few in number, occur preferentially with base atoms and with the major-groove atoms of cytosine, in particular. Cytosine also interacts with cationic amino acids but there is a decided bias for contacts with anionic amino acids at distances of 3.2 Å or less and interactions with cationic amino acids at distances of 3.8 Å or more, a result suggestive of the presence of ion pairs in the surrounding protein microenvironment. The donation of amino protons from cytosine to aspartic or glutamic acid, originally anticipated by Suzuki[37] and considered in earlier structural surveys,[38-41] is apparently much easier than that from adenine.

The collective data show that polar amino acids interact preferentially with A·T over G·C base pairs, and build up in greater numbers on the surface of adenine compared to thymine, *i.e.*, within distances of 3.4 Å or less. By

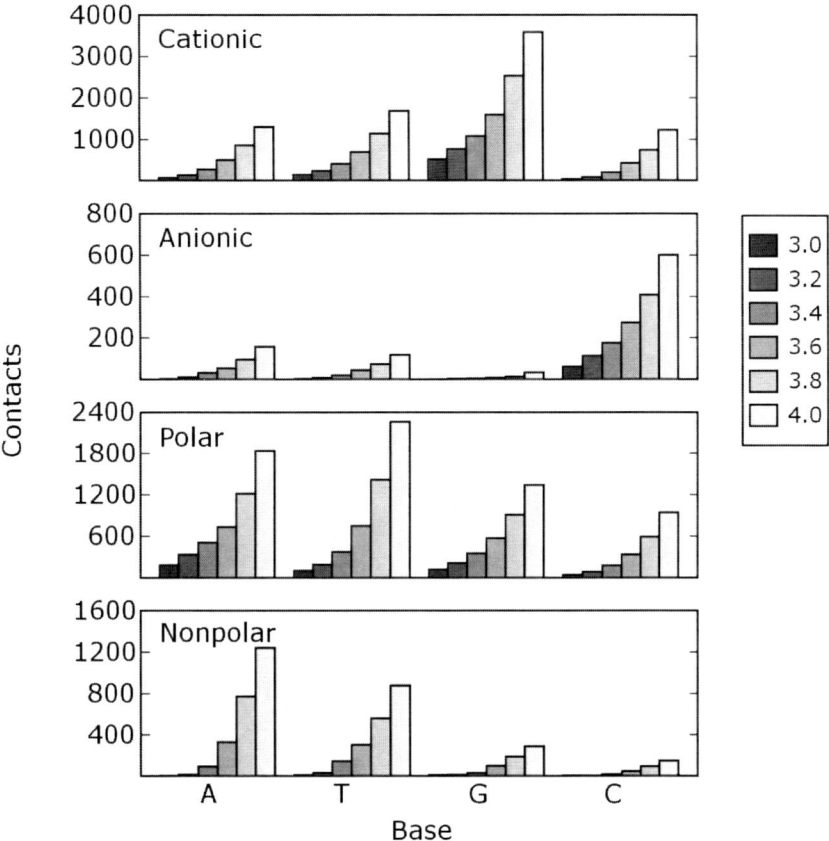

Figure 9-7. Histograms of the number of pairwise contacts of heavy (non-hydrogen) atoms, one from DNA and the other from protein, within specified distance cutoffs. Contacts to DNA are limited to atoms on the bases. Amino acids are grouped according to standard chemical properties (see Methods).

contrast, there are more polar amino acid atoms in the vicinity of thymine if the contact limit is increased to 3.8 Å or more. Whereas adenine associates preferentially with polar amino acids, thymine interacts with more cationic than polar amino acid atoms on its immediate surface. The atoms of nonpolar amino acids do not lie as close to the bases as the atoms of other amino acid types, and make roughly fourfold more contacts with A·T than G·C base pairs. The number of nonpolar contacts to thymine exceed those to adenine within distances up to 3.4 Å, but the relative accumulation of nonpolar atoms is reversed if the cutoff is increased to 3.8 Å or more.

4.   CONCLUDING REMARKS

The positions of individual atoms, which constitute the independent variables of conventional molecular simulations, are needed to decipher the structural "codes" that govern the sequence-dependent recognition and deformability of DNA. All-atom computer simulations of short duplexes are beginning to account for basic structural features of the double helix, including its sequence-dependent bending and solvent-induced conformational changes. The future promise of these methods in deciphering protein-induced distortions of DNA rests on continuing improvements of the force fields which underlie the calculations. A number of critical properties of double helical DNA structures described herein provide useful new checks of computational predictions.

The rapidly accumulating crystallographic data reveal long anticipated[23,27] sequence-dependent variability of DNA chain conformation. Nucleotides in particular trimeric contexts exhibit a greater tendency to undergo changes of rotational state, such as the S to N repuckering of the sugar ring associated with the B to A transition and the *gauche* to *trans* (BI to BII) change of the phosphodiester backbone torsion ζ linked to the B to C transition. The sugar puckering preferences are tied, in turn, to changes of complementary base-pair structure. The observed orientation and displacement of complementary bases and sequential base pairs accordingly depend on sequence. There are now sufficient data to obtain reliable estimates of dimeric structure and deformability in terms of sequence context. Recent atomic-level simulations[15] overestimate the deformability and enhance the A-like character of DNA models compared to the conformational states observed in high-resolution protein-DNA structures, even when states of extreme distortion are included in the X-ray dataset. With few exceptions, *e.g.*, Ref. 46, large-scale, atomic-level computations of DNA in solution are not generally checked against the known binding properties of water and solute molecules captured from crystallographic structures. The sequence-dependent positioning of water and the build-up of amino acid residues around the DNA bases are suggestive of mechanisms, such as economy of hydration[36] or modulation of phosphate repulsion,[47] which underlie the sequence-dependent structure and deformability of DNA. The inferences that can be drawn from known DNA structures provide new stimuli for improvement of nucleic acid force fields and fresh ideas for exploration of the sequence-dependent properties of DNA at the all-atom level.

ACKNOWLEDGMENTS

This work has been generously supported by the U.S. Public Health Service under research grant GM20861. Computations were carried out at the Rutgers University Center for Computational Chemistry and through the facilities of the Nucleic Acid Database project (NSF grant DBI-0110076).

REFERENCES

1. Brooks, B. R., Bruccoleri, R. E., Olafson, B. D., States, D. J., Swaminathan, S. & Karplus, M. (1983). CHARMM: a program for macromolecular energy, minimization, and dynamics calculations. *J. Comp. Chem.* **4**, 187-217.
2. Weiner, S. J., Kollman, P. A., Case, D. A., Singh, U. C., Ghio, C., Alagona, G., Profeta, J. S. & Weiner, P. (1984). A new force field for molecular mechanical simulation of nucleic acids and proteins. *J. Am. Chem. Soc.* **106**, 765-784.
3. Foloppe, N. & MacKerell, Jr., A. D. (2000). All-atom empirical force field for nucleic acids: I. Parameter optimization based on small molecule and condensed phase macromolecular target data. *J. Comp. Chem.* **21**, 86-104.
4. Cieplak, P., Caldwell, J. & Kollman, P. (2001). Molecular mechanical models for organic and biological systems going beyond the atom centered two body additive approximation: aqueous solution free energies of methanol and N-methyl acetamide, nucleic acid base, and amide hydrogen bonding and chloroform/water partition coefficients of the nucleic acid bases. *J. Comp. Chem.* **22**, 1048-1057.
5. Soares, T. A., Hunenberger, P. H., Kastenholz, M. A., Krautler, V., Lenz, T., Lins, R. D., Oostenbrink, C. & van Gunsteren, W. F. (2005). An improved nucleic acid parameter set for the GROMOS force field. *J. Comp. Chem.* **26**, 725-737.
6. Olson, W. K. (1976). The spatial configuration of ordered polynucleotide chains. I. Helix formation and stacking. *Biopolymers* **15**, 859-878.
7. Yathindra, N. & Sundaralingam, M. (1976). Analysis of the possible helical structures of nucleic acids and polynucleotides. Application of (n-h) plots. *Nucleic Acids Res.* **3**, 729-748.
8. Olson, W. K. (1982). Theoretical studies of nucleic acid conformation: potential energies, chain statistics, and model building. In *Topics in Nucleic Acid Structure* (Neidle, S. E., ed.), pp. 1-79, Macmillan Press, London.
9. Fratini, A. V., Kopka, M. L., Drew, H. R. & Dickerson, R. E. (1982). Reversible bending and helix geometry in a B-DNA dodecamer: CGCGAATTCGCG. *J. Biol. Chem.* **257**, 14686-14707.
10. Cruse, W. B. T., Salisbury, S. A., Brown, T., Cosstick, R., Eckstein, F. & Kennard, O. (1986). Chiral phosphorothioate analogues of B-DNA: the crystal structure of Rp-d(Gp(S)CpGp(S)CpGp(S)C). *J. Mol. Biol.* **192**, 891-905.
11. Heinemann, U. & Alings, C. (1989). Crystallographic study of one turn of G/C-rich B-DNA. *J. Mol. Biol.* **210**, 369-381.
12. van Dam, L. & Levitt, M. H. (2000). BII nucleotides in the B and C forms of natural-sequence polymeric DNA: a new model for the C form of DNA with 40° helical twist. *J. Mol. Biol.* **304**, 541-561.

13. Feig, M. & Pettitt, B. M. (1998). Structural equilibrium of DNA represented with different force fields. *Biophys. J.* **75**, 134-149.

14. Olson, W. K. & Zhurkin, V. B. (2000). Modeling DNA deformations. *Curr. Opin. Struct. Biol.* **10**, 286-297.

15. Beveridge, D. L., Barreiro, G., Byun, K. S., Case, D. A., Cheatham III, T. E., Dixit, S. B., Giudice, E., Lankaš, F., Lavery, R., Maddocks, J. H., Osman, R., Seibert, E., Sklenar, H., Stoll, G., Thayer, K. M., Varnai, P. & Young, M. A. (2004). Molecular dynamics simulations of the 136 unique tetranucleotide sequences of DNA oligonucleotides. I. Research design and results on d(CpG) steps. *Biophys. J.* **87**, 3799-3813.

16. Berman, H. M., Olson, W. K., Beveridge, D. L., Westbrook, J., Gelbin, A., Demeny, T., Hsieh, S.-H., Srinivasan, A. R. & Schneider, B. (1992). The Nucleic Acid Database: a comprehensive relational database of three-dimensional structures of nucleic acids. *Biophys. J.* **63**, 751-759.

17. Dickerson, R. E., Bansal, M., Calladine, C. R., Diekmann, S., Hunter, W. N., Kennard, O., von Kitzing, E., Lavery, R., Nelson, H. C. M., Olson, W. K., Saenger, W., Shakked, Z., Sklenar, H., Soumpasis, D. M., Tung, C.-S., Wang, A. H.-J. & Zhurkin, V. B. (1989). Definitions and nomenclature of nucleic acid structure parameters. *J. Mol. Biol.* **208**, 787-791.

18. Lu, X.-J. & Olson, W. K. (2003). 3DNA: a software package for the analysis, rebuilding, and visualization of three-dimensional nucleic acid structures. *Nucleic Acids Res.* **31**, 5108-5121.

19. Olson, W. K., Gorin, A. A., Lu, X.-J., Hock, L. M. & Zhurkin, V. B. (1998). DNA sequence-dependent deformability deduced from protein-DNA crystal complexes. *Proc. Natl. Acad. Sci., USA* **95**, 11163-11168.

20. Ge, W., Schneider, B. & Olson, W. K. (2005). Knowledge-based elastic potentials for docking drugs or proteins with nucleic acids. *Biophys. J.* **88**, 1166-1190.

21. Olson, W. K., Bansal, M., Burley, S. K., Dickerson, R. E., Gerstein, M., Harvey, S. C., Heinemann, U., Lu, X.-J., Neidle, S., Shakked, Z., Sklenar, H., Suzuki, M., Tung, C.-S., Westhof, E., Wolberger, C. & Berman, H. M. (2001). A standard reference frame for the description of nucleic acid base-pair geometry. *J. Mol. Biol.* **313**, 229-237.

22. Schneider, B., Cohen, D. M., Schleifer, L., Srinivasan, A. R., Olson, W. K. & Berman, H. M. (1993). A systematic method for studying the spatial distribution of water molecules around nucleic acid bases. *Biophys. J.* **65**, 2291-2303.

23. Sundaralingam, M. (1969). Stereochemistry of nucleic acids and their constituents. IV. Allowed and preferred conformations of nucleosides, nucleoside mono-, di-, tri-, tetraphosphates, nucleic acids and polynucleotides. *Biopolymers* **7**, 821-860.

24. Schneider, B., Neidle, S. & Berman, H. M. (1997). Conformations of the sugar-phosphate backbone in helical DNA crystal structures. *Biopolymers* **42**, 113-124.

25. IUPAC-IUB Joint Commission on Biochemical Nomenclature (JCBN). (1983). Abbreviations and symbols for the description of conformations of polynucleotide chains. Recommendations 1982. *Eur. J. Biochem.* **131**, 9-15.

26. Altona, C. & Sundaralingam, M. (1972). Conformational analysis of the sugar ring in nucleosides and nucleotides. A new description using the concept of pseudorotation. *J. Am. Chem. Soc.* **94**, 8205-8212.

27. Leslie, A. G., Arnott, S., Chandrasekaran, R. & Ratliff, R. L. (1980). Polymorphism of DNA double helices. *J. Mol. Biol.* **143**, 49-72.

28. Timsit, Y. (1999). DNA structure and polymerase fidelity. *J. Mol. Biol.* **293**, 835-853.

29. Gorin, A. A., Zhurkin, V. B. & Olson, W. K. (1995). B-DNA twisting correlates with base pair morphology. *J. Mol. Biol.* **247**, 34-48.

30. Wu, Z., Delaglio, F., Tjandra, N., Zhurkin, V. B. & Bax, A. (2003). Overall structure and sugar dynamics of a DNA dodecamer from homo- and heteronuclear dipolar couplings and ^{31}P chemical shift anisotropy. *J. Biomol. NMR* **26**, 297-315.

31. Ivanov, V. I. & Minchenkova, L. E. (1995). The A-form of DNA: in search of biological role (a review). *Mol. Biol.* **28**, 780-788.

32. Kamath, S., Sarma, M. H., Zhurkin, V. B., Turner, C. J. & Sarma, R. H. (2000). DNA bending and sugar switching. In *Proceedings of the Eleventh Conversation in Biomolecular Stereodynamics. J. Biomol. Struct. Dynam., Conversation 11, No. 2* (Sarma, R. H. & Sarma, M. H., eds.), pp. 317-325, Adenine Press, Schenectady, NY.

33. Zhurkin, V. B., Tolstorukov, M. Y., Xu, F., Colasanti, A. V. & Olson, W. K. (2005). Sequence-dependent variability of B-DNA: an update on bending and curvature. In *DNA Conformation and Transcription* (Ohyama, T., ed.), pp. 18-34, Landes Bioscience, Georgetown, TX and Springer, New York.

34. Heinemann, U., Alings, C. & Bansal, M. (1992). Double helix conformation, groove dimensions and ligand binding potential of a G/C stretch in B-DNA. *EMBO J.* **11**, 1931-1939.

35. Lu, X.-J., Shakked, Z. & Olson, W. K. (2000). A-form conformational motifs in ligand-bound DNA structures. *J. Mol. Biol.* **300**, 819-840.

36. Saenger, W., Hunter, W. N. & Kennard, O. (1986). DNA conformation is determined by economics in the hydration of phosphate groups. *Nature* **324**, 385-388.

37. Suzuki, M. (1994). A framework for the DNA-protein recognition code of the probe helix in transcription factors: the chemical and stereochemical rules. *Structure* **2**, 317-326.

38. Mandel-Gutfreund, Y., Schueler, O. & Margalit, H. (1995). Comprehensive analysis of hydrogen bonds in regulatory protein DNA-complexes: in search of common principles. *J. Mol. Biol.* **253**, 370-382.

39. Mandel-Gutfreund, Y., Margalit, H., Jernigan, R. L. & Zhurkin, V. B. (1998). A role for CH···O interactions in protein-DNA recognition. *J. Mol. Biol.* **277**, 1129-1140.

40. Kono, H. & Sarai, A. (1999). Structure-based prediction of DNA target sites by regulatory proteins. *Proteins: Struct. Funct. Genet.* **35**, 114-131.

41. Luscombe, N. M., Laskowski, R. A. & Thornton, J. M. (2001). Amino acid-base interactions: a three-dimensional analysis of protein-DNA interactions at an atomic level. *Nucleic Acids Res.* **29**, 2860-2874.

42. Ohlendorf, D. H. & Matthew, J. B. (1985). Electrostatics and flexibility in protein-DNA interactions. *Advan. Biophys.* **20**, 137-151.

43. Chirgadze, Y. N. & Larionova, E. A. (2003). Definitive role of polar residue clusters in B-DNA major groove recognition by protein factors. *Mol. Biol. (Engl. Tr.)* **37**, 232-239.

44. Jones, S., Shanahan, H. P., Berman, H. M. & Thornton, J. M. (2003). Using electrostatic potentials to predict DNA-binding sites on DNA-binding proteins. *Nucleic Acids Res.* **31**, 7189-7198.

45. Stawiski, E. W., Gregoret, L. M. & Mandel-Gutfreund, Y. (2003). Annotating nucleic acid-binding function based on protein structure. *J. Mol. Biol.* **326**, 1065-1079.

46. Feig, M. & Pettitt, B. M. (1999). Modeling high-resolution hydration patterns in correlation with DNA sequence and conformation. *J. Mol. Biol.* **286**, 1075-1095.

47. Elcock, A. H. & McCammon, J. A. (1996). The low dielectric interior of proteins is sufficient to cause major structural changes in DNA on association. *J. Am. Chem. Soc.* **118**, 3787-3788.

48. Kielkopf, C. L., Ding, S., Kuhn, P. & Rees, D. C. (2000). Conformational flexibility of B-DNA at 0.74 Å resolution: d(CCAGTACTGG)$_2$. *J. Mol. Biol.* **296**, 787-801.

Chapter 10

RNA: THE COUSIN LEFT BEHIND BECOMES A STAR

Tamar Schlick

Department of Chemistry and Courant Institute of Mathematical Sciences, New York University, 251 Mercer Street, New York, New York 10012.
Phone: (212) 998-3116 Fax: (212) 995-4152 Email: schlick@nyu.edu

Abstract: A brief introduction to RNA structure and function is offered, including recent exciting discoveries concerning RNA's starring roles in gene regulation. Challenges in RNA research are outlined, and the role of modeling and bioinformatics approaches to these problems is suggested. Applications of a graph-theory representation of RNA secondary structures to RNA analysis and design are illustrated.

Key words: RNA structure and function, secondary structure, RNA graphs, RNA structural repertoire, RNA design, RNA folding

Even after the completion of many genome sequences, both the number and diversity of ncRNA genes remain largely unknown.[1]

—Eddy, 2001

New evidence suggests, however, that this junk DNA may encode RNA molecules that perform a variety of regulatory functions. This new theory may explain why the structural and developmental complexity of organisms does not parallel their numbers of protein-coding genes.[2]

—Mattick, 2004

1. INTRODUCTION

While proteins are household words and DNA is an icon—in science as well as art (e.g., Ref.[3])—their biomolecular cousin, RNA, has largely been left behind until recently. Indeed, RNA's starring role in the cell is emerging with

J. Šponer and F. Lankaš (eds.), Computational Studies of RNA and DNA, 259–281.

new discoveries concerning RNA's vital regulatory roles. We now appreciate that RNA molecules are integral components of the cellular machinery for protein synthesis and transport, RNA editing, chromosome replication and regulation, catalysis, and many other functions (see Table 1 for RNA's diverse roles). In this chapter, some of the current excitement in the field of RNA biology and chemistry is described, along with pressing challenges concerning RNA structure and function. Novel computational approaches, including molecular modeling and simulation and mathematical represent-tations of secondary RNA structures, have great potential for impacting the field of RNA research, including genome-wide initiatives concerning RNA structure and function, or *ribonomics*. Such applications are illustrated for RNA structure analysis and design using graph theory representation of RNA secondary structures.

Table10.1.   Some classes of non-coding RNA (ncRNA).

RNA	Function
transfer RNA (tRNA)	protein synthesis
ribosomal RNA (rRNA)	protein synthesis
small nucleolar RNA (snoRNA)	rRNA modification
micro RNA (miRNA)	translation regulation
transfer-messenger RNA (tmRNA)	protein stability in ribosome
telomerase RNA	replication
guide RNA (gRNA)	mRNA editing
spliced leader RNA (SL RNA)	mRNA trans-splicing
small nuclear RNA (snRNA)	RNA splicing
hammerhead ribozyme	self-cleavage
hepatitis delta virus ribozyme	self-cleavage
Group I intron	self-splicing
Group II intron	self-splicing
RNase P	pre-tRNA processing
23S rRNA	peptide bond formation

2.   RNA AT ATOMIC RESOLUTION

RNA is a single-stranded polynucleotide chain which can fold upon itself to form double-stranded segments stabilized by complementary hydrogen bonds such as adenine with uracil, cytosine with guanine, as well as thermo-dynamically stable guanine-uracil wobble pairs. These folded structures are

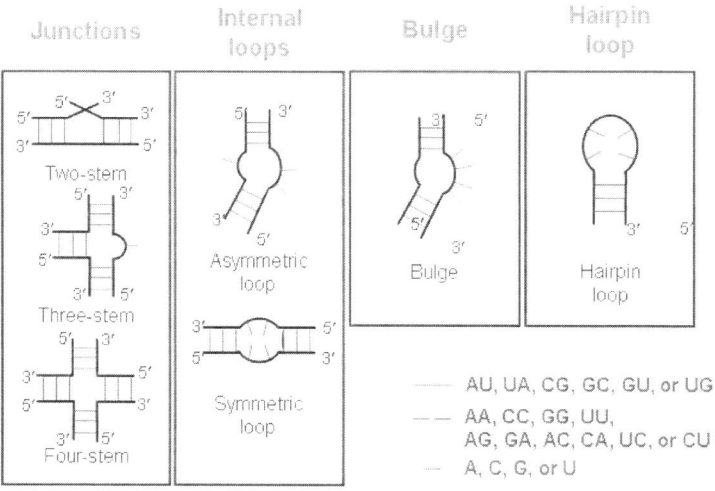

Figure 10.1. RNA Secondary Structural Motifs.

imperfect due to non-complementary base pairs and unpaired bases and thus form bulges, loops, junctions, and other motifs, as shown in Figure 10.1, stabilized by various stacking interactions, hydrogen bonding, and intra-molecular networks between distant regions in the linear sequence[4].

The motifs shown in Figure 10.1 are known as *secondary structures*; these can lead to compact and complex *tertiary interactions*, as shown in Figure 10.2 for the hammerhead ribozyme[5] and Figure 10.3 for the hepatitis delta virus (HDV) ribozyme[6]. The latter RNA reveals a common feature of RNA that can be considered as a super-secondary structural element: a *pseudoknot*. RNA pseudoknots have a stretch of nucleotides within a hairpin loop that pairs with nucleotides external to that loop. In other words, hydrogen bonding occurs between alternating regions (e.g., **a with c and b with d** to produce an inter-twined geometry), as illustrated in Figure 10.4.

The clover-leaf structure of the tRNA molecule has been known for over 25 years, and for a long time was the only well-characterized major structure of an RNA molecule; see Ref.[7] for a perspective following the high-resolution determination of yeast phenylalanine tRNA in 2000. However, with vast improvements in crystallization procedures for RNA (e.g., RNA structure determination through crystallization with a protein that would not interfere with the enzyme's activity[8]), as well as alternative approaches for studying RNAs such as high-resolution NMR, spectroscopy, cross-linking relations and phylogenic data analysis, our knowledge of RNA structure has increased dramatically.

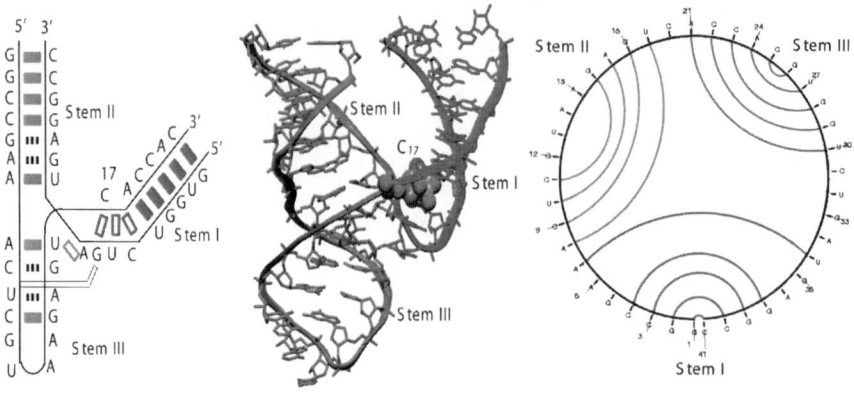

Figure 10.2. Hammerhead Ribozyme Secondary and Tertiary Structures: two types of base-pairing diagrams (left and right) illustrate the secondary structure, and the middle illustration shows the folded, three-dimensional configuration.

The high-resolution ribosome structures have added dramatically to our library of known RNA structures[9,10,11,12,13,14,15,16,17]. The ribosome is the cell's protein synthesis factory, a complex of many proteins and several RNA molecules, which are folded as many stable secondary motifs; the ribosome's small and large subunits cooperate tightly to coordinate the interplay between tRNA, mRNA, and proteins in the process of protein synthesis and catalyze the peptide bond formation. As of Summer 2005, there are about 825 known structures of RNA in the public databases (see http://www.rcsb.org), but many entries are duplicates of the same molecule or motif.

3. RNA's DIVERSITY

The wonderful capacity of RNA to form complex, stable tertiary structures has been exploited by evolution.

Traditionally, RNA is known for its key role as mediator between the agent of heredity — DNA — and the cell's workhorses — proteins (see Figure 10.5). For example, tRNA molecules carry amino acids and deposit them in the correct order, mRNAs mediate translation of the hereditary information from DNA into protein, and rRNAs are involved in protein biosynthesis (within a complex of ribosomal RNA and numerous proteins). However, work in the 1980s established that RNA, like protein, can act as a catalyst in living cells. Thomas Cech and Sidney Altman received the 1989 Nobel Prize for chemistry for this discovery of RNA enzymes or *ribozymes*. More than 500 ribozyme types have been found in a diverse range of organisms. Many ribozymes make or break phosphodiester bonds in nucleic acid backbones, but other biological and chemical functions are continuously being discovered.

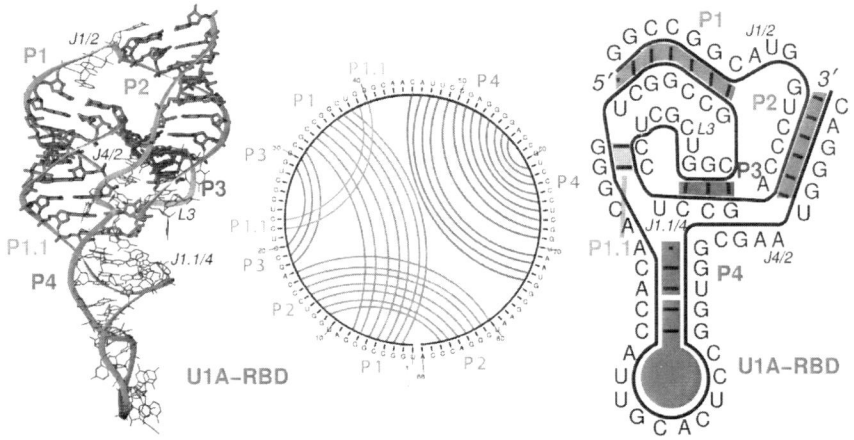

Figure 10.3. Hepatitis Delta Virus (HDV) ribozyme structure. The secondary and tertiary structures are shown, along with the pseudoknot details. Two pseudoknots are present: between regions **P1** and **P2** (red) and **P1** and **P3** (green). This intertwined base-pairing is evident from the middle image, which shows crossings in the paired bases (compare to the hammerhead ribozyme in Figure 10.2, which shows no crossings).

Ribozymes have also been designed (e.g., Refs.[18,19,20]), as spare in composition as two base building-block units (rather than four)[21]. In fact, the 83-nucleotide ribozyme composed only of two different building blocks — uracil and 2,6-diaminopurine — was shown to catalyze the ligation of two RNA molecules with a rate 36,000 times faster than the uncatalyzed reaction[21]. That RNA's genetic code may be simpler than today's four bases lends further support to the "RNA world" hypothesis.

4. RECENT DISCOVERIES CONCERNING RNA's STARRING ROLE

Exciting recent discoveries concerning RNA came in the end of 2002, when a high-quality draft sequence of the mouse genome was published and analyzed (see the 5 December 2002 issue of *Nature*, volume 420). The 2.5-billion size of the mouse genome is slightly smaller than the human genome (3 billion base pairs in length), and the number of estimated mouse genes, around 30,000, is roughly similar to the number approximated for humans. Intriguingly, the various analyses reported in December 2002 revealed that only a small percentage (1%) of the mouse's genes has no obvious human counterparts. This similarity makes the mouse genome an excellent organism for studying human diseases and novel treatments. But the obvious dissimilarity between mice and men also begs for further comparative investigations: why are we not more like mice? Part of this question may be explained through an understanding of how mouse and human genes might be regulated differently.

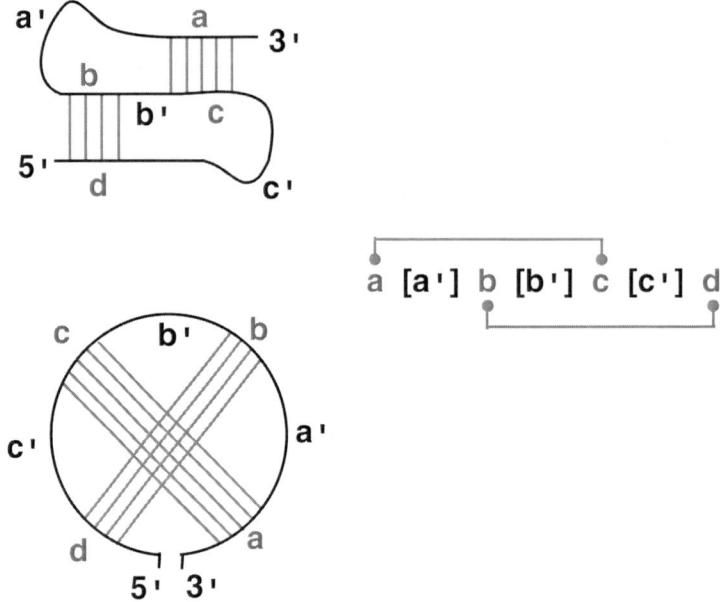

Figure 10.4. RNA 3D Folds Involve Pseudoknots: RNA Pseudoknots have an intertwined form of base pairing, which can be evident from a circular representation of base pairing, as shown in Figure 10.2 (no pseudoknot) and Figure 10.3 (2 pseudoknots).

Aside from complex *networks* of genes rather than *single* genes that are responsible for controlling phenotypes, another factor for this difference between humans and mice traits is related to control of gene activation and function by a novel class of genes called *RNA genes* — RNA transcripts that do not code for proteins (ncRNA for non-coding RNAs). These genes have essential regulatory functions that may play significant roles in each organism's survival[22,23,24,25]. In fact, scientists are amazed at how RNA's critical activities have eluded them so long. As early as 1961, suggestions that RNA can control gene activity were mentioned in Jacob and Monod's classic paper[26]; but only recently have some of these mechanisms and immense applications been discovered[2].

Such newly found roles for RNAs, especially concerning tiny RNAs that do not encode proteins (e.g., micro-RNAs) but can influence gene action, won DNA's cousin the venerable trophy of "Breakthrough of The Year" by

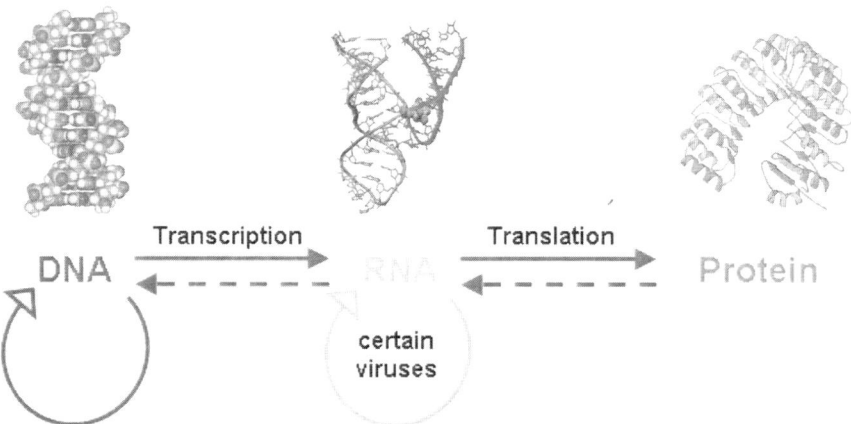

Figure 10.5. The Central Dogma of Biology. In the traditional flow of information, DNA makes RNA through transcription and RNA makes protein in translation. However, discoveries that RNA can act as an enzyme, and can carry actual instructions for protein synthesis (other than via the RNA code) enrich RNA's cellular roles dramatically.

Science editors in 2002 (see the 20 December 2002 issue of *Science*, volume 298). Non-protein coding stretches of mRNAs range in size from only 20 nucleotides to over 10000 nucleotides[27]; they are required to control the translation from the mRNA transcript into protein.

The 2002 award recognized a large group of papers that unraveled various fascinating features of small RNAs (affectionately termed *nanoRNAs*). Micro-RNAs (miRNAs), generally 21 to 25 nucleotides long, form a regulatory class of ncRNAs[28]. These RNAs, encoded in genomes, control gene expression by repressing translation of target genes through, for example, binding to 3′ untranslated regions of the messenger-RNA targets or by destroying the messenger-RNA targets. Thus, regions in the genome which were previously denoted as "junk-DNA" may actually define a gold-mine of biological information.

Such small RNAs in animals, plants, and fungi are collectively termed RNAi (for RNA interference). The agents that initiate RNAi in a sequence-specific manner are double-stranded RNA segments (siRNAs, for small interfering RNAs) that silence gene expression. For example, they may seek out messenger RNA (corresponding to them) and destroy it, or they may bind to chromatin and/or modify chromatin structure[29,30,31]. Though initially regarded as anomalies, work is revealing that such siRNAs regulate gene expression in a variety of organisms. SiRNAs can also be synthetic, designed to target specific genes; they have therapeutic applications.

Such interference mechanisms by RNA silencing through chromatin structure can provide an organism a natural defense against invading viruses and transposons (DNA segments that migrate within and across organisms and are associated with bacterial pathogenicity)[32]. Consequently, this natural protection is being exploited by scientists using siRNAs to target viral genes that can inhibit the replication of HIV-1, polio, or other viruses (e.g., Ref.[33]). Moreover, RNA interference mechanisms are being investigated by several companies who are applying them to discover the functions of genes by turning them off to determine the effect on the plant or the animal. For example, a landmark study on obesity employed RNA interference[34] to inactivate about 85% of the roundworm's predicted 19,757 genes that code for proteins in a single experiment[35]. Results of the experiment helped identify the genes that play a role in an organism's tendency for obesity.

These fascinating discoveries regarding RNA's interference with gene activity are associated with many *epigenetic* phenomena — changes in gene expression that do not involve alterations in the genome and persist across at least one generation.

A different kind of epigenetic control was also discovered by Breaker and coworkers[36,37] in bacterial messenger RNAs containing sequences that sense small molecules directly to control translation of mRNA into protein. Namely, specific control regions of mRNA can bind directly to metabolites, such as associated with vitamin B synthesis and import, and induce a conformational change in RNA's folding state; this metabolite-triggered conformational change acts as part of the signal transduction pathway that senses vitamin level and controls enzyme production. Conformational switches have also shown to be important for the catalytic activity of the hepatitis delta virus (HDV) ribozyme[38].

Such a switch of RNA conformation between two states in a ligand-dependent manner (a *riboswitch*) also opens new avenues for thinking about RNA design in a variety of contexts[39]. Like the *Paracelsus challenge* for proteins[40], one can formulate a similar challenge for RNA design: describe minimal changes in the nucleotide sequence to trigger a conformational rearrangement in the folding of a given RNA molecule. Such a challenge in the RNA world will be appropriately approached by a combination of computational and experimental wizardry. In fact, *Science* editor Jennifer Couzin[41] exclaims: *"Having exposed RNAs' hidden talents, scientists now hope to put them to work.".*

Already, numerous applications can be envisioned for RNAs as regulators of gene expression, therapeutic agents, molecular switches, and molecular sensors. This is because therapeutic agents could be designed to exploit

RNA's functional sites as potential drug targets[42]. The construction of ligands that bind RNA and interfere with protein synthesis, transcription, or viral replication may lead to new antibiotic/antiviral drugs, for example. Together with the design of novel RNAs and of RNA sequences called *aptamers* — RNAs that are *apt* to bind to specific molecular targets or perform desired catalytic functions by design (see Refs.[43,44] for example) — RNA offers a great molecular machinery with potential benefits to biomedicine and nanotechnology. Many such novel synthetic RNAs have been found by random RNA sequence pool ("in vitro") experiments[27,45,46,47,48]. With rapidly growing interest in RNA structure and function and its applications in biomedical research, enhancing the repertoire of both natural and synthetic RNA is a central goal.

5. MAJOR CHALLENGES IN RNA RESEARCH

At least five key challenges concerning RNA naturally arise: finding novel RNA genes, identifying the biological roles of these RNA genes, determining the structural repertoire of RNA, determining RNA tertiary folds from sequence, and designing novel RNAs.

5.1 RNA Gene Location

Identifying locations of RNA genes in genome sequences is much more difficult than protein genes, since the start and stop codons for protein transcripts do not apply. Thus, searching for RNA genes in intron and intergenic regions, which comprise over 90% of the genomes, presents a challenge (see Ref.[49] for example). Current programs like tRNAscan-SE, FAStRNA, and Snoscan for identifying RNA genes are based on existing sequences of functional RNA, conservation of RNA secondary structure in identified sequences, and comparative genome analysis[50,51,52]. However, these programs often lead to many false identifications and are not successful for non-conserved ncRNA sequences and as high-throughput discoveries. Most methods instead rely on biochemical and molecular biology techniques. In the case of mRNAs, their associated functions are inferred from complementary target mRNAs where possible[53].

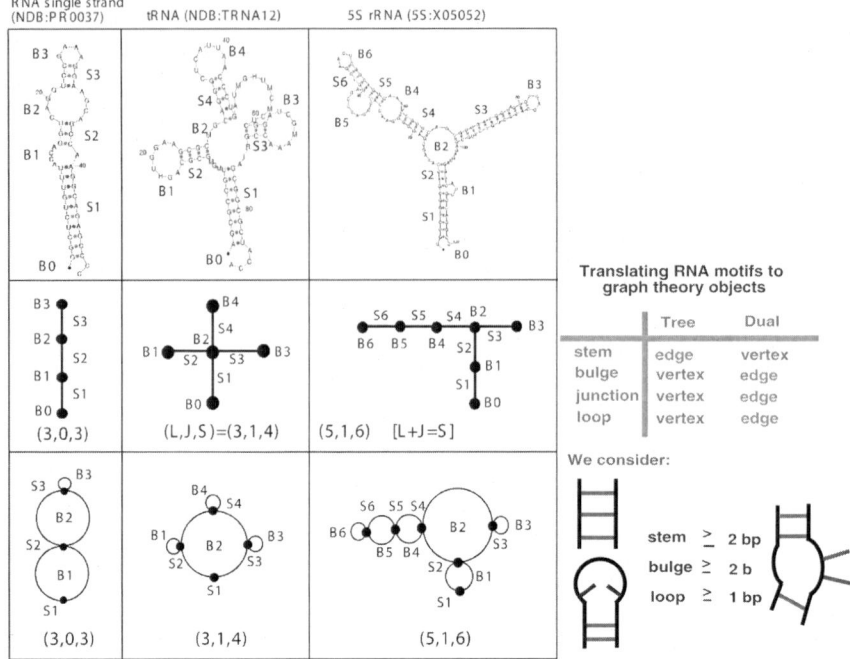

Figure 10.6. Graphical representations of RNA secondary structures as trees (middle row) and dual graphs (bottom row).

5.2   RNA Gene Function

Once potential sequences that correspond to RNA genes are identified, experiments are required to verify the expression of these transcripts in cells. Typically, genes are assayed by Northern blotting and microarray techniques[54], and real-time PCR is also used to verify RNA transcripts. Comparative genome analysis to indicate conservation of secondary structures also aids in the verification process of RNA gene candidates. The functional characterization is a laborious and time-consuming process, and thus predictions by modeling and computation need to be as reliable or as discriminating as possible.

5.3   RNA's Structural Repertoire

Defining the structural repertoire of RNA involves delineating all possible folded motifs of functional RNAs. For proteins, catalogues of known folds are collected in many databases, such as SCOP, CATH, PROSITE, or

PFAM, and organized by classes, folds, superfamilies, families, and domains (see illustrations in Ref.[55], Chapter 4). The enormous interest in the *protein folding problem* has led to structural genomics initiatives that seek to define and design new protein sequences that will fold into novel motifs. Funding opportunities from the National Institutes of Health have already led to important discoveries in this area[56,57].

However, experience has also taught structural genomicists that Nature is tricky. While some pairs of disparate sequences lead to similar folds, despite expectations to be different, other sequence pairs thought to lead to similar architectures produce new unanticipated folds. These cause-and-effect patterns are likely explained by various scenarios of structural evolution of protein active sites. Besides protein folding and protein structural genomics, scientists in the protein field are also speculating on the total number of existing protein folds; this number is likely in the range of several thousands. Discussions on to how best to employ experimental technology and computation to find all Nature's folds have led to concentrated initiatives by large teams of structural genomicists.

In this regard, the RNA field lags behind considerably. The number of known RNA folds is an order-of-magnitude less than that available for proteins (e.g., 825 vs. ~25,000 as of Summer 2005), and the best way to organize RNA motifs is not known. Possibly, the number of functional RNA motifs may grow with sequence size rather than approach a limit.

The application of graph theory to describe secondary structural motifs of RNA (e.g., Refs.[58,59,60]) holds great promise in this area of RNA structure analysis. Together with many reports of newly discovered RNA motifs, our group's RNA-As-Graphs database (RAG) (http://monod.biomath.nyu.edu/rna/rna.html)[61,62,63] suggests that the currently-known RNAs represent only a *small fraction* of the possible stable and functional RNAs.

5.4 The RNA Folding Problem

Our graph theory application involves defining RNA secondary structures as two types of graphical objects: trees and dual graphs (see Figure 10.6)[64]. These classic representations, coupled with linear algebra tools, such as the Laplacian second eigenvalue corresponding to a graph (see Figure 10.7), and graph enumeration theorems allow us to enumerate the possible RNA secondary motifs (see Figure 10.8 and also Figure 10.7, bottom). The color coding (red/blue/black) in the library segment distinguishes motifs corresponding to existing RNAs (red) from those that are hypothetical and RNA-like (blue) and hypothetical but non-RNA-like (black). This classi-fication of the hypothetical motifs has been accomplished by statistical clustering methods[63]. In fact, clustering suggested ten novel RNA motifs (Figure 10.9a and 10.9b) which contain sub-components corresponding to

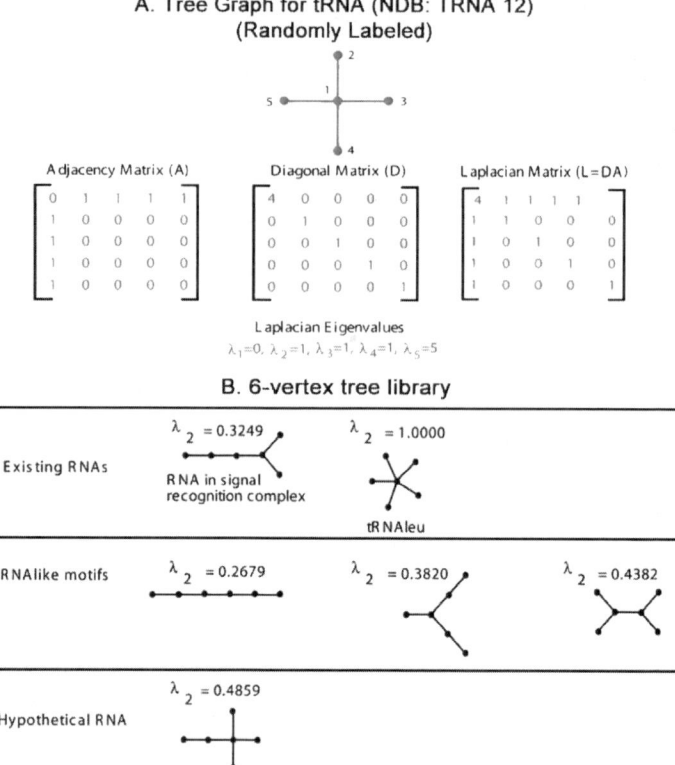

A. Tree Graph for tRNA (NDB: TRNA 12) (Randomly Labeled)

Adjacency Matrix (A)   Diagonal Matrix (D)   Laplacian Matrix (L=D–A)

Laplacian Eigenvalues
$\lambda_1=0$, $\lambda_2=1$, $\lambda_3=1$, $\lambda_4=1$, $\lambda_5=5$

B. 6-vertex tree library

Existing RNAs — $\lambda_2 = 0.3249$ RNA in signal recognition complex — $\lambda_2 = 1.0000$ tRNAleu

RNAlike motifs — $\lambda_2 = 0.2679$ — $\lambda_2 = 0.3820$ — $\lambda_2 = 0.4382$

Hypothetical RNA — $\lambda_2 = 0.4859$

C. Dual-Graph Motif Libraries

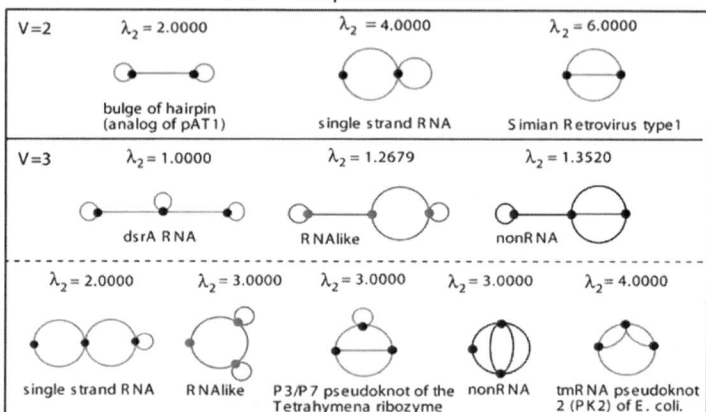

V=2 — $\lambda_2 = 2.0000$ bulge of hairpin (analog of pAT1) — $\lambda_2 = 4.0000$ single strand RNA — $\lambda_2 = 6.0000$ Simian Retrovirus type1

V=3 — $\lambda_2 = 1.0000$ dsrA RNA — $\lambda_2 = 1.2679$ RNAlike — $\lambda_2 = 1.3520$ nonRNA

$\lambda_2 = 2.0000$ single strand RNA — $\lambda_2 = 3.0000$ RNAlike — $\lambda_2 = 3.0000$ P3/P7 pseudoknot of the Tetrahymena ribozyme — $\lambda_2 = 3.0000$ nonRNA — $\lambda_2 = 4.0000$ tmRNA pseudoknot 2 (PK2) of E. coli.

Figure 10.7. Top: Quantifying the tRNA graph using Laplacian matrix and eigenvalues. Middle and bottom: Tree (B) and dual graph (C) motif library segments from the RAG database. They are categorized into functional RNA, RNA-like, and hypothetical RNA (black graphs in C) using motif clustering. RNA-like tree and dual graphs can be used to search for novel ncRNAs.

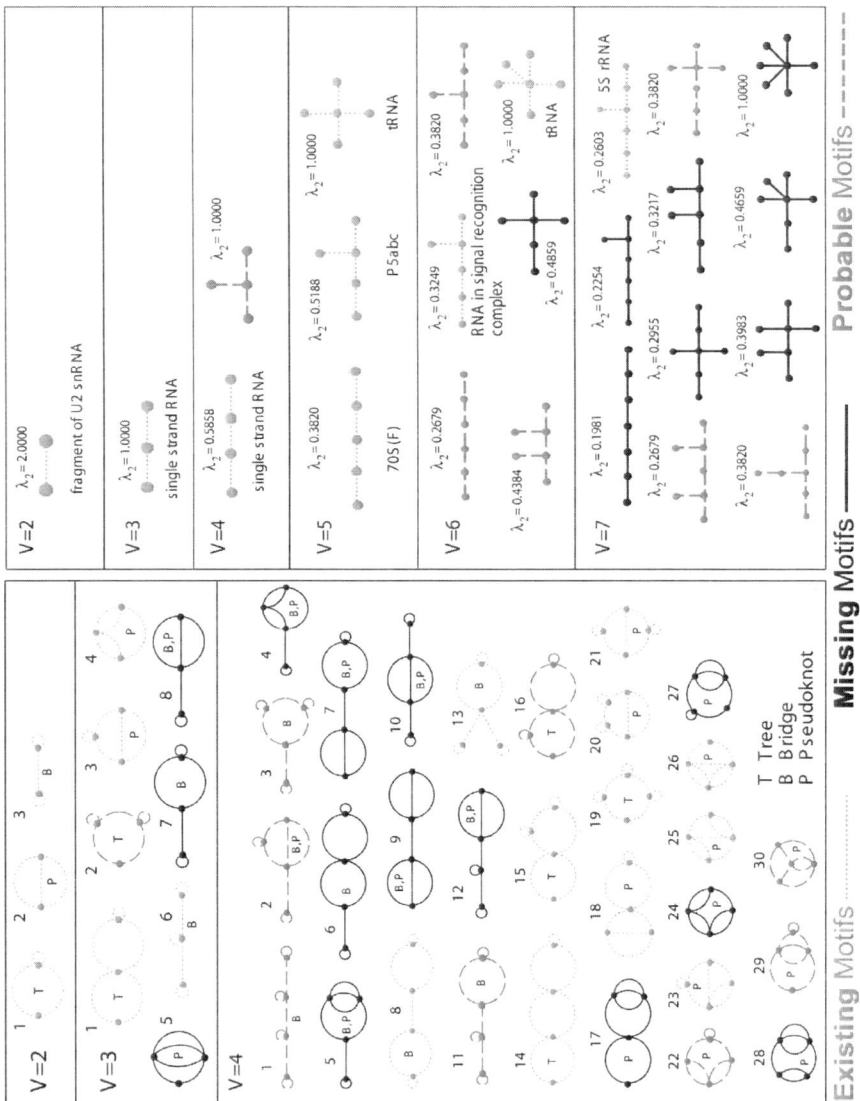

Figure 10.8. Segments of the RAG (http://monod.biomath.nyu.edu/rna) library for dual graphs (bottom) and tree graphs (top).

existing RNA motifs. This allowed us to use build-up procedures in combination with existing 2D folding packages to propose candidate sequences that will generate these target RNA motifs (Figure 10.7)[63]. Such cataloging of RNA structures, both existing and hypothetical, and their applications are important for the development of the RNA field on a large scale, *ribonomics*.

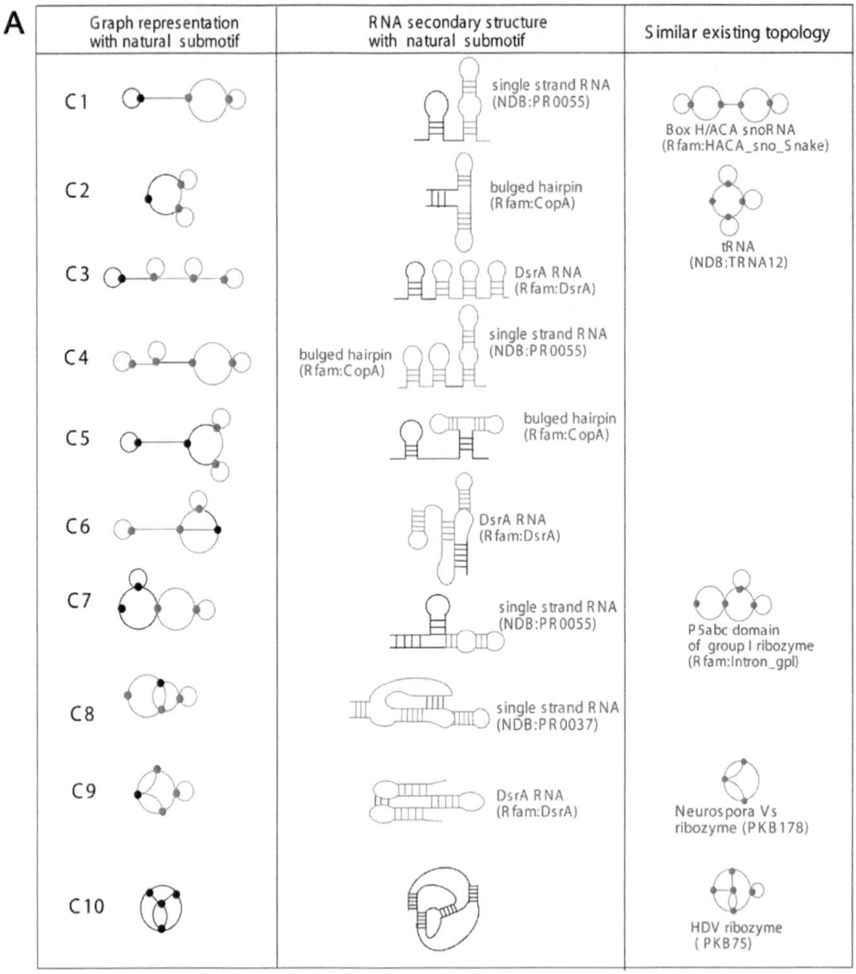

Figure 10.9a. Ten examples of predicted novel RNA-like (dual graph) topologies (1st column, labeled C1, ...,C10) shown with their secondary structures (2nd column) and natural submotifs (red lines) that occur in known RNAs. The sequences of those submotifs are used in a build-up procedure to generate candidate sequences for motifs C1 through C10, as shown in Figure 10.9b[63].

B

ID	Designed sequence	Novel RNA structure
C1	AACACAUCAGAUUUCCUGGUGUAA CGCCAAUGAGGUUUAUCCGAGGC	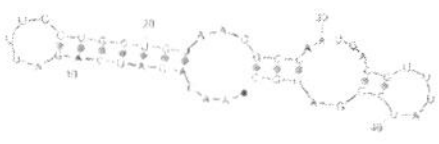
C2	AGCGCCGUGGCAGGGCUCAUAACC CUGAUGUCCUCGGAUCGAAACCGA GCGGCGCUACCA	
C3	AACACUCAGAUUUCCUGGUGUAAC GAAUUUUUUAAGUGCUUCUUGCUU AAGCAAGUUUCUACCCGACCCCCU CAGGGUCGGGAUUUUGGACCUCCA UGACGUUAUGGUCC	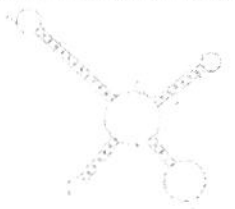
C4	AACACUCAGAUUGGACCUCAUGAC GUUAUGGUCCUUCCUGGUGUAACG AAUUUUUUAAGUGCUUCUUGCUUA AGCAAGUUUCUACCCGACCCCCUC AGGGUCGGGAUUU	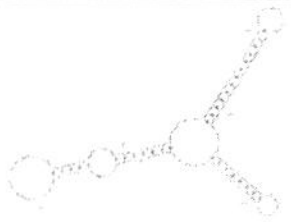
C5	CCUGGUAUUGCAGUACCUCCAGGU AGCGCCGUGGCAGGGCUCAUAACC CUGAUGUCCUCGGAUCGAAACCG AGCGGCGCUACCA	
C6	AGACCGUCAAACACAGACUAAAUGU CGGUCGGGGAAGAUGUAUUCUUCU CAUAAGAUAUAGUCGGCCUGGUAU UGCAGUACCUCCAGGU	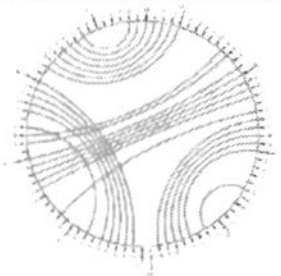
C7	GGCAGUACCAAGUCGCGAAAGCGA UGAUGGUAAGCCUUGCAAAGGGUU AAGCUGCC	
C8	not yet found	
C9	CUUCUUAUAUGAUUAGGUUGUCAU UUAGAAUAAGAAAACCUGGUAUUG CAGUACCUCCAGGUUAACCUG	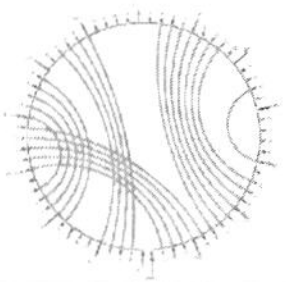
C10	not yet found	

Figure 10.9b. Designed candidate sequences that "fold" into the target RNA-like motifs (see Figure 10.9a) using a modular assembly approach where fragments from existing RNAs are assembled and folded[63]. Note that motifs C6 and C9 are pseudoknots.

While the protein folding problem (of course to *us*, not to Nature...) has received much attention — due to the enormous intellectual challenge, not to speak of practical applications to drug design — there is an analogous problem in RNA. In fact, because RNA folding is hierarchical, with secondary-structural elements forming first and then forming tertiary interactions independently, deducing RNA tertiary folds from the primary sequence might be simpler [65,66]. Unfortunately, only a tiny fraction of the number of protein folding aficionados are addressing the analogous challenge for RNA.

In this context, a challenge in RNA folding is to understand how strong electrostatic repulsions between closely packed phosphates in RNA are alleviated. Indeed, the stability of compact RNA forms is strongly maintained through interactions with both monovalent and divalent cations and by pseudoknotting.

Predicting the secondary and tertiary folding of RNA is a difficult and ongoing enterprise[67,68,69]. Various 2D algorithms like MFOLD[70] predict the patterns of base pairing, the presence of base pair mismatches, and regions of unpaired bases (loops, bulges, junctions, etc.). Other programs have been developed[71,72,73,74]. Secondary structural elements are easier to identify through modeling combined with evolutionary and database relation-ships[75,76]. Though imperfect, especially for long RNAs, these pre-dictions provide opportunities for learning what works, as well as what fails, in structure prediction for RNA.

Emerging themes in RNA structure include the importance of metal ions and loops for structural stability, various groove binding motifs, architectural motifs tailored for intermolecular interactions[77], hierarchical folding, fast establishment of secondary structural elements, and extreme flexibility of the molecule as a whole[78,69,6,66].

At present, relatively successful algorithms are available to predict secondary structure of RNA molecules up around 200nt by calculating the most energetically favorable base-pairing schemes. However, discriminating among the possible tertiary interactions to obtain the final folded state remains a challenge[66].

A direct measurement of the complete folding pathway of the *Tetrahymena* ribozyme suggests that tertiary structure of the P4–P6 domain forms cooperatively within three seconds, but several minutes are required for complete folding of the catalytic center of the enzyme[79]. The folding pathways of large functional RNAs may also take minutes or longer.

Still, findings concerning the folding kinetics of *Tetrahymena* ribozyme[79] have suggested that, as thermodynamic data on tertiary structure interactions become available[66], the RNA folding problem might be easier to solve than protein folding[65,66]. Therefore, with advances in RNA synthesis and structure determination[80] and in the availability of thermodynamic data on tertiary interactions[66], it is likely that our understanding of RNA structure, RNA folding, and RNA's role in enzyme evolution will dramatically increase in the coming decade. These advances will undoubtedly propel *ribonomics*[81] and RNA design (e.g., Refs.[20,18,46,48]).

5.5 Designing Novel RNAs

Because of the many potential applications of RNA in biomedicine and technology[82], designing novel functional RNAs is a promising enterprise.

Typically, *in vitro* selection experiments are used to explore systematically the ability of large sequence pools of nucleic-acid building blocks to form RNAs with desired function[47,45,46,48]. Essentially, huge pools (of order 10^{15}) of substrate nucleotides are mixed in special apparati and amplified by PCR (polymerase chain reaction) to enhance the success of producing functional molecules with respect to desired function (catalysis) or binding activity. The process is iterated many times, thereby mimicking an evolutionary process by which the "fittest molecules" (i.e., those with high binding affinities for ligands) survive. While such *in vitro* technology has produced RNA *aptamers* — synthetic RNAs that bind to target biomolecules, antibiotics, or viruses — the success rate is quite low. This is because the sequence space 4^N, where N is the number of nucleotides in the RNA sequence, has a very low function to information ratio[83]. In other words, only a small fraction of the theoretically possible RNA sequences leads to actual functional RNAs. Thus, random sequence pools tend to produce a biased pattern of resulting products, and these may not encourage structural diversity[84,85].

Thus, enhanced technology using targeted sequence pools can potentially improve this yield. Novel RNAs could also be designed by novel computational techniques under development[63,74,50,54,51]. Such techniques may involve structured RNA libraries[86,87]. Recent work has also shown that ribozyme engineering[88] and molecular design of RNAs by combining self-folding building blocks and optimizing connecting regions has great promise[89,90,63]. New mathematical tricks to analyze *in vitro* pools (e.g., Ref.[85]) and additional engineering tools to describe targeted libraries may prove fruitful as complementary techniques.

6. INVITATION TO COMPUTATIONAL BIOLOGISTS

Given the exciting new discoveries concerning RNA's starring role in gene regulation and the endless possibilities of engineering and utilizing RNA structure for medicine and technology, it is not difficult to understand the sharp rise of experimental efforts to discover and exploit RNA's diversity. Indeed, numerous companies focusing on RNA technology and its biomedical applications, for example for antibiotic and antiviral agents.

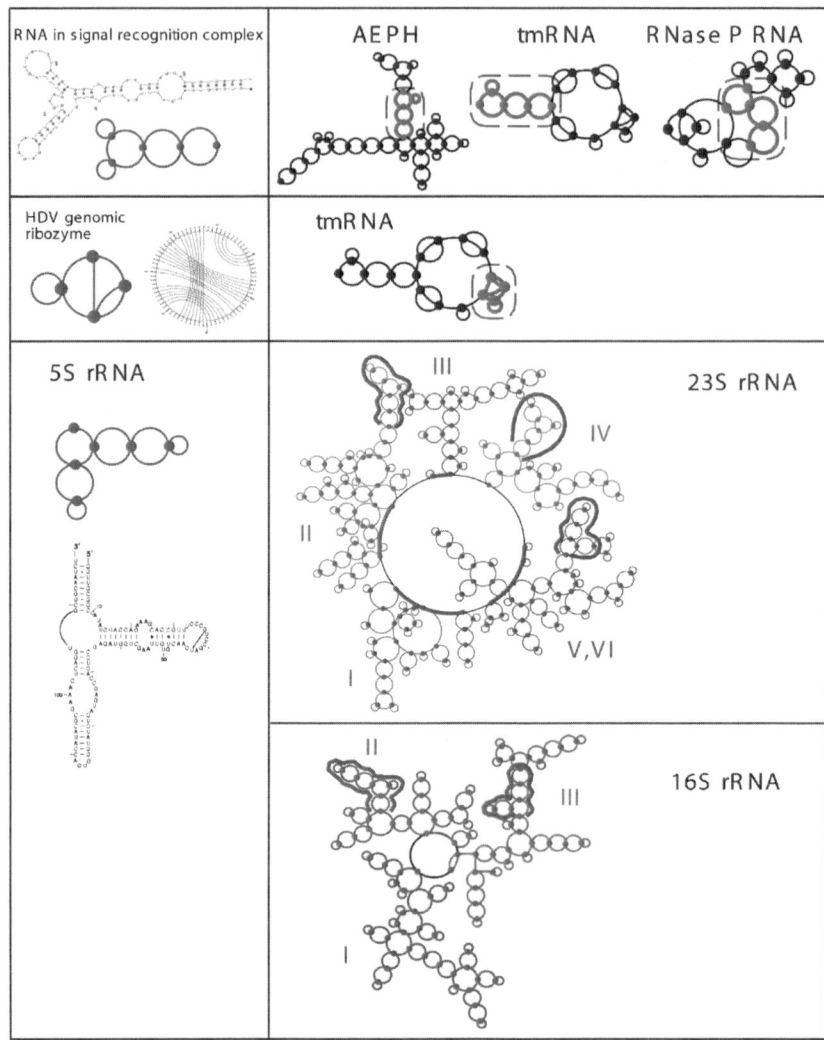

Figure 10.10. RNA topologies within larger RNAs. The search for submotifs can be performed automatically using the concept of graph isomorphism[91].

At the same time, numerous new opportunities become available to computational scientists who can apply and develop tools in dynamic programming, molecular modeling, cluster analysis, statistics, and network theory to pressing problems in RNA structure and function.

For example, work in the RAG group has shown that graph theory can be used to catalogue and enumerate RNA motifs[62,61]; identify RNA submotifs within larger RNAs and find new structural and functional similarity between pairs of RNAs[62,91] (see Figure 10.10); predict and design novel

RNAs by a build-up procedure that targets the most RNA-like motifs among all the candidate, hypothetical RNAs[63]; enhance *in vitro* selection by a targeted design approach[85]; and search for small functional RNAs in genomes[92,93]. These applications rely on the critical advantages of graph theory: the much smaller motif space (enumerated via graphical objects) compared to the sequence space available for RNA, and the automation that such compact mathematical descriptions allow. These critical differences allow exhaustive analysis of the RNA pattern space, subject to the usual modeling limitations.

For now, predicting the *tertiary* folds of RNA, even given the secondary structure, lurks at a distance, but there is no doubt that the dedicated RNA community will make great advances in this challenge too in the coming decade.

ACKNOWLEDGEMENTS

I thank Carse Ramos, Melanie Hill-Cantey, Yanli Wang, and Karunesh Arora for their excellent technical assistance and Hin Hark Gan for his critical reading of this chapter and contribution to Table 1. This work is supported by Human Frontier Science Program (HFSP), by the National Institute of Health (NIGMS award R01-GM055164), and by NSF/NIGMS Initiative in Mathematical Biology (DMS-0201160).

REFERENCES

[1]   Eddy, S. R. (2001). Non-coding RNA genes and the modern RNA world. *Nat. Rev. Genet.* **2**, 919–929.

[2]   Mattick, J. S. (2004). The hidden genetic program of complex organisms. *Sci. Amer.* **291**, 61–67.

[3]   Schlick, T. (2005). The critical collaboration between art and science: *Applying an Experiment on a Bird in an Air Pump to the Ramifications of Genomics on Society. Leonardo*, **38**, 323–329.

[4]   Moore, P. B. (1999). Structural motifs in RNA. *Ann. Rev. Biochem.* **68**, 287–300.

[5]   Scott, W. G., Finch, J. T. & Klug, A. (1995). The crystal structure of an all-RNA hammerhead ribozyme: A proposed mechanism for RNA catalytic cleavage. *Cell*, **81**, 991–1002.

[6]   Ferré-D'Amaré, A. R. & Doudna, J. A. (1999). RNA folds: Insights from recent crystal structures. *Ann. Rev. Biophys. Biomol.* Struc. **28**, 57–73.

[7]   Shi, H. & Moore, P. (2000). The crystal structure of yeast phenylalanine tRNA at 1.93 Å resolution: A classic structure revisited. *RNA*, **6**, 1091–1105.

[8]   Ferré-D'Amaré, A. R., Zhou, K. & Doudna, J. A. (1998). Crystal structure of a hepatitis delta virus ribozyme. *Nature*, **395**, 567–574.

[9]   Cate, J. H., Yusupov, M. M., Yusupova, C. Z., Earnest, T. N. & Noller, H. F. (1999). X-ray crystal structure of 70S ribosome functional complexes. *Science*, **285**, 2095–2104.

[10] Cech, T. R. (2000). The ribosome is a ribozyme. *Science*, **289**, 878–879.

[11] Ban, N., Nissen, P., Hansen, J., Moore, P. B. & Steitz, T. A. (2000). The complete atomic structure of the large ribosomal subunit at 2.4 Å resolution. *Science*, **289**, 905–920.

[12] Nissen, P., Hansen, J., Ban, N., Moore, P. B. & Steitz, T. A. (2000). The structural basis of ribosome activity in peptide bond synthesis. *Science*, **289**, 920–930.

[13] Williamson, J. R. (2000). Small subunit, big science. *Nature*, **407**, 306–307.

[14] Wimberly, B. T., Brodersen, D. E., Clemons Jr., W. M., Morgan-Warren, R. J., Carter, A. P., Vonrhein, C., Hartsch, T. & Ramakrishnan, V. (2000). Structure of the 30S ribosomal subunit. *Nature*, **407**, 327–339.

[15] Carter, A. P., Clemons, W. M., Brodersen, D. E., Morgan-Warren, R. J., Wimberly, B. T. & Ramakrishnan, V. (2000). Functional insights from the structure of the 30S ribosomal subunit and its interactions with antibiotics. *Nature*, **407**, 340–348.

[16] Schluenzen, F., Tocilj, A., Zarivach, R., Harms, J., Gluehmann, M., Janell, D., Bashan, A., Bartels, H., Agmon, I., F. Franceschi & Yonath, A. (2000). Structure of functionally activated small ribosomal subunit at 3.3 Å resolution. *Cell*, **102**, 615–623.

[17] Yusupov, M. M., Yusupova, G. Z., Baucom, A., Lieberman, K., Earnest, T. N., Cate, J. H. D. & Noller, H. F. (2001). Crystal structure of the ribosome at 5.5 Å resolution. *Science*, **292**, 883–896.

[18] Schultes, E. A. & Bartel, D. B. (2000). One sequence, two ribozymes: Implications for the emergence of new ribozyme folds. *Science*, **289**, 448–452.

[19] Soukup, G. A. & Breaker, R. R. (2000). Allosteric nucleic acid catalysts. *Curr. Opin. Struct. Biol.* **10**, 318–325.

[20] Tang, J. & Breaker, R. R. (2001). Structural diversity of self-cleaving ribozymes. *Proc. Natl. Acad. Sci. USA*, 97, 5784–5789.

[21] Read, J. S. & Joyce, G. F. (2002). A ribozyme composed of only two different nucleotides. *Nature*, **420**, 841–844.

[22] Lau, N. C. & Bartel, D. P. (2003). Censors of the genome. *Sci. Amer.* **289**, 34–41.

[23] Gregory, R. I. & Shiekhattar, R. (2005). MicroRNA biogenesis and cancer. *Cancer Res.* **65**, 3509–3512.

[24] Lu, J., Getz, G., Miska, E. A., Alvarez-Saavedra, E., Lamb, J., Peck, D., Sweet-Cordero, A., Ebert, B. L., Mak, R. H., Ferando, A. A., Downing, J. R., Jacks, T., Horvitz, H. R. & Golub, T. R. (2005). MicroRNA expression profiles classify human cancers. *Nature*, **435**, 834–838.

[25] McManus, M. T. (2003). MicroRNAs and cancer. *Semin. Cancer Biol.* **13**, 253–258.

[26] Jacob, F. & Monod, J. (1961). Genetic regulatory mechanisms in the synthesis of proteins. *J. Mol. Biol.* **3**, 318–356.

[27] Storz, G. (2002). An expanding universe of noncoding RNAs. *Science*, **296**, 1260–1263.

[28] Carrington, J. C. & Ambros, V. (2003). Role of microRNAs in plant and animal development. *Science,* **301**, 336–338.

[29] Plasterk, R. H. A. (2002). RNA silencing: the genome's immune system. *Science*, **296**, 1263–1265.

[30] Zamore, P. D. (2002). Ancient pathways programmed by small RNAs. *Science*, **296**, 1265 1269.

[31] Felsenfeld, G. & Groudine, M. (2003). Controlling the double helix. *Nature*, **421**, 448–453.

[32] Ahlquist, P. (2002). RNA-dependent RNA polymerases, viruses, and RNA silencing. *Science*, **296**, 1270–1273.

[33] Li, H., Li, W. X. & Ding, S. W. (2002). Induction and suppression of RNA silencing by an animal virus. *Science*, 296, 1319–1321.

[34] Kamath, R. S., Fraser, A. G., Dong, Y., Poulin, G., Durbin, R., Gotta, M., Kanapin, A., Bot, N. L., Moreno, S., Sohrmann, M., Welchman, D. P., Zipperlen, P. & Ahringer, J. (2003). Systematic functional analysis of the Caenorhabditis elegans genome using RNAi. *Nature*, 421, 231–237.

[35] Ashrafi, K., Chang, F. Y., Watts, J. L., Fraser, A. G., Kamath, R. S., Ahringer, J. & Ruvkun, G. (2003). Genome-wide RNAi analysis of Caenorhabditis elegans fat regulatory genes. *Nature*, 421, 268–272.

[36] Winkler, W., Nahvi, A. & Breaker, R. R. (2002). Thiamine derivatives bind messenger RNAs directly to regulate bacterial expression. *Nature*, 419, 952–956.

[37] Nahvi, A., Sudarsan, N., Ebert, M. S., Zou, X., Brown, K. L. & Breaker, R. R. (2002). Genetic control by a metabolite binding mRNA. *Chem. Biol.* 9, 1043–1049.

[38] Ke, A., Zhou, K., Ding, F., Cate, J. & Doudna, J. (2004). A conformational swithc controls hepatitis delta virus ribozyme catalysis. Nature, 429, 201–205.

[39] Szostak, J. W. (2002). RNA gets a grip on translation. *Nature*, 419, 890–891.

[40] Rose, G. (1997). Protein folding and the Paracelsus challenge. *Nature Struc. Biol.* 4, 512–514.

[41] Couzin, J. (2002). Small RNAs make big splash. *Science*, 298, 2296–2297.

[42] Pearson, N. D. & Prescott, C. D. (1997). RNA as a drug target. *Chem. Biol.* 4, 409–414.

[43] Tereshko, V., Skripkin, E. & Patel, D. J. (2003). Encapsulating streptomycin withing a small 40-mer RNA. *Chem. Biol.* 10, 175–187.

[44] Piganeau, N. & Schroeder, R. (2003). Aptameter structures: a preview into regulatory pathways? *Chem. Biol,* 10, 103–104.

[45] Wilson, D. & Szostak, J. (1999). in vitro selection of functional nucleic acids. Ann. Rev. *Biochem.* 68, 611–647.

[46] Ellington, A. D. & Szostak, J. W. (1990). in vitro selection of RNA molecules that bind specific ligands. *Nature*, 346, 818–822.

[47] Jäschke, A. (2001). Artificial ribozymes and deoxyribozymes. *Curr. Opin. Struct. Biol.* 11, 321–326.

[48] Tuerk, C. & Gold, L. (1990). Systematic evolution of ligands by exponential enrichment: RNA ligands to bacteriophage T4 DNA polymerase. *Science*, 249, 505–570.

[49] Hershberg, R., Altuvia, S. & Margalit, H. (2003). A survey of small RNA-encoding genes in Escherichia Coli. *Nucl. Acids Res.* 31, 1813–1820.

[50] Rivas, E., Klein, R., Jones, T. & Eddy, S. (2001). Computational identification of noncoding RNAs in E-coli by comparative genomics. *Curr. Biol.* 11, 1369–1373.

[51] Carter, R., Dubchak, I. & Holbrook, S. (2001). A computational approach to identify genes for functional RNAs in genomic sequences. *Nucl. Acids Res.* 29, 3928–3938.

[52] Macke, T. J., Ecker, D. J., Gutell, R. R., Gautheret, D., Case, D. A. & Sampath, R. (2001). RNA motif, an RNA secondary structure definition and search algorithm. *NAR*, 29, 4724–4735.

[53] Rhoades, M., Reinhart, B. J., Lim, L. P., Burge, C. B., Bartel, B. & Bartel, D. P. (2002). Prediction of plant microRNA targets. *Cell*, 110, 513–520.

[54] Wassarman, K., Repoila, F., Rosenow, C., Storz, G. & Gottesman, S. (2001). Identification of novel small RNAs using comparative genomics and microarrays. *Genes Dev.* 15, 1637–1651.

[55] Schlick, T. (2002). *Molecular Modeling and Simulation: An Interdisciplinary Guide.* Springer-Verlag, New York, NY.

[56] Vitkup, D., Melamud, E., Moult, J. & Sander, C. (2001). Completeness in structural genomics. *Nat. Struc. Biol.* **8**, 559–565.

[57] Burley, S. & Bonanno, J. (2004). Structural genomics. *Methods Biochem. Anal.* **44**, 591–612.

[58] Waterman, M. & Smith, T. (1978). RNA secondary structure: a complete mathematical analysis. *Math. Biosci.* **42**, 257–266.

[59] Le, S., Nussinov, R. & Maizel, J. (1989). Tree graphs of RNA secondary structures and their comparisons. *Comput. Biomed. Res.* **22**, 461–473.

[60] Benedetti, G. & Morosetti, S. (1996). A graph-topological approach to recognition of pattern and similarity in RNA secondary structures. *Biophys. Chem.* **59**, 179–184.

[61] Fera, D., Kim, N., Shiffeldrim, N., Zorn, J., Laserson, U., Kim, N. & Schlick, T. (2004). RAG: RNA-As-Graphs web resource. *BMC Bioinformatics*, **5**, 88.

[62] Gan, H., Fera, D., Zorn, J., Tang, M., Shiffeldrim, N., Laserson, U., Kim, N. & Schlick, T. (2004). RAG: RNA-As-Graphics database – concepts, analysis, and features. *Bioinformatics*, **20**, 1285–1291.

[63] Kim, N., Shiffeldrim, N., Gan, H. & Schlick, T. (2004). Candidates for novel RNA topologies. *J. Mol. Biol.* **341**, 1129–1144.

[64] Gan, H. H., Pasquali, S. & Schlick, T. (2003). A survey of existing RNAs using graph theory with implications to RNA analysis and design. *Nucl. Acids Res.* **31**, 2926–2943.

[65] Batey, R. T. & Doudna, J. A. (1998). The parallel universe of RNA folding. *Nature Struc. Biol.* **5**, 337–340.

[66] Tinoco, Jr., I. & Bustamante, C. (1999). How RNA folds. *J. Mol. Biol.* **293**, 271–281.

[67] Pyle, A. M. & Green, J. B. (1995). RNA folding. *Curr. Opin. Struct. Biol.* **5**, 303–310.

[68] Schuster, P., Stadler, P. F. & Renner, A. (1997). RNA structures and folding: From conventional to new issues in structure predictions. *Curr. Opin. Struct. Biol.* **7**, 229–235.

[69] Brion, P. & Westhof, E. (1997). Hierarchy and dynamics of RNA folding. *Ann. Rev. Biophys. Biomol. Struc.* **26**, 113–137.

[70] Zuker, M. M. (2003). Mfold web server for nucleic acid folding and hybridization prediction. *Nucl. Acids Res.* **31**, 3406. http://www.bioinfo.rpi.edu/zukerm.

[71] McCaskill, J. (1990). The equilibrium partition function and base pair binding probabilities for RNA secondary structure. *Biopolymers*, **29**, 1105–1119.

[72] Rivas, E. & Eddy, S. (1999). A dynamic programming algorithm for RNA structure prediction including pesudoknots. *J. Mol. Biol.* **285**, 2053–2068.

[73] Xayaphoummine, A., Bucher, T., Thalmann, F. & Isambert, H. (2003). Prediction and statistics of pseudoknots in RNA structures using exactly clustered stochastic simulations. *Proc. Natl. Acad. Sci. USA*, **100**, 15310–15315.

[74] Andronescu, M., Fejas, A., Hutter, F., Hoos, H. & Condon, A. (2004). A new algorithm for RNA secondary structure design. *J. Mol. Biol.* **336**, 607–624.

[75] Zuker, M. & Stiegler, P. (1981). Optimal computer folding of large RNA sequences using thermodynamics and auxiliary information. *Nucl. Acids Res.* **9**, 133–148.

[76] Mathews, D. H., Sabina, J., Zuker, M. & Turner, D. H. (1999). Expanded sequence dependence of thermodynamic parameters improves prediction of RNA secondary structure. *J. Mol. Biol.* **288**, 911–940.

[77] Hermann, T. & Patel, D. J. (2000). Adaptive recognition by nucleic acid aptamers. *Science*, **287**, 820–825.

[78] Hagerman, P. J. (1997). Flexibility of RNA. *Ann. Rev. Biophys. Biomol. Struc.* **26**, 139–156.

[79] Sclavi, B., Sullivan, M., Chance, M. R., Brenowitz, M. & Woodson, S. A. (1998). RNA folding at millisecond intervals by synchrotron hydroxyl radical footprinting. *Science*, **279**, 1940–1943.

[80] Holbrook, S. R. & Kim, S.-H. (1997). RNA crystallography. *Biopolymers*, **44**, 3–21.

[81] Doudna, J. A. (2000). Structural genomics of RNA. *Nature Struc. Biol.* **7**, 954–956. (Structural Genomics Supplement).

[82] Puerta-Fernández, E., Romero-López, C., Barroso-delJesus, A. & Berzal-Herranz, A. (2003). Ribozymes: recent advances in the development of RNA tools. *FEMS Microbiol. Rev.* **27** (1), 75–97.

[83] Szostak, J. (2003). Molecular messages. *Nature*, **423**, 689.

[84] Carothers, J., Oestreich, S., Davis, J. & Szostak, J. (2004). Informational complexity and functional activity of RNA structures. *J. Amer. Chem. Soc.* **126**, 5130–5137.

[85] Gevertz, J., Gan, H. & Schlick, T. (2005). in *vitro* random rools are not structurally diverse: a computational analysis. *RNA*, **11**, 853–863.

[86] Davis, J. H. & Szostak, J. W. (2002). Isolation of high-affinity gtp aptamers from partially structured RNA libraries. *Proc. Natl. Acad. Sci.* **99**, 11616–11621.

[87] Lau, M. W., Cadieux, K. E. & Unrau, P. J. (2004). Isolation of fast purine nucleotide synthase ribozymes. *J. Amer. Chem. Soc.* **126**, 15686–15693.

[88] Cech, T. (1992). Ribozyme engineering. Curr. Opin. Struct. Biol. 2, 605–609.

[89] Ikawa, Y., Fukada, K., Watanabe, S., Shiraishi, S. & Inoue, T. (2002). Design, construction, and analysis of a novel class of self-folding RNA. *Structure*, **10**, 527–534.

[90] Chworos, A., Severcan, I., Koyfman, A., Weinkam, P., Oroudjev, E., Hansma, H. & Jaeger, L. (2004). Building programmable jigsaw puzzles with RNA. *Science*, **306**, 2068–2072.

[91] Pasquali, S., Gan, H. & Schlick, T. (2005). Modular RNA architecture revealed by computational analysis of existing pseudoknots and ribosomal RNAs. *Nucl. Acids Res.* **33**, 1384–1398.

[92] Laserson, U., Gan, H. & Schlick, T. (2004). Searching for 2D RNA geometries in bacterial genomes. *In Proceedings of the Twentieth Annual ACM Symposium on Computational Geometry, June 9–11* pp. 373–377. ACM Press.

[93] Laserson, U., Gan, H. & Schlick, T. (2005). Exploring the connection between synthetic and natural RNAs in genomes via a novel computational approach. In New Algorithms for Macromolecular Simulation, *Proceedings of the Fourth International Workshop on Algorithms for Macromolecular Modelling, Leicester, UK, August 2004*, (Chipot, C., Elber, R., Laaksonen, A., Leimkuhler, B., Mark, A., Schlick, T., Skeel, R. D. & Schuette, C., eds), vol. 49, of *Lecture Notes in Computational Science and Engineering* Springer-Verlag, New York.

Chapter 11

MOLECULAR DYNAMICS SIMULATIONS OF RNA SYSTEMS: IMPORTANCE OF THE INITIAL CONDITIONS

RNA molecular dynamics

Pascal Auffinger

Institut de Biologie Moléculaire et Cellulaire du CNRS, Modélisations et Simulations des Acides Nucléiques, UPR9002, 15 rue René Descartes, 67084 Strasbourg Cedex, France

Abstract: Several factors can lead to the generation of accurate and informative molecular dynamics simulations. Among them, the choice of an appropriate starting structure is primordial. Unfortunately, experimental structures are not void of inaccuracies and errors. The aim of this review is to provide practical guidelines for detecting and correcting such inaccuracies in order to start molecular dynamics simulations of nucleic acid systems under the best possible conditions.

Key words: Molecular dynamics simulation; RNA; nucleic acids; interatomic interactions

1.   INTRODUCTION

Since 1995, we have witnessed a significant raise in the number of molecular dynamics (MD) simulations of RNA systems (see Table 11-1), essentially driven by: *(i)* a rapid increase in available computational power, *(ii)* the constant improvement of MD simulation protocols and *(iii)* the accumulation of experimental data steered by a growing interest of the community of biologists for these biomolecular systems. Yet, the history of MD simulations of nucleic acids has made clear that the generated models are often only informative on a time scale that is directly linked with our current structural and chemical knowledge of these systems. For instance, the first

J. Šponer and F. Lankaš (eds.), Computational Studies of RNA and DNA, 283–300.
© 2006 *Springer.*

Table 11-1. List of MD simulations of RNA systems (up to October 2005) using an explicit representation of the solvent and Ewald summation methods for the treatment of the long-range electrostatic interactions (simulations using truncation[43,49,60] or density functional methods[61] are not listed in the table).

Starting structures	nt	length (ns)	ions	modified nucleotide	forcefield	water model	ref.
MODEL STRUCTURES							
Single strand A$_3$, U$_3$	3	>60	K$^+$		AMBER	TIP3P	62
Single strand A$_6$, U$_6$	6						
HNA.RNA duplex	16	1.1	Na$^+$	HNA	AMBER	TIP3P	63
MOE.RNA duplex	20	1.3	Na$^+$	MOE	AMBER	TIP3P	64
PNA.RNA duplex	12	2.5	Na$^+$	PNA	AMBER	TIP3P	65
Amide-3 DNA/RNA hydrids	12	10.0	NH$_4^+$	Amide-3 linkage; Am; Gm	CHARMM	TIP3P	66
r{(ApU)$_{12}$}$_2$	48	2.4	≈ 0.2 M KCl		AMBER	SPC/E	67
r{(CpG)$_{12}$}$_2$	48	2.4	≈ 0.2 M KCl		AMBER	SPC/E	67-69
r{(CmpGm)$_{12}$}$_2$	48	4.4	≈ 0.2 M KCl	Cm; Gm	AMBER	SPC/E	69
r(CCAACGUUGG)$_2$	20	2.0	Na$^+$		AMBER	TIP3P	47
r(CGCGAAUUCGCG)$_2$	24	11.0	Na$^+$		AMBER	TIP3P	70
r(CGCGCG)$_2$	12	0.7	Na$^+$		CHARMM	Flexible SPC	71
r(CGCUGCG)$_2$	16	5.0	NaCl	F (fluorinated);	AMBER	TIP3P	72
r(CGUUACG)$_2$				P (phenyl)			
r(GAGUACUC)$_2$	16	5.0	≈ 0.3, ≈ 1.0 M NaCl		CHARMM	Modified TIP3P	73
r(GCGAGUACUCGC)$_2$	24						
r(CGCGAUCGCG)$_2$	20						
r(CCUUUCGAAAGG)$_2$	24						
r(GGCUGGCC)$_2$	16	5.0	≈ 0.3, ≈ 1.0 M NaCl		CHARMM	Modified TIP3P	74
r(GGCGUGCC)$_2$							
r(GACUGGUC)$_2$							
r(GACGUGUC)$_2$							
r(GGAUGUCC)$_2$							
r(GGAGUUCC)$_2$							
r(GGCUAGCC)$_2$							
r(GGCAUGCC)$_2$							
r(GACUAGUC)$_2$							
r(GACAUGUC)$_2$							
r(RNAA) hairpins	26	3.0	≈ 0.1 M NaCl		AMBER	TIP3P	13
Base pair opening study	18	-	≈ 0.1 M NaCl		AMBER	TIP3P	75
	10	1.4	Na$^+$	I	CHARMM	TIP3P	76
GAAG hairpin (folding)	8	2.0	≈ 0.3, ≈ 0.5 M NaCl		AMBER	TIP3P	77
U1A hairpin	21	5.0	NaCl		AMBER	TIP3P	78
Human U4 snRNA	62	3.0	K$^+$		AMBER	TIP3P	79
X-RAY STRUCTURES							
ApU and GpC steps (in crystal)	8	2.0	Na$^+$		AMBER	TIP3P	80
r(CmGmCmGmCmGm)$_2$	12	0.7	Na$^+$	Cm; Gm	CHARMM	Flexible SPC	71
r(GGACUUCGGUCC)$_2$	24	4.0	≈ 0.1 M NaCl		AMBER	TIP3P	81
r(UAAGGAGGUGAU)$_2$	24	5.0	≈ 0.3, ≈ 1.0 M NaCl		CHARMM	Modified TIP3P	73
r(GCCAGUUCGCUGGC)$_2$	28	3.0	Na$^+$, ≈ 0.1 M NaCl	br^{5}C	AMBER	TIP3P	82
tRNA^{Asp}	76	0.5	NH$_4^+$	D; Ψ; m^{1}G; m^{5}C; m^{5}U	AMBER	SPC/E	21,44
tRNA^{Asp} anticodon hairpin	17	0.5	NH$_4^+$	Ψ; m^{1}G	AMBER	SPC/E	35,44,83
5S rRNA loop E	24	10.0	Na$^+$ & Mg^{2+}		AMBER	TIP3P	84
	24	11.5	≈ 0.2, ≈ 1.0 M KCl & Mg^{2+}		AMBER	SPC/E	35,40,85
HIV (lai-mal) kissing loops	46	2.0-5.5	Na$^+$ & Mg^{2+}		AMBER	TIP3P	86
HIV (lai-mal) kissing loops	46	7.5	Na$^+$ & Mg^{2+}		AMBER	TIP3P	36

16S rRNA	81	5.5	≈ 0.1 M NaCl		AMBER	TIP3P	87
Pseudoknot	26	5.0	Na⁺	C⁺	AMBER	TIP3P	22
Hammerhead ribozyme	41	1.1	Na⁺ & Mg²⁺		AMBER	TIP3P	88
	41	0.8	≈ 0.1 M NaCl & Mg²⁺		AMBER	SPC/E	37,89
HDV ribozyme	78	8.0-16.0	Na⁺, Mg²⁺	C⁺	AMBER	TIP3P	90
Kink-turns	47	74.0	Na⁺		AMBER	TIP3P	95
RNA/ligand (x-ray)							
16S rRNA site A / neomycin B	21	10.0	Na⁺		AMBER	TIP3P	26
RNA/protein (x-ray)							
U1A RNA/protein complex	21	1.0	Na⁺		AMBER	TIP3P	91
	21	1.8	≈ 0.1 & ≈ 1.0 M NaCl		AMBER	SPC/E	92
	21	5.0	Na⁺ & Cl⁻		AMBER	TIP3P	78
	20	5.0	Na⁺, Cl⁻, Mg²⁺		AMBER	TIP3P	93
U2 snRNA/protein complex	23	2.2	K⁺		AMBER	TIP3P	94
U4 snRNA/protein complex	47	10.0	Na⁺		AMBER	TIP3P	51
5S rRNA loopE-HelixIV/L25	32	24.0	Na⁺ & Mg²⁺		AMBER	TIP3P	29
NMR STRUCTURES							
GCAA hairpin	12	0.2	Na⁺		OPLS	SPC/E	96
UUCG hairpin	12	1.0-2.0	Na⁺		AMBER	TIP3P	48,97
UUUU hairpin	10	10.0	Na⁺		AMBER	TIP3P	53
tRNA^Ala acceptor stem hairpin	22	2.5	NH₄⁺		CHARMM	Modified TIP3P	98
	22	2.0	Na⁺	I; 2AA; 2AP; IsoC; dU; Z; M; 7DAA	AMBER	TIP3P	99
HIV (lai) SL1 hairpin	23	10.0	Na⁺		AMBER	TIP3P	100
HIV TAR RNA hairpin	29	1.6	Na⁺	2AP	CHARMM	TIP3P	101
	30	2.0	Mg²⁺		AMBER	TIP3P	102
HIV (lai) kissing loop	46	5.5-20.0	Na⁺		AMBER	TIP3P	54,86
H3 kissing loop	28	16.0	Na⁺		AMBER	TIP3P	36
HCV IRES IIId domain	28	2.6	Na⁺ & Mg²⁺		AMBER	TIP3P	103
Telomerase RNA hairpin	30	20.0	≈ 0.1 M KCl		AMBER	TIP3P	104
RNA/ligand (NMR)							
FMN aptamer	35	1.7	Na⁺		AMBER	TIP3P	105
BIV tat-TAR complex	28	1.2	Na⁺		AMBER	TIP3P	106
TAR-aptamer complex	29	3.0	Na⁺		CHARMM	Modified TIP3P	107
RNA/protein (NMR)							
U1A RNA/protein complex	30	1.0	Na⁺		AMBER	TIP3P	91
dsRBD/dsRNA	30	2.0	Na⁺ & Cl⁻		AMBER	TIP3P	108

MD simulations of biomolecular systems, limited to the picosecond timescales without explicit consideration of the solvent, were certainly accurate enough for describing the fastest intermolecular motions.[1] However, they failed dramatically in reproducing the stability of simple nucleic acid duplexes over the 100 ps timescale.[2] Hence, modelers improved the informative power of their simulations by integrating in their models more and more subtle interactions such as long-range electrostatic interactions, at first considered as negligible mainly because the available computer power was not sufficient to take them into account. Indeed, the quality of a given MD simulation is often strongly correlated with the precision with which the initial conditions are known. This includes the starting structure, but also the manner in which all the important intermolecular interactions are taken into account by the force-field that is used. In that respect, an analogy can be made between MD simulations and weather forecasting methods for which a precise knowledge of the initial conditions is of similar paramount importance.[3] This review is therefore devoted to the examination of the various issues and traps that pave the road of a modeler when setting up a MD simulation of an RNA system with the additional aim of providing useful practical guidelines. Specific issues related to MD simulations of RNA systems using implicit solvation models can be found in [4,5] and will not be addressed here. The reader is also referred to other reviews on nucleic acid MD simulations.[4,6-9]

2.   VARIOUS ISSUES RELATED TO MD SETUPS

A large number of methodological issues related to the setup of MD simulations of RNA systems have already been addressed in a preceding review.[9] These issues are:

(i)   the choice of an adequate starting structure (model-built or derived from x-ray or NMR data);

(ii)   the importance of checking thoroughly the starting structures with respect to their protonation states, conformation and, more specifically for x-ray structures, interpretation of the solvent density maps;

(iii)   the programs and web sites that can be used for adding hydrogen atoms to an RNA system;

(iv)   the choice of an appropriate thermodynamic ensemble, environment (crystal, liquid, ...), ionic conditions, simulation packages and force-fields including water models;

(v)   parametrization issues of modified nucleotides and ligands;

(vi)   the treatment of long-range electrostatic interactions;

(vii)   equilibration and sampling problems;

(viii) multiple molecular dynamics simulations;

In the following, I will develop some important aspects associated with recent methodological advances and particularly insist on the importance of choosing and checking thoroughly the initial structure used to initiate an MD study.

2.1   Choosing an Appropriate Starting Structure ...

Among all possible choices, model-built structures are generally considered as the least precise. Yet, their degree of precision is strongly correlated to the complexity of the system that is investigated. For example, it is relatively straightforward to construct double-helices or simple stem-loops while it is certainly much more difficult to elaborate an accurate three-dimensional model of the largest RNA structures such as the ribosome or of other RNA/protein complexes.[10] Thus, one has occasionally to accept severe local or even global deviations from the "true" conformational space of the investigated system. Such inescapable deviations are difficult to pinpoint for the simple reason that the experimental structures are not known. The ways through which three-dimensional models can be made are multiple. They range from the hierarchical assembly of known three-dimensional building blocks[11] through the use of more complex methods based on the biopolymer chain elasticity (BCE) approach that are currently under development.[12] In some occurrences, it is also possible to consider the assembly of model-built parts with fragments extracted from experimental structures.[13]

NMR structures are commonly more precise than model-built structures and are often used when no crystallographic data are available.[14] One major drawback associated with this structural source is that it provides only very little information related to the hydration and the ionic environment of biomolecular systems.[15]

When derived from moderate to high resolution data, x-ray experiments certainly provide the most accurate source of starting structures since, besides precise coordinates for the solute, they also supply important information related to its hydration and ionic environment.[16] NMR and crystallographic structures as well as some model-built structures of nucleic acids are deposited in the NDB (Nucleic acid Database; http://ndbserver. rutgers.edu/)[17] and in the PDB (Protein Data Bank; http://www.rcsb.org/pdb/).[18]

2.2 … and Checking it

Yet, even with very precise starting structures such as those derived from high quality crystals, one has to be aware that various types of errors or inaccuracies can result from the interpretation of the experimental data that may subsequently affect to various degrees the quality of the generated MD trajectories. Some of these problems are listed below.

2.2.1 Protonation Issues

In specific structural contexts, nucleic acid residues can be found in their protonated form. This is true for standard nucleotides like A or C, but also for several naturally or synthetically modified nucleotides. Unfortunately, the direct detection of protonation states is beyond the ability of most experimental techniques, with the exception of neutron[19] and high resolution x-ray diffraction experiments.[20] This may lead to locally inaccurate starting structures. For instance, in a 500 ps long MD simulation of a tRNA^{Asp} molecule based on a crystal structure,[21] a reordering of its tertiary core has been observed. This reordering was first attributed to the absence of Mg^{2+} cations in the model. Later on it was proposed that an adenine located in the core of the tRNA was probably protonated and that this difference induced the calculated structural deviations.[22]

Working at different pH conditions can also modify the experimentally observed conformational equilibria. For example, the NMR structure of a tRNA^{lys} anticodon stem-loop obtained at an acidic pH is different from the canonical structures derived from x-ray data primarily because of the formation, at the beginning of the stem, of an A$^+$-C pair[23] instead of a more standard bifurcated A-C base pair.[24] Thus, caution has to be exerted in the choice of a starting structure that must reflect the chemical conditions in which one desires to investigate the properties of a biomolecular system.

Evidently, protonation states of charged amino acids and of drugs such as aminoglycosides[25] have also to be clearly defined before starting an MD simulation.[26]

2.2.2   Conformational Issues

In some instances, especially for medium to low-resolution crystal structures, the correct orientation of specific residues is difficult to derive from experimental data. This is true for the HIS, ASN, and GLU amino acids.[27] Hence, some groups have proposed automated methods based on the generation of "contact maps" (see, e.g. MolProbity; http://kinemage. biochem.duke.edu) for detecting incorrect or at least suspicious orientations of amino acid residues in experimental structures and, subsequently, proposing corrected orientations. In the case of a His residue, the correct orientation can be inferred from the number and type of hydrogen bond these residues might establish with their environment. These methods can also be used to detect misorientations of modified nucleotides like pseudouridines (Ψ) that are present in a large number of natural RNA systems.[28] Importantly, special residue names are provided by the PDB when the experimental data are not precise enough to distinguish unambiguously between the ASP/ASN and the GLU/GLN residues (http://www.rcsb. org/pdb/). These residue names are "ASX" and "GLX". Similarly, "UNK" and "A" mark undetermined residues and atoms. The code "A" applies genrally only to the terminal atoms of ASP and GLU and to the ring atoms of the HIS residues. It is obvious that starting an MD simulation with wrongly oriented residues may have serious consequences on the generated trajectories and on their interpretation.[29]

The methods based on "contact maps" are also very useful for detecting steric clashes that may be present in the experimental structures of nucleic acids. In these structures, the respective position of the bases is generally well determined while, for the sugar-phosphate backbone, there are generally too many degrees of freedom relative to the number of observable data.[30] Locating such steric clashes before starting a simulation is often required in order to avoid miss- and/or over-interpretations of the generated trajectories.

One has also to be aware that occasionally, experimental structures, due to a lack of observable data, can be locally wrong. A well-documented example is provided by the first NMR structures of an r(UUCG) tetraloop.[31] The first structure proposed an incorrect base-pairing pattern for the GU pair of the loop that was subsequently corrected. Another illustration of the difficulty of interpreting correctly experimental data is provided by a crystal structure of a DNA polymerase bound to DNA obtained at 2.8 Å. For this complex the authors proposed the formation of Hoogsteen pairs for the DNA duplex substrate.[32] However, an ensuing report suggests that a Watson-Crick base-pairing pattern is more appropriate.[33] The controversy is not yet resolved and points to important issues that can be encountered when dealing with experimental structures.

2.2.3 Solvent Density Maps

Solvent density maps derived from x-ray measurements provide data related to regions of high concentration of electrons located around the solutes. Unfortunately, such density maps do not contain any information related to the chemical type of the atoms located at those positions. Thus, based on current knowledge, crystallographers have to decide which atom type has to be associated with a given region of high electron density. This experimentally underdetermined problem leaves the door open to possible misinterpretations. A recent survey of the NDB has revealed that it is not unusual for water molecules to be assigned to sites with very large electron densities (above those expected for a water molecule). Such large densities are rather indicative of the presence of a metal ion.[34] Among other factors, temperature factors of water molecules lower than those of neighboring solvent molecules should incite to caution.

It has also been emphasized that some density spots resulting from the presence of anions (SO_4^{2-}, Cl^-, ...) in the vicinity of nucleic acids have possibly been wrongly assigned to divalent cations such as Mg^{2+} or Mn^{2+}.[35] It is therefore advised to consider such an option when evaluating a starting structure. Help can be provided by comparing anion binding maps with the crystallographic position of electron density spots. These maps were obtained through a thorough survey of nucleic acid structure.[35] Indeed, incorrectly assigned metal ions represent a major source of errors in MD simulations.[35-37]

As noted above, crystallographers should be encouraged to use the "A" atom code when there is an ambiguity associated with the type of solvent molecule that should be assigned to a particular region of high electron density. Such precautions of an "ethic" type may have profound implications on the quality of future MD studies.

On the other side, it has also been shown through a careful inspection of deposited structure factors (through the use of the SFCHECK program)[38] that a sizeable fraction of the position of crystallographic water molecules has poorly defined electron density. Consequently, these positions are not reliable.[34] Yet, the authors of this study concluded that the nucleic acid structures they analyzed were of rather uniform quality, with only very few structures exhibiting unusual values for the indicators they checked. Another useful way to assess the validity of crystallographic water molecules is to compare their position with those derived from stereochemical maps derived from statistical surveys of deposited crystal structures.[39] Indeed, there is hope that statistical surveys cancel out wrong and inaccurate data.

2.2.4   "Inappropriate" Ionic Concentrations

As stated before, experimental structures derived from crystallographic data are often used as starting points for simulating biomolecular systems in solution. Very frequently, it is necessary to take into account information related to the position of divalent cations. This is mainly due to the fact that MD simulations conducted on a nanosecond timescale are not long enough to ensure a correct sampling of the configurational space associated with these cations. Furthermore, Mg^{2+} cations are usually found in either a penta- or hexa-hydrated form. The associated desolvation process has a relaxation time in the order of microseconds and can consequently not be reproduced by current MD simulations. Hence, Mg^{2+} cations must be placed at crystallographically well defined positions. Yet, for the purpose of obtaining crystals that diffract well, crystallographers are sometimes inclined to raise the concentration of divalent cations in mother liquors that is no longer close to "*in vivo*" conditions. Consequently, some x-ray structures identify probably many more divalent cations than are needed in "*in vivo*" conditions and caution must be exerted in the interpretation of simulations that are based on structures containing a large number of divalent cations. For instance, five Mg^{2+} cations have been detected in the 5S rRNA loop E crystal structure while probably only one or two divalent cations are necessary "*in vivo*" for maintaining the structure of this internal loop.[40] A further example is provided by a crystal structure of an hydrolase where four Mg^{2+} cations were located close to the active site while it has been inferred through other experiments that only two Mg^{2+} cations are necessary for catalysis.[41] Hence, the authors concluded that the additional divalent cations seen around the catalytic site might result from crystallization artifacts.

2.2.5 Is Our Chemical Knowledge Always Appropriate?

A further "trap" in the interpretation of starting structures is related to our "limited" knowledge of the chemical properties of the atoms composing biochemical systems and of the type of interatomic interactions they can establish. For instance, besides classical hydrogen bonds, it is now accepted that: (i) C-H...O/N or C-H...π interactions, that enter into the category of weak hydrogen bonds and (ii) cation...π interactions that are observed in many crystal structures of proteins, play a significant structural role in biological systems.[42-44] Yet, the structural importance of these "parallel" types of interatomic interactions is certainly less well appreciated by the biological community. This is surely the case for the halogen bonds.[45] This unusual type of chemical bond (C-X...O with X = Cl,Br,I) that involves an electrostatic interaction between a polarized halogen atom and the lone pair electrons of an oxygen atom, although known since the 50s in organic chemistry,[45] has only been recently surveyed in structural biology.[46] Such type of "exotic" interactions are not integrated in classical force-fields rendering the simulations of systems containing bonded halogen atoms sometimes hazardous. With respect of these "exotic" interactions, it seems that structural biologists have only slightly opened the "treasure chest" of interatomic interactions already widely explored by the chemist community. Hence, many more specific and significant interatomic interactions remain to be uncovered in biological systems.

2.2.6 Can MD "Correct" Inaccurate Starting Conditions?

In the preceding section, several issues related to experimental imprecision in the starting structures have been raised. Yet, the modelers carry always the hope that during the course of an MD simulation, the simulation protocol and the associated force-field will be able to correct small deficiencies in the starting structure. This is may be true for the smallest components of biomolecular systems such as water molecules and monovalent cations (Na^+, K^+, ...). However, the complexity of the problem already increases when one considers Mg^{2+} cations that have very slow diffusion constants and for which, as noted above, current simulation lengths are too short. This is even more true when large conformational transitions are considered. For example, early attempts were made to convert an artificial B-RNA duplex into a more standard A-RNA structure. On the investigated nanosecond time scale, the expected conformational transition was not observed.[47] MD simulations starting from an incorrect NMR structure of an r(UUCG) tetraloop could not lead to the "correct" experimental structure.[48] Indeed, significant conformational changes can generally only be obtained by imposing artificial constraints to the system or by using various protocols that have been developed for achieving such goals like umbrella sampling,[49,50] local enhanced sampling (LES),[51-53] targeted molecular dynamics[54] or replica-exchange techniques.[55]

On the other hand, strong force-field dependences have been reported. For nucleic acids, early simulations made clear that starting DNA structure drifted, over the nanosecond time range, toward an A-form with the AMBER force-field and toward a B-form with the CHARMM force-field.[56] Subsequent versions of these force-fields have addressed this issue. Such considerations led to the proposal that, at least in the realm of MD simulations, the well known nucleic acid polymorphism coexists with a more difficultly apprehended force-field polymorphism.[7] Thus, one important goal of theoretical studies is to deconvolute these two effects. This is without doubt a complex task since it is expected that more such "force-field" manifestations will become manifest on the longer time scales. A recent report of a 50 ns long MD simulation of a B-DNA duplex[57] described an accumulation of non-canonical α/γ backbone torsion angles towards the end of the simulation that may indicate that the structure is evolving toward a "non-natural" state (see also [58]). Further studies will support or infirm this statement. There is also some concern about the parameters currently used for monovalent ions. Some studies reported the formation of unnatural ionic clusters in the bulk surrounding nucleic acid solutes (Auffinger & Vaiana, unpublished results).[8,59]

3.   CONCLUSIONS

With this review, it was my aim to stress the importance of carefully checking and understanding the initial conditions that are used both: *(i)* from a structural point of view but also *(ii)* from those associated with the energetics of intermolecular interactions. Indeed, many traps will have to be carefully identified and neutralized. It is only through such a long, often tedious but also fascinating process that progress can be achieved in the field.

ACKNOWLEDGEMENTS

Andrea Vaiana is acknowledged for numerous discussions and his help in "checking" this manuscript.

REFERENCES

1. Prabhakaran, M., Harvey, S. C., Mao, B. & McCammon, J. C. (1983). Molecular dynamics of phenylalanine transfer RNA. *J Biomol Struct Dyn* **1**, 357-369; McCammon, J. A. & Harvey, S. C. (1987 *Dynamics of proteins and nucleic acids.* (McCammon, J. A. & Harvey, S. C., Eds.), Cambridge University Press, New-York.
2. Swaminathan, S., Ravishanker, G. & Beveridge, D. L. (1991). Molecular dynamics of B-DNA including water and counterions: a 140 ps trajectory for d(CGCGAATTCGCG) based on the GROMOS force field. *J Am Chem Soc* **111**, 5027-5040.
3. Palmer, T. N. (2000). Predicting uncertainty in forecasts of weather and climate. *Rep Prog Phys* **63**, 71-116.
4. Zacharias, M. (2000). Simulation of the structure and dynamics of nonhelical RNA motifs. *Curr Op Struct Biol* **10**, 311-317.
5. Orozco, M. & Luque, F. J. (2000). Theoretical methods for the description of the solvent effect in biomolecular systems. *Chem Rev* **100**, 4187-4226.
6. Auffinger, P. & Westhof, E. (1998). Molecular dynamics of nucleic acids. In *Encyclopedia of Computational Chemistry* (Schleyer, P. v. R., Allinger, N. L., Clark, T., Gasteiger, J., Kollman, P. A., Schaefer III, H. F. & Schreiner, P. R., eds.), Vol. 5, pp. 1629-1640. Wiley & Sons, Chichester; Auffinger, P. & Westhof, E. (1999). Conformational preferences in RNA molecules as inferred from molecular dynamics simulations. In *Perspective in Structural Biology* (Vijayan, M., Yathindra, N. & Kolaskar, A. S., eds.), pp. 545-555. University Press, Hyderabad; Norberg, J. & Nilsson, L. (2002). Molecular dynamics applied to nucleic acids. *Acc Chem Res* **35**, 465-472; Giudice, E. & Lavery, R. (2002). Simulation of nucleic acids and their complexes. *Acc Chem Res* **35**, 350-357; Cheatham, T. E., 3rd & Young, M. A. (2000). Molecular dynamics simulation of nucleic acids: successes, limitations, and promise. *Biopolymers* **56**, 232-256; Cheatham, T. E. & Kollman, P. A. (2000). Molecular dynamics simulation of nucleic acids. *Annu Rev Phys Chem* **51**, 435-471; Cheatham, T. E. & Kollman, P. A. (1998). Molecular dynamics simulation of nucleic acids in solution: how sensitive are the results to small perturbations in the force field and environment? In *Structure, Motion, Interaction and Expression of Biological Macromolecules Proceedings of the Tenth Conversation, State University of New York, Albany, 1997* (Sarma, R. H. & Sarma, M. H., eds.), pp. 99-116. Adenine Press, New-York; Auffinger, P. & Vaiana, A. C. (2005). Hydrogens in RNA as visualized by molecular dynamics simulations. In *Hydrogen- and hydration-sensitive structural biology* (Niimura, N., Mizuno, H., Helliwell, J. & Westhof, E., eds.), pp. 227-234. Kubapro Co, Ltd, Tokyo.
7. Auffinger, P. & Westhof, E. (1998). Simulation of the molecular dynamics of nucleic acids. *Curr Op Struct Biol* **8**, 227-236.
8. Cheatham, T. E. (2004). Simulations and modeling of nucleic acid stucture, dynamics and interactions. *Cur Op Struct Biol* **14**, 360-367.
9. Auffinger, P. & Vaiana, A. C. (2005). Molecular dynamics simulations of RNA systems. In *Handbook of RNA biochemistry* (Westhof, Bindereif, Schön & Hartmann, eds.), pp. 560-576. Willey-VCH, Manheim.

10. Knight, W., Hill, W. & Lodmell, S. J. (2005). Ribosome builder: a software project to simulate the ribosome. *Comput Biol and Chem* **29**, 163-174; Tung, C. S. & Sanbonmatsu, K. Y. (2004). Atomic model of the *Thermus thermophilus* 70S ribosome developped *In Silico. Proc Natl Acad Sci USA* **87**, 2714-2722.

11. Massire, C. & Westhof, E. (1998). MANIP: An interactive tool for modelling RNA. *J Mol Graph* **16**, 197-205; Westhof, E. & Auffinger, P. (2000). RNA tertiary structure. In *Encyclopedia of analytical chemistry* (Meyers, R. A., ed.), pp. 5222-5232. John Wiley & Sons, Ltd, Chichester; Jossinet, F. & Westhof, E. (2005). Sequence to structure (S2S): display, manipulate and interconnect RNA data from sequence to structure. *Bioinformatics* **21**, 3320-3321.

12. Santini, G. P., Pakleza, C. & Cognet, J. A. (2003). DNA tri- and tetra-loops and RNA tetra-loops hairpins fold as elastic biopolymer chains in agreement with PDB coordinates. *Nucleic Acids Res* **31**, 1086-1096; Pakleza, C. & Cognet, J. A. (2003). Biopolymer Chain Elasticity: a novel concept and a least deformation energy principle predicts backbone and overall folding of DNA TTT hairpins in agreement with NMR distances. *Nucleic Acids Res* **31**, 1075-1085.

13. Schneider, C. & Sühnel, J. (2000). A molecular dynamics simulation study of coaxial stacking in RNA. *J Biomol Struct Dyn* **18**, 345-352.

14. Wüthrich, K. (1995). NMR-This other method for protein and nucleic acid structure determination. *Acta Cryst* **D51**, 249-270.

15. Halle, B. & Denisov, V. P. (1998). Water and monovalent ions in the minor groove of B-DNA oligonucleotides as seen by NMR. *Biopolymers* **48**, 210-233.

16. Sundaralingam, M. & Pan, B. (2002). Hydrogen and hydration of DNA and RNA oligonucleotides. *Biophys Chem* **95**, 273-282; Egli, M. (2002). DNA-cation interactions: quo vadis? *Chem Biol* **9**, 277-286.

17. Berman, H. M., Westbrook, J., Feng, Z., Iype, L., Schneider, B. & Zardecki, C. (2002). The nucleic acid database. *Acta Cryst* **D58**, 889-898.

18. Berman, H. M., Battistuz, T., Bhat, T. N., Bluhm, W. F., Bourne, P. E., Burkhardt, K., Feng, Z., Gilliland, G. L., Iype, L., Jain, S., Fagan, P., Marvin, J., Padilla, D., Ravichandran, V., Schneider, B., Thanki, N., Weissig, H., Westbrook, J. D. & Zardecki, C. (2002). The protein data bank. *Acta Cryst* **D58**, 899-907.

19. Niimura, N. (1999). Neutrons expand the field of structural biology. *Curr Opin Struct Biol* **9**, 602-608; Niimura, N., Chatake, T., Kurihara, K. & Maeda, M. (2004). Hydrogen and hydration in proteins. *Cell Biochem and Biophys* **40**, 351-369.

20. Petrova, T. & Podjarny, A. (2004). Protein crystallography at subatomic resolution. *Rep Prog Phys* **67**, 1565-1605.

21. Auffinger, P., Louise-May, S. & Westhof, E. (1999). Molecular dynamics simulations of the solvated yeast tRNA^{Asp}. *Biophys J* **76**, 50-64.

22. Csaszar, K., Spackova, N., Stefl, R., Šponer, J. & Leontis, N. B. (2001). Molecular dynamics of the frame-shifting pseudoknot from beet western yellow virus: the role of non-Watson-Crick base-pairing, ordered hydration, cation binding and base mutations on stability and unfolding. *J Mol Biol* **313**, 1073-1091.

23. Durant, P. C. & Davis, D. R. (1999). Stabilization of the anticodon stem-loop of tRNA^{Lys,3} by an A$^+$-C base pair and by pseudouridine. *J Mol Biol* **1999**, 115-131.

24. Auffinger, P. & Westhof, E. (2001). An extended structural signature for the tRNA anticodon loop. *RNA* **7**, 334-341.

25. Kaul, M., Barbieri, C. M., Kerrigan, J. E. & Pilch, D. S. (2003). Coupling of drug protonation to the specific binding of aminoglycosides to the A site of 16S rRNA: elucidation of the number of drug amino groups involved and their identities. *J Mol Biol*

326, 1373-1387; Pilch, D. S., Kaul, M., Barbieri, C. M. & Kerrigan, J. E. (2003). Thermodynamics of aminoglycoside-rRNA recognition. *Biopolymers* **70**, 58-79.

26. Asensio, J. L., Hidalgo, A., Cuesta, I., Gonzalez, C., Canada, J., Vicent, C., Chiara, J. L., Cuevas, G. & Jimenez-Barbero, J. (2002). Experimental evidence for the existence of non-exo-anomeric conformations in branched oligosaccharides: NMR analysis of the structure and dynamics of aminoglycosides of the neomycin family. *Chemistry* **8**, 5228-5240.

27. McDonald, I. K. & Thornton, J. M. (1995). The application of hydrogen bonding analysis in X-ray crystallography to help orientate asparagine, glutamine and histidine side chains. *Prot Eng* **8**, 217-224; Word, J. M., Lovell, S. C., Richardson, J. S. & Richardson, D. C. (1999). Asparagine and glutamine: using hydrogen atom contacts in the choice of side-chain amide orientation. *J Mol Biol* **285**, 1735-1747.

28. Auffinger, P. & Westhof, E. (1998). Effects of pseudouridylation on tRNA hydration and dynamics: a theoretical approach. In *Modification and editing of RNA* (Grosjean, H. & Benne, R., eds.), pp. 103-112. American Society for Microbiology, Washington, DC 2005.

29. Reblova, K., Spackova, N., Koca, J., Leontis, N. B. & Šponer, J. (2004). Long-residency hydration, cation binding, and dynamics of loop E/helix IV rRNA-L25 protein complex. *Biophys J* **87**, 3397-3412.

30. Davis, I. W., Murray, L. W., Richardson, J. S. & Richardson, D. C. (2004). MOLPROBITY: structure validation and all-atom contact analysis for nucleic acids and their complexes. *Nucleic Acids Res* **32**, W615-619; Murray, L. J., Arendall, W. B., 3rd, Richardson, D. C. & Richardson, J. S. (2003). RNA backbone is rotameric. *Proc Natl Acad Sci U S A* **100**, 13904-13909; Murray, L. J., Richardson, J. S., Arendall, W. B. & Richardson, D. C. (2005). RNA backbone rotamers--finding your way in seven dimensions. *Biochem Soc Trans* **33**, 485-487.

31. Varani, G., Cheong, C. & Tinoco, I., Jr. (1991). Structure of an unusually stable RNA hairpin. *Biochemistry* **30**, 3280-3289; Varani, G. (1995). Exceptionally stable nucleic acid hairpins. *Annu Rev Biophys Biomol Struct* **24**, 379-404.

32. Nair, D. T., Johnson, R. E., Prakash, S., Prakash, L. & Aggarwal, A. K. (2004). Replication by human DNA polymerase-iota occurs by Hoogsteen base-pairing. *Nature* **430**, 377-380.

33. Wang, J. (2005). DNA polymerases: Hoogsteen base-pairing in DNA replication? *Nature* **437**, E6-7.

34. Das, U., Chen, S., Fuxreiter, M., Vaguine, A. A., Richelle, J., Berman, H. M. & Wodak, S. J. (2001). Checking nucleic acid crystal structures. *Acta Cryst* **D57**, 813-828.

35. Auffinger, P., Bielecki, L. & Westhof, E. (2004). Anion binding to nucleic acids. *Structure* **12**, 379-388.

36. Reblova, K., Spackova, N., Šponer, J. E., Koca, J. & Šponer, J. (2003). Molecular dynamics simulations of RNA kissing-loop motifs reveal structural dynamics and formation of cation-binding pockets. *Nucleic Acids Res* **31**, 6942-6952.

37. Hermann, T., Auffinger, P. & Westhof, E. (1998). Molecular dynamics investigations of the hammerhead ribozyme RNA. *Eur J Biophys* **27**, 153-165.

38. Vaguine, A. A., Richelle, J. & Wodak, S. J. (1999). SFCHECK: a unified set of procedures for evaluating the quality of macromolecular structure-factor data and their agreement with the atomic model. *Acta Cryst* **D55**, 191-205.

39. Schneider, B., Cohen, D. & Berman, H. M. (1992). Hydration of DNA bases: analysis of crystallographic data. *Biopolymers* **32**, 725-750; Schneider, B., Cohen, D. M., Schleifer, L., Srinivasan, A. R., Olson, W. K. & Berman, H. M. (1993). A systematic method for studying the spatial distribution of water molecules around nucleic acid bases. *Biophys J*

65, 2291-2303; Schneider, B. & Berman, H. M. (1995). Hydration of the DNA bases is local. *Biophys J* **69**, 2661-2669; Schneider, B., Patel, K. & Berman, H. M. (1998). Hydration of the phosphate group in double-helical DNA. *Biophys J* **75**, 2422-2434; Auffinger, P. & Westhof, E. (1998). Hydration of RNA base pairs. *J Biomol Struct Dyn* **16**, 693-707.

40. Auffinger, P., Bielecki, L. & Westhof, E. (2004). Symmetric K^+ and Mg^{2+} ion binding sites in the 5S rRNA loop E inferred from molecular dynamics simulations. *J Mol Biol* **335**, 555-571.

41. Bailey, S., Sedelnikova, S. E., Blackburn, G. M., Abdelghany, H. M., Baker, P. J., McLennan, A. G. & Rafferty, J. B. (2002). The crystal structure of diadenosine tetraphosphate hydrolase from *caenorhabditis elegans* in free and binary complex forms. *Structure* **10**, 589-600.

42. Desiraju, G. & Steiner, T. (1999). *The weak hydrogen bond.*, Oxford University Press, New-York; Steiner, T. (2002). The hydrogen bond in the solid state. *Angew Chem Int Ed* **41**, 48-76; Dougherty, D. A. (1996). Cation-π interactions in chemistry and biology: a new view of benzene, Phe, Tyr, and Trp. *Science* **271**, 163-168; Gallivan, J. P. & Dougherty, D. A. (1999). Cation-p interactions in structural biology. *Proc Natl Acad Sci* **96**, 9459-9464; Nishio, M., Hirota, M. & Umezawa, Y. (1998). *The CH/π interaction. Evidence, nature, and consequences.*, Wiley-VCH, New York.

43. Auffinger, P., Louise-May, S. & Westhof, E. (1996). Molecular dynamics simulations of the anticodon hairpin of tRNA^{Asp}: structuring effects of C-H.O hydrogen bonds and of long-range hydration forces. *J Am Chem Soc* **118**, 1181-1189.

44. Auffinger, P., Louise-May, S. & Westhof, E. (1996). Hydration of C-H groups in tRNA. *Farad Discuss* **103**, 151-174.

45. Metrangolo, P., Neukirch, H., Pilati, T. & Resnati, G. (2005). Halogen bonding based recognition processes: a world parallel to hydrogen bonding. *Acc Chem Res* **38**, 386-395.

46. Auffinger, P., Hays, F. A., Westhof, E. & Ho, P. S. (2004). Halogen bonds in biological molecules. *Proc Natl Acad Sci* **101**, 16789-16794.

47. Cheatham, T. E. & Kollman, P. A. (1997). Molecular dynamics simulations highlights the structural differences among DNA:DNA, RNA:RNA, and DNA:RNA hybrid duplexes. *J Am Chem Soc* **119**, 4805-4825.

48. Miller, J. & Kollman, P. A. (1997). Theoretical studies of an exceptionally stable RNA tetraloop: observation of convergence from an incorrect NMR structure to the correct one using unrestrained molecular dynamics. *J Mol Biol* **270**, 436-450.

49. Hart, K., Nystrom, B., Ohman, M. & Nilsson, L. (2005). Molecular dynamics simulations and free energy calculations of base flipping in dsRNA. *RNA* **11**, 609-618.

50. Giudice, E., Varnai, P. & Lavery, R. (2003). Base pair opening within B-DNA: free energy pathways for GC and AT pairs from umbrella sampling simulations. *Nucleic Acids Res* **31**, 1434-1443.

51. Cojocaru, V., Nottrott, S., Klement, R. & Jovin, T. M. (2005). The snRNP 15.5K protein folds its cognate K-turn RNA: a combined theoretical and biochemical study. *RNA* **11**, 197-209.

52. Cojocaru, V., Klement, R. & Jovin, T. M. (2005). Loss of G-A base pairs is insufficient for achieving a large opening of U4 snRNA K-turn motif. *Nucleic Acids Res* **33**, 3435-3446.

53. Koplin, J., Mu, Y., Richter, C., Schwalbe, H. & Stock, G. (2005). Structure and dynamics of an RNA tetraloop: a joint moleular dynamics and NMR study. *Structure* **13**, 1255-1267.

54. Aci, S., Mazier, S. & Genest, D. (2005). Conformational pathway for the kissing complex -> extended dimer transition of the SL1 stem-loop from genomic HIV-1 RNA as monitored by targeted molecular dynamics techniques. *J Mol Biol* **351**, 520-530.

55. Sanbonmatsu, K. Y. & Garcia, A. E. (2002). Structure of Met-enkephalin in explicit aqueous solution using replica exchange molecular dynamics. *Proteins* **46**, 225-234.

56. Feig, M. & Pettitt, B. M. (1997). Experiment vs force fields: DNA conformation from molecular dynamics simulations. *J Chem Phys B* **101**, 7361-7363; Feig, M. & Pettitt, B. M. (1998). Structural equilibrium of DNA represented with different force-fields. *Biophys J* **75**, 134-149.

57. Varnai, P. & Zakrzewska, K. (2004). DNA and its counterions: a molecular dynamics study. *Nucleic Acids Res* **32**, 4269-4280.

58. Dixit, S. B., Beveridge, D. L., Case, D. A., Cheatham, T. E., Giudice, E., Lankaš, F., Lavery, R., Maddocks, J. H., Osman, R., Sklenar, H., Thayer, K. M. & Varnai, P. (2005). Molecular dynamics simulations of the 136 unique tetranucleotide sequences of DNA oligonucleotides. II. Sequence context effects on the dynamical structures of the 10 unique dinucleotides steps. *Biophys J* **89**, 3721-3740.

59. Mazur, A. K. (2003). Titration in silico of reversible B <=> A transitions in DNA. *J Am Chem Soc* **125**, 7849-7859.

60. Tang, Y. & Nilsson, L. (1999). Molecular dynamics simulations of the complex between human U1A protein and hairpin II of U1 small nuclear RNA and of free RNA in solution. *Biophys J* **77**, 1284-1305; Lahiri, A. & Nilsson, L. (2000). Molecular dynamics of the anticodon domain of yeast tRNA(Phe): codon- anticodon interaction. *Biophys J* **79**, 2276-2289; Sarzynska, J., Kulinski, T. & Nilsson, L. (2000). Conformational dynamics of a 5S rRNA hairpin domain containing loop D and a single nucleotide bulge. *Biophys J* **79**, 1213-1227; Norberg, J. & Nilsson, L. (1996). Constant pressure molecular dynamics simulations of the dodecamers d(GCGCGCGCGCGC)$_2$ and r(GCGCGCGCGCGC)$_2$. *J Chem Phys* **104**, 6052-6057; Auffinger, P., Louise-May, S. & Westhof, E. (1995). Multiple molecular dynamics simulations of the anticodon loop of tRNA^{Asp} in aqueous solution with counterions. *J Am Chem Soc* **117**, 6720-6726.

61. Hutter, J., Carloni, P. & Parrinello, M. (1996). Nonempirical calculations of a hydrated RNA duplex. *J Am Chem Soc* **118**, 8710-8712.

62. Yeh, I. C. & Hummer, G. (2004). Diffusion and electrophoretic mobility of single-stranded RNA from molecular dynamics simulations. *Biophys J* **86**, 681-689.

63. De Winter, H., Lescrinier, E., Van Aerschot, A. & Herdewijn, P. (1998). Molecular dynamics simulation to investigate differences in minor groove hydration of HNA/RNA hybrids as compared to HNA/DNA complexes. *J Am Chem Soc* **120**, 5381-5394.

64. Lind, K. E., Mohan, V., Manoharan, M. & Ferguson, D. M. (1998). Structural characteristics of 2'-O-(2-methoxyethyl)-modified nucleic acids from molecular dynamics simulations. *Nucleic Acids Res* **26**, 3694-3699.

65. Soliva, R., Sherer, E., Luque, F. J., Laughton, C. A. & Orozco, M. (2000). Molecular dynamics simulations of PNA.DNA and PNA.RNA duplexes in aqueous solution. *J Am Chem Soc* **122**, 5997-6008.

66. Nina, M., Fonne-Pfister, R., Beaudegnies, R., Chekatt, H., Jung, P. M., Murphy-Kessabi, F., De Mesmaeker, A. & Wendeborn, S. (2005). Recognition of RNA by amide modified backbone nucleic acids: molecular dynamics simulations of DNA-RNA hybrids in aqueous solution. *J Am Chem Soc* **127**, 6027-6038.

67. Auffinger, P. & Westhof, E. (2001). Water and ion binding around r(UpA)$_{12}$ and d(TpA)$_{12}$ oligomers - Comparison with RNA and DNA (CpG)$_{12}$ duplexes. *J Mol Biol* **305**, 1057-1072.

68. Auffinger, P. & Westhof, E. (2000). Water and ion binding around RNA and DNA (C,G)-oligomers. *J Mol Biol* **300**, 1113-1131; Auffinger, P. & Westhof, E. (2002). Melting of the solvent structure around a RNA duplex: a molecular dynamics simulation study. *Biophys Chem* **95**, 203-210.

69. Auffinger, P. & Westhof, E. (2001). Hydrophobic groups stabilize the hydration shell of 2'-O-methylated RNA duplexes. *Angew Chem Int Ed* **40**, 4648-4650.

70. Noy, A., Perez, A., Lankaš, F., Javier Luque, F. & Orozco, M. (2004). Relative flexibility of DNA and RNA: a molecular dynamics study. *J Mol Biol* **343**, 627-638.

71. Kulinska, K., Kulinski, T., Lyubartsev, A., Laaksonen, A. & Adamiak, R. W. (2000). Spatial distribution functions as a tool in the analysis of ribonucleic acids hydration - molecular dynamics studies. *Comp & Chem* **24**, 451-457.

72. Zacharias, M. & Engels, J. W. (2004). Influence of a fluorobenzene nucleobase analogue on the conformational flexibility of RNA studied by molecular dynamics simulations. *Nucleic Acids Res* **32**, 6304-6311.

73. Pan, Y. & MacKerell, A. D. (2003). Altered structural fluctuations in duplex RNA versus RNA: a conformational switch involving base pair opening. *Nucleic Acids Res* **31**, 7131-7140.

74. Pan, Y., Priyakumar, U. D. & MacKerell, A. D., Jr. (2005). Conformational determinants of tandem GU mismatches in RNA: insights from molecular dynamics simulations and quantum mechanical calculations. *Biochemistry* **44**, 1433-1443.

75. Varnai, P., Canalia, M. & Leroy, J. L. (2004). Opening mechanism of G.T/U pairs in DNA and RNA duplexes: a combined study of imino proton exchange and molecular dynamics simulation. *J Am Chem Soc* **126**, 14659-14667.

76. Sarzynska, J., Nilsson, L. & Kulinski, T. (2003). Effects of base substitutions in an RNA hairpin from molecular dynamics and free energy simulations. *Biophys J* **85**, 3445-3459.

77. Li, W., Ma, B. & Shapiro, B. (2001). Molecular dynamics simulations of the denaturation and refolding of an RNA tetraloop. *J Biomol Struct Dyn* **19**, 381-396.

78. Pitici, F., Beveridge, D. L. & Baranger, A. M. (2002). Molecular dynamics simulation studies of induced fit and conformational capture in U1A-RNA binding: Do molecular substates code for specificity? *Biopolymers* **65**, 424-435.

79. Guo, J., Daizadeh, I. & Gmeiner, W. H. (2000). Structure of the Sm binding site from human U4 snRNA derived from a 3 ns PME molecular dynamics simulation. *J Biomol Struct Dyn* **18**, 335-344.

80. Lee, H., Darden, T. & Pedersen, L. (1995). Accurate crystal molecular dynamics simulations using particle-mesh-Ewald: RNA dinucleotides - ApU and GpC. *Chem Phys Lett* **243**, 229-235.

81. Schneider, C., Brandl, M. & Sühnel, J. (2001). Molecular dynamics simulation reveals conformational switching of water-mediated uracil-cytosine base-pairs in an RNA duplex. *J Mol Biol* **305**, 659-667.

82. Sherer, E. C. & Cramer, C. J. (2002). Internal loop-helix coupling in the dynamics of the RNA duplex (GC*C*AGUUCGCUGGC)(2). *J Phys Chem B* **106**, 5075-5085.

83. Auffinger, P. & Westhof, E. (1996). H-bond stability in the tRNA[Asp] anticodon hairpin: 3 ns of multiple molecular dynamics simulations. *Biophys J* **71**, 940-954; Auffinger, P. & Westhof, E. (1997). RNA hydration: three nanoseconds of multiple molecular dynamics simulations of the solvated tRNA[Asp] anticodon hairpin. *J Mol Biol* **269**, 326-341; Auffinger, P. & Westhof, E. (1997). Rules governing the orientation of the 2'-hydroxyl group in RNA. *J Mol Biol* **274**, 54-63.

84. Reblova, K., Spackova, N., Stefl, R., Csaszar, K., Koca, J., Leontis, N. B. & Šponer, J. (2003). Non-Watson-Crick basepairing and hydration in RNA motifs: molecular dynamics of 5S rRNA loop E. *Biophys J* **84**, 3564-3582.

85. Auffinger, P., Bielecki, L. & Westhof, E. (2003). The Mg^{2+} binding sites of the 5S rRNA loop E motif as investigated by molecular dynamics simulations. *Chem Biol* **10**, 551-561.

86. Aci, S., Gangneux, L., Paoletti, J. & Genest, D. (2004). On the stability of different experimental dimeric structures of the SL1 sequence from the genomic RNA of HIV-1 in solution: a molecular dynamics simulation and electrophoresis study. *Biopolymers* **74**, 177-188.

87. Li, W., Ma, B. & Shapiro, B. A. (2003). Binding interactions between the core central domain of 16S rRNA and the ribosomal protein S15 determined by molecular dynamics simulations. *Nucleic Acids Res* **31**, 629-638.

88. Torres, R. A. & Bruice, T. C. (1998). Molecular dynamics study displays near in-line attack conformations in the hammerhead ribozyme self-cleavage reaction. *Proc Natl Acad Sci* **95**, 11077-11082; Torres, R. A. & Bruice, T. C. (2000). The mechanism of phosphodiester hydrolysis - Near in-line attack conformations in the hammerhead ribozyme. *J Am Chem Soc* **122**, 781-791.

89. Hermann, T., Auffinger, P., Scott, W. G. & Westhof, E. (1997). Evidence for a hydroxide ion bridging two magnesium ions at the active site of the hammerhead ribozyme. *Nucleic Acids Res* **25**, 3421-3427.

90. Krasovska, M. V., Sefcikova, J., Spackova, N., Šponer, J. & Walter, N. G. (2005). Structural dynamics of precursor and product of the RNA enzyme from the hepatitis delta virus as revealed by molecular dynamics simulations. *J Mol Biol* **351**, 731-748.

91. Reyes, C. M. & Kollman, P. A. (1999). Molecular dynamics studies of U1A-RNA complexes. *RNA* **5**, 235-244.

92. Hermann, T. & Westhof, E. (1999). Simulations of the dynamics at an RNA-protein interface. *Nat Struct Biol* **6**, 540-544.

93. Showalter, S. A. & Hall, K. B. (2005). Correlated motions in the U1 snRNA stem/loop 2: U1A RBD1 complex. *Biophys J* **89**, 2046-2058.

94. Guo, J. & Gmeiner, W. H. (2001). Molecular dynamics simulation of the human U2B protein complex with U2 snRNA hairpin IV in aqueous solution. *Biophys J* **81**, 630-642.

95. Razga, F., Koca, J., Šponer, J. & Leontis, N. B. (2005). Hinge-like motions in RNA kink-turns: the role of the second a-minor motif and nominally unpaired bases. *Biophys J* **88**, 3466-3485.

96. Zichi, D. A. (1995). Molecular dynamics of RNA with the OPLS force field. Aqueous simulation of a hairpin containing a tetranucleotide loop. *J Am Chem Soc* **117**, 2957-2969.

97. Cheatham, T. E., Miller, J. L., Fox, T., Darden, T. A. & Kollman, P. A. (1995). Molecular dynamics simulations on solvated biomolecular systems: the particle mesh Ewald method leads to stable trajectories of DNA, RNA and proteins. *J Am Chem Soc* **117**, 4193-4194.

98. Nina, M. & Simonson, T. (2002). Molecular dynamics of the tRNA(Ala) acceptor stem: Comparison between continuum reaction field and particle-mesh Ewald electrostatic treatments. *J Phys Chem B* **106**, 3696-3705.

99. Nagan, M. C., Beuning, P., Musier-Forsyth, K. & Cramer, C. J. (2000). Importance of discriminator base stacking interactions: molecular dynamics analysis of A73 microhelix[Ala] variants. *Nucleic Acids Res* **28**, 2527-2534; Nagan, M. C., Kerimo, S. S., Musierforsyth, K. & Cramer, C. J. (1999). Wild-type tRNA microhelix(Ala) and 3:70 variants: molecular dynamics analysis of local helical structure and tightly bound water. *J Am Chem Soc* **121**, 7310-7317; Beuning, P. J., Nagan, M. C., Cramer, C. J., Musier-Forsyth, K., Gelpi, J. L. & Bashford, D. (2002). Efficient aminoacylation of the tRNA(Ala) acceptor stem: dependence on the 2:71 base pair. *RNA* **8**, 659-670.

100.Kieken, F., Arnoult, E., Barbault, F., Paquet, F., Huynh-Dinh, T., Paoletti, J., Genest, D. & Lancelot, G. (2002). HIV-1(Lai) genomic RNA: combined use of NMR and molecular dynamics simulation for studying the structure and internal dynamics of a mutated SL1 hairpin. *Eur Biophys J* **31**, 521-531; La Penna, G., Genest, D. & Perico, A. (2003). Modeling the dynamics of the solvated SL1 domain of HIV-1 genomic RNA. *Biopolymers* **69**, 1-14.

101.Kulinski, T., Olejniczak, M., Huthoff, H., Bielecki, L., Pachulska-Wieczorek, K., Das, A. T., Berkhout, B. & Adamiak, R. W. (2003). The apical loop of the HIV-1 TAR RNA hairpin is stabilized by a cross-loop base pair. *J Biol Chem* **278**, 38892-38901.

102.Golebiowski, J., Antonczak, S., Fernandez-Carmona, J., Condom, R. & Cabrol-Bass, D. (2004). Closing loop base pairs in RNA loop-loop complexes: structural behavior, interaction energy and solvation analysis through molecular dynamics simulations. *J Mol Model* **10**, 408-417.

103.Golebiowski, J., Antonczak, S., Di-Giorgio, A., Condom, R. & Cabrol-Bass, D. (2004). Molecular dynamics simulation of hepatitis C virus IRES IIId domain: structural behavior, electrostatic and energetic analysis. *J Mol Model* **10**, 60-68.

104.Yingling, Y. G. & Shapiro, B. A. (2005). Dynamic behavior of the telomerase RNA hairpin structure and its relationship to dyskeratosis congenita. *J Mol Biol* **348**, 27-42.

105.Schneider, C. & Sühnel, J. (1999). A molecular-dynamics simulation of the flavin mononucleotide-RNA aptamer complex. *Biopolymers* **50**, 287-302.

106.Reyes, C. M., Nifosi, R., Frankel, A. D. & Kollman, P. A. (2001). Molecular dynamics and binding specificity analysis of the bovine immunodeficiency virus biv tat-tar complex. *Biophys J* **80**, 2833-2842.

107.Beaurain, F., Di Primo, C., Toulme, J. J. & Laguerre, M. (2003). Molecular dynamics reveals the stabilizing role of loop closing residues in kissing interactions: comparison between TAR-TAR* and TAR-aptamer. *Nucleic Acids Res* **31**, 4275-4284.

108.Castrignano, T., Chillemi, G., Varani, G. & Desideri, A. (2002). Molecular dynamics simulation of the RNA complex of a double-stranded RNA-binding domain reveals dynamic features of the intermolecular interface and its hydration. *Biophys J* **83**, 3542-3552.

Chapter 12

MOLECULAR DYNAMICS SIMULATIONS OF NUCLEIC ACIDS
MD simulations of G-DNA and functional RNAs

Naďa Špačková,[1] Thomas E. Cheatham, III[2] and Jiří Šponer[1]

[1]*Institute of Biophysics, Academy of Sciences of the Czech Republic, Královopolská 135, 612 65 Brno, Czech Republic, spackova@ncbr.chemi.muni.cz, sponer@ncbr.chemi.muni.cz;* [2]*Departments of Medicinal Chemistry and of Pharmaceutics and Pharmaceutical Chemistry and of Bioengineering, 2000 East, 30 South, Skaggs Hall 201, University of Utah, Salt Lake City, UT 84112, USA, tec3@utah.edu.*

Abstract: In the first part, structures and dynamic behavior of a representative set of guanine quadruplex (G-DNA) molecules studied by explicit solvent molecular dynamics (MD) simulations are described. These MD studies represent one of the most successful applications of MD to nucleic acids. Most important results summarized in this review are the following: the unique stability and rigidity of G-DNA stem, importance of cations bound in its central channel, ability of G-DNA to catch ions from the solvent, flexibility of loops surrounding the G-DNA stem, and prediction of hypothetical formation pathway of G-DNA stem through free energy analysis of possible intermediates. In the second part, the most important results of MD simulations of various RNA molecules (kissing-loop complexes, loop E and K-turn ribosomal motifs, HDV ribozyme) are briefly mentioned. Limitations of empirical force fields are also discussed.

Key words: molecular dynamics simulations; guanine quadruplex; thymine loops; channel ions; quadruplex formation; kissing-loop complex; loop E; hepatitis delta virus ribozyme; kink-turn motif

301

J. Šponer and F. Lankaš (eds.), Computational Studies of RNA and DNA, 301–325.
© 2006 *Springer.*

1.   GUANINE QUADRUPLEX MOLECULES

Guanine quadruplex (G-DNA) molecules represent a prominent group of non-canonical multistranded DNAs. G-DNA is involved in a variety of biological and biochemical processes including transcription, translation and replication. Guanine-rich DNA sequences with capability to form the quadruplex structures are present at the ends of eukaryotic chromosomes, the telomeres, and G-DNA formed by these telomeric sequences has gained an attention as a target for cancer therapy.[1,2] Sequences with stretches of guanine residues occur not only at the telomeres, but also in many other regions in the genome of cells, including the regulatory region of the insulin gene, fragile X-syndrome triplet repeats, and the promoter region of the c-myc oncogene.[3]

G-DNA molecules have been characterized by X-ray crystallography[4-10] as well as NMR spectroscopy[11-26] and also studied in detail by large-scale MD simulations[27-36] and other theoretical methods[37-42] providing interesting insight into the structural and dynamical properties of guanine quadruplex molecules. This chapter demonstrates the ability of MD simulation methods (AMBER) to successfully describe the unique properties of the guanine quadruplex stem. On the other hand, it also shows limitations of the current force field leading to major inaccuracies in description of loop regions.

1.1   Quadruplex Structure

The basic structural unit of the quadruplex stem is a guanine quartet which represents a planar set of four cyclically arranged guanine bases connected by Hoogsteen-like hydrogen bonds (Figure 12-1a).

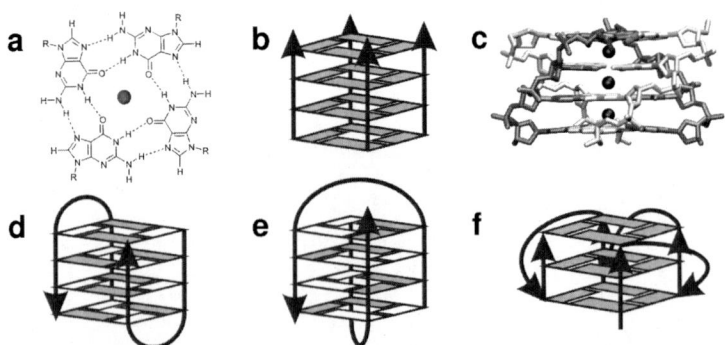

Figure 12-1. Guanine quadruplex molecules. a) Guanine quartet stabilized by the monovalent ion present in its center, b) schematic representation of parallel quadruplex, c) parallel quadruplex with a channel occupied by ions, d)-f) schematic representations of various quadruplexes: with lateral, diagonal, and groove loop arrangement, respectively. Open and grey boxes indicate guanines in syn and anti conformation, respectively.

The central hole of the quartet is characterized by a deep minimum of electrostatic potential due to the presence of guanine carbonyl groups oriented into the center of the quartet. The quadruplex stem is formed by stacking of 2-4 consecutive quartets and stabilized by monovalent cations present in the central channel (Figure 12-1). Majority of G-DNA structures has been determined by atomic resolution techniques in the presence of Na^+ and/or K^+ ions. In the crystal, position of the Na^+ ion is preferentially in-plane of the guanine quartet, while the K^+ ion is localized in the space between two adjacent quartets. Although there also are studies dealing with RNA quadruplexes,[43-46] majority of structural studies on quadruplexes have been carried out for DNA structures. Quadruplex structures are highly variable.[47] There are unimolecular and bimolecular quadruplexes (folded quadruplexes) which contain connecting loop regions as well as structures formed by four independent DNA strands (linear quadruplexes) (Figure 12-1b,c). The strands forming the guanine quadruplex can be parallel or antiparallel. Parallel orientation is typical for four-stranded linear quad-ruplexes while the combination of parallel and antiparallel orientation is necessary in folded G-DNA structures. There are also differencies in the glycosidic bond orientation of guanines in one strand - while *syn* and *anti* conformations alternate along the sequence in the folded quadruplexes, only *anti* conformation has been observed in the linear quadruplex.

The quadruplex stem is predominantly formed by all-guanine quartets, but participation of other bases in the quartet arrangement has also been observed, leading to, e.g., mixed guanine/cytosine quartets,[13,16,17] all-thymine,[48] all-cytosine[49] and all-adenine quartets.[50,51] Quadruplex stem can also incorporate guanine analogs like inosine, thioguanine and others.[52,53]

In antiparallel folded quadruplexes, the strands are connected by single-stranded loop regions preferentially formed by several thymine residues. In case of bimolecular G-DNA, two types of loop arrangements were observed (Figure 12-1d,e). The diagonal type is characterized by the loop arrangement across the quadruplex channel interconnecting the opposite strands. The lateral type of the loop connects two adjacent strands forming a hairpin. To complete the loop arrangement listing, it is necessary to mention the "groove" or external type of loops observed in the unimolecular parallel quadruplex formed by the human telomeric repeat $d(TTAG_3)$.[9] The groove loops are positioned alongside the grooves leading to an asymmetric propeller shape of the G-DNA structure (Figure 12-1f).

Here we report MD results obtained for the following quadruplex systems. The main model system is the parallel quadruplex $d(G_4)_4$ based on the highly resolved crystal structure of $d(TG_4T)_4$ in the presence of Na^+.[10]

Also, several types of d(G₄T₄G₄)₂ antiparallel folded quadruplexes have been studied: diagonal-looped quadruplexes in presence of Na$^+$ solved by NMR[20] and X-ray,[6] diagonal-looped quadruplexes in presence of K$^+$ (X-ray),[4] lateral-looped quadruplex in presence of K$^+$ (X-ray),[7] as well as lateral-looped quadruplexes containing mixed GCGC quartets like d(GCGGT₃GCGG)₂ in presence of Na$^+$ ions[16] and two structures of d(G₃CT₄G₃C)₂ in presence of K$^+$ and/or Na$^+$ (NMR).[13,17] Further, variety of hypothetical structures containing guanine analogs have been designed using previously mentioned structures as templates. Detailed information of studied quadruplex systems can be found in our preceding articles.[27-31,35] (Note, that X-ray structure reported in ref.[7] is seriously flawed and shows an entirely incorrect overall topology of the G-DNA – cf. ref.[4]).

1.2   Behavior of the Quadruplex Stem

G-DNA stem represents a rigid molecular assembly with essential contribution of monovalent cations to its stabilization. Simulations of the parallel d(G₄)₄ quadruplex yielded stable trajectories on a nanosecond scale, with exceptionally low fluctuations and close agreement between the experimental and theoretical structures with RMSd values below 1 Å.[28] The simulations thus reveal that G-DNA stem is, once formed, extremely rigid. Highest deviations were found for the more flexible sugar phosphate backbone, where bistability was also noticed in the X-ray structure. The four-quartet stem was stabilized by three Na$^+$ cations positioned between adjacent quartets. In contrast to crystal data, the Na$^+$ cations did not occupy a position in-plane of a guanine quartet layer. This discrepancy is caused by the inaccuracy in the force field description of the interaction between guanine quartets and cations. Polarization effects, which are important in cation-containing systems, are not included in the pairwise additive force field. The simulations also did not show significant differences for a G-DNA stem simulated with Na$^+$ and K$^+$, respectively.

Simulations with a reduced number of ions in the channel were made to obtain insights which are beyond the experimental data. Our simulations showed that the parallel quadruplex structure is entirely stable even with reduced number of cations in the central channel. This is an important lead regarding the smooth equilibrium exchange of cations between the quadruplex channel and bulk solvent experimentally observed on the hundreds of microseconds to milisecond time scale.[14]

Simulations without any ion in the channel demonstrated the central role of cations for the stability of the guanine quadruplex stem. The quadruplex channel was fully hydrated when left vacant by cations. However, water molecules were not able to sufficiently stabilize the stem. Rearrangements involving vertical slippage of one strand ("strand walking") were observed (Figure 12-2), leading to temporary formation of a "slipped" molecule with only three guanine quartets, one guanine triad at one end of the quadruplex stem, and an stacked guanine residue at the other pole. The slipped quadruplex molecule was stabilized by placing cations back into the central channel. This final structure was very stable and showed a behavior very similar to the original native quadruplex.

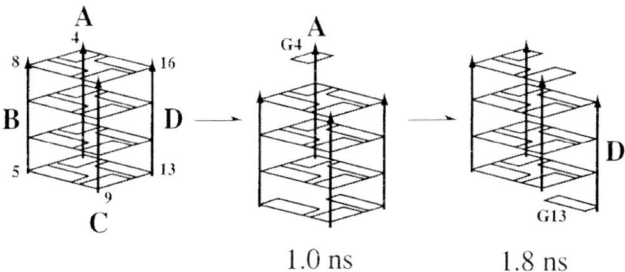

1.0 ns   1.8 ns

Figure 12-2. Schematic representation of a vertical strand slippage observed during the simulation of the parallel quadruplex without ions in the channel. Strands are labeled A-D, guanines are shown as open boxes, first and last residues are denoted. The starting structure is shown on the left.

We suggested that the slipped structure represents one of the possible intermediates occurring in the last stages of G-DNA stem formation (see below). Actually, slipped structures were recently detected in solution for interlocked G-DNA.[54] The quadruplex with the vacant channel is also able to spontaneously intercept and incorporate an initially remote cation from the bulk.[29,34] Typically, it is a two-step process (Figure 12-3): first, the ion arrives to the channel entrance while the channel cavity is occupied by water molecule. Then, the positions of the water and the ion are exchanged. On the basis of our simulations we estimated the lifetime of a G-DNA stem with a vacant channel to be ca. 10 ns. This observation further supported the genuine view that G-DNA stems were never left vacant by cations.

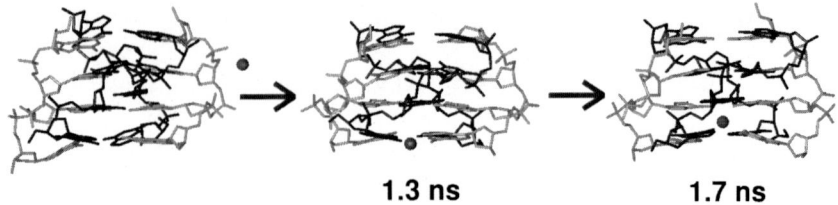

1.3 ns   **1.7 ns**

Figure 12-3. Spontaneous interception of a bulk cation by the initially vacant stem.

The ns-scale simulations did not show any distinct differences in stem behavior depending on the strand orientation.[28,29] As noted above, in contrast to crystal data, Na^+ cations in our simulations evade in-plane position in the guanine quartet layer, thus they effectively look to be oversized. Reduction of the radius of Na^+ cations improved the mobility of cations in the channel and increased the probability of the in-plane position of the ion. However, it did not eliminate another artifact observed, namely, a bifurcated type of hydrogen bonding of the inner guanine quartets.[28,29] In case of K^+ ions, these ions are unambiguously too large to be placed in the plane of the guanine quartet. However, both Na^+ and K^+ ions were capable of passing through the plane of the quartet without causing any structural destabilization. The global distribution of the cations was captured well, since it is primarily determined by the long-range electrostatic interactions properly included in the current force field.

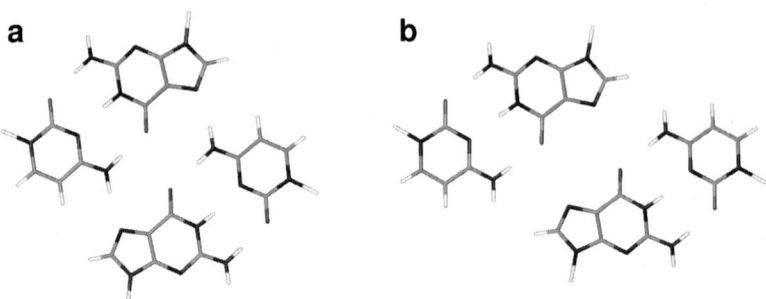

Figure 12-4. a) "closed" and b) "sheared" geometries of mixed GCGC quartets observed during MD simulations.

Mixed quadruplexes containing guanine/cytosine GCGC quartets (localized at the center or at the ends of the quadruplex stem) were studied by NMR.[13,16,17] In simulations,[29] the GCGC quartets adopt two distinct conformations, termed "closed" and "sheared" (Figure 12-4), in very good agreement with the NMR.[13,16,17] The "closed" quartet is stabilized by H-bonds connecting the two G/C base pairs forming the quartet. The "sheared" quartet has no H-bonds between the G/C base pairs and is formed upon a close contact with a cation. When a large K^+ ion is positioned in the channel cavity, the adjacent GCGC quartet adopts the "sheared" geometry while the presence of Na^+ ion leads to the "closed" arrangement. The "sheared" geometry in presence of Na^+ ions was observed only if the ion closely contacted the mixed quartet. However, interactions between monovalent ions and mixed quartets are less attractive than interactions between cations and all-guanine quartets. Our simulations suggested that stability of mixed quadruplexes originates in the presence of cation-stabilized guanine quartets while GCGC quartets are just tolerated in the stem.

We also tested the ability of the stem to incorporate non-standard guanine analog bases like 6-oxoguanine (inosine), 6-thioguanine and 6-thiopurine.[31] Our simulations showed that inosine is capable to form cation-stabilized all-inosine as well as mixed inosine/cytosine quadruplex stems with almost identical properties as the native all-guanine quadruplex. The loss of all amino groups implies the loss of four hydrogen bonds per quartet, although no apparent destabilization is observed. This clearly indicates that the stems are primarily stabilized by the quartet-cation interactions, which are very similar for inosine and guanine due to the similarity of their electronic structures. On the other hand, the inosine quadruplex with a vacant channel exhibits a different behavior than the similar all-guanine structure (Figure 12-5).

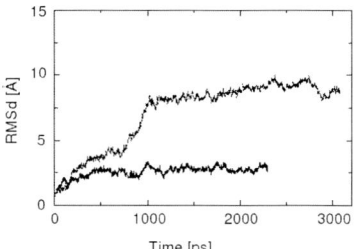

Figure 12-5. Evolutions of the RMS deviations along the trajectory in simulations of parallel quadruplexes with the vacant channel formed by guanine (solid line) and inosine (dotted line).

The inosine quadruplex stem disintegrates within the first nanosecond of simulation having no time to catch any cation. We suggested that the difference between guanine and inosine H-bonding capability may affect the folding process of the respective quadruplexes leading to different intermediates in the formation pathway before the ions start to dominate the stabilization (see below).

Presence of thioguanine in the quadruplex structure is strongly destabilizing. Experiments clearly demonstrate that thioguanine inhibits the G-DNA formation.[55] In our MD simulations, the presence of the all-thioguanine quartet was associated with a large perturbation of the G-DNA stem and replacement of all guanines by thioguanines led to imminent collapse of the quadruplex structure.[31] Although the thiogroup has a very favorable interaction with Na$^+$ cations, this group is too bulky and its presence leads to expulsion of the cations from the channel, triggering the collapse. Classical MD simulations were completed by Thermodynamic Integration (TI) free energy calculations. It was found that even a single guanine → thioguanine mutation has a substantial impact on the stability of the G-DNA stem.[30]

1.3   Behavior of the Quadruplex Loops

Diagonal and/or lateral loops represent an important part of the antiparallel folded quadruplex molecules. Although there is a high degree of sequence and length variability of quadruplex loops, our studied systems typically contained four thymine bases. The flexibility of the loops is in a sharp contrast with the rigidity of the G-DNA stem. The MD simulations starting from experimental geometries revealed multiple nanosecond-scale stable conformations of the four-thymine loops.[28,29,35] Thus, ns-scale simulations are not able to fully characterize the conformational space of the loops and to identify the global minimum of the loop region. Therefore, the conventional MD simulations were combined with other techniques and approaches like the Locally Enhanced Sampling (LES, see another part of the book), the systematic CICADA conformational search, and the MM-PBSA free energy analysis to provide (within the approximations of the applied force field) a qualitatively complete analysis of the available loop conformational space. LES simulations launched from different starting structures appeared to localize the correct global minimum of loop topology, and LES and MM-PBSA data were mutually consistent.[35] That means, the LES loop geometries had the best free energies. The results further suggested that optimal lateral and diagonal geometries (predicted by LES) are close in energy (Table 12-1).[35] The difference is only on scale of several kcal/mol which is within the error margin of the MM-PBSA method.

Table 12-1. Estimated free energies (calculated using the MM-PBSA method) for the antiparallel quadruplex structure (G_{total}), and relative free energy compared with the predicted optimal geometry of the diagonal loop ($\Delta G_{diagonal}$). Entropy components are not included. Different initial structures used for calculations are distinguished by superscripts: [a]the lowest energy lateral-looped structure, [b]the lowest-energy diagonal-looped structure, [c]the NMR diagonal-looped structure,[20] [d]the X-ray diagonal-looped structure.[4] Table adapted from ref.[35]

Structure	G_{total} (kcal/mol)	$\Delta G_{diagonal}$ (kcal/mol)
G-DNA[a] with lateral loops	−5206	+4
(LES, MD)		
G-DNA[b] with diagonal loops	−5210	0
(LES, MD; incorrect geometry)		
G-DNA[c] with diagonal loops	−5181	+29
(MD; correct geometry)		
G-DNA[d] with diagonal loops	−5189	+21
(MD; correct geometry)		

Unfortunately, the LES-predicted optimal geometry of the diagonal loop arrangement differed substantially from a geometry which is notoriously known from multiple independent NMR and X-ray experiments[4,6,20,24] (Figure 12-6), with the free energy difference between the LES MD (incorrect) diagonal loop and the correct (experimental) loop geometries being ca. 20-30 kcal/mol in favor of the incorrect one (Table 12-1). We suggest that the inability of the computations to predict the correct diagonal loop geometry originates in some previously unknown but substantial imbalances of the force field. Thus, stability of the experimental loop structure in the ns-scale conventional simulations is not an ultimate indicator of the correctness of the force field.

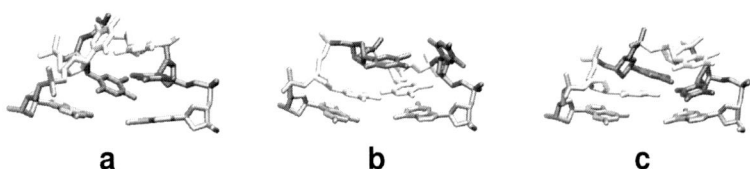

a   **b**   **c**

Figure 12-6. Comparison of different geometries of four-thymine diagonal loops attached to guanines from the adjacent quartet. a) correct geometry observed in NMR and X-ray experiments, b) and c) incorrect geometries predicted by LES.

The crystal structure of the diagonally looped quadruplex showed a stable in-loop coordination of K[+] ions[4] and also the NMR structure of mixed quadruplex suggested the presence of K[+] ions in the lateral loops.[13] Our simulations did not reveal any stable coordination of cations within the loops. However, three transient binding sites were identified: above the channel entry, in the loop center, and at the loop apex.[29] It has been suggested that these positions represent a possible route for cation exchange between the quadruplex channel and solvent. On the other hand, initially loop-coordinated Na^+ and/or K^+ ions were unstable and left the loops within several nanoseconds (1 ns in case of K^+, 3 ns in case of Na^+).[29,35] The cations did not return back to the loops and were not replaced by other cations. It also indicates that the force field description of the G-DNA loops is imperfect.

1.4   Quadruplex Stem Formation

Formation of G-DNA structures is an exceptionally slow process taking from hours to days, with a complex kinetics. Experimental insight into G-DNA folding is very limited and hypothetical models of folding are mainly based on spectroscopic studies. There are three mechanisms suggested:[56] stepwise strand addition, duplex dimerization, and triplex-triplex disproportionation (Figure 12-7). Experimental data also indicate that G-DNA formation proceeds in discrete steps progressing through stable intermediate states.[56] Most questions remain unsolved, such as at what stage are the monovalent cations integrated into the structure or what is the atomic-level structure of the intermediates? The formation process also likely varies with different sequences and experimental conditions. The anticancer drugs targeting G-DNA should speed up G-DNA formation.[1]

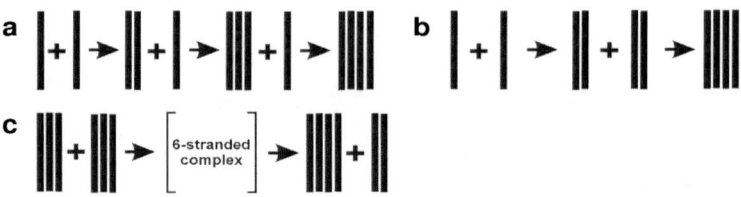

Figure 12-7. Hypothetical models proposed for quadruplex formation from experiments. a) stepwise strand addition, b) duplex dimerization, and c) disproportionation.

A spontaneous aggregation of the four strands and formation of the quadruplex stem using MD simulations is not realistic. However, the MD approach can be successfully applied to probe the stability of possible intermediates on the formation pathway.[27] Relatively simple structure of all-parallel quadruplex without loops was selected as the model system and various four-stranded, three-stranded, and duplex assemblies were designed based on this molecule (Figure 12-8).

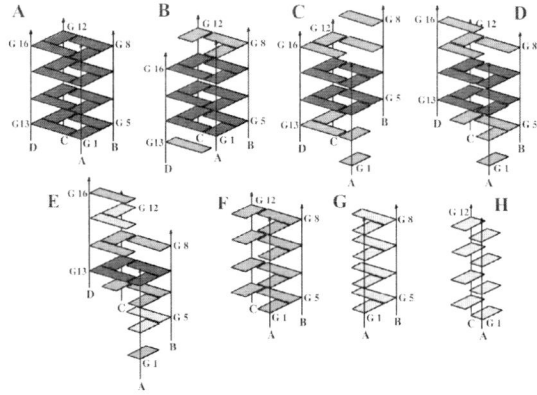

Figure 12-8. Molecular structures of the parallel quadruplex and its possible intermediates: a) native G-DNA, b) G-DNA with one strand shifted, c) and d) G-DNA with two adjacent strands shifted down and up, e) spiral stem, f) triplex, g) parallel edge duplex, and h) parallel diagonal duplex.

Structural stability of all proposed intermediates was tested using classical MD simulations with subsequent relative free energy evaluation using the MM-PBSA method. The simulated structures were divided into two groups. The first group comprised structures that were completely stable (some with minor initial rearrangements) during MD simulations. These structures are supposed to be possible intermediates of the quadruplex formation. The second group of structures disintegrated on the nanosecond scale and thus is not likely involved in G-DNA formation.

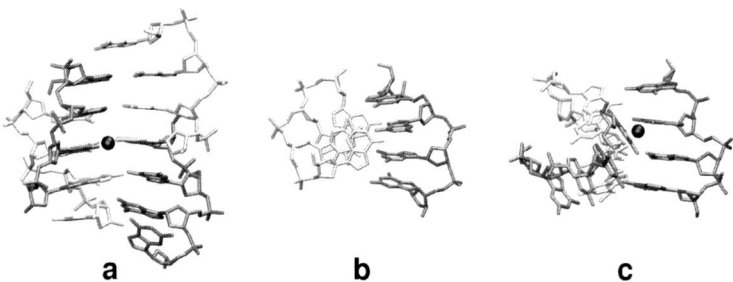

Figure 12-9. Averaged structures of key simulated intermediates. a) averaged structure of the spiral stem with one ion coordinated at the guanine quartet (structure E on Figure 12-8), b) "cross-like" structure observed in simulations of duplex structures, and c) "coil-like" structure observed in simulation of triplex structure.

The group of the direct intermediates is represented by a number of four-stranded structures with one, two or three mutually shifted strands. These "slipped" structures contained a reduced number of guanine quartets compared to the native quadruplex. An extreme case of the "slipped" structures is a "spiral stem" with all strands shifted by one step and only one intact guanine quartet in the center of the structure. Despite the higher flexibility leading to modest structural changes, the structure remained stable on the ns-scale. The structure was stabilized by one ion positioned in the plane of the central quartet (Figure 12-9a). Simulations also indicated that the spiral stem intermediate is able to further rearrange towards the native G-DNA stem. Note, that the final rearrangements in quadruplex stem formation may be quite slow since cation stabilized intermediates with several quartets are assumed to be rather stable and rigid and their further geometrical changes toward the final G-DNA structure are likely to occur during the exchange of cations between solute and bulk solvent.

Table 12-2 presents relative free energies calculated for selected four-stranded structures with inclusion of the bound ions. The experimentally observed native cation-stabilized G-DNA is the free-energy minimum and any strand slippage leads to the free energy increase. The vacant quadruplex is significantly destabilized compared to the native quadruplex. The spiral structure represents the least stable assembly and it is significantly less stable than the vacant quadruplex. Even the four-stranded structures with a single strand slippage are significantly less stable than the native stem and thus they are not supposed to be observed once the formation process is completed and equilibrium is reached.

Table 12-2. Estimated free energies (calculated using MM-PBSA method) for the parallel quadruplex structure including 3 ions (G_{total}), and differences in the free energy to the native G-DNA stem (ΔG_{native}). Entropy components are not included. Table adapted from ref.[27]

structure	G_{total} (kcal/mol)	ΔG_{native} (kcal/mol)
native G-DNA stem (3 ions in the channel)	−3722.4	0.0
G-DNA stem with vacant channel	−3696.6	+25.8
G-DNA stem with 1 strand slippage (3 ions in the channel)	−3710.3	+12.1
spiral stem	−3672.8	+49.6

The second group of structures is represented by assemblies with reduced number of strands, namely by three-stranded and duplex structures. Both "edge" and "diagonal" duplexes are unstable and convert to virtually identical "cross-like" structures formed by two perpendicularly oriented strands and stabilized by extensive non-planar interbase hydrogen bonding (Figure 12-9b). Free energy analysis suggests that formation of the "cross-like" structure is favorable and that ions do not stabilize this duplex structure. The three-stranded structure also disintegrates during the simulation, leading to "coil-like" structure formed by the "cross-like" structure interacting with the third strand via a Na^+ cation (Figure 12-9c). Thermodynamics calculations reveal that the "coil-like" structure is likely weakly stabilized by the presence of this cation, whereas the "cross-like" duplex itself is not stabilized by cation coordination.

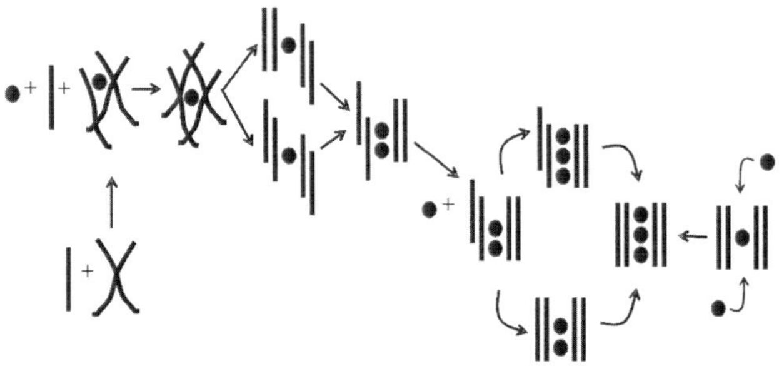

Figure 12-10. Sketch of a possible folding pathway as suggested by the simulations.

Both "cross-like" and "coil-like" structures are stable in ns-scale simulations. They contain unpaired guanines with capability to intercept other single strands and/or "cross-like" structures via hydrogen bonding of bases, serving as "nucleation" centers for a further buildup of the four-stranded stem (Figure 12-10). Using a combination of "cross-like" and "coil-like" intermediates we are hypothetically able to construct all three formation pathways (Figure 12-7). Although these structures are not suggested as the only possible three- and two-stranded intermediates, they likely are representative examples of plausible structures.

In conclusion, the G-DNA simulation studies provide an excellent example of the successes and pitfalls of contemporary MD studies of nucleic acids and the way how this technique should be applied and assessed. On one side, when a qualitative task was properly formulated, the simulations were capable to provide unique qualitative insights into the complex process of the G-DNA stem formation which is currently beyond the applicability of experimental methods. On the other side, attempts to carry out an in depth study of the loop region of G-DNA molecules resulted into a major, frustrating failure of the force field to describe this region properly.

2. MD SIMULATIONS OF RNA MOLECULES

The last decade is characterized by intense structural investigations of functional RNA molecules. While in the middle of nineties first X-ray structures of compactly folded medium sized RNA molecules were obtained,[57] since 2000 atomic resolution structures of ribosomal subunits[58,59] and the whole ribosome were reported.[60] RNAs are organized on hierarchical principles that are very different from the long B-DNA double helix. Key structural, dynamical and biological roles are played by versatile non-Watson-Crick base pairs (base pairs different from the canonical ones), often utilizing the 2'OH group of ribose.[61] RNA molecules became also a target for simulation studies (See the complete survey of RNA simulations in the chapter 11 by P. Auffinger). MD simulations of functional RNAs are not easy, since the analysis of compactly folded molecules with a complex network of intricate molecular interactions is much more demanding than reporting a simulation of a simple B-DNA duplex. At the same time, the RNA simulations appear to be more robust, since we are rarely interested in

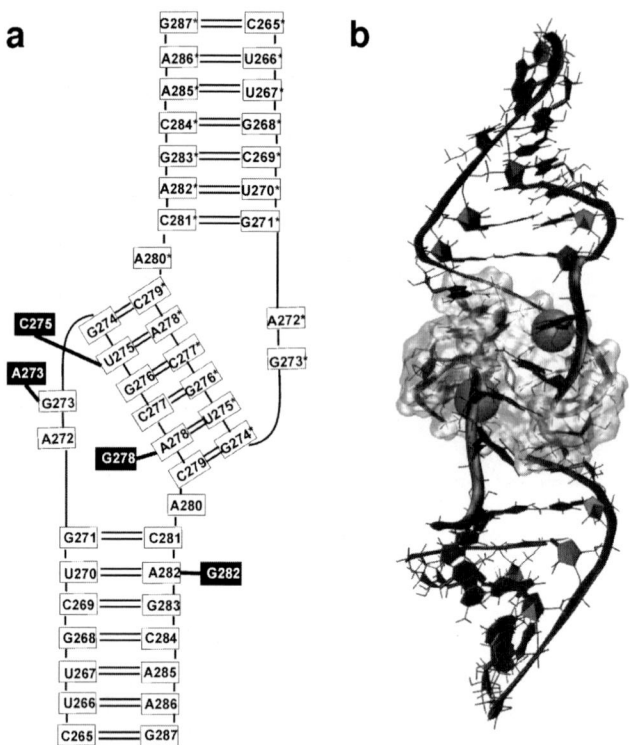

Figure 12-11. a) schematic representation of 6 bp kissing-loop complex of HIV-1 DIS subtype A. Black boxes correspond to changes in loop sequence of HIV-1 subtype B. b) structure of the same complex (subtype A) from the MD simulation with the highlighted central pocket occupied by Na⁺ ions (balls).

such subtle structural details as in typical B-DNA studies (modest local conformational variations of B-DNA double helix and elusive cation binding in B-DNA grooves may be well below the accuracy of the force fields). Our experience so far is that the performance of the RNA force field is superior to the DNA one (unpublished results) and the RNA simulations appear to be surprisingly realistic, considering the enormous complexity of backbone topologies and molecular interactions in functional RNAs.

Well-executed and carefully analyzed simulations can provide useful data complementing the structural and biochemical experiments. For example, in simulations of HIV-1 dimerization initiation site RNA kissing complexes we found that their central part is stabilized by a unique cation binding pocket permanently occupied by 2-3 monovalent cations[62] (Figure 12-11). Despite the site occupancy, the ions exchange with the bulk solvent on a time scale 2-3 ns and thus we do not deal with trapped cations. Further, the ions are delocalized in the pocket, i.e., they interact with multiple RNA atoms. Such dynamical cations are essentially invisible to any atomic-resolution experiments. Specifically, such ion binding sites can easily be missed by crystallography. Crystallography deals well with static binding sites but not with the dynamical ones. Note that it is entirely unjustified to a priory claim that delocalized binding sites are weak or structurally unimportant. The ion delocalization in the RNA kissing complex is not caused by their weak binding, but by flatness of the very deep electrostatic potential minimum. Besides G-DNA, this is the most prominent cation-binding site reported in nucleic acids simulations so far. Compared to this, the ion binding events around B-DNA are negligible. Major monovalent cation binding sites with occupancy 50% or more for inner shell (direct solute-cation) binding should be well reproduced by simulations even considering the force field approximations and sampling limitations.

Major cation binding sites are not rare in RNAs. 5S rRNA Loop E is a unique RNA duplex motif consisting of seven consecutive non-Watson-Crick base pairs[63-66] (Figure 12-12a).

Its deep major groove represents a major binding site for monovalent and divalent ions. Further, the simulations of 5S rRNA Loop E revealed formation of highly specific hydration sites involving long-residency water molecules with binding times 1-5 ns (Figure 12-12b-d). Such water binding times are considerably longer than water binding times in common hydration sites in nucleic acids (50-500 ps). When simulating a rather large molecular complex between 5S rRNA Loop E/Helix IV segment and protein L25 (Figure 12-13), we detected several key long-residency waters, some of them bound for the entire 25 ns trajectory.[64]

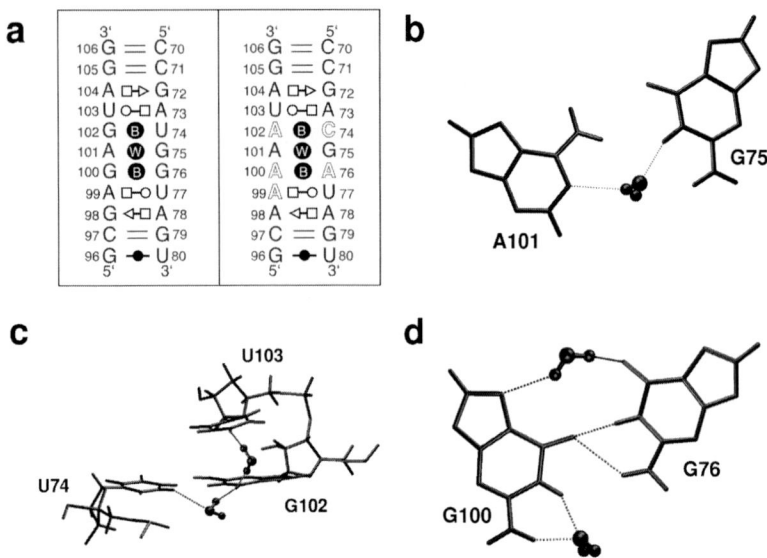

Figure 12-12. a) schematic representation of studied Loop E systems: bacterial E. coli 5S rRNA Loop E (right panel) and spinach chloroplast 5S rRNA Loop E (left panel); the specific symbols mark distinct non-Watson-Crick base pairs, b) water-mediated cis Watson-Crick A101/G75 base pair from average MD structure, c) long residency water molecules in the hydration pocket near G102, d) water-mediated cis bifurcated G100/G76 base pair from average MD structure.

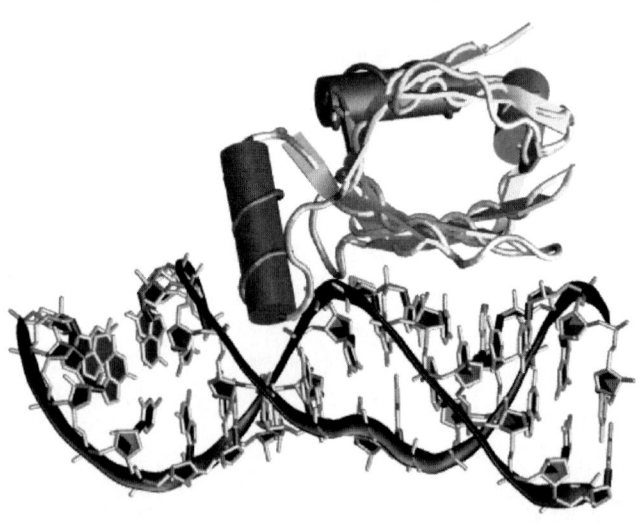

Figure 12-13. Loop E/Helix IV rRNA – L25 Protein complex.

Thus, we suggested that long-residency water molecules seen in small systems could actually convert into close to permanent waters inside big molecular assemblies such as the ribosome. MD simulation technique is uniquely poised to analyze and detect specific structural long-residency hydration patterns, as the crystallography provides no dynamical information while NMR is limited when detecting hydration. Unfortunately, many recent MD studies on nucleic acids failed to analyze specific hydration. Thus, one of the most valuable pieces of data that can be gathered by simulations is lost. We have also successfully used MD simulations to predict the 3D structure of Loop E from spinach chloroplast, based on X-ray structure of *E. coli* Loop E, phylogenetic analysis and isostericity principle.[61] The spinach chloroplast sequence has, compared to the bacterial one, 5 mutated positions out of 14 (40% mutation) including 4 G→A mutations. Despite this, MD simulations show that the spinach chloroplast sequence forms a Loop E structure with close to identical 3D shape as the bacterial one. It is because the original bacterial non-Watson-Crick base pairs are replaced by new non-Watson-Crick base pairs which are isosteric, i.e., they have the same shape. This finding was independently confirmed by solution NMR measurement.[67]

Several major cation binding sites were also reported in extended simulations of Hepatitis Delta Virus ribozyme (HDVr)[68] (Figure 12-14). HDVr is a self-cleaving catalytic RNA. The self-cleavage is a transesterification reaction and is important for the virus replication. It is well established that cytosine C75 is absolutely required for the catalysis.

The simulations were utilized to investigate structural aspects of product and precursor of HDVr, assuming several mutations at the 75 position. The simulations support the role of C75 acting as the general base rather than general acid in the catalysis, provided the ground state seen in X-ray structures of HDVr precursor is sufficiently close to the transition state. (Simulation of the catalytic process itself would require to use combined quantum chemical/molecular mechanical approach, which presently is not available for nucleic acids).

Simulations of Kink-turns[69] perhaps best illustrate the power of the simulation technique in the RNA field.[70-72] RNA Kink-turn (K-turn) motifs are asymmetric internal loops characterized by a sharp bend in the

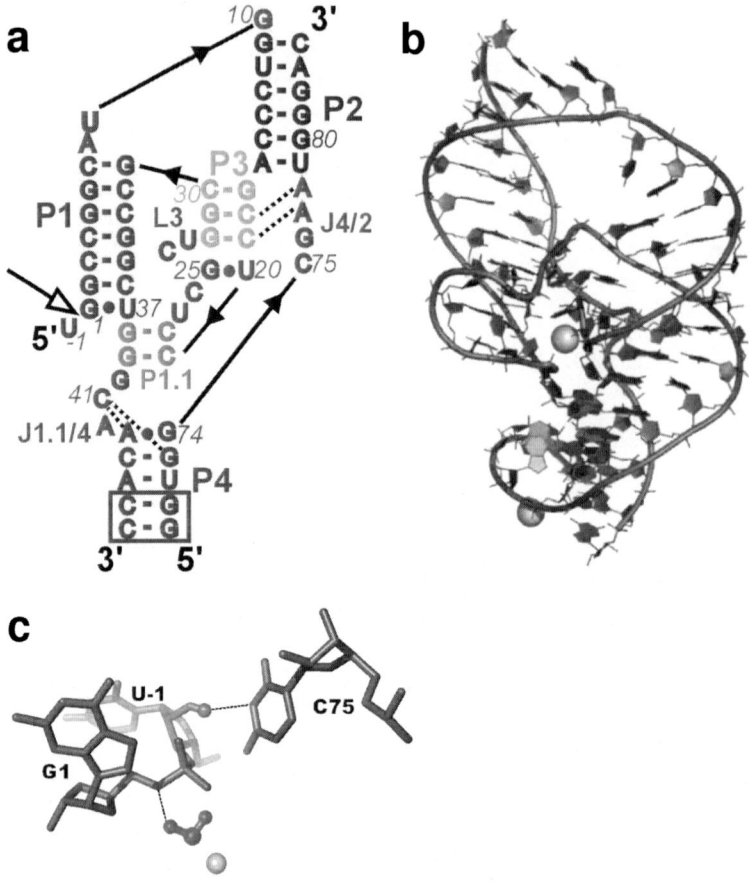

Figure 12-14. a) schematic representation of the simulated genomic HDV ribozyme (precursor form). The product form lacks U-1. The cleavage site is represented by open arrow. b) 3D structure of HDV ribozyme with two Mg^{2+} ions (balls) positioned in the catalytic pocket (upper) and in the major groove of P4 stem (lower). c) The cleavage site (MD substate) with H-bonded O2'H of U^{-1} and N3 of C75. In this position C75 could act as the general base and hydrated Mg^{2+} (gray ball) acts as the general acid.

phosphodiester backbone resulting in "V"-shaped structures (Figure 12-15). K-turns are composed of three distinct structural elements. **NC** is the non-canonical helix (**NC**-stem) while **C** represents the canonical helix (**C**-stem). The **C**- and **NC**-stems flank the internal loop (**K**), which comprises nominally unpaired nucleotides and produces a sharp bend of the RNA helix by ca. 120°. **K** thus forms the tip of the "V", with **NC** and **C** being the attached arms. The crucial interactions between **NC** and **C** are mediated through pairs of A-minor motifs. A-minor motifs are key RNA tertiary interactions involving adenosines that are usually involved in non-Watson-Crick base pairs, leaving their Watson-Crick and Sugar edges available for further interactions.[73] The adenosines then interact with shallow grooves of Watson-Crick base pairs. This leads to insertion of the smooth, minor groove edges of adenines into the minor groove of neighboring helices, preferentially at C = G base pairs, where they form hydrogen bonds with one or both of the 2'OHs of those pairs. The A-minor motif is by far the most abundant tertiary structure interaction in the large ribosomal subunit and apparently also in other RNAs. The X-ray structures of ribosomal subunits reveal multiple K-turns present at universally conserved regions mainly of 23S ribosomal RNA.[69] The K-turns thus can be considered as the most important non-Watson-Crick molecular building blocks (RNA motifs) of the whole ribosome LEGO.

We have reported series of carefully analyzed explicit solvent AMBER simulations (more than 200 ns) of selected K-turns, utilizing their X-ray geometries as starts. The simulations revealed an unprecedented dynamical flexibility of the K-turns around their X-ray geometries, making them strikingly different from other RNAs that are quite rigid in similar simulations. The sampled geometries are essentially isoenergetic and separated by modest energy barriers. Based on MD simulations we have suggested that the K-turns represent unique RNA-based molecular elbows or floppy systems. The ns-scale dynamics of isolated K-turns can be qualitatively considered as motion of two rigid helical stems controlled by a flexible internal loop. This leads to hinge-like motions between the two stems (Figure 12-15c).

Subtle local variations at the kink can thus produce substantial modification of the inter-segment distances (up to 10 Å or more) at the ends of the flanking stems. Note that in contrast to the solution experiments the simulations investigated finer motions that are not accompanied by any marked destabilization of the kink region. The most striking local motion regulating the elbow dynamics of certain K-turns is insertion of a long-residency (2-3 ns) water molecule into the second A-minor interaction between the C and NC stems (Figure 12-15d-f).

Viewed in light of the phylogenetic conservation and location of K-turns at the base of functionally significant protuberances in the ribosome, the dynamical results support the notion that some K-turns could serve as flexible hinges to allow biologically significant motions in protein synthesis.

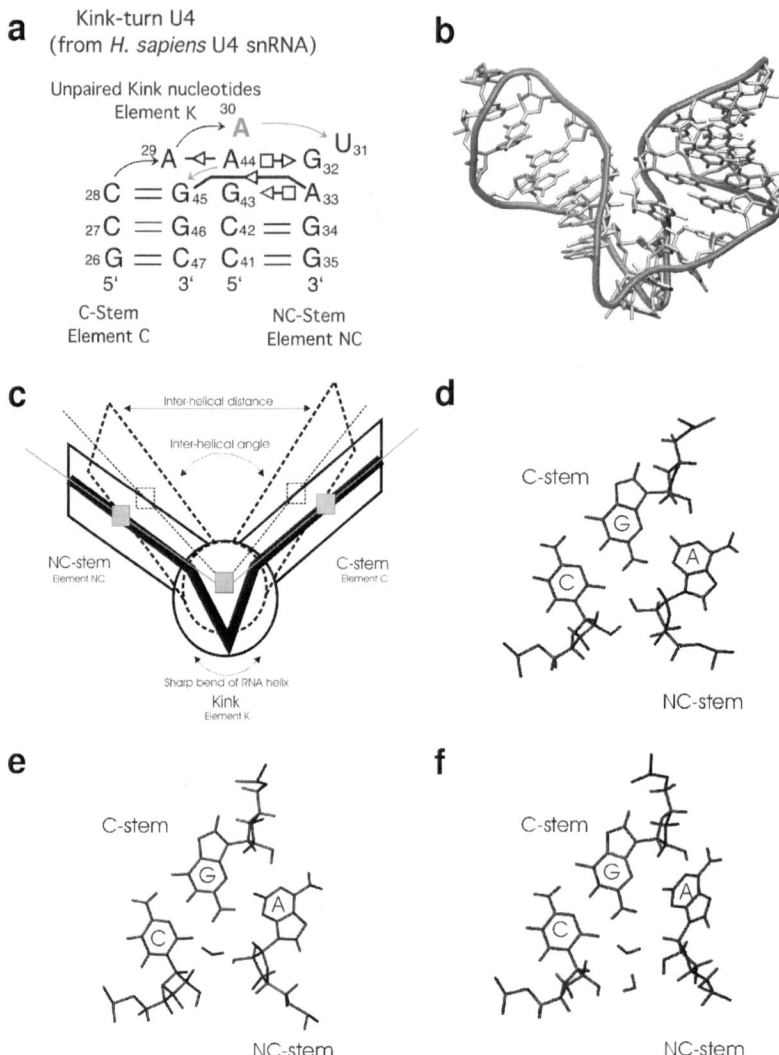

Figure 12-15. a) scheme of K-turn U4 structure using Leontis et al.[61] nomenclature, b) 3D "V" shaped structure of K-turn 58 from *H. marismortui* 23S rRNA c) modular hinge-like motion of K-turns (centers of mass of the particular elements are represented as squares), dynamics of A-minor type I: the closed substate (d) - direct H-bonds mediate the interaction between helical stems, open substate (e) - single water molecule bridges the interacting C and A nucleotides and fully open substate (f) - insertion of two water molecules between C and A nucleotides. Water insertion causes the large-scale motion of adjacent helices.

ACKNOWLEDGEMENTS

This research was supported by Wellcome Trust International Senior Research Fellowship in Biomedical Science in Central Europe GR067507, grants GA203/05/0388 and GA203/05/0009, Grant Agency of the Czech Republic, and research projects AVOZ50040507 and MSM0021622413, Ministry of Education of the Czech Republic.

REFERENCES

1. Neidle, S. & Parkinson, G. N. (2003). The structure of telomeric DNA. *Curr. Opin. Struct. Biol.* **13**, 275-283.
2. Neidle, S. & Parkinson, G. (2002). Telomere maintenance as a target for anticancer drug discovery. *Nat. Rev. Drug Discov.* **1**, 383-393.
3. Shafer, R. H. & Smirnov, I. (2000). Biological aspects of DNA/RNA quadruplexes. *Biopolymers* **56**, 209-227.
4. Haider, S., Parkinson, G. N. & Neidle, S. (2002). Crystal structure of the potassium form of an Oxytricha nova G-quadruplex. *J. Mol. Biol.* **320**, 189-200.
5. Haider, S. M., Parkinson, G. N. & Neidle, S. (2003). Structure of a G-quadruplex-ligand complex. *J. Mol. Biol.* **326**, 117-125.
6. Horvath, M. P. & Schultz, S. C. (2001). DNA G-quartets in a 1.86 angstrom resolution structure of an Oxytricha nova telomeric protein-DNA complex. *J. Mol. Biol.* **310**, 367-377.
7. Kang, C., Zhang, X. H., Ratliff, R., Moyzis, R. & Rich, A. (1992). Crystal-Structure of 4-Stranded Oxytricha Telomeric DNA. *Nature* **356**, 126-131.
8. Laughlan, G., Murchie, A. I. H., Norman, D. G., Moore, M. H., Moody, P. C. E., Lilley, D. M. J. & Luisi, B. (1994). The High-Resolution Crystal-Structure of a Parallel-Stranded Guanine Tetraplex. *Science* **265**, 520-524.
9. Parkinson, G. N., Lee, M. P. H. & Neidle, S. (2002). Crystal structure of parallel quadruplexes from human telomeric DNA. *Nature* **417**, 876-880.
10. Phillips, K., Dauter, Z., Murchie, A. I. H., Lilley, D. M. J. & Luisi, B. (1997). The crystal structure of a parallel-stranded guanine tetraplex at 0.95 angstrom resolution. *J. Mol. Biol.* **273**, 171-182.
11. Aboulela, F., Murchie, A. I. H. & Lilley, D. M. J. (1992). NMR-Study of Parallel-Stranded Tetraplex Formation by the Hexadeoxynucleotide d(TG₄T). *Nature* **360**, 280-282.
12. Aboulela, F., Murchie, A. I. H., Norman, D. G. & Lilley, D. M. J. (1994). Solution Structure of a Parallel-Stranded Tetraplex Formed by d(TG₄T) in the Presence of Sodium-Ions by Nuclear-Magnetic-Resonance Spectroscopy. *J. Mol. Biol.* **243**, 458-471.
13. Bouaziz, S., Kettani, A. & Patel, D. J. (1998). A K cation-induced conformational switch within a loop spanning segment of a DNA quadruplex containing G-G-G-C repeats. *J. Mol. Biol.* **282**, 637-652.
14. Hud, N. V., Schultze, P., Sklenar, V. & Feigon, J. (1999). Binding sites and dynamics of ammonium ions in a telomere repeat DNA quadruplex. *J. Mol. Biol.* **285**, 233-243.

15. Keniry, M. A., Strahan, G. D., Owen, E. A. & Shafer, R. H. (1995). Solution Structure of the Na+ Form of the Dimeric Guanine Quadruplex d($G_3T_4G_3$)$_2$. *Eur. J. Biochem.* **233**, 631-643.

16. Kettani, A., Kumar, R. A. & Patel, D. J. (1995). Solution Structure of a DNA Quadruplex Containing the Fragile-X Syndrome Triplet Repeat. *J. Mol. Biol.* **254**, 638-656.

17. Kettani, A., Bouaziz, S., Gorin, A., Zhao, H., Jones, R. A. & Patel, D. J. (1998). Solution structure of a Na cation stabilized DNA quadruplex containing G.G.G.G and G.C.G.C tetrads formed by G-G-G-C repeats observed in adeno-associated viral DNA. *J. Mol. Biol.* **282**, 619-636.

18. Marathias, V. M. & Bolton, P. H. (1999). Determinants of DNA quadruplex structural type: Sequence and potassium binding. *Biochemistry* **38**, 4355-4364.

19. Marathias, V. M. & Bolton, P. H. (2000). Structures of the potassium-saturated, 2 : 1, and intermediate, 1 : 1, forms of a quadruplex DNA. *Nucleic Acids Res.* **28**, 1969-1977.

20. Schultze, P., Smith, F. W. & Feigon, J. (1994). Refined Solution Structure of the Dimeric Quadruplex Formed from the Oxytricha Telomeric Oligonucleotide d(GGGGTTTTGGGG). *Structure* **2**, 221-233.

21. Schultze, P., Hud, N. V., Smith, F. W. & Feigon, J. (1999). The effect of sodium, potassium and ammonium ions on the conformation of the dimeric quadruplex formed by the Oxytricha nova telomere repeat oligonucleotide d($G_4T_4G_4$). *Nucleic Acids Res.* **27**, 3018-3028.

22. Sket, P., Crnugelj, M. & Plavec, J. (2004). d($G_3T_4G_4$) forms unusual dimeric G-quadruplex structure with the same general fold in the presence of K+, Na+ or NH4+ ions. *Bioorg. Med. Chem.* **12**, 5735-5744.

23. Sket, P., Crnugelj, M. & Plavec, J. (2005). Identification of mixed di-cation forms of G-quadruplex in solution. *Nucleic Acids Res.* **33**, 3691-3697.

24. Smith, F. W. & Feigon, J. (1992). Quadruplex Structure of Oxytricha Telomeric DNA Oligonucleotides. *Nature* **356**, 164-168.

25. Smith, F. W. & Feigon, J. (1993). Strand Orientation in the DNA Quadruplex Formed from the Oxytricha Telomere Repeat Oligonucleotide d($G_4T_4G_4$) in Solution. *Biochemistry* **32**, 8682-8692.

26. Smith, F. W., Schultze, P. & Feigon, J. (1995). Solution Structures of Unimolecular Quadruplexes Formed by Oligonucleotides Containing Oxytricha Telomere Repeats. *Structure* **3**, 997-1008.

27. Stefl, R., Cheatham, T. E., Spackova, N., Fadrna, E., Berger, I., Koca, J. & Šponer, J. (2003). Formation pathways of a guanine-quadruplex DNA revealed by molecular dynamics and thermodynamic analysis of the substates. *Biophys. J.* **85**, 1787-1804.

28. Spackova, N., Berger, I. & Šponer, J. (1999). Nanosecond molecular dynamics simulations of parallel and antiparallel guanine quadruplex DNA molecules. *J. Am. Chem. Soc.* **121**, 5519-5534.

29. Spackova, N., Berger, I. & Šponer, J. (2001). Structural dynamics and cation interactions of DNA quadruplex molecules containing mixed guanine/cytosine quartets revealed by large-scale MD simulations. *J. Am. Chem. Soc.* **123**, 3295-3307.

30. Spackova, N., Cubero, E., Šponer, J. & Orozco, M. (2004). Theoretical study of the guanine -> 6-thioguanine substitution in duplexes, triplexes, and tetraplexes. *J. Am. Chem. Soc.* **126**, 14642-14650.

31. Stefl, R., Spackova, N., Berger, I., Koca, J. & Šponer, J. (2001). Molecular dynamics of DNA quadruplex molecules containing inosine, 6-thioguanine and 6-thiopurine. *Biophys. J.* **80**, 455-468.

32. Strahan, G. D., Keniry, M. A. & Shafer, R. H. (1998). NMR structure refinement and dynamics of the K+- $d(G_3T_4G_3)_2$ quadruplex via particle mesh Ewald molecular dynamics simulations. *Biophys. J.* **75**, 968-981.

33. Chowdhury, S. & Bansal, M. (2001). A nanosecond molecular dynamics study of antiparallel $d(G)_7$ quadruplex structures: Effect of the coordinated cations. *J. Biomol. Struct. Dyn.* **18**, 647-669.

34. Chowdhury, S. & Bansal, M. (2001). G-quadruplex structure can be stable with only some coordination sites being occupied by cations: A six-nanosecond molecular dynamics study. *J. Phys. Chem. B* **105**, 7572-7578.

35. Fadrna, E., Spackova, N., Stefl, R., Koca, J., Cheatham, T. E. & Šponer, J. (2004). Molecular dynamics simulations of guanine quadruplex loops: Advances and force field limitations. *Biophys. J.* **87**, 227-242.

36. Hazel, P., Huppert, J., Balasubramanian, S. & Neidle, S. (2004). Loop-length-dependent folding of G-quadruplexes. *J. Am. Chem. Soc.* **126**, 16405-16415.

37. Suhnel, J. (2001). Beyond nucleic acid base pairs: From triads to heptads. *Biopolymers* **61**, 32-51.

38. Ross, W. S. & Hardin, C. C. (1994). Ion-Induced Stabilization of the G-DNA Quadruplex - Free-Energy Perturbation Studies. *J. Am. Chem. Soc.* **116**, 6070-6080.

39. Mohanty, D. & Bansal, M. (1994). Conformational Polymorphism in Telomeric Structures - Loop Orientation and Interloop Pairing in $d(G_4T_nG_4)$. *Biopolymers* **34**, 1187-1211.

40. Meyer, M., Schneider, C., Brandl, M. & Suhnel, J. (2001). Cyclic adenine-, cytosine-, thymine-, and mixed guanine-cytosine-base tetrads in nucleic acids viewed from a quantum-chemical and force field perspective. *J. Phys. Chem. A* **105**, 11560-11573.

41. Meyer, M. & Suhnel, J. (2003). Interaction of cyclic cytosine-, guanine-, thymine-, uracil- and mixed guanine-cytosine base tetrads with K+, Na+ and Li+ ions - A density functional study. *J. Biomol. Struct. Dyn.* **20**, 507-517.

42. Gu, J. D. & Leszczynski, J. (2000). A remarkable alteration in the bonding pattern: An HF and DFT study of the interactions between the metal cations and the Hoogsteen hydrogen-bonded G-tetrad. *J. Phys. Chem. A* **104**, 6308-6313.

43. Pan, B. C., Xiong, Y., Shi, K., Deng, J. P. & Sundaralingam, M. (2003). Crystal structure of an RNA purine-rich tetraplex containing adenine tetrads: Implications for specific binding in RNA tetraplexes. *Structure* **11**, 815-823.

44. Pan, B. C., Xiong, Y., Shi, K. & Sundaralingam, M. (2003). An eight-stranded helical fragment in RNA crystal structure: Implications for tetraplex interaction. *Structure* **11**, 825-831.

45. Liu, H., Matsugami, A., Katahira, M. & Uesugi, S. (2002). A dimeric RNA quadruplex architecture comprised of two G : G(: A): G : G(: A) hexads, G : G : G : G tetrads and UUUU loops. *J. Mol. Biol.* **322**, 955-970.

46. Cheong, C. J. & Moore, P. B. (1992). Solution Structure of an Unusually Stable RNA Tetraplex Containing G-Quartet and U-Quartet Structures. *Biochemistry* **31**, 8406-8414.

47. Keniry, M. A. (2000). Quadruplex structures in nucleic acids. *Biopolymers* **56**, 123-146.

48. Patel, P. K. & Hosur, R. V. (1999). NMR observation of T-tetrads in a parallel stranded DNA quadruplex formed by Saccharomyces cerevisiae telomere repeats. *Nucleic Acids Res.* **27**, 2457-2464.

49. Patel, P. K., Bhavesh, N. S. & Hosur, R. V. (2000). NMR observation of a novel C-tetrad in the structure of the SV40 repeat sequence GGGCGG. *Biochem. Biophys. Res. Commun.* **270**, 967-971.

50. Searle, M. S., Williams, H. E. L., Gallagher, C. T., Grant, R. J. & Stevens, M. F. G. (2004). Structure and K+ ion-dependent stability of a parallel-stranded DNA quadruplex containing a core A-tetrad. *Org. Biomol. Chem.* **2**, 810-812.

51. Patel, P. K. & Hosur, R. V. (1998). NMR observation of a novel A-tetrad in a telomeric DNA segment in aqueous solution. *Curr. Sci.* **74**, 902-906.

52. Risitano, A. & Fox, K. R. (2005). Inosine substitutions demonstrate that intramolecular DNA quadruplexes adopt different conformations in the presence of sodium and potassium. *Bioorg. Med. Chem. Lett.* **15**, 2047-2050.

53. Mergny, J. L., De Cian, A., Ghelab, A., Sacca, B. & Lacroix, L. (2005). Kinetics of tetramolecular quadruplexes. *Nucleic Acids Res.* **33**, 81-94.

54. Krishnan-Ghosh, Y., Liu, D. S. & Balasubramanian, S. (2004). Formation of an interlocked quadruplex dimer by d(GGGT). *J. Am. Chem. Soc.* **126**, 11009-11016.

55. Marathias, V. M., Sawicki, M. J. & Bolton, P. H. (1999). 6-Thioguanine alters the structure and stability of duplex DNA and inhibits quadruplex DNA formation. *Nucleic Acids Res.* **27**, 2860-2867.

56. Hardin, C. C., Perry, A. G. & White, K. (2000). Thermodynamic and kinetic characterization of the dissociation and assembly of quadruplex nucleic acids. *Biopolymers* **56**, 147-194.

57. Ferre-d'Amare, A. R. & Doudna, J. A. (1999). RNA FOLDS: Insights from recent crystal structures. *Annu. Rev. Biophys. Biomolec. Struct.* **28**, 57-73.

58. Ogle, J. M., Carter, A. P. & Ramakrishnan, V. (2003). Insights into the decoding mechanism from recent ribosome structures. *Trends Biochem. Sci.* **28**, 259-266.

59. Moore, P. B. & Steitz, T. A. (2003). The structural basis of large ribosomal subunit function. *Annu. Rev. Biochem.* **72**, 813-850.

60. Schuwirth, B. S., Borovinskaya, M. A., Hau, C. W., Zhang, W., Vila-Sanjurjo, A., Holton, J. M. & Cate, J. H. D. (2005). Structures of the bacterial ribosome at 3.5 angstrom resolution. *Science* **310**, 827-834.

61. Leontis, N. B., Stombaugh, J. & Westhof, E. (2002). The non-Watson-Crick base pairs and their associated isostericity matrices. *Nucleic Acids Res.* **30**, 3497-3531.

62. Reblova, K., Spackova, N., Šponer, J. E., Koca, J. & Šponer, J. (2003). Molecular dynamics simulations of RNA kissing-loop motifs reveal structural dynamics and formation of cation-binding pockets. *Nucleic Acids Res.* **31**, 6942-6952.

63. Reblova, K., Spackova, N., Stefl, R., Csaszar, K., Koca, J., Leontis, N. B. & Šponer, J. (2003). Non-Watson-Crick basepairing and hydration in RNA motifs: Molecular dynamics of 5S rRNA loop E. *Biophys. J.* **84**, 3564-3582.

64. Reblova, K., Spackova, N., Koca, J., Leontis, N. B. & Šponer, J. (2004). Long-residency hydration, cation binding, and dynamics of loop E/helix IV rRNA-L25 protein complex. *Biophys. J.* **87**, 3397-3412.

65. Auffinger, P., Bielecki, L. & Westhof, E. (2003). The Mg2+ binding sites of the 5S rRNA loop E motif as investigated by molecular dynamics simulations. *Chem. Biol.* **10**, 551-561.

66. Auffinger, P., Bielecki, L. & Westhof, E. (2004). Symmetric K+ and Mg2+ ion-binding sites in the 5 S rRNA loop E inferred from molecular dynamics simulations. *J. Mol. Biol.* **335**, 555-571.

67. Vallurupalli, P. & Moore, P. B. (2003). The solution structure of the loop E region of the 5 S rRNA from spinach chloroplasts. *J. Mol. Biol.* **325**, 843-856.

68. Krasovska, M. V., Sefcikova, J., Spackova, N., Šponer, J. & Walter, N. G. (2005). Structural dynamics of precursor and product of the RNA enzyme from the hepatitis delta virus as revealed by molecular dynamics simulations. *J. Mol. Biol.* **351**, 731-748.

69. Klein, D. J., Schmeing, T. M., Moore, P. B. & Steitz, T. A. (2001). The kink-turn: a new RNA secondary structure motif. *Embo J.* **20**, 4214-4221.

70. Razga, F., Spackova, N., Reblova, K., Koca, J., Leontis, N. B. & Šponer, J. (2004). Ribosomal RNA kink-turn motif - A flexible molecular hinge. *J. Biomol. Struct. Dyn.* **22**, 183-193.

71. Razga, F., Koca, J., Šponer, J. & Leontis, N. B. (2005). Hinge-like motions in RNA kink-turns: The role of the second A-minor motif and nominally unpaired bases. *Biophys. J.* **88**, 3466-3485.

72. Cojocaru, V., Klement, R. & Jovin, T. M. (2005). Loss of G-A base pairs is insufficient for achieving a large opening of U4 snRNA K-turn motif. *Nucleic Acids Res.* **33**, 3435-3446.

73. Nissen, P., Ippolito, J. A., Ban, N., Moore, P. B. & Steitz, T. A. (2001). RNA tertiary interactions in the large ribosomal subunit: The A-minor motif. *Proc. Natl. Acad. Sci. USA* **98**, 4899-4903.

Chapter 13

USING COMPUTER SIMULATIONS TO STUDY DECODING BY THE RIBOSOME

Kevin Y. Sanbonmatsu
Theoretical Biology and Biophysics Dept.; Theoretical Division, Los Alamos National Laboratory

Abstract: The ribosome is the central information-processing unit of the cell. Decoding is the step during protein synthesis where information is transferred from protein to nucleic acid. The decoding problem is reviewed, including recent kinetic, x-ray and cryo-EM data. Molecular dynamics simulations are described which investigate the role of hydrogen bond networks, "base-flipping" events, and changes in the conformation of transfer RNA.

Key words: Ribosome, Molecular Dynamics Simulations, High Performance Computing, RNA, Base-flipping, Translation, tRNA Selection, Decoding

The ribosome is the universal information-processing unit responsible for the synthesis of proteins in the cell. Composed of a handful of long strands of RNA and more than fifty proteins, this ribonucleoprotein complex is ancient in origin and lies at the heart of every biological system. The ribosome is so highly conserved across all kingdoms of life that its sequence is extensively used for the construction of phylogenetic trees. The importance of the ribosome is underscored by the amount of material and infrastructure devoted to ribosome biosynthesis. For example, in yeast, the ribosome constitutes approximately 80% of the RNA. When combined with transfer RNAs (tRNAs), which account for 15% of the RNA, a total of 95% of the RNA is designated to ribosome function (protein synthesis). Lastly, ribosomal RNA transcription accounts for approximately 60% of the total transcription performed by the cell.[1] In light of the fact that much of cellular metabolism is devoted to the manufacture of ribosomes, it is not outlandish to use the perspective of the 'selfish ribosome' to describe the evolution of single-celled organisms.

J. Šponer and F. Lankaš (eds.), Computational Studies of RNA and DNA, 327–342.
© 2006 *Springer.*

As stated in the central dogma of molecular biology, genetic information generally flows from DNA to RNA (transcription) and from RNA to protein (translation). During transcription, the RNA polymerase complex binds to DNA, interacts with a certain gene on the DNA, and manufactures RNA molecules whose sequences are exactly complementary (in terms of Watson-Crick base-pairing) to the gene on the DNA. During translation, or protein biosynthesis, the ribosome reads genetic information encoded in the RNA molecules (messenger RNAs) and synthesize the exact protein specified by this information. Translation has three major steps: (1) initiation, where the ribosomal complex is formed, (2) elongation where the message is decoding and the protein is manufactured, and (3) termination, where the ribosomal complex dissociates. Elongation itself has three steps: (1) decoding, where the message is read, (2) peptidyl transferase, where an amino acid is added to the new protein, and (3) translocation, where the ribosome moves exactly three nucleotides along the messenger RNA.

While there are many so-called molecular machines which perform catalysis and many which bind to nucleic acid molecules, there are only four which process long strings of information: DNA polymerase, RNA polymerase, reverse transcriptase, and the ribosome. The first three of these perform the trivial lookup operation specified by Watson-Crick base-pairing. The ribosome is the only information-processing complex, which uses a non-trivial lookup table.

To understand the meaning of "non-trivial" in this context, the lookup operations must first be formally defined. Consider the 4-letter "alphabet" of RNA nucleotides,

$$\Omega = \{a, c, g, u\ \},$$

and the 20-letter alphabet of amino acids,

$$\Sigma = \{A, C, D, E, F, G, H, I, K, L, M, N, P, Q, R, S, T, V, W, Y\ \}.$$

The ribosome performs the many-to-one mapping, $\eta: xyz \rightarrow U$, where x, y, and z are members of Ω and U is a member of Σ. The mapping is further complicated by the so-called degeneracy of the genetic code where several codons map to one amino acid. In contrast to the operation performed by the ribosome, the polymerases perform trivial one-to-one mappings. For example, DNA polymerase performs the mapping $\zeta: U \rightarrow V$, where U is a member of Σ and V is a member of $\Sigma' = \{a, c, g, t\ \}$. The non-trivial map used by the ribosome requires the suite of transfer RNAs (tRNAs) that are not required by the polymerases, where the information transfer is achieved directly by single nucleotides. Because the ribosome is the only molecular machine that performs a non-trivial information processing operation that is central to the operation of the cell, the ribosome is analogous to a central processing unit (CPU) of the cell.

How does this operation take place physically? With the help several initiation factors, the ribosomal complex forms and is positioned at the start codon on the mRNA molecule. Once the initiation complex has formed, the ribosome exposes the codon following the start codon (*i.e.*, the aminoacyl site or 'A' site) to the suite of tRNAs present in the cell. The ribosome must then distinguish between tRNAs whose anticodons are complementary to this codon (cognate) and tRNAs whose anticodons are not complementary to the codon (non-cognate).[2-4] The process of selecting the correct tRNA molecule is known as decoding. This step allows the transfer of information from nucleic acid to protein: without it, polypeptides with random sequences would be produced. During decoding, a certain fraction of tRNAs selected are actually incorrect (the error rate in bacteria is approximately 5×10^{-3}).[5] Watson-Crick pairing, of course, plays an important role in this process. However, Watson-Crick pairing alone cannot explain the observed error rates. This fact, in combination with the observation that ribosomal ambiguity mutations alter the error rate, strongly suggests that the ribosome plays a role in decoding.[4]

In the kinetic proofreading model of decoding, tRNA selection occurs in two stages: initial selection and proofreading.[6-9] During initial selection, the candidate tRNA (in an aminoacyl-tRNA♦GTP♦EF-Tu ternary complex) binds in a partially bound state (the A/T state). The initial selection and proofreading stages are separated by hydrolysis of the GTP in the ternary complex. Following the proofreading step, EF-Tu♦GDP dissociates and the candidate tRNA moves from the partially bound (A/T) to the fully bound (A/A) state (accommodation).[6] The ribosome must adjust its conformation to accommodate this change in the state of the candidate tRNA.

The recent x-ray structure of the full 70S ribosome has uncovered the configuration of A/A state tRNA along with its ribosomal interactions.[10] Furthermore, the x-ray structure of the small subunit (SSU) of the ribosome has illuminated the mechanism of cognate codon-anticodon pair recognition as well as the recognition of near-cognate codon-anticodon pairs which occurs during misreading induced by paromomycin.[11] X-ray structures have also suggested an open-to-closed change in SSU conformation that differs for cognate and near-cognate codon-anticodon interactions.[12] More recently, cryo electron microscopy (cryo-EM) data have shown that the tRNA in the A/T state is curved with respect to the A/A state, suggesting that the accommodation change involves the relaxation of the curved tRNA to the native linear tRNA structure.[13] Several effects may be important in the mechanism of decoding, including the formation and reorganization of hydrogen bond networks, the "flipping" of ribosomal RNA bases (rRNA) in the decoding center and the kinetic rates for certain transitions.

The quite beautiful structure of the SSU has shown that an intricate network of hydrogen bonds exists between the tRNA anticodon, the mRNA codon, and the universally conserved SSU rRNA bases G530, A1492 and A1493 (Figure 13-1). In addition to the Watson-Crick hydrogen bonds between the codon and anticodon, hydrogen bonds including the 2'-hydroxyl groups occur between the ribosomal RNA and codon-anticodon pairs. To predict the difference between the cognate and near-cognate networks, we have performed all-atom molecular dynamics simulations of the mRNA and tRNA, complexed with the decoding center of the ribosome, based on the high resolution x-ray structure.[11]

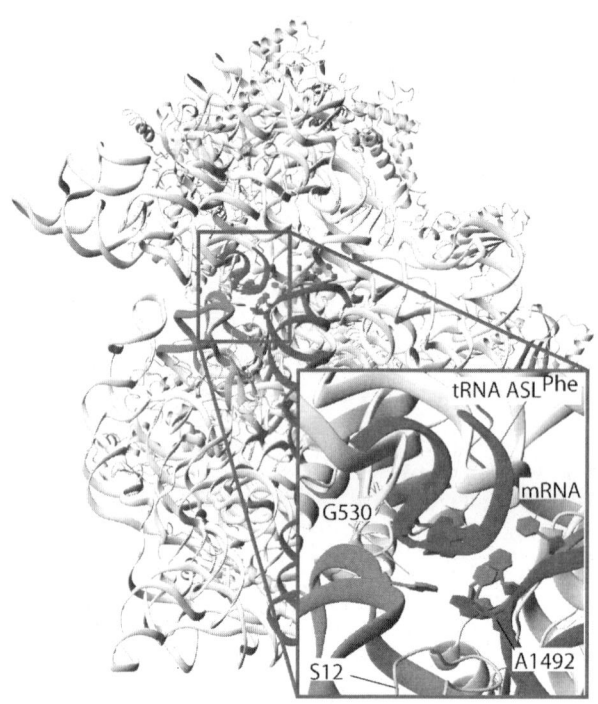

Figure 13-1. The structure of the decoding center of the ribosome of Ramakrishnan. A1493 is not shown but lies next to A1492. The 3-base codon is shown in green.

The importance of studying the local interactions in the decoding center is underscored by the identical foot prints in SSU rRNA of tRNA bound to the ribosome in the A/A and A/T states.[14] Furthermore, cryo-EM data suggests the local codon-anticodon-ribosome interactions at the decoding site in the

A/T and A/A states are similar.[13] Finally, A site anticodon stem loops (ASLs) soaked into SSU crystals have been successfully used as a model system to study the selection of cognate tRNA.[11] We use this model system as a basis for our work, which is relevant to both A/T and A/A states. While large scale conformational changes are clearly important in the decoding process (examined below), the interaction between codon and anticodon in the SSU A site must also play an important role, in light of the fact that the specificity is, in large part, determined by the codon-anticodon interaction.

The method of all-atom explicit solvent molecular dynamics simulations allows us to observe the structural stability of near-cognate tRNAs bound to the ribosome over very short time scales. The high resolution structure of the SSU•ASL^{Phe}•mRNA complex[11] was used as a basis for the simulations.[15] The focus of the simulation is the three base-pair codon-anticodon "mini-helix" and its interactions with the ribosome. For the cognate case, the codon-anticodon mini-helix consists of U:A, U:A, and U:G base pairs for the UUU codon on the mRNA and GAA anticodon on the tRNA. The simulation domain consists of those stretches of small subunit rRNA and small subunit ribosomal proteins which directly interact with the A site codon-anticodon pair. This amounted to 54 nucleotides and 9 amino acid residues. Each system was solvated with approximately 7×10^3 water molecules.

We have performed twenty 3.0 ns molecular dynamics simulations (60.0 ns total sampling time of the decoding problem), including all codons differing from the cognate codon (UUU) by one base. As control simulations, we examined the cognate cases (UUU and UUC) as well as a non-cognate case with three mismatches (AAG). To test the sensitivity of these results to the size of the simulation domain, additional simulations were performed including small subunit rRNA bases 912-916 and 1395-1404 in addition to the domain described above for cognate and near-cognate cases. The simulation methods were described previously.[15]

The control case (cognate UUU mRNA codon) produced stable codon-anticodon-ribosome hydrogen bonds, consistent with those observed in the crystal structure (Figure 13-2). The average number of codon-anticodon-ribosome hydrogen bonds involving the tRNA-ASL is $N_H E 8.70$, according to the criteria used by Ogle, et al.,[11] similar to the value $N_H = 9$ observed in the crystal structure. Simulations with identical protocol were performed for the cognate UUC codon. Here, the third position G:C pair in UUC forms three hydrogen bonds compared to the two bonds formed by the third position G:U wobble for UUU. The total number of codon-anticodon-ribosome hydrogen bonds involving the tRNA-ASL is greatest for the cognate cases (UUU and UUC) and the average root-mean-squared-deviation (RMSD) away from the crystal structure is lowest for the cognate cases.

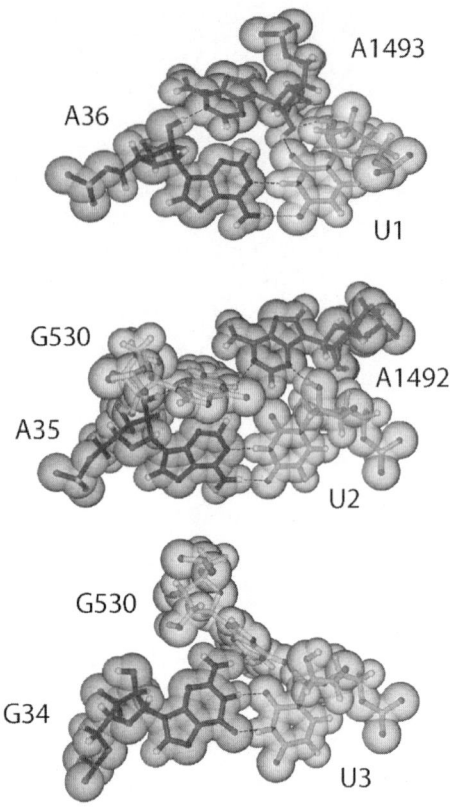

Figure 13-2. The hydrogen bond network in the decoding center of the ribosome for cognate codon-anticodon pairs. The three codon-anticodon base pairs are shown with their ribosome hydrogen bond partners for a structure at late times during the cognate simulation. The network includes the mRNA codon (green), the tRNA anticodon (purple), the ribosome SSU H18(cyan) and H44(red). The simulations produced stable codon-anticodon-ribosome hydrogen bonds similar to those observed in the x-ray structure.

In the cognate case, SSU rRNA bases G530 and A1492 interact with each other by forming a hydrogen bond across the minor groove of the codon-anticodon mini-helix, shielding the codon-anticodon minor groove from solvent (Figure 13-3). In the near-cognate UUA case, the tRNA-mRNA G34:A3 purine-purine mismatch sterically hinders the rRNA-rRNA G530:A1492 interaction, preventing the proper (*i.e.*, cognate) hydrogen bond system from forming, while increasing the solvent exposure of the codon-anticodon pair.

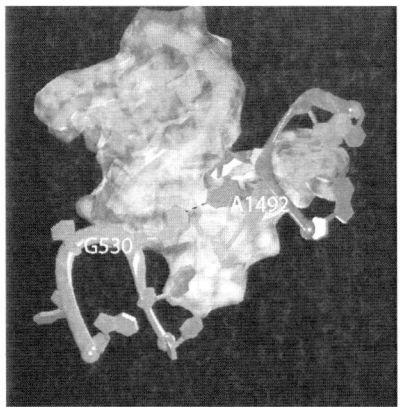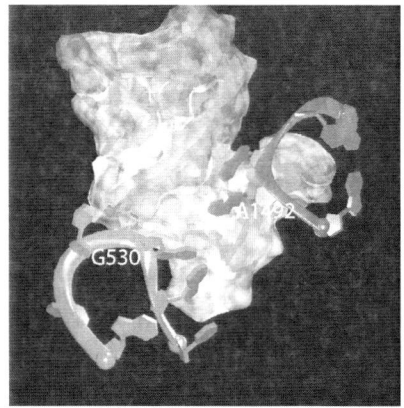

Figure 13-3. Cognate (left) and near-cognate (right) simulations show the G530-A1492 occurs for cognate but does not occur for near-cognate anticodons. Translucent surface displays the minor groove of the codon-anticodon mini-helix formed by the tRNA anticodon (purple) and mRNA codon (green). SSU H18 (cyan) and H44 (red) appear in the foreground.

To quantify the minor groove measurement mechanism, we examined the groove dimensions and hydrogen bond distances of the cognate UUU and near-cognate UUA systems. The average groove depth for UUA is approximately 1.69 Å, significantly shallower than the UUU groove depth of 4.1 Å. The shallower minor groove prevents the G530:A1492 N1-H...N1 hydrogen bond from forming, giving an average distance of 6.2 Å and N1-H...N1 angle of 75° for UUA, compared to an average distance of 2.0 Å and N1-H...N1 angle of 150° for UUU. It is clear that the shallower minor groove of the UUA codon-anticodon mini-helix contributes to the steric hindrance of the G530:A1492 interaction by the near-cognate codon-anticodon mini-helix.

The steric hindrance of the rRNA G530:A1492 interaction results in a loss of codon-anticodon-ribosome hydrogen bonds, due to the repositioning of G530 and A1492. In our simulations, steric effects and hydrogen bonding work together to distinguish between cognate and near-cognate interactions. The use of the G530:A1492 interaction in discrimination is consistent with the lethal mutation A1492G and the lethal mutation G530A, whose G:G and A:A clashes, respectively, would disrupt the G530:A1492 interaction, preventing proper discrimination. From our simulations we were able to correctly predict that the double mutant A1492G:G530A is viable because it preserves the N1-H...N1 interaction between SSU bases 530 and 1492.

To investigate the role of the ribosome, simulations for the UCU case were repeated without any ribosomal RNA or protein. The ASL^{Phe} and mRNA were simulated with the same geometry and restraints, enabling us to compare the codon-anticodon stability with and without ribosomal bases. In the absence of the ribosome, the base pair fluctuations were markedly lower.

The relative stability of the codon-anticodon mini-helix for UCU in absence of the ribosome contrasts with the disruption of the first and second position base pairs in the presence of the ribosome, suggesting that the ribosome may interfere with near-cognate UCU codon-anticodon pairing. In contrast, for cognate cases, the ribosome enhances the stability of the codon-anticodon interactions by forming hydrogen bonds with both the codon and anticodon. Our simulations suggest that steric effects, in combination with competing hydrogen bond configurations, result in different hydrogen bond network configurations for cognate, near-cognate, and non-cognate codon-anticodon-ribosome interactions.

The flipping of universally conserved SSU rRNA bases A1492-A1493 is also thought to play a role in decoding. The nature of this role is not understood. While it is known that the bases flip out of SSU helix 44 upon cognate tRNA binding, it is not known whether these bases flip as a result of near-cognate tRNA binding alone. The effect of this base flipping on the conformation of helix 44 and the overall conformation of the SSU subunit is also poorly understood. As a first step to understand the role of base flipping, we have performed molecular dynamics simulations of the decoding helix.

While the above study is effective in examining the stabilities of hydrogen bonds and subtle conformational changes (*e.g.,* the movement of G530 and A1492 shown in Figure 13-3), the 3 ns sampling per simulation is not sufficient to observe the flipping in and out of A1492 and A1493, which is a larger conformational change than the local rearrangement of hydrogen bonds. Replica exchange molecular dynamics simulations enhance the sampling by 5-20 fold over conventional molecular dynamics simulations.[16,17] The method is characterized by performing simulations of a large number of copies of the original system. Copies of the system, identical except for temperature, exchange temperatures after a given time interval, avoiding kinetic traps by sampling high temperatures. This temperature sampling facilitates barrier crossings on the energy landscape (*i.e.,* transitions between stable configurations), creating trajectories well suited for detailed investigation of the mechanism of conformational changes.

The initial structure consisted of small ribosomal subunit bases 1404-1411 and 1489-1497 from the structures of Ramakrishnan (PDB accession code 1J5E and 1IBM).[18] Four initial configurations were used: (1) both A1492-A1493 flipped in; (2) both A1492-A1493 flipped out; (3) A1492 flipped in and A1493 flipped out; and (4) A1492 flipped out and A1493 flipped in. Excess ions were placed randomly in a box of $(55 \text{ Å})^3$ at concentrations of 150 mM KCl, and 7 mM $MgCl_2$. Ions were then equilibrated with the AMBER force field using a continuum water model with a 5 Å radius for the ions for 1 ns to ensure electrostatic energy convergence. Next the system was solvated with TIP3P water ($N_{atoms} = 16,389$), minimized by steepest descent and equilibrated again with the AMBER force field and particle mesh Ewald electrostatics. A distribution of 48 replicas was used for $312.0 < T < 544.5$ K (Figure 13-4, right). The system was run in production exchange mode for 900 ps per replica, with exchange attempts every 0.25 ps, giving a total sampling of 43.2 ns. We have observed several base flip transitions. The flip-in transitions occur with A1492 and A1493 flipping into helix 44 simultaneously (Figure 13-5). Of the 12 replicas beginning in configuration (1) with both bases flipped out of helix 44, 5 replicas display flip-in conformational changes. Of the 12 replicas in configuration (2), 2 replicas display flip-out conformational changes. Configurations (3) and (4) each show 1 replica flip-in event. Figure 13-4 (right) shows the potential-of-mean force ($-\ln (n_i/N_0)$ where n_i is the number of configurations in a particular bin and N_0 is the total number of configurations) in units of kT, where blue represents minima and red represents regions of low probability. The order parameters, θ_{1492} and θ_{1493}, describe the flipped in/out conformation change of small subunit rRNA A1492 and A1493, respectively. Because the torsional parameters and local helical parameters do not uniquely describe the flipping-in and flipping-out of SSU rRNA bases A1492-A1493, we have chosen the angle $\theta_{A1492/3} = N1_{A1492/3} - C1'_{A1492/3} - N1_{A1408}$ as our order parameter. While this is certainly not the only parameter that uniquely describes base flipping, it correctly categorizes structures as flipped-out, flipped-in, or partially flipped-out/in. The landscape is dominated by populations with both bases flipped in or both bases flipped out, with some structures displaying one base flipped in.

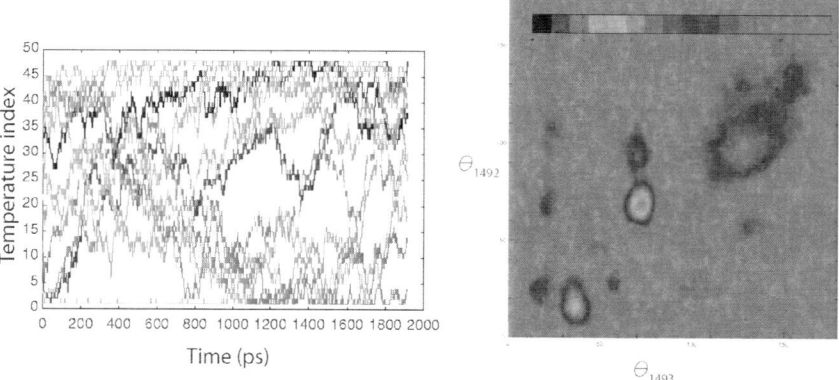

Figure 13-4. Temperature index of replicas vs. time for selected replicas (left) and potential-of mean-force for order parameters describing the decoding bases flipping out of SSU H44.

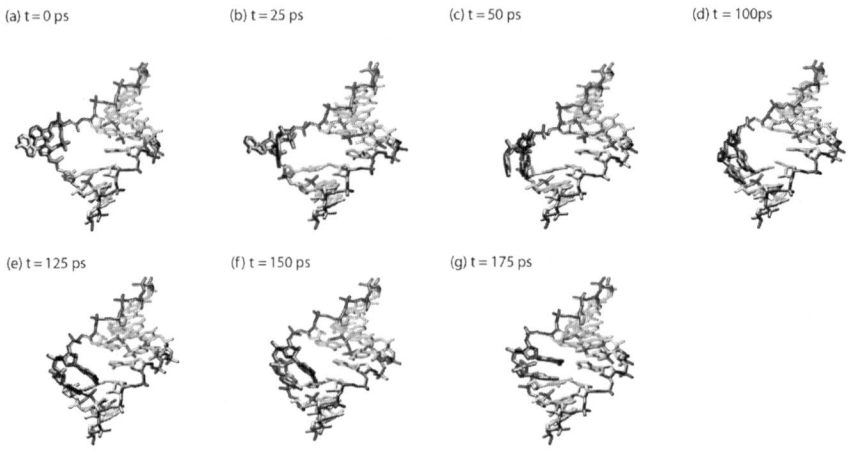

(a) t = 0 ps   (b) t = 25 ps   (c) t = 50 ps   (d) t = 100ps

(e) t = 125 ps   (f) t = 150 ps   (g) t = 175 ps

Figure 13-5. A flipping in event during the replica simulation. The decoding bases are shown in red.

In addition to local effects such as hydrogen bond stability and base-flipping, larger scale conformational changes have recently been shown to be important for decoding. In particular, the rate-limiting step of decoding has been shown to be the accommodation step for the case of cognate-tRNAs.[19] We have performed targeted molecular dynamics simulations of the accommodation step to understand which parts of the ribosome are important for the cognate tRNA accommodation.

Because the accommodation rate of cognate tRNAs is ≈7/s (8), we have implemented the targeted MD algorithm[20-22] in explicit solvent, which gradually reduces the root-mean-squared distance (RMSD) of the simulated A/T ribosome complex to the A/A state while allowing thermal fluctuations of the structure at any given RMSD. Rather than providing an exact energy landscape of the transition, our simulations produce stereochemically feasible pathways that can be tested experimentally. In particular, our goal is to predict large subunit (LSU) rRNA bases and ribosomal protein residues that are important for accommodation.

Because it is not clear which regions of the ribosome are essential for accommodation, we have taken the rather conservative approach of simulating the whole ribosome in explicit solvent. During accommodation, the tRNA moves from the partially bound, or A/T state, to the fully bound, or A/A state. The A/A state model structure is similar to the previously published model.[23] The A/T state model structure is similar to the A/A state, with the exception of the tRNA and the GTPase activation center (LSU rRNA helices 43-44 and L11), which were the only portions of the complex that displayed differences between the A/A and A/T state cryo-EM maps.[23] Excess ions were placed randomly and equilibrated (NAMD2.5)[24] in a box around the solute at concentrations of 0.1 M KCl and 7 mM MgCl$_2$ to approximate the Wintermeyer/Rodnina and Puglisi buffers.[6,25] The ion-solute systems were embedded in an explicit TIP3P water solvent box with a buffer of water around the solute of at least 16 Å in all directions, yielding a system of Natoms = 2,640,030, minimized, and equilibrated (NAMD2.5) for 1.6 ns. Targeted molecular dynamics (TMD) was implemented. Eight simulations of 2 ns were performed with different initial conditions. To test the dependence of the simulation results on the simulation time, one 4 ns and two 1 ns simulations were performed giving a total production simulation sampling of 22 ns. Performance studies demonstrated near-linear scaling to 768 processors.

The accommodation of the aminoacyl-tRNA in all simulations displays a bulk motion of the tRNA body into the large subunit A site (Figure 13-6). The motion involves the relaxation of the A/T kink about tRNA positions 44-45 and 26, combined with a ≈45° rotation around the anticodon stem loop axis. Each simulation also shows the subsequent accommodation of the universally conserved 3'-CCA end into the peptidyl transferase center and convergence to the target structure. The entrance of the 3'-CCA end into the peptidyl transferase center is impeded with respect to the motion of the tRNA body due to the interaction with the LSU A-loop (2552-2561). The codon-anticodon interactions, as well as the interactions of SSU rRNA G530, A1492 and A1493 with the tRNA and mRNA, remained intact throughout the accommodation simulations. The interactions between the tRNA and the ribosome are divided into four stages (Figure 13-7), describing A/T interactions (stage 1), relaxation of the tRNA body (stage 2), relaxation of the tRNA acceptor stem (stage 3) and A/A interactions (stage 4).

During stage 3, both the 3'-CCA end and the A-loop flex slightly to allow accommodation of the 3'-CCA end through the A-loop towards the peptidyl transferase center. The mutual flexibility between the tRNA 3'-CCA end and the LSU rRNA A-loop resolves the apparent steric clash present during accommodation. The A-loop constitutes a corridor through which the amino acid may pass if the tRNA and the A-loop are sufficiently flexible and the tRNA is properly aligned. The large available volume in the corridor makes this pathway a candidate for the accommodation of all other amino acids.

By stage 4, the tRNA body has been accommodated by the ribosome and the 3'-CCA end enters the peptidyl transferase center. The phosphate RMSD values between the final simulated structure and the A/A state x-ray structures are 1.21 Å, 1.27 Å, and 2.41 Å for the SSU aminoacyl (A) and peptidyl (P) sites,[11] the LSU A and P sites,[26] and the 70S complex,[10] respectively (A and P sites are defined by all rRNA nucleotides and protein residues within 4.5 Å of the tRNA analogs in the higher resolution x-ray structures).

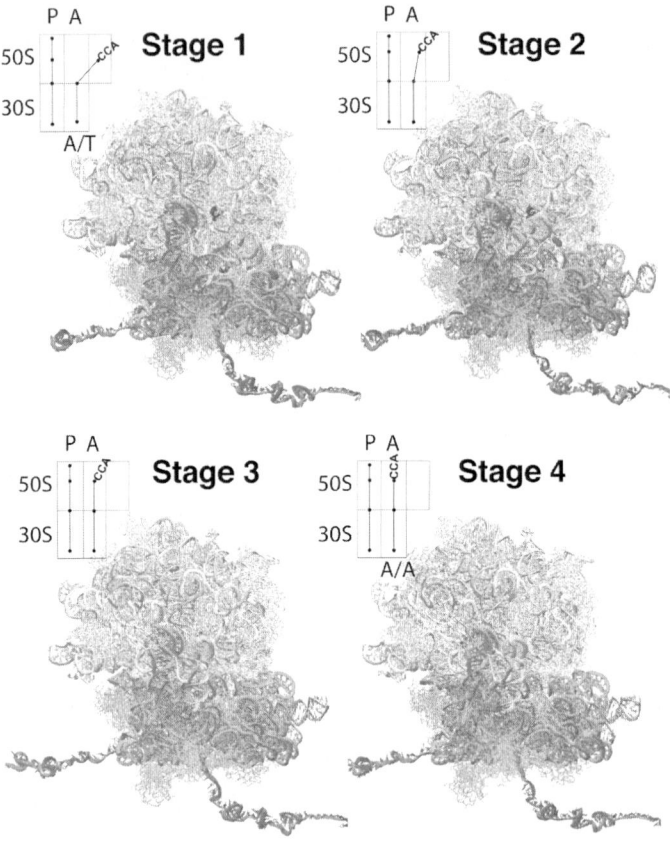

Figure 13-6. Movement of the aminoacyl-tRNA into the ribosome during accommodation simulated by targeted molecular dynamics.

The targeted MD simulations have shown explicitly that the codon-anticodon-ribosome interactions can persist throughout accommodation. The successful movement of the system from the A/T to the A/A states for 8 sets of initial velocities demonstrates the stereochemical feasibility of maintaining similar codon-anticodon-SSU geometries throughout accommodation, as proposed by the Ramakrishnan[27] and Frank groups.[28]

Of the 68 nucleotides in the LSU rRNA that were observed to interact with the aminoacyl-tRNA during our simulations, 28 nucleotides are ≥95% conserved. Eighteen nucleotides are 'universally conserved' in the sense that they are as highly conserved (~99%) or more highly conserved than SSU rRNA base A1493.[29]

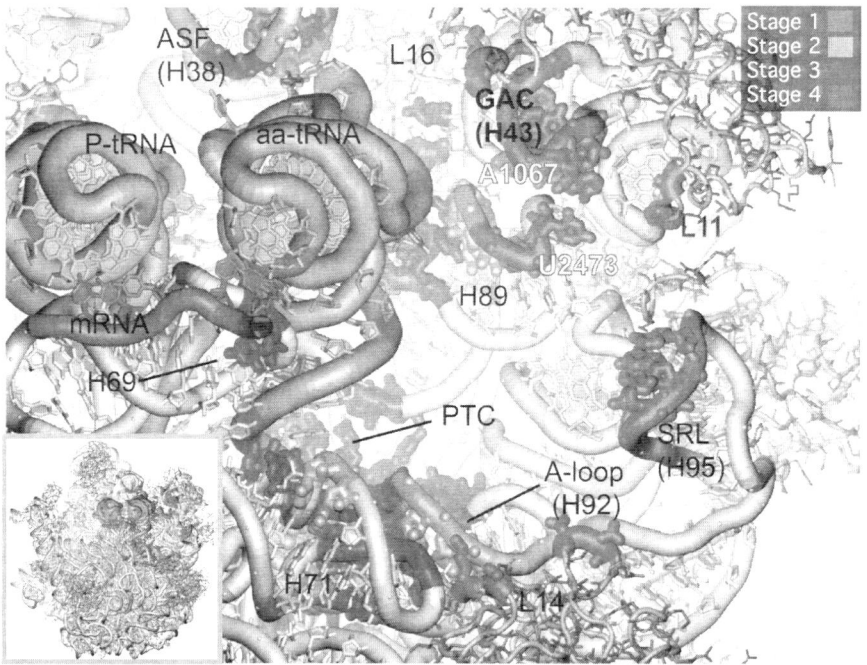

Figure 13-7. Stages of accommodation. Aminoacyl-tRNA (yellow), peptidyl-tRNA (cyan), mRNA (green), large subunit RNA (grey) and large subunit proteins (green) are shown.

Within the context of the decoding process, the favorable interactions of the ternary complex with the ribosome likely overcome the energetic cost of inducing the A/T kink in the aminoacyl-tRNA. Once EF-Tu dissociates, the tRNA is allowed to relax to the A/A state (which more closely resembles the native tRNA state) while its anticodon stem loop remains tightly bound to the SSU A site. The 3'-CCA end, however, is deterred mainly by the A-loop. The flexibility of the tRNA 3'-CCA end facilitates its movement into the peptidyl transferase center. Because proper accommodation requires a properly aligned anticodon stem loop, a near-cognate tRNA with near-cognate codonanticodon interaction may significantly affect the accommodation trajectory, potentially decreasing the chances of entry of the 3'-CCA end into the peptidyl transferase center. If the 3'-CCA end does not enter the peptidyl transferase center, the tRNA dissociation rate will likely increase.

Kinetic effects have recently been shown to dominate decoding.[19] The mRNA•tRNA•rRNA hydrogen bond network has the potential to play an important role in discrimination; however, the relationship between the hydrogen bond network and kinetic discrimination remains poorly understood.

One possible role of the hydrogen bond network may be the proper aligning of cognate tRNAs in the ribosome. This has the potential to affect both the GTPase activation step used to reject near-cognate tRNAs and the accommodation of cognate tRNAs.

With regard to signal communication via the tRNA, although the tRNA is quite flexible, the orientation of the acceptor stem is not independent of the orientation of the ASL. There must be an optimal orientation for the ASL portion of the tRNA at the decoding site to achieve the most efficient rates of GTP hydrolysis. In particular, the equilibrium conformation of the CCA end of the tRNA is coupled to the equilibrium conformation of the anticodon through the body of the tRNA. That is, in the rigid body limit (*i.e.,* infinitely strong covalent bonds), a "misalignment" of the anticodon stem-loop caused by a change in the mRNA•tRNA•rRNA hydrogen bond network will certainly cause a misalignment of the CCA end of the tRNA in either the A/T state or the accommodated A/A state. A misalignment of the CCA end can change the GTPase activation rate and the accommodation rate. In the opposite limit (weak covalent bonds), a misaligned anticodon stem-loop would have no effect on the CCA end. Obviously, the tRNA cannot be rigid since it must exist in the A/T and A/A conformations. Conversely, the tRNA cannot be completely flexible since this would eliminate any structural integrity. Clearly, the actual tRNA lies between these two limits in a *quasi-rigid* regime. Because the tRNA must have some degree of rigidity, a misalignment of the anticodon stem-loop will cause some degree of misalignment of the CCA end, resulting in both a change in the GTPase activation rate and a change in the accommodation rate, which were shown to be the rate-limiting steps in tRNA selection for near-cognate and cognate tRNAs, respectively. Thus, the mechanism of "communication" through the tRNA may be a *quasi-rigid* body alignment and misalignment for cognate and near-cognate tRNAs, respectively.

REFERENCES

1. Warner, J. R. (1999). The economics of ribosome biosynthesis in yeast. *TIBS* **24**, 437-440.
2. Gilbert, W. (1963). Polypeptide synthesis in Escherichia coli. *J.Mol.Biol.* **6**, 389-403.
3. Watson, J. D. (1963). Involvement of RNA in the synthesis of proteins. *Science* **140**, 17-26.
4. Davies, J., Gilbert, W., Gorini, L. (1964). Streptomycin, suppression and the code. *Biochemistry* **51**, 883-890.
5. Kurland, C. G. (1992). Translational accuracy and the fitness of bacteria. *Annu.Rev.Gen.* **26**, 29-50.

6. Rodnina, M. V. & Wintermeyer, W. (2001). Fidelity of aminoacyl-tRNA selection on the ribosome: Kinetic and structural mechanisms. *Annu.Rev. Biochem.* **70**, 415-435.

7. Ninio, J. (1974). A semi-quantitative treatment of missense and nonsense suppression in the strA and ram ribosomal mutants of Escherichia coli: Evaluation of some molecular parameters of translation in vivo. *J.Mo. Biol.* **84**, 297-313.

8. Hopfield, J. J. (1974). Kinetic Proofreading: New Mechanism for Reducing Errors in Biosynthetic Processes Requiring High Specificity. *Proc.Nat.Acad.Sci.USA* **71**, 4135-4139.

9. Thompson, R. C. & Stone, P. J. (1977). Proofreading of Codon-Anticodon Interaction on Ribosomes. *Proc.Nat.Acad.Sci.USA* **74**, 198-202.

10. Yusupov, M. M., Yusupova, G. Z., Baucom, A., Lieberman, K., Earnest, T. N., Cate, J. H. D. & Noller, H. F. (2001). Crystal structure of the ribosome at 5.5 angstrom resolution. *Science* **292**, 883-896.

11. Ogle, J. M., Brodersen, D. E., Clemons, W. M., Tarry, M. J., Carter, A. P. & Ramakrishnan, V. (2001). Recognition of cognate transfer RNA by the 30S ribosomal subunit. *Science* **292**, 897-902.

12. Ogle, J. M., Murphy, F. V., Tarry, M. J. & Ramakrishnan, V. (2002). Selection of tRNA by the ribosome requires a transition from an open to a closed form. *Cell* **111**, 721-732.

13. Frank, J., Sengupta, J., Gao, H., Li, W., Valle, M., Zavialov, A. & Ehrenberg, M. (2005). The role of tRNA as a molecular spring in decoding, accommodation, and peptidyl transfer. *FEBS Lett.* **579**, 959-962.

14. Moazed, D. & Noller, H. F. (1989). Intermediate States in the Movement of Transfer-RNA in the Ribosome. *Nature* **342**, 142-148.

15. Sanbonmatsu, K. Y. & Joseph, S. (2003). Understanding Discrimination by the Ribosome: Stability Testing and Groove Measurement of Codon-Anticodon Pairs. *J.Mol.Biol.* **328**, 33-47.

16. Sugita, Y. & Okamoto, Y. (1999). Replica-exchange molecular dynamics method for protein folding. *Chem.Phys.Lett.* **314**, 141-151.

17. Sanbonmatsu, K. & Garcia, A. (2002). Structure of Met-enkephalin in explicit aqueous solution using replica exchange molecular dynamics. *Proteins* **46**, 225-234.

18. Ramakrishnan, V. (2002). Ribosome structure and the mechanism of translation. *Cell* **108**, 557-572.

19. Gromadski, K. B. & Rodnina, M. V. (2004). Kinetic determinants of high-fidelity tRNA discrimination on the ribosome. *Mol.Cell* **13**, 191-200.

20. Ma, J., Sigler, P., Xu, Z. & Karplus, M. (2000). A dynamic model for the allosteric mechanism of GroEL. *J.Mol.Biol.* **302**, 303-313.

21. Schlitter, J., Engels, M. & Kruger, P. (1994). Targeted molecular dynamics: a new approach for searching pathways of conformational transitions. *J.Mol.Graph.* **12**, 84-89.

22. Young, M. A., Gonfloni, S., Superti-Furga, G., Roux, B. & Kuriyan, J. (2001). Dynamic coupling between the SH2 and SH3 domains of c-Src and Hck underlies their inactivation by C-terminal tyrosine phosphorylation. *Cell* **105**, 115-126.

23. Valle, M., Zavialov, A., Li, W., Stagg, S. M., Sengupta, J., Nielsen, R. C., Nissen, P., Harvey, S. C., Ehrenberg, M. & Frank, J. (2003). Incorporation of aminoacyl-tRNA into the ribosome as seen by cryo-electron microscopy. *Nat.Struct.Biol.* **10**, 899-906.

24. Phillips, J., Gengbin, Z., Kumar, S. & Kale, L. (2002). NAMD: Biomolecular simulation on thousands of processors. *Proceedings of the SuperComputing 2002 annual meeting*.

25. Blanchard, S. C., Kim, H. D., Gonzalez, R. L., Jr., Puglisi, J. D. & Chu, S. (2004). tRNA dynamics on the ribosome during translation. *Proc.Natl.Acad.Sci.USA* **101**, 12893-12898.

26. Hansen, J. L., Schmeing, T. M., Moore, P. B. & Steitz, T. A. (2002). Structural insights into peptide bond formation. *Proc.Natl.Acad.Sci.USA* **99**, 11670-11675.
27. Ramakrishnan, V. (2002). Ribosome structure and the mechanism of translation. *Cell* **108**, 557-572.
28. Valle, M., Sengupta, J., Swami, N. K., Grassucci, R. A., Burkhardt, N., Nierhaus, K. H., Agrawal, R. & Frank, J. (2002). Cryo-EM reveals an active role for aminoacyl-tRNA in the accommodation process. *Embo J.* **21**, 3557-3567.
29. Cannone, J. J., Subramanian, S., Schnare, M. N., Collett, J. R., D'Souza, L. M., Du, Y., Feng, B., Lin, N., Madabusi, L. V., Muller, K. M., Pande, N., Shang, Z., Yu, N. & Gutell, R. R. (2002). The comparative RNA web (CRW) site: an online database of comparative sequence and structure information for ribosomal, intron, and other RNAs. *BMC Bioinformatics* **3**, 2.

Chapter 14

BASE STACKING AND BASE PAIRING
Advanced quantum chemical studies

Jiří Šponer[1,2], Petr Jurečka[1,2] and Pavel Hobza[2]

[1]*Institute of Biophysics, Academy of Sciences of the Czech Republic, Královopolská 135, 612 65 Brno, Czech Republic* [2]*Institute of Organic Chemistry and Biochemistry, Academy of Sciences of the Czech Republic, Flemingovo náměstí 2, 166 10 Prague 6, Czech Republic*

*Corresponding authors: e-mail: sponer@ncbr.chemi.muni.cz, pavel.hobza@uochb.cas.cz

Abstract: This chapter summarizes molecular interactions of nucleic acid bases as revealed by advanced ab initio quantum chemical (QM) calculations. We explain advantages and limitations of modern QM calculations of nucleobase interactions. Then we give a detailed overview of the basic methodological issues and present selected recent results.

Key words: Base stacking and pairing, accurate calculations, molecular interactions, DNA

1. INTRODUCTION, METHODS AND PRINICPLES

Nucleobases are involved in two qualitatively different mutual interactions: hydrogen bonding and aromatic base stacking. Further, nucleobases interact with solvent, cations, drugs, etc. Below, we summarize results obtained by QM studies of base stacking and base pairing.[1-3]

1.1 The Advantage of Ab Initio Studies

The major advantage of ab initio (first principle) QM calculations is absence of any empirical parameters. In contrast, all empirical potential and semiempirical QM methods rely on parameterizations. Semi-empirical QM approaches were often used in older studies of base-base interactions. However, even the most recent methods of semiempirical nature are not

343

J. Šponer and F. Lankaš (eds.), Computational Studies of RNA and DNA, 343–388.
© 2006 *Springer.*

suitable to study molecular interactions.[4] Accuracy of ab initio calculations is determined by the size of the basis set of atomic orbitals which is used to construct the molecular orbitals and the inclusion of electron correlation effects. *Improving both factors in a balanced way leads to a systematic improvement of the results while solid evaluations of the error margins are possible even before reaching the convergence. This is a unique feature of high-level ab initio calculations. Obviously, even the best QM calculations are influenced by some approximations. Fundamental approximation is the Born-Oppenheimer approximation, i.e., separation of electronic and nuclear motions. Also, the calculations are non-relativistic. For heavier elements, at least some relativistic effects can be included quite easily using relativistic pseudopotentials. Nevertheless, accuracy of high-level ab initio QM methods is comparable to the best physico-chemical experiments. The advanced QM calculations can be applied for systems where no relevant experiments exist, such as H-bonding and stacking of nucleic acid bases. Fundamental advantage is that QM calculations reveal a direct relation between structures and energies, which is close to impossible to achieve via experimental approaches. In addition, the QM calculation can be done on any selected geometry, even in regions, which can not be studied by experiments.* Base pairs and stacks are closed shell systems and can be safely studied with a very high accuracy using standard single determinant QM techniques. There obviously are some basic rules to be followed in order to obtain meaningful results. These rules nevertheless are rather simple and essentially identical for all studies of closed-shell molecular clusters.

Before further explanations, let us briefly illustrate the development of computational chemistry. Table 14-1 summarizes the dependence of base pairing and stacking energy on the computational level. AT WC, mAmT H, AT S2 and mAmT S2 stand for Adenine Thymine Watson-Crick base pair, Adenine Thymine Hoogsteen pair with N9 methylated adenine and N1 methylated thymine, a stacked complex (S2) of adenine and thymine, and finally the same stacking complex with methylated bases. The subsequent four columns represent similar numbers for Guanine – Cytosine dimer (Figure 14-1). As shown below, the aim of QM calculations is to derive geometries of the complexes, and then the interaction energies.

The first row, HF/6-31G**, brings the result of Hartree-Fock gradient optimization with the standard 6-31G** basis set, with inclusion of the deformation energies and with correction for the basis set superposition error[5] (see below for details). This is the first meaningful level of theory and could be applied at the beginning of nineties, using, e.g. large nitrogen-cooled supercomputers such as Cray Y-MP and later SGI workstations such as R4000 and higher, with the use of the Gaussian suit of programs. Gradient optimization of a base pair took ca 3 weeks on the SGI workstation.

The second row shows interaction energy calculated within the MP2 approximation (the lowest-level electron correlation method, see below) which includes a significant portion of the electron correlation, using the same HF/6-31G** optimized geometry. The abbreviation 6-31G(*0.25) means that 6-31G* basis set with one set of diffuse d-polarization functions was used, in order to improve the description of the dispersion energy.[5] This electron correlation evaluation took ca 10 cpu hours on Cray Y-MP at the beginning of nineties (1993-1994) and with Gaussian 90 code. At around 1994, evaluation of a single point of stacked guanine dimer with the 6-31G*(0.25) basis set required ca. 14 cpu hours on Cray Y-MP supercomputer. Such calculation could be today completed within 2-3 hours on 2 GHz xeon processor. Considerably more accurate MP2/aug-cc-pVDZ evaluation (using RI-MP2 method, see below) can be nowadays done within few hours on the xeon machine while such task was not imaginable in 1994.

Table 14-1. Interaction energies of selected A...T and G...C DNA complexes. Cf. Figure 14-1.

Metod/Basis set	AT WC	mAm T H	AT S2	mAm T S2	GC WC	mGmC WC	GC S	mGm C S
HF/6-31G**	–9.6	-	-	-	–23.4	-	-	-
MP2//HF[a]	–12.6	-	-	-	–23.8	-	-	-
RI-MP2/TZVPP	–14.3	–14.8	–12.1	–14.4	–25.8	–25.6	–16.3	–17.7
RI-MP2/aDZ[b]	–13.8	–15.2	–12.8	–14.9	–25.6	–25.4	–16.9	–18.3
RI-MP2/aTZ[b]	–14.7	–15.9	–13.8	–16.2	–27.0	–26.8	–18.1	–19.6
RI-MP2/aQZ[b]	–15.1	–16.2	–14.1	–16.4	–27.7	–27.5	–18.5	–20.2
D→T	–15.0	–16.2	–14.3	–16.8	–27.5	–27.4	–18.6	–20.1
T→Q	–15.4	–16.4	–14.4	–16.6	–28.2	–27.9	–18.8	–20.5
D→T Truhlar	–15.3	–16.4	–14.6	–17.3	–27.9	–27.7	–19.0	–20.6
MP2→CCSD(T)	0.0	0.1	2.8	3.5	–0.6	–0.6	1.9	2.5
ΔEtot	–15.4	–16.3	–11.6	–13.1	–28.8	–28.5	–16.9	–18.0
ΔH[00]	–14.6	–15.8	–11.0	–12.2	–27.5	–27.2	–15.7	–16.8
ΔH[0]T	–14.0[c]	–15.1[c]	–10.3[c]	–11.9[c]	–27.0[d]	–27.0[d]	–15.0[d]	–16.5[d]

[a]MP2/6-31G*(0.25)//HF/6-31G**, [b]aug-cc-pVXZ basis sets (X = D,T,Q) were abbreviated as aXZ, [b]323 K, [c]381 K

The subsequent rows illustrate data that started to emerge since ca 2001. First, (see also below), the costly standard MP2 procedure was replaced by exceptionally efficient RI-MP2 (resolution of identity MP2). The suitability of RI-MP2 for base pairing was demonstrated in 2001.[6] Thus, the RI-MP2 row shows interaction energies obtained with the RI-MP2 method using the TZVPP basis set for geometries obtained at the same level of theory.[6-8] The TZVPP basis set is close to equivalent to more common cc-pVTZ basis set. Then, the subsequent three rows show interaction energies calculated with three systematically improved augmented correlation consistent basis sets.[7,8] The aug-cc-pVDZ basis set contains (besides s and p functions) two sets of d-functions, the aug-cc-pVTZ basis set three sets of d and two sets of f functions and finally the aug-cc-pVQZ basis set four sets of d, three sets of f and two sets of g functions on the second row elements. One set of each type is always very diffuse. The MP2/aug-cc-pVQZ calculations are considered to be very close to the saturation. D,T,Q stands for double, triple and quadruple-zeta basis set.

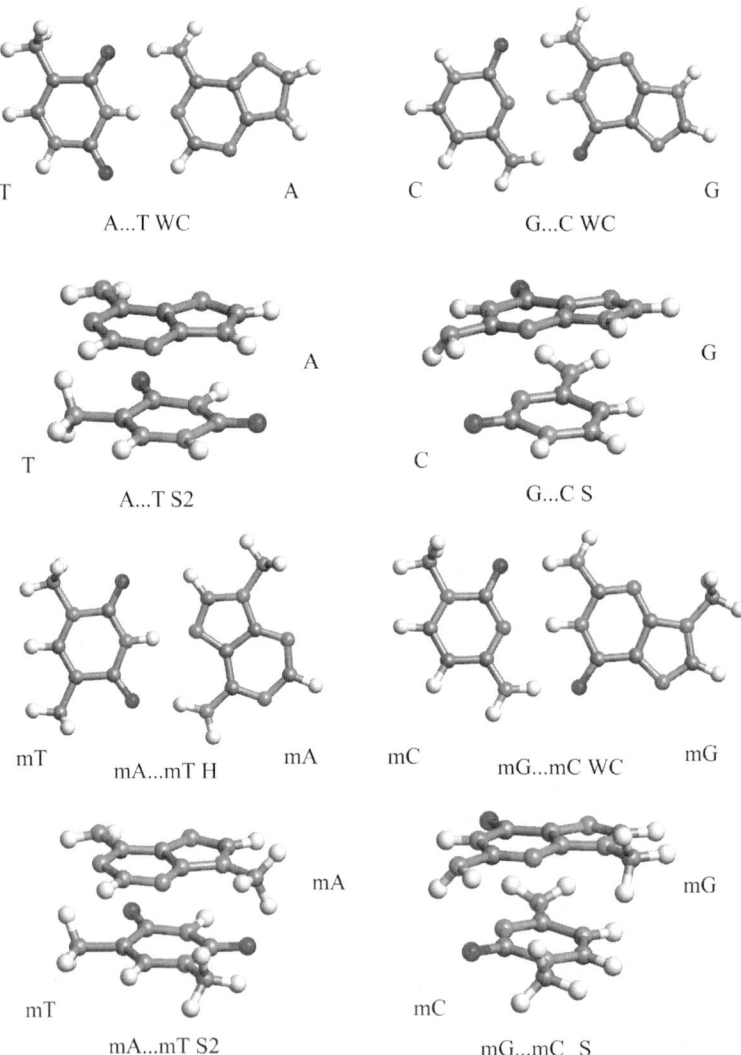

Figure 14-1. Structures studied in Table 14-1.

The aug-cc-pVXZ calculations show a clear trend towards convergence. This allows to carry out so called extrapolation to the complete basis set (CBS) limit at the MP2 level.[7,8] The Table 14-1 presents three methods. D->T is extrapolation based on method by Helgaker and co-workers,[9,10] utilizing aug-cc-pVXZ for X = 2,3 = D,T data and is actually predicting close-to aug-cc-pVQZ values without calculating them directly. In general, Helgaker's N, N + 1 extrapolation usually predicts values close to the N+2 basis set results. The T->Q extrapolation thus yields results of almost aug-cc-pV5Z quality, which are assumed to be very close to the basis set limit.[8] D -> T Truhlar's extrapolation is also based on the X = 2,3 data,[11] but different parametrization causes that the extrapolation is a bit less robust compared to the Helgaker's method and may little randomly deviate from the target (unknown) CBS values. Nevertheless, using the RI-MP2 calculations and basis sets as large as aug-cc-pVTZ or even aug-cc-pVQZ, we can today reach MP2 interaction energies that are considered as the final MP2 values.[12,13]

An open question is direction and magnitude of the higher-order electron correlation effects not included in the MP2 method.[7,8,13,14] This is presently estimated using CCSD(T) calculations with medium-sized basis sets. The difference between the CCSD(T) and MP2 values with the 6-31G*(0.25) basis set is given in the Table 14-1 as MP2->CCSD(T) correction (sometimes abbreviated as Δ(T)). This correction is almost negligible for hydrogen bonded base pairs but substantially repulsive (positive) for base stacking. The next row of the Table 14-1 gives the total interaction energies based on combining the most accurate T->Q and MP2->CCSD(T) data. The last two rows, finally, show 0K and room temperature interaction enthalpies, based on zero point vibrational analysis and thermal corrections.[7]

1.2   The Electron Correlation

The basic method used in QM calculations is the Hartree-Fock (HF) approximation, which solves the time-independent Schroedinger equation by assuming that an electron moves in an averaged field of the other electrons. (Equivalent abbreviation is SCF, Self-Consistent Field). More sophisticated electron correlation "post-HF" methods consider explicitly that electron motions are correlated. The London dispersion attraction is exclusively a consequence of intermolecular correlation of electrons (induced dipole – induced dipole interactions in the classical sense). Inclusion of electron correlation is also important to obtain accurate values of charge distributions, as the HF method overestimates the dipole moments of DNA bases.

The cheapest wave-function based electron correlation method is the second-order Moeller-Plesset perturbational method (MP2) considering double electron excitations to the second order of the perturbation approach. This method mostly provides (with a reasonable basis set of atomic orbitals) very reliable results. The third method to be noted here is coupled-cluster method with noniterative inclusion of triple electron excitations, abbreviated as CCSD(T). This method is of spectroscopic accuracy, but is prohibitively costly. Note that additional expansions of the MP method are not advised. MP3 is imbalanced while the MP4 method is good but inferior to CCSD(T). A cheaper variant of full MP4 is MP4SDQ, neglecting the triple electron excitations. This method is severely imbalanced.

Density functional theory (DFT) methods are considerably more economical compared with the MP2 method while they also include major portion of electron correlation effects. In the wave function methods, the wave function of an N-electron molecule depends for closed shell systems on 3N spatial coordinates. Instead, the DFT methods are based on evaluation of electron densities that depend only on three coordinates. While complexity of the wave function increases with the number of electrons, the electron density has the same number of variables, independently of the number of electrons. Unlike wave function based methods, the DFT functionals often contain empirical parameters. There are also non-empirical functionals such as PBE and PW91, and it is encouraging that their performance is usually only slightly worse than that of the empirical ones. The most accurate DFT approaches, like B3LYP, are based on the combination (hybrid) of the gradient corrected (GGA) exchange-correlation functionals with certain portion of the exact exchange form Hartree-Fock calculation. Such hybrids represent considerable improvement in accuracy with respect to the HF method, and for certain types of molecules they are even superior to MP2. However, the relatively good performance of the DFT is not systematic and often reflects error cancellation between the exchange and the correlation parts of the functional combined with the basis set superposition error (see below). The most serious drawback of the currently available DFT functionals is that they are in principle unable to capture the London dispersion attraction.[15-21] It means that for all molecular complexes with important dispersion contribution (such as the base stacking) DFT techniques fail. DFT methods can provide meaningful results for H-bonded systems. However, *indiscriminate application of DFT methods to molecular clusters with significant participation of dispersion interactions belongs to the worst abuses of QM methods currently seen in the literature.* The inclusion of the intermolecular correlation effects is still only partial, even for H-bonding, where dispersion interaction contributes usually by (much) more than 10% to the total interaction energy. In addition, accuracy of the results can be systematically improved neither by increasing the basis set size nor by a rational choice of the functional. If the set of studied structures contains H-bonded systems of different types and strengths, or systems that are borderline between H-bonded and stacked complexes, results are biased. Density functionals are typically parametrized employing the training sets of molecules which contain very few molecular complexes, therefore they are not optimized for intermolecular interactions. Even inclusion of the rare gas dimers into training set brings no improvements. We especially warn against studies claiming improvement of DFT for molecular interactions while failing to test a single extended dispersion-controlled system, or failing to simultaneously test the performance for H-bonded and stacked systems. Promising option is to introduce the dispersion interaction either by proposing a suitable non-local functional or, e.g., by combining the standard functional with empirical dispersion term.[17,20]

1.3   Fast Variants of MP2

Two approximations can speed up the key MP2 method, resolution of Identity MP2 (RI-MP2)[22] and Local MP2 (LMP2).[23] Both approximations are of entirely different origin and are complementary. Whereas RI-MP2 is particularly effective for very large atomic basis sets and medium sized molecules, LMP2 is efficient for very large molecules with medium basis sets. RI-MP2 (or alternatively Density Fitting MP2 – DF-MP2) method is based on approximate evaluation of coulomb two-electron, four-centered integrals in perturbation treatment. As a result, memory and disk space requirements are substantially reduced. Time savings are usually one order of magnitude for a basis set of triple-ζ quality and efficiency rapidly increases with the basis set size, making HF calculation the by far time-dominating step. Therefore, main applications are accurate benchmark calculations on medium-size molecules up to about 80 atoms. According to our test calculations, loss in accuracy is negligible (error in the interaction energy is typically smaller than 0.02 kcal/mol). Drawbacks are that scaling with system size remains unchanged (only prefactor is smaller) and additional optimized basis set (so called fitting or auxiliary) is needed.[24] LMP2 speedup is achieved in two ways. First, virtual space of each correlated electron pair is drastically reduced by localization of occupied and virtual orbitals. Number of configuration state functions (CSFs) is therefore much smaller and scaling is reduced from $O(N^4)$ to $O(N^2)$ (dependence of the computer demands on the number of atoms or basis functions). Second, while close electron pairs are treated at full MP2 level, distant pairs can be treated approximately, and interaction of very distant pairs can be completely neglected. Such scheme can serve as a basis for linear-scaling of computational cost with the system size. Certainly welcomed side-effect of the local approximation is almost complete elimination of BSSE (see below) which allows for efficient BSSE-free geometry optimizations. Setup of

LMP2 calculation is currently not trivial and requires some level of experience. Special care must be paid if smooth intermolecular potential energy surfaces are needed or if a molecule contains intrinsically delocalized orbitals, which obviously is the case of nucleic acid bases. A non-negligible portion of the intermolecular correlation energy of hydrogen-bonded clusters (10-15%) is lost due to neglect of certain kind of ionic excitations.[25] Consequent underestimation of intermolecular interaction energy is perhaps small, nevertheless, extensive testing would be desirable in future. While RI-MP2 is currently routinely used for nucleic acid bases, the applicability of LMP2 remains to be established. Just recently, idea of combining RI-MP2 and LMP2 methods appeared, resulting in a method called DF-LMP2.[26]

1.4   Basis Set of Atomic Orbitals

Variational QM methods (such as HF) guarantee that quality of the results is systematically improved by increasing the size of the one-electron basis set used. Also for non-variational approaches, such as Møller-Plesset theory, the values obtained with very large basis sets are superior. Most of contemporary calculations rely on basis sets of relatively low quality, such as cc-pVDZ or 6-31G**. The results are usually good estimates of properties desired although very often based on a fortunate cancellation of errors. The largest errors are the basis set superposition error (BSSE – see below) and the basis set incompleteness error (BSIE).[27,28]

BSIE affects mainly the dispersion energy which originates in the "empty" space between the monomers and thus is only included when the basis set is large and diffuse enough to cover this region. For H-bonded base pairs, the intermonomer space is quite well covered by basis functions localized on the hydrogens involved in the H-bonding. In contrast, much larger intermononer gap needs to be filled in stacked complexes. Our experience is that while for H-bonded systems the standard higher-angular momentum polarization functions (f, g....) on second-row (non-hydrogen) atoms are important, diffuse polarization functions (with reduced exponents) are key for stacking. Diffuse sp shells (such as in 6-31 + G* basis set) bring marginal improvement.[12] To improve the description of the dispersion interaction series of aug-cc-pVXZ basis sets augmented with diffuse functions were designed[29]. Alternatively, small 6-31G*(0.25) basis set can be used. It is a modified 6-31G* basis set with exponents of d-polarization functions on C, N and O reduced from the standard value of 0.8 to 0.25.

Typical magnitudes of the above discussed errors can be inferred from the Figure 14-2, which shows basis set dependence of the uncorrected and BSSE-corrected interaction energies of H-bonded and stacked complexes between adenine and thymine. Sum of BSSE and BSIE constitutes the difference between the uncorrected value for a given basis set and the CBS (complete basis set) limit. BSSE is much larger than BSIE for the stacked complex, whereas in the case of H-bonded AT pair they are of similar magnitude. The CBS values represent the "correct" result and are free of BSSE and BSIE.

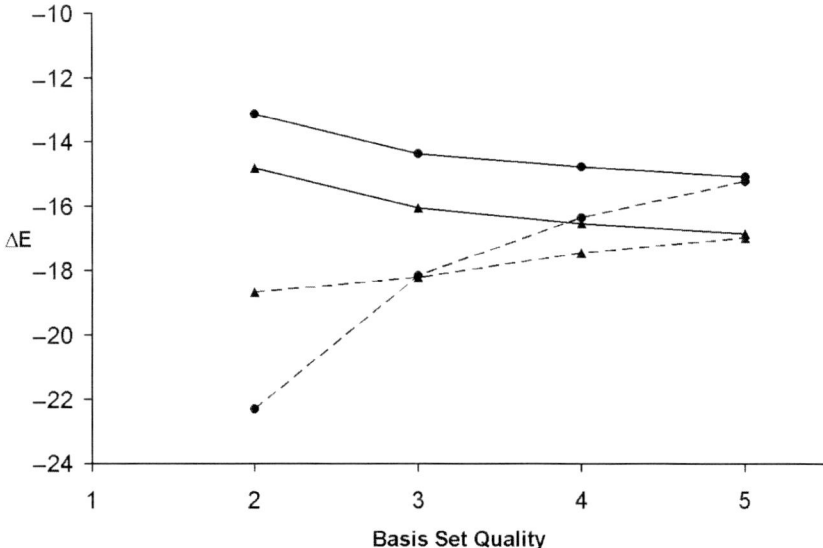

Figure 14-2. Basis set dependence of the interaction energy (ΔE) of Watson-Crick H-bonded (triangles) and stacked (circles) adenine…thymine base pair. Solid lines correspond to counterpoise corrected values (without BSSE), dashed to uncorrected ones (affected by both BSSE and BSIE). On the Basis Set Quality axis 2, 3 and 4 correspond to aug-cc-pVDZ, aug-cc-pVTZ and aug-cc-pVQZ basis sets, and 5 is the extrapolated value, i.e., the CBS limit.

The best way to reach quantitatively reliable QM interaction energies is to estimate the complete basis set (CBS) limit via extrapolation. Let us give the typical forms of the extrapolation equations for the Hartree-Fock energy and the correlation energy[9,10] (eq. 2 and 3, respectively).

$$E_X^{HF} = E_{CBS}^{HF} + Ae^{-\delta X} \tag{2}$$

$$E_X^{Corr} = E_{CBS}^{Corr} + AX^{-3} \tag{3}$$

X is the highest angular momentum quantum number in the basis set given (2 for DZ, 3 for TZ, …), A and δ are constants. The HF energy converges with the basis set size swiftly. The correlation interaction energy (its dispersion part) converges very slowly and the same is true about the BSSE fraction that originates in electron correlation. It is important to use a consecutive sequence of basis sets yielding systematic improvements, such as the correlation-consistent basis sets of Dunning and co-workers.[30,31]

Table 14-2. Composition of some common Gaussian basis sets. M is the number of atomic orbitals, s,p,d,f.... shells contain 1, 3, 5, 7 independent atomic orbitals. Thus, 4s2p1d consists of 4s + 2x3p + 5d = 15 basis functions. "/" separates second-row elements and hydrogens. Contracted basis function are composed of 2 or more primitive (uncontracted) basis functions linked by fixed coefficients.

Basis set	Number and type of basis functions per atom		Number and type of uncontracted basis functions per atom	
	M_{CNO}/M_H	Composition	M_{CNO}/M_H	Composition
STO-3G[a]	5/1	2s1p/1s	15/3	6s3p/3s
6-31(d,p)[b]	15/5	3s2p1d/2s1p	28/7	11s4p1d/4s1p
6-311G(2df,2pd)	30/14	4s3p2d1f/3s2p1d	43/16	11s5p2d1f/5s2p1d
6-311++G(3df,3pd)	39/18	5s4p3d1f/4s3p1d	52/20	12s6p3d1f/6s3p1d
cc-pVDZ[c]	14/5	3s2p1d/2s1p	26/7	9s4p1d/4s1p
cc-pVTZ	30/14	4s3p2d1f/3s2p1d	42/16	10s5p2d1f/5s2p1d
cc-pVQZ	55/30	5s4p3d2f1g/ 4s3p2d1f	68/42	12s6p3d2f1g/ 6s3p2d1f
cc-pV5Z	91/55	6s5p4d3f2g1h/ 5s4p3d2f1g	108/58	14s8p4d3f2g1h/ 8s4p3d2f1g
cc-pV6Z	140/91	7s6p5d4f3g2h1i/ 6s5p4d3f2g1h	161/95	16s10p5d4f3g2h1i/ 10p5d4f3g2h1i
aug-cc-pVDZ	23/9	4s3p2d/3s2p	35/11	10s5p2d/5s2p
aug-cc-pVTZ	46/23	5s4p3d2f/4s3p2d	58/25	11s6p3d2f/6s3p2d
aug-cc-pVQZ	80/46	6s5p4d3f2g/ 5s4p3d2f	93/58	13s7p4d3f2g/ 7s4p3d2f
aug-cc-pV5Z	127/80	7s6p5d4f3g2h/ 6s5p4d3f2g	144/83	15s9p5d4f3g2h/ 9s5p4d3f2g
aug-cc-pV6Z	189/127	8s7p6d5f4g3h2i/ 7s6p5d4f3g2h	210/131	17s11p6d5f4g3h2i/ 11s6p5d4f3g2h
SVP[d]	14/5	3s2p1d/2s1p	24/7	7s4p1d/4s1p
TZVPP	31/14	5s3p2d1f/3s2p1d	46/16	11s6p2d1f/5s2p1d
QZVPPP	57/30	7s4p3d2f1g/ 4s3p2d1f	77/33	15s8p3d2f1g/ 7s3p2d1f

[a] STO-3G minimal basis sets
[b] set of Pople's basis sets, 6-31G(d) and 6-31G(d,p) are often abbreviated as 6-31G* and 6-31G**, 6-31G*(0.25) is diffuse variant of the 6-31G* basis set suitable for stacking calculations. + indicates that diffuse sp shells are added.
[c] correlation consistent basis sets; "aug" basis sets are augmented by very diffuse functions.
[d] Basis sets by Ahlrichs and co-workers.

The basis set choice follows two requirements. An economy requirement, since time requirements of the calculation are usually proportional to a power of the number of basis functions (usually the power varies from 2 to 7, depending on the QM method; especially the higher-order electron correlations methods scale poorly). The other requirement is accuracy. The most common Atomic Orbital forms are the exponentials, Gaussians or plane waves. The AO sets built with the exponential functions are the most natural basis sets with correct asymptotic behavior both near the nucleus and far away from it. Nevertheless, most of the current calculations employ Gaussian basis functions due to their computational efficiency. Still in use are the Pople style basis sets STO-3G, 6-31G(d,p) and 6-311G(2df,2dp) of minimum, double- and triple-ζ quality.[32] Better alternatives usually represent correlation consistent basis sets cc-pVXZ[29-31] designed for methods such as MP2 or CCSD(T). To improve the description of the intermolecular interactions, anions and transition states, aug-cc-pVXZ series augmented by diffuse functions were introduced.

The basis sets generally differ in their size and quality. By the size we understand the number of the basis functions. Quality of the basis set depends on the type of the basis functions: their exponents, angular momentum quantum numbers and contractions (see below and Table 14-2). The smallest possible basis set is the minimum basis set such as STO-3G. For the carbon atom with electronic configuration (1s 2s 2p) the minimum basis set contains 2 s-type functions and 1 p-type shell with three functions p_x, p_y and p_z, which is abbreviated as 2s1p; for hydrogen atom only 1s function is necessary, so the overall abbreviation for CNO/H minimum basis set is 2s1p/1s. Larger basis sets can be obtained by multiplying the number of the functions in the minimum basis set by 2, 3, 4, ..., thus yielding so called double zeta (DZ), triple zeta (TZ), ..., basis sets. The Z or zeta stands for the greek ζ, which denotes exponent of the function. For the carbon/hydrogen the DZ basis set is 4s2p/2s, the TZ basis set 6s3p/3s, etc. Since the chemists are interested mainly in the processes in the valence shells, split valence basis sets were introduced. In the split valence basis set only the number of the valence basis functions is multiplied by 2,3,4, ..., and the number of the inner functions remains unchanged. Resulting VDZ basis set contains 3s2p/2s (CNO/H) basis functions and VTZ set 4s3p/3s. The exponents of the basis set functions have to be carefully optimized.

Polarization functions are those with angular momenta higher than in the mimimum basis. They are essential to describe (usually non-symmetrical) redistributions of the electron density upon bonding or in the external field. Also, decisive portion of the correlation energy is recovered by utilizing high angular momentum polarization functions. The convergence of the correlation energy to its basis set limit is very slow and at least f-type functions should be included. In case we are interested in some special properties like accurate polarizabilities or hyperpolarizabilities, or when dealing with excited states, anions or weak van der Waals complexes, the basis set should be augmented with diffuse functions, characterized by very small exponents. Evaluation of the NMR shielding parameters and couplings requires accurate knowledge of the electron density very close to the nucleus and therefore additional large-exponent functions. Note, that the number of shells decreases when going to higher angular momentum quantum numbers (e.g., 4s3p2d1f in the cc-pVTZ) in a pyramidal way. This is one of the requirements for balanced basis sets, ensuring minimum of artifacts in polarization and other properties. Another useful concept is contraction of several different Gaussian functions into one. The contracted Gaussian is a fixed (preoptimized) linear combination of usually 2 -10 primitive functions. In the variational calculation it is treated with single variational coeffient, thus saving computer time. DFT calculations may have distinct design of basis sets, since DFT energy converges to the basis set limit swiftly.[33]

1.5   Energetics of Molecular Interactions – The Main Task.

The ab initio technique can be used to determine optimal structures of molecular clusters and to calculate energies for any single geometry of the cluster. QM calculations provide molecular wave functions, which can be used to derive physicochemical properties, such as vibrational spectra, dipole and higher multipole moments, polarizabilites, proton affinities, NMR parameters and others. *Nevertheless, the main achievement of QM calculations was description of the nature and energetics of nucleobase interactions,* because it is difficult for experimental approaches.

1.6   What QM Calculations Tell About DNA and RNA?

QM calculations reveal the binding energy between two bases in the gas phase, i.e., in *complete isolation*. Such calculations thus reveal the *intrinsic* interactions of the systems with no perturbation by external effects such as solvent. The intrinsic intermolecular stabilities are directly linked to molecular structures and can be derived in any selected geometry. The gas phase interaction energies, however, do *not* reveal the stability of the interactions in nucleic acids. At first glance, the gas phase calculations may appear to be rather far fetched from physiological systems. However, one needs to correctly understand the *intrinsic* interactions to understand the role of nucleobases in nucleic acids. Nevertheless, it is not possible to easily correlate the QM calculations with measured base pairing and stacking stabilities in nucleic acids. The apparent (measured) strength of the base-

base interactions in nucleic acids in various experiments[34-38] is determined by a complex interplay of many factors and the intrinsic base–base term is only one of them. Many researchers incorrectly believe that the experiments reflect the "true" stabilities of base–base interactions and vice versa.

1.7 Interplay Between Intrinsic and Environmental Effects

Stacking of consecutive protonated cytosines is a highly repulsive interaction in the gas phase due to a charge – charge repulsion.[39] Nevertheless, in intercalated i-DNA quadruplex, stacking of a number of consecutive closely spaced protonated cytosines occurs (Figure 14-3).[40, 41] I-DNA is paired via cytosine – protonated (N3) cytosine base pairs. Both cytosines are equivalent in X-ray and NMR experiments suggesting rather fast intra-basepair proton switches. Two duplexes intercalate to form the tetraplex. Stability of i-motif is due to this massive cumulation of closely spaced protonated base pairs.

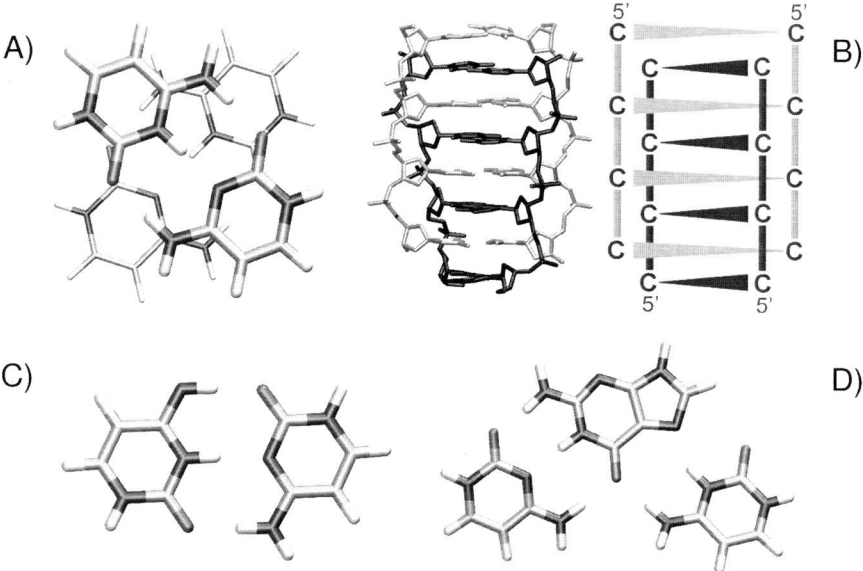

Figure 14-3. A – stacking of two hemiprotonated CCH$^+$ base pairs in i-DNA, B – i-DNA structure and scheme, C – hypothetical neutral iminoC-C base pair, D – CH$^+$-G.C triple.

The experimental groups postulated that the intrinsic i-DNA stacking should be attractive, being stabilized by some unusual dipole-dipole molecular interactions involving base exocyclic groups. Then QM calculations revealed a substantial stacking repulsion in the i-DNA stacking geometry and absence of the hypothetical "dipole-dipole" interactions.[39] We then suggested to reduce the degree of protonation via temporary formation of cytosine imino tautomers, still having proton at the N3 position (Figure 14-3 C). However, it did not agree with the NMR picture. The i-DNA puzzle was ultimately solved by our explicit solvent MD simulations[42] proving that the vertical repulsion between consecutive protonated base pairs in i-DNA is counterbalanced by solvent screening effects and backbone arrangements. Thus i-DNA indeed has, in contrast to other DNAs, *repulsive intrinsic stacking energy terms*. Simply, one cannot predict i-DNA stability based solely on the gas phase data. At the same time, one cannot evaluate the *intrinsic* stacking energetics based on studies of i-DNA. Sometimes we can make no extrapolation from gas phase to nucleic acids while, conversely, studies of nucleic acids bring no unambiguous information about the intrinsic base–base terms. To show the full complexity of molecular interactions, let us underline that the screening is specific for i-DNA.[42,43] Strikingly contrasting i-DNA is behaviour of consecutive protonated cytosines in C+–G.C triples (Figure 14-3) of Pyr-Pur.Pyr triplexes.[44] Consecutive protonated cytosines would be needed to recognize consecutive guanines in the second strand. The vertical position of protonated cytosines in triplex would adopt arrangement closely resembling i-DNA and also planar H-bonding of the third-strand protonated cytosines to N7 of second-strand guanines resembles the i-DNA base pairing. However, this sharply destabilizes the DNA triplex, and even two or three consecutive CH+ are not tolerated, because in triplex the screening of the vertical electrostatic repulsion by the backbone and environment is different than in i-DNA. The vertical repulsion between protonated cytosines is a major problem in a design of antisense triplexes leading to efforts to find appropriate neutral analogs of cytosine.[45] Thus in this particular case of i-DNA and triplexes we cannot transfer experience concerning nucleobase interactions between two DNA forms. Each case should be studied separately. In other words, a given type of base pairing and base stacking may have entirely opposite roles in different NA forms. A given interaction may be a crucial stabilizing factor for one type of nucleic acid architecture (protonation of consecutive cytosines in the i-motif) while it may be even not tolerated in another architecture. In polar solutions ion-pairs with equal charges may attract each other while ion-pairs with opposite charge may be repulsing, completely reversing the gas phase trend.[46] It means that there is no way to design an ultimate experiment to decide about the *common nature* of base stacking in nucleic acids. In order to understand the interactions in nucleic acids, we need to consider wide range of systems and the gas phase data represent an important part of the overall picture.

Many striking cases of major differences between gas phase and condensed phase stabilities were studied for metal-ion containing systems; see the chapter on metal binding to NA.[47-53]

Cytosine and guanine easily form rare tautomers in the gas phase and there have been numerous QM studies of the nucleobase tautomerism in the gas phase.[54,55,56] In contrast, X-ray, NMR and spectroscopic studies have firmly established decades ago that in biologically relevant environments guanine and cytosine are overwhelmingly in the major keto tautomeric forms as the environment stabilizes the major forms.[57] It actually is fully confirmed by QM studies modeling the aqueous solution via continuum solvent models.[54, 55] As an exception, participation of guanine tautomer is possibly indicated by the electron density of the G.U base pair in the second position of near-cognate codon - anticodon helix in one of the x-ray structures of small ribosomal subunit.[58] This guanine is buried rather deeply inside the ribosomal subunit and thus participation of the tautomer is not entirely ruled out. Elsewhere, occurrence of tautomers would bias the genetic code and RNA folding. This obviously does not rule out an occasional spontaneous temporary formation of tautomers which can be promoted for example by an action of metal cations,[59,60] Isoguanine, a base that very well pairs with cytosine, forms tautomers even in aqueous solution and DNA crystals. However, isoguanine plays no substantial biological roles, most likely due to its tautomeric promiscuity.[61,62]

As another example, it is quite common in biochemistry to carry out base mutation studies. They introduce single-base mutations to extract the effect of individual bases and even single H-bonds on the interactions, folding and function of NA. However, results of such studies are typically inconclusive, due to non-additive nature of the changes introduced by the mutations. I.e., the single-base substitution affects simultaneously too many contributions to allow unambiguous conclusion.[63] In B-DNA double helix, the guanine-cytosine base pair can be substituted for by inosine-cytosine (or 5-methyl-cytosine) one (Figure 14-4). Both base pairs are isosteric, as regarding their size and orientation of glycosidic bonds. Since the inosine lacks the N2 aminogroup and the pair has only two hydrogen bonds, the substituted base pair is considered to be a good mimic of A.U or A.T base pair.[64-66] However, this expectation is not always valid. The I.C base pair is intrinsically much stronger than A.U one and its electron distribution (dipole moment, electrostatic potential and charge distribution) is very similar to G.C. This leads to G.C-like electrostatic part of stacking interactions, major groove hydration and cation binding properties.[64,66] It then depends on the individual experimental conditions and the type of nucleic acid whether steric or electronic "similarity" prevails. Similar explanation can be used for the behavior of 2-amino-adenine-thymine (or uracil) base pair, which is an isosteric mimic of guanine – 5methyl-cytosine (or cytosine) base pair, namely due to the N2 amino group in the minor groove and three H-bonds. However, the pair remains electrostatically similar to the A.T one.[64]

An appropriate comparison with gas phase data helps to interpret experiments and to understand the interplay between the intrinsic base – base terms, the environment effects, and the structure and stability of nucleic acids. E.g., experimental studies of the nature of aromatic stacking are inconclusive, some of them attributing the stacking forces to electrostatic effects while others to dispersion or hydrophobic contributions. Instructive computational analysis has been recently published, dealing with ambiguous interpretation of condensed phase experimental data on aromatic stacking.[67] Another experimental study claimed, based on NMR data, that there is a marked

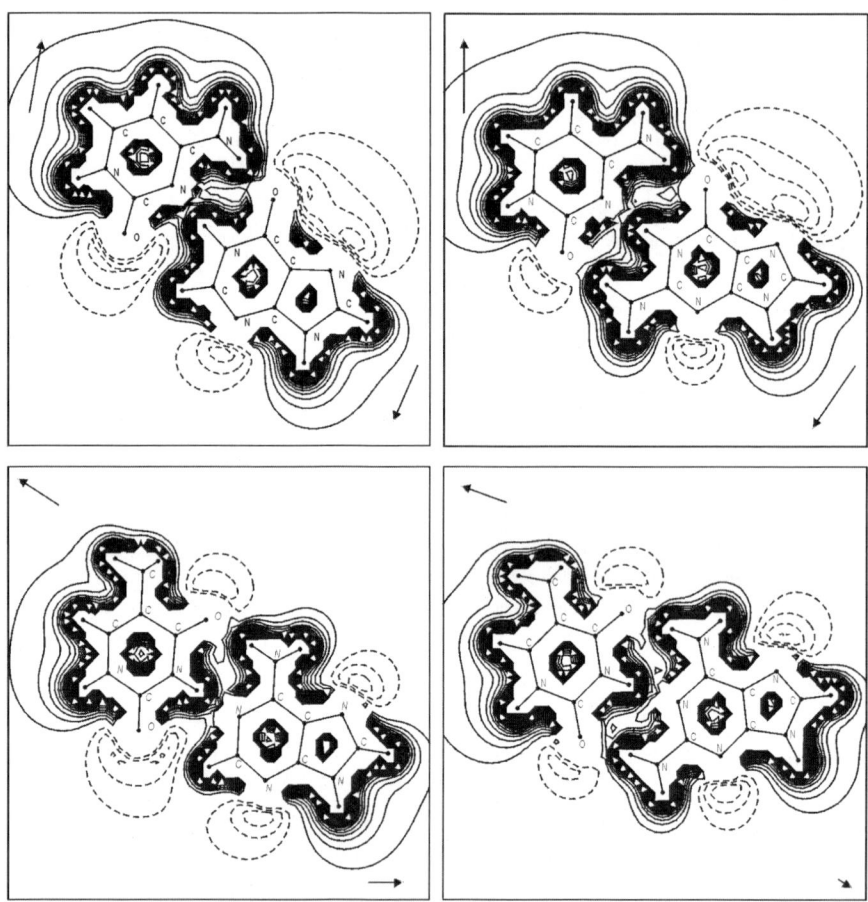

Figure 14-4. Electrostatic potentials and steric contours: inosine-cytosine, guanine-cytosine (top, left and right), adenine – thymine and 2-amino adenine – thymine (bottom left and right) base pairs. Nucleobase dipole directions and magnitudes are indicated by the arrows.

difference in the intrinsic strength of RNA A.U and DNA A.T base pairs.[68] QM, MD and database studies pinpointed a major flaw in interpretation of the above experimental data.[69,70] Recent crystallographic study suggested a presence of water-mediated cation – π interaction between hydrated magnesium and a cytosine in a B-DNA duplex.[71] Quantum-chemical analysis ruled out this claim;[72] see also the chapter devoted to metal binding to nucleic acids. The CpA(TpG) step in some B-DNA crystals adopts specific high-twist, high-slide, low-roll geometry. The widely accepted view in the literature is that this step has weak stacking since the overlap of bases (judged by visual inspection) is rather poor.[73] Calculations show that stacking and overlap are quite good.[74] *Thus, we can summarize that QM calculations provide very valuable data that are of primary importance for everyone who wants to obtain correct qualitative insights into the basic principles of molecular interactions in nucleic acids.*

The natural counterpart of QM calculations is gas phase experiment. Many years ago mass field spectroscopy data provided crude estimates of gas phase enthalpies of selected base pairs,[75] in good agreement with modern QM data.[7] The experiments did not provide any clues about the structures of the studied species but it was believed that H-bonded base pairs were studied. Gas phase QM and MD studies suggest that the gas phase experiments in fact deal with a *mixture of simultaneously populated structures*, involving various H-bonded dimers, T-shaped and stacked structures.[76-78] Another experiment on the intrinsic energetics of base pairs in the gas phase failed because of the lack of thermodynamic equilibrium in the molecular beam expansion, and should be disregarded entirely.[79] Several groups are currently advancing in experimental gas phase studies of nucleobase pairing.[80-84] These experiments will provide excellent data mainly regarding the spectroscopy of base pairs. Gas phase experimental studies have also been reported on interactions between nucleobases and monovalent metal cations.[85] Nevertheless, there are currently no experiments that could show the energies of base – base interactions.

QM calculations *represent a major source of quantitative data for parametrization and verification of other computational tools, including biomolecular force fields for simulations of nucleic acids. Indirectly, in this way QM calculations have a substantial impact on our understanding of the nucleic acids.* Corrections (scaling) are, however, often needed to achieve a balance of distinct parts of the force fields for condensed phase simulations. Future polarization potentials should be able to reproduce gas phase and condensed phase data simultaneously.[86]

Inclusion of solvent effects into QM calculations is difficult. One option is to extend the studied system by explicit water molecules.[87-89] Such calculations still deal with a gas phase molecular cluster. The hydration pattern will thus differ from those in water where the first shell waters interact with the second shell etc. In a small cluster, individual water molecules will form bridges and zippers between bases to maximize the number of H-bonds. The hydration picture in solution (MD data) and x-ray structures typically reveals simple non-cooperative in-plane hydration of the polar nucleobase sites (the nitrogens and oxygens).[90-93] Water binding sites in common hydration sites around nucleic acids have water binding times ca. 50-500 ps.[91-94] However, especially in complex RNA molecules, some hydration sites may be occupied by tightly bound water molecules.[92,93] In the cluster approach, the potential energy surface contains a large number of minima and without an efficient sampling technique it is virtually impossible to verify the true global minimum. Gas phase MD studies have demonstrated that inclusion of very few water molecules (two to six) into the cluster is sufficient to convert the H-bonded base pairs into stacked structures.[87]

The other option is to include the solvent as a polarizable continuum. QM methods[95-97] consider effects of the continuum on the electronic structure of the solute molecules, in contrast to classical continuum approaches. These techniques are, however, based on substantial approximations. The outcomes are quite sensitive to the choice of parameters such as the atomic radii used to define the "solute" cavity; no universal accurate set of "true" radii can be established. Older methods such as the Onsanger spherical cavity should be avoided. Modern methods such as the Polarization Continuum Model and COSMO can provide pretty good results. However, tests of the accuracy including adjustments of parameters for a given type of systems are advised. The fundamental problem of continuum models is difficulty to calculate accurate solvation energies. It then obviously transfers into inaccuracies of the calculated binding energies. The classical continuum solvent models such as GB/SA and PB/SA are facing the same limitation (see other parts of this book). The continuum models are considerably more successful in predictions of the effects of the solvation on electronic structure and the wave functions (calculations of NMR parameters, excited states and other properties). The continuum calculations may be combined with cluster calculations, where the first hydration shell is treated explicitly.[98] Even if QM continuum solvent calculation is properly performed, the numbers would differ from experiment. In practice we are neglecting loss of degrees of freedom upon moving to the average DNA conformation and the differential flexibility due to the difference in twist-roll flexibility of different base pair steps, etc. Also, all effects associated with the presence of sugar-phosphate backbone are neglected. There is no way in which these terms ca be computed for DNA unless simulation is made.

1.8 Definition of Interaction Energy and its Components

Interaction energy of two bases A and B in a given geometry, $\Delta E^{A...B}$, is the energy difference between the total electronic energy of the dimer $E^{A...B}$ and the electronic energies E^A and E^B of isolated bases.

$$\Delta E^{AB} = E^{AB} - E^A - E^B \qquad (4)$$

Often it can be useful to decompose the total interaction energy into its physically meaningful components, i.e., the electrostatic, induction, exchange-repulsion and dispersion contributions. Intuitively (but not rigorously) the induction contribution can be further divided into the polarization and charge transfer. Many different decomposition schemes have been suggested, e.g., Kitaura-Morokuma[99] or SAPT (Symmetry Adapted Perturbation Theory)[100] decompositions. The cheap Kitaura-Morokuma decomposition is based on series of supermolecular calculations carried out with certain restrictions regarding the wave function. Its major drawback is that the electron correlation is not taken into account. On the other hand, SAPT evaluates the interaction energy components directly, without performing supermolecular calculations. The electron correlation is respected and the accuracy at the highest implemented level (approximately MP4) is comparable to the accuracy of the CCSD(T) calculations. Unfortunately the computational demand of SAPT calculations is comparable to the CCSD(T) demands.

Individual components evaluated with different decomposition schemes can differ, also due to the differences in definition (e.g., the sum of the SAPT polarization terms should be approximately equal to the sum of Kitaura-Morokuma induction and charge transfer terms). *In other words, decomposition schemes are method-dependent.* This can lead to confusion and comparisons of different results should be always carried with care. However, if properly understood, knowledge of interaction energy components can improve our understanding of intermolecular interactions. For example, the sum of the exchange-repulsion and dispersion attraction roughly corresponds to the van der Waals term of empirical force fields (neglecting contributions such as anisotropy of atoms, higher order terms in the dispersion expansion, induction, and others). As we will argue below, H-bonding of base pairs is clearly dominated by the electrostatic term, followed by the dispersion attraction. Charge transfer/polarization effects are also quite important but not dominating. Simple coulombic + Lennard-Jones potential with no polarization/charge transfer term can reproduce fairly well interaction energies of base pairs in stability range 5-50 kcal/mol. The only way to explain this result is to consider the electrostatic term as the dominating one. Nevertheless, some QM studies suggest dominant contribution of charge transfer/polarization to H-bonding. This could be related to ambiguity in decompositions and the way how are the individual terms individual terms grouped. Another problem can be inherent deficiency of the DFT methods for molecular interactions, if DFT is applied.

The interaction energy (eg. 4) reflects a hypothetical dimerization process at 0K and is not measurable. If to be related to experimental dissociation energies D_0 and enthalpies of formation, the deformation energy of monomers and the zero-point vibration energy must be included. The zero-point energies and enthalpy and entropy contributions at nonzero temperature are usually calculated in the harmonic approximation.[3,7] Since base-pair complexes are weak, anharmonicity can play an important role, especially for stacked systems and particularly at higher temperatures. Nevertheless, except of direct comparison with gas phase experiment, interaction energy evaluation is the sufficient outcome of QM analysis.

1.9   What is Basis Set Superposition Error (BSSE)?

There are several advantages of the variation determination of interaction energy (e.g. easy applicability, any QM code can be used, high accuracy) but also an important drawback, which is the basis set inconsistency leading to BSSE. The BSSE originates from different descriptions of the supersystem (complex) and subsystems. The basis set of the supersystem is larger than that of subsystems. The A...B supersystem has m + n basis functions while subsystems A and B have only m and n basis functions, respectively. Because of the variation principle the total energy of the supersystem is larger than would be when basis set of supersystem and subsystems were comparable. The subsystem A in the complex is getting an artefactual additional stabilization (compared with isolated A) since it can utilize atomic orbitals of B, and vice versa. BSSE is a pure mathematical artifact and should be subtracted otherwise the stabilization (dimerization) energy is too large (in absolute value). Elimination of BSSE is surprisingly easy. The function counterpoise method suggested independently by Jansen and Ross and Boys and Bernardi solved the problem by using the supersystem basis set also for all subsystems.[102] This is done by defining „ghost" atoms which do not contain electrons and nuclei but do contain basis functions. Ghost atoms are placed at positions of atoms of the other subsystem. The monomers are computed first in the basis set of the complex and then in the monomer basis set. The energy difference is the BSSE to be subtracted from the uncorrected data. In practice, the binding energy is evaluated as the difference between energies of the dimer minus the energies of monomers (cf. eq. 4), but all calculated in the dimer centered basis set. (Evaluations in the monomer centered basis set are not needed unless we want to know the magnitude of BSSE). With highly symmetric complexes the BSSE calculation might bring a considerable computational burden since passing from a complex to subsystem the symmetry is reduced.

The corrected stabilization energies are much less basis set dependent. The BSSE is large for small basis sets and it can reach even 50% or more of the stabilization energy (Figure 14-2). If extended basis sets are used, the BSSE converges to zero. The convergence is, however, very slow, and basis sets containing higher polarization functions (g and h) are required for keeping the BSSE of a rather small molecular cluster below about 10% limit of the stabilization energy. Although the elegant counterpoise procedure was sometimes questioned in the past it brings a reasonable solution of the BSSE problem, and many studies criticizing its use were shown to be biased, for example since they were done with poor basis sets.[102] Then lack of BSSE correction or only a partial correction may appear to improve the agreement with experimental data. That means, the stabilization energy is under-estimated by poor basis set and this deficiency is compensated for by the artifact stabilization by BSSE. This, however, is inconsistent and not transferable approach to improve quality of low-level calculations. Although the CP approach is fairly coarse and the BSSE correction may be overestimated to some extent, experience shows that the BSSE corrected values are superior to the uncorrected ones (see above Figure 14-2).

Whereas in single point calculations BSSE can be removed by the CP, in the course of geometry optimization CP correction is needed in each step of optimization. Major trouble is that BSSE affects intramolecular interactions, where no correction is possible since we deal with a monomer. This occurs, e.g., when comparing two conformers of one molecule, one with and the other without an intramolecular H-bond, or a dinucleotide stack, etc.

1.10   What are Deformation Energies of Monomers?

When gradient optimization is carried out, the monomer geometries are changed upon complexation. First, there are real deformations (mutual adaptations) of the monomers that improve the intermolecular interaction (binding) in expense of the intramolecular energy terms. Some of the deformations can be directly related to the binding strength. However, for larger systems, some monomer rearrangements reflect rather long-range effects. For example, there could be a substantial adaptation of flexible sugar-phosphate backbone upon complex formation. Thus, real deformations consist of two distinct contributions. Direct deformations (always present) reflect the force of the binding and may be complemented by various indirect larger-scale conformational changes and adaptations. The H-bonding usually leads to X-H stretching (elongation) which is manifested as red shift in IR vibrational spectra, although there might be blue shifts in so-called improper H-bonds. Besides real deformations, the BSSE contributes to the deformation when standard gradient optimization is applied.

In our base pair calculations, the interaction energies of the optimized complexes are a posteriori corrected for the BSSE, as explained above (using the geometry of the complex and dimer-centerd basis set). Then we separately correct for the deformation energy in such a way that, using the monomer basis sets, we evaluate the monomer energies in the deformed (complex) and optimized (isolated) monomer geometries.

$$E_{Def}{}^{A} = E^{A(\text{dimer geometry})} - E^{A(monomer\ geometry)} \qquad (5)$$

Thus, the interaction energy of a dimer is defined in the following way:

$$\Delta E^{A\cdots B} = E^{A\cdots B} - (E^{A} + E^{B}) + E_{Def}{}^{A} + E_{Def}{}^{B}. \qquad (6)$$

The first three energies are calculated in dimer-centered basis set. The intramolecular deformation energy actually cancels a major part of the intermolecular energy improvement caused by mutual monomer adaptations. Some studies formally include the deformation energy as a part of the BSSE correction.[103] However, although it might look more consistent mathematically, we and others[104] vigorously oppose this approach, especially for larger systems such as base pairs and other fragments of biopolymers. First, the integrated expression is, after formal rearrangements, *entirely indentical* to the above definition, which in addition is older. I.e., the correction was commonly known *before* being rearranged and "re-discovered".[103] Second, while BSSE is a mathematical artifact, monomer deformations represent real effects related to fundamental properties of the studied clusters including their vibrational spectra and polarization/charge transfer effects. Thus, it is quite useful to evaluate the magnitude of the monomer deformations explicitly. Further, for flexible systems (for example many non-Watson Crick interactions involving also the sugar-phosphate atoms, see below) the changes of geometries upon complex formation include substantial conformational changes unrelated to the force between the monomers. Then any formal inclusion of the deformation term into the BSSE correction would be meaningless. Although the approach by Xantheas[103] was immediately criticized,[104] it firmly entered the "small-molecule" literature. For base pairs and larger systems of chemical and biological interest, it is confusing. We have recently tested the relative magnitude (unpublished data) of "real" and "BSSE" contributions to monomer deformations and concluded, that the deformation energies are dominated by "real" geometry adaptations. An alternative approach is to use recently introduced counterpoise-corrected gradient optimization where the BSSE is removed in each gradient iteration, although with a substantial increase of the computer requirements.[105,106] It eliminates the BSSE part of the deformation energies while true deformations persist.

Note that in base pairing studies the deformation energy can be calculated either with respect to the planar monomers, thus neglecting the amino group nonplanarity, or with respect to nonplanar bases. These two numbers differ simply by the difference between energies of planar and noplanar monomers and can be easily compared when needed. The reader should always check the method section how is the deformation defined.

1.11   Nonplanarity of Amino Groups

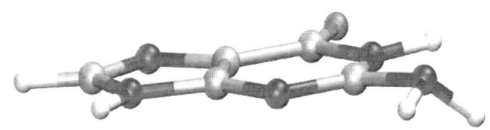

Figure 14-5. Nonplanarity of guanine amino group in complete isolation.

QM studies discovered a decade ago that, in the gas phase, nucleobases have nonplanar amino groups due to a partial sp^3 pyramidalization of their amino nitrogens.[107-111] The 0K electronic energy difference between planar and nonplanar structures is ca 0.1 kcal/mol for adenine and cytosine and 1.0 kcal/mol for guanine. The pyramidalization is weaker than for aniline aminogroup.[111] Intrinsic nonplanarity of the guanine amino group enveloped by two ring nitrogens is substantially larger compared to adenine and cytosine, where there is one carbon atom adjacent to the $C-NH_2$ group. Further, pyramidalization of guanine amino group is nonsymmetrical, i.e., one of the hydrogens is substantially more nonplanar (N1-C2-N2-H21 dihedral angle of ca 40°) than the other (N3-C2-N2-H22, ca 10°) (Figure 14-5). It is because of its repulsion with the adjacent polar (N1)-H1 ring hydrogen; actually the H1 ring hydrogen is bent in the opposite direction by about 6° [107] (pyramidal-rotated geometry). Actually, the sum of the three valence angles around the amino group nitrogen (its deviation from 360°) is a better measure of the nonplanarity than the individual dihedrals. Pyramidal-rotated geometries of amino groups can be also induced by intermolecular interactions. Note that rotation of the amino group weakens the $C=N$ double bond and facilitates further pyramidalization. The nonplanarity is important in the gas phase but rarely expressed in nucleic acids, since in-plane H-bonds (including common base pairing and hydration) shift the electronic structure of the amino groups to sp². Nevertheless, there are interactions in nucleic acids likely stabilised or facilitated by the pyramidalization.[107-114]

One may argue that the amino groups of bases must be planar because they, in contrast to aniline amino group, do not undergo protonation under solution experiments. This judgement based on basicity (pKa) studies is fundamentally wrong because the protonation of the amino group nitrogen (even considering the sp³ hybridisation) is considerably less favorable compared to bare ring nitrogen sites of the nucleobases. Thus, the ring nitrogens are protonated first, and once any of them is protonated, the amino group becomes planar, due to electronic structure changes. The lack of ring nitrogens permits a protonation of the aniline amino group. There is no reason to relate pyramidalization and basicity. Note that the guanine amino group showing the largest degree of pyramidalization has in fact the worst amino nitrogen proton affinity (–205 kcal/mol).[1] Amino proton affinity of adenine is substantially better (–219 kcal/mol) and this explains why guanine appears to be more efficient in establishing out-of-plane H-bonds while adenine is more often seen in amino-acceptor contacts.[112] Amino group nonplanarity was recently confirmed by gas phase experiments.[81] Common force fields lack the ability to consistently describe the sp² <-> sp³ balance of the amino nitrogens and mostly assume the sp² planar hybridization.

1.12   Can Strength of Individual H-Bonds be Dissected?

When there are more H-bonds, their effect is cooperative and the strength of the individual H-bonds cannot be calculated. The degree of cooperativity depends on the system. The relative strength of H-bond can be indirectly assessed (but not exactly evaluated) via vibrational analysis. Crystallographers occasionally speculated that the C2-H2(adenine)... O2(thymine or uracil) contact in Watson-Crick base pairs forms a third, albeit weak H-bond. Let us first re-iterate that the C2 and O2 heteroatoms are evidently well separated, and thus there is no reason to introduce the C-H...O H-bond. Nevertheless, a QM study was made to elucidate the strength of the C2-H2 ... O2 H-bond, concluding that it is a weak H-bond contributing ca 10% (1.6 kcal/mol) to the base pair strength.[115] The authors first evaluated the base pair and then its analog lacking the C-H...O interaction. Such calculations are misleading since deletion of a single group changes the overall electronic structure and thus also the strength of the other H-bonds. Thorough electron topology and vibrational analysis confirmed that C2-H2...O2 contact in the A.T and A.U base pairs is inactive.[116] The mutual position of the two groups is determined by the other two H-bonds and formation of active C2H2...O2 interaction would disturb the true base pair H-bonds. C-H...O H-bond, however, exists in U.U RNA base pair known as the Calcutta base pair (see below).[116]

1.13   What are the Many Body Effects?

When the studied system consists of three monomers (A,B,C), the interaction energy of a trimer, $\Delta E^{A...B...C}$, can be expressed in two ways:[117,118]
   (i) as energy difference of the complex and the monomers:

$$\Delta E^{A...B...C} = E^{A...B...C} - (E^A + E^B + E^C) \tag{7}$$

or (ii) as a sum of pair interaction energies and the three-body term ΔE^3:

$$\Delta E^{A...B...C} = \Delta E^{A...B} + \Delta E^{A...C} + \Delta E^{B...C} + \Delta E^3, \tag{8}$$

$$\Delta E^3 = E^{A...B...C} - (E^{A...B} + E^{A...C} + E^{B...C}) + (E^A + E^B + E^C). \tag{9}$$

These equations can be easily extended for complexes consisting of N subsystems.[119] (ΔE stands for the interaction energies and E for the total electronic energies.) Many body effects arise only when the complex is assembled and cannot be attributed to the individual monomers. Major empirical force fields neglect all many-body effects. This is a major problem of contemporary molecular modelling, since one of the leading nonadditivities is polarization of solute by polar solvent. To partly compensate for this the charge distributions in common force fields are derived at the HF level, exaggerating the permanent dipole moments. Many body terms in systems with multiple neutral nucleobases are mostly 1-2 kcal/mol.[118] Many-body effects are of key importace when cations interact with the cluster. Total amount of nonadditivities in the first ligand shell of a divalent cation is on a scale of 50-100 kcal/mol. Such charge-transfer effects cannot be treated even by polarization force fields except of specialised ones.[120]

1.14   Gradient Optimization vs. Single Points

Modern quantum chemical programs allow easy gradient optimizations of base pairs, where all coordinates (or parameters) are optimized. Standard optimizations are not corrected for BSSE. Since the optimization itself is more computer-demanding than the subsequent interaction energy evaluation, very often a better level of theory (level X) is used for interaction energy calculation than for optimization (level Y). This is abbreviated as X//Y. For example, abbreviation MP2/aug-cc-pVTZ//MP2/cc-pVDZ tells that the optimized structure was obtained at the MP2/cc-pVDZ level while the energies were derived for this optimized geometry at the MP2/aug-cc-pVTZ level (MP2 method with aug-cc-pVTZ set of atomic orbitals).

Nevertheless, not always is the gradient optimization a method of choice. It is good for systems with well-defined local minima, such as H-bonded base pairs. However, it is often of no use for stacking, since stacking seen in nucleic acids does not correspond to minima on the potential energy surface of isolated stacked dimers and conformational scanning is preferred.[121] Further, gradient optimizations of dispersion-controlled clusters (such as stacking of bases) are affected by huge BSSE, unless a very large basis set is used. With lower quality methods the structures are unstable and convert to H-bonded ones. In addition, gradient optimization of stacked dimers leads to puckering of the aromatic ring.[122,123] This is usually not desirable since in real environments the nucleobases have some interactions at both their sides, preventing such puckering. Thus, stacking calculations are mostly carried out as series of single points with fixed geometries and rigid monomers.[121]

It is not advisable to directly use monomer geometries from PDB files of NA x-ray structures. Due to limited resolution of biomolecular x-ray structures the monomer geometries carry no experimental information while the bases are often deformed after the refinement. Such deformed monomer geometries frustrate the electronic structure. We advise to replace (for example via overlay) the monomers from the PDB files via QM-optimized monomers. Also intermolecular x-ray geometries may cause substantial errors in calculations.[124] Especially drastic distorsion of the calculated energies can be caused by artifactual steric clashes in the refined crystal structures. A real nightmare occurs when the x-ray base stacks are apparently vertically compressed or extended due to inaccurate determination of the interbase angles (propeller twist, base pair roll, etc).[124] This requires a case by case judgement and some experience with crystallography. Note that a rather small error in the interbase distances (which may still be tolerable from the geometry point of view) can lead to a considerable energy artifact.[124] This is when the geometry falls into a region of interatomic distances where the short-range repulsion starts to dominate. In this region the calculated energy is a highly nonlinear function of the interatomic distance.[124] Similarly, H-bonded base pairs are sensitive to experimental geometry errors due to the genuine close contact between H-bond partners. Besides data and refinement errors a bad geometry can for example result from presence of two or more local substates. Substates cannot be distinguished except of having nominal resolution better than ca 1 Å. The refined geometry reflects an averaged geometry which may have very poor energy. Fiber diffraction models cannot be recommended for direct calculations. We think that use of a fiber model was responsible for rather poor (compared to B-DNA) calculated stacking energy of A-DNA in one earlier study.[125] When high-resolution B-A intermediate was used no difference was seen.[74] Often it is useful to freeze the x-ray intermolecular geometry with partial internal relaxation of monomers.[108, 112-114]

1.15   Atomic Charges

Due to basic principles of quantum mechanics, atomic charge distributions are arbitrary. There is no quantum mechanical operator for atomic charges and thus there is no experiment (even a hypothetical one) to determine atomic charges. Partial atomic charges (even those derived from experimental electron densities) do not correspond to any real physicochemical quantity. Thus charge distributions can be greatly manipulated by changing the way they are derived and there exist vast of possibilities how to define them. It makes not much sense to compare individual atomic charges in different force fields.

One of the most reasonable ways to derive charge distributions is fitting to molecular electrostatic potential (MEP, ESP charges) around the monomers.[126-129] *The molecular electrostatic potential, dipole and higher moments are well-defined, measurable quantities.* Thus the individual charges still remain arbitrary, nevertheless, the electrostatic field created by all of them together closely approximates the actual electrostatic potential around the nucleobases. The charges effectively include higher multipole moments and there is no need to consider point dipoles, quadrupoles, etc. ESP charge distributions are very useful for calculations of electrostatic interaction energies and the popular AMBER force field uses this way to derive charges.[129] Disadvantage of the ESP approach is that for flexible molecules such as amino acids and sugar-phosphate backbone, the charges derived in one conformer do not reproduce the electrostatic potential in other conformers. This is one of the major problems of current molecular modeling. In addition, the sugar-phosphate backbone has anionic nature and anions are inherently very difficult to be described by simple pair additive force fields.

The Natural Bond Orbital method, NBO[130] aims to characterize the electronic structure and charge transfer. There are other approaches, which under certain circumstances may be used to assess interactions and charge transfer. However, *the popular Mulliken charge distribution is extremely method and basis set dependent.* When investigating binding of hydrated metal cations to bases we observed that an entirely negligible adjustment of an exponent of a single inner s- atomic orbital on Mg^{2+} (with 6-31G* basis set) changed the calculated Mulliken charge on the cation by as much as 0.4 e.[131] Two similar basis sets gave an opposite trend when comparing the partial Mulliken charges on S6(O6) of (thio)guanine.[132] Mulliken analysis can lead to wrong conclusions even when assessing the relative trends.

1.16   Advance of High-Level Ab Initio Calculations

QM studies of DNA base pairs have been attempted for more than 30 years. Before ca 1985, ab initio methods could not be applied to large systems such as the base pairs. Thus, old semiempirical methods had to be used. Such methods are incapable to treat molecular complexes.[4]

Then, till 1994, ab initio Hartree-Fock (HF) approximation was applied. Such calculations inherently neglect all electron correlation effects. This has a significant implication for studies of molecular clusters, since the dispersion attraction originates in intermolecular electron correlation. The dispersion energy is thus completely neglected. The calculations were in addition carried out with very small basis sets of atomic orbitals. Modern specialized pair additive force fields such as the Cornell et al.[129] provide much better results than such low-level QM data.[4] Notably, there have been several pioneering attempts to include the dispersion forces into the computations.[133,134]

A breakthrough in QM studies of base-base interactions was achieved in the period 1994-1996, when first calculations with inclusion of electron correlation effects via the MP2 theory were made, with medium-sized basis sets.[1-5,17,39,52,74,109,110,117,118,121,132,135,136] The calculations provided close to converged values of interaction energies, except of the dispersion energy. Proper evaluation of the dispersion energy requires huge basis sets of atomic orbitals while for base stacking calculations the higher electron correlation effects need to be considered.

The last chapter in QM studies of base-base interactions has been written at around 2003-2005 due to two major improvements.[7,8,13,137] i) MP2 (RI-MP2) calculations can be carried out with series of large basis sets of atomic orbitals, allowing extrapolations to the basis set limit (CBS). ii) Higher-order electron correlation effects can be estimated, although still with small basis sets of atomic orbitals, using CCSD(T) method. It is assumed that modern calculations are within $0.5 - 1.0$ kcal/mol value away the exact numbers.

2. SELECTED RECENT RESULTS

2.1 H-bonded Base Pairs

Table 14-3 summarizes the latest energy data on base pairing (Figure 14-6). The first three columns show the RI-MP2/aug-cc-pVDZ (aDZ), RI-MP2/aug-cc-pVTZ (aTZ) and aDZ→aTZ interaction energies extrapolated according to Helgaker's extrapolation CBS method.[9,10] All numbers include the monomer deformation energies. The next two columns give the RI-MP2/aug-cc-pVQZ (aQZ) and aTZ→aQZ Helgaker's extrapolated data for seven base pairs. The values in parenthesis in the aTZ→aQZ column are obtained with Truhlar's D→T (aDZ→aTZ^T) extrapolation for all base pairs.[11] The aDZ→aTZ Helgaker's extrapolation stops short of reaching the infinite basis set value and predicts the aQZ values. The Truhlar's aDZ→aTZ (aDZ→aTZ^T) extrapolation is aimed to directly predict the infinite basis set values. The range of differences of relative base pairing energies between the aTZ→aQZ and aDZ→aTZ^T methods is 0.6 kcal/mol. There are no coefficients available to extend the Truhlar's extrapolation for the aTZ→aQZ case. The extrapolation scheme of Truhlar leads to more negative interaction energies by 0.1 to 1.1 kcal/mol compared to the aDZ→aTZ extrapolations of Helgaker. The aTZ→aQZ extrapolation (predicting values close to the a5Z results) are close to converged. The aDZ→aTZ^T extrapolations by Truhlar compare with the Helgaker's aTZ→aQZ extrapolation within the range of –0.3 to + 0.3 kcal/mol.

The seventh column shows the ΔCCSD(T) correction. The differences between the aDZ→aTZ (Helgaker) values and the aTZ→aQZ data corrected for CCSD(T) are in the range of –0.3 to –1.3 kcal/mol. The next two columns decompose the interaction energy (the aDZ→aTZ values) into the SCF and MP2 contributions.

To give an example of the BSSE artifact at the RI-MP2/cc-pVTZ level, the pure BSSE correction is 3.5 kca/mol for the thioG.C WC base pair and 2.7 kcal/mol for the G.C WC base pair.

Table 14-4 compares the reference aDZ->aTZ *ab initio* data with other methods. (Note that both DFT methods fail for base stacking.) The last column of Table 14-4 shows the Cornell et al. (AMBER) force field values.[129] The force field data are close to the reference aDZ→aTZ values, in the range of –1.0 to + 2.8 kcal/mol (except of the C.CH$^+$ base pair). The force field performs better for stronger base pairs with considerable electrostatic interaction. Its stabilization sometimes exceeds the reference QM data. It is because the AMBER charges are derived at the HF level and thus overestimate the dipole moments (polarity) of bases.

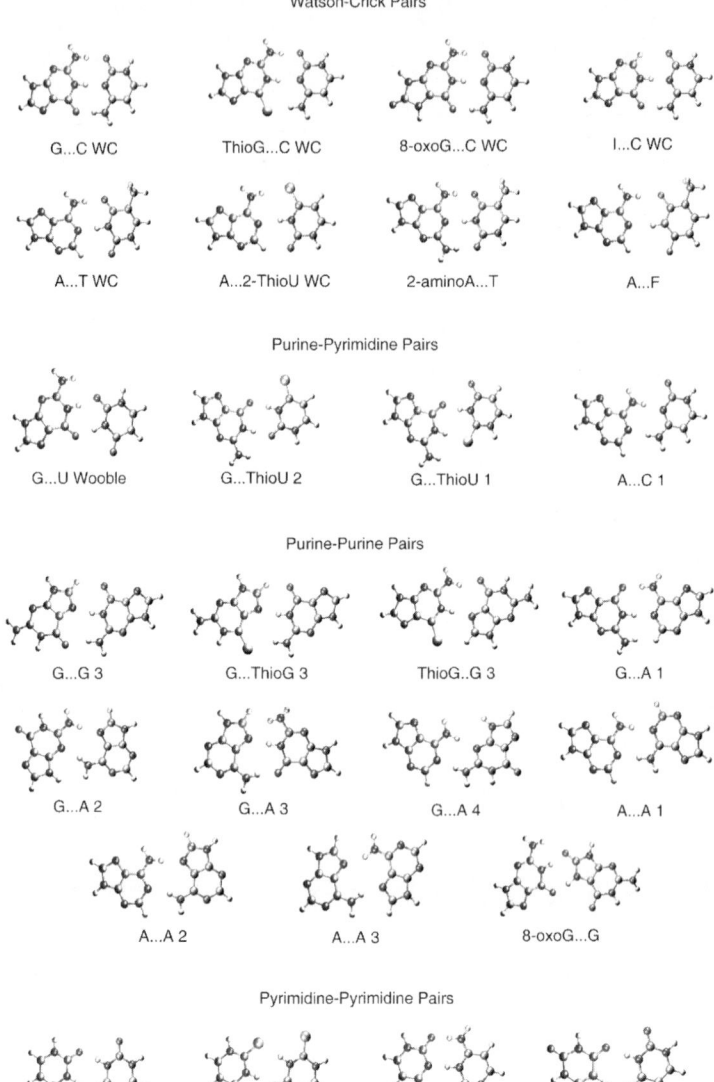

Figure 14-6. H-bonded base pairs.

Table 14-3. Interaction energies of base pairs (kcal/mol).

Structure[a]	ΔE_{MP2}[b]					$\Delta CCSD(T)$[c]	aDZ→aTZ[d]		
	aDZ	aTZ	aDZ→ aTZ	aQZ	aTZ→aQZ (aDZ→aTZ^T)		ΔE^{SCF}	ΔE^{cor}	E^{def}
Watson-Crick Pairs									
G.C WC	−25.6	−27.0	−27.5	−27.7	−28.2(−27.9)	−0.6	−20.0	−7.4	3.6
6-thioG.C WC pl	−23.6	−25.0	−25.5		(−25.9)		−18.3	−7.0	4.0
8-oxoG.C WC pl	−26.3	−27.7	−28.2	−28.4	−29.0(−29.3)	−0.4	−21.3	−6.9	3.9
I.C WC pl	−20.6	−21.6	−22.0	−22.1	−22.5(−22.7)	−0.2	−15.4	−6.6	2.2
A.T WC	−13.8	−14.7	−15.0	−15.1	−15.4(−15.3)	0.0	−7.0	−8.0	1.5
A.4-thioU WC	−12.8	−13.1	−13.2		(−13.4)		−6.9	−6.3	1.0
2-aminoA.T	−16.1	−17.1	−17.6		(−17.9)		−7.7	−9.9	1.9
2-aminoA.T pl	−15.8	−16.9	−17.3		(−17.6)		−8.8	−8.5	2.4
A.F	−4.75	−4.9	−4.9		(−5.1)		−1.0	−4.0	0.3
Purine-Pyrimidine Pairs									
G.U Wobble	−14.4	−15.4	−15.8	−15.7	−15.8(−16.0)	−0.3	−9.7	−6.1	3.0
G.4-thioU2	−14.7	−15.6	−15.9		(−16.2)		−9.3	−6.6	1.9
G.2-thioU1	−13.1	−14.2	−14.6		(−14.9)		−7.1	−7.5	2.0
A.C 1 pl	−14.9	−15.6	−15.9		(−16.1)		−8.1	−7.7	1.7
Purine-Purine Pairs									
G.G 3 pl	−17.4	−18.1	−18.4		(−18.5)		−14.1	−4.3	2.9
G.-6-thioG 3 pl	−17.5	−18.6	−19.0		(−19.2)		−14.3	−4.8	2.8
6-thioG.G 3 pl	−18.3	−19.3	−19.6		(−19.9)		−14.5	−5.2	3.1
G.A 1	−16.4	−17.2	−17.5		(−17.8)		−8.2	−9.3	1.9
G.A 1 pl	−14.8	−15.8	−16.1		(−16.3)		−8.7	−7.4	2.8
G.A 2	−9.7	−10.5	−10.9		(−11.3)		−3.1	−7.7	3.5
G.A 2 pl	−9.5	−10.2	−10.5		(−10.6)		−4.4	−6.1	2.3
G.A 3	−15.8	−16.5	−16.8		(−17.0)		−7.6	−9.1	2.0
G.A 4	−11.0	−11.8	−12.1		(−12.2)		−4.8	−7.3	1.4
A.A 1 pl	−12.2	−12.8	−13.1		(−13.2)		−5.1	−8.0	1.4
A.A 2 pl	−11.5	−12.0	−12.3		(−12.4)		−4.5	−7.8	1.4
A.A 3 pl	−10.2	−10.7	−10.9		(−11.0)		−3.5	−7.4	1.3
8-oxoG.G	−18.3	−19.2	−19.6		(−19.9)		−11.1	−8.5	3.2
Pyrimidine-Pyrimidine Pairs									
U.U 1 pl	−11.4	−12.1	−12.4		(−12.7)	−0.2	−6.8	−5.7	1.1
2-thioU.2-thioU 1 pl	−10.2	−11.2	−11.6		(−12.0)		−4.0	−7.6	1.0
C.CH+ pl	−44.3	−45.7	−46.4	−46.4	−47.0(−47.2)	−0.1	−36.5	−9.9	4.9
U.U Calcutta pl	−8.9	−9.3	−9.5	−9.6	−9.7(9.8)	−0.1	−4.9	−4.5	0.5

[a] see Figure 14-6; pl means that the base pair has been optimized under C_s symmetry. Attempts to locate nonplanar geometries for I...C (I – inosine), 8-oxoG...C, G...G 3, A...A 1, A...C 1 A...A 2 and A...T WC resulted in planarization. G...C WC is weakly nonplanar while the planar structure is essentially isoenergetic. All studied pyrimidine...pyrimidine base pairs and some others are assumed to be intrinsically planar. For all base pairs that are significantly nonplanar with provide the true minimum and in most cases also the planar structure.

[b] aXZ stands for the RI-MP2 interaction energy values with aug-cc-pVXZ (X = 2,3,4) basis set with inclusion of extrapolated (X = 2,3) deformation energies, aDZ→aTZ and aTZ→aQZ are the CBS extrapolations by Helgaker while aDZ→aTZ^T is the extrapolation by Truhlar.

[c] The difference between CCSD(T) and MP2 values with the 6-31G* basis set.

[d] ΔE^{SCF} and ΔE^{cor} stand for SCF and correlation parts of the aDZ→aTZ interaction energies including the respective deformation terms, E^{def} is the extrapolated deformation energy.

Table 14- 4. Comparison of QM data (aDZ→aTZ) with other methods (kcal/mol).[a]

Structure	QM	DFT1	DFT2	MP2//HF	AMBER
G.C WC	−27.5	−25.5 (4.0) / 2.0	−27.7/−0.2	−23.4 (2.5)/4.1	−28.0/−0.5
6-thioG.C WC	−25.3	−23.4 (4.4) /1.9	−25.8 /−0.5	−22.5 (2.5)/2.8	−25.1/0.2
8-oxoG.C WC	−28.2	−26.8 (3.4) /1.4	−28.7 /−0.5	−24.0 (2.3)/4.2	-
I.C WC	−22.0	−19.3 (2.5) /2.7	−21.0 /1.0	−18.0 (1.4)/4.0	−22.0/0.0
A.T WC	−15.0	−12.3 (1.7) /2.7	−14.5 /0.8	−11.8 (0.7)/3.2	−12.8/2.2
A.4-thioU WC	−13.2	−11.0 (1.6) /2.2	-	−11.2 (0.6)/2.1	-
2-aminoA.T WC	−17.6	−14.9 (2.2) /2.7	−17.5 /−0.1	−13.7 (1.4)/3.9	−15.8/1.8
A.F WC	−4.9	−3.1 (0.2) /1.8	−4.1 /0.8	-	-
G.U Wobble	−15.8	−13.4 (2.3) /2.4	−14.8 /1.0	−12.7 (1.6)/2.9	−16.0/−0.2
G.4-thioU Wobble	−15.9	−12.3 (2.2) /3.6	−13.7 /1.8	−12.1 (1.3)/3.8	-
G.2-thioU Wobble	−14.6	−11.4 (2.1) /3.2	−13.4/1.2	−12.1 (1.2)/2.5	-
A.C 1	−15.9	−13.7 (1.8) /2.2	−14.9/1.0	−13.5 (0.9)/2.4	−13.5/2.4
G.G 3	−18.4	−15.8(2.6) /2.6	−17.4 /1.0	−16.3 (1.6)/2.1	−19.4/−1.0
6-thioG.G 3	−19.0	−16.4 (2.3) /2.6	−18.2 /0.8	-	-
G.6-thioG 3	−19.6	−17.1(2.9) /2.5	−19.1 /0.5	-	-
G.A 1	−17.5	−14.5 (1.9)/ 3.0	−16.8 /0.7	−14.2 (1.1)/3.3	−14.7/2.8
G.A 2	−10.9	−9.4 (1.7) /1.5	−11.7 /−0.8	−9.7 (0.9)/1.2	−11.4/1.2
G.A 3	−16.8	−14.4 (2.0) /2.4	−15.8 /1.0	−13.5 (1.2)/3.3	−15.2/1.6
G.A 4	−12.1	10.5 (1.6) /1.6	−12.8 /−0.7	−10.3 (0.8)/1.8	−10.7/1.4
A.A 1	−13.1	−10.6 (1.6) /2.5	−12.3/0.8	−11.0 (0.6)/2.1	−10.8/2.3
A.A 2	−12.3	−10.1 (1.6) /2.2	−11.5/0.8	−10.3 (0.6)/2.0	−10.9/1.4
A.A 3	−10.9	−9.1 (1.4) /1.8	−11.4 /−0.5	−9.2 (1.0)/1.7	−10.9/0.0
8-oxoG.G	−19.6	−16.2 (3.0) / 3.4	18.0 /1.6	-	-
U.U 1	−12.4	−10.2 (1.3) /2.2	-	−10.0 (0.6)/2.4	−12.1/0.3
2-thioU.2-thioU 1	−11.6	−8.1 (1.0) /3.5	-	−8.8 (0.5)/2.8	-
C.CH+	−46.4	−44.3 (5.6) /2.1	−46.6 /−0.2	−41.8 (3.1)/4.6	−41.7/ −4.7
U.U Calcutta	−9.5	−7.5 (0.7) /2.0	−8.7 /0.8	7.4 (0.4 2.1	-

[a]DFT1 = Becke3LYP/6-31G**, DFT2 = PW91/6-31G**//Becke3LYP/6-31G**, MP2//HF = MP2/6-31G*(0.25)//HF/6-31G**, AMBER = Cornell et al force field. The first number is the binding energy with inclusion of the monomer deformation and the second number, following "/", is the difference with respect to the aDZ→aTZ data. The values in parentheses separately list the deformation energies where relevant. DFT2 has the same deformation energy as DFT1 while the AMBER values are small.

The AMBER force field by definition neglects all polarization and charge transfer effects. The excellent correlation between the AMBER and QM values confirms that the main stabilizing contribution of H-bonded base pairs originates in the electrostatic interactions well reproduced by the atom-centered electrostatic-potential fitted charges. There is no other way to explain the agreement between AMBER and QM data for base pairs with stability in the range –5 to –50 kcal/mol. This rules out dominant role of charge transfer.

2.2 RNA Base Pairing

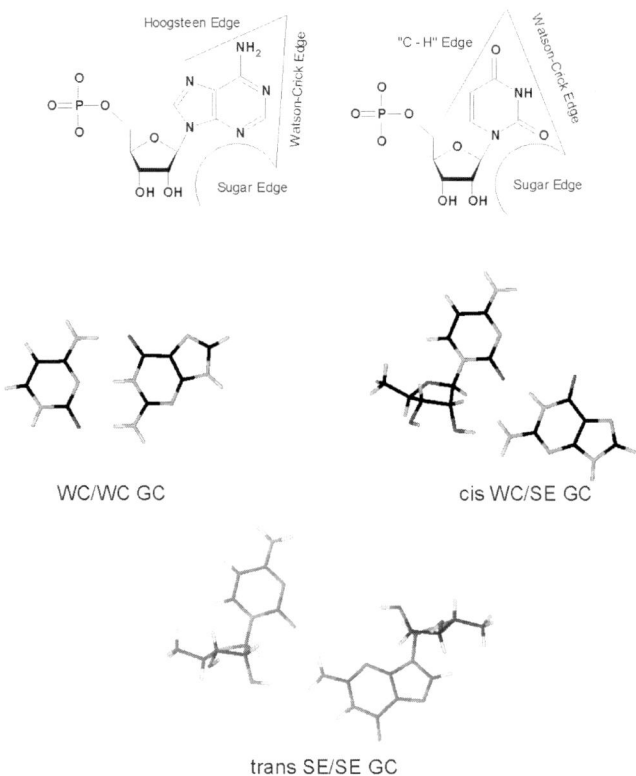

WC/WC GC

cis WC/SE GC

trans SE/SE GC

Figure 14-7. In RNA, the nucleotides can base pair via three edges.[138]

The interaction patterns of nucleic acid bases in RNA are complex. Standard Watson–Crick (WC) A.U and G.C base pairs account for only slightly above 50% of base pairs in large RNAs. The G.U wobble base pair is the third standard RNA base pair. In a sharp contrast to DNA, large fraction of RNA H-bonding interactions is represented by highly variable non-canonical (non-Watson-Crick, non-WC) base pairs.[92,93,114,138-150] They are absolutely essential for building up the complex three-dimensional architectures of large RNAs. Each nucleobase possesses three edges (Figure 14-7).[138] The WC edge, the Hoogsteen (H) or 'C-H' edge, and the Sugar edge (SE). The SE involves the ribose and its 2'-hydroxyl group is capable of forming efficient H-bonds, in contrast to deoxyribose in DNA. Two nucleobases can interact with each other through any of the three edges. In addition, the nucleobase can be either in cis or trans orientation with respect to its sugar. This leads to a total of 12 distinct families with 168 basic possible base pairing patterns.[138] Actually, ca. 110 of them were already identified via x-ray. Many non-WC base pairs are well paired with two or more hydrogen bonds. Systematic quantum chemical analysis of RNA pairing is currently under way. [149-151]

2.3   Nature of Base Stacking

A decade ago the nature of base stacking was not fully clear. There were incorrect alternative models, assuming specific interactions between delocalised π electrons of the aromatic rings and polar exocyclic groups of the bases.[151] Another unusual model is the sandwich model, postulating that description of stacking would require set of explicit charges (besides some van der Waals term) localized below and above the aromatic planes.[153] Validity of the sandwich model would be a disaster for molecular modeling, as none of the force fields includes such out of plane charges. In order to clarify the nature of base stacking, we evaluated hundreds of stacked complexes at the MP2/6-31G*(0.25) level, i.e., with diffuse d-orbitals.[17,39,74,109,119,121] This greatly improved the description of dispersion energy; standard 6-31G* basis set would fail for stacking. As verified ca 8 years later, the MP2/6-31G*(0.25) method is qualitatively correct.[13,154] The hundreds of points were compared with other models of stacking.[4,119,121] We found that *base stacking is basically determined by three contributions: dispersion attraction, short-range exchange repulsion, and electrostatic interaction. No specific "π-π" interactions have been revealed (by "π-π" interactions we mean specific terms that would make the aromatic stacking different in nature from nonaromatic stacking).*

The stabilization of base stacking is dominated by the dispersion attraction, which is rather isotropic and proportional to the geometrical overlap of the bases. The vertical distance between stacked bases is determined by balance between dispersion attraction and short-range exchange repulsion. Finally, the mutual orientation of bases and their displacement are primarily determined by the electrostatic interactions. In water, the response of polar solvent to the electric field of the bases would lead to a destabilization of the most stable gas phase arrangements.[74,96]

Simple empirical potential consisting of a Lennard-Jones van der Waals term and Coulombic term with atom-centered point charges (Eq. 8) is able to reproduce the *ab initio* stacking energy within the range of +/– 1.5 kcal/mol per stacked dimer over the major portion of the conformational space:[121]

$$\Delta E(kcal/mol) = \quad \Sigma 332 q_i q_j / r_{ij} + \Sigma - A_{ij} / r_{ij}^{\,n} \quad + \Sigma B_{ij} / r_{ij}^{\,m}, \tag{8}$$

where $n = 6$, $m = 9\text{--}12$, q are atom-centered point charges, A, B are constants of the van der Waals term and r is the interatomic distance between atoms i and j. The summation is performed over all atoms of both interacting monomers. The surprisingly good agreement of the simple empirical potential with the *ab initio* calculations bolster the current molecular modeling, since this analytical potential function is employed in almost all force fields. Already the comprehensive empirical potential characterization of stacked base dimers reported by Poltev and Shyliupina was correct.[155] However, this does not mean that each empirical potential is accurate. Differences of force fields used in the past in studies of nucleic acids are rather large. We have also shown that inclusion of out-of-plane charges derived via ESP-fitting does not improve the results.[17] We also reported failure of DFT[17] and semiempirical[4] methods for stacking, and substantial problems of method called distributed multipole analysis (DMA).[17]

QM calculations have been used to characterise base stacking in standard and high-resolution B-DNA and Z-DNA geometries.[74,119] The local conformational variations seen in crystals of duplexes do not improve the intrinsic stacking energy terms.[74] When QM calculations were compared with the sandwich model[153] for idealized B-DNA geometries very large difference was noticed. The sandwich model for base stacking is incompatible with QM calculations (see Figure 3 in ref. 119). As for DMA, there is no need to introduce out-of-plane charges to calculate stacking, since in-plane charges provide enough accuracy.[17] The intercalator – base interactions are similar to base stacking.[156,157] Recently, we started reevaluation of older data using the MP2/CBS+CCSD(T) level. Although the results are qualitatively the same, there are quantitative changes. The most systematic data are available for stacked cytosine dimer[13] and B-DNA.[158]

2.4 Future Directions: Combined Quantum Mechanical and Molecular Mechanical Approaches

In the preceding paragraphs, we have shown that modern quantum-chemical methods can provide exceptionally accurate data regarding the intrinsic energetics of molecular interactions for model systems such as base pairs. The main challenge for future quantum chemistry is development of QM-based methods that would be applicable for large systems of biological and biochemical relevance. The most promising direction is certainly the development of combined QM/MM methods. In this book, we do not have any specialized chapter dealing with QM/MM techniques, as these methods were, for a variety of reasons and limitations, not yet systematically applied to nucleic acids. Nevertheless, their application in future is likely, provided some key limitations are successfully addressed.

QM/MM is an acronym for a wide variety of methods combining accurate quantum mechanics (QM) calculations with fast molecular mechanics (MM) methods. A large molecular system (S) is split into an inner part (I) and remaining large outer region (O). In this way a small part of a large and complex molecular system, which is for some reason difficult to describe by means of molecular mechanics only, can be modeled appropriately by a sophisticated QM (high-level) method. The remaining part is usually satisfactorily described by an empirical force-field (low-level). Fast QM methods (including semiempirical approaches) can possibly also be used for the QM core with the aim to perform QM/MM molecular dynamics, but such cheap QM methods will always rely on substantial specific parametrization efforts and their accuracy may be very limited.

A typical application of the QM/MM method is a chemical reaction or a photochemical process taking place in a complex environment, when the environment significantly interferes electrostatically or mechanically. In such a case the QM/MM technique is usually the only option. As interest in modeling complex molecules and environments increases, also the QM/MM methods are becoming more and more popular and constitute a part of almost every modern *ab initio* code. Most codes offer standard tools to search for the ground and transition states and to perform molecular dynamics simulations. In the (still rare) case when also constrained dynamics is implemented, free energy reaction profiles can be obtained and the dynamic effects of the environment motion on the reactive process can be studied. QM/MM methods are commonly used in studies of enzyme catalysis, so their application to nucleic acids should primarily concern the RNA enzymes (ribozymes) including ribosome. Their success in future remains to be seen.

The QM/MM method is a powerful but also a very complex tool. A great deal of attention must be paid to a chemically reasonable choice of the high-level part of the system and to a proper choice of the QM and MM methods, but also – and perhaps mainly – to their interconnection, the QM/MM interface. Widely discussed are questions of appropriate treatment of the inner-outer electrostatic interactions (and polarization)[159] and treatment of the frontier bonds and link atoms,[160] which interconnect the inner and outer regions. A number of solutions have been suggested (for reviews see, e.g., refs [161-163]). For instance, in the mechanical embedding[159] the electrostatic interaction of the inner and outer part is treated at the force-field level only. This is the simplest way, however, it may not be adequate when inner part is significantly polarized by a charge distribution of the surroundings. In the electrostatic embedding method[159] polarization of the cluster by the outer MM charges is accounted for while polarization of the outer region is considered in the polarized embedding[159] scheme. Different problems may require different solutions and certain experience is necessary to yield satisfactory results. Another difficulty inherent to the QM/MM studies is related to the complicated potential energy surfaces. Since the QM/MM typically deals with very complex systems, characterized by a large number of degrees of freedom, exploration of the potential energy surface can become extremely difficult. Even relatively simple problems may require a huge number of single point and gradient evaluations and computational cost can be consequently enormous.

Despite its complexity, the QM/MM approach can be expected to dominate the field of chemical reactions and photochemical processes in the complex environments in the near future.[164] This can be documented by rapidly growing number of applications to enzyme catalysis. Studies on the nucleic acids are rare up to now.[165,166]

REFERENCES

1. Šponer, J., Leszczynski, J. & Hobza, P. (2001). Electronic properties, hydrogen bonding, stacking, and cation binding of DNA and RNA bases. *Biopolymers*, **61**, 3-31.

2. Šponer, J., Berger, I., Spackova, N., Leszczynski, J. & Hobza, P. (2000). Aromatic base stacking in DNA: From ab initio calculations to molecular dynamics simulations. *J. Biomol. Struct. Dyn.*, Special issue, 383-407.

3. Hobza, P. & Šponer, J. (1999). Structure, energetics, and dynamics of the nucleic acid base pairs: Nonempirical ab initio calculations. *Chem. Rev.* **99**, 3247-3276.

4. Hobza, P., Kabelac, M., Šponer, J., Mejzlik, P. & Vondrasek, J. (1997). Performance of empirical potentials (AMBER, CFF95, CVFF, CHARMM, OPLS, POLTEV), semiempirical quantum chemical methods (AM1, MNDO/M, PM3), and ab initio Hartree-Fock method for interaction of DNA bases: Comparison with nonempirical beyond Hartree-Fock results. *J. Comput. Chem.* **18**, 1136-1150.

5. Šponer, J., Leszczynski, J. & Hobza, P. (1996). Structures and energies of hydrogen-bonded DNA base pairs. A nonempirical study with inclusion of electron correlation. *J. Phys. Chem.*, **100**, 1965-1974.

6. Jurecka, P., Nachtigall, P. & Hobza, P. (2001). RI-MP2 calculations with extended basis sets - a promising tool for study of H-bonded and stacked DNA base pairs. *Phys. Chem. Chem. Phys.*, **3**, 4578-4582.

7. Jurecka, P. & Hobza, P. (2003). True stabilization energies for the optimal planar hydrogen-bonded and stacked structures of guanine center dot center dot center dot cytosine, adenine center dot center dot center dot thymine, and their 9-and 1-methyl

derivatives: Complete basis set calculations at the MP2 and CCSD(T) levels and comparison with experiment. *J. Am. Chem. Soc.*, **125**, 15608-15613.

8. Šponer, J., Jurecka, P. & Hobza, P. (2004). Accurate interaction energies of hydrogen-bonded nucleic acid base pairs. *J. Am. Chem. Soc.*, **126**, 10142-10151.

9. Halkier, A., Helgaker, T., Jorgensen, P., Klopper, W., Koch, H., Olsen, J. & Wilson, A. K. (1998). Basis-set convergence in correlated calculations on Ne, N-2, and H2O. *Chem. Phys. Lett.*, **286**, 243-252.

10. Halkier, A., Helgaker, T., Jorgensen, P., Klopper, W. & Olsen, J. (1999). Basis-set convergence of the energy in molecular Hartree-Fock calculations. *Chem. Phys. Lett.*, **302**, 437-446.

11. Truhlar, D. G. (1998). Basis-set extrapolation. *Chem. Phys. Lett.*, **294**, 45-48.

12. Šponer, J. & Hobza, P. (2000). Interaction energies of hydrogen-bonded formamide dimer, formamidine dimer, and selected DNA base pairs obtained with large basis sets of atomic orbitals. *J. Phys. Chem. A*, **104**, 4592-4597.

13. Jurecka, P., Šponer, J. & Hobza, P. (2004). Potential energy surface of the cytosine dimer: MP2 complete basis set limit interaction energies, CCSD(T) correction term, and comparison with the AMBER force field. *J. Phys. Chem. B*, **108**, 5466-5471.

14 Šponer, J. & Hobza, P. (1997). MP2 and CCSD(T) study on hydrogen bonding, aromatic stacking and nonaromatic stacking. *Chem. Phys. Lett.* **267**, 263-270.

15. Kristyan, S. & Pulay, P. (1994). Can (Semi)Local Density-Functional Theory Account for the London Dispersion Forces. *Chem. Phys. Lett.*, **229**, 175-180.

16. Hobza, P., Šponer, J. & Reschel, T. (1995). Density-Functional Theory and Molecular Clusters. *J. Comput. Chem.*, **16**, 1315-1325.

17. Šponer, J., Leszczynski, J. & Hobza, P. (1996). Base stacking in cytosine dimer. A comparison of correlated ab initio calculations with three empirical potential models and density functional theory calculations. *J. Comput. Chem.*, **17**, 841-850.

18. Wesolowski, T. A., Parisel, O., Ellinger, Y. & Weber, J. (1997). Comparative study of benzene center dot center dot center dot X (X = O-2, N-2, CO) complexes using density functional theory: The importance of an accurate exchange-correlation energy density at high reduced density gradients. *J. Phys. Chem. A*, **101**, 7818-7825.

19. Tran, F., Weber, J. & Wesolowski, T. A. (2001). Theoretical study of the benzene dimer by the density-functional-theory formalism based on electron-density partitioning. *Helv. Chim. Acta*, **84**, 1489-1503.

20. Elstner, M., Hobza, P., Frauenheim, T., Suhai, S. & Kaxiras, E. (2001). Hydrogen bonding and stacking interactions of nucleic acid base pairs: A density-functional-theory based treatment. *J. Chem. Phys.*, **114**, 5149-5155.

21. van Mourik, T. & Gdanitz, R. J. (2002). A critical note on density functional theory studies on rare-gas dimers. *J. Chem. Phys.*, **116**, 9620-9623.

22. Kendall, R. A., Früchtl, H.A. (1997). The impact of the resolution of the identity approximate integral method on modern ab initio algorithm development. *Theoret. Chem. Acta*, **97**, 158-163.

23. Saebø, S., Pulay, P. (1993). Local treatment of electron correlation. *Annu. Rev. Phys. Chem.*, **44**, 213-236.

24. Weigend, F., Köhn, A., Hättig, C. (2002). Efficient use of the correlation consistent basis sets in resolution of the identity MP2 calculations. *J. Chem. Phys.*, **116**, 3175-3183.

25. Schutz, M., Rauhut, G. & Werner, H. J. (1998). Local treatment of electron correlation in molecular clusters: Structures and stabilities of (H2O)(n), n = 2-4. *J. Phys. Chem. A*, **102**, 5997-6003.

26. Werner, H. J., Manby, F. R. & Knowles, P. J. (2003). Fast linear scaling second-order Moller-Plesset perturbation theory (MP2) using local and density fitting approximations. *J. Chem. Phys.*, **118**, 8149-8160.

27. Paizs, B., Suhai, S. (1996). Extension of SCF and DFT versions of chemical Hamiltonian approach to N interacting subsystems and an algorithm for their efficient implementation. *J. Comput. Chem.*, 18, 694-701.

28. Boys, S.F., Bernardi, F. (1970). The calculation of small molecular interactions by thedifferences of separate total energies. Some procedures with reduced errors. *Mol. Phys.*, 19, 553–566.

29. Kendall, R.A., Dunning, T.H., Jr., Harrison, R.J. (1992). Electron-affinities of the 1-st row atoms revisited – systematic basis-sets and wave-functions. *J. Chem. Phys.*, 96, 6796-6806.

30. Dunning, T.H, Jr. (1989). Gaussian-basis sets for use in correlated molecular calculations.1. The atoms boron through neon and hydrogen. *J. Chem. Phys.*, 90, 1007-1023.

31. Dunning, T.H, Jr. (2000). A road map for the calculation of molecular binding energies. *J. Phys. Chem. A*, 104, 9062-9080.

32. Frisch, M.J., Pople, J.A., Binkley, J.S. (1984). Self-consistent molecular-orbital methods. 25. Supplementary functions for Gaussian-basis sets. *J. Phys. Chem*, 80, 3265-3269. And references therein.

33. Jensen, F. (2002). Polarization consistent basis sets. II. Estimating the Kohn-Sham basis set limit. *J. Chem. Phys.*, 116, 7372-7379.

34. Mathews, D. H., Sabina, J., Zuker, M. & Turner, D. H. (1999). Expanded sequence dependence of thermodynamic parameters improves prediction of RNA secondary structure. *J. Mol. Biol.*, **288**, 911-940.

35. SantaLucia, J. (1998). A unified view of polymer, dumbbell, and oligonucleotide DNA nearest-neighbor thermodynamics. *Proc. Natl. Acad. Sci. U.S.A*, **95**, 1460-1465.

36. Lai, J. S., Qu, J. & Kool, E. T. (2003). Fluorinated DNA bases as probes of electrostatic effects in DNA base stacking. *Angew. Chem.-Int. Ed.*, **42**, 5973-5977.

37. Sartorius, J. & Schneider, H. J. (1997). Supramolecular chemistry .71. Intercalation mechanisms with ds-DNA: binding modes and energy contributions with benzene, naphthalene, quinoline and indole derivatives including some antimalarials. *J. Chem. Soc.-Perkin Trans. 2*, 2319-2327.

38. Protozanova, E., Yakovchuk, P. & Frank-Kamenetskii, M. D. (2004). Stacked-unstacked equilibrium at the nick site of DNA. *J. Mol. Biol.*, **342**, 775-785.

39. Šponer, J.; Leszczynski, J.; Vetterl, V.; Hobza, P. (1996) Base stacking and hydrogen bonding in protonated cytosine dimer: The role of molecular ion-dipole and induction interactions. *J. Biomol. Struct. Dyn.* **13**, 695-706

40. Gehring, K., Leroy, J. L. & Gueron, M. (1993). A Tetrameric DNA-Structure with Protonated Cytosine.Cytosine Base-Pairs. *Nature*, **363**, 561-565.

41. Berger, I., Kang, C., Fredian, A., Ratliff, R., Moyzis, R. & Rich, A. (1995). Extension of the 4-Stranded Intercalated Cytosine Motif by Adenine-Center-Dot-Adenine Base-Pairing in the Crystal-Structure of D(Cccaat). *Nat. Struct. Biol.*, **2**, 416-429.

42. Spackova, N., Berger, I., Egli, M. & Šponer, J. (1998). Molecular dynamics of hemiprotonated intercalated four-stranded i-DNA: Stable trajectories on a nanosecond scale. *J. Am. Chem. Soc.*, **120**, 6147-6151.

43. Gallego, J., Golden, E. B., Stanley, D. E. & Reid, B. R. (1999). The folding of centromeric DNA strands into intercalated structures: A physicochemical and computational study. *J. Mol. Biol.*, **285**, 1039-1052.

44. Soliva, R., Laughton, C. A., Luque, F. J. & Orozco, M. (1998). Molecular dynamics simulations in aqueous solution of triple helices containing d(G center dot C center dot C) trios. *J. Am. Chem. Soc.*, **120**, 11226-11233.

45. Doronina, S. O. & Behr, J. P. (1997). Towards a general triple helix mediated DNA recognition scheme. *Chem. Soc. Rev.*, **26**, 63-71.

46. No, K. T., Nam, K. Y. & Scheraga, H. A. (1997). Stability of like and oppositely charged organic ion pairs in aqueous solution. *J. Am. Chem. Soc.*, **119**, 12917-12922.

47. Šponer, J. E., Leszczynski, J., Glahe, F., Lippert, B. & Šponer, J. (2001). Protonation of platinated adenine nucleobases. Gas phase vs condensed phase picture. *Inorg. Chem.*, **40**, 3269-3278.

48. Burda, J. V., Šponer, J., Hrabadkova, J., Zeizinger, M. & Leszczynski, J. (2003). The influence of N-7 guanine modifications on the strength of Watson-Crick base pairing and guanine N-1 acidity: Comparison of gas-phase and condensed-phase trends. *J. Phys. Chem. B*, **107**, 5349-5356.

49. Denhartog, J. H. J., Vandenelst, H. & Reedijk, J. (1984). Coordination of 9-Methyladenine to [Cis-Pt(Nh3)2]2+ and [Pt(Dien)]2+ as Studied by Proton NMR. *J. Inorg. Biochem.*, **21**, 83-92.

50. Arpalahti, J., Klika, K. D., Sillanpaa, R. & Kivekas, R. (1998). Dynamic processes in platinum(II)-adenosine complexes. Preparation, NMR spectroscopic characterisation and crystal structure of isomeric Pt-II(dien)-adenosine complexes. *J. Chem. Soc.-Dalton Trans.*, 1397-1402.

51. Sigel, R. K. O., Freisinger, E. & Lippert, B. (2000). Effects of N7-methylation, N7-platination, and C8-hydroxylation of guanine on H-bond formation with cytosine: platinum coordination strengthens the Watson-Crick pair. *J. Biol. Inorganic Chem.*, **5**, 287-299.

52. Šponer, J., Sabat, M., Gorb, L., Leszczynski, J., Lippert, B. & Hobza, P. (2000). The effect of metal binding to the N7 site of purine nucleotides on their structure, energy, and involvement in base pairing. *J. Phys. Chem. B*, **104**, 7535-7544.

53. Navarro, J. A. R. & Lippert, B. (1999). Molecular architecture with metal ions, nucleobases and other heterocycles. *Coord. Chem. Rev.*, **186**, 653-667.

54. Trygubenko, S. A., Bogdan, T. V., Rueda, M., Orozco, M., Luque, F. J., Šponer, J., Slavicek, P. & Hobza, P. (2002). Correlated ab initio study of nucleic acid bases and their tautomers in the gas phase, in a microhydrated environment and in aqueous solution - Part 1. Cytosine. *Phys. Chem. Chem. Phys.*, **4**, 4192-4203.

55. Colominas, C., Luque, F. J. & Orozco, M. (1996). Tautomerism and protonation of guanine and cytosine. Implications in the formation of hydrogen-bonded complexes. *J. Am. Chem. Soc.*, **118**, 6811-6821.

56. Leszczynski, J. (1998). The potential energy surface of guanine is not flat: An ab initio study with large basis sets and higher order electron correlation contributions. *J. Phys. Chem. A*, **102**, 2357-2362.

57. Voet, D., Voet, J.G. & Pratt, C.W., Fundamentals of Biochemistry, John Wiley and Sons Ltd, 1999.

58. Ogle, J. M., Murphy, F. V., Tarry, M. J. & Ramakrishnan, V. (2002). Selection of tRNA by the ribosome requires a transition from an open to a closed form. *Cell*, **111**, 721-732.

59. Šponer, J., Šponer, J. E., Gorb, L., Leszczynski, J. & Lippert, B. (1999). Metal-stabilized rare tautomers and mispairs of DNA bases: N6-metalated adenine and N4-metalated cytosine, theoretical and experimental views. *J. Phys. Chem. A*, **103**, 11406-11413.

60. Zamora, F., Kunsman, M., Sabat, M. & Lippert, B. (1997). Metal-stabilized rare tautomers of nucleobases .6. Imino tautomer of adenine in a mixed-nucleobase complex of mercury(II). *Inorg. Chem.*, **36**, 1583-1587.

61. Blas, J. R., Luque, F. J. & Orozco, M. (2004). Unique tautomeric properties of isoguanine. *J. Am. Chem. Soc.*, **126**, 154-164.

62. Robinson, H., Gao, Y. G., Bauer, C., Roberts, C., Switzer, C. & Wang, A. H. J. (1998). 2'-deoxyisoguanosine adopts more than one tautomer to form base pairs with thymidine observed by high-resolution crystal structure analysis. *Biochemistry*, **37**, 10897-10905.

63. Bevilacqua, P. C., Brown, T. S., Nakano, S. & Yajima, R. (2004). Catalytic roles for proton transfer and protonation in ribozymes. *Biopolymers*, **73**, 90-109.

64. Lankaš, F., Cheatham, T. E., Spackova, N., Hobza, P., Langowski, J. & Šponer, J. (2002). Critical effect of the N2 amino group on structure, dynamics, and elasticity of DNA polypurine tracts. *Biophys. J.*, **82**, 2592-2609.

65. Mollegaard, N. E., Bailly, C., Waring, M. J. & Nielsen, P. E. (1997). Effects of diaminopurine and inosine substitutions on A-tract induced DNA curvature. Importance of the 3'-A-tract junction. *Nucl. Acids Res.*, **25**, 3497-3502.

66. Crothers, D. M., Haran, T. E. & Nadeau, J. G. (1990). Intrinsically Bent DNA. *J. Biol. Chem.*, **265**, 7093-7096.

67. Luo, R., Gilson, H. S. R., Potter, M. J. & Gilson, M. K. (2001). The physical basis of nucleic acid base stacking in water. *Biophys. J.*, **80**, 140-148.

68. Vakonakis, L. & LiWang, A. C. (2004). N1(...)N3 hydrogen bonds of A : U base pairs of RNA are stronger than those of A : T base pairs of DNA. *J. Am. Chem. Soc.*, **126**, 5688-5689.

69. Swart, M., Guerra, C. F. & Bickelhaupt, F. M. (2004). Hydrogen bonds of RNA are stronger than those of DNA, but NMR monitors only presence of methyl substituent in uracil/thymine. *J. Am. Chem. Soc.*, **126**, 16718-16719.

70. Pérez, A., Šponer, J., Jurecka, P., Hobza, P., Luque, F. J. & Orozco, M. (2005) Are the Hydrogen Bonds of RNA (A·U) Stronger Than those of DNA (A·T)? A Quantum Mechanics Study *Chem. Eur. J.* **11**, 5062-5066.

71. McFail-Isom, L., Shui, X. Q. & Williams, L. D. (1998). Divalent cations stabilize unstacked conformations of DNA and RNA by interacting with base Pi systems. *Biochemistry*, **37**, 17105-17111.

72. Šponer, J., Šponer, J. E. & Leszczynski, J. (2000). Cation-pi and amino-acceptor interactions between hydrated metal cations and DNA bases. A quantum-chemical view. *J. Biomol. Struct. Dyn.*, **17**, 1087-1096.

73. Prive, G. G., Heinemann, U., Chandrasegaran, S., Kan, L. S., Kopka, M. L. & Dickerson, R. E. (1987). Helix Geometry, Hydration, and G.A Mismatch in a B-DNA Decamer. *Science*, **238**, 498-504.

74. Šponer, J., Florian, J., Ng, H. L., Šponer, J. E. & Spackova, N. (2000). Local conformational variations observed in B-DNA crystals do not improve base stacking: computational analysis of base stacking in a d(CATGGGCCCATG)(2) B <-> A intermediate crystal structure. *Nucl. Acids Res.*, **28**, 4893-4902.

75. Sukhodub, L. F. (1987). Interactions and Hydration of Nucleic-Acid Bases in a Vacuum - Experimental-Study. *Chem. Rev.*, **87**, 589-606.

76. Kratochvil, M., Engkvist, O., Šponer, J., Jungwirth, P. & Hobza, P. (1998). Uracil dimer: Potential energy and free energy surfaces. Ab initio beyond Hartree-Fock and empirical potential studies. *J. Phys. Chem. A*, **102**, 6921-6926.

77. Kratochvil, M., Šponer, J. & Hobza, P. (2000). Global minimum of the adenine center dot center dot center dot thymine base pair corresponds neither to Watson-Crick nor to Hoogsteen structures. Molecular dynamic/quenching/AMBER and ab initio beyond Hartree-Fock studies. *J. Am. Chem. Soc.*, **122**, 3495-3499.

78. Kabelac, M. & Hobza, P. (2001). At nonzero temperatures, stacked structures of methylated nucleic acid base pairs and microhydrated nonmethylated nucleic acid base

pairs are favored over planar hydrogen-bonded structures: A molecular dynamics simulations study. *Chem. Eur. J.*, **7**, 2067-2074.

79. Dey, M., Grotemeyer, J. & Schlag, E. W. (1994). Pair Formation of Free Nucleobases and Mononucleosides in the Gas-Phase. *Z. Naturfors. Sect. A. - J. Phys. Sci.*, **49**, 776-784.

80. Nir, E., Kleinermanns, K. & de Vries, M. S. (2000). Pairing of isolated nucleic-acid bases in the absence of the DNA backbone. *Nature*, **408**, 949-951.

81. Dong, F. & Miller, R. E. (2002). Vibrational transition moment angles in isolated biomolecules: A structural tool. *Science*, **298**, 1227-1230.

82. Roscioli, J. R. & Pratt, D. W. (2003). Base pair analogs in the gas phase. *Proc. Natl. Acad. Sci. U.S.A.*, **100**, 13752-13754.

83. Abo-Riziq, A., Grace, L., Nir, E., Kabelac, M., Hobza, P. & de Vries, M. S. (2005). Photochemical selectivity in guanine-cytosine base-pair structures. *Proc. Natl. Acad. Sci. U.S. A.*, **102**, 20-23.

84. Abo-Riziq, A., Crews, B., Grace, L. & de Vries, M. S. (2005). Microhydration of guanine base pairs. *J. Am. Chem. Soc.*, **127**, 2374-2375.

85. Rodgers, M. T. & Armentrout, P. B. (2000). Noncovalent interactions of nucleic acid bases (uracil, thymine, and adenine) with alkali metal ions. Threshold collision-induced dissociation and theoretical studies. *J. Am. Chem. Soc.*, **122**, 8548-8558.

86. Halgren, T. A. & Damm, W. (2001). Polarizable force fields. *Curr. Opinion Struct. Biol.* **11**, 236-242.

87. Kabelac, M., Ryjacek, F. & Hobza, P. (2000). Already two water molecules change planar H-bonded structures of the adenine center dot center dot center dot thymine base pair to the stacked ones: a molecular dynamics simulations study. *Phys. Chem. Chem. Phys.*, **2**, 4906-4909.

88. Gorb, L. & Leszczynski, J. (1998). Intramolecular proton transfer in mono- and dihydrated tautomers of guanine: An ab initio post Hartree-Fock study. *J. Am. Chem. Soc.*, **120**, 5024-5032.

89. Brandl, M., Meyer, M. & Suhnel, J. (2000). Water-mediated base pairs in RNA: A quantum-chemical study. *J. Phys. Chem. A*, **104**, 11177-11187.

90. Schneider, B. & Berman, H. M. (1995). Hydration of the DNA bases is local. *Biophys. J.*, **69**, 2661-2669.

91. Auffinger, P. & Westhof, E. (2000). Water and ion binding around RNA and DNA (C,G) oligomers. *J. Mol. Biol.*, **300**, 1113-1131.

92. Reblova, K., Spackova, N., Stefl, R., Csaszar, K., Koca, J., Leontis, N. B. & Šponer, J. (2003). Non-Watson-Crick basepairing and hydration in RNA motifs: Molecular dynamics of 5S rRNA loop E. *Biophys. J.*, **84**, 3564-3582.

93. Reblova, K., Spackova, N., Koca, J., Leontis, N. B. & Šponer, J. (2004). Long-residency hydration, cation binding, and dynamics of loop E/helix IV rRNA-L25 protein complex. *Biophys. J.*, **87**, 3397-3412.

94. Phan, A. T., Leroy, J. L. & Gueron, M. (1999). Determination of the residence time of water molecules hydrating B'-DNA and B-DNA, by one-dimensional zero-enhancement nuclear Overhauser effect spectroscopy. *J. Mol. Biol.*, **286**, 505-519.

95. Cramer, C. J. & Truhlar, D. G. (1999). Implicit solvation models: Equilibria, structure, spectra, and dynamics. *Chem. Rev.*, **99**, 2161-2200.

96. Florian, J., Šponer, J. & Warshel, A. (1999). Thermodynamic parameters for stacking and hydrogen bonding of nucleic acid bases in aqueous solution: Ab initio/Langevin dipoles study. *J. Phys. Chem. B*, **103**, 884-892.

97. Orozco, M. & Luque, F. J. (2000). Theoretical methods for the description of the solvent effect in biomolecular systems. *Chem. Rev.*, **100**, 4187-4225.

98. Sychrovsky, V., Šponer, J. & Hobza, P. (2004). Theoretical calculation of the NMR spin-spin coupling constants and the NMR shifts allow distinguishability between the specific direct and the water-mediated binding of a divalent metal cation to guanine. *J. Am. Chem. Soc.*, **126**, 663-672.

99. Kitaura, K., Morokuma, K. (1976). *Int. J. Quantum Chem.*, **10**, 325-344..

100. Jeziorski, B., Moszynski, R., Szalewicz, K. (1994) Perturbation-theory approach to intermolecular potential-energy surfaces of van-der-Waals complexes. *Chem. Rev.*, **94**, 1887-1930.

101. Hobza, P. & Šponer, J. (1996). Thermodynamic characteristics for the formation of H-bonded DNA base pairs. *Chem. Phys. Lett.*, **261**, 379-384.

102. Chalasinski, G. & Szczesniak, M. M. (2000). State of the art and challenges of the ab initio theory of intermolecular interactions. *Chem. Rev.*, **100**, 4227-4252.

103. Xantheas, S. S. (1996). On the importance of the fragment relaxation energy terms in the estimation of the basis set superposition error correction to the intermolecular interaction energy. *J. Chem. Phys.*, **104**, 8821-8824.

104. Szalewicz, K., Jeziorski, B. (1998) Comment on "On the importance of the fragment relaxation energy terms in the estimation of the basis set superposition error correction to the intermolecular interaction energy" [J Chem Phys 104, 8821 (1996)] *J. Chem. Phys.* **109**, 1198-1200.

105. Hobza, P., Bludsky, O. & Suhai, S. (1999). Reliable theoretical treatment of molecular clusters: Counterpoise-corrected potential energy surface and anharmonic vibrational frequencies of the water dimer. *Phys. Chem. Chem. Phys.*, **1**, 3073-3078.

106. Simon, S., Duran, M. & Dannenberg, J. J. (1996). How does basis set superposition error change the potential surfaces for hydrogen bonded dimers? *J. Chem. Phys.*, **105**, 11024-11031.

107. Šponer, J. & Hobza, P. (1994). Nonplanar Geometries of DNA Bases - Ab-Initio 2nd-Order Moller-Plesset Study. *J. Phys. Chem.*, **98**, 3161-3164.

108. Šponer, J. & Hobza, P. (1994). Bifurcated Hydrogen-Bonds in DNA Crystal-Structures - an Ab-Initio Quantum-Chemical Study. *J. Am. Chem. Soc.*, **116**, 709-714.

109. Šponer, J., Leszczynski, J. & Hobza, P. (1996). Hydrogen bonding and stacking of DNA bases: A review of quantum-chemical ab initio studies. *J. Biomol. Struct. & Dyn.*, **14**, 117-135.

110. Šponer, J., Florian, J., Hobza, P. & Leszczynski, J. (1996). Nonplanar DNA base pairs. *J. Biomol. Struct. & Dyn.*, **13**, 827-833.

111. Bludsky, O., Šponer, J., Leszczynski, J., Spirko, V. & Hobza, P. (1996). Amino groups in nucleic acid bases, aniline, aminopyridines, and aminotriazine are nonplanar: Results of correlated ab initio quantum chemical calculations and anharmonic analysis of the aniline inversion motion. *J. Chem. Phys.*, **105**, 11042-11050.

112. Luisi, B., Orozco, M., Šponer, J., Luque, F. J. & Shakked, Z. (1998). On the potential role of the amino nitrogen atom as a hydrogen bond acceptor in macromolecules. *J. Mol. Biol.*, **279**, 1123-1136.

113. Vlieghe, D., Šponer, J. & Van Meervelt, L. (1999). Crystal structure of d(GGCCAATTGG) complexed with DAPI reveals novel binding mode. *Biochemistry*, **38**, 16443-16451.

114. Šponer, J., Mokdad, A., Šponer, J. E., Spackova, N., Leszczynski, J. & Leontis, N. B. (2003). Unique tertiary and neighbor interactions determine conservation patterns of Cis Watson-Crick A/G base-pairs. *J. Mol. Biol.*, **330**, 967-978.

115. Starikov, E. B. & Steiner, T. (1997). Computational support for the suggested contribution of C-H center dot center dot center dot O = C interactions to the stability of nucleic acid base pairs. *Acta Cryst. Sect. D-Biol. Crystall.*, **53**, 345-347.

116. Hobza, P., Šponer, J., Cubero, E., Orozco, M. & Luque, F. J. (2000). C-H center dot center dot center dot O contacts in the adenine center dot center dot center dot uracil Watson-Crick and uracil center dot center dot center dot uracil nucleic acid base pairs: Nonempirical ab initio study with inclusion of electron correlation effects. *J. Phys. Chem. B*, **104**, 6286-6292.

117. Burda, J. V., Šponer, J., Leszczynski, J. & Hobza, P. (1997). Interaction of DNA base pairs with various metal cations (Mg2+, Ca2+, Sr2+, Ba2+, Cu+, Ag+, Au+, Zn2+, Cd2+, and Hg2+): Nonempirical ab initio calculations on structures, energies, and nonadditivity of the interaction. *J. Phys. Chem. B*, **101**, 9670-9677.

118. Šponer, J., Burda, J. V., Mejzlik, P., Leszczynski, J. & Hobza, P. (1997). Hydrogen-bonded trimers of DNA bases and their interaction with metal cations: Ab initio quantum-chemical and empirical potential study. *J. Biomol. Struct. Dyn.*, **14**, 613-628.

119. Šponer, J., Gabb, H. A., Leszczynski, J. & Hobza, P. (1997). Base-base and deoxyribose-base stacking interactions in B-DNA and Z-DNA: A quantum-chemical study. *Biophys. J.*, **73**, 76-87.

120. Gresh, N., Šponer, J. E., Spackova, N., Leszczynski, J. & Šponer, J. (2003). Theoretical study of binding of hydrated Zn(II) and Mg(II) cations to 5'-guanosine monophosphate. Toward polarizable molecular mechanics for DNA and RNA. *J. Phys. Chem. B*, **107**, 8669-8681.

121. Šponer, J., Leszczynski, J. & Hobza, P. (1996). Nature of nucleic acid-base stacking: Nonempirical ab initio and empirical potential characterization of 10 stacked base dimers. Comparison of stacked and H-bonded base pairs. *J. Phys. Chem.*, **100**, 5590-5596.

122. Hobza, P. & Šponer, J. (1998) Significant structural deformation of nucleic acid bases in stacked base pairs: an ab initio study beyond Hartree-Fock. *Chem. Phys. Lett.* **288**, 7-14.

123. Shishkin, O. V., Šponer, J. & Hobza, P. (1999). Intramolecular flexibility of DNA bases in adenine-thymine and guanine-cytosine Watson-Crick base pairs. *J. Mol. Struct.*, **477**, 15-21.

124. Šponer, J. & Kypr, J. (1993). Theoretical-Analysis of the Base Stacking in DNA - Choice of the Force-Field and a Comparison with the Oligonucleotide Crystal-Structures. *J. Biomol. Struct. Dyn.*, **11**, 277-292.

125. Alhambra, C., Luque, F. J., Gago, F. & Orozco, M. (1997). Ab initio study of stacking interactions in A- and B-DNA. *J. Phys. Chem. B*, **101**, 3846-3853.

126. Chirlian, L. E. & Francl, M. M. (1987). Atomic Charges Derived from Electrostatic Potentials - a Detailed Study. *J. Comput. Chem.*, **8**, 894-905.

127. Besler, B. H., Merz, K. M. & Kollman, P. A. (1990). Atomic Charges Derived from Semiempirical Methods. *J. Comput. Chem.*, **11**, 431-439.

128. Breneman, C. M. & Wiberg, K. B. (1990). Determining Atom-Centered Monopoles from Molecular Electrostatic Potentials - the Need for High Sampling Density in Formamide Conformational-Analysis. *J. Comput. Chem.*, **11**, 361-373.

129. Cornell, W. D., Cieplak, P., Bayly, C. I., Gould, I. R., Merz, K. M., Ferguson, D. M., Spellmeyer, D. C., Fox, T., Caldwell, J. W. & Kollman, P. A. (1995). A 2nd Generation Force-Field for the Simulation of Proteins, Nucleic-Acids, and Organic-Molecules. *J. Am. Chem. Soc.*, **117**, 5179-5197.

130. Reed, A. E., Curtiss, L. A. & Weinhold, F. (1988). Intermolecular Interactions from a Natural Bond Orbital, Donor-Acceptor Viewpoint. *Chem. Rev.*, **88**, 899-926.

131. Šponer, J., Sabat, M., Burda, J. V., Leszczynski, J. & Hobza, P. (1999). Interaction of the adenine-thymine Watson-Crick and adenine-adenine reverse-Hoogsteen DNA base pairs with hydrated group IIa (Mg2+, Ca2+, Sr2+, Ba2+) and IIb (Zn2+, Cd2+, Hg2+) metal

cations: Absence of the base pair stabilization by metal-induced polarization effects. *J. Phys. Chem. B*, **103**, 2528-2534.

132. Šponer, J., Leszczynski, J. & Hobza, P. (1997). Thioguanine and thiouracil: Hydrogen-bonding and stacking properties. *J. Phys. Chem. A*, **101**, 9489-9495.

133. Hobza, P. & Sandorfy, C. (1987). Nonempirical Calculations on All the 29 Possible DNA-Base Pairs. *J. Am. Chem. Soc.*, **109**, 1302-1307.

134. Aida, M. & Nagata, C. (1986). An Abinitio Molecular-Orbital Study on the Stacking Interaction between Nucleic-Acid Bases - Dependence on the Sequence and Relation to the Conformation. *Int. J. Quantum Chem.*, **29**, 1253-1261.

135. Gould, I. R. & Kollman, P. A. (1994). Theoretical Investigation of the Hydrogen-Bond Strengths in Guanine Cytosine and Adenine Thymine Base-Pairs. *J. Am. Chem. Soc.*, **116**, 2493-2499.

136. Hobza, P., Šponer, J. & Polasek, M. H-Bonded and stacked DNA-base pairs – cytosine dimer – an ab initio 2nd-order Moller-Plesset study. *J. Am. Chem. Soc.* **117**, 792-798.

137. Leininger, M.L., Nielsen, I.M.B, Colvin, M.E. & Janssen, C.L. (2002) Accurate structures and binding energies for stacked uracil dimers. *J. Phys. Chem.* A 106, 3850-3854.

138. Leontis, N. B., Stombaugh, J. & Westhof, E. (2002). The non-Watson-Crick base pairs and their associated isostericity matrices. *Nucl. Acids Res.*, 30, 3497-3531.

139. Leontis, N. B. & Westhof, E. (1998). The 5S rRNA loop E: Chemical probing and phylogenetic data versus crystal structure. *RNA*, 4, 1134-1153.

140. Moore, P. B., Steitz T. A .(2003) The structural basis of large ribosomal subunit function. *Annu. Rev. Biochem.* **72**, 813-850

141. Ban, N., Nissen, P., Hansen, J., Moore, P. B., Steitz, T. A. (2000) The complete atomic structure of the large ribosomal subunit at 2.4 angstrom resolution. *Science* **289**, 905-920.

142. Wimberly, B. T., Brodersen, D. E., Clemons, W. M., Morgan-Warren, R. J., Carter, A. P., Vonrhein, C., Hartsch, T., Ramakrishnan, V. (2000) Structure of the 30S ribosomal subunit *Nature* **407**, 327-339.

143. Klein, D. J., Schmeing, T. M., Moore, P. B., Steitz, T. A. (2001) The kink-turn: a new RNA secondary structure motif *EMBO J.* **20**, 4214-4221

144. Nissen, P., Ippolito, J. A., Ban, N., Moore, P. B., Steitz, T. A. (2001) RNA tertiary interactions in the large ribosomal subunit: The A-minor motif. *Proc. Natl. Acad. Sci. USA* **98**, 4899-4903

145. Razga, F., Koca, J., Šponer, J., Leontis, N. B. (2005) Hinge-like motions in RNA kink-turns: The role of the second A-minor motif and nominally unpaired bases. *Biophys. J.* **88**, 3466-3485.

146. Leontis, N. B., Westhof, E. (2003) Analysis of RNA motifs. *Curr. Opin. Struct. Biol.* **13**, 300-308.

147. Adams, P. L., Stahley, M. R., Gill, M. L., Kosek, A. B., Wang, J. M., Strobel, S. A. (2004) Crystal structure of a group I intron splicing intermediate: *RNA* **10**, 1867-1887.

148. Ke, A. L., Zhou, K. H., Ding, F., Cate, J. H. D., Doudna, J. A. (2004) A conformational switch controls hepatitis delta virus ribozyme catalysis *NATURE,* **429**, 201-205.

149. Šponer, J. E., Spackova, N., Kulhanek, P., Leszczynski, J., Šponer, J. (2005) Non-Watson-Crick base pairing in RNA. Quantum chemical analysis of the cis Watson-Crick/sugar edge base pair family. *J. Phys. Chem. A,* **109**, 2292-2301.

150. Šponer, J. E., Spackova, N., Leszczynski J. & Šponer, J. (2005) Principles of RNA base pairing: Structures and energies of the trans Watson-Crick/sugar edge base pairs. *J. Phys. Chem. B* **109**, 11399-11410.

151. Šponer, J. E., Leszczynski, J., Sychrovsky, V. & Šponer, J (2005) Sugar edge/sugar edge base pairs in RNA: Stabilities and structures from quantum chemical calculations *J. Phys. Chem. B* **109**, 18680-18689.

152. Bugg, C. E., Thomas, J. M., Sundaralingam, M., Rao, S. T. (1971) Stereochemistry of nucleic acids and their constituents. X. Solid-state base-stacking patterns in nucleic acid constituents and polynucleotides. *Biopolymers* **10**, 175-219.

153. Hunter, C. A. (1993). Sequence-Dependent DNA-Structure - the Role of Base Stacking Interactions. *J. Mol. Biol.*, **230**, 1025-1054.

154. Hobza, P. & Šponer, J. (2002). Toward true DNA base-stacking energies: MP2, CCSD(T), and complete basis set calculations. *J. Am. Chem. Soc.*, **124**, 11802-11808.

155. Poltev, V. I. & Shulyupina, N. V. (1986). Simulation of Interactions between Nucleic-Acid Bases by Refined Atom-Atom Potential Functions. *J. Biomol. Struct. Dyn.*, **3**, 739-765.

156. Bondarev, D. A., Skawinski, W. J. & Venanzi, C. A. (2000). Nature of intercalator amiloride-nucelobase stacking. An empirical potential and ab initio electron correlation study. *J. Phys. Chem. B*, **104**, 815-822.

157. Reha, D., Kabelac, M., Ryjacek, F., Šponer, J., Šponer, J. E., Elstner, M., Suhai, S. & Hobza, P. (2002). Intercalators. 1. Nature of stacking interactions between intercalators (ethidium, daunomycin, ellipticine, and 4',6-diaminide-2-phenylindole) and DNA base pairs. Ab initio quantum chemical, density functional theory, and empirical potential study. *J. Am. Chem. Soc.*, **124**, 3366-3376.

158. Šponer, J., Jurecka, P., Marchan, I., Luque, F.J., Orozco, M., Hobza, P. (2006) Nature of base stacking. Reference quantum chemical stacking energies in ten unique B-DNA base pair steps. *Chem. Eur. J.* **12**, 2854-2865.

159. Bakowies, D., Thiel, W. (1996). Hybrid models for combined quantum mechanical and molecular mechanical approaches. *J. Phys. Chem.* **100**, 10580-10594.

160. Reuter, N., Dejaegere, A., Maigret, B.,Karplus, M. (2000). Frontier bonds in QM/MM methods: A comparison of different approaches. *J. Phys. Chem. A*, **104**, 1720-1735.

161. Sherwood, P. (2000). Hybrid Quantum Mechanics/Molecular Mechanics approaches. Modern Methods and Algorithms of Quantum Chemistry. Proceedings, Second Edition, J. Grotendorst (Ed.), John von Neumann Institute for Computing, Jülich, NIC Series, Vol. 3, ISBN 3-00-005834-6, 285-305.

162. Lin, H., Truhlar, D. G. (2005). QM/MM: What have we learned, where are we, and where do we go from here? *Theor. Chem. Acc.*, in press.

163. Gao, J., York, D. (2003). Preface to proceedings of the symposium on methods and applications of combined quantum mechanical and molecular mechanical potentials (2001). *Theor. Chem. Acc.*, **109**, 99-99.

164. Friesner, R. A., Guallar, V. (2005). Ab initio quantum chemical and mixed quantum mechanics/molecular mechanics (QM/MM) methods for studying enzymatic catalysis. *Annu. Rev. Phys. Chem.*, **56**, 389-427.

165. Zoete, V., Meuwly, M. (2004). Double proton transfer in the isolated and DNA-embedded guanine-cytosine base pair. *J. Chem. Phys.* **121**, 4377-4388.

166. Spiegel, K., Rothlisberger, U., Carloni, P. (2004). Cisplatin binding to DNA oligomers from hybrid Car-Parrinello/molecular dynamics simulations. *J. Phys. Chem. B* **108**, 2699-2707.

Chapter 15

INTERACTION OF METAL CATIONS WITH NUCLEIC ACIDS AND THEIR BUILDING UNITS

A comprehensive view from quantum chemical calculations

Judit E. Šponer,[1] Jaroslav V. Burda,[2] Jerzy Leszczynski,[3] and Jiří Šponer[1,4]

[1]*Institute of Biophysics, Academy of Sciences of the Czech Republic, Královopolská 135, 612 65 Brno, Czech Republic;* [2]*Department of Chemical Physics and Optics, Faculty of Mathematics and Physics, Charles University, Ke Karlovu 3, 121 16 Prague 2, Czech Republic;* [3]*Jackson State University, 1325 J. R. Lynch Street, Jackson, Mississippi 39217-0510;* [4]*Institute of Organic Chemistry and Biochemistry, Academy of Sciences of the Czech Republic, Flemingovo náměstí 2, 166 10 Prague 6, Czech Republic*

Abstract: This chapter summarizes the main achievements in quantum chemical modeling of metal-nucleic acid interactions. The topics discussed cover a wide range of problems taken from bio-inorganic chemistry starting from the nature of metal-nucleobase and metal-phosphate interactions, through joint binding modes to the phosphate and nucleobase moieties in nucleosides, up to the mechanism of action of platinum-based metallodrugs. The examples shown are of biological relevance and are aimed at understanding experimentally known phenomena.

Key words: DNA, RNA, nucleic acids, metals, quantum chemistry, DFT

1. INTRODUCTION

After the four basic elements, O, C, H, N, sodium and potassium are the most frequent components of the human organism.[1] Under physiologic conditions, these two metals serve as charge compensating species and provide with the proper osmotic pressure in the living cell.[2]

Nucleic acids (NA) are negatively charged polyelectrolytes whose charge is compensated mainly by cations condensed on their surface. While monovalent cations mostly bind in a non-specific manner, divalent metal cations are apt to occupy structurally well-defined binding positions in the

389

J. Šponer and F. Lankaš (eds.), Computational Studies of RNA and DNA, 389–410.
© 2006 *Springer.*

RNA and DNA architectures and, thereby, to fulfill important structural and biochemical functions. They, for example, participate in the stabilization of the three-dimensional architecture or provide with bio-catalytic properties.

Although metal-NA interactions belong to an advanced branch of experimental bio-inorganic chemistry, accurate theoretical studies in this field have yet become possible in the last decade, especially due to the development of DFT-based quantum chemical techniques.[3]

In this chapter we summarize those achievements in the quantum chemical investigations of metal-NA interactions which pointed beyond the level attainable by common experimental methods. Computational studies are able to provide a simultaneous picture of the structure, stability and electronic properties of the studied systems, which is important to understand the biochemical functions. As with all modeling approaches, one has to keep in mind the limitations introduced into the model. It is pretty challenging to relate gas phase data to properties observable in physiological DNA and RNA. In addition, site-specific binding of metal ions obeys the rules of coordination chemistry. Thus, metal cations have to be surrounded with properly situated ligand field to provide a realistic picture of their interactions. This adds another level of complexity to the models used for these systems and shows that metal cations are more sophisticated entities than to treat them as charged spheres. A proper description of metal cations must account for both the electrostatic and molecular orbital effects. As a consequence, quantum chemistry is superior to the force field techniques at describing interaction of metal cations with NAs.

2.   METAL-NUCLEOBASE INTERACTIONS

The nominal charge of the polynucleotide chain is contributed by the phosphate group. Nevertheless, nucleobases themselves are excellent N and O donor ligands, forming well-defined binding sites primarily for divalent cations. Under physiological conditions, i.e. in aqueous solution, metal cations are always present in a hydrated form. Thus, the relevant model for the site-specific binding of metal cations to nucleobases must always include the hydration shell of the metal cation. Then, the hydrated metal cation may interact with the nucleobase either in the (i) outer shell, (ii) inner shell, and occasionally, (iii) bidentate fashions (Figure 15-1).

While the geometry is largely determined by the coordination chemistry of the cation, the binding strength depends on the ligand type. Nucleobases offer several binding sites of various interaction strengths. Thus, one of the main questions is to determine the binding strength depending on the

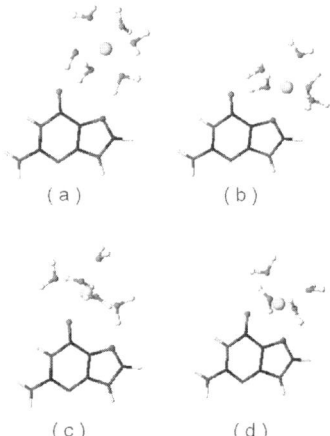

Figure 15-1. Binding modes of octahedral Mg^{2+} cations to guanine. Structures were obtained from optimizations at Becke3LYP/6-31G** level of theory. (a) Outer-shell binding; (b) inner shell binding to N7; (c) inner shell binding to O6; (d) bidentate binding mode.

metalation site for a given base and cation. Moreover, various cations exhibit different affinity towards a given type of binding.

This issue was systematically studied for the inner and outer shell binding of hydrated Mg^{2+} and Zn^{2+} cations to guanine.[4] These cations have the same charge and almost the same radius, yet, they play very different roles in chemistry and biology. The thermodynamical driving force of cation binding was estimated on the basis of three different thermodynamic cycles. As for the inner shell complexes, the computed results reveal that the N7 position of guanine has a greater affinity towards Zn^{2+} than Mg^{2+}, while the O6 position exhibits a similar propensity towards both cations. For the outer shell binding the gas-phase results were not accurate enough to unambiguously show preference towards either cations. Decomposition of the interaction energies into pairwise terms (see below) has revealed that inner shell binding of hydrated zinc to guanine can be viewed as a process in which the base is first metalated and then hydrated, as the dominant part of the total stabilization is due to the interaction between the bare cation and the nucleobase. In contrast, for the analogous complex with magnesium the metal-base term is reduced with respect to zinc, and, therefore, the complex is shifted more towards a complex of nucleobase with hydrated cation.

Electronic background of the difference between the complexes of group IIb and IIa cations is examined in Ref. 5. While interaction of bare IIa group cations with nucleobases is dominated by Coulombic interactions, the covalent d-orbital-lone pair interaction strengthens binding of IIb metals.

A more elaborated approach has pointed at the necessity to include solvation models into the calculations when binding affinities are addressed in solution. One recent study[6] compares geometrical parameters of an outer shell complex of guanine with $[Mg(H_2O)_6]^{2+}$ optimized in three different approximations: (i) with the AMBER force field in vacuo, (ii) by B3LYP/ 6-31G(d,p) optimizations in the gas-phase and (iii) at the same level as (ii) but within the framework of the CPCM (COSMO polarized continuum) solvent model. The force field calculation results in overly long Mg - nucleobase distance and in a change of the experimentally suggested H-bonding pattern. Inclusion of the solvation model into the DFT-optimizations pushes the nucleobase away from the metal because the solvent considerably screens the electric field of the hydrated cation. Finally, this study suggests that in solution relatively accurate interaction energies can be obtained when using gas-phase optimized geometries from Becke3LYP/6-31G(d,p) calculations and computing binding energies by the CPCM method (cf. also section 3).

Other aspects of metal-nucleobase interactions will be addressed in section 6.1., in connection with platinated nucleobases.

NMR shifts and spin-spin coupling constants for the hydrated magnesium-guanine system have been reported recently.[7] This study illustrates, that the intermolecular spin-spin coupling constants $^1J(X,O6)$ and $^1J(X,N7)$ ($X = Mg^{2+}$, Zn^{2+}) can be successfully applied to differentiate between outer shell and inner shell binding modes.

In contrast to Mg^{2+}, which has a clear preference towards the hexa-coordinated form, in biological systems Zn^{2+} might be present also in tetra-coordinated complexes. An in-depth database analysis reveals that, in general, Mg^{2+} is more susceptible towards O-donor ligands while Zn^{2+} rather favors N- and S-donors. Mn^{2+} represents a borderline between Zn^{2+} and Mg^{2+}.[8]

There is a branch of studies considering interaction of bare metal cations with nucleobases. Although these contributions are of high value from methodological point of view, they are not discussed in the frame of this review. For details, refer to Ref. 9.

3.   METAL-PHOSPHATE INTERACTIONS

The nominally -1 negative charge of each nucleotide unit in NAs is carried by the phosphate moiety. Therefore, the phosphate groups form the primary metal binding sites in NAs. To a large extent the charge of NAs is compensated by monovalent cations partly condensed on the surface of the macromolecule. Note, however, that since polar solvent efficiently screens the negative charge of phosphates, the individual phosphates for most of the time do not interact directly with an ion. In addition to this, (divalent) metal cations specifically bound to the phosphate groups may have important catalytic functions: they can catalyze the hydrolysis of the phophodiester linkage in RNA.

Crystal database analyses show, that in contrast to the analogous carboxylates, metal binding to the phosphinyl ($-PO_2^-$) fragment of the phosphate group does not take place in a symmetric fashion: i.e. in most published crystal structures the electron density maxima are outside the bisector of the $O = P = O$ angle.[10] A thorough ab initio analysis of variously charged mono- and diphosphate derivatives with mono- and divalent cations has shown that the asymmetric coordination of the metal cation is likely to be caused by the direct participation of polar particles, such as water, in the cation binding.[11] This again calls attention on the fact, that biologically relevant models must include the proper number of ligated water molecules to represent the hydration sphere of the cation.

Petrov et al. analyzed interaction of hexahydrated magnesium with dimethyl-phosphate anion and also concluded that to provide reliable estimates of the binding affinities in solution, it is absolutely essential to include sufficiently accurate solvation models into the calculations.[6]

Metal catalyzed hydrolysis of the phosphodiester bonds in NAs is a long-disputed problem of bio-inorganic chemistry. One study suggested, that the process involves the cleavage of the C5'-O bonds of the sugar rather than that of the P-O bonds of the phosphate.[12] Quantum chemical studies of hydrated dimethyl-phosphate models with alkali and alkaline earth cations point at the role of metal cations in these processes. It has been found that catalytic activity of a metal cation is due to its ability to (i) shield the negative charge of the phosphate group and (ii) simultaneously weaken the C-O bonds in the model compounds. This can be correlated with the charge abstracting and polarizing properties of the cations which change in the Mg^{2+} > Li^+ > Ca^{2+} ~ Na^+ > K^+ order. The entire mechanism of the metal-assisted phosphodiester hydrolysis in the hammerhead ribozyme was investigated by Torres et al.[13] The reaction proceeds according to an $S_N2(P)$ scenario and is initiated by deprotonation of the 2'-OH by a hydrated $Mg^{2+}OH^-$ cofactor bound to the phosphate moiety. In the forthcoming steps of the reaction the role of the metal center is twofold: (i) via its hydration shell it stabilizes the intermediates and transition state complexes formed, and (ii) donates a proton from its hydration shell to facilitate the loss of the leaving group in the rate determining step of the reaction.

Let us note here, however, that the mechanism of phosphodiester hydrolysis is strongly dependent on the local RNA architecture and often requires cooperation of remote segments of the macromolecule.[14] Sometimes, even two cations might be active in the process. Thus, exploring the mechanism of phosphodiester hydrolysis in biologically relevant systems is a very intricate and computationally challenging task. Indisputably, combined

QM/MM techniques are the hot candidates to solve this problem. However, no QM/MM techniques suited for reliable studies of NAs were presented so far.

4.   METAL-NUCLEOTIDE INTERACTIONS

Combination of the N- and O-donor sites of nucleobases with the negatively charged phosphate oxygens gives rise to versatile metal-binding modes in nucleotide structures. Due to their distinct biological relevance, characterization of metal-nucleotide interactions in the literature is restricted to the N7-binding in purine nucleotides.

Hydrated zinc and magnesium group divalent cations bind to the N7 position of purine nucleotides to form very strong H-bonds between the cation and anionic oxygen atoms of the phosphate group (see Figure 15-2a). Thereby the water shell exerts a screening of the negative charge concentrated on the phosphate moiety and at the same time influences the backbone geometry. The pairwise nucleotide-cation interaction provides the major contribution to the stabilization of these complexes, which can be characterized with a high non-additivity (17-28% of the total interaction energies). The non-additivity term expresses the shielding of the metal-nucleotide electrostatic attraction by the water shell and is larger for Zn^{2+} than for Mg^{2+}.[15]

As non-additivity and polarization effects are absented from standard force fields, the force fields are not suitable to capture the energetics of the metal - nucleotide interactions with a satisfactory accuracy. The SIBFA polarizable molecular mechanics force field of Gresh provides a partial remedy for this problem. It is parametrized to describe interaction of hydrated Zn^{2+} and Mg^{2+} cations with 5'-guanosine-monophosphate and properly includes polarization and charge transfer terms necessary to treat this kind of systems.[16]

Due to the high non-additivities, the pairwise interaction energies are not representative enough for the strength of the metal-nucleobase and metal-phosphate interactions. In contrast, the geometrical parameters well reflect the balance of all contributing effects and can be used to indirectly evaluate binding selectivity in these systems. From systematic changes in the interatomic distances, electronic energies and interaction energies the binding selectivity of divalent cations towards N7-binding varies in the following order: $Cu^{2+} \gg Zn^{2+} = Cd^{2+} > Mg^{2+}$. In fact, the binding selectivity

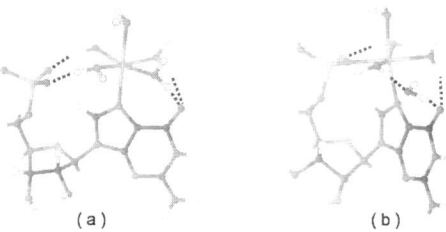

(a)   (b)

Figure 15-2. Binding of penthydrated Zn^{2+} to the N7 site of 5'-guaninosine-monophosphate. Optimized structures obtained from HF calculations with the 6-31G* basis set and Christiansen's pseudopotential on Zn. (a) Inner-shell binding to N7(G), outer shell binding to the phosphate; (b) inner shell binding both to the nucleobase and phosphate. The dotted lines indicate H-bonding contacts. Note that the highly polarized water molecules from the cation's hydration shell form extremely strong H-bonds.[15] Binding of cations in NAs would be affected by the overall topology of NAs which destabilize the water bridges to phosphates.[15]

is the result of the balance between the water - cation and cation - nucleobase contributions. Let us mention here, that the binding selectivity in biological (solution) systems is determined by pretty tiny energy differences (on the scale of 3-10 kcal/mol), while the absolute gas phase binding energies are around 300 kcal/mol. This underlines the accuracy of the methods needed to evaluate such systems.[17] The gas phase data are typically dominated by the electrostatics which is eliminated in polar environment. However, specific non-electrostatic effects well reflected by the QM calculations are often fully transferred into solution.[15,17]

Simultaneous inner-shell binding to N7(G) and the phosphate oxygen (see Figure 15-2b) attenuates the water-phosphate contacts. On the other hand, inner shell binding to the phosphate does not change the above established ranking of cations related to their selectivity towards N7-binding.[17]

It has been found that inner shell binding of guanine to hexahydrated cation-phosphate complexes substantially weakens the metal-phosphate outer shell binding. Thus, the phosphate group recognizes when water is replaced by guanine in the hydration shell of the cation. However, it is unable to recognize the Mg^{2+} to Zn^{2+} substitution, assuming the metals adopt the same coordination mode.[17] These findings are of key importance at unraveling the principles of metal-phosphate recognition in NAs.

5.   INTERACTION OF METAL CATIONS WITH BASE PAIRS

It has been suggested long ago, that coordination of metal cations to the N7 position of the purine base enhances the stability of GC and AT base pairs.[18]

The enhancement is due to polarization effects and classical electrostatic attraction between the cation and the remote base. A very convenient approach has been found to asses the polarization contribution to the base pair enhancement for the interaction of cations with base pairs.[19] The following formulae decomposes the total interaction energy:

$$\Delta E_{total} = \Delta E_{Me\text{-}B1} + \Delta E_{B1\text{-}B2} + \Delta E_{Me\text{-}B2} + \Delta E_3 \tag{1}$$

where $\Delta E_{Me\text{-}B1}$ is the pairwise interaction energy between the metal (Me) and the nucleobase directly bonded to it (B1); $\Delta E_{B1\text{-}B2}$ stands for the pairwise base-base (B1 - B2) interaction energy (see Figure 15-3), and $\Delta E_{Me\text{-}B2}$ expresses interaction between the hydrated cation and the remote base (mostly long-range electrostatics). Finally ΔE_3 is the three-body (nonadditivity) term that reveals how the three interacting species cooperate when forming the complex, thus, polarization of the base pairing by the N7 metal binding. ΔE_3 and $\Delta E_{Me\text{-}B2}$ term contribute to the enhancement of the base pairing.

Table 15-1 compares the strength of various intermolecular contacts in the metalated GC pair, for bare and pentahydrated cations. While the pairwise GC interaction is not affected by the presence of the cation's first hydration shell, the three-body term becomes strongly attenuated in the hydrated systems due to screening and polarization re-distribution (compare the entries of bare and hydrated Ca^{2+} and Ba^{2+}). Further, the pairwise $\Delta E_{Me\text{-}B1}$ terms of analogous systems with Mg^{2+} and Zn^{2+} illustrate the differences (section 2) in the cation binding affinity of the N7(G) position. The polarization enhancement (ΔE_3) depends also on the polarity of the base

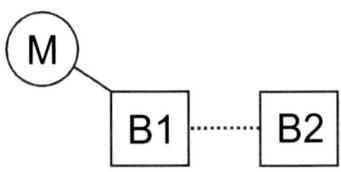

Figure 15-3. Schematic model of metalated base pairs. B1, B2 = nucleobases, M = pentahydrated metal.

Table 15-1. Interaction energies (kcal/mol) for various metalated base pairs computed at the MP2/6-31G(d)//HF/6-31G(d) level*.

Me	B1	B2	$\Delta E_{\text{Me-B1}}$	$\Delta E_{\text{Me-B2}}$	$\Delta E_{\text{B1-B2}}$	ΔE_3	ΔE_{total}
-	G	C	–	–	–26.2	–	–
Mg^{2+}	G	C	–198.7	–	–26.0	–	–243.8
$[Mg(H_2O)_5]^{2+}$	G	C	–89.3	–1.5	–26.4	–8.1	–125.4
$[Mg(H_2O)_5]^{2+}$	PG	C	–237.7	–0.7	–23.5	–5.9	–303.9
Zn^{2+}	G	C	–237.2	–	–	–	–285.4
$[Zn(H_2O)_5]^{2+}$	G	C	–93.8	–1.5	–26.4	–8.7	–130.4
Ca^{2+}	G	C	–133.9	–3.0	–25.8	–10.1	–172.7
$[Ca(H_2O)_5]^{2+}$	G	C	–82.6	–1.7	–26.3	–5.2	–115.8
Ba^{2+}	G	C	–118.3	–2.0	–25.6	–9.6	–156.1
$[Ba(H_2O)_5]^{2+}$	G	C	–71.2	–7.7	–23.2	–2.1	–104.1
-	G	G	–	–	–17.7	–	–
$[Mg(H_2O)_5]^{2+}$	G	G	–89.8	–9.5	–19.9	–10.4	–129.6
$[Mg(H_2O)_5]^{2+}$	PG	G	–274.3	–8.5	–12.9	–7.0	–302.7
-	A	U	–	–	–12.3	–	–
$[Mg(H_2O)_5]^{2+}$	A	U	–59.6	–9.8	–9.3	–0.4	–79.1
$[Mg(H_2O)_5]^{2+}$	PA	U	–244.3	–9.0	–8.1	+1.9	–259.5

* Me = metal entity; B1 = proximal base interacting with the cation (PN stands for nucleotide), B2 = remote nucleobase, $\Delta E_{\text{Me-B1}}$, $\Delta E_{\text{Me-B2}}$ = pairwise metal - nucleobase interaction energy terms; $\Delta E_{\text{B1-B2}}$ = pairwise nucleobase-nucleobase interaction energy; ΔE_3 = three-body term; ΔE_{total} = total interaction energy as defined in equation 1. Data are taken from Refs. 19 and 20.

pair, thus it is quite substantial for the GC and GG base pairs and pretty small for the AU pair. When the proximal base is replaced by a nucleotide, the metal binding is improved because of the favorable electrostatic contacts between the cation and the anionic phosphate group. On the contrary, this reduces the three-body term.

No significant differences were observed in the electron topology of the GC, GG and AU base pairs metalated with pentahydrated Mg^{2+} at N7.[20] NBO was not able to detect any significant charge transfer effects caused by the cation coordination. Therefore, the polarization base pair enhancement upon cation binding is best interpreted on the basis of "the electrostatic response of the complexes to the presence of the cation".[20]

6. NUCLEIC ACIDS AND METALLODRUGS

Introduction of cisplatin in cancer treatment ignited a series of experimental[21] and, later, theoretical studies aimed at describing its mechanism of action. Clinical application of platinum compounds is restricted due to their high cytotoxicity and the experimental research was recently shifted towards other less toxic DNA-binder metal-based drugs. These studies also benefit from findings of theoretical investigations on square planar platinum complexes, which are also of interest for the basic bioinorganic chemistry.

6.1   Platinated Nucleobases

A systematic study on the proton affinities of platinated adenines has revealed that the *gas-phase* protonation energies depend primarily on the overall charge and to a much lesser extent on the location of the metal.[22] The charge-dependence is enormous, yet it appears to be completely abolished in *aqueous solution,* where the corresponding pKa values are essentially independent on the charge of the metal adduct in the range of charges +3 to −1. In contrast, the subtle variations in the interplay between different cation binding and protonation sites stem from molecular orbital effects and appears to be translated from the gas phase to solution. Figure 15-4 summarizes the computed gas-phase proton affinities as a function

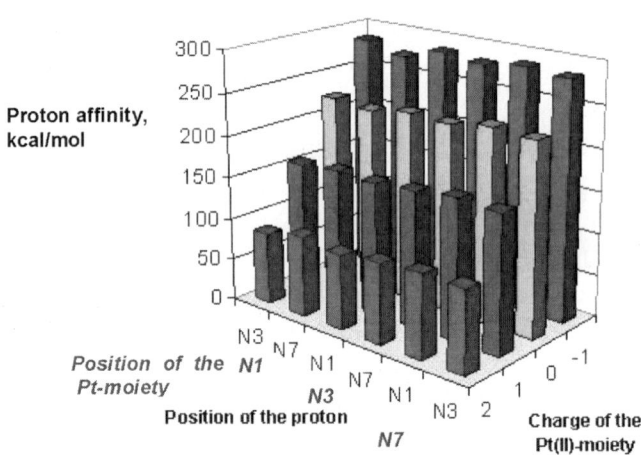

Figure 15-4. Gas phase proton affinity of platinated adenines for different combinations of platinum binding and protonation sites, as a function of the total charge of the metal adduct. Computed results were obtained at Becke3LYP level of theory using the 6-31G* basis set and the lanl2dz pseudopotential on Pt.

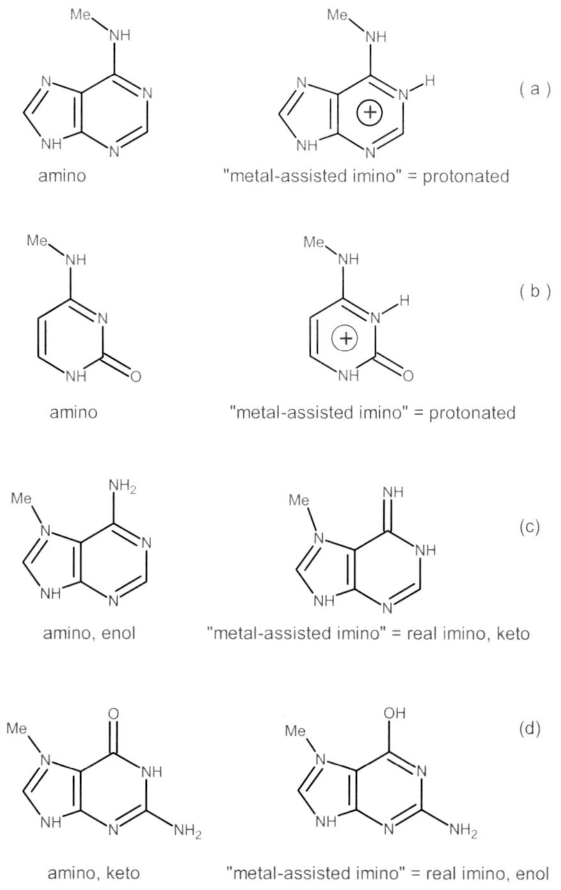

Figure 15-5. Metal-assisted tautomeric forms of (a) N6-metalated adenine, (b) N4-metalated cytosine, (c) N7-metalated adenine, and (d) N7-metalated guanine. The amino metalation affects the tautomerism by non-electrostatic effects and the tautomerism is thus seen in bioinorganic experiments. The N7 metalation results in primarily electrostatic influence on tautomerism which is likely fully eliminated in polar environments.

of the total charge as well as the position of the proton and platinum moiety. For a given platinum and proton position one can recognize an almost linear relationship between the proton affinity and the charge of the platinum adduct. On the other hand, for a given total charge of the metal adduct the maximum is reached if the platinum is at N7 and the proton binds to N1. This is in qualitative agreement with experimental observations from solution studies and at the same time illustrates that site-dependence of the protonation energy stems from molecular orbital effects, because it is independent on the total charge of the metal-adduct.

The improvement of the protonation energy is larger for the N4-platinated cytosine than for N6-mercurated adenine (Figure 15-5a and b) and can be correlated with the shift of the π-electron density from the nucleobase to the metal center in the major tautomeric forms. Thus, metal-assisted tautomerism (protonation) of these otherwise charge-neutralized systems is unambiguously dominated by molecular orbital (nonelectrostatic) effects and this is why it is seen in all solution and x-ray experiments. Metal-induced shift of the protonation equilibria and thereby stabilization of the rare tautomeric forms may give rise to stabilization of mispairs in DNA.[23] This was studied for amino-metalated base pairs. For example, stabilization of the N1-protonated form of adenine by N6-metalation with Hg^{2+} gives rise to the formation of stable AH^+.C and AH^+.G base pairs. Even more considerable stabilizing effect is proposed for the CH+.G Hoogsteen base pair platinated at N4(C), where metalation enhances the stability of the mispair by ca. 20 kcal/mol. Interbase proton transfer processes may also be accelerated by amino-platination. For example, proton transfer from the N1 position of guanine to N3 of cytosine results in the formation of a CH^+.G$^-$ ion pair. The energy difference between the ion pair and canonical forms of this base pair is ca. 24 kcal/mol in the gas-phase. Due to the stabilization of the protonated (i.e. "rare tautomeric") form in N4-platinated cytosines, however, this energy difference can be reduced to the half by metalation.[23]

N7-metalation also changes the gas-phase tautomeric equilibria of purine bases. However, the effect vanishes for neutral metal adducts.[24] Charged metal-adducts stabilize the major (amino) and minor (imino) forms of guanine and adenine, respectively (Figure 15-5c, d). As the shift of tautomeric equilibria in these gas-phase structures is due to electrostatic effects, they are not expected to occur in solution or crystal.

Catalytic scenarios (cytosine deamination) brought about by platinum in N3-metalated cytosines are discussed in Ref. 25.

6.2   Interaction Strength of Platinated Base Pairs

Enhancement of the base pairing by N7 binding of $[Pt(NH_3)_3]^{2+}$ is much larger than that caused by the inner shell binding of hydrated Mg^{2+} or Zn^{2+} cations.[26] The electronic changes stimulated by Pt-coordination in the structure of nucleobases implicate even more pronounced changes in the stability of platinated base pairs.

A correlation has been found between the N1-acidity of N7 platinated guanines and the strength of the GC base pairs in the gas-phase.[27] The enhancement of the base pair is caused primarily by polarization effects, while the N1-acidity is mostly influenced by ionic-electrostatic forces with

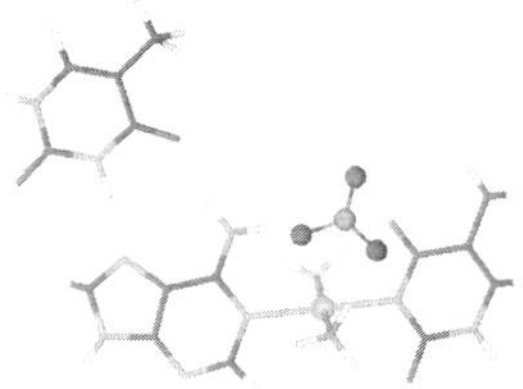

Figure 15-6. Counterion effects in platinated systems. Note, that the nitrate ion is coordinated to the platinum(II)-center in a pseudoaxial position and entirely biases the experimental outcome, since it belongs to the buffer.

some additional polarization effects. While the computed gas-phase N1-deprotonation energies can be correlated with the solution pK_a values,[28] there is no correlation with the $K_{GC,DMSO}$ association constants measured in solution. This led to the conclusion that the deviation (in the condensed phase experiment) from the expected trend is likely to be caused by environmental effects, such as unwanted (biasing) presence of counterions in the experimentally studied system.

In fact, similar observations were made for the AT Hoogsteen pair platinated at N1(A)-position.[29] While the computed interaction energy for the AT Hoogsteen pair is –14.1 kcal/mol, platination enhances its strength to –18.8 kcal/mol, primarily due to the long-range electrostatic attraction between the charged Pt^{2+} cation and thymine. In contrast, no base pairing was detected by experiment. When including, however, a nitrate anion coordinated in pseudoaxial position to Pt^{2+} (see Figure 15-6) the base pair stability dropped down to –11.6 kcal/mol, because of the compensation of the charge on Pt^{2+}. This suggests that the buffer anion determines the experimental outcome.

6.3 Studies on Cisplatin Binding to Nucleobases

Cisplatin is the first anticancer drug successfully employed in the clinical treatment of various kinds of cancers, such as small cell lung, ovarian, testicular, head and neck tumors. From chemical point of view it is a square planar molecule (see Figure 15-7) whose hydrolysis results in the formation of the active form suitable to bind to DNA.[30,31] Cisplatin binds to the G-reach segments of DNA in such a way to form intrastrand cross-links between two adjacent guanines. The antitumor activity of the drug is associated with the bending of the DNA as a consequence of the cross-link formation. The binding was modeled by classical MD simulation.[32] In fact, the binding is a ligand exchange reaction between the hydrated form of cisplatin and guanine, in which the water ligand is replaced by the nucleobase. The ligand exchange reaction proceeds via a trigonal bipyramidal transition state and is kinetically controlled. A comparative study on cisplatin binding to guanine and adenine has shown, that for guanine the binding process is energetically more favored both from kinetic and thermodynamic points of view.[33] The extra stabilization of the transition state with guanine is attributed to (i) a very strong H-bond between the ammine ligands of the Pt-moiety and O6 of guanine as well as (ii) the significantly stronger electronic interaction between the Pt and the guanine ligand as compared to that between Pt and adenine. In addition, the diaqua complex has been found to exhibit a higher selectivity towards guanine.

There are controversial views on the role of π-back-donation in cisplatin-nucleobase binding.[34] A study on hypothetical square planar Pt-complexes containing CO-ligands concludes that, even with the strongest π-acids, π-back-donation may operate exclusively in the zero oxidation state. Thus, it hardly has any role in stabilizing the cisplatin-nucleobase adducts. In addition, it is suggested, that non-coplanarity of the cisplatin and nucleobase moieties also impedes the π-interactions in these systems. On the contrary, a recent orbital analysis of guanine complexes cross-linked with cisplatin reports on a rare binding combination between the d_{xy} atomic orbital of Pt and the antibonding π*-orbital of guanine.[35]

In complex biological matrices, such as the cell, there are several potential targets of cisplatin binding. In fact, Pearson's theory predict a larger binding affinity of cisplatin towards S-donor ligands.[36] Considering the high concentration of cystein and methionine in the cell one can thus almost rule out that cisplatin reaches DNA. A comprehensive theoretical rationale on the competition of S- and N-donor ligands in cisplatin binding has been given by Deubel.[36-38] From an analysis of the Pt-L binding energies in various $[Pt(NH_3)_3L]^{2+}$ (L = ligand) complexes he has inferred that intrinsic binding affinity of cisplatin is higher towards N-donors than S-donors in the gas-phase. However, solvation effects, particularly in polar solvents, strongly alter the gas-phase trends, and eventually in water S-donors become more favorable targets than the N-donor ligands. At $\varepsilon = 78.4$ (water) only

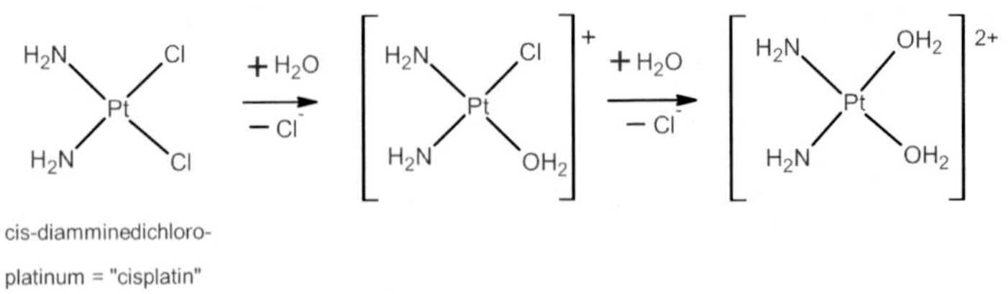

Figure 15-7. Cisplatin and its hydrolysis products.

methyl-guanine remains competitive with neutral S-donors.[36] The kinetic control of cisplatin binding is determined by three factors: (i) the nucleophile, (ii) the substituents of the nucleophile, (iii) the environment, i.e. solvation effects. Computed gas-phase activation energy of the nucleophilic substitution reaction of cis-$[Pt(NH_3)_2(OH_2)Cl]^+$ with NH_3 has been found to be lower than that of H_2S, showing that N-donors may be intrinsically better nucleophiles than S-donors. However, this trend can be easily reverted by substituent as well as environmental effects. While in a more polar environment, i.e. water, the scenarios with sulfur nucleophiles are kinetically more favorable in the low ε regions of chromatin binding to the guanine-reach regions of DNA is the decisive reaction mechanism.[37] Amine loss of cisplatin derivatives via nucleophilic attack of S-donors at the platinum center is a possible way of cisplatin inactivation in the cell. Thermodynamic and kinetic aspects of this process were studied for a series of substituted $[Pt(NH_3)_4]^{2+}$ derivatives both in gas-phase as well as in aqueous solution. The substrate selectivity of the reaction is determined by the kinetic and thermodynamic trans-effect of the ligands connected to the platinum entity. The following order is established for the trans effect of ligand L in complexes with composition of $[Pt(NH_3)_3L]^{2+/+}$: S-donors > N-donors > water = 0. Similarly, cis-$[Pt(NH_3)_3MeS]^+$ has been selected as a substrate to model the effect of various nucleophiles on the kinetic and thermodynamic control of the deactivation process, i.e. ammine loss reaction. It has been shown that adenine is unable to replace ammine from the model substrate both for kinetic and thermodynamic reasons. Similarly, thiolates can be rejected due to the unfavorable activation energy of the substitution reaction. In contrast, nucleophiles like Met, Cys, and Gua may play a more significant role in the deactivation of cisplatin.[38]

Cisplatin induced depurination reaction of nucleosides has been studied by Baik et al.[39] They have described the cleavage of the glycosidic bond in N7 platinated as well as protonated forms including solvent effects in the SCRF (Self-Consistent-Reaction-Field) approach. The lack of reactivity in the platinated systems is explained by the very similar activation energy obtained for non-metalated guanosine and its platinated adduct. In contrast, protonation causes a significant stabilization of the transition state and thereby accelerates the depurination reaction.

7. CATION-Π INTERACTIONS

In principle, not only the heteroatoms of nucleobases may form binding sites for metal cations. It is for example well known that metal cations may interact with the π-electron system of aromatic compounds. The prototype of these interactions is depicted in Figure 15-8a, which shows the optimized geometry of magnesium-hexahydrate with benzene, obtained from optimization at the HF/6-31G* level of theory. To catch a similar minimum for the cytosine - magnesium-hexahydrate system (see Figure 15-8b), however, one has to introduce geometrical constraints, as full optimization will always lead to the considerably more stable in-plane binding.[40,41] As nucleobases have no intrinsic propensity towards participating in cation-π interactions, occurrence of cation-π-like geometries in NA crystals is accidental and they are fixed by other interactions.

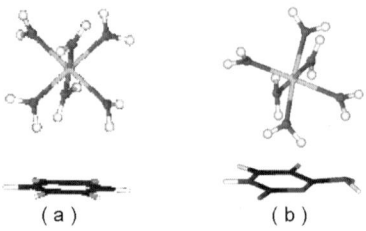

(a)   (b)

Figure 15-8. Cation - π interaction between magnesium-hexahydrate and (a) benzene, (b) cytosine. Geometry optimizations were performed at the HF/6-31G* level of theory

8. SITE-SPECIFIC BINDING OF CATIONS TO NUCLEIC ACIDS

Interaction of metal cations with NAs is, in fact, a very complex network of numerous contributing effects. All studies summarized above capture only selected aspects of this very intricate process, and, therefore their applicability to project to larger systems is limited.

The recently published approach by Petrov et al represents an interesting step towards understanding the binding of metal cations to NAs in its complexity.[42] They divide the total binding free energy into four contributions on the basis of a thermodynamic cycle approach. At first, interaction energy of the gas-phase system, describing the binding site is determined. In the second step the gas-phase system is immersed into water using e.g. a COSMO-polarized continuum model (CPCM). Interaction between the RNA and the metal cation is considered in the next step, adapting the approach introduced by Bashford and Karplus[43] for the calculation of ionization constants of titratable sites in proteins. Finally, the free energy change upon addition of diffuse electrolyte ions is calculated using non-linear Poisson-Boltzmann equations. The summation of these four contributions gives the total binding free energy in solution.

9. COMMENT ON THE ACCURACY OF FORCE FIELD CALCULATIONS FOR CATIONS

Force field calculations enable an explicit treatment of a very complex matrix consisting of nucleic acids, cations, counterions as well as water molecules. Systems up to 100+ nucleotides can be routinely treated with explicit solvent molecular dynamics method.[44, 45] However, a caution should be taken when evaluating results of MD simulations due to severe force field and sampling limitations.

The most remarkable shortages of force field calculations are the following: (i) they are not able to account for polarization and charge transfer effects, (ii) metal ions are treated as van der Waals spheres with a point charge in the center, disregarding their electronic structure, (iii) they fail to fully capture the competition between water molecules, monovalent and divalent cations for the binding to NAs. The profound effect of the ion on the first-shell ligand waters (which are very strong H-bond donors compared to the bulk waters) is neglected.[15-17,42,46] To overcome these problems specially tuned polarization potentials have been developed, yet they have not so far been introduced in common MD simulations.[16,47] Note that simple polarization force fields would still not capture charge transfer.

In general, performance of MD technique is much better for monovalent than divalent cations.[46,48] Monovalent ions sample rather reasonably on the time scale 10-100 ns, at least for strong binding sites. One has to be more careful with anions such as Cl⁻, since anions are polarizable species. Description of divalent cations suffers from large force field imbalances while they sample entirely insufficiently in contemporary simulations. Incorrectly placed divalent ion may behave like an unguided missile in the simulation and cause solute perturbation that is not reparable on a common simulation time scale.[48]

Keeping in mind its limitations the MD technique can still be successfully exploited to study interaction of cations with NAs. For example, MD simulations were applied to examine the role of monovalent ions in stabilizing guanine quadruplexes.[49] Simulations can also be used to detect major ion binding sites in RNA, including key binding sites with very wide free energy minima. These are characterized by delocalized (dynamical) cation binding and thus difficult to capture by X-ray technique.[50]

10. CONCLUSIONS

The uniqueness of quantum chemical methods to simultaneously capture the structure, energy and electronic properties can be fruitfully exploited at studying the interaction of metal cations with NAs.

When applying quantum chemistry to metal ions, it is critically important to distinguish between ionic electrostatic effects and (molecular orbital, polarization) non-electrostatic effects. While the ionic effects are usually not visible in polar environment due to solvent screening the non-electrostatic effects are typically well expressed. Consideration of gas phase experimental or QM data without at least a qualitative inclusion of the solvent effects leads to conclusions that are not relevant outside the gas phase.

In NAs, a great variety of binding sites are available for the cations, represented by the heteroatoms of nucleobases and the charged oxygens of the phosphate groups. Electronic effects control the intrinsic propensity of a cation towards a given type of binding, which cannot be accounted for by standard force field calculations. Site-specific coordination of metal cations causes sensible changes in the protonation and tautomeric equilibria of nucleobases and base pair strengths. QM calculations represent a dedicated tool to study such physical properties. Gas-phase properties can be projected into solution either by using continuum solvent techniques or by following the charge dependence of the gas-phase results.

To understand mechanism of action of DNA-binder metallodrugs, beyond the NAs, one has to consider interaction with other components of the biological matrix.

Assessing the strength of metal-NA interactions calls for combining information from gas-phase quantum chemical calculations with other theoretical approaches used to describe solvation effects as well as polymer-cation interactions. The state-of-the-art approach[42] elaborated by Petrov et al illustrates the way of computational quantum chemistry towards more extended systems, such as RNA and DNA.

In general, the nature of metal-nucleobase, metal-phosphate and metal-nucleotide interactions is a well-explored field of computational quantum chemistry. Thus, the new directions of research should concentrate either on the metal-NA interactions or on mechanistic studies. In the later subject there are a lot of intriguing problems to investigate, e.g. the cleavage of the phosphodiester bond in catalytic RNAs and mechanism of action of metallodrugs. To describe these issues, however, one needs to employ computational methods suited to tackle extended systems, such as properly parametrized QM/MM techniques. QM/MM approaches could also be very useful in future, albeit the methods are currently used in practice only for proteins. Therefore, there is an urgent need to continue development of plausible computational platforms for this purpose.

ACKNOWLEDGEMENTS

This contribution was supported by the Grant Agency of the Academy of Sciences of the Czech Republic grant No. 1QS500040581 (J. E. Š., J. Š.), by grants No. 203/05/0009 and 203/05/0388 (J. E. Š., J. Š.) from the Grant Agency of the Czech Republic, as well as by the Wellcome Trust International Senior Research Fellowship GR067507 (J. Š.). J. L., J. Š. and J. E. Š. acknowledge for the financial support from NIH grant S06 GM008047, NSF-CREST grant HRD-0318519 and ONR grant N00034-03-1-0116. J. V. B. is very grateful for the financial support from MSMT-NSF grant 1P05ME784. IBP and IOBC are supported by the grants AVO Z5 004 0507 and Z4 055 0506, Ministry of Education of the Czech Republic.

REFERENCES

1. Frieden, E. J. (1985). New perspectives on the essential trace-elements. *J. Chem. Ed.* **62**, 917-923.
2. Bloomfield, V. A., Crothers, D. M. and Tinoco, Jr., I. (1999). *Nucleic Acids: Structures, Properties and Functions*, pp. 475-534, University Science Books, Sausalito, California.
3. Koch, W. and Holthausen, M. C. (2002). *A Chemist's Guide to Density Functional Theory*, 2nd ed., Wiley-VCH, Weinheim, New York, Chicester.
4. Šponer, J. E., Sychrovský, V., Hobza, P., and Šponer, J. (2004). Interactions of hydrated divalent metal cations with nucleic acid bases. How to relate the gas phase data to solution situation and binding selectivity in nucleic acids. *Phys. Chem. Chem. Phys.* **6**, 2772-2780.
5. Burda, J. V., Šponer, J. and Hobza, P. (1996). Ab initio study of the interaction of guanine and adenine with various mono- and bivalent cations (Li^+, Na^+, K^+, Rb^+, Cs^+; Cu^+, Ag^+, Au^+; Mg^{2+}, Ca^{2+}, Sr^{2+}, Ba^{2+}; Zn^{2+}, Cd^{2+}, and Hg^{2+}). *J. Phys. Chem.* **100**, 7250-7255.
6. Petrov, A. S., Pack, G. R. and Lamm, G. (2004). *J. Phys. Chem. B* **108**, 6072-6081.
7. Sychrovský, V., Šponer, J. and Hobza, P. (2004). Theoretical calculation of the NMR spin-spin coupling constants and the NMR shifts allow distinguishability between the specific direct and the water-mediated binding of a divalent metal cation to guanine. *J. Am. Chem. Soc.* **126**, 663-672.
8. Bock, C. W., Katz, A. K., Markham, G. D. and Glusker, J. P. (1999). Manganese as a replacement for magnesium and zinc: functional comparison of the divalent ions. *J. Am. Chem. Soc.* **121**, 7360-7372.
9. Bertran, J., Sodupe, M., Šponer, J. and Šponer, J. E. (2005). Metal cation-nucleic acids interactions. In: *Encyclopedia of Computational Chemistry (online edition)*, (von Ragué Schleyer, P., Schaefer, III, H. F., Schreiner, P. R., Jorgensen, W. L., Thiel, W. and Glen, R. C., eds), John Wiley & Sons, Chichester, doi:10.1002/0470845015.cn0094.

10. Alexander, R. S., Kanyo, Z. F., Chirlian, L. E. and Christianson, D. W. (1990). Stereochemistry of phosphate-Lewis acid interactions - implications for nucleic-acid structure and recognition. *J. Am. Chem. Soc.* **112**, 933-937.

11. Schneider, B., Kabeláč, M. and Hobza, P. (1996). Geometry of the phosphate group and its interactions with metal cations in crystals and ab initio calculations. *J. Am. Chem. Soc.* **118**, 12207-12217.

12. Murashov, V. V. and J. Leszczynski (1999). Theoretical study of complexation of phosphodiester linkage with alkali and alkaline-earth cations. *J. Phys. Chem. B* **103**, 8391-8397.

13. Torres, R. A., Himo, F., Bruice, T. C., Noodleman, L. and Lowell, T. (2003). Theoretical examination of Mg^{2+}-mediated hydrolysis of a phophodiester linkage as proposed for the hammerhead-ribozyme. *J. Am. Chem. Soc.* **125**, 9861-9867.

14. Fedor, M. J. (2002). The role of metal ions in RNA catalysis. *Curr. Opin. Struct. Biol.* **12**, 289-295.

15. Šponer, J., Sabat. M., Gorb, L., Leszczynski, J., Lippert, B. and Hobza, P. (2000). The effect of metal binding to the N7 site of purine nucleotides on their structure, energy, and involvement in base pairing. *J. Phys. Chem. B* **104**, 7535-7544.

16. Gresh, N., Šponer, J. E., Špačková, N. and Šponer, J. (2003). Theoretical study of binding of hydrated Zn(II) and Mg(II) cations to 5'-guanosine monophosphate. Toward polarizable molecular mechanics for DNA and RNA. *J. Phys. Chem. B* **107**, 8669-8681.

17. Rulíšek, L. and Šponer, J. (2003). Outer-shell and inner-shell coordination of phosphate group to hydrated metal ions (Mg^{2+}, Cu^{2+}, Zn^{2+}, Cd^{2+}) in the presence and absence of nucleobase. The role of nonelectrostatic effects. *J. Phys. Chem. B* **107**, 1913-1923.

18. Anwander, E. H. S., Probst, M. M. and Rode, B. M. (1990). The influence of Li^{+}, Na^{+}, Mg^{2+}, Ca^{2+}, and Zn^{2+} ions on the hydrogen-bonds of the Watson-Crick base-pairs, *Biopolymers* **29**, 757-769.

19. Šponer, J., Burda, J. V., Sabat, M., Leszczynski, J. and Hobza, P. (1998). Interaction between the guanine-cytosine Watson-Crick DNA base pair and hydrated group IIa (Mg^{2+}, Ca^{2+}, Sr^{2+}, Ba^{2+}) and group IIb (Zn^{2+}, Cd^{2+}, Hg^{2+}) metal cations. *J. Phys. Chem. A* **102**, 5951-5957.

20. Munoz, J., Šponer, J., Hobza, P., Orozco, M. and Luque, F. J. (2001). Interactions of hydrated Mg^{2+} cation with bases, base pairs, and nucleotides. Electron topology, natural bond orbital, electrostatic and vibrational study. *J. Phys. Chem. B* **105**, 6051-6060.

21. Lippert, B. ed (1999). *Cisplatin: chemistry and biochemistry of a leading anticancer drug.* Wiley-VCH, Weinheim, New York, Chichester.

22. Šponer, J. E., Leszczynski, J., Glahe, F., Lippert, B. and Šponer, J. (2001). Protonation of platinated adenine nucleobases. Gas phase vs condensed phase picture. *Inorg. Chem.* **40**, 3269-3278.

23. Šponer, J., Šponer, J. E., Gorb, L., Leszczynski, J. and Lippert, B. (1999). Metal-stabilized rare tautomers and mispairs of DNA bases: N6-metalated adenine and N4-metalated cytosine. Theoretical and experimental views. *J. Phys. Chem. A* **103**, 11406-11413.

24. Burda, J. V., Šponer, J. and Leszczynski, J. (2000). The interactions of square platinum(II) complexes with guanine and adenine: a quantum-chemical ab initio study of metalated tautomeric forms. *J. Biol. Inorg. Chem.* **5**, 178-188.

25. Šponer, J. E., Sanz Miguel, P. J., Rodríguez-Santiago, L., Erxleben, A., Krumm, M., Sodupe, M., Šponer, J. and Lippert, B. (2004). Metal-mediated deamination of cytosine:experiment and DFT calculations, *Angew. Chem. Int. Ed.* **43**, 5396-5399.

26. Burda, J. V., Šponer, J. and Leszczynski, J. (2001). The influence of square planar platinum complexes on DNA base pairing. An ab initio DFT study. *Phys. Chem. Chem. Phys.* **3**, 4404-4411.

27. Burda, J. V., Šponer, J., Hrabáková, J., Zeizinger, M. and Leszczynski, J. (2003). The influence of N₇ guanine modifications on the strength of Watson-Crick base pairing and guanine N₁ acidity: comparison of gas-phase and condensed-phase trends. *J. Phys. Chem. B* **107**, 5349-5356.

28. Siegel, R. K. O., Freisinger, E. and Lippert, B. (2000). Effects of N7-methylation, N7-platination, and C8-hydroxylation of guanine on H-bond formation with cytosine: platinum coordination strengthens the Watson-Crick pair. *J. Biol. Inorg. Chem.* **5**, 287-299.

29. Schmidt, K. S., Reedijk, J., Weisz, K., Janke, E. M. B., Šponer, J. E., Šponer, J. and Lippert, B. (2002). Loss of Hoogsteen pairing ability upon N1 adenine platinum binding. *Inorg. Chem.* **41**, 2855-2863.

30. Burda, J. V., Zeizinger, M. and Leszczynski, J. (2005). Hydration process as an activation of trans- and cisplatin complexes in anticancer treatment. DFT and ab initio computational study of thermodynamic and kinetic parameters. *J. Comput. Chem.* **26**, 907-914.

31. Costa, L. A. S., Rocha, W. R., De Almeida, W. B. and Dos Santos, H. F. (2004). The solvent effect on the aquation processes of the cis-dichloro(ethylenediammine)platinum(II) using continuum solvation models. *Chem. Phys. Lett.* **387**, 182-187.

32. Elizondo-Riojas, M. A. and Kozelka, J. (2001). Unrestrained 5 ns molecular dynamics simulation of a cisplatin-DNA 1,2-GG adduct provides a rationale for the NMR features and reveals increased conformational flexibility at the platinum binding site. *J. Mol. Biol.*, **314**, 1227-1243.

33. Baik, M.-H., Friesner, R. A. and Lippard, S. J. (2003). Theoretical study of cisplatin binding to purine bases: Why does cisplatin prefer guanine over adenine? *J. Am. Chem. Soc.* **125**, 14082-14092.

34. Baik, M.-H., Friesner, R. A. and Lippard, S. J. (2003). cis-{Pt(NH₃)₂(L)}²⁺/⁺ (L = Cl, H₂O, NH₃) binding to purines and CO: Does π-back-donation play a role? *Inorg. Chem.* **42**, 8615-8617.

35. Burda, J. V. and Leszczynski, J. (2003). How strong can the bend be on a DNA helix from cisplatin? DFT and MP2 quantum chemical calculations of cisplatin-bridged DNA purine bases. *Inorg. Chem.* **42**, 7162-7172.

36. Deubel, D. V. (2002). On the competition of the purine bases, functionalities of peptide side chains, and protecting agents for the coordination sites of dicationic cisplatin derivatives. *J. Am. Chem. Soc.* **124**, 5834-5842.

37. Deubel, D. V. (2004). Factors governing the kinetic competition of nitrogen and sulfur ligands in cisplatin binding to biological targets. *J. Am. Chem. Soc.* **126**, 5999-6004.

38. Lau, J. K.-C. and Deubel, D. V. (2005). Loss of ammine from platinum(II) complexes: implications for cisplatin inactivation, storage and resistance. *Chem. - Eur. J.* **11**, 2849-2855.

39. Baik, M.-H., Friesner, R. A. and Lippard, S. J. (2002). Theoretical study on the stability of N-glycosyl bonds: Why does N7-platination not promote depurination? *J. Am. Chem. Soc.* **124**, 4495-4503.

40. Šponer, J., Šponer, J. E. and Leszczynski, J. (2000). Cation - π and amino-acceptor interactions between hydrated metal cations and DNA bases. A quantum chemical view. *J. Biomol. Struct. Dyn.* **17**, 1087-1096.

41. Magnuson, E. C., Koehler, J., Lamm, G. and Pack, G. R. (2002). $Mg(H_2O)_6^{2+}$ - π (cytosine) interactions in a DNA dodecamer. *Int. J. Quantum Chem.* **88**, 236-243.

42. Petrov, A. S., Lamm, G. and Pack, G. R. (2005). Calculation of the binding free energy for magnesium-RNA interactions. *Biopolymers*, **77**, 137-154.

43. Bashford, D. and Karplus, M. (1990). pK_as of ionizable groups in proteins - atomic detail from a continuum electrostatic model. *Biochemistry*, **29**, 10219-10225.
44. Cheatham, T. E. and Young, M. A. (2000). Molecular dynamics simulation of nucleic acids: Successes, limitations, and promise. *Biopolymers*, **56**, 232-256.
45. Orozco, M., Perez, A., Noy, A. and Luque, F. J. (2003). Structure, recognition properties, and flexibility of the DNA-RNA hybrid. *Chem. Soc. Rev.*, **32**, 350-364.
46. Reblová, K., Špačková, N., Štefl, R., Csaszar, K., Koča, J., Leontis, N. B., Šponer, J. (2003). Non-Watson-Crick basepairing and hydration in RNA motifs: Molecular dynamics of 5S rRNA loop E. *Biophys. J.*, **84**, 3564-3582.
47. Halgren, T. A. and Damm., W. (2001). Polarizable force fields. *Curr. Opin. Struct. Biol.*, **11**, 236-242.
48. Reblová, K., Špačková, N., Koča, J., Leontis, N. B., Šponer, J. (2004). Long-residency hydration, cation binding, and dynamics of loop E/helix IV rRNA-L25 protein complex. *Biophys. J.*, **87**, 3397-3412.
49. Štefl, R., Cheatham, T. E., Špačková, N., Fadrná, E., Berger, I., Koča, J., Šponer, J. (2003). Formation pathways of a guanine-quadruplex DNA revealed by molecular dynamics and thermodynamic analysis of the substates. *Biophys. J.*, **85**, 1787-1804.
50. Reblová, K., Špačková, N., Šponer, J. E., Koča, J., Šponer, J. (2003). Molecular dynamics simulations of RNA kissing-loop motifs reveal structural dynamics and formation of cation-binding pockets. *Nucleic Acids Res.*, **31**, 6942-6952.

Chapter 16

PROTON TRANSFER IN DNA BASE PAIRS
Potential mutagenic processes

J. Bertran,[1] L. Blancafort,[2] M. Noguera,[1] M. Sodupe[1]

[1]Departament de Química,Universitat Autònoma Barcelona, Bellaterra 08193 Spain; [2]Institut de Química Computacional, Universitat de Girona, Girona E-17071 Spain

Abstract: This chapter reviews the theoretical studies performed on single and double proton transfer reactions in guanine-cytosine and adenine-thymine base pairs. The influence of excitation, ionization, protonation and metal cation binding of base pairs in these processes is explored from the analysis of the potential energy surfaces.

Key words: DNA, proton transfer, oxidation, excited states, metal cation interaction

1.   INTRODUCTION

Interactions between hydrogen-bonded nucleobases of DNA and RNA are responsible for the storage and transfer of biological information. In the double helix model,[1] these interactions are thought to be in terms of the keto-amine tautomers, the canonical forms, but, as suggested by Watson and Crick, the genetic code may be perturbed by the presence of nucleotide bases appearing in "rare" tautomeric forms. A few years after Watson and Crick proposed their double helix model, Lowdin[2] pointed out that "rare" tautomers may be produced in pairs by proton rearrangement within the hydrogen bonds connecting a base pair, and that replications involving such rearranged base pairs would perpetuate genetic coding alterations. Nevertheless, this spontaneous double proton transfer rearrangement must be a rare event in neutral systems.

Larger effects may be obtained by means of induced proton transfer processes, caused by the fact that the two bases within the pair have obtained unequal charges.[2] When one of the bases takes an extra charge, the

411

J. Šponer and F. Lankaš (eds.), Computational Studies of RNA and DNA 411–432.
© 2006 *Springer.*

probability of a single proton tunneling increases due to the changes in the energy profile. This way the genetic information is lost and the deletion of a base in one strand leads to an irreversible mutation. An unequal charge between bases may be obtained through electron donor-acceptor reactions with other molecules, by ionization, or by the addition of a proton to one of the purine bases. In addition the double-well potential may be disturbed through additional electrostatic potentials or by UV-radiation.[2-4] Rein and Ladik[5] found that a charge-transfer excited state of guanine-cytosine base pair (GC) seemed relatively favorable for a proton rearrangement.

For a long time it was not possible to check these ideas theoretically due to the size of the system. Nowadays, methodology developments and the increase of computational power have allowed testing these hypotheses by means of accurate quantum chemical calculations. Nevertheless, a certain caution must be taken in some particularly difficult cases such as excited states, radical cations or metal-ligand systems. Moreover, one should be careful when comparing theoretical results of simplified models with experimental results, since for charged systems electrostatic effects are largely screened by the environment. One should keep in mind that the selected procedure always results from a compromise between the computational cost of the calculations and the accuracy of the level of theory plus the complexity of the system considered. In all cases, it is important to check the reliability of the level of theory chosen by performing a calibration study in small models.

In this chapter, the spontaneous double proton transfer in neutral base pairs, and the induced single proton transfer in excited and ionized states, and in protonated and cationized nucleobases will be explored from the analysis of the potential energy surfaces.

2. NEUTRAL BASE PAIRS

Many theoretical studies have been devoted to check Lowdin's hypotheses. First studies[6-12] considered the double proton transfer process on the ground state of neutral pairs. However, due to of the size adenine-thymine (AT) and guanine-cytosine (GC) base pairs, semiempirical[6-9] and low level initio[10-12] methods were used. In addition, these studies, were performed using fixed geometries for the monomers during the proton transfer process. Consequently, both the single and double proton transfer reactions were found to be too unfavorable. More recent studies,[13-15] in which the geometries were fully optimized, found smaller energy barriers.

Florian and coworkers[14] studied the double proton transfer reaction in AT within the Hartree-Fock (HF) approach in conjunction with MINI-1 and

MIDI-1 basis. At the HF/MINI-1 level a barrier of 9.7 kcal/mol was calculated for the lowest energy reaction path from the canonical AT structure to the "rare" tautomeric A^*T^* one. They also showed that the "rare" tautomer A^*T^* lies below the barrier by only 0.22 kcal/mol which makes it rather metastable. Furthermore, the dissociation of A^*T^* was found to be 10 kcal/mol higher than that of AT and no minima corresponding to the ion pair structures A^+T^- and A^-T^+, produced by a single proton transfer reaction, were located. Florian and Leszczynski[15] studied the whole potential energy surface corresponding to several proton transfer processes in GC base pair at the HF level. At this level of theory, both the transition structure of the N1-N3 and O6-N4 double proton transfer reaction and the stationary points of the two step mechanism were located on the potential energy surface. Among them, the two-step mechanism was computed to be the most favorable process, the transition state structure lying 3.2 kcal/mol higher than the ion pair intermediate. However, this situation changed dramatically when electron correlation was considered by single point calculations at the MP2 level, since the one-step transition state became 4.0 kcal/mol more stable that the ion pair intermediate. Furthermore, this intermediate disappeared at the MP2 level of calculation, since both transition states, before and after the intermediate, became lower in energy than the ion pair structure.

Figure 16-1 presents the double proton transfer energy profiles for GC[16,17] and AT base pairs. As expected, hydrogen bond distances at the transition structure are shortened compared to the reactant, and the $O_6...H-N_4$ transfer is more advanced than the $N_1-H...N_3$ one. The pathway of the double proton transfer is concerted and asynchronous. All attempts to optimize the single proton transferred ion pair collapsed to the initial non transferred structure, showing that electron correlation is essential to properly describe the topology of the potential energy surface. The B3LYP energy profile is very similar to the one presented previously[15] from MP2 single point calculations and to those recently obtained by performing full optimizations at the B3LYP/6-31G(d), MP2/6-31G(d) and B3LYP/6-311++G(d,p) levels of theory.[18, 19]

At present, all studies agree that the single proton transfer reaction is less favorable than the double one, because the single transfer process implies a charge separation when forming the ion-pair complex, while in the double proton transfer process the electroneutrality is maintained. An interesting result is that the barrier of the double proton transfer in an isolated GC base pair does not change significantly when a larger system is considered, that is, when the central GC base pair is embedded in a DNA double stranded GCGCG helix using mixed quantum/classical methods.[19]

For isolated GC, the ΔG_{298} for the double proton transfer reaction is 9.8 kcal/mol.[17] The equilibrium between the canonical base pair and the double proton transferred one, with a calculated equilibrium constant of 6.2×10^{-8}, will be largely displaced to reactants and so the production of "rare" tautomers by this process is expected to occur rarely. Furthermore, solvent stabilizes the canonical base pair more than the "rare" tautomer one,[17, 18] so that this equilibrium is disfavored in solution. In contrast to this, the solvent stabilizes the ion pair coming from a single proton transfer process, in such a way that the intermediate may exist in solution and play a role in mutagenic processes. It should be noted that in solution the N_1 position of guanine presents the higher pK_a[20] while the proton affinity is maximum at the N_3 site of cytosine.[21]

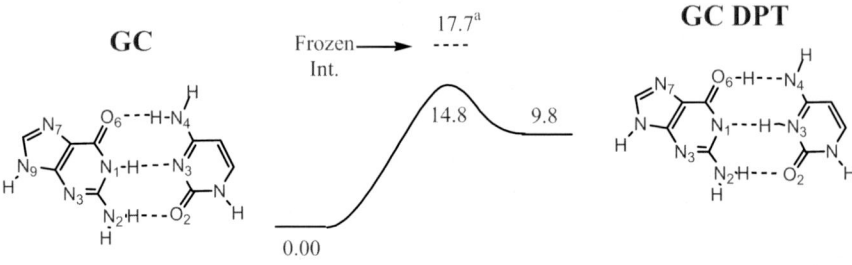

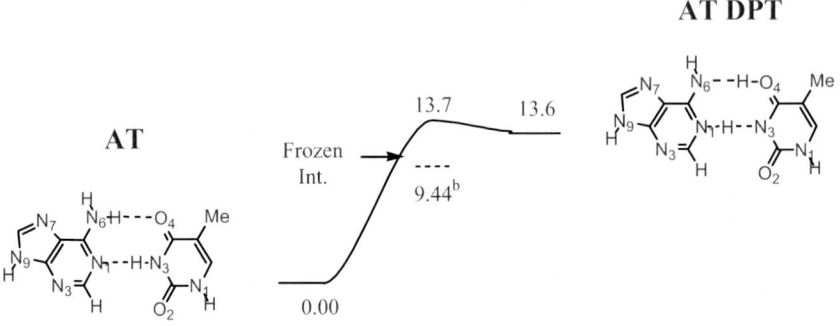

Figure 16-1. B3LYP energy profile for the double proton transfer reaction in neutral guanine-cytosine (GC) and adenine-thymine (AT). In kcal/mol. [a] H-N3 value fixed at 1.049 Å. [b] N1-H fixed at 1.095 Å.

Figure 16-1 also shows the energy profile of the double proton transfer process in AT base pair at the B3LYP/6-311++G(d,p).[22] The main difference with respect to GC is that the double proton transfer product is found to correspond to a very shallow minimum, in agreement with previous work at the HF level.[14] In the transition structure the proton transfer is more advanced in $N_1...H-N_3$ than in $N_6-H...O_4$, and no ion pair intermediate on the potential energy surface is located. Thus, the process is concerted and asynchronous. Similar to GC, the single proton transferred ion pair intermediate of AT is not stable and thus, an estimation of its relative energy has been obtained by freezing the N_1-H distance. This intermediate is only localized at the HF level,[23, 24] which again shows that electron correlation is essential for a correct description of the topology of the potential energy surface. On the other hand, it has been shown, at the B3LYP and MP2 levels, that the shallow minimum found for the double proton transferred product is not real when Gibbs free energy values are considered.[18] Furthermore, AT dimerization implies a lower association energy than that between the A* and T* "rare" tautomers.[18] Thus, the double proton transfer reaction in AT base is not expected to have mutagenic effects. The solvent does not seem to have any effect in this process,[25] but it can stabilize the ion pair of the single proton transfer reaction in such a way that this last process must not be disregarded.

Rare forms of DNA bases can also be obtained by means of intramolecular proton transitions. The biological importance of such a proton transfer may become apparent during the catalytic incorporation of new nucleotides into the growing DNA strand when a "rare" form of a nucleic acid base forms a pair with an incorrect base. In adenine and thymine the corresponding equilibria are shifted towards the canonical forms more significantly than in guanine and cytosine[18] for which the "rare" tautomeric forms are observed experimentally.[26, 27] Theoretical calculations have shown that in guanine[20, 28, 29] and cytosine[30-33] the equilibrium between different tautomers highly depends on whether the solvent is taken into account or not. Furthermore, along with the thermodynamics of the process, one should also consider the kinetics of the proton transfer reaction, since by the time needed for the synthesis of new DNA, the equilibrium between tautomers may not be reached. At this respect several theoretical works[34-38] have shown that the potential energy barrier for the intramolecular proton transfer reaction decreases dramatically when the process takes place through one or more water molecules, the water molecules playing a similar role than the second nucleobase in base pairs double proton transfer processes.

3.   EXCITED BASE PAIRS

The exposure of DNA to UV radiation lies behind the interest in hydrogen transfer processes in electronically excited nucleobase pairs. This reaction is one of the possible responses of DNA to electronic excitation and is generally assumed to be energetically easier than in the ground state.

Experimental evidence on excited-state hydrogen transfer in isolated base pairs has recently appeared.[39, 40] There are many quantum chemical calculations on models such as the 7-azaindole[41-44] and 2-aminopyridine dimers[45, 46] (a discussion on the validity of these models can be found in Ref. [45]), but here we centre on the calculations on base pairs.[47-49] However one should not forget that the role of excited states of single bases or base pairs in the excited-state behavior of DNA is a matter of discussion.[50]

We focus on single tautomerization because theoretical evidence indicates that it is favored in the excited state with respect to the double tautomerization. Single tautomerization is also the process that has been detected experimentally, although the double tautomerization would not be easily detected in the experiments referred to here, since they are based on detection by mass spectroscopy.

To put the computations in context, the hydrogen transfer is described as a (coupled or sequential) proton and electron transfer. The two main possible pathways are presented with a semiquantitative correlation diagram (Figure 16-2) which considers the ground and two excited states for the canonic GC base pair and its single proton transferred form (GCSPT). The diagram for AT is similar. The two excited states are a locally excited (LE) one, where the local excitation only affects one of the bases, and a charge transfer (CT) one that involves an electron transfer between the two monomers (the number of electrons on each base is shown in brackets). Calculations show that the LE and CT states are the lowest ones for GC and GCSPT forms, respectively.[48, 49, 51] Two possible pathways, called CT path and adiabatic path here, are shown in Figure 16-2. Thus, dashed lines correlate the non-transferred structures with the transferred ones, and the hydrogen transfer can be induced "directly" by excitation to the CT state of the canonic form, which should cause an almost spontaneous proton transfer (i.e. sequential electron and proton transfer). On the other hand, the LE state of GC is connected to the CT state of GCSPT by an adiabatic hydrogen transfer pathway going through a transition structure, associated to an energy barrier (coupled electron and proton transfer). Only the cytosine LE and the guanine to cytosine CT states are shown in the diagram because this state is the lowest CT state of the pair, in agreement with the oxidation and reduction potentials of the bases. However the same picture comes up from interaction between the guanine LE and the cytosine to guanine CT states.

The possibility of excited state hydrogen transfer in the base pairs can be considered from two points of view: for the adiabatic case the height of the barriers (relative to the excited state minimum of the canonical pair) can be estimated, while the points of interest for the CT path are the accessibility of the charge transfer state, i.e. its excitation energy and oscillator strength, and the existence of the almost barrierless path suggested by Figure 16-2.

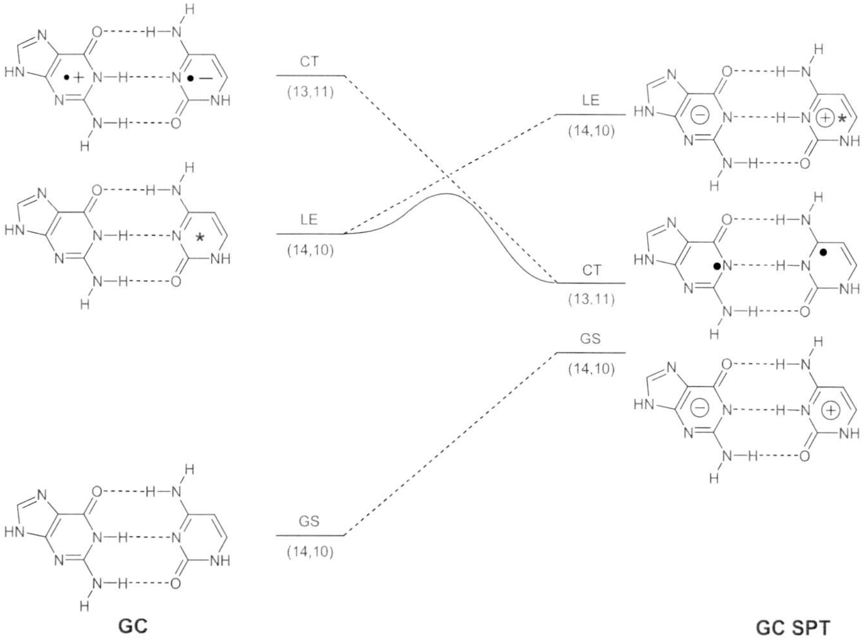

Figure 16-2. Semiquantive correlation diagram for excited state hydrogen transfer (proton and electron transfer) in the GC base pair.

Optimization of the relevant structures and calculation of the barrier heights for the adiabatic path has been the subject of an earlier CIS/6-31G study.[49] For the AT pair, two transition structures for the concerted, asynchronous double hydrogen transfer were located, with the excitation localized on either of the bases. The analogous structure was found for the GC pair, with excitation localized on the guanine moiety. In these cases, the barriers were 14–22 kcal/mol. The single hydrogen transfer, with a barrier of approximately 20 kcal/mol, was only found for the cytosine LE state, but the corresponding SPT intermediate was only a shallow minimum and the barrier for the second hydrogen transfer was less than 1 kcal/mol. However the SPT intermediate does not correspond to the CT state shown in Figure 16-2, and the reported path is not the adiabatic one shown in the Figure. Moreover, a CT-SPT structure for GC is reported in the paper as the lowest energy one on the excited state. This suggests that the energetically favored hydrogen transfer actually follows the adiabatic mechanism sketched in Figure 16-2. However the corresponding transition structure could not be located, presumably because of the limitations of the CIS method. Similar results are described for AT, and it appears that the role of the CT-SPT structures and the associated coupled proton and electron path requires further clarification.

A more recent study for the GC pair[48] uses a better level of theory, namely CAS-PT2//CIS (single point CAS-PT2 calculations on geometries determined at the CIS level) and focuses on the role of the guanine to cytosine CT state. The authors have traced an almost barrierless decay coordinate from the canonical base pair to the SPT intermediate. This corresponds to the dashed line of Figure 16-2 and is an exothermic process by about 0.8 eV or 22 kcal/mol. Moreover, the results indicate that for the canonical form there are distinct minima for the LE and CT states separated by a barrier. However this could not be confirmed due to limitations of the methodology. The adiabatic path was not considered in this study.

The vertical excitation energy of the CT state and its oscillator strength, which gives an estimate of the probability of populating the corresponding state by excitation, can be used to assess the experimental relevance of the CT path. The reported CAS-PT2 excitation energy is 4.75 eV, which lies very close to the energy of the two guanine- and cytosine-localized excited states (4.35 and 4.62 eV). However, the oscillator strength, calculated from the CASSCF transition dipole moment and the CAS-PT2 excitation energy, is only 0.003,[48] which predicts a small probability of populating the CT path of Figure 16-2. In contrast to this, a different CIS/6-31++G** study gives a significant probability with a reported oscillator strength of 0.27.[51] Further clarification of this point would be important for a correct interpretation of gas phase experiments with base pairs, although the relevance for the DNA case is unclear in view of the uncertainty about the role of the single base or base pair chromophores in the stacked polymer.

For the SPT product (Figure 16-2), transfer of the second hydrogen is possible, but the barrier has not been reported. However, at the geometry of the biradical SPT minimum, the state is close in energy to the ion pair ground state.[45, 47, 48] A near lying conical intersection between the two states has been found, separated from the minimum by a negligible barrier. It provides a path to return to the ground state. MP2/6-31G* calculations indicate that there is no stable ground state intermediate in the SPT region,[18] and one can expect that the base pair will decay further, either back to the canonical form or further to the double tautomer.

The SPT structure also provides a connection between the singlet and triplet excited state manifolds. It is a "pure" biradical coincident with the SPT minimum on the triplet surface. In fact triplet states must be relevant in DNA in view of the known photoreactivity of the thymine and cytosine triplets,[52] and a path to access the triplet state from the singlet excited one has been calculated for cytosine.[53] The barrier for the adiabatic hydrogen transfer in the GC triplet (Figure 16-2) has been calculated at the CAS-PT2//CASSCF level.[47] There are two different paths which start from the cytosine- and guanine-locally excited triplet minima and have sizeable barriers of 17 kcal/mol and 19 kcal/mol, respectively, and the process is exothermic by approximately 11 kcal/mol. The transfer of the second hydrogen to form the double tautomer has a large barrier of 25 kcal/mol in the triplet state. Therefore intersystem crossing to the degenerate singlet state, or other reactive processes of the triplet biradical which have not been explored, will be the favored reactive routes.

The resulting overall picture is that in the singlet excited state, transfer of a single hydrogen will lead to the SPT form, followed preferentially by its decay to the ground state and transfer of the second hydrogen or return to the canonical pair. Moreover, the SPT biradical connects the singlet and triplet excited states and makes intersystem crossing possible.

4.   IONIZED BASE PAIRS

One-electron oxidations in DNA are important processes due to their connection to the DNA damage caused by ionizing radiation[54] or oxidizing agents.[55, 56] One of the induced reactions is the intermolecular proton transfer in the base pair, which has been discussed to stop the migration of holes in DNA.[57-60] Because of that, the reactivity of base pair radical cations has received considerable attention from a theoretical point of view.[55, 57, 61-68]

The ground electronic state of $GC^{\cdot+}$ and $AT^{\cdot+}$ is $^2A''$, the open shell orbital being of π nature.[64] Because purines present lower ionization energies than pyrimidines,[69] the oxidation of GC and AT is mainly localized at G and A, respectively. Since ionized monomers increase their acidity, those hydrogen bonds in which G or A act as proton donor strengthen while those in which they act as proton acceptor weaken. Nevertheless, both for GC and AT, the dimerization energy increases significantly upon ionization[64] due to the presence of a positive charge, which largely enhances the electrostatic interaction. For $GC^{\cdot+}$ the computed D_e value (44.3 kcal/mol) is about 19 kcal/mol larger than for the neutral system (25.5 kcal/mol). For AT the increase of the binding energy upon ionization is of 10 kcal/mol, from 12.3 kcal/mol for neutral AT to 22.3 for $AT^{\cdot+}$ radical cation.[64]

The energy profiles of the single proton transfer process in ionized GC and AT are shown in Figure 16-3. Strong and weak hydrogen bonds are indicated with the letters s and w, respectively.

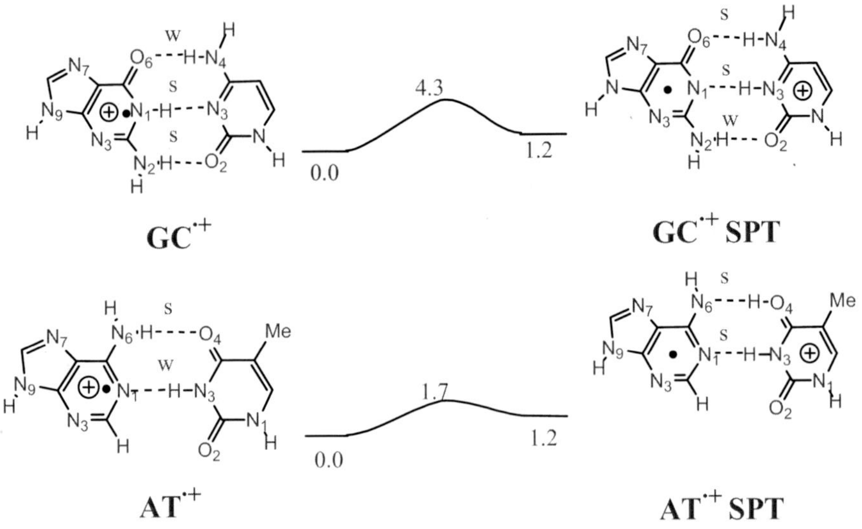

Figure 16-3. B3LYP energy profile for the single proton transfer reaction in GC and AT radical cations. In kcal/mol.

GC has three hydrogen bonds: two in which G acts as proton donor and one in which it acts as proton acceptor. Thus, upon ionization the N_1-N_3 and N_2-O_2 strengthen and the O_6-N_4 hydrogen bond weakens; that is, $GC^{\cdot+}$ shows a w-s-s pattern. The N_1-N_3 proton transfer moves the positive charge to C and so, the two hydrogen bonds in which C acts as proton donor strengthen and we get a similar s-s-w pattern with two neighbor hydrogen bonds. In contrast to neutral systems, the proton transfer reaction is much more favorable, the relative energy of $GC^{\cdot+}$ and $GC^{\cdot+}SPT$ being only 1.2 kcal/mol.[64] This is not surprising considering that i) the proton transfer reaction does not imply a creation of charges but a transfer of a positive charge, ii) the two centers involved in the transfer present similar proton affinities and iii) both the sequence and number of strong and weak hydrogen bonds is similar in $GC^{\cdot+}$ and $GC^{\cdot+}SPT$. The double proton transfer structure would have again the positive charge at the G moiety, but this structure is not found to be stable because now the weak hydrogen bond (N_1-N_3) is the central one; that is, it presents an alternate pattern (s-w-s), which does not allow a simultaneous optimal interaction of the three hydrogen bonds.

For $AT^{\cdot+}$ we have one strong and one weak hydrogen bond. After the single proton transfer, the positive charge moves to T and thus, both hydrogen bonds become strong. As for GC, the non-transferred and the single proton transferred structures, $AT^{\cdot+}$ and $AT^{\cdot+}SPT$, lie close in energy (1.2 kcal/mol) even though in this case the proton affinity of O_4 is about 25 kcal/mol smaller than that of N_6. Note, however, that in this case the proton transfer implies a change of pattern, from s-w to s-s. The double proton transfer reaction leads to a situation similar to the initial one but with the two hydrogen bonds reversed. This structure is about 7 kcal/mol less stable than $AT^{\cdot+}$ due to the stability of the two tautomers involved in the pairing and the strength of the interaction.

Both for $GC^{\cdot+}$ and $AT^{\cdot+}$ radical cations the single proton transfer reaction presents a low energy barrier (4.3 and 1.6 kcal/mol, respectively),[64] the $AT^{\cdot+}$ behaving more as a strong low barrier hydrogen bond than $GC^{\cdot+}$. In contrast the double proton transfer reaction is not a favorable process. It has to be mentioned that recent experimental studies have found evidences for the proton transfer in $GC^{\cdot+}$.[39, 70] Finally, it should be mentioned that the proton transferred structures $GC^{\cdot+}SPT$ and $AT^{\cdot+}SPT$ present the radical character at the purine moiety. These species have been invoked to play an important role in DNA damage caused by ionization.[71] Thus, in addition to the potentially mutagenic proton transfer reaction, other unwanted important processes can be induced from the formation of purine radicals.

5.   PROTONATED BASE PAIRS

Although DNA is immersed in a physiological pH solution it is possible to find local regions of low pH on the DNA surface.[72] Therefore, in addition to acid-catalyzed reactions of the DNA backbone, basic centers in the minor and major groves of the double helix are sensitive to protonation. Similar to oxidation, protonation of the base pair introduces a positive charge in the system. Because of that, base pairing is reinforced and significant changes on hydrogen bond distances are observed.[16, 17, 73] The general trend is that those hydrogen bonds in which the moiety holding the charge acts as proton donor are strengthened while those in which it acts as proton acceptor are weakened. These effects become especially important when the positive charge is delocalized at the six membered ring of purine; that is, upon protonation of N_3 of guanine and adenine. Moreover, due to the presence of an extra charge, the single proton transfer reaction no longer generates an ion-pair structure, as in the neutral system, and only displaces the charge from one moiety to another. Protonation may occur at the main basic centers of each base pair, preferably at the purine moiety. For GC protonation may occur at the N_7, O_6 and N_3 centers of G, while for AT basic sites are found at N_7 and N_3 of A and O_2 and O_4 of T. These basic sites have been considered for protonation when studying the different proton transfer processes (see Figure 16-4).

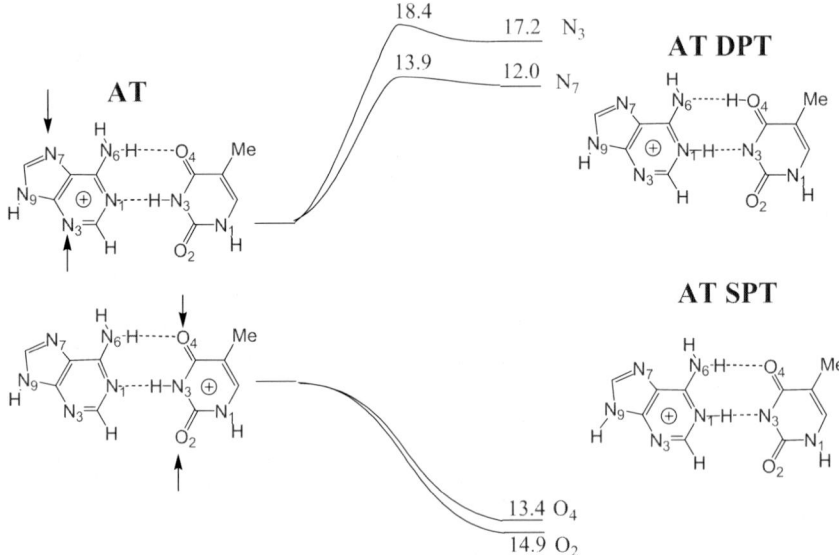

Figure 16-4. B3LYP energy profiles for single and double proton transfer processes in protonated GC and AT base pairs. In kcal/mol.

As mentioned earlier, the single proton transfer product for neutral GC is not found as a minimum on the potential energy surface, the frozen intermediate lying 17.7 kcal/mol above reactants. For protonated systems, however, the single proton-transferred products are largely stabilized.

In the case of $H^+G_{N3}C$, stabilization leads to a near-degeneracy for reactants and products, the reaction energy and energy barrier of this process being only 0.6 kcal/mol and 4.6 kcal/mol, respectively. In fact, among the three different protonated GC systems, $H^+G_{N3}C$ shows the largest pairing energy increase (about 30% more than $H^+G_{N7}C$ and $H^+G_{O6}C$) and the shorter hydrogen bond distances. Effects induced by protonation at the O_6 and N_7

sites are less pronounced, the reaction energies being 2.0 and 4.2 kcal/mol, and the energy barriers 6.3 and 6.5 kcal/mol, respectively. These results were to be expected considering that protonation at the 6-membered ring has a larger influence on the acidity of the atoms involved on the hydrogen bonds than protonation at the five-membered ring of the purine.

For AT the situation is somewhat more complex because basic centers are not only found in A but also in T. Moreover, no single proton transfer products are found for A protonated systems. These differences are mainly due to the poorer basicity of O_4 of T compared to that of the imino nitrogen atom of A. In fact, the double proton transfer transition state lies very close in energy to the products, indicating the high instability of such products.

When protonation occurs at the T moiety important differences are observed. Since N_1 adenine's basicity is much larger than that of thymine's O_2 and O_4, the N_3 proton of T spontaneously transfers to N_1 of A, and no minimum is found for the non-transferred structure.

In summary, the GC base pair is sensitive to locally acid pH because of the feasibility of single proton transfer process when the base pair is protonated at the three basic centers. Products arising from such processes are found to be surprisingly stable and can be involved in mutagenic phenomena. For AT, however, the picture is different. If A is the protonated moiety the single proton transfer reaction does not occur and the double proton transfer process would hardly take place considering the energetics of the reaction. On the other hand, if T is the protonated species the N_3-N_1 proton transfer reaction becomes exothermic without energy barrier; that is, the process is spontaneous.

6. METAL CATION BINDING

Another possible way to induce intermolecular proton transfer reactions in Watson-crick base pairs is by the interaction of metal cations with the purine bases, which would stabilize the formed ion pair. Although the primary influence of metal cations is to neutralize the negative charges on the backbone phosphate groups, they can also interact, to a lesser extent, with the DNA bases, modifying the strength and structure of the H-bonded pairs. Several theoretical and experimental works have determined the metal cation affinity of the common DNA and RNA nucleobases.[74] The influence of the metal cation (naked and hydrated) on the structure and binding energy of the base pair has also been considered from a theoretical point of view.[75-79] However, few works have considered the influence of the metal cation on intermolecular proton transfer reactions.[79-81]

Among the various binding sites, the N_7 of A and the N_7 and O_6 of G are the preferred ones. For A, metal cation binding to N_7 forces the amino group to be highly pyramidal and rotated,[74] which induces important geometry distortion of the base pair. Because of that, in this section we will only expose the results obtained for GC interacting with three metal cations: $Cu^{2+}(d^9)$, an open shell metal cation with important electrostatic and oxidant effects, and $Cu^+(d^{10})$ and Ca^{2+}, two closed shell metal cations for which M-GC interaction is mainly due to electrostatic and polarization effects.

Metal cation binding at N_7 of G enhances the GC interaction energy.[75, 79] In addition to the electrostatic and polarization effects, this is also due to the enhancement of the N_1-H and N_2-H acidity by metal cation coordination.[82, 83] Consequently, the N_1-N_3 and N_2-O_2 hydrogen bonds strengthen whereas the O_6-N_4 weakens. These effects are more pronounced for the divalent cations that for the monovalent one.[81] However, comparison between the two divalent cations shows that the strengthening of the hydrogen bonds is significantly larger for Cu^{2+}. This is due to the oxidant character of Cu^{2+} which leads to the formation of GC radical cation.[81]

Figure 16-5 shows the energy profiles of the N_1-N_3 proton transfer reaction for GC coordinated to Cu^+, Ca^{2+} and Cu^{2+}. The products resulting from double proton transfer reaction (N_1-N_3 and N_4-O_6) were not found to be stable and any attempt to optimize them collapsed to the single proton transferred structure.[81] Not surprisingly, the ion-pair structure derived from the single proton transfer reaction is stabilized by metal cation coordination, the reaction energy decreasing from 17.7 kcal/mol for the uncationized system to 5.1 for Cu^+, -3.8 kcal/mol for Ca^{2+} and -10.1 kcal/mol for Cu^{2+}. Moreover, for the dications, the proton transferred asymptote lies much lower in energy due to the important electrostatic repulsion between the two positively charged fragments. Although both divalent dications show a similar behavior, the process is especially efficient for Cu^{2+} due to its oxidant character, which leads to the formation of a Cu^+G^+C complex with the open shell orbital localized at guanine.

These results correspond to the gas phase situation, and they can be quite different from the ones in real living systems since in these cases the metal cation is solvated by water molecules and interacting with the negatively charged backbone. Therefore, the electrostatic effects will be largely screened, which will make the proton-transfer reaction less favourable. In fact, calculations for hydrated Mg^{2+} cation binding at N_7 of guanine show that the canonical and ion-pair structures are equal in energy. However, if the negative phosphate is taken into account the ion pair lies 12.9 kcal/mol higher.

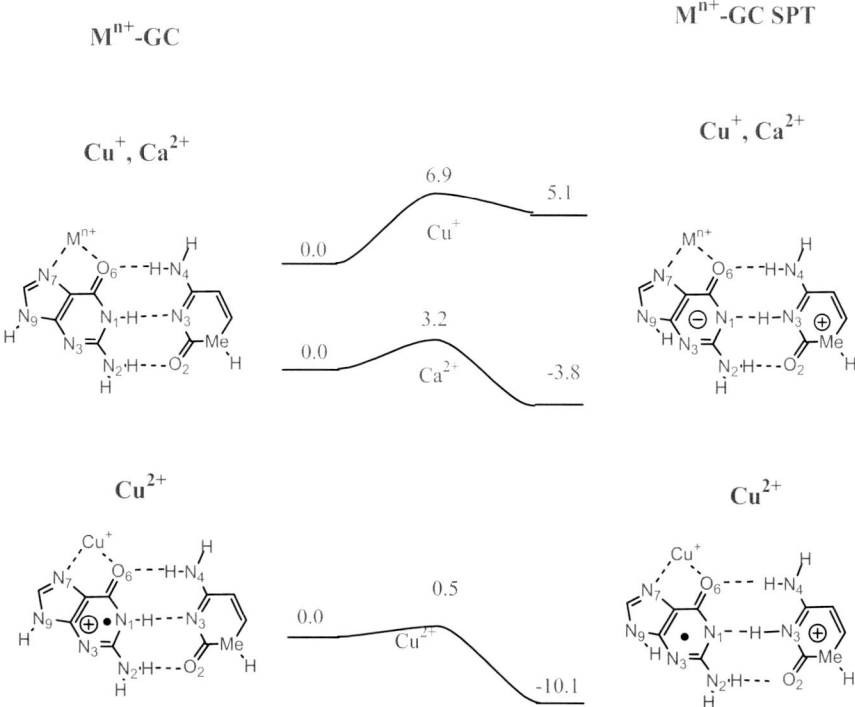

Figure 16-5. B3LYP energy profile for metal cationized GC base pair. In kcal/mol.

Focusing on Cu^{2+}, the effect of the hydrated cation coordinated to N_7 has been investigated. It has been found that the degree of metal oxidation depends on the coordination environment. For the pentacoordinated $(H_2O)_4Cu^{2+}GC$ system the population analysis shows that copper behaves as Cu^{2+}. The spin density mainly lies on this atom, as expected for a d^9 open-shell cation. However, in the tetracoordinated $H_2O(H_2O)_3Cu^{2+}GC$ complex, the spin density is highly delocalised between the metal cation and G, and for situations in which the metal cation presents a lower coordination number the spin density is mainly localized at G. Thus, the ability of Cu^{2+} to oxidize guanine depends on the number of water molecules directly interacting with the metal cation and on the kind of coordination.[81] This different behaviour can be understood considering the metal-ligand interactions.

As expected, the proton-transfer reaction is disfavoured by the hydration of the metal cation, and the reaction energy becomes more negative the smaller the number of water molecules directly interacting with Cu^{2+}.[81] That is, for pentacoordinated Cu^{2+} the reaction energy is –0.6 kcal/mol, for tetracoordination –2.2 kcal/mol and for a tricoordination –3.0 kcal/mol.

These variations are due to the changes both on the electrostatic and oxidant effects. However, one observed trend is that the proton transfer becomes more favourable the larger the degree of oxidation of guanine, which points out the importance of oxidative effects of Cu^{2+}. Thus, it is observed that certain local environments of the metal can induce the oxidation of guanine and, as a consequence, favour the mutagenic proton transfer to cytosine.

7.   CONCLUSIONS

In a pioneering work,[2] Lowdin introduced the hypothesis that intermolecular proton transfer reactions in DNA base pairs could produce spontaneous mutations since rare tautomers could be formed. This chapter reviews the quantum chemical studies performed on intermolecular proton transfer processes in DNA base pairs. We have considered the neutral systems both in their ground and excited states as well as oxidized, protonated and cationized systems. The main conclusions obtained for gas phase isolated base pairs are that for the ground state neutral systems the double proton transfer reaction is more favorable than the single one. Nevertheless, the energy barrier is high, and the double tautomer is thermodynamically unstable. Thus, no important mutagenic effects are expected from this double proton transfer reaction. In contrast to this, an excited hydrogen transferred structure arising from a proton plus an electron transfer has been found to be quite stable, although it can easily deactivate to the ground state through a conical intersection. For the ionized GC and AT systems, the single proton transfer reaction appears to be very favorable since ionization of the purine bases increases their acidity and the reaction does not imply a creation of charges, but the transfer of a positive charge. Finally, protonation and cationization of guanine also favors the single proton transfer reaction, especially for divalent cations for which the process is exothermic and the energy barrier is small. These effects are very important for Cu^{2+} due to its oxidant character. Although general trends are observed for GC and AT neutral, excited or oxidized base pairs, they can show important differences upon metal cationization or protonation due to the different nature of the hydrogen bonds. For instance, cationization and protonation of adenine does not induce a single proton transfer reaction to the pyrimidine moiety, in contrast to what is observed for GC.

Finally, it should be noted that although these studies provide interesting results to analyze the intrinsic reactivity (in absence of solvent) of these systems, a step forward for understanding these processes in real living systems requires the introduction of the environment. For instance, the formation of an ion-pair by a single proton transfer process in neutral systems will be favored by a polar medium. On the other hand, solvation of metal cations will screen the electrostatic effects described and modify the oxidant character of the metal cation. Although some steps have been performed in this direction, there still remains much work to be done to fully understand these complex biological processes.

A further important point is that the current development of powerful experimental techniques often puts a challenge for theoreticians to support the interpretation of the experiments. An example is the observation of hydrogen transfer in the spectroscopy of GC, where it is not clear if the process takes place in the excited state or in the ionic ground state.[39] The calculations described in this review are steps towards answering such questions.

REFERENCES

1. Watson, J. D. & Crick, F. H. C. (1953). Molecular structure of nucleic acids. *Nature* **171**, 737-738.
2. Löwdin, P. O. (1963). Proton tunneling in DNA [deoxyribonucleic acid] and its biological implications. *Rev. Mod. Phys.* **35**, 724-732.
3. Löwdin, P. O. (1963). Quantum genetics and the aperiodic solid. Some aspects of the biological problems of heredity, mutations, aging, and tumors in view of the quantum theory of the DNA molecule. Technical Report.
4. Rein, R. & Harris, F. E. (1964). Proton tunneling in radiation-induced mutation. *Science* **146**, 649-650.
5. Rein, R. & Ladik, J. (1964). Semiempirical SCF-LCAO-MO Calculation of the electronic structure of the guanine-cytosine base pair: Possible interpretation of the mutagenic effect of radiation. *J. Chem. Phys.* **40**, 2466-2470.
6. Rein, R. & Harris, F. E. (1964). Studies of hydrogen-bonded systems. I. The electronic structure and the double well potential of the N-H...N hydrogen bond of the guanine-cytosine base pair. *J. Chem. Phys.* **41**, 3393-3401.
7. Lunell, S. & Sperber, G. (1967). Hydrogen bonding in the adenine-thymine, adenine-cytosine, and guanine-thymine base pairs. *J. Chem. Phys.* **46**, 2119-2124.
8. Scheiner, S. & Kern, C. W. (1978). Theoretical study of proton transfers between base pairs of DNA. *Chem. Phys. Lett.* **57**, 331-333.
9. Scheiner, S. & Kern, C. W. (1979). Molecular Orbital Investigation of Multiply Hydrogen Bonded Systems. Formic Acid Dimer and DNA Base Pairs. *J. Am. Chem. Soc.* **101**, 4081-4085.
10. Clementi, E., Mehl, J. & von Niessen, W. (1971). Electronic structure of molecules. XII. Hydrogen bridges in the guanine-cytosine pair and in the dimeric form of formic acid. *J. Chem. Phys.* **54**, 508-520.
11. Clementi, E. (1972). Computation of large molecules with the Hartree-Fock model. *Proc. Natl. Acad. Sci.* **69**, 2942-2944.
12. Kong, Y. S., John, M. S. & Löwdin, P. O. (1987). Studies on proton transfers in water clusters and DNA base pairs. *Int. J. Quantum Chem. Quant. Biol. Symp.* **14**, 189-209.

13. Hrouda, V., Florian, J. & Hobza, P. (1993). Structure, Energetics, and Harmonic Vibrational Spectra of the Adenine-Thymine and Adenine*-Thymine* Base Pairs: Gradient Nonempirical and Semiempirical Study. *J. Phys. Chem.* **97**, 1542-1557.

14. Florian, J., Hrouda, V. & Hobza, P. (1994). Proton Transfer in the Adenine-Thymine Base Pair. *J. Am. Chem. Soc.* **116**, 1457-1460.

15. Florián, J. & Leszczynski, J. (1996). Spontaneous DNA Mutations Induced by Proton Transfer in the Guanine-Cytosine Base Pairs: An Energetic Perspective. *J. Am. Chem. Soc.* **118**, 3010-3017.

16. Bertran, J., Noguera, M. & Sodupe, M. (2002). Protonation vs. ionization on intermolecular proton transfer processes in Guanine-Cytosine Watson-Crick base pair. *Afinidad* **59**, 470-478.

17. Noguera, M., Sodupe, M. & Bertran, J. (2004). Effects of protonation on proton-transfer processes in guanine-cytosine Watson-Crick base pairs. *Theor. Chem. Acc.* **112**, 318-326.

18. Gorb, L., Podolyan, Y., Dziekonski, P., Sokalski, W. A. & Leszczynski, J. (2004). Double-Proton Transfer in Adenine-Thymine and Guanine-Cytosine Base Pairs. A Post-Hartree-Fock ab Initio Study. *J. Am. Chem. Soc.* **126**, 10119-10129.

19. Zoete, V. & Meuwly, M. (2004). Double proton transfer in the isolated and DNA-embedded guanine-cytosine base pair. *J. Chem. Phys.* **121**, 4377-4388.

20. Jang, Y. H., Goddard III, W. A., Noyes, K. T., Sowers, L. C., Hwang, S. & Chung, D. S. (2003). pKa Values of Guanine in Water: Density Functional Theory Calculations Combined with Poisson-Boltzmann Continuum-Solvation Model. *J. Phys. Chem. B* **107**, 344-357.

21. Colominas, C., Luque, F. J. & Orozco, M. (1996). Tautomerism and Protonation of Guanine and Cytosine. Implications in the formation of Hydrogen-Bonded Complexes. *J. Am. Chem. Soc.* **118**, 6811-6821.

22. Noguera, M., Sodupe, M. & Bertran, J. (2005). Effects of protonation on proton transfer processes in Watson-Crick Adenine-Thymine base pair. *To be submitted.*

23. Kryachko, E. S. & Sabin, J. R. (2003). Quantum Chemical Study of the Hydrogen-Bonded Patterns in A - T Base Pair of DNA: Origins of Tautomeric Mispairs, Base Flipping, and Watson–Crick to Hoogsteen Conversion. *Int. J. Quant. Chem.* **91**, 695-710.

24. Shimizu, N., Kawano, S. & Tachikawa, M. (2005). Electron correlated and density functional studies on hydrogen-bonded proton transfer in adenine-thymine base pair of DNA. *J. Mol. Struct.* **735**, 243-248.

25. Marañon, J., Fantoni, A. & Grigera, J. R. (1999). Molecular dynamics simulation of double proton transfer: adenine-thymine base pair. *J. Theor. Biol.* **201**, 93-102.

26. Nir, E., Janzen, C., Imhof, P., Kleinermanns, K. & de Vries, M. S. (2001). Guanine tautomerism revealed by UV–UV and IR–UV hole burning spectroscopy. *J. Chem. Phys.* **115**, 4604-4611.

27. Mons, M., Dimicole, I., Piuzzi, F., Tardivel, B. & Elhanine, M. (2002). Tautomerism of the DNA Base Guanine and Its Methylated Derivatives as Studied by Gas-Phase Infrared and Ultraviolet Spectroscopy. *J. Phys. Chem. A* **106**, 5088-5094.

28. Leszczynski, J. (1998). The potential Energy Surface of Guanine Is Not Flat: An ab Initio Study with Large Basis Sets and higher Order Electron Correlation Contributions. *J. Phys. Chem. A* **102**, 2357-2362.

29. Hanus, M., Ryjâcek, F., Kabeläc, M., Kubar, T., Bogdan, T. V., Trygubenko, S. A. & Hobza, P. (2003). Correlated ab Initio Study of Nucleic Acid Bases and Their Tautomers

in the Gas Phase, in a Microhydrated Environment and in Aqueous Solution. Guanine: Surprising Stabilization of Rare Tautomers in Aqueous Solution. *J. Am. Chem. Soc.* **125**, 7678-7688.

30. Kobayashi, R. (1998). A CCSD(T) Study of the Relative Stabilities of Cytosine Tautomers. *J. Phys. Chem. A* **102**, 10813-10817.

31. Alemán, C. (2000). The keto-amino/enol tautomerism of cytosine in aqueous solution. A theoretical study using combined discrete/self consistent reaction field models. *Chem. Phys.* **253**, 13-19.

32. Fogarasi, G. (2002). Relative Stabilities of Three Low-Energy Tautomers of Cytosine: A Coupled-Cluster Electron Correlation Study. *J.Phys.Chem. A* **106**, 1381-1390.

33. Trygubenko, S. A., Bogdan, T. V., Rueda, M., Orozco, M., Luque, F. J., Šponer, J., Slavicek, P. & Hobza, P. (2002). Correlated ab initio study of nucleic acid bases and their tautomers in the gas phase, in a microhydrated environment and in aqueous solution. Part 1. Cytosine. *Phys. Chem. Chem. Phys.* **4**, 4192-4203.

34. Gorb, L. & Leszczynski, J. (1998). Intramolecular Proton Transfer in Monohydrated Tautomers of Cytosine: An Ab Initio Post-Hartree-Fock Study. *Int. J. Quant. Chem.* **70**, 855-862.

35. Gorb, L. & Leszczynski, J. (1998). Intramolecular Proton transfer in Mono- and Dihydrated tautomers of Guanine: An ab initio Post Hartree-Fock Study. *J. Am. Chem. Soc.* **120**, 5024-5032.

36. Gu, J. & Leszcynski, J. (1999). A DFT Study of the Water-Assisted Intramolecular Proton Transfer in the Tautomers of Adenine. *J. Phys. Chem. A* **103**, 2744-2750.

37. Ahn, D.-S., Lee, S. & Kim, B. (2004). Solvent-Mediated tautomerization of Purine: single to quadruple proton transfer. *Chem. Phys. Lett.* **390**, 384-388.

38. Hu, X., Li, H., Liang, W. & Han, S. (2004). Theoretical Study of the proton transfer of Uracil and (Water)n (n = 0-4) Water stabilization and Mutagenicity of uracil. *J. Phys. Chem. B* **108**, 12999-13007.

39. Nir, E., Kleinermanns, K. & de Vries, M. S. (2000). Pairing of isolated nucleic-acid bases in the absence of the DNA backbone. *Nature* **408**, 949-951.

40. Hunig, I., Plutzer, C., Seefeld, K. A., Lowenich, D., Nispel, M. & Kleinermanns, K. (2004). Photostability of isolated and paired nucleobases: N-H dissociation of adenine and hydrogen transfer in its base pairs examined by laser spectroscopy. *ChemPhysChem* **5**, 1427-1431.

41. Douhal, A., Guallar, V., Moreno, M. & Lluch, J. M. (1996). Theoretical study of molecular dynamics in model base pairs. *Chem. Phys. Lett.* **256**, 370-376.

42. Guallar, V., Batista, V. S. & Miller, W. H. (1999). Semiclassical molecular dynamics simulations of excited state double-proton transfer in 7-azaindole dimers. *J. Chem. Phys.* **110**, 9922-9936.

43. Moreno, M., Douhal, A., Lluch, J. M., Castano, O. & Frutos, L. M. (2001). Ab initio based exploration of the potential energy surface for the double proton transfer in the first excited singlet electronic state of the 7-azaindole dimer. *J. Phys. Chem. A* **105**, 3887-3893.

44. Serrano-Andrés, L., Merchan, M., Borin, A. C. & Stalring, J. (2001). Theoretical studies on the spectroscopy of the 7-azaindole monomer and dimer. *Int. J. Quantum Chem.* **84**, 181-191.

45. Sobolewski, A. L. & Domcke, W. (2003). Ab initio study of the excited-state coupled electron-proton-transfer process in the 2-aminopyridine dimer. *Chem. Phys.* **294**, 73-83.

46. Schultz, T., Samoylova, E., Radloff, W., Hertel, I. V., Sobolewski, A. L. & Domcke, W. (2004). Efficient deactivation of a model base pair via excited-state hydrogen transfer. *Science* **306**, 1765-1768.

47. Blancafort, L., Bertran, J. & Sodupe, M. (2004). Triplet (π,π*) reactivity of the guanine-cytosine DNA base pair: Benign deactivation versus double tautomerization via intermolecular hydrogen transfer. *J. Am. Chem. Soc.* **126**, 12770-12771.

48. Sobolewski, A. L. & Domcke, W. (2004). Ab initio Studies on the photophysics or the Gu-Cy base pair. *Phys. Chem. Chem. Phys.* **6**, 2763-2771.

49. Guallar, V., Douhal, A., Moreno, M. & Lluch, J. M. (1999). DNA mutations induced by proton and charge transfer in the low-lying excited singlet electronic states of the DNA base pairs: A theoretical insight. *J. Phys. Chem. A* **103**, 6251-6256.

50. Crespo-Hernández, C. E., Cohen, B., Hare, P. M. & Kohler, B. (2004). Ultrafast Excited-State Dynamics in nucleic acids. *Chem. Rev.* **104**, 1977-2019.

51. Shukla, M. K. & Leszczynski, J. (2002). A theoretical study of excited state properties of adenine-thymine and guanine-cytosine base pairs. *J. Phys. Chem. A* **106**, 4709-4717.

52. Cadet, J. & Vigny, P. (1990). The Photochemistry of Nucleic Acids. In *Bioinorganic Photochemistry* (Morrison, H., ed.). John Wiley & Sons, Inc., New York.

53. Merchán, M., Serrano-Andrés, L., Robb, M. A. & Blancafort, L. (2005). Triplet-state formation along the ultrafast decay of excited singlet cytosine. *J. Am. Chem. Soc.* **127**, 1820-1825.

54. Sevilla, M. D., Becker, D., Yan, M. & Summerfield, S. R. (1991). Relative Abundances of Primary Ion Radicals in γ-Irradiated DNA: Cytosine vs Thymine Anions and Guanine vs Adenine Cations. *J. Phys. Chem.* **95**, 3409-3415.

55. Steenken, S. (1989). Purine Bases, Nucleosides, and Nucleotides: Aqueous Solution Redox Chemistry and Transformation Reactions of Their Radical Cations and e$^-$ and OH adducts. *Chem. Rev.* **89**, 503-520.

56. Steenken, S., Telo, J. P., Novais, H. M. & Candeias, L. P. (1992). One-Electron-Reduction Potentials of Pyrimidine Bases, Nucleosides, and Nucleotides in Aqueous Solution. Consequences for DNA Redox Chemistry. *J. Am. Chem. Soc.* **114**, 4701-4709.

57. Steenken, S. (1997). Electron transfer in DNA? Competition by ultra-fast proton transfer? *Biol. Chem.* **378**, 1293-1297.

58. Murphy, C. J., Arkin, M. R., Jenkins, Y., Ghatlia, N. D., Bossmann, S. H., J. Turro, N. & Barton, J. K. (1993). Long-range photoinduced electron transfer through a DNA helix. *Science* **262**, 1025-1029.

59. Kelley, S. O. & Barton, J. K. (1999). Electron transfer between bases in double helical DNA. *Science* **283**, 375-381.

60. Bhattacharya, P. K. & Barton, J. K. (2001). Influence of intervening mismatches on long-range guanine oxidation in DNA duplexes. *J. Am. Chem. Soc.* **123**, 8649-56.

61. Hutter, M. & Clark, T. (1996). On the Enhanced stability of the Guanine-Cytosine Base-Pair Radical Cation. *J. Am. Chem. Soc.* **118**, 7574-7577.

62. Li, X., Cai, Z. & Sevilla, M. D. (2001). Investigation of Proton Transfer within DNA Base Pair Anion and Cation Radicals by Density Functional Theory (DFT). *J. Phys. Chem. B* **105**, 10115-10123.

63. Colson, A.-O., Besler, B. & Sevilla, M. D. (1992). Ab Initio Molecular Orbital Calculations on DNA Base Pair Radical Ions: Effect of Base Pairing on Proton-Transfer Energies, Electron Affinities, and Ionization Potentials. *J. Phys. Chem.* **96**, 9787-9794.

64. Bertran, J., Oliva, A., Rodríguez-Santiago, L., & Sodupe, M. (1998). Single Versus Double Proton-Transfer Reactions in Watson-Crick Base Pair Radical Cations. A Theoretical Study. *J. Am. Chem. Soc.* **120**, 8159-8167.

65. Reynisson, J. & Steenken, S. (2002). DFT studies on the pairing abilities of the one-electron reduced or oxidized adenine-thymine base pair. *Phys. Chem. Chem. Phys.* **4**, 5353-5358.

66. Reynisson, J. & Steenken, S. (2002). DNA-base radicals. Their base pairing abilities as calculated by DFT. *Phys. Chem. Chem. Phys.* **4**, 5346-5352.

67. Improta, R., Scalmani, G. & Barone, V. (2000). Radical cations of DNA bases: some insights on structure and fragmentation patterns by density functional methods. *Int. J. Mass. Spectrom.* **201**, 321-336.

68. Colson, A. O. & Sevilla, M. D. (1995). Elucidation of primary radiation damage in DNA through application of ab initio molecular orbital theory. *Int. J. Radiat. Biol.* **67(6)**. 627-45.

69. (2005). *NIST Chemistry WebBook, NIST Standard Reference Database Number 69* (Mallard, W. G., Ed.), National Institute of Standards and Technology, Gaithersburg MD, 20899 (http://webbook.nist.gov).

70. Kobayashi, K. & Tagawa, S. (2003). Direct Observation of Guanine Radical Cation Deprotonation in Duplex DNA Using Pulse Radiolysis. *J. Am. Chem. Soc.* **125**, 10213-10218.

71. Melvin, T., Botchway, S. W., Parker, A. W. & O'Neill, P. (1996). Induction of Strand Breaks in Single-Stranded Polyribonucleotides and DNA by Photoionization: One Electron Oxidized Nucleobase Radicals as Precursors. *J. Am. Chem. Soc.* **118**, 10031-10036.

72. Lamm, G. & Pack, G. P. (1990). Acidic domains around nucleic acids. *Proc. Natl. Acad. Sci.* **87**, 9033-9036.

73. Ford, G. P. & Wang, B. (1992). Prototropic Changes in Cationic Base-Pair Adducts. I. Guanine Protonation. *Int. J. Quantum. Chem.* **44**, 587-603.

74. Bertran, J., Sodupe, M., Šponer, J. & Šponer, J. E. (2005). Metal Cation-Nucleic Acids Interactions. In *Encyclopedia of ComputationalChemistry (online edition)* (Glen, R. C., ed.), Vol. online posting date: 15th March 2005, DOI: 10.1002/0470845015.cn0094. John Wiley & Sons, Ltd, Chichester, UK.

75. Burda, J. V., Šponer, J., Leszczynski, J. & Hobza, P. (1997). Interaction of DNA Base Pairs with Various Metal Cations (Mg^{2+}, Ca^{2+}, Sr^{2+}, Ba^{2+}, Cu^+, Ag^+, Au^+, Zn^{2+}, Cd^{2+}, and Hg^{2+}): Nonempirical ab Initio Calculations on Structures, Energies, and nonadditivity of the Interaction. *J. Phys. Chem. B* **101**, 9670-9677.

76. Muñoz, J., Šponer, J., Hobza, P., Orozco, M. & Luque, F. J. (2001). Interactions of Hydrated Mg^{2+} Cation with Bases, Base Pairs, and Nucleotides. Electron Topology, Natural Bond Orbital, Electrostatic, and Vibrational Study. *J. Phys. Chem. B* **105**, 6051-6060.

77. Šponer, J., Burda, J. V., Sabat, M., Leszczynski, J. & Hobza, P. (1998). Interaction between the guanine-cytosine Watson-Crick DNA Base Pair and Hydrated group IIa

(Mg^{2+}, Ca^{2+}, Sr^{2+}, Ba^{2+}) and group IIb (Zn^{2+}, Cd^{2+}, Hg^{2+}) metal cations. *J. Phys. Chem. A* **102**, 5951-5957.

78. Šponer, J., Sabat, M., Burda, J. V., Leszczynski, J. & Hobza, P. (1999). Interaction of the Adenine-Thymine Watson-Crick and Adenine-Adenine Reverse-Hoogsten DNA Base Pairs with Hydrated Group IIa (Mg^{2+}, Ca^{2+}, Sr^{2+}, Ba^{2+}) and IIb (Zn^{2+}, Cd^{2+}, Hg^{2+}) Metal Cations: Absence of the Base Pair Stabilization by Metal-Induced Polarization effects. *J. Phys. Chem. B* **103**, 2528-2534.

79. Šponer, J., Sabat, M., Gorb, L., Leszczynski, J., Lippert, B. & Hobza, P. (2000). The Effect of Metal Binding to the N7 Site of Purine Nucleotides on Their Structure, Energy, and Involvement in Base Pairing. *J. Phys. Chem. B* **104**, 7535-7544.

80. Bertran, J., Noguera, M. & Sodupe, M. (2003). Effects of ionization and cationization on intermolecular proton transfer reactions in DNA base pairs. *Fundamental World of Quantum Chemistry* **2**, 557-581.

81. Noguera, M., Bertran, J. & Sodupe, M. (2004). A Quantum Chemical Study of Cu^{2+} Interacting with Guanine-Cytosine Base Pair. Electrostatic and Oxidative Effects on Intermolecular Proton-Transfer Processes. *J. Phys. Chem. A* **108**, 333-341.

82. Burda, J. V., Šponer, J., Hrabáková, J., Zeizinger, M. & Leszczynski, J. (2003). The influence of N7 Guanine Modifications on the Strength of Watson-Crick Base Pairing and Guanine N1 Acidity: Comparison of Gas-Phase and Condensed-Phase Trends. *J. Phys. Chem. B* **107**, 5349-5356.

83. Song, B., Zhao, J., Griesser, R., Meiser, C., Sigel, H. & Lippert, B. (1999). Effects of (N7)-Coordinated Nickel(II), Copper(II) or Platinum (II) on the Acid-Base Properties of Guanine Derivatives and other Related Purines. *Chem. Eur. J.* **5**, 2374-2387.

Chapter 17

COMPARATIVE STUDY OF QUANTUM MECHANICAL METHODS RELATED TO NUCLEIC ACID BASES: ELECTRONIC SPECTRA, EXCITED STATE STRUCTURES AND INTERACTIONS

M.K. Shukla and Jerzy Leszczynski
Computational Center for Molecular Structure and Interactions, Department of Chemistry, Jackson State University, Jackson, Mississippi 39217 (USA)

Abstract: This review is devoted to the comprehensive analysis of theoretical and experimental investigations of excited state properties of nucleic acid bases (NAB), base pairs, and their interactions with water molecules. The ground state geometries of these molecules were determined long ago using X-ray crystallographic and neutron diffraction techniques. However, the geometries of such complex molecules cannot be determined experimentally in the excited state. Fortunately, some limited information can be obtained in this context experimentally. Experimental methods, which have been also validated in recent theoretical studies, in some cases have suggested the nonplanar geometries of NABs in the excited state. The mode of interaction of water molecules with NABs has been found to be significantly different in the $n\pi^*$ excited state compared to those in the ground state. Though, in this review article some attention pertains to the ground state properties of these molecules; however, the primary emphasis is on electronic singlet excited state properties, such as electronic transition energies, excited state geometries and interaction with water molecules in electronic singlet excited states.

Key words: Nucleic acid bases, base pairs, electronic transitions, excited state geometries, ultrafast nonradiative decay, ab initio, CIS method

J. Šponer and F. Lankaš (eds.), Computational Studies of RNA and DNA, 433–461.

1.   INTRODUCTION

The sequences of *hydrogen bonding patterns* in nucleic acids which contain genetic information of living organisms depend on the nucleic acid bases (NABs) existing in the canonical tautomeric forms.[1] Adenine and guanine belong to purine bases, while cytosine and thymine belong to pyrimidine bases (Figure 17-1). In RNA, the thymine base is replaced with uracil which has a hydrogen atom at the C5 site instead of a methyl group which is present in thymine (Figure 17-1). In DNA and RNA, adenine is hydrogen-bonded with thymine (uracil in the case of RNA), while guanine is hydrogen-bonded with cytosine (Figure 17-2). Minor tautomers of nucleic acid bases can be formed under certain environmental conditions and that may lead to mispairing of bases.[1] Generally, such deformations are repaired by cellular self-defense mechanisms. However, sometimes due to the malfunctioning of the repair system, such deformations may exist and may yield to mutations.[1]

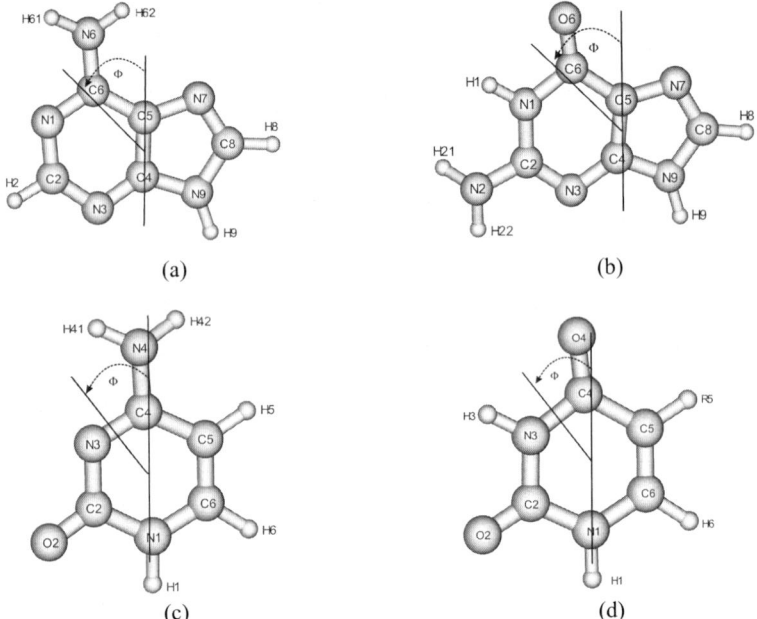

Figure 17-1. Structure and atomic numbering schemes of nucleic acid bases, (a) adenine (N9H), (b) guanine (keto-N9H), (c) cytosine, (d) uracil (R5=H), and thymine (R5=Ch3) The Φ represents the transition moment direction according to the DeVoe-Tinoco convention [9].

Fortunately, the number of *minor tautomers* in free nucleic acid bases is reduced in nucleic acid polymers due to the presence of sugar at the N9 site of purines and the N1 site of pyrimidine. Therefore, the possible tautomerism in nucleic acid polymers is restricted to the keto-enol and the amino-imino forms of bases. It has been shown by our group that the presence of a water molecule in the proton transfer reaction path of the keto-enol tautomerization reaction of nucleic acid bases and analogs drastically reduces the barrier height of tautomerization.[2-6] Further, the transition states of such water-assisted proton transfer reactions have a zwitterionic structure.[3,4]

It is well known that the *fluorescence quantum yields* for all NABs are very low in aqueous solution at the room temperature and most of the excitation energy is lost in the form of nonradiative decays.[7-10] Recent state-of-the-art experimental studies have demonstrated that an internal conversion process occurs on a subpicosecond time scale and that the inclusion of bulky groups in the NABs increases the time scale.[10-17] Different possible mechanisms for the ultrafast nonradiative decay in nucleic acid bases have been suggested.[10,18-41] Theoretical calculations have suggested the prominant role of nonplanar geometries of NABs in the excited state in providing possible route for different nonradiative decay channels.[32-41]

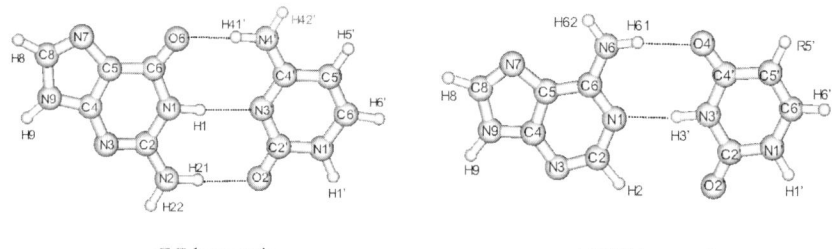

GC base pair   AT(U) base pair

Figure 17-2. Structure and atomic numbering schemes of the GC, AT (R5′=CH3) and AU (R5′=H) base pairs.

With the advent of *state-of-the-art hardware and advanced algorithms*, quantum chemical methods are now routinely used to study ground state properties of nucleic acid bases and related molecules at a high level of accuracy.[2-6,42-46] The ground state geometries of these molecules were determined long ago using X-ray crystallographic and neutron diffraction techniques. However, the geometries of such complex molecules cannot be determined experimentally in the excited state. Fortunately, some limited information can be obtained in this context experimentally. For example, supersonic jet-cooled[47] and resonance Raman[48] studies have indicated non-planar excited state geometries and are supported by theoretical studies.[49] Although remarkable progress has been made to study the ground state properties of molecules, the situation is not as good for the study of the excited states of molecular systems. Recently, ab initio calculations of the electronic spectra, transition moments and excited state geometries of selected nucleic acid bases and related molecules were reported.[19,34,35,49-74]

However, excited state studies are far fewer than those dealing with the ground state properties of nucleic acid bases.

The last substantial review on excited state properties of nucleic acid systems was done by Callis in 1983.[9] In this review, he performed an excellent analysis of the experimental and theoretical results of electronic transitions of nucleic acid bases and related analogues. But it should be noted that theoretical results were limited to semiempirical methods (the ab initio calculation for this class of molecules were almost impossible at that time). An excellent review article on nucleic acid bases has also appeared very recently, but this article is mostly devoted to the ultrafast excited state dynamics of bases.[10] The aim of the present review is to analyze the results of theoretical and experimental studies of the electronic singlet excited state properties of natural nucleic acid bases and base pairs in order to examine the performance of the different ab initio theoretical results in explaining the excited state properties and in resolving the existing ambiguity of different transitions. The major emphasis has been devoted to electronic singlet transition energies, excited state geometries and interactions with water molecules.

2.   GROUND STATE PROPERTIES OF NUCLEIC ACID BASES AND BASE PAIRS

Depending on the environmental conditions, NABs can coexist in several tautomeric forms. Although prototropic tautomerism (N9 ↔ N7 in adenine and guanine and N1↔N3 in cytosine) is blocked in nucleic acid polymers, the possibility for the formation of other tautomeric forms (enol and imino) exists. A rare tautomer in DNA can yield a base pair by forming bonds with a noncomplementary partner which may yield spontaneous mutation. A significantly large number of theoretical and experimental investigations are devoted to studying ground state properties of nuclecic acid bases and related molecules.[2-6,42-46] Different experimental and high level ab initio theoretical calculations show that adenine exists mainly in the N9H and N7H tautomeric forms, the relative distribution depends upon the environment.[2,75,76] Guanine shows both keto-enol and prototropic (N7↔N9) tautomerisms.[2,5,42-44,77] A recent jet-cooled spectroscopic investigation has suggested the existence of four tautomeric forms of guanine, namely keto-N9H, keto-N7H, enol-N9H, and enol-N7H, in the gas phase.[78] In low temperature argon matrices, both keto and enol forms of guanine exist in equal propotions, while in a polar solvent the keto-N9H form dominates.[2,5,77] The pyrimidine bases uracil and thymine are generally believed to exist mainly in the keto form. Different tautomeric forms of cytosine, a pyrimidine base, are also known to exist in various environmental conditions.[2,42-44,79] It exists as a mixture of the amino-hydroxy and amino-oxo (N1H) tautomeric forms with the equilibrium being shifted slightly towards the former tautomeric form in the argon and nitrogen matrices.[79,80] In aqueous solutions both of the amino-oxo forms (N1H and N3H) are present.[81]

The amino groups of NABs are pyramidal due to the partial sp^3 hybridization of the amino nitrogen. The amino group pyramidalization of guanine is highest among the nucleic acid bases.[2,42-44,49,59,60] Experimental evidence for the nonplanarity of adenine and cystosine has been recently indicated in the vibrational transition moment direction measurement study by Dong and Miller.[82] Further, it has also been revealed recently from the different levels of computations that the pyrimidine ring in the NABs possesses high conformational flexibility.[2] The Watson-Crick (WC) base pairs (GC, AT and AU) geometries are planar including the amino group at the HF and DFT levels.[2,53,54] However, at the MP2 level the Watson Crick GC base pair has a nonplanar while the AT base pair has a planar structure.[45,46] Detailed information about different ground state properties of NABs can be found in some of the recent review articles.[2,42-44]

3. EXCITED STATE PROPERTIES OF NUCLEIC ACID BASES

The molecular geometries of nucleic acid bases are planar (except the amino group which is pyramidal) in the ground state,[2,43,44] but excited state geometries are generally nonplanar.[19,47-49,55,59-63,83] Such nonplanarity has been suggested to provide a possible route for the ultrafast nonradiative decay channels.[10,19,32-41] The modes of interaction of NABs with water molecules are also found to be different in the electronic excited states compared to the ground state, and such changes are more pronounced in the nπ* excited states.[84-87] In the nπ* excited states hydrogen bond accepting sites, which are involved in the electronic excitations, provide repulsive potential. Consequently base pairs are destabilized under such excitations.[53,54] In a recent femtosecond spectroscopic study of adenine-water clusters, the adenine-water hydrogen bonds were found to be dissociated on the nπ* potential energy surfaces of adenine.[84,85]

3.1   Electronic Transitions

3.1.1   Adenine

The main absorption band of adenine observed near 260 nm (4.77 eV) is composed of two electronic transitions with different intensity and transition moment directions. In the aqueous medium, the main absorption transition of adenine located near 261 nm (4.75 eV) is short axis-polarized, while the other transition appearing in the form of a weak shoulder near 267 nm (4.64 eV) is long axis-polarized.[88] Similar results were also found in the linear dichroism (LD) spectra of 9-methyladenine,[64] and in the photoacoustic spectra of adenine.[89] But, the splitting of the 260 nm band is not found in the CD spectra.[90-92] In the vapor phase and in the trimethyl phosphate (TMP) solution the 260 nm band is not resolved.[93] The results of the transition moment directions (according to the DeVoe—Tinoco convention, (Figure 17-1a) differ in the various experiments. An elegant work was performed by Clark[95,96] to model the electronic spectra of adenine by measuring the polarized spectra of crystals of 9-methyladenine and 6-(methylamino)purine. The strong transition (near 265 nm) was shown to be polarized at 25° with respect to the C4C5 bond, while the weaker transition (near 275 nm) was found to be polarized close to the long molecular axis. The transition moment directions of several transitions of 9-methyl and 7-methyl adenine samples (9MA and 7MA) oriented in stretched polymer films were also measured.[64] Clark[97] has tentatively assigned the existence of nπ* transitions near 244 and 204 nm (5.08 and 6.08 eV) in the crystal of 2'-deoxyadenosine. The possibility of the existence of such nπ* transitions is also supported from a recent theoretical study.[53] There are also some investigations suggesting the existence of an nπ* transition close to the first singlet ππ* transition.[62,64,84,85] Experimental electronic transitions of adenine and its derivatives are summarized in Table 17-1.

The multi-reference configuration interaction (MRCI) and random phase approximation (RPA) levels of electronic transition calculations on adenine using the ground state self-consistent field orbitals yielded transition energies higher by 1.48–1.86 eV compared to the corresponding experimental data.[65] Roos and coworkers[66] have used the complete active space self-consistent field (CASSCF) method and the complete active space multiconfigurational second-order perturbation theory (CASPT2) to study the electronic transitions of the planar form of adenine. As expected, the CASPT2 correlation correction to the CASSCF energies yielded significant improvements in the CASSCF excitation energies.[66] It was suggested that the absorption of adenine will be dominated by the N9H tautomer, and some contribution from the N7H tautomer was also suggested.[66]

Table 17-1. Summary of experimental transition energies (ΔE, eV) of adenine, guanine, thymine, uracil, cytosine and their derivatives. The 'f' represents oscillator strength, and Φ represents transition moment direction (°) according to the Devoe-Tinoco convention (Figure 17-1).

Molecule/Transitions								References
Adenine								
Absorption Spectra								
ΔE		4.92		5.99				Adenine, vapor [93]
ΔE		4.77		5.96				Adenine, TMP [93]
ΔE		4.81		5.85				9MA, MCH [93]
ΔE		4.77		5.90				9MA, TMP [93]
ΔE		4.77		5.99				Adenine, water [94]
ΔE	4.59	4.77		5.90				Adenine, water [98]
ΔE	4.63	4.77		6.05				Adenine, water [99]
ΔE		4.77		6.02				Adenosine, water [99]
ΔE	4.59			5.90		6.81	7.75	Adenine sublimed film [100]
ΔE	4.51	4.68		5.82	6.08	6.81	7.75	9MA, crystal[96]
f	0.1	0.2		0.25	0.11	0.30	0.23	
Φ	83	25		−45	15	72	6	
LD Spectra								
ΔE	4.55	4.81	5.38	5.80	5.99			9MA, stretched film [64]
f	0.047	0.24	0.027	0.14	0.12			
Φ	66	19	−15	−21	−64			
CD Spectra								
ΔE		4.63		5.93	6.36			Adenines, water [90]
ΔE		4.77		5.74	6.36	6.63		Adenines, water [91]
ΔE		4.68		5.51				Adenosine, water [99]
MCD Spectra								
ΔE	4.59	4.92		5.90				Adenine, water [99]
ΔE	4.56	4.90		5.77				Adenosine, water [99]
Photo Acoustic Spectra								
ΔE	4.28	4.59			6.20	6.89		Adenine, film [89]
Electron Scattering								
ΔE	4.53			5.84		6.50	7.71	Adenine, film [91]
Guanine								
Absorption Spectra								
ΔE	4.46	5.08	6.20	6.57				Guanine, model [104]
f	0.15	0.24	0.40	0.48				
Φ	−12	80	70	−10				
ΔE	4.51	5.04	6.33					Guanine, water [94]
ΔE	4.56	5.04	6.19	6.67				Guanosine, water [104]
f	0.15	0.24	0.40	0.48				
Φª	−24	88	86	-8 to 44				
ΔE	4.56	4.98	6.02	6.63				9EtG, water [103]

f	0.14	0.21		0.38	0.42		
ΔE	4.51	4.84		6.11	6.52		9EtG, TMP [102]
ΔE	4.51	4.92		6.05	6.59		9EtG, water [102]
ΔE	4.46	4.88	5.46	6.08	6.56		9EtG, crystal [103]
f	0.16	0.25	<0.05	0.41	0.48		
Φ	–4	–75	–75	–9			
ΔE	4.35	5.00		6.23	6.70		Guanine, sublimed film [100]
LD Spectra							
ΔE	4.43	5.00					Guanine, stretched film [107]
Φ	4		–88				
CD Spectra							
ΔE	4.51	4.92	5.51	6.20	6.63		dGMP, water [91]
ΔE		5.06	5.77				Guanosine, water [99]
MCD Spectra							
ΔE	4.46	5.00					Guanosine, water [99]
Uracil							
Absorption Spectra							
ΔE	5.08			6.05	6.63		Uracil, vapor [93]
ΔE	4.84			6.05	6.63		1,3-Dimethyluracil,vapor [93]
ΔE	4.68			6.08	6.63		1,3-Dimethyluracil,water [93]
ΔE	4.81			6.11			Uracil, water [99]
ΔE	4.75			6.05			Uridine, water [99]
ΔE	4.81			6.11	6.85		Uracil, TMP [102]
ΔE	4.79			6.14	6.85		Uracil, water [102]
ΔE	4.70			6.02	6.74		1,3-Dimethyluracil, TMP [102]
ΔE	4.73			6.11	6.81		1,3-Dimethyluracil,MCH [102]
ΔE	4.51		5.82				1-Methyluracil, crystal [118]
Φ	–9		59				
ΔE	4.66			6.08	6.97	7.90	Uracil, sublimed film [100]
CD Spectra							
ΔE	4.73		5.77	6.36	7.00		Uridine, water [91]
ΔE	4.63		5.71				Uridine, water [99]
ΔE	4.68		5.82	6.26			Uridine, water [90]
MCD Spectra							
ΔE	4.86		5.85				Uracil, water [99]
ΔE	4.77		5.71				Uridine, water [99]
Electron Scattering							
ΔE	4.70		5.93		6.93		Uracil, film [115]
Thymine							
Absorption Spectra							
ΔE	4.68			6.08			Thymine, water [99]
ΔE	4.64			6.05			Thymidine, water [99]
ΔE	4.54			5.99			1-methylthymine, water [88]
f	0.19			0.28			
ΔE	4.64		5.88		7.04		Thymine, sublimed film [100]
CD Spectra							
ΔE	4.68		5.77	6.36	7.00		Thymidine, water [91]
ΔE	4.54		5.69				Thymidine, water [99]
ΔE	4.63		5.85	6.42			Thymidine, water [90]
MCD Spectra							

ΔE	4.71	5.77					Thymine, water [99]
ΔE	4.73	5.64					Thymidine, water [99]
Photo Acoustic Spectra							
ΔE	4.59	5.90		7.08			Thymine, film [89]
Electron Scattering							
ΔE	4.66	5.94		7.08		8.82	Thymine, film [115]
Cytosine							
Absorption Spectra							
ΔE	4.66	5.39	5.85	6.29			Cytosine, water [121]
f	0.14	0.03	0.13	0.36			
Φ[b]	6	–46	76	-27 or 86			
ΔE	4.64		6.31				Cytosine, water [94]
ΔE	4.57	5.39	6.26				Cytidine, water [94]
ΔE	4.48	5.23	6.08	6.63			Cytosine, TMP [102]
ΔE	4.59	5.28	5.74	6.26			dCMP, water [91]
ΔE	4.64	5.21	5.83	6.46			Cytosine, water [123]
f	0.096	0.100	0.211	0.639			
ΔE	4.57	5.34	5.77	6.26			Cytidine, water [122]
ΔE	4.57			6.17			Cytosine, sublimed film [100]
ΔE	4.54	5.40	6.07	6.67		7.35	Cytosine, sublimed film [124]
f	0.058	0.073	0.115	0.072		0.072	
LD Spectra							
ΔE	4.63	5.17					Cytosine, polymer film [107]
Φ	25±3	6±4					
	or	or					
	–46±4	–27±3					
CD Spectra							
ΔE	4.59	5.27	5.74	6.14	6.56	7.38	dCMP, water [91]
ΔE	4.59	5.02	5.64	6.36			Cytosine nucleosides [125]
ΔE	4.59	5.17	5.64	6.36			Cytidine[c] [125]

TMP: trimethylphosphate, 9MA: 9-methyladenine, MCH: methylcyclohexane, adenines: adenine derivatives; for details see relevant references; [a]Based on polarized absorption spectra of crystalline guanosine [104], 9EtG: 9-ethylguanine, TMP: trimethylphosphate; for details see relevant reference; [b]Based on polarized spectra of cytosine crystal [121], TMP: trimethylphosphate, for details see relevant reference, [c]Based on CD and absorption measurements of cytosine nucleosides in different solvents (water, acetonitrile, dioxane, 1,2-dichloroethane).

We have computed electronic vertical singlet $\pi\pi^*$ and $n\pi^*$ transition energies of adenine at the CIS/6-311G(d,p) level using the HF/6-311G(d,p) geometry.[50] The super molecular approach considering three water molecules in the first solvation shell of the adenine was used to model aqueous solvation. The CIS transition energies shown in the Table 17-2 were scaled by a factor of 0.72.[19,50,53,64] The experimental transition energies of adenine shown in Table 17-1 can be generally explained within an accuracy of 0.2 eV in terms of the scaled CIS computed transition energies of the hydrated adenine.

Table 17-2. Scaled vertical singlet ππ* and nπ* excitation energies (ΔE, eV), oscillator strengths (f), transition moment directions (Φ,°) and dipole moments (μ, Debye) of isolated and hydrated adenine at the CIS/6-311G(d,p)//HF/6-311G(d,p) level.[a,50]

CIS							CASPT2/CASSCF[b]
Isolated				Hydrated			
ΔE	f	Φ	μ[c]	ΔE	f	Φ	ΔE^1/ΔE^2/f/Φ/μ
ππ* Transitions							
4.76	0.394	60	2.85	4.76	0.440	50	5.20/6.48/0.37/37/2.30
4.79	0.024	−6	3.40	4.74	0.038	−66	5.13/5.73/0.07/23/2.37
5.90	0.398	−38	0.83	5.82	0.342	−31	6.24/7.80/0.851/-57/2.13
6.18	0.447	15	2.02	6.12	0.423	19	
6.21[d]				6.15[e]			6.72/8.30/0.159/40/4.60
6.24	0.547	−87	3.14	6.17	0.589	−77	
6.76	0.232	29	2.65	6.78	0.375	23	6.99/8.77/0.565/27/3.42
nπ* Transitions							
5.18	0.001	–	2.47	5.38	0.000	–	6.15/6.43/0.001/-/2.14
5.52	0.002	–	0.93	5.75	0.001	–	6.86/7.16/0.001/-/1.93
5.74	0.014	–	1.62	5.97	0.015	–	

[a]Scaling factor 0.72, [b]ΔE^1 corresponds to CASPT2 and ΔE^2 corresponds to CASSCF transition energies [66], [c]ground state dipole moments of adenine at the HF/6-311G(d,p) level is 2.51 Debye, [d]Average of transitions at 6.18 and 6.24 eV, [e]Average of transitions at 6.12 and 6.17eV.

The CIS calculation suggests that the first ππ* transition of adenine in the gas phase is stronger, while the second ππ* transition is much weaker. After hydration the transition energy of the weaker transition is decreased; therefore, the stronger transition becomes the second transition (Table 17-2). Experimentally, a weak shoulder near 270 nm (4.59 eV) and a strong peak near 260 nm (4.77 eV) in the water solution are observed.[98] Thus the calculated transitions of the hydrated adenine are in qualitative agreement with the experimental data, although the computed splitting between the two transitions is too small. The CIS calculation predicts that the two transitions computed at 6.18 and 6.24 eV for the isolated adenine and at 6.12 and 6.17 eV for its hydrated form (Table 17-2) would contribute to the 6.2 eV experimental region of the molecule (Table 17-1). The calculation predicts that the transition moment directions of these transitions would be approximately perpendicular to each other (Table 17-2). The MCD results suggest that the UV-absorption band in the 200 nm (6.2 eV) region is composed of two transitions with non-parallel transition dipole moments.[101] Thus, the CIS results may correspond to the MCD observation in this regard.

The CIS computed first and third nπ* transitions of adenine can be explained in terms of the experimental transitions near 244 nm (5.08 eV) and 204 nm (6.08 eV) of 2'-deoxyadenosine.[97]

3.1.2   Guanine

The first electronic transition of guanine lies near 275 nm (4.51 eV), and the second appears near 250 nm (4.96 eV). The intensity of the second transition is stronger than the first one.[9,66,99,102-105] The existence of the third transition which is very weak and generally lies near 225 nm (5.51 eV) is not very often observed.[91,92,103] The fourth and fifth transitions of the guanine are intense and located near 204 nm (6.08 eV) and 188 nm (6.59 eV), respectively.[9,102-104] The existence of three transitions of the nπ* type near 238, 196, and 175 nm (5.21, 6.32, and 7.08 eV) in guanine has been tentatively suggested by Clark.[104] The precise measurement of transition moment directions of guanine is complicated by a crystal field which was shown to have a significant effect on the transition moment directions.[88,106] Clark[104] has suggested that directions in guanine for transitions near 4.46, 5.08, 6.20 and 6.57 eV would be −12, 80, 70 and −10 degrees, respectively. Experimental transition energies and transition moment directions of guanine and its derivatives are summarized in Table 17-1.

Recently, some advanced spectroscopic studies have been performed on guanine and substituted analogs.[78,108-114] In these studies, the spectral origin (0-0 transition) of the S_1 excited state and some lower vibrational frequencies corresponding to the same were determined, and the existence of different tautomers of guanine was investigated. The existence of upto four tautomers of guanine, namely enol-N7H (32864 cm^{-1}), keto-N7H (33269 cm^{-1}), keto-N9H (33910 cm^{-1}) and enol-N9H (34755 cm^{-1}); the values in parentheses correspond to the 0-0 transition of the corresponding tautomer, in the gas phase has been suggested.[78,108,114]

In guanine also the the random phase approximation (RPA) and multireference configuration interaction (MRCI) methods yielded electronic transition energies to be higher by 1.48–1.86 eV compared to the experimental ones, and linear scaling was used for comparison with the experimental data.[65] Roos and coworkers[66] have used the CASSCF and CASPT2 methods to study electronic transitions of guanine. The effect of the aqueous solvent on electronic transitions was considered using the self-consistent reaction field (SCRF) model. The computed CASPT2 transition energies were found to be in good agreement (with an accuracy of 0.3 eV) with the experimental data. The CIS computed vertical electronic singlet ππ* and nπ* transition energies of guanine and hydrated guanine with three water molecules to model the effect of aqueous solvation are shown in the

Table 17-3. It should be noted that these transition energies were scaled by a factor of 0.72.[19,50,53,64] The ground state geometries optimized at the HF/6-311G(d,p) level were used in the transition energy calculations. The CIS/6-311G(d,p) and CASPT2/CASSCF results are in agreement with respect to the assignment of the first $n\pi^*$ transition as being due to the excitation of the carbonyl group lone pair electron.

Table 17-3. Scaled vertical singlet $\pi\pi^*$ and $n\pi^*$ excitation energies (ΔE, eV), oscillator strengths (f), transition moment directions (Φ,°) and dipole moments (μ, Debye) of isolated and hydrated guanine at the CIS/6-311G(d,p)//HF/6-311G(d,p) level.[a,50]

CIS								
Isolated				Hydrated			CASPT2/CASSCF[b]	
ΔE	f	Φ	μ^c	ΔE	f	Φ	$\Delta E^1/\Delta E^2/f/\Phi/\mu$	$\Delta E^3/f/\Phi$
$\pi\pi^*$ *Transitions*								
4.60	0.282	–42	5.96	4.64	0.245	–25	4.76/6.08/0.113/–15/7.72	4.73/0.154/–4
5.22	0.516	66	7.74	5.17	0.567	70	5.09/6.99/0.231/73/6.03	5.11/0.242/75
5.99	0.104	51	6.18	5.95	0.089	59	5.96/7.89/0.023/7/5.54	5.98/0.021/6
6.66[d]	0.113	79	5.52					
6.67[e]				6.57	0.512	–89	6.65/8.60/0.161/–80/10.17	6.49/0.287/–85
6.67	0.356	81	6.17					
$n\pi^*$ *Transitions*								
5.02	0.001	–	4.71	5.24	0.001	–	5.79/6.22/10^{-4}/–/4.31	
5.63	0.010	–	5.84	5.77	0.010	–	6.60/8.05/0.013/–/4.63	
6.18	0.003	–	7.24	6.40	0.002	–	6.63/7.97/0.002/–/2.64	

[a]Scaling factor 0.72, [b]ΔE^1 corresponds to CASPT2, and ΔE^2 corresponds to CASSCF transition energies in the gas phase, $\Delta E^3/f/\Phi$ corresponds to results in water [66], [c]ground state dipole moments of guanine at the HF/6-311G(d,p) level is 6.77 Debye, [d]Rydberg contamination, [e]Average of transitions at 6.66 and 6.67 eV.

The CIS computed transition intensity order agrees with the solution spectra of guanine and its derivatives in which the first transition (near 275 nm) appears as a weak peak in comparison to the stronger peak near the 250 nm region.[9] Further, there is a good correspondence between the computed transitions (scaled) of the guanine (and its hydrated form) and the CASPT2 results (solvation included), in particular when comparison is made with the transition of the hydrated tautomer (Table 17-3). However, the agreement is better in the lower energy region than in the higher energy region. The third $\pi\pi^*$ transition computed at 5.95 eV of the hydrated guanine can be considered for an explanation of the 5.5 eV band in the experimental data.[91,92,103] Therefore, the CIS calculation favors this transition as a weak $\pi\pi^*$ type. Computed $n\pi^*$ transitions of the hydrated guanine are found to be at 5.24, 5.77 and 6.40 eV (Table 17-3). These results support the findings of Clark[104] with regard to the existence of the $n\pi^*$ transitions near 5.21 and 6.32 eV in guanine.

3.1.3   Uracil and Thymine

The electronic spectra of uracil and thymine are generally similar showing three absorption bands near 260, 205 and 180 nm (4.77, 6.05 and 6.89 eV). The first and third bands in thymine are generally slightly red- and blue-shifted, respectively, with respect to the corresponding bands in uracil.[9,89,93,94,99,102,115,116] The CD spectra have shown that the 205 nm band is composite in nature with peaks near 215 and 195 nm (5.77 and 6.36 eV).[91] The CD spectra indicated the presence of a band near 240 nm (5.17 eV) that was assigned as $n\pi^*$ type.[9,91] Hug and Tinoco[117] have predicted an $n\pi^*$ transition at 250 nm (4.96 eV) which has been suggested as the possible source of the 240 nm band observed in the CD spectra.[9,91] The transition moment direction for the first band is found to be close to 0° for uracil and -20° for thymine with respect to the N1C4 direction (Figure 17-1d).[9,88] Novros and Clark[118] have suggested two values –53° or +59° with respect to the N1C4 direction for the second band. But they have favored +59° since it is in agreement with the LD spectra of uracil.[107] Holmen et al.[119] have found an angle of 35° with respect to the N1C4 direction for the second transition of 1,3-dimethyluracil. Several investigations of uracil, thymine and analogs have suggested the existence of an $n\pi^*$ transition within the 260 nm envelope.[9,49,89,116,120] The relative positions of the $n\pi^*$ transition is found to be solvent dependent. In the gas phase and in an aprotic solvent, the $n\pi^*$ state lies the lowest, while in a protic environment it lies higher than the $\pi\pi^*$ state. Consequently, the $n\pi^*$ state becomes the second state in a protic environment.[9,49,116,120] Summary of the experimental transitions obtained for uracil and thymine is shown in Table 17-1.

The MRCI and RPA methods[68] and CASSCF and CASPT2 methods[67] have been used the to compute electronic transition energies of uracil and thymine. The CIS method was also used in the study.[49,50,87] Singlet vertical $\pi\pi^*$ and $n\pi^*$ transition energies of uracil and thymine were computed at the CIS/6-311G(d,p) level using the HF/6-311G(d,p) optimized geometries.[50,87] The three water molecules in the first solvation shell of uracil and thymine were used to model aqueous solvation. The transition energies were scaled by a factor of 0.72. The MRCI, RPA,[68] CASSCF/CASPT2[67] and CIS[50,87] methods agree with the assignment that the first $n\pi^*$ transition of uracil and thymine is localized at the C4O4 group and the second $n\pi^*$ transition is localized at the C2O2 group. The CIS results[50,87] agree with experimental observations of uracil and thymine that, in the gas phase or in an aprotic solvent, the nπ* state is the lowest singlet excited state, while in a protic solvent the ππ* state is the lowest.[9,116] A good correlation between the scaled CIS-computed excitation energies[50,87] and the CASPT2 excitation energies[67] was revealed except for the second transition in uracil and the second and third transitions in thymine. The CIS computed scaled transition energies of uracil and thymine[50,87] were generally found to be within the range of experimental transition energies of the corresponding molecules shown in Table 17-1.

3.1.4   Cytosine

The aqueous absorption spectrum of cytosine shows two broad peaks near 266 and 197 nm (4.66 and 6.29 eV) and two weak peaks or shoulders near 230 and 212 nm (5.39 and 5.85 eV).[9,91,94,100,102,121-124] However, the general features of the cytosine spectrum are found to be solvent dependent.[9,102] The 197 nm band of cytosine is found to be composed of 202 nm (6.14 eV) and 189 nm (6.56 eV) transitions as suggested by the CD spectra.[91,125,126] Clark's group[121] have assigned the transition moment direction of the first three transitions of cytosine (I = 6°, II = –46°, III = 76°) explicitly while two values (–27° or 86°) were suggested for the fourth transition of the molecule. With regard to the nπ* transitions, the experimental[99,121,122,127] and theoretical[49,69] evidences suggest the existence of such a transition in the 5.3 eV (232 nm) region in cytosine. Zaloudek et al.[121] have suggested the existence of another nπ* transition near 5.6 eV (220 nm). Table 17-1 contains a summary of the experimental transitions of cytosine and its derivatives in different environments.

The MRCI, RPA,[68] CASSCF, CASPT2[69] and CIS[86] methods have been used to compute the electronic transition energies of cytosine. In the CIS/6-311G(d,p) calculation[86] the HF/6-311G(d,p) optimized geometries of the isolated and hydrated cytosine were used for the transition energy calculations. In the hydration of cytosine, three water molecules were taken in the first solvation shell to model the aqueous environment. The CIS computed transition energies were scaled by a factor of 0.72. The CIS computed transition energies (after scaling) of the hydrated cytosine were found to be in good agreement (within an accuracy of 0.2 eV) with the experimental data of cytosine in water, except for the second transition.[86] Since the general features of the cytosine spectrum are found to be solvent dependent,[9,102] differences between the computed and experimental data is not unexpected. The CIS computed nπ* transitions of hydrated cytosine[86] near 5.37 and 5.71 eV were also found to be in good agreement with the observed nπ* transition near 5.3 and 5.6 eV.[9,99,121,122,127]

3.2 Geometries

The ground state geometry of adenine is planar except for the amino group which is pyramidal. Some selected geometrical parameters of adenine in the ground and excited states obtained at the HF/6-311G(d,p) and CIS/6-311G(d,p) levels, respectively are presented in Table 17-4. It should be noted that all geometries belong to minima at their respective potential energy surface.

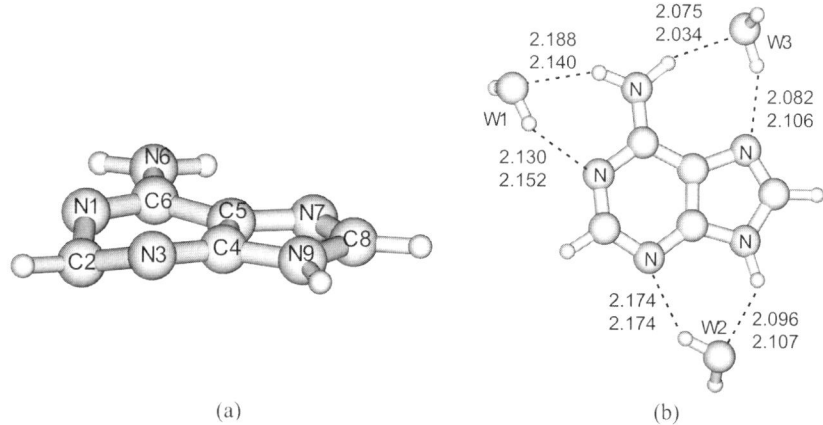

(a)   (b)

Figure 17-3. Structure of adenine: (a) in the $S_3(n\pi^*)$ state, (b) hydrogen bond lengths of hydrated form in the ground state (top indices) and in the $S_1(\pi\pi^*)$ state (bottom indices).

Table 17-4. Selected ground and singlet excited state optimized bond lengths (Å), bond angles (°) and dihedral angles (°) of adenine.[a,50]

Parameters	S_0	$S_1(\pi\pi^*)$	$S_3(n\pi^*)$	Parameters	S_0	$S_1(\pi\pi^*)$	$S_3(n\pi^*)$
N1C2	1.327	1.320	1.377	H61N6H62	119.4	118.0	119.6
C2N3	1.312	1.402	1.399	N3C2N1C6	0.2	0.0	38.4
N3C4	1.327	1.278	1.320	C4N3C2N1	0.1	0.4	−39.2
C4C5	1.375	1.428	1.377	C5C4N3C2	−0.1	−0.3	21.8
C6N1	1.325	1.310	1.306	N9C4N3C2	179.7	179.5	−162.5
C5C6	1.400	1.459	1.427	C6C5C4N3	−0.1	−0.2	0.8
C5N7	1.383	1.318	1.374	N7C5C4N3	179.8	179.8	177.4
N7C8	1.278	1.335	1.276	C5C6N1C2	−0.4	−0.6	−17.0
C4N9	1.362	1.383	1.356	N6C6N1C2	178.7	177.2	163.6
C8N9	1.371	1.366	1.384	N1C6C5C4	0.4	0.7	−3.9
N6C6	1.344	1.348	1.340	H8C8N7C5	180.0	179.8	179.7
N1C2N3	128.7	126.9	111.6	C8N9C4N3	179.7	−179.8	177.4
C2N3C4	111.6	112.6	116.4	H61N6C6N1	10.1	11.7	−11.2
C2N1C6	118.6	120.4	122.4	H61N6C6C5	−170.8	−170.6	169.3
H61N6C6	118.2	117.6	119.2	H62N6C6N1	170.4	163.3	174.2
H62N6C6	119.4	118.1	119.0	H62N6C6C5	−10.5	−18.9	6.4

[a] Ground and excited state parameters are at the HF/6-311G(d,p) and CIS/6-311G(d,p) levels, respectively. The S_1, S_3 represent singlet excited states in the ascending energy order (Table 17-2).

The adenine in the $S_1(\pi\pi^*)$ excited state has an almost planar geometry except the amino group which is slightly more pyramidal than that in the ground state. In the $S_3(n\pi^*)$ excited state, the ring geometry is nonplanar especially around the N1C2N3 fragment of the ring (Figure 17-3a), and the amino group nonplanarity is similar to that in the ground state. The geometries of the hydrated form of adenine, in which three water molecules were considered in the first solvation shell, were also optimized in the ground and the lowest singlet $\pi\pi^*$ excited states. The hydrated structures showing hydrogen bond distances in the ground and the lowest singlet $\pi\pi^*$ excited states are shown in (Figure 17-3b). Hydration generally induces planarity in the ground and lowest singlet $\pi\pi^*$ excited states. Consequently, the tautomers were found to be almost planar in the ground and lowest singlet $\pi\pi^*$ excited states.

Selected geometrical parameters of the guanine in the ground and lowest singlet $\pi\pi^*$ and $n\pi^*$ excited states obtained at the HF/6-311G(d,p) and CIS/6-311G(d,p) levels, respectively are shown in Table 17-5. The geometries represent global minima at their respective potential energy surfaces. The ground state geometry is planar except for the amino group which is pyramidal. The geometry of guanine in the $S_1(\pi\pi^*)$ excited state is strongly nonplanar around the C6N1C2N3 fragment of the ring (Figure 17-4a). The amino group pyramidalization is also increased appreciably in this state. In the $S_2(n\pi^*)$ excited state, the length of the C6O6 bond is increased by about 0.1 Å compared to that in the ground state. The ring geometry is slightly nonplanar and the amino group is also more pyramidal than that in the ground state. The O6 and H1 atoms are displaced away from the ring plane on the opposite side to each other in the $S_2(n\pi^*)$ state. The geometries of the hydrated form of guanine in which three water molecules were considered in the first solvation shell were also optimized in the ground and lowest singlet $\pi\pi^*$ excited states. It was revealed that hydration generally induces planarity in the system. The hydrogen bond lengths of the complex in the ground and excited states are also depicted in (Figure 17-4b). A similar trend of change in the geometrical parameters of guanine in the excited state was found in another study[63] where the geometries of the molecules were also optimized in the ground and excited states in an aqueous medium with the effect of solvation being considered using the integral equation formalism of the polarizable continuum model. The restricted open-shell Kohn-Sham level of calculation on the first singlet $\pi\pi^*$ excited state of guanine[83] also shows structural deformation similar to that obtained using the CIS method as discussed previously.

Table 17-5. Selected ground and singlet excited state optimized bond lengths (Å), bond angles (°) and dihedral angles (°) of guanine.[a,50]

Parameters	S_0	$S_1(\pi\pi^*)$	$S_2(n\pi^*)$	Parameters	S_0	$S_1(\pi\pi^*)$	$S_2(n\pi^*)$
N1C2	1.357	1.397	1.364	H21N2C2	117.9	115.3	116.3
C2N3	1.286	1.393	1.282	H22N2C2	113.8	112.7	112.9
N3C4	1.355	1.284	1.367	H21N2H22	115.0	111.4	114.1
C4C5	1.367	1.431	1.361	N3C2N1C6	−0.6	−64.0	−1.6
N1C6	1.417	1.429	1.434	N3C2N1H1	176.0	170.9	157.7
C5C6	1.436	1.455	1.463	C4N3C2N1	0.8	44.2	−5.5
C5N7	1.378	1.354	1.371	C5C4N3C2	−1.0	−2.4	5.9
N7C8	1.276	1.293	1.283	N9C4N3C2	179.5	174.5	170.8
C4N9	1.352	1.376	1.356	C6C5C4N3	0.9	−18.5	0.9
C8N9	1.375	1.360	1.368	N7C5C4N3	179.6	175.9	178.3
O6C6	1.188	1.184	1.282	C5C6N1C2	0.3	36.2	8.1
N1C2N3	123.9	118.4	125.7	O6C6N1C2	179.4	−145.1	−127.2
C2N3C4	112.7	109.1	112.8	N1C6C5C4	−0.4	−0.6	−7.4
N3C4C5	128.9	125.1	128.6	N1C6C5N7	179.8	162.0	171.6
C2N1C6	126.3	111.2	123.7	O6C6C5C4	179.3	179.2	127.8
N3C4N9	125.7	130.9	126.0	H21N2C2N1	30.4	42.3	−39.6
O6C6N1	119.2	121.7	116.2	H22N2C2N1	169.4	171.8	174.3

[a]Ground and excited state parameters are at the HF/6-311G(d,p) and CIS/6-311G(d,p) levels, respectively. The S_1, S_2 represent singlet excited state in the ascending energy order (Table 17-3).

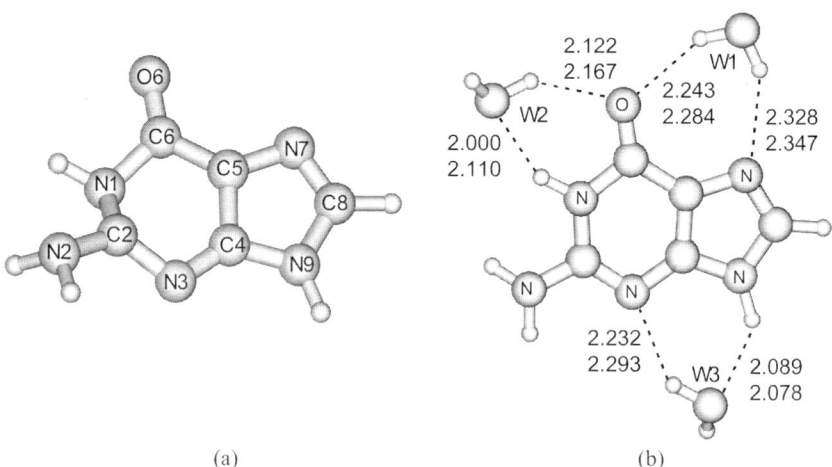

(a)   (b)

Figure 17-4. Geometry of guanine: (a) in the $S_1(\pi\pi^*)$ state, (b) hydrogen bond lengths of hydrated form in the ground state (top indices) and in the $S_1(\pi\pi^*)$ state (bottom indices).

Selected geometrical parameters of uracil and thymine in the ground and singlet excited states obtained at the HF/6-311G(d,p) and CIS/6-311G(d,p) levels, respectively are presented in Table 17-6. It should be noted that geometries correspond to minima at the respective potential energy surfaces.[50,87] The ground state geometries are planar, while the lowest singlet nπ* excited state geometries are slightly nonplanar. Further, in the lowest singlet nπ* excited state the C4O4 bond length is increased by about 0.1 Å compared to that in the ground state, and the C4O4 bond lies appreciably out of the approximate ring plane. However, such deformation is consistent with the fact that the excited state is assigned to the excitation of lone pair electron of the C4O4 group. Thymine in the lowest singlet ππ* excited state adopts a boat type of structure with the N1, C2, C4 and C5 atoms being approximately in a plane, while the N3 and C6 atoms are out of the same plane (Figure 17-5a). The geometries of the hydrated forms of thymine and uracil in which three water molecules were considered in the first solvation shell were also optimized in the ground and singlet excited states. The structure and hydrogen bond lengths of hydrated thymine are shown in (Figure 17-5b), while those for hydrated uracil are shown in (Figure 17-5c). It was revealed that hydration generally induces planarity in the system; however, structural deformations in the excited state are generally similar to that in isolated species as discussed earlier. In jet-cooled studies, the geometrical deformation of the ππ* excited states of thymine and uracil has been suggested to be responsible for the diffuseness of spectra of these compounds.[47,48]

Table 17-6. Selected ground and singlet excited state optimized bond lengths (Å), bond angles (°) and dihedral angles (°) of uracil and thymine.[a,50,87]

Parameters	Uracil		Thymine		
	S_0	$S_1(n\pi^*)$	S_0	$S_1(n\pi^*)$	$S_2(\pi\pi^*)$
C2N1	1.373	1.371	1.368	1.366	1.403
N3C2	1.370	1.367	1.371	1.367	1.371
C4N3	1.391	1.417	1.387	1.417	1.417
C5C4	1.463	1.465	1.472	1.474	1.441
C6N1	1.371	1.388	1.377	1.393	1.346
C6C5	1.328	1.326	1.328	1.326	1.439
O2C2	1.188	1.191	1.189	1.192	1.184
O4C4	1.188	1.280	1.190	1.279	1.196
N3C2N1C6	0.0	7.0	0.0	−7.5	5.5
C4N3C2N1	0.0	0.8	0.0	−0.3	28.5
C5C4N3C2	0.0	−6.3	0.0	5.9	−24.8
C6C5C4N3	0.0	4.3	0.0	−3.7	−10.8
C5C6N1C2	0.0	−9.1	0.0	9.9	−39.0
N1C6C5C4	0.0	2.8	0.0	−3.5	41.3
O4C4N3C2	180.0	137.1	180.0	−179.2	−155.0

[a]Ground and excited state parameters are at the HF/6-311G(d,p) and CIS/6-311G(d,p) levels, respectively.

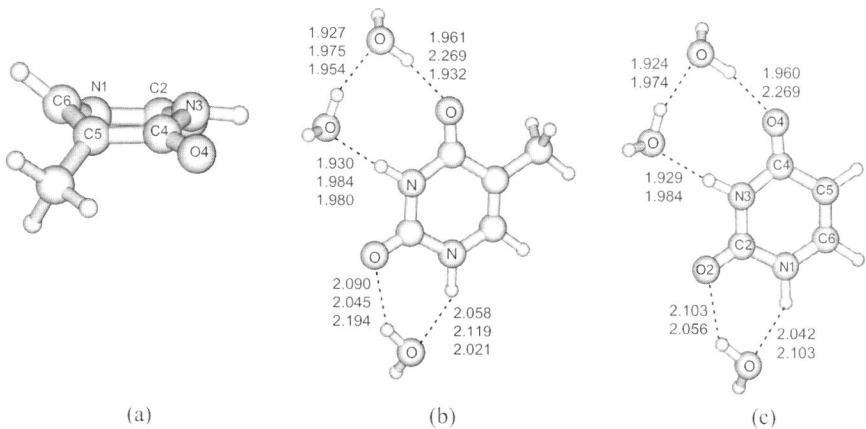

(a)   (b)   (c)

Figure 17-5. Geometry of thymine (a) in the singlet $S_2(\pi\pi^*)$ excited state, (b) hydrogen bond lengths of the hydrated thymine in the ground, $S_1(n\pi^*)$ and $S_2(\pi\pi^*)$ states (from top to bottom indices, respectively). For hydrated thymine in the $S_2(\pi\pi^*)$ excited state, the 6-311++G(d,p) basis set was used. Hydrogen bond lengths of hydrated uracil in the ground (top indices) and lowest singlet $n\pi^*$ excited state (bottom indices) is in (c).

Ground and singlet excited state geometries of the isolated and hydrated cytosine were also studied at the HF/6-311G(d,p) and CIS/6-311G(d,p) level.[86] In the hydrated form, three water molecules were considered in the first solvation shell. The structure and hydrogen bond lengths of the hydrated forms are shown in (Figure 17-6). All geometries were found to be minima at their respective potential energy surfaces. It should be noted that there was a convergence problem for the geometry optimizations of the cytosine and its hydrated form in the lowest singlet $\pi\pi^*$ excited state with the 6-311G(d,p) basis set. Therefore, geometry optimizations were performed using the 4-31G basis set for all atoms except the amino nitrogen for which the 6-311+G(d) basis set was employed.[86] Further, geometry optimization of the lowest singlet $\pi\pi^*$ excited states of the hydrated form was not successful. As the geometry optimizations proceeded, the excitation energy went down to about 0.02 eV, and the calculation aborted due to the lack of the convergence.[86] The ground state geometry of cytosine is planar, while the $S_1(\pi\pi^*)$ excited state geometry is nonplanar mainly around the N1C6C5C4 fragment. The amino group is pyramidal in both states. In the lowest singlet $n\pi^*$ excited state ($S_2(n\pi^*)$), the amino group, is considerably rotated, and the N3 atom lies appreciably out-of-plane.[49,86] Such deformation was explained as being a result of the excitation of the N3 lone pair. The rotation of the amino group was speculated due to the partial mixing in excitation due to the amino lone pair electron.[49,86] Under hydration with three water molecules, the amino group is more planar (Figure 17-6a). Further, the mode hydration in the $S_2(n\pi^*)$ excited state is completely modified (Figure 17-6b). The N3 site provides a repulsive potential to hydrogen bonding. However, the repulsive nature of the N3 atom in this state is well justified since the lone pair electron is promoted to the antibonding orbitals.

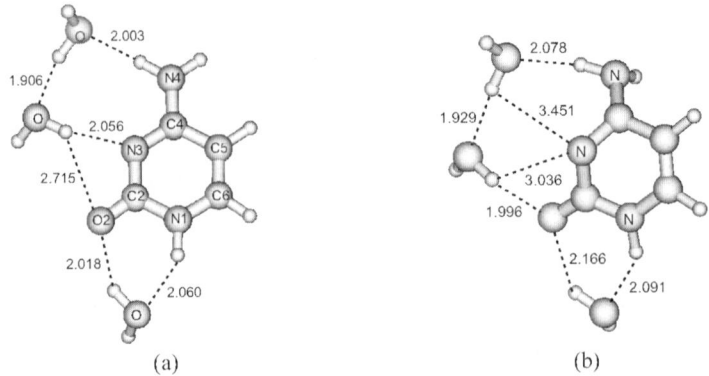

(a)   (b)

Figure 17-6. Structure and hydrogen bond lengths of hydrated cytosine (a) in the ground state, and (b) in the $S_2(n\pi^*)$ excited state.

4. EXCITED STATE PROPERTIES OF WATSON-CRICK BASE PAIRS

A theoretical evaluation of the singlet excited state property of the Watson-Crick AU, AT and GC base pairs was recently performed.[53,54] The ground state geometries were optimized at the HF level, while the excited states were generated at the CIS level. The excited state geometries were optimized at the same level under the constraint of planarity. The 6-31++G(d,p) basis set was used in the study. The computed vertical singlet $\pi\pi^*$ and $n\pi^*$ transitions were assigned to the excitation of the monomeric unit. This result is in agreement with the experimental observation of the AT and GC polymers and natural DNA bases where the electronic transitions were assigned to the corresponding monomer bases.[128] Some states of the charge transfer type were also found lying slightly higher in energy. These types of states (charge transfer type) are characterized by the excitation of electrons from the occupied orbitals of one base to the virtual orbitals of the complementary base of the base pair. It was found that the base pair formation does not have a significant effect on the singlet $\pi\pi^*$ transition energies of the constituent bases. However, the $n\pi^*$ transition energies were found to be appreciably blue-shifted. Such a blue shift is in accordance with the experimental findings concerning $n\pi^*$ transitions in hydrogen-bonded environments.[8,9,129] The change in the excited state geometry was found to be localized to the monomer which was involved in the excitation of the complex. The basis set superposition error (BSSE)-corrected interaction energies of the base pair formation were also calculated in the ground and excited state.[53,54]

The hydrogen bond parameters and interaction energies of the base pairs in the ground and excited states shown in Table 17-7 suggest that the hydrogen bond properties and interaction energies for the AT and AU base pairs are similar. The hydrogen bonds are more nonlinear in the singlet $n\pi^*$ states than those in the ground and singlet $\pi\pi^*$ states. It follows from the results of the calculations that the base pairs are appreciably destabilized due to $n\pi^*$ excitations.

Table 17-7. Hydrogen bond lengths (Å), hydrogen bond angles (°) and interaction energies (E_{int}, kcal/mol) of the AT, AU and GC base pairs in the ground and different singlet excited states[a, 53, 54]

AT base pair	S_0	$S_1(\pi\text{-}\pi^*)$	$S_2(\pi\text{-}\pi^*)$	$S_4(n\text{-}\pi^*)$	$S_6(n\text{-}\pi^*)$
N6...O4'	3.083	3.011	3.027	3.778	3.094
N1...N3'	3.023	3.071	3.025	3.090	3.152
H61(N6)...O4'	2.090	2.016	2.029	2.778	2.106
N1...H3'	2.010	2.063	2.014	2.086	2.149
O4'H61N6	172.6	171.1	174.5	161.8	170.2
N3'H3'N1	177.7	174.8	177.2	183.2	174.3
E_{int}	−9.9	−10.0	−10.6	−5.8	−7.1
AU base pair	S_0	$S_1(\pi\pi^*)$	$S_3(\pi\pi^*)$	$S_4(n\pi^*)$	$S_6(n\pi^*)$
N6...O4'	3.082	3.011	3.029	3.742	3.092
N1...N3'	3.019	3.066	3.022	3.087	3.146
H61(N6)...O4'	2.088	2.016	2.032	2.786	2.104
N1...H3'	2.007	2.059	2.010	2.083	2.143
O4'H61N6	172.6	171.0	174.2	161.9	170.2
N3'H3'N1	177.6	174.6	177.0	183.0	174.1
E_{int}	−10.1	−10.2	−10.7	−5.9	−7.3
GC base pair	S_0	$S_1(\pi\text{-}\pi^*)$	$S_2(\pi\text{-}\pi^*)$		
O6...N4'	2.932	2.937	2.958		
N1...N3'	3.053	3.023	3.066		
N2...O2'	3.028	3.037	3.098		
O6...H41'(N4')	1.926	1.929	1.954		
H1(N1)...N3'	2.048	2.021	2.059		
H21(N2)...O2'	2.028	2.034	2.100		
O6H41'N4'	175.7	177.7	177.3		
N1H1N3'	175.3	176.3	176.4		
N2H21O2'	176.9	177.4	177.7		
E_{int}	−24.8	−22.9	−20.4		

[a] States are given in the ascending energy order (see refs. 53, 54).

5.   CONCLUDING REMARKS

The structure and properties of molecular systems in the electronic excited states are different than the corresponding ground states. In order to obtain information about the excited state properties of complex molecular systems, different techniques should be used. An ideal situation would be a comprehensive experimental and theoretical study of a given system. However, there are several problems with this. It is difficult to model exact experimental conditions and is formidable in the study of excited states of complex biomolecular systems like nucleic acid bases. Experimental methods on the other hand also have limitations. For example, the experimental determination of excited state geometries of large complex systems not possible currently; however, only some limited information about the nature of geometry can be acquired. Reliable ab initio theoretical methods on the other hand can give somewhat precise information about excited state geometries. In fact theoretical and experimental methods are complementary to each other. Although excellent progress has been made from a theoretical point of view regarding the study of ground state properties of molecules, such advancement is still lacking for the excited state. The application of the MCSCF method for the evaluation of excited state properties and especially geometries are limited to smaller systems; other relatively less expensive methods such as CIS do not include electron correlation. It is evident from the present article that molecular geometries in excited states are generally nonplanar compared to the corresponding ground state geometries. Theoretical methods are becoming more relied upon among techniques used to study excited states of DNA fragments. In the near future we hope that tremendous development of state-of-the-art algorithms to study excited states of complex systems such as NABs and large fragments of nucleic acids (oligonucleic acids) will make possible such investigations in a reliable routine manner.

ACKNOWLEDGEMENTS

Authors are thankful to NSF-CREST grant No. 9805465 & 9706268, ONR grant No. N00014-03-1-0116 and the NIH-SCORE grant No. 3-S06 GM008047 31S1 for financial support. Authors are especially thankful to Prof. M. Kasha, Institute of Molecular Biophysics, Department of Chemistry and Biochemistry, Florida State University, Tallahassee, Florida, who has devoted his extremely valuable time in reading the manuscript and provided very encouraging and stimulating suggestions.

REFERENCES

1. Saenger, W. *Principles of Nucleic Acid Structures*, Springer-Verlag, New York, 1984.
2. Leszczynski, J. (2000). Isolated, Solvated, and Complexed Nucleic Acid Bases: Structures and Properties. In *Advances in Molecular Structure and Research*, **6**, Hargittai, M., Hargittai, I. (eds.,) JAI Press Inc., Stamford, Connecticut, 209-265.
3. Shukla, M.K., Leszczynski, J. (2000). A Theoretical Study of Proton Transfer in Hypoxanthine Tautomers: Effects of Hydration. *J. Phys. Chem. A* **104**, 3021-3027.
4. Shukla, M.K., Leszczynski, J. (2000). A DFT Investigation on Effects of Hydration on the Tautomeric Equilibria of Hypoxanthine *J. Mol. Struct. (THEOCHEM)* **529**, 99-112.
5. Gorb, L., Leszczynski, J. (1998). Intramolecular Proton Transfer in Mono- and Dihydrated Tautomers of Guanine: An ab Initio Post Hartree—Fock Study. *J. Am. Chem. Soc*, **120**, 5024-5032.
6. Gorb, L., Podolyan, Y., Leszczynski, J., Siebrand, W., Fernandez-Ramos, A., Smedarchina, Z. (2002). A Quantum-Dynamics Study of the Prototropic Tautomerism of Guanine and Its Contribution to Spontaneous Point Mutations in Escherichia Coli. *Biopolymers (Nuc. Acid Sci)*. **61**, 77-83.
7. Eisinger, J., Shulman, R.G. (1968). Excited Electronic States of DNA. *Science* **161**, 1311-1319.
8. Eisinger, Lamola, A.A. (1971). The Excited States of Nucleic Acids. In *Excited State of Proteins and Nucleic Acids*, Steiner, R.F., Weinryb, I. (eds.,) Plenum Press, New York-London, 107-198.
9. Callis, P.R. (1983). Electronic States and Luminescence of Nucleic Acid Systems. *Ann. Rev. Phys. Chem. 34*, 329-357.
10. Crespo-Hernandez, C.E., Cohen, B., Hare, P.M., Kohler, B. (2004). Ultrafast Excited-State Dynamics in Nucleic Acids. *Chem. Rev.* 2004, **104**, 1977-2019.
11. Kang, H., Lee, K.T., Jung, B., Ko, Y.J., Kim. S.K. (2002). Intrinsic Lifetimes of the Excited State of DNA and RNA Bases. *J. Am. Chem. Soc.* **124**, 12958-12959.
12. Peon, J., Zewail, A.H. (2001). DNA/RNA Nucleotides and Nucleosides: Direct Measurement of Excited-State Lifetimes by Femtosecond Fluorescence Up-Conversion. *Chem. Phys. Lett.* **348**, 255-262.
13. Cohen, B., Hare, P.M., Kohler, B. (2003). Ultrafast Excited-State Dynamics of Adenine and Monomethylated Adenines in Solution: Implications for the Nonradiative Decay Mechanism. *J. Am. Chem. Soc.* **125**, 13594-13601.
14. Gustavsson, T., Sharonov, A., Markovitsi, D. (2001). Thymine, Thymidine and Thymidine 5'-Monophosphate Studied by Femtosecond Fluorescence Upconversion Spectroscopy. *Chem. Phys. Lett.* **351**, 195-200.
15. Onidas, D., Markovitsi, D., Marguet, S., Sharonov, A., Gustavsson, T. (2002). Fluorescence Properties of DNA Nucleosides and Nucleotides: A Refined Steady-State and Femtosecond Investigation. *J. Phys. Chem. B* **106**, 11367-11374.
16. Sharonov, A., Gustavsson, T., Carre, V., Renault, E., Markovitsi, D. (2003). Cytosine Excited State Dynamics Studied by Femtosecond Fluorescence Upconversion and Transient Absorption Spectroscopy. *Chem. Phys. Lett.* **380**, 173-180.
17. Malone, R.J., Miller, A.M., Kohler, B. (2003). Singlet Excited-state Lifetimes of Cytosine Derivatives Measured by Femtosecond Transient Absorption. *Photochem. Photobiol.* **77**, 158-164.
18. Lim, E.C. (1986). Proximity Effect in Molecular Photophysics: Dynamical Consequences of Pseudo-Jahn-Teller Interaction. *J. Phys. Chem.* **90**, 6770-6777.

19. Broo, A. (1998). A Theoretical Investigation of the Physical Reason for the Very Different Luminescence Properties of the Two Isomers Adenine and 2-Aminopurine. *J. Phys. Chem. A* **102**, 526-531.

20. Sobolewski, A.L., Domcke, W. (2002). On the Mechanism of Nonradiative Decay of DNA Bases: ab initio and TDDFT Results for the Excited States of 9H-Adenine. *Eur. Phys. J. D* **20**, 369-374.

21. Ebata, T., Minejima, C., Mikami, N. (2002). A New Electronic State of Aniline Observed in the Transient IR Absorption Spectrum from S_1 in a Supersonic Jet. *J. Phys. Chem. A* **106**, 11070-11074.

22. Kang, H., Jung, B., Kim, S.K. (2003). Mechanism for Ultrafast internal Conversion of Adenine. *J. Chem. Phys.* **118**, 6717-6719.

23. He, Y., Wu, C., Kong, W. (2003). Decay Pathways of Thymine and Methyl-Substituted Uracil and Thymine in the Gas Phase. *J. Phys. Chem. A* **107**, 5145-5148.

24. Pancur, T., Schwalb, N.K., Renth, F., Temps, F. (2005). Femtosecond Fluorescence Up-conversion Spectroscopy of Adenine and Adenosine: Experimental Evidence for the πσ* State? *Chem. Phys.* **313**, 199-212.

25. Hunig, I., Plutzer, C., Seefeld, K.A., Lowenich, D., Nispel, M., Kleinermanns, K. (2004). Photostability of Isolated and Paired Nucleobases: N-H Dissociation of Adenine and Hydrogen Transfer in its Base Pairs Examined by Laser Spectroscopy. *ChemPhysChem* **5**, 1427-11431.

26. Samoylova, E., Lippert, H., Ullrich, S., Hertel, I.V., Radloff, W., Schultz, T. (2005). Dynamics of Photoinduced Processes in Adenine and Thymine Base Pairs. *J. Am. Chem. Soc.* **127**, 1782-1786.

27. Ullrich, S., Schultz, T., Zgierski, M.Z., Stolow, A. (2004). Electronic Relaxation Dynamics in DNA and RNA Bases Studied by Time-Resolved Photoelectron Spectroscopy. *Phys. Chem. Chem. Phys.* **6**, 2796-2801.

28. Canuel, C., Mons, M., Piuzzi, F., Tardivel, B., Dimicoli, I., Elhanine, M. (2005). Excited States Dynamics of DNA and RNA Bases: Characterization of a Stepwise Deactivation Pathway in the Gas Phase. *J. Chem. Phys.* **122**, 74316-74321.

29. Zierhut, M., Roth, W., Fischer, I. (2004). Dynamics of H-atom Loss in Adenine. *Phys. Chem. Chem. Phys.* **6**, 5178-5183.

30. Cohen, B., Crespo-Hernandez, C.E., Kohler, B. (2004). Strickler-Berg Analysis of Excited Singlet State Dynamic in DNA and RNA Nucleosides. *Faraday Discuss.* **127**, 137-147.

31. Kim, N.J., Kang, H., Park, Y.D., Kim, S.K. (2004). Dispersed Fluorescence Spectroscopy of Jet-Cooled Adenine. *Phys. Chem. Chem. Phys.* **6**, 2802-2805.

32. Perun, S., Sobolewski, A.L., Domcke, W. (2005). Photostability of 9H-Adenine: Mechanisms of the Radiationless Deactivation of the Lowest Excited Singlet States. *Chem Phys.* **313**, 107-112.

33. Perun, S., Sobolewski, A.L., Domcke, W. (2005). Ab Initio Studies on the Radiationless Decay Mechanisms of the Lowest Excited Singlet States of 9H-Adenine. *J. Am. Chem. Soc.* **127**, 6257-6265.

34. Merchan, M., Serrano-Andres, L. (2003). Ultrafast Internal Conversion of Excited Cytosine via the Lowest ππ* Electronic Singlet State. *J. Am. Chem. Soc.* **125**, 8108-8109.

35. Blancafort, L., Robb, M.A. (2004). Key Role of a Threefold State Crossing in the Ultrafast Decay of Electronically Excited Cytosine. *J. Phys. Chem. A* **108**, 10609-10614.

36. Blancafort, L., Cohen, B., Hare, P.M., Kohler, B., Robb, M.A. (2005). Singlet Excited-State Dynamics of 5-Fluorocytosine and Cytosine: An Experimental and Computational Study. *J. Phys. Chem. A* **109**, 4431-4436.

37. Matsika, S. (2004). Radiationless Decay of Excited States of Uracil through Conical Intersections. *J. Phys. Chem. A* **108**, 7584-7590.
38. Langer, H., Doltsinis, N.L. (2004). Nonradiative Decay of Photoexcited Methylated Guanine. *Phys. Chem. Chem. Phys.* **6**, 2742-2748.
39. Woutersen, S., Cristalli, G. (2004). Strong Enhancement of Vibrational Relaxation by Watson-Crick Base Pairing. *J. Chem. Phys.* **121**, 5381-5386.
40. Schultz, T., Samoylova, E., Radloff, W., Hertel, I.V., Sobolewski, A.J., Domcke, W. (2004). Efficient Deactivation of a Model Base Pair via Excited-State Hydrogen Transfer. *Science* **306**, 1765-1768.
41. Sobolewski, A.L., Domcke, W. (2004). Ab Initio Studies on the Photophysics of the Guanine–Cytosine Base Pair. *Phys. Chem. Chem. Phys.* **6**, 2763-2771.
42. Šponer, J., Hobza, P., Leszczynski, J. (2000). Aromatic DNA Base Stacking and H-Bonding. In *Computational Chemistry: Reviews of Current Trends*, **5**, Leszczynski, J., (ed.,) World Scientific, 171-210.
43. Hobza, P., Šponer, J. (1999). Structure, Energetics, and Dynamics of the Nucleic Acid Base Pairs: Nonempirical *Ab Initio* Calculations. *Chem. Rev.* **99**, 3247-3276.
44. Šponer, J., Hobza, P., Leszczynski, J. (1999). Computational Approaches to the Studies of the Interactions of Nucleic Acid Bases. In *Computational Molecular Biology, Theoretical and Computational Chemistry Book Series*, **8**, Leszczynski, J., (ed.,) Elsevier, 85-117.
45. Gorb, L., Podolyan, Y., Dziekonski, P., Sokalaski, W.A., Leszczynski, J. (2004). Double-Proton Transfer in Adenine-Thymine and Guanine-Cytosine Base Pairs. A Post-Hartree-Fock ab Initio Study. *J. Am. Chem. Soc.* **126**, 10119.
46. Kurita, N., Danilov, V.I., Anisimov, V.M. (2005). The Structure of Watson–Crick DNA Base Pairs obtained by MP2 Optimization. *Chem. Phys. Lett.* **404**, 164-170.
47. Brady, B.B., Peteanu, L.A., Levy, D.H. (1988). The Electronic Spectra of the Pyrimidine Bases Uracil and Thymine in a Supersonic Molecular Beam. *Chem. Phys. Lett.* **147**, 538-547.
48. Chinsky, L., Laigle, A., Peticolas, L., Turpin, P.-Y. (1982). Excited State Geometry of Uracil from the Resonant Raman Overtone Spectrum using a Kramers–Kronig Technique. *J. Chem. Phys.* **76**, 1-5.
49. Shukla, M.K., Mishra, P.C. (1999). A Gas Phase ab initio Excited State Geometry Optimization Study of Thymine, Cytosine and Uracil *Chem. Phys.* **240**, 319-329.
50. Shukla, M.K., Leszczynski, J. (2003). Excited States of Nucleic Acid Bases. In *Computational Chemistry: Reviews of Current Trends*, **8**, Leszczynski, J., (ed., World Scientific, 249-344.
51. Mishra, S.K., Mishra, P.C. (2001). An ab initio Study of Electronic Structure and Spectra of 8-Bromoguanine: A Comparative Study with Guanine. *Spectrochim. Acta A* **57**, 2433-2450.
52. Shukla, M.K., Leszczynski, J (2003). Electronic Spectra, Excited State Geometries and Molecular Electrostatic Potentials of Hypoxanthine: A Theoretical Investigation. *J. Phys. Chem. A* **107**, 5538-5543.
53. Shukla, M.K., Leszczynski, J. (2002). Theoretical Investigation on Excited State Properties of Adenine-Uracil Base Pair *J. Phys. Chem. A* **106**, 1011-1018.
54. Shukla, M.K., Leszczynski, J. (2002). A Theoretical Study of Excited State Properties of Adenine-Thymine and Guanine-Cytosine Base Pairs. *J. Phys. Chem. A* **106**, 4709-4717.
55. Shukla, M.K., Leszczynski, J. (2000). Investigations of the Excited State Properties of Isocytosine: An ab initio Approach. *Int. J. Quantum Chem.* **77**, 240-254.

56. Shukla, M.K., Kumar, A., Mishra, P.C. (2001). An Ab Initio Study of Excited State Molecular Electrostatic Potential Maps and Other Related Properties of 5-Fluorouracil: A Comparative Study with Uracil. *J. Mol. Struct. (THEOCHEM)* **535**, 269-277.

57. Shukla, M.K., Mishra, P.C. (1998). Excited State Molecular Electric Properties of Some Biologically Important Purines, Pyrimidines and Azines: An ab initio Study. *J. Chem. Infor. Computer Sci.* **38**, 678-684.

58. Shukla, M.K., Mishra, P.C. (1998). Electronic Structure, Spectra and Excited State Properties of N,N-Dimethylaminoguanine. *Spectrochim. Acta* **54A**, 937-946.

59. Shukla, M.K., Mishra, S.K., Kumar, A., Mishra, P.C. (2000). An ab initio Study of Excited States of Guanine in the Gas Phase and Aqueous Media: Electronic Transitions and Mechanism of Spectral Oscillations. *J. Comput. Chem.* **21**, 826-846.

60. Mishra, S.K., Shukla, M.K., Mishra, P.C. (2000). Electronic Spectra of Adenine and 2-Aminopurine: An ab initio Study of Energy Level Diagrams of Different Tautomers in Gas Phase and Aqueous Solution. *Spectrochim. Acta* **56A**, 1355-1384.

61. Salter, L.M., Chaban, G.M. (2002). Theoretical Study of Gas Phase Tautomerization Reactions for the Ground and First Excited Electronic States of Adenine. *J. Phys. Chem. A* **106**, 4251-4256.

62. Mennucci, B., Toniolo, A., Tomasi, J. (2001). Theoretical Study of the Photophysics of Adenine in Solution: Tautomerism, Deactivation Mechanisms, and Comparison with the 2-Aminopurine Fluorescent Isomer. *J. Phys. Chem. A* **105**, 4749-4757.

63. Mennucci, B., Toniolo, A., Tomasi, J. (2001). Theoretical Study of Guanine from Gas Phase to Aqueous Solution: Role of Tautomerism and Its Implications in Absorption and Emission Spectra. *J. Phys. Chem. A* **105**, 7126-7134.

64. Holmen, A., Broo, A., Albinsson, B., Norden, B. (1997). Assignment of Electronic Transition Moment Directions of Adenine from Linear Dichroism Measurements. *J. Am. Chem. Soc.* **119**, 12240-12250.

65. Petke, J.D., Maggiora, G.M., Christoffersen, R.F. (1990). Ab initio Configuration Interaction and Random Phase Approximation Calculations of the Excited Singlet and Triplet States of Adenine and Guanine. *J. Am. Chem. Soc.* **112**, 5452-5460.

66. Fulscher, M.P., Serrano-Andres, L., Roos, B.O. (1997). A Theoretical Study of the Electronic Spectra of Adenine and Guanine. *J. Am. Chem. Soc.* **119**, 6168-6176.

67. Lorentzon, J., Fulscher, M.P., Roos, B.O. (1995). Theoretical Study of the Electronic Spectra of Uracil and Thymine. *J. Am. Chem. Soc.* **117**, 9265-9273.

68. Petke, J.D., Maggiora, G.M., Christoffersen, R.E. (1992). Ab initio Configuration Interaction and Random Phase Approximation Calculations of the Excited Singlet and Triplet States of Uracil and Cytosine. *J. Phys. Chem.* **96**, 6992-7001.

69. Fulscher, M.P., Roos, B.O. (1995). Theoretical Study of the Electronic Spectrum of Cytosine. *J. Am. Chem. Soc.* **1995**, *117*, 2089-2095.

70. Rachofsky, E.L., Roos, J.B.A., Krauss, M., Osman, R. (2001). CASSCF Investigation of Electronic Excited States of 2-Aminopurine. *J. Phys. Chem. A* **105**, 190-197.

71. Borin, A.C., Serrano-Andres, L., Fulscher, M.P., Roos, B.O. (1999). A Theoretical Study of the Electronic Spectra of N_9 and N_7 Purine Tautomers. *J. Phys. Chem. A* **103**, 1838-1845.

72. Ohrn, A., Christiansen, O. (2001). Electronic Excitation Energies of Pyrimidine Studied using Coupled Cluster Response Theory. *Phys. Chem. Chem. Phys.* **3**, 730-740.

73. Catalan, J. (2002). Feasibility of Adenine Photoinduced Mispairing of the Watson-Crick Pairing in DNA. *J. Phys. Chem. B* **106**, 11384-11390.

74. Fischer, G., Cai, Z.-L., Reimers, J.R., Wormell, P. (2003). Singlet and Triplet Valence Excited States of Pyrimidine. *J. Phys. Chem. A* **107**, 3093-3103.

75. Chenon, M.T., Pugmire, R.J., Grant, D.M., Panzica, R.P., Townsend, L.B. (1975). Carbon-13 Magnetic Resonance. XXVI. A Quantitative Determination of the Tautomeric Populations of Certain Purines. *J. Am. Chem. Soc.* **97**, 4636-4642.

76. Nowak, M.J., Lapinski, L., Kwiatkowski, J.S., Leszczynski, J. (1996). Molecular Structure and Infrared Spectra of Adenine. Experimental Matrix Isolation and Density Functional Theory Study of Adenine 15N Isotopomers. *J. Phys. Chem.* **100**, 3527-3534.

77. Sheina, G.G., Stepanian, S.G., Radchenko, E.D., Blagoi, Yu.P. (1987). IR Spectra of Guanine and Hypoxanthine isolated Molecules. *J. Mol. Struct.* **158**, 275-292.

78. Mons, M., Dimicoli, I., Piuzzi, F., Tardivel, B., Elhanine, M. (2002). Tautomerism of the DNA Base Guanine and Its Methylated Derivatives as Studied by Gas-Phase Infrared and Ultraviolet Spectroscopy. *J. Phys. Chem. A* **106**, 5088-5094.

79. Szczesniak, M., Szczepaniak, K., Kwiatowski, J.S., KuBulat, K., Person, W.B. (1988). Matrix Isolation Infrared Studies of Nucleic Acid Constituents. 5. Experimental Matrix-Isolation and Theoretical ab Initio SCF Molecular Orbital Studies of the Infrared Spectra of Cytosine Monomers. *J. Am. Chem. Soc.* **110**, 8319-8330.

80. Kwiatkowski, J.S., Leszczynski, J. (1996). Molecular Structure and Vibrational IR Spectra of Cytosine and Its Thio and Seleno Analogues by Density Functional Theory and Conventional ab Initio Calculations. *J. Phys. Chem.* **100**, 941-953.

81. Drefus, M., Bensaude, O., Dodin, G., Dubois, J.E. (1976). Tautomerism in Cytosine and 3-Methylcytosine. A Thermodynamic and Kinetic Study. *J. Am. Chem. Soc.* **98**, 6338-6349.

82. Dong, F., Miller, R.E. (2002). Vibrational Transition Moment Angles in Isolated Biomolecules: A Structural Tool. *Science* **298**, 1227-1230.

83. Langer, H., Doltsinis, N.L. (2003). Excited State Tautomerism of the DNA Base Guanine: A Restricted Open-Shell Kohn–Sham Study. *J. Chem. Phys.* **118**, 5400-5407.

84. Kang, H., Lee, K.T., Kim, S.K. (2002). Femtosecond Real Time Dynamics of Hydrogen Bond Dissociation in Photoexcited Adenine–Water Clusters. *Chem. Phys. Lett.* **359**, 213-219.

85. Kim, N.J., Jeong, G., Kim, Y.S., Sung, J., Kim, S.K., Park, Y.D. (2000). Resonant Two-Photon Ionization and Laser Induced Fluorescence Spectroscopy of Jet-Cooled Adenine. *J. Chem. Phys.* **113**, 10051-10055.

86. Shukla, M.K., Leszczynski, J. (2002). Interaction of Water Molecules with Cytosine Tautomers: An Excited State Investigation. *J. Phys. Chem. A* **106**, 11338-11346.

87. Shukla, M.K., Leszczynski, J. (2002). Phototautomerism in Uracil: A Quantum Chemical Investigation. *J. Phys. Chem. A* **106**, 8642-8650.

88. Stewart, R.F., Davidson, J. (1963). Polarized Absorption Spectra of Purines and Pyrimidines. *J. Chem. Phys.* **39**, 255-266.

89. Inagaki, T., Ito, A., Heida, K., Ho, T. (1986). Photoacoustic Spectra of Some Biological Molecules Between 300 and 130 nm. *Photochem. Photobiol.* **44**, 303-306.

90. Brunner, W.C., Maestre, M.F. (1975). Circular Dichroism of Some Mononucleosides. *Biopolymers* **14**, 555-565.

91. Sprecher, C.A., Johnson, Jr., W.C. (1977). Circular Dichroism of the Nucleic Acid Monomers. *Biopolymers* **16**, 2243-2264.

92. Miles, D.W., Hann, S.J., Robins, R.K., Eyring, H. (1968). Vicinal Effects on the Optical Activity of Some Adenine Nucleosides. *J. Phys. Chem.* **72**, 1483-1491.

93. Clark, L.B., Peschel, G.G., Tinoco, Jr., I. (1965). Vapor Spectra and Heats of Vaporization of Some Purine and Pyrimidine Bases. *J. Phys. Chem.* **69**, 3615-3618.

94. Voet, D., Gratzer, W.B., Cox, R.A., Doty, P. (1963). Absorption Spectra of Nucleotides, Polynucleotides, and Nucleic Acids in the far Ultraviolet. *Biopolymers* **1**, 193-208.

95. Clark, L.B.(1989). Polarization Assignments in the 270-nm Band of the Adenine Chromophore. *J. Phys. Chem.* **93**, 5345-5347.

96. Clark, L.B. (1990). Electronic Spectrum of the Adenine Chromophore. *J. Phys. Chem.* **94**, 2873-2879.

97. Clark, L.B. (1995). Transition Moments of 2'-Deoxyadenosine. *J. Phys. Chem.* **99**, 4466-4470.

98. Santhosh, C., Mishra, P.C. (1990). Electronic Spectra of Adenine: Interaction with Dissolved Oxygen in Solution, Oscillation and Intensification of n-π^* Transition. *J. Mol. Struc.* **220**, 25-41.

99. Voelter, W., Records, R., Bunnenberg, E., Djerassi, C. (1968). Magnetic Circular Dichroism Studies. VI. Investigation of Some Purines, Pyrimidines, and Nucleosides. *J. Am. Chem. Soc.* **90**, 6163-6170.

100. Yamada, T., Fukutome, H. (1968). Vacuum Ultraviolet Absorption Spectra of Sublimed Films of Nucleic Acid Bases. *Biopolymers* **6**, 43-54.

101. Sutherland, J.C., Griffin, K. (1984). Magnetic Circular Dichroism of Adenine, Hypoxanthine, and Guanosine 5'-Diphosphate to 180 nm. *Biopolymers* **23**, 2715-2724.

102. Clark, L.B., Tinoco, I. (1965). Correlations in the Ultraviolet Spectra of the Purine and Pyrimidine Bases. *J. Am. Chem. Soc.* **87**, 11-15.

103. Clark, L.B. (1977). Electronic Spectra of Crystalline 9-Ethylguanine and Guanine Hydrochloride. *J. Am. Chem. Soc.* **99**, 3934-3938.

104. Clark, L.B. (1994). Electronic Spectra of Crystalline Guanosine: Transition Moment Directions of the Guanine Chromophore. *J. Am. Chem. Soc.* **116**, 5265-5270.

105. Santhosh, C., Mishra, P.C. (1989). Complexation of Guanine with Dissolved Oxygen in Solution and Study of Excited-State Lifetime. *J. Mol. Struct.* **198**, 327-337.

106. Theiste, D., Callis, P.R., Woody, R.W. (1991). Effects of the Crystal Field on Transition Moments in 9-Ethylguanine. *J. Am. Chem. Soc.* **113**, 3260-3267.

107. Matsuoks, Y., Norden, B. (1982). Linear Dichroism Studies of Nucleic Acid Bases in Stretched Poly(vinyl alcohol) Film. Molecular Orientation and Electronic Transition Moment Directions. *J. Phys. Chem.* **86**, 1378-1386.

108. Nir, E., Janzen, Ch., Imhof, P., Kleinermanns, K., de Vries, M.S. (2001). Guanine Tautomerism Revealed by UV–UV and IR–UV Hole Burning Spectroscopy. *J. Chem. Phys.* **115**, 4604-4611.

109. Nir, E., Kleinermanns, K., Grace, L., de Vries, M.S. (2001). On the Photochemistry of Purine Nucleobases. *J. Phys. Chem. A* **105**, 5106-5110.

110. Nir, E., Kleinermanns, K., de Vries, M.S. (2000). Pairing of Isolated Nucleic-Acid Bases in the Absence of the DNA Backbone. *Nature* **408**, 949-951.

111. Nir, E., Imhof, P., Kleinermanns, K., de Vries, M.S. (2000). REMPI Spectroscopy of Laser Desorbed Guanosines. *J. Am. Chem. Soc.* **122**, 8091-8092.

112. Nir, E., Muller, M., Grace, L.I., de Vries, M.S. (2002). REMPI spectroscopy of cytosine. *Chem. Phys. Lett.* **355**, 59-64.

113. Nir, E., Hunig, I., Kleinermanns, K., de Vries, M.S. (2004). Conformers of Guanosines and their Vibrations in the Electronic Ground and Excited States, as Revealed by Double-Resonance Spectroscopy and Ab Initio Calculations. *ChemPhysChem* **5**, 131-137.

114. Piuzzi, F., Mons, M., Dimicoli, I., Tardivel, B., Zhao, Q. (2001). Ultraviolet Spectroscopy and Tautomerism of the DNA Base Guanine and its Hydrate Formed in a Supersonic Jet. *Chem. Phys.* **270**, 205-214.

115. Isaacson, M. (1972). Interaction of 25 keV Electrons with the Nucleic Acid Bases, Adenine, Thymine, and Uracil. I. Outer Shell Excitation. *J. Chem. Phys.* **56**, 1803-1812.

116. Becker, R.S., Kogan, G. (1980). Photophysical Properties of Nucleic Acid Components-1. The Pyrimidines: Thymine, Uracil, N,N-Dimethyl Derivatives and Thymidine, *Photochem. Photobiol.* **31**, 5-13.

117. Hug, W., Tinoco, I. (1973). Electronic Spectra of Nucleic Acid Bases. I. Interpretation of the in-plane Spectra with the Aid of All Valence Electron MO-CIA [configuration interaction] Calculations. *J. Am. Chem. Soc.* **95**, 2803-2813.

118. Novros, J.S., Clark, L.B. (1986). On the Electronic Spectrum of 1-Methyluracil. *J. Phys. Chem.* **90**, 5666-5668.

119. Holmen, A., Broo, A., Albinsson, B. (1994). IR Transition Moments of 1,3-Dimethyluracil: Linear Dichroism Measurements and ab Initio Calculations. *J. Phys. Chem.* **98**, 4998-5009.

120. Fujii, M., Tamura, T., Mikami, N., Ito, M. (1986). Electronic Spectra of Uracil in a Supersonic Jet. *Chem Phys. Lett.* **126**, 583-587.

121. Zaloudek, F., Novros, J.S., Clark, L.B. (1985). The Electronic Spectrum of Cytosine. *J. Am. Chem. Soc.* **1985**, *107*, 7344-7351.

122. Johnson, Jr., W.C., Vipond, P.M., Girod, J.C. (1971). Tautomerism on Cytidine. *Biopolymers* **10**, 923-933.

123. Kaito, A., Hatano, M., Ueda, T., Shibuya, S. (1980). The Application of Magnetic Circular Dichroism to the Study of the Tautomerism of Cytosine and Isocytosine. *Bull. Chem. Soc. Jpn.* **53**, 3073-3078.

124. Raksanyi, K., Foldvary, I. (1978). The Electronic Structure of Cytosine, 5-Azacytosine, and 6-Azacytosine. *Biopolymers* **17**, 887-896.

125. Miles, D.W., Robins, M.J., Robins, R.K., Winkley, M.W., Eyring, H. (1969). Circular Dichroism of Nucleoside Derivatives. V. Cytosine Derivatives. *J. Am. Chem. Soc.* **91**, 831-838.

126. Inskeep, W.H., Miles, D.W., Eyring, H.(1970). Circular Dichroism of Nucleoside Derivatives. VIII. Coupled Oscillator Calculations of Molecules with Fixed Structure. *J. Am. Chem. Soc.* **92**, 3866-3872.

127. Miles, D.W., Robins, R.K., Eyring, H. (1967). Optical Rotatory Dispersion, Circular Dichroism, and Absorption Studies on Some Naturally Occurring Ribonucleosides and Related Derivatives. *Proc. Natl. Acad. Sci.* **57**, 1138-1145.

128. Chou, P.-J., Johnson, Jr., W.C. (1993). Base Inclinations in Natural and Synthetic DNAs. *J. Am. Chem. Soc.* **115**, 1205-1214.

129. Brealey, G.J., Kasha, M.(1955). The Role of Hydrogen Bonding in the n→π* Blue-shift Phenomenon. *J. Am. Chem. Soc.* **77**, 4462-4468.

Chapter 18

SUBSTITUENT EFFECTS ON HYDROGEN BONDS IN DNA
A Kohn-Sham DFT approach

Célia Fonseca Guerra and F. Matthias Bickelhaupt*

Afdeling Theoretische Chemie, Scheikundig Laboratorium der Vrije Universiteit, De Boelelaan 1083, NL-1081 HV Amsterdam, The Netherlands, Fax: +31 - 20 - 59 87629, E-mail: FM.Bickelhaupt@few.vu.nl

Abstract: In this Chapter, we discuss how the hydrogen bonds in Watson-Crick base pairs can be tuned both structurally and in terms of bond strength by exposing the DNA bases to different kinds of substitutions: (1) substitution in the $X–H\cdots Y$ hydrogen bonding moiety, (2) remote substitution, i.e., introducing substituents in the DNA base at positions not directly involved in hydrogen bonding; and (3) environment effects which can be conceived as supramolecular substitution. The hydrogen bonds in DNA Watson-Crick base pairs have long been considered predominantly electrostatic phenomena. Here, we show with state-of the art calculations that this is not true and that electrostatic interactions and covalent (i.e. orbital interaction) contributions in these hydrogen bonds are in fact of the same order of magnitude. We also discuss the role of electrostatic and orbital interactions in the mechanism behind the substituent effects on hydrogen bonds in DNA.

Key words: covalent interaction, density functional theory, DNA, hydrogen bonding, nucleic acids, substituent effects, Watson-Crick base pairs

1. INTRODUCTION

Hydrogen bonding plays an important role in the formation of deoxyribonucleic acid (DNA) base pairs[1] and has, therefore, been the subject of several theoretical studies.[2–4] DNA is macromolecule that consists of two helical strands of nucleotides that are twisted around each other thus rendering the well-known double helix.[1] It carries the genetic information of

J. Šponer and F. Lankaš (eds.), Computational Studies of RNA and DNA, 463–484.
© 2006 *Springer.*

living organisms, which is encoded by means of the order in which the four nucleic acids adenine (A), thymine (T), guanine (G) and cytosine (C) occur in a DNA strand. The two strands of the DNA double helix are held together by the hydrogen bonds of the Watson-Crick (WC) base pairs that form selectively between A and T, and between G and C (see Scheme 1). The strength of these hydrogen bonds, according to gas-phase experiments, is – 12.1 and –21.0 kcal/mol for AT and GC, respectively (bond enthalpies at 298 K).[5] This is an important contribution to the intrinsic stability of the DNA molecule.[6] It was shown that the experimental AT and GC hydrogen-bond strength are accurately, i.e., within 1 – 2 kcal/mol, reproduced with the generalized-gradient approximation of density functional theory (DFT) at BP86/TZ2P. This DFT approach is in fact suitable for describing the even weaker hydrogen bond in the neutral water dimer for which it yields an O–H•••O distance of 2.90 Å and a bond energy of –4.3 kcal/mol which agrees within a few hundredths of an Å and a few tenths of a kcal/mol with the corresponding CCSD(T) benchmark values.[2b] Thus, the forthcoming discussions will be based on BP86/TZ2P computations.

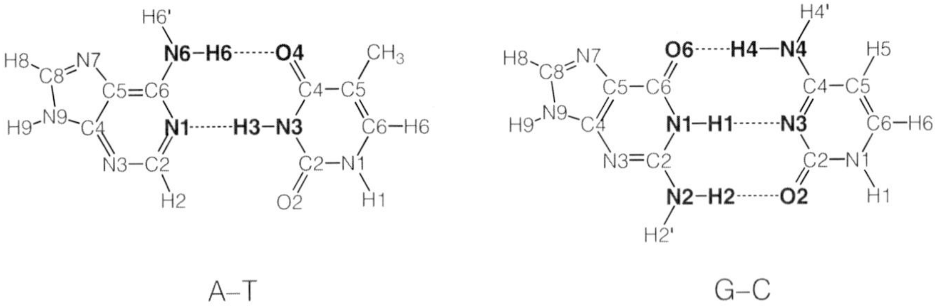

Scheme 1. AT and GC Watson-Crick pairs of Natural DNA.

In addition to providing stability to the carrier of genetic information, the cohesion of the two DNA strands through Watson-Crick pairs also has another important consequence: it causes the two strands to be complementary copies of each other in terms of the order of the composing nucleotides. This principle is the basis of enzyme-catalyzed DNA replication and transcription, two chemical reactions that play a key role in the highly complex biological processes of cell growth and gene expression. In these reactions, the strands of the double helix dissociate, and the genetic information is copied by using one of them as a template for making a new complementary strand. In the standard model, the immensely high accuracy with which this process occurs is ascribed to the hydrogen bonds between the complementary bases.[1] Recently, Kool et al.[7] found experimental evidence for shape complementarity also contributing to the accuracy of the replication.[6] Recently, we have shown that, proceeding from Kool's idea that the DNA base of the template strand and that of the incoming nucleotide must fit like the peaces of a puzzle into the active site of the DNA polymerase enzyme, Watson-Crick hydrogen bonding is still crucial for the process of biomolecular recognition.[2e,f] Note that this is in fact a synthesis of the standard and the steric model.

In this Chapter, we discuss how the hydrogen bonds in Watson-Crick pairs can be affected or tuned both structurally and in terms of bond strength by exposing the DNA bases to different kinds of substitutions. Our starting point, in Section 2, is the structure and bonding in the natural Watson-Crick pairs AT and GC (see Scheme 1) which, at variance with textbook-knowledge, are shown to have substantial covalent (i.e., donor–acceptor orbital interaction) character. Thereafter, in Sections 3 – 5, the following categories of substitutions and there effects on the hydrogen bonds are discussed: (i) substitution in the X–H•••Y hydrogen bonding moiety (see Scheme 2); (ii) remote substitution, i.e., introducing substituents in the DNA base at positions not directly involved in hydrogen bonding (see Scheme 3); and (iii) environment effects which, we argue, can be conceived as "supramolecular" substitution (see Scheme 4). Understanding how Watson-Crick hydrogen bonds react to modifications in or around the constituting DNA bases serves to deepen our understanding of their nature and behavior. However, in addition to this more fundamental aspect, such knowledge is also of direct relevance for achieving a more rational design of artificial DNA bases with the purpose of tuning the bonding capabilities, i.e., making them stronger or weaker as desired. On the long term, this is of relevance for applications in, for example, supramolecular chemistry[8] and antisense technology.[9]

2. HYDROGEN BONDS IN NATURAL WATSON-CRICK BASE PAIRS

2.1 Structure and Strength of Watson-Crick Hydrogen Bonds

The hydrogen bond enthalpies, computed at the BP86/TZ2P level of DFT, amount to –13.0 kcal/mol for AT and –26.1 kcal/mol for GC (see Table 18-1).[2a-c] This, as already pointed out in the introduction, agrees excellently with the experimental values[5] of –11.8 and –23.8 kcal/mol for AT and GC, respectively, but also with theoretical values of others. For example, the corresponding values obtained by Bertran et al.[4b] at B3LYP are –10.9 and –24.0 kcal/mol and -those obtained by Šponer et al.[4d] at MP2 are –9.5

and –20.8 kcal/mol. The corresponding hydrogen bond distances computed at BP86/TZ2P are 2.85 and 2.81 Å for N6–O4 and N1–N3 in AT, and 2.73, 2.88 and 2.87 Å for O6–N4, N1–N3 and N2–O2 in GC (see Scheme 1). Discrepancies with existing experimental data[10] are mainly due to the fact that the latter refer to the solid state in which DNA bases experience an environment that affects their structure. This is discussed in more detail in Section 5 (see also Figure 18-6).

2.2   Nature of Watson-Crick Hydrogen Bonds

Next, we address the nature of the forces that hold together the two bases and establish the importance of the different components in the interaction.[2a,c-g,11,12] To reveal this nature, the bond energy ΔE can be separated, at first, into the preparation energy ΔE_{prep} associated with deforming the individual DNA bases from their equilibrium structure into the geometry they adopt in the Watson-Crick pair and the actual interaction energy ΔE_{int} between these deformed bases in the base pair (eq. 1).[3b]

$$\Delta E = \Delta E_{prep} + \Delta E_{int} \qquad (1)$$

As the DNA bases undergo only minor geometrical changes in the Watson-Crick pairs, the corresponding preparation energies are small, 4 kcal/mol or less (see Table 18-1), and will not be further discussed. The interaction energy ΔE_{int} is however important and it has been analyzed in the framework of the Kohn-Sham MO model by quantitatively decomposing it into (i) the electrostatic attraction ΔV_{elstat} between the unperturbed charge distributions of the bases, (ii) Pauli repulsive orbital interactions ΔE_{Pauli} between occupied orbitals (c.f. 2-orbital 4-electron repulsion), and (iii) bonding orbital interactions ΔE_{oi} that account for charge transfer (i.e., donor–acceptor inter-actions between occupied orbitals on one moiety with unoccupied orbitals of the other, including the HOMO–LUMO interactions) and polarization (eq. 2).[3b]

$$\Delta E_{int} = \Delta V_{elstat} + \Delta E_{Pauli} + \Delta E_{oi} \qquad (2)$$

The orbital interaction term can be further decomposed into contributions from the σ and the π electron system, i.e., $\Delta E_{oi} = \Delta E_{\sigma} + \Delta E_{\pi}$.

In parallel to the bond energy decomposition, the charge redistribution associated with forming the hydrogen bonds can be analyzed on a per-atom basis by computing the corresponding gain or loss of charge of an atom, ΔQ_A, with the Voronoi deformation density method (VDD).[3c,d,2c,f] The changes ΔQ_A in atomic charge associated with ΔE_{int} can be further decomposed into contributions $\Delta Q_{A,Pauli}$, associated with the Pauli repulsive orbital interactions, and $\Delta Q_{A,oi}$, associated with the bonding orbital interactions (eq. 3).[2f]

$$\Delta Q_A = \Delta Q_{A,Pauli} + \Delta Q_{A,oi} \qquad (3)$$

Both, $\Delta Q_{A,Pauli}$ and $\Delta Q_{A,oi}$, can be further decomposed into contributions from the σ and the π electron system, i.e., $\Delta Q_{A,Pauli} = \Delta Q^{\sigma}_{A,Pauli} + \Delta Q^{\pi}_{A,Pauli}$ and $\Delta Q_{A,oi} = \Delta Q^{\sigma}_{A,oi} + \Delta Q^{\pi}_{A,oi}$.[2c]

Table 18-1. Hydrogen-bond energy decomposition (in kcal/mol) for selected DNA base pairs.[a]

Energy terms[b]	AT	GC	$AT_{FC'H}$	$G_{FC'N}C_{N'NF}$	AT^{F6}	GC^{F6}
ΔE_{Pauli}	39.2	52.1	7.2	8.9	40.1	51.5
ΔV_{elstat}	−32.1	−48.6	−6.5	−9.1	−32.8	−47.5
$\Delta E_{Pauli} + \Delta V_{elstat}$	7.1	3.5	0.7	−0.2	7.3	4.0
ΔE_{σ}	−20.7	−29.3	−3.4	−4.1	−21.4	−29.1
ΔE_{π}	−1.7	−4.8	−0.2	−0.5	−1.8	−4.7
ΔE_{oi}	−22.4	−34.1	−3.6	−4.6	−23.2	−33.8
ΔE_{int}	−15.3	−30.6	−2.9	−4.8	−15.9	−29.8
ΔE_{prep}	2.3	4.1	0.2	0.9	2.5	4.5
ΔE	−13.0	−26.5	−2.7	−3.9	−13.4	−25.3
$\Delta E(C_1)$	−13.0	−26.1	−2.7	−4.1		

[a]Computed at BP86/TZ2P (see Schemes 1 - 4 for definition of species). ΔE_{int} and its components refer to DNA base pairs optimized in C_s symmetry and the corresponding DNA bases in the geometry they adopt in the base pair. ΔE and ΔE_{prep} refer to DNA base pairs and isolated DNA bases both optimized in C_s symmetry. $\Delta E(C_1)$ refers to DNA base pairs optimized in C_s symmetry and isolated DNA bases optimized without any symmetry restriction in C_1 symmetry. Note the small discrepancy between ΔE and the more accurate $\Delta E(C_1)$ in the case of GC-type pairs. This is due to the pronounced pyramidalization of the amino group in the isolated guanine-type base.[2b,g]
[b]See equations 1 and 2.

The most striking result of our analysis is that charge-transfer or donor–acceptor orbital interactions are not at all a negligible or minor component in the hydrogen-bond energy of Watson-Crick base pairs (see Table 18-1). On the contrary, charge transfer is of the same order of magnitude as the electrostatic interaction! For AT, the orbital interactions ΔE_{oi} are −22.4 kcal/mol and the electrostatic attraction ΔV_{elstat} is −32.1 kcal/mol, and for

GC, ΔE_{oi} is –34.1 kcal/mol and ΔV_{elstat} is –48.6 kcal/mol (see Table 18-1). A favorable electrostatic attraction is in line with the charge distribution in the isolated DNA bases: all proton-acceptor atoms have a negative charge whereas the corresponding protons they face are positively charged. This can be seen in Figure 18-1 which shows the atomic charges of all four DNA bases[2c] computed with the VDD method.[3c,d] The orbital interaction term ΔE_{oi} can be further divided into a σ-component and a π-component. The former, ΔE_σ, consists mainly of donor–acceptor interactions between (i) a lone pair σ_{LP} on the proton-acceptor atom (nitrogen or oxygen) of one base pointing toward and donating charge into, (ii) an unoccupied σ* orbital of an N–H group of the other base. This leads to the formation of $\sigma_{LP} + \sigma^*$ donor–acceptor bonds. The π-component, ΔE_π, accounts basically for the polarization in the π-system which partly compensates the local build-up of charge caused by the charge-transfer interactions in the Σ-system. Figure 18-2 shows the relevant frontier-orbital interactions for AT as they emerge from our quantitative Kohn-Sham MO analysis. As one can see, the lone pair on the N1 atom of adenine interacts with the σ* orbitals on H3–N3 of thymine and that the lone pair on the O4 atom of thymine is involved in donor-acceptor interactions with the σ* orbitals on N6–H6 of adenine. A similar picture emerges for the Watson-Crick pair GC.[2c]

The prominent role of orbital interactions that evolves from both the bond energy decomposition and the Kohn-Sham orbital electronic structure are confirmed also by the VDD analyses (see eq 3 and corresponding text) of the charge distribution in the Watson-Crick pairs. The results of the VDD analyses for AT are displayed in Figure 18-3. In the formation of this Watson-Crick pair from adenine + thymine, Pauli repulsion causes a depletion of charge density away from the central region of overlap and toward the periphery of the N6–H6•••O4 and N1•••H3–N3 hydrogen bonds. This causes a build-up of positive charge on the central hydrogen atom and of negative charge on the terminal atoms on either side of the hydrogen bond (see ΔQ^σ_{Pauli} in Figure 18-3). The changes in the atomic charges caused by the bonding orbital interaction in the σ-system (see ΔQ^σ_{oi} in Figure 18-3) reveal that the electron-donor atoms of the N6–H6•••O4 and N1•••H3–N3 hydrogen bonds lose 24 and 46 mili-a.u. while the N–H bonds gain up to 54 mili-a.u.. The charge redistribution in the π-electron system polarizes in such a way that the built-up of positive or negative atomic charges in the Σ-electron system is cancelled or even overcompensated (see ΔQ^π_{oi} in Figure 18-3). Again, a similar picture emerges for the Watson-Crick pair GC.[3b,2c,f]

Interestingly, the electrostatic interaction alone is not capable of providing a net bonding interaction at the equilibrium geometry at which it can only partly compensate the Pauli-repulsive orbital interactions ΔE_{Pauli}. Without the bonding orbital interactions, the net interaction energies of AT and GC at their equilibrium structures would be repulsive by 7.1 and 3.5 kcal/mol, respectively (Table 18-1). This can be understood such that the orbital interactions cause the DNA bases to approach each other significantly more closely than if there were only electrostatic attraction. This results in hydrogen bond lengths at which the "electrostatic-attraction-only" energy curve (i.e., $\Delta E_{int} - \Delta E_{oi} = \Delta E_{Pauli} + \Delta V_{elstat}$) is already repulsive.

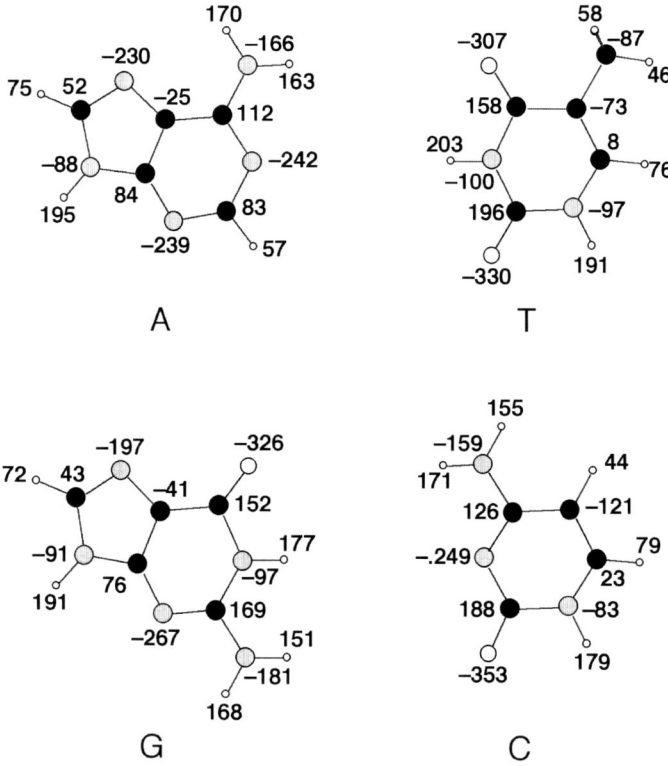

Figure 18-1. VDD atomic charges (in mili-a.u.) of the isolated DNA bases A, T, G and C, computed at BP86/TZ2P.

2.3 Orbital Interactions versus Electrostatic Attraction

Thus, our analyses disprove the still widespread believe that hydrogen bonding in DNA base pairs is a predominantly electrostatic phenomenon. In both Watson-Crick pairs, the orbital interactions ΔE_{oi} provide a significant portion, namely, 41%, of all attractive forces (i.e., $\Delta E_{oi} + \Delta V_{elstat}$), while electrostatic attraction contributes the remaining 59% (see Table 18-1). The ΔE_{oi} term can be further split into 38% ΔE_σ and 3% ΔE_π for AT, and 35% ΔE_σ and 6% ΔE_π for GC. Very similar ratios between orbital interactions and electrostatic attraction (with large contributions of the former, see, for example, in Table 18-1) were also found for AT and GC mimics in which N–H is substituted by C–H and oxygen by fluorine (see Section 3),[2c-f] for remotely halogen-substituted (i.e., at purine-H8 and pyrimidine-H6) Watson-Crick pairs (see Section 4),[11] for Watson-Crick pairs exposed to micro-solvation and a crystal environment (see Section 5),[2a,b,d] and, finally, for mismatches of DNA bases.[2d] We conclude that, at variance with current belief, charge-transfer or donor–acceptor orbital interactions play an essential role in the hydrogen bonds of various types of DNA base pairs. One may expect that they also play an important role in other classes of hydrogen-bonded complexes.

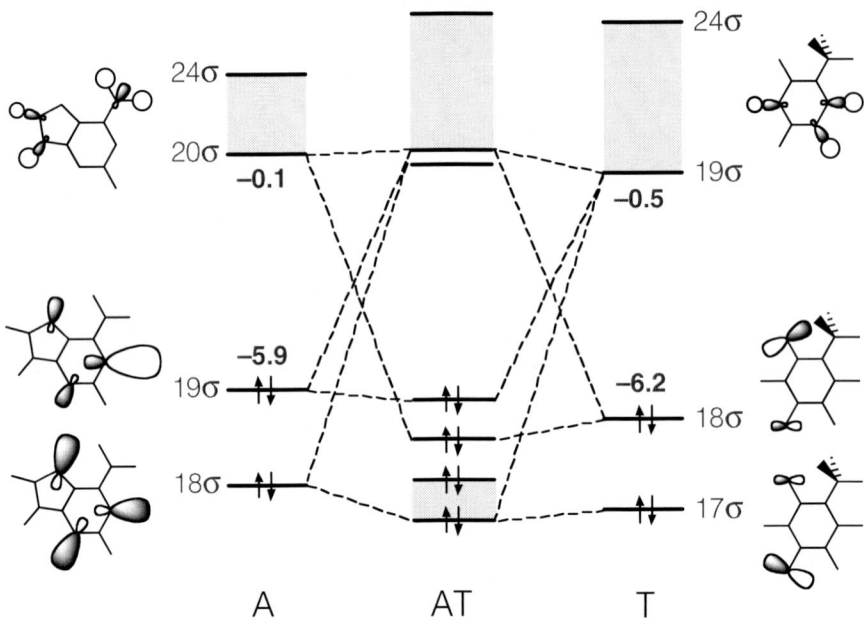

Figure 18-2. Orbital interactions in the σ-electron system between adenine and thymine in AT with σ_{HOMO} and σ_{LUMO} energies in eV, computed at BP86/TZ2P. The group of lowest unoccupied orbitals involved is represented as a grey block. Selected orbitals of adenine (18σ to 20σ) and thymine (17σ - 19σ) are schematically represented.

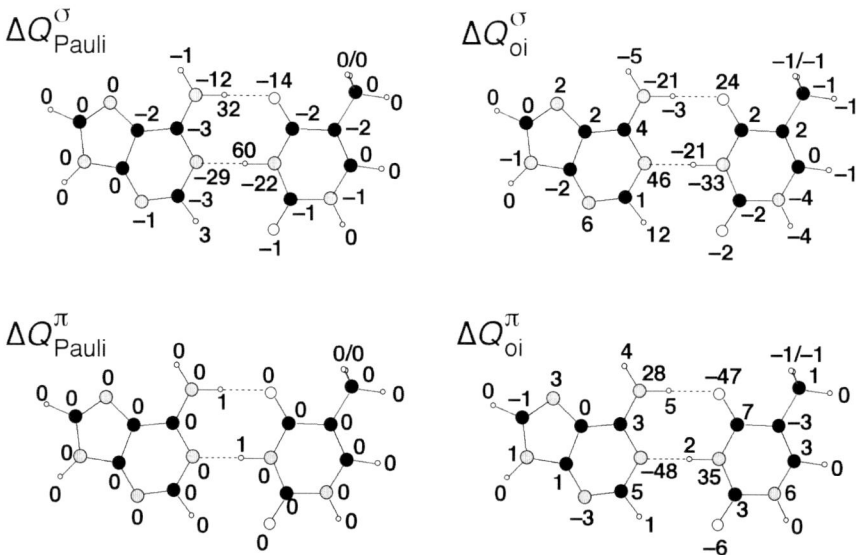

Figure 18-3. VDD decomposition of atomic charges ΔQ (in mili-a.u.) associated with the formation of the Watson-Crick AT pair from A and T into contributions $\Delta Q^{\sigma}{}_{Pauli}$ and $\Delta Q^{\pi}{}_{Pauli}$, caused by Pauli repulsion in the σ and π electron system, and $\Delta Q^{\sigma}{}_{oi}$ and $\Delta Q^{\pi}{}_{oi}$, caused by bonding orbital interactions in the Σ and π electron system, computed at BP86/TZ2P.

3. SUBSTITUTIONS IN X–H•••Y HYDROGEN BONDS

The effect of substituting N–H•••O and N•••H–N hydrogen bonds in natural Watson-Crick pairs by N–H•••F and N•••H–C, as shown in Scheme 2, is a weakening of the bond energy and an elongation of the hydrogen bond that has undergone the replacement (see Figure 18-4). Note however that, as mentioned above, the nature of the hydrogen bonds, in particular, the pronounced covalent character is not much affected. This is illustrated in Table 18-1 by a hydrogen-bond energy decomposition for two representative examples, $AT_{FC'H}$ and $G_{FC'N'}C_{N'NF}$. As the effects are very similar for AT and GC, we focus in the following mainly on substitutions in AT. The above-mentioned behavior agrees well with the idea that C–H is a poor hydrogen bond donor and F a poor hydrogen bond acceptor compared to N–H and O, respectively. This is also confirmed by our analyses. They show that the

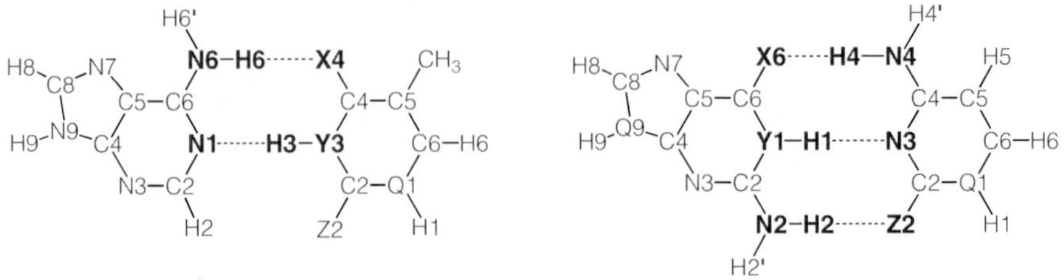

Scheme 2. AT and GC Watson-Crick pairs with Substitutions in Hydrogen Bonds: X, Z = O, F, H and Y = N, C (also Q = N, C).

reduced bonding capabilities of C–H and F do not only derive from a lower polarity (i.e., smaller positive and negative atomic charges on C–H and C–F compared to N–H and C=O). Rather, an important factor is the increase of ca. 0.5 eV in σ^* orbital energy from N–H to C–H and the substantial decrease of ca. 2.5 eV of the σ_{LP} orbital energy from O to F.[2c-g] This makes the unoccupied σ^*_{C-H} a worse acceptor orbital and the σ_{LP} a worse donor orbital in the donor–acceptor orbital interactions of the Watson-Crick hydrogen bonds, see Figure 18-2 (see also Table 18-1).

In case of AT and its mimics, both the first and the second replacement lead to a reduction in bond energy of 4 - 7 kcal/mol (see Figure 18-4). In case of GC, the corresponding effect is a reduction in bond energy of ca 9 kcal/mol for each of the first two substitutions and a further reduction of 5 kcal/mol for the third replacement (see Figure 18-4; for more details on GC, the reader is referred to the original literature[2g]). For example, the replacement of thymine O4 by F4 followed by the replacement of N3–H3 by C3–H3 along $AT_{ON'O}$ (= AT), $AT_{FN'O}$ and $AT_{FC'O}$ causes the bond energy of the complex to decrease from –13.0 to –8.8 to –2.0 kcal/mol. This indicates that turning one of the two hydrogen bonds in natural AT into a weak hydrogen bond has little effect on the strength of the second hydrogen bond, which preserves approximately its original strength.

This is in line with the fact that substitution of one of the natural hydrogen bonds by a weaker one has little effect on the bond distance of the other hydrogen bond. For example, replacing thymine O4 by F4 along $AT_{ON'O}$ (= AT) and $AT_{FN'O}$ causes hydrogen bond N6–H6•••X4 to elongate by 0.33 Å whereas hydrogen bond N1•••H3–N3 becomes only slightly longer, i.e., 0.03 Å (see Figure 18-4). Likewise, changing thymine N3–H3 by C3–H3 along $AT_{ON'H}$ and $AT_{FC'H}$ leads to an elongation of the N1•••H3–Y3

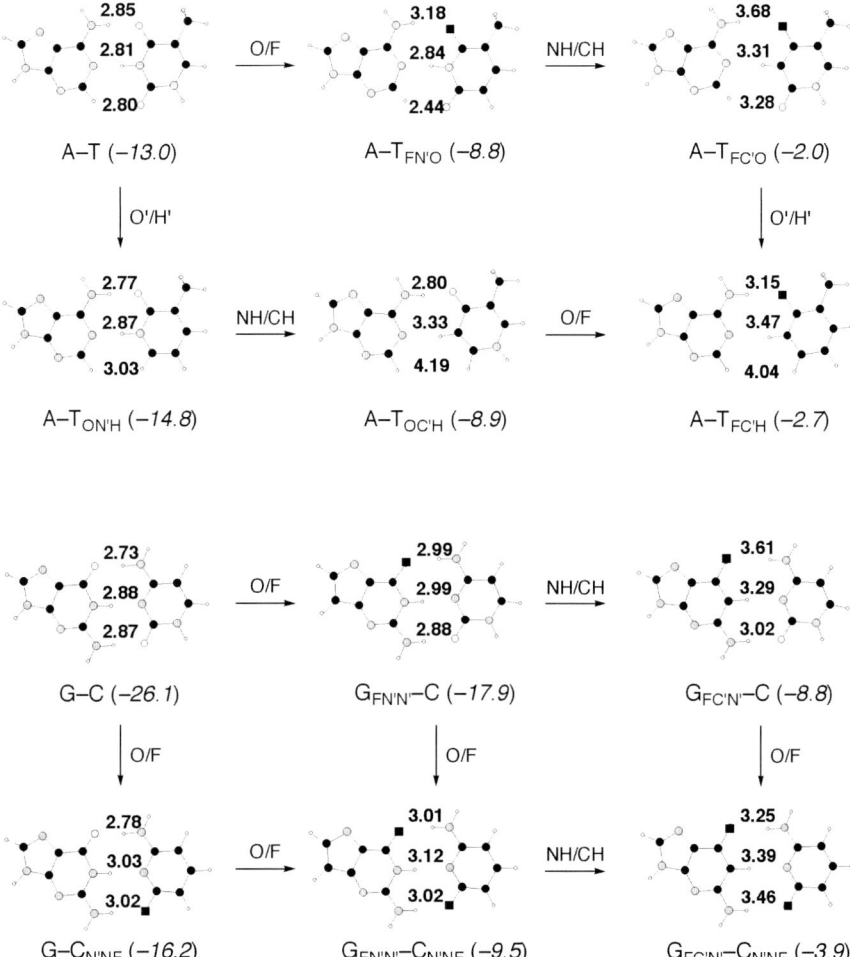

Figure 18-4. Hydrogen bond distances (in Å), N6–X4 and N1–Y3 in AT (upper) and X6–N4, Y1–N3 and N2–Z2 in GC derivatives (lower), and hydrogen-bond energies (in kcal/mol, in parentheses) in Watson-Crick pairs and artificial mimics thereof, computed at BP86/TZ2P (see Scheme 2). Arrows indicate base pairs related by a substitution at X6, Y1H1 or Z2 (e.g., O/F or NH/CH).

of 0.46 Å whereas the N6–H6•••O4 bond elongates only very slightly, i.e., by 0.03 Å. This is schematically represented in **1** and **2** below:

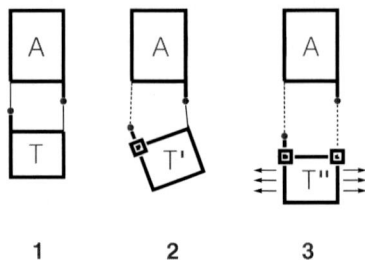

Here, the bases are represented with bold lines, natural hydrogen bonds by plain lines and weak hydrogen bonds by dashed lines; the little squares in **2** and **3** represent substitution in the hydrogen bonding moiety. Interestingly, the two bases of the complex hinge with respect to each around a pivot through the central hydrogen atom of the strong (unsubstituted) hydrogen bond if one weak hydrogen bond is introduced. This is because the latter must bend to allow, without elongating by itself, for the elongation of the other hydrogen bond that is substituted for a weaker one. The bases hinge back to a situation of more linear (and again somewhat longer) hydrogen bonds if the remaining strong hydrogen bond is replaced by a weaker one, for example, if one goes from $AT_{FN'O}$ to $AT_{FC'O}$ (see also Figure 18-4). A schematic representation of this behavior is provided by **2** and **3**.

Substitution of thymine O2 by H2 (or F2, not discussed here, but in Ref. 2g) has little effect on the bond energy. This is not unexpected because atom Z2 is not involved in any stabilizing C2–H2•••O2 interaction as pointed out before.[2c] Note however that the bond energy increases, although only slight, from Z2 = O2 to H2, for example, from –2.0 to –2.7 kcal/mol along $AT_{FC'O}$ and $AT_{FC'H}$ (see Figure 18-4) The above substitution of thymine O2 has also little effect on the geometry of the two hydrogen bonds N6–H6•••X4 and N1•••H3–Y3 if the latter are strong, i.e., N6–H6•••O4 and N1•••H3–N3, as from $AT_{ON'O}$ (= AT) to $AT_{ON'H}$. This is further evidence for the absence on any significant C2–H2•••O3 interaction in AT. However, fluctuations in bond distances of up to 0.53 Å can occur in the much weaker N6–F4 and N1–C3 hydrogen bonds, as can be seen in Figure 18-4 along $AT_{FC'O}$ and $AT_{FC'H}$. This is indicative for the much softer geometry associated with the shallow potential energy surface of the more weakly bound mimics of AT. In illustration **3**, this has been indicated by the arrows pointing left and right.

4.   REMOTE SUBSTITUTIONS AT DNA BASES

The effects of remotely substituting hydrogen by halogen atoms (F, Cl, Br) at X8 and Y6 in Watson-Crick pairs, as shown in Scheme 3, consist of relatively subtle changes in hydrogen bond distances and energies (see Table 18-2) if compared with the much larger effects that occur if N–H•••O and N•••H–N hydrogen bonds are replaced by N–H•••F and N•••H–C (see Figure 18-4).[11] The former affects hydrogen bond distances and strengths by only a few hundredths of an Ångstrøm and by less than a kcal/mol, respectively, whereas the latter goes with hydrogen bond expansions of nearly half an

$A^{X8}{-}T^{Y6}$ $G^{X8}{-}C^{Y6}$

Scheme 3. $A^{X8}T^{Y6}$ and $G^{X8}C^{Y6}$ WC pairs with remote substituents X8, Y6.

Ångstrøm and bond weakening of up to 7 kcal/mol per replacement. Again, we focus in the following mainly on the effects of remote substitution in AT, as the effects in GC are very similar. For substituent effects on NMR parameters in Watson-Crick pairs, see Ref. 12.

We recall that the hydrogen bond distances N6–H6•••O4 and N1•••H3–N3 in the natural Watson-Crick pair AT are 2.85 and 2.81 Å and provide a hydrogen bond energy ΔE of –13.00 kcal/mol (see Table 18-2). Introducing a halogen atom at adenine X8 (see Scheme 3) has hardly any effect: hydrogen bond distances remain essentially unaffected and hydrogen bond energies change by only a few hundredths of a kcal/mol. At variance, introducing a halogen atom at thymine Y6 has more pronounced effects: the hydrogen bonds N6–H6•••O4 and N1•••H3–N3 are elongated and contracted, respectively, by up to 0.03 Å, and overall Watson-Crick hydrogen bonding is stabilized by –0.30, –0.34 and –0.43 kcal/mol if thymine H6 is replaced by Br, Cl and F, respectively (see Table 18-2).[11] This is shown schematically for AT and AT^{F6} manner in illustrations **4** and **5** below:

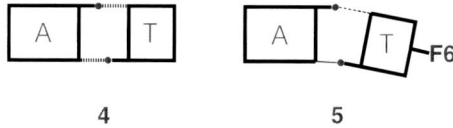

4   5

Here, bases are represented with bold lines, natural hydrogen bonds by broad dashes, weakened hydrogen bonds by narrow dashes, and strengthened hydrogen bonds by plain lines. Simultaneously introducing halogen atoms at adenine X8 and thymine Y6 in $A^{X8}T^{Y6}$ has essentially the same effect, only slightly less pronounced, as introducing only one halogen atom in AT^{Y6}: the hydrogen bonds N6–H6•••O4 and N1•••H3–N3 (see Scheme 3) are elongated and contracted, respectively, by up to 0.02 Å, and the hydrogen bond is strengthened by up to –0.31 kcal/mol (see Table 18-2). An equivalent behavior is found for $G^{X8}C^{Y6}$.[11] Substituting, for example, cytosine H6 by a halogen atom causes the outer hydrogen bonds O6•••H4–N4 and N2–H2•••O2 to be contracted and elongated, respectively, by up to 0.03 Å, whereas the central hydrogen bond N1–H1•••N3 is less affected, while the overall hydrogen bond strength is reduced by 0.74 - 0.82 kcal/mol. This is shown for GC and GC^{F6} in a schematic manner in **6** and **7** below:

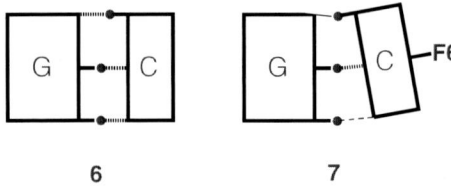

6   **7**

Table 18-2. Hydrogen bond lengths (in Å) and energies (in kcal/mol) in $A^{X8}T^{Y6}$.[a]

X8	Y6	N6–H6•••O4	N1•••H3–N3	ΔE
H	H	2.85	2.81	–13.00
	F	2.87	2.78	–13.43
	Cl	2.86	2.79	–13.34
	Br	2.88	2.78	–13.30
F	H	2.85	2.81	–12.98
	F	2.87	2.79	–13.31
Cl	H	2.85	2.81	–13.01
	Cl	2.87	2.79	–13.25
Br	H	2.85	2.81	–13.01
	Br	2.86	2.79	–13.27

[a] Computed at BP86/TZ2P. See also Scheme 3.

Thus, a halogen atom at adenine X8 has little effect whereas a halogen atom at thymine Y6 weakens hydrogen bond N6–H6•••O4 and strengthens hydrogen bond N1•••H3–N3 (see Scheme 1 and Table 18-1). The latter effect apparently dominates (and is strongest in case of Y6 = fluorine) as follows from the fact that the net hydrogen bond strength increases. The substituent effects can be understood on the basis of the electron density distribution and the orbital electronic structure of the DNA bases in combination with a quantitative decomposition of the hydrogen bond energy (see Section 2.1). Figure 18-5 shows the VDD atomic charges[3c,d] of isolated (i.e., noninteracting) DNA bases A^{X8} and T^{Y6} with substituents X8, Y6 = F, Cl, Br (for X8, Y6 = H, see Figure 18-1). Replacing hydrogen by halogen causes the substituent to become significantly more negatively charged, in line with the higher electronegativity of the halogen atom: the VDD atomic charge of adenine X8 is +77, –5, –21 and –42 mili-a.u. and that of thymine Y6 is +73, –8, –22 and –49 mili-a.u. along H, Br, Cl and F, thus following the trend in electronegativity of the substituent. This causes both the build-up of a net positive charge and a stabilization of the molecular orbitals in the DNA base. Consequently, the carbon atom to which the substituent is connected becomes significantly more positively charged (see Figure 18-5). But also the more remote front-atoms, i.e., the atoms involved in hydrogen bonding, are affected, although to a lesser extent (see Figure 18-5): hydrogen front atoms become slightly *more positive* by up to 5 mili-a.u., which *strengthens* the electrostatic attraction in the corresponding hydrogen bond, whereas nitrogen or oxygen front atoms become slightly *less negative* by up to 9 mili-a.u., which *weakens* the electrostatic attraction in the corresponding hydrogen bond. The stabilization of the orbitals in the halogen-substituted DNA base is in the order of a few tenths of an eV (not shown in tables or figures) and has the same effect on hydrogen bond lengths and strengths: the *stabilization of* σ^*_{N-H} *acceptor orbitals strengthens* the donor–acceptor orbital interaction that occurs in the corresponding hydrogen bond (because this reduces the orbital energy gap with lone-pair donor orbitals of the other base) whereas the *stabilization of N or O lone-pair orbitals weakens* the orbital interactions that occur in the corresponding hydrogen bond (because this increases the orbital energy gap with σ^*_{N-H} acceptor orbitals of the other base); see Figure 18-2 for the orbital interactions in AT. Thus, halogen substitution at adenine X8 promotes strengthening and contraction of N6–H6•••O4 and weakening and expansion of N1•••H3–N3. The opposite happens in case of halogen substitution at thymine Y6, which promotes weakening and expansion of N6–H6•••O4 and strengthening and contraction of N1•••H3–N3.

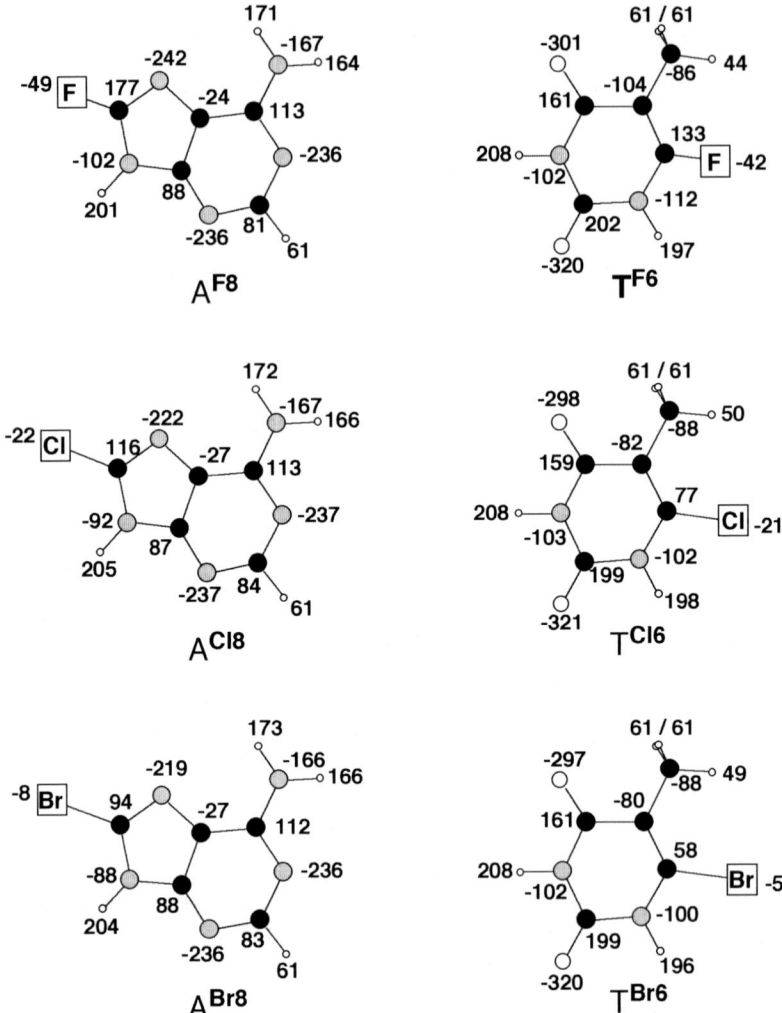

Figure 18-5. VDD atomic charges Q_A (in mili-a.u.) in *isolated* DNA bases A^{X8} and T^{Y6} with the geometries they adopt in the corresponding $A^{X8}T^{Y6}$ pair, computed at BP86/TZ2P (see Figure 18-1 for values in unsubstituted A and T).

Hydrogen bonding receives in all $A^{X8}T^{Y6}$ pairs an important stabilizing contribution from occupied–virtual orbital interactions, which are in the same order of magnitude as electrostatic interaction (see Table 18-1 for two examples: AT^{F6} and GC^{F6}). The percentage contribution of the orbital interactions ΔE_{oi} to all bonding forces (i.e., $\Delta E_{oi} + \Delta V_{elstat}$) between the bases in Watson-Crick $A^{X8}T^{Y6}$ and $G^{X8}C^{Y6}$ pairs amounts to 41-42% (see also

Table 18-1).[2g] The remaining 59-58% is provided the electrostatic attraction ΔV_{elstat}.

The substituent effects in thymine are more pronounced than those in adenine. This is in line with the fact that the pyrimidine base thymine is smaller than the purine base adenine and that the charges of its front atoms as well as its orbital energies are more strongly affected by the substituent. Introducing, for example, fluorine at X8 in adenine has no effect on the charge of H6 and causes N1 to become 4 mili-a.u. less negative, whereas introducing fluorine at Y6 in thymine causes both H4 and N3 to become 5 mili-a.u. more positive/less negative (Figure 18-5). Likewise, the fluorine substituent stabilizes the LUMO and LUMO+1 of adenine by not more than 0.2 eV whereas it stabilizes the LUMO and LUMO+1 of thymine by up to 0.4 eV. Furthermore, in terms of overall bond strength, the effect on hydrogen bond N1•••H3–N3 overrules that on hydrogen bond N6–H6•••O4. This can be rationalized on the basis of the fact that hydrogen bond N1•••H3–N3 is associated with a smaller orbital energy gap between the electron-donating lone pair on one base and the electron-accepting σ* orbital on the other base (see Ref. 2c): it is the stronger of the two hydrogen bonds but it also reacts more strongly to changes in the orbital energy gap caused by introducing the halogen substituent. Eventually, if hydrogen bond N1•••H3–N3 is weakened by introducing a halogen atom at adenine X8, both the A–T orbital interactions ΔE_{oi} and the electrostatic attraction ΔV_{elstat} are slightly reduced (e.g., from –22.00 to –21.56 and from –31.78 to –31.42 kcal/mol, respectively, if one goes from AT to $A^{F8}T$); the net effect on the overall bond strength ΔE is negligible (see Table 18-2). On the other hand, both ΔE_{oi} and ΔV_{elstat} are reinforced to a larger extent (e.g., from –22.00 to –23.18 and from –31.78 to –32.76 kcal/mol, respectively, if one goes from AT to AT^{F6}) if N1•••H3–N3 is strengthened by introducing the same halogen atom at thymine Y6 (see Table 18-2); the net effect on the overall bond strength ΔE is a pronounced stabilization.

5.   SUPRAMOLECULAR SUBSTITUENT EFFECTS

Not only substituents bound to the DNA bases through regular chemical bonds (*vide supra*) can affect the hydrogen bonds in Watson-Crick pairs. Similar effects can also be achieved when the substituents are attached more loosely, through hydrogen bonds. We designate this supramolecular substituents. Scheme 4 shows various conceivable positions for supramolecular substituents in AT and GC Watson-Crick pairs; the substituents, **S**, are represented by spheres.

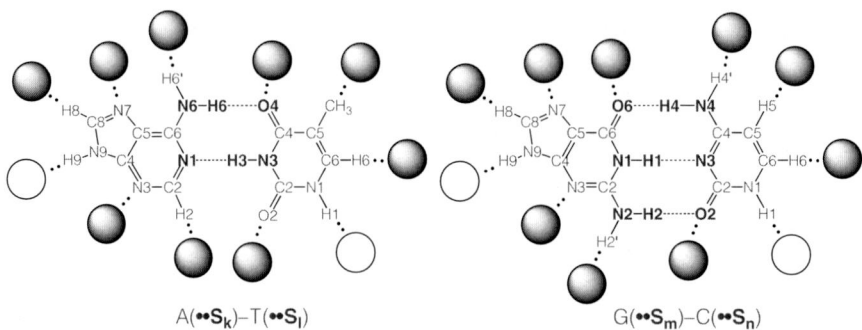

Scheme 4. AT and GC WC pairs with Supramolecular Substituents **S**.

An example of such supramolecular substituents is the environment that Watson-Crick pairs experience in (physiological) solution or in crystals, which consists of water molecules, counterions and/or other DNA bases.[2a,b,d] Note that the simple AT and GC models shown in Scheme 4 may bind supramolecular substituents (white spheres) at purine-H9 and pyrimidine-H1. This is of course not possible if the DNA bases are part of a DNA strand in which the indicated hydrogens are replaced by bulky 2'-deoxyribose-5'-phosphate residues which form the backbone of the DNA strand. All other positions for supramolecular substituents (shaded spheres) remain available also in the DNA double helix in which they are located either in the major groove (upper side of base pairs) or in the minor groove (lower side, see Scheme 4).

The effects of supramolecular substitution can be as large as those of introducing regular, chemically bound substituents described in Section 4. In Figures 18-6, two examples are shown, one for AT and one for GC. Introducing two water molecules at AT, i.e., going to $[AT] \cdot (H_2O)_2$, causes the N6–O4 bond to expand significantly, from 2.85 to 2.92 Å. The N1–N3 bond is not so much affected; it shortens only slightly from 2.81 to 2.80 Å. Likewise, introducing four water molecules and a sodium cation at GC, i.e., going to $[GC] \cdot (H2O)_4 Na^+$, causes the bond-length pattern of the three hydrogen bonds O6–N4, N1–N3 and N2–O2 to change even qualitatively, namely, from short-long-long (2.73, 2.88 and 2.87Å) to long-long-short (2.88, 2.95 and 2.85Å, see Figure 18-6).

Interestingly, this brings the computed and experimental geometries into close agreement (compare $[AT] \cdot (H_2O)_2$ and $[GC] \cdot (H_2O)_4 Na^+$ with AT(exp) and GC(exp) in Figure 18-6). This is no coincidence! Our examples of supramolecular substitutions closely model the molecular environment that the Watson-Crick pairs experience in the crystals used to carry out the X-ray diffraction experiments.[10]

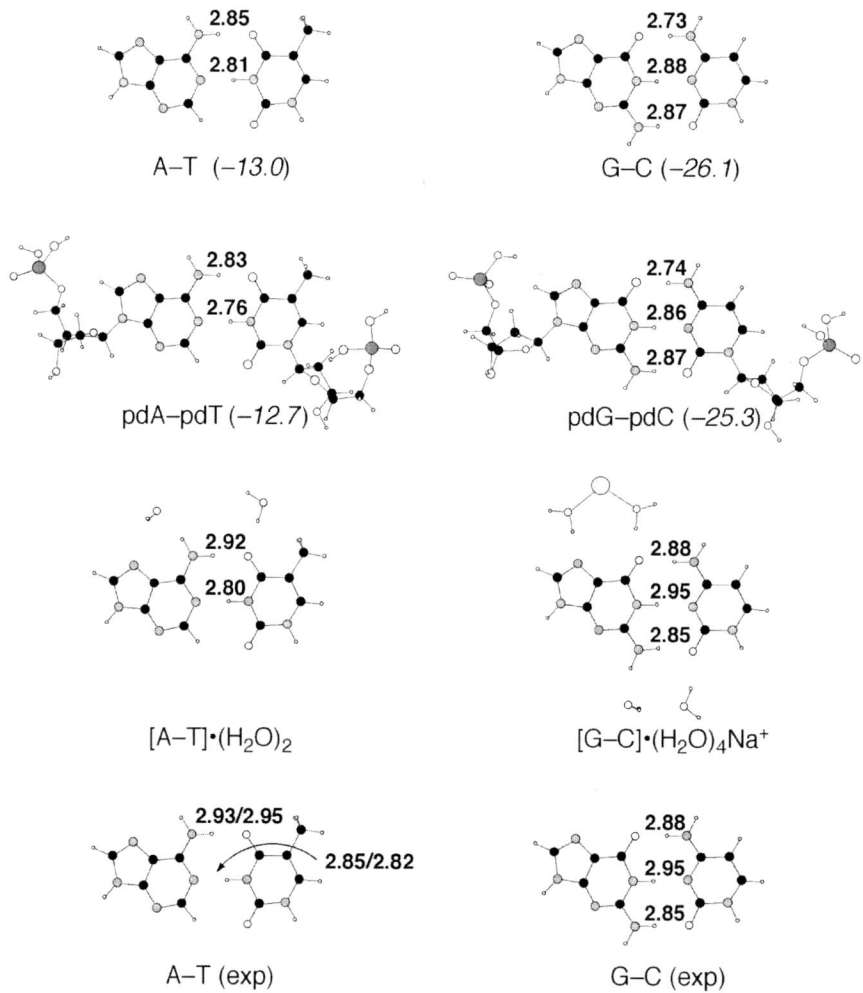

Figure 18-6. Hydrogen bond distances (in Å), N6–O4 and N1–N3 in AT and O6–N4, N1–N3 and N2–O2 in GC pairs, and bond energies (in kcal/mol, in parentheses) in selected Watson-Crick base pairs, computed at BP86 (for details see Ref. 2b): A–T and G–C (with hydrogen atoms at purine-H9 / pyrimidine-H1); pdA–pdT and pdG–pdC (with deoxyribose-5'-phosphate substituents); [A–T]•(H₂O)₂ and [G–C]•(H₂O)₄Na⁺ (with "supramolecular substituents", water and sodium); AT (exp) and GC (exp)[10b] (experimental X-ray crystal structures of sodium adenylyl-3',5'-uridine hexahydrate[10a] and sodium guanylyl-3',5'-cytidine nonahydrate[10b]). Note that for AT (exp), there are two experimental values for N6–O4 and N1–N3 because the two AU pairs in the crystal experience different environments.

Neglecting this environment in the computations of plain AT and GC model systems leads to a relatively large discrepancy between theory and experiment. This discrepancy is not caused by neglecting the substituent effect of deoxyribose-5'-phosphate residues of the backbone at purine-N9 and pyrimidine-N1 positions: introducing the latter in pdA–pdT and pdG–pdC has hardly any effect on the hydrogen bond distances and the discrepancy with experiment (see Figure 18-6). These results also show that DFT yields results which agree excellently with experiment if the most important hydrogen bonding interactions of a base pair with its environment in the crystal are incorporated into the model system. It also justifies tackling the computationally extremely demanding problems of biological molecules involving hydrogen bonding with DFT as an efficient alternative to traditional ab initio methods.

6.   CONCLUSIONS

There is a sizeable covalent or orbital-interaction component in the hydrogen bonds of Watson-Crick base pairs that, in fact, is of the same order of magnitude as the electrostatic attraction. This disproves the still widespread believe that hydrogen bonding in DNA base pairs is a predominantly electrostatic phenomenon.

     This insight into the nature of hydrogen bonding is not just an academic issue that only fascinates the theoretician, it can also lead to more practical applications. We have shown that it enables one to understand substituent effects in Watson-Crick pairs. The effect of, for example, remote substitution (i.e., at purine-H8 and pyrimidine-H6) on the Watson-Crick hydrogen bonds is largely determined by how the substituents influence the orbital energies of the DNA bases and, in this way, the donor–acceptor orbital interactions between DNA bases in a Watson-Crick pair. Such understanding on the origin of substituent effects may also facilitate a more rational and efficient design of new artificial DNA bases and other molecular building blocks with the potential of molecular recognition and selforganization toward tailor-made supramolecular structures.

     Finally, density functional theory (DFT) is an adequate and efficient tool for accurately describing systems involving (multiple) hydrogen bonding. In particular, a long-standing discrepancy between theory (ab initio and hybrid DFT) and experiment (X-ray crystal structures) could be traced to effects of the molecular environment in the crystal that were missing in previous theoretical model systems. Inclusion of these effects in DFT computations leads to reconciliation of theory and experiment. An important implication is that DFT can be applied as an efficient alternative to conventional ab initio methods for tackling large biomolecular systems in which hydrogen bonds play a prominent role.

ACKNOWLEDGEMENT

We thank the National Research School Combination - Catalysis (NRSC-C) and the Netherlands organization for Scientific Research (NWO) for financial support.

REFERENCES

1. (a) Stryer, L., *Biochemistry* (W.H. Freeman and Company, New York, 1988). (b) Jeffrey, G.A., Saenger, W., *Hydrogen Bonding in Biological Structures* (Springer-Verlag, Berlin, 1991). (c) Jeffrey, G.A., *An Introduction to Hydrogen Bonding* (Oxford University Press, New York, 1997).
2. (a) Fonseca Guerra, C., Bickelhaupt, F.M. (1999). Charge Transfer and Environment Effects Responsible for Characteristics of DNA Base Pairing, *Angew. Chem. Int. Ed.* **38**, 2942–2945. (b) Fonseca Guerra, C., Bickelhaupt, F.M., Snijders, J.G., Baerends, E.J. (2000). Hydrogen Bonding in DNA Base Pairs: Reconciliation of Theory and Experiment, *J. Am. Chem. Soc.* **122**, 4117–4128. (c) Fonseca Guerra, C., Bickelhaupt, F.M., Snijders, J.G., Baerends, E.J. (1999). The Nature of the Hydrogen Bond in DNA Base Pairs: the Role of Charge Transfer and Resonance Assistance, *Chem. Eur. J.* **5**, 3581–3594. (d) Fonseca Guerra, C., Baerends, E.J., Bickelhaupt, F.M. (2002). Orbital Interactions in Hydrogen Bonds Important for Cohesion in Molecular Crystals and Mismatched Pairs of DNA Bases, *Crystal Growth & Design* **2**, 239–245. (e) Fonseca Guerra, C., Bickelhaupt, F.M. (2002). Orbital Interactions in Strong and Weak Hydrogen Bonds are Essential for DNA Replication, *Angew. Chem. Int. Ed.* **41**, 2092–2095. (f) Fonseca Guerra, C., Bickelhaupt, F.M., (2003). Orbital interactions and charge redistribution in weak hydrogen bonds: The Watson–Crick AT mimic adenine-2,4-difluorotoluene, *J. Chem. Phys.* **119**, 4262–4273. (g) Fonseca Guerra, C., Bickelhaupt, F.M., Baerends, E.J. (2004). Hydrogen Bonding in Mimics of Watson-Crick Base Pairs Involving CH Proton Donor and F Proton Acceptor Groups: A Theoretical Study, *ChemPhysChem* **5**, 481–487.
3. For theory and methodology used in our work in Ref. 2, see: (a) te Velde, G., Bickelhaupt, F.M., van Gisbergen, S.J.A., Fonseca Guerra, C., Baerends, E.J., Snijders, J.G., Ziegler, T. (2001). Chemistry with ADF , *J. Comput. Chem.* **22**, 931–967. (b) Bickelhaupt, F.M., Baerends, E.J., Kohn-Sham Density Functional Theory: Predicting and Understanding Chemistry, in: *Rev. Comput. Chem, Vol 15.* edited by K.B. Lipkowitz, D.B. Boyd (Wiley-VCH, New York, 2000), p. 1-86. (c) Fonseca Guerra, C., Handgraaf, J.-W., Baerends, E.J., Bickelhaupt, F. M. (2004). Voronoi deformation density (VDD) charges: Assessment of the Mulliken, Bader, Hirshfeld, Weinhold, and VDD methods for charge analysis, *J. Comput. Chem.* **25**, 189–210. (d) Bickelhaupt, F.M., van Eikema Hommes, N.J.R., Fonseca Guerra, C., Baerends, E.J. (1996). The Carbon-Lithium Electron Pair Bond in $(CH_3Li)_n$ (n = 1, 2, 4) *Organometallics* **15**, 2923–2931.
4. (a) Hobza, P., Šponer, J. (1999). Structure, energetics, and dynamics of the nucleic acid base pairs: Nonempirical ab initio calculations, *Chem. Rev.* **99**, 3247–3276. (b) Bertran, J., Oliva, A., Rodríguez-Santiago, L., Sodupe, M. (1998). Single versus double proton-transfer reactions in Watson-Crick base pair radical cations. A theoretical study, *J. Am. Chem. Soc.* **121**, 8159–8167. (c) Brameld, K., Dasgupta, S., Goddard III, W.A. (1997).

Distance dependent hydrogen bond potentials for nucleic acid base pairs from ab initio quantum mechanical calculations (LMP2/cc-pVTZ), *J. Phys. Chem. B* **101**, 4851–4859. (d) Šponer, J., Leszczynski, J., Hobza, P. (1996). Structures and energies of hydrogen-bonded DNA base pairs. A nonempirical study with inclusion of electron correlation, *J. Phys. Chem.* **100**, 1965–1974. (e) Gould, I.R., Kollman, P.A. (1994). Theoretical Investigation of the Hydrogen Bond Strengths in Guanine-Cytosine and Adenine-Thymine Base Pair, *J. Am. Chem. Soc.* **116**, 2493–2499. (f) Santamaria, R., Vázquez, A. (1994). Structural and electronic property changes of the nucleic-acid bases upon base-pair formation, *J. Comp. Chem.* **15**, 981–996. (g) Šponer, J., Hobza, P. (2000). Interaction energies of hydrogen-bonded formamide dimer, formamidine dimer, and selected DNA base pairs obtained with large basis sets of atomic orbitals, *J. Phys. Chem. A* **104**, 4592–4597. (h) Hobza, P., Šponer, J., Cubero, E., Orozco, M., Luque, F.J., (2000). C-H center dot center dot center dot O contacts in the adenine center dot center dot center dot uracil Watson-Crick and uracil center dot center dot center dot uracil nucleic acid base pairs: Nonempirical ab initio study with inclusion of electron correlation effects, *J. Phys. Chem. B* **104**, 6286–6292. (i) Poater, J., Fradera, X., Solà, M., Duran, M., Simon, S. (2003). On the electron-pair nature of the hydrogen bond in the framework of the atoms in molecules theory, *Chem. Phys. Lett.* **369**, 248–255. (j) Gilli, G., Bellucci, F., Ferretti, V., Bertolasi, V. (1989). Evidence for resonance-assisted hydrogen-bonding from crystal-structure correlations on the enol form of the beta-diketone fragment, *J. Am. Chem. Soc.* **111**, 1023–1028.

5. (a) Yanson, I.K., Teplitsky, A.B., Sukhodub, L.F. (1979). Experimental studies of molecular-interactions between nitrogen bases of nucleic-acids, *Biopolymers* **18**, 1149–1170 (with corrections by reference 4c).

6. Other factors that play a role for the intrinsic stability of DNA and for DNA replication are, for example, the aromatic stacking interactions between the bases within a strand and the solvent effects that occur under physiological conditions. See, for example: (a) Summerer, D., Marx, A. (2001) DNA polymerase selectivity: Sugar interactions monitored with high-fidelity nucleotides, *Angew. Chem. Int. Ed.* **40**, 3693–3695. (b) Berger, M., Ogawa, A.K., McMinn, D.L., Wu, Y., Schultz, P.G., Romesberg, F.E. (2000). Stable and selective hybridization of oligonucleotides with unnatural hydrophobic bases, *Angew. Chem. Int. Ed.* **39**, 2940–2942.

7. Kool, E.T., Morales, J.C., Guckian, K.M. (2000). Mimicking the structure and function of DNA: Insights into DNA stability and replication, *Angew. Chem. Int. Ed.* **39**, 990–1009 and references cited therein.

8. See, for example: Vögtle, F., *Supramolecular Chemistry* (Wiley, Chichester, 1993).

9. See, for example: *Antisense Research and Application*, edited by Crooke, S. T. (Springer Verlag, Berlin, 1998).

10. (a) Seeman, N.C., Rosenberg, J.M., Suddath, F.L., Kim, J.J.P., Rich, A. (1976). RNA double-helical fragments at atomic resolution – .1. crystal and molecular-structure of sodium adenylyl-3',5'-uridine hexahydrate, *J. Mol. Biol.* **104**, 109–144. (b) Rosenberg, J.M., Seeman, N.C., Day, R.O., Rich, A. (1976). RNA double-helical fragments at atomic resolution – .2. crystal and molecular-structure of sodium guanylyl-3',5'-cytidine nonahydrate, *J. Mol. Biol.* **104**, 145–167.

11. Fonseca Guerra, C., van der Wijst, T., Bickelhaupt, F.M. (2005). Substituent Effects on Hydrogen Bonding in Watson-Crick Base Pairs. A Theoretical Study, *Struct. Chem.* **16**, 211–221.

12. Swart, M., Fonseca Guerra, C., Bickelhaupt, F.M. (2004). Hydrogen Bonds of RNA Are Stronger than Those of DNA, but NMR Monitors Only Presence of Methyl Substituent in Uracil/Thymine, *J. Am. Chem. Soc.* **126**, 16718–16719.

Chapter 19

COMPUTATIONAL MODELING OF CHARGE TRANSFER IN DNA

Alexander A. Voityuk

Institució Catalana de Recerca i Estudis Avançats (ICREA), Institute of Computational Chemistry, Universitat de Girona,17071 Girona, Spain, alexander.voityuk@icrea.es

Abstract: During the past decade, charge transfer (CT) through DNA has been an area of extensive experimental and theoretical studies. The migration of excess charge plays an important role in DNA damage and repair and it is of great potential for molecular electronics. The CT process mediated by DNA is essentially determined by the structural dynamics of the π stack and its surroundings. Theoretical calculations provide microscopic insights into CT characteristics which are difficult to analyze by experimental techniques, and therefore, they are essential for understanding how and why charge transport occurs. In this chapter we deal with various aspects of computational modeling of CT in DNA. We consider estimation of key parameters- the driving force ΔG°, the electronic coupling V_{da} and the reorganization energy λ, that govern CT efficiency in DNA. The effects of molecular motions on these quantities are described. We discuss excess charge delocalization over adjacent base pairs. The mechanistic details derived from theoretical calculations are used to analyze the consistency and limitations of mechanisms for CT. In the last section, we consider some perspectives of theoretical studies on charge movement in DNA based systems and formulate some requirements to computational tools for microscopic modeling of the electron transfer process.

Key words: DNA, hole transfer, excess electron transfer, electronic coupling, quantum mechanical calculations

J. Šponer and F. Lankaš (eds.), Computational Studies of RNA and DNA, 485–511.

1.   INTRODUCTION

Soon after Watson and Crick discovered the double-helix structure of DNA in 1953,[1] Eley and Spivey suggested that DNA could be an electronic conductor.[2] However, only 30 years later, Barton and co-workers experimentally demonstrated rapid photoinduced electron transfer in DNA duplex over a distance of greater than 40 angstroms.[3-5] In the past decade, long-distance charge transfer mediated by DNA has received considerable experimental and theoretical attention and the main aspects of this process are now well understood.[6] Several experiments have demonstrated that a radical-cation state in DNA stack can migrate over distances up to ~200 Å.

Although DNA is not primarily an electron transfer species, the array of stacked base pairs in the interior of the double helix makes DNA a medium for CT. The overlapping π orbitals of the bases provide a pathway for migration of charge carriers generated on the stack. Besides the phenomenon has fundamental physical interest, attention to this problem is also motivated by various applications ranging from DNA damage and repair to design of nanoelectronic devices.

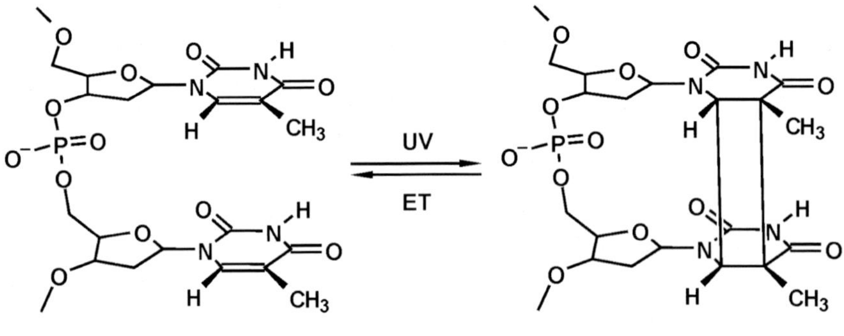

Figure 19-1. Ultraviolet light induces the formation of the thymine photodimer. Electron transfer initiates the repair of the damage.

Absorption of ultraviolet light may cause neighboring pyrimidine bases in a DNA strand to form cyclobutane-type dimers (see Figure 19-1). Such lesions have mutagenic effects and block the replication and the translation of DNA. The enzyme photolyase can repair these lesions. Quantum chemical calculations showed that splitting of photodimers can be triggered by ET.[7,8] Furthermore, the formation of an intermediate radical cation of guanine by oxidative stress may lead to DNA damage. Long-distance ET through DNA might protect genes from the oxidative mutation.[9]

The structural features and molecular interactions which make DNA efficient by transferring genetic information can be used to design new molecular devices. On the nano-scale (10-100Å), DNA fragments demonstrate intermolecular recognition and form predictable structures when they associate. Self-assembly and charge transfer properties make DNA a crucial construction material for molecular electronics. It appears to be a long way until the key biological molecule will play an important role in nanotechnology, however, the first steps to design electronic circuits and photoelectronic devices have been already made.[10,11]

Two issues are of special interest- the distance over which charge can be transferred through a DNA π stack and the time-scale of the CT process. There are two experimental approaches to measure charge transfer in DNA. In experiments on the electrical conductivity, one applies voltage to induce an electrical current between electrodes connected by DNA fragments. Recent measurements have provided contradictory results. In different studies, DNA ranges from an insulator to a semiconductor and even to a highly conducting wire.[12-14] Thus, the fundamental question of whether DNA is a wire or not remains unsettled because of a number of factors that are very difficult to control. In particular, the nature of the contacts of DNA to metallic electrodes presents a considerable challenge.

Reliable data on the CT rate in DNA have been obtained in photochemical experiments. Using a range of spectroscopic and biochemical methods, it has been shown that DNA is a medium for the efficient charge transfer.[6,15-18] In such studies one deals with structurally well-defined systems consisting of electron donor and acceptor covalently attached to an oligonucleotide. Charge transfer in DNA includes the following elementary steps: charge injection, hopping, trapping or/and recombination (Figure 19-2). The characteristic time of these steps is essentially dependent on the nature of donor and acceptor as well as on the length and structure of the DNA bridge.[6] By photoexcitation of the chromophore, an elementary charge is injected into DNA π-stack. Then, one observes migration of the charge along the chain. The efficiency of charge transport is determined, either by measuring the fluorescence decay of the chromophore or by analysis of the relative yield of strand damages at different positions in the DNA stack. Such experiments make it possible to examine the effects of structural and electronic factors on the CT rate.

Electron transfer can occur both in oxidized and in reduced DNA. In oxidized strands, an electron hole (radical cation state) moves through the stack. This process is called *hole transfer* (HT). Guanine bases (G) have the lowest oxidation potential among four nucleobases. As a result, HT implies the formation of intermediate radical cation states G^+. The majority of the experimental information has been obtained for hole transport. Schematically (Figure 19-2), irradiation of an intercalated chromophore leads to injection of a radical cation into adjacent guanine G_1. Then, another guanine, G_2 next to G_1, can donate an electron to G_1^+ resulting in G_1 and G_2^+. The charge hops through DNA using guanine bases as stepping stones. The sequential hole transfer between G bases, G-hopping, allows rapid transport of the positive charge in DNA.

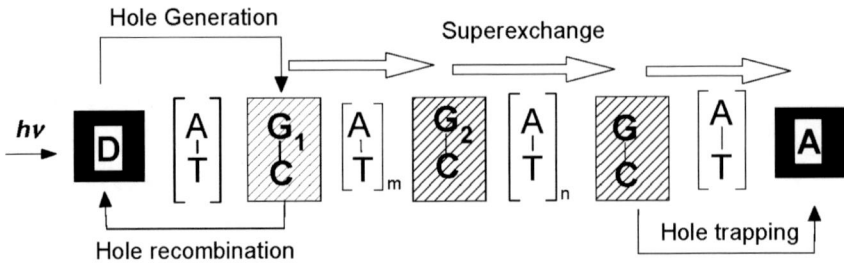

Figure 19-2. Schematic representation of photoinduced hole transfer in DNA.

The radical cation G^+ migrates through DNA before it reacts irreversibly with water. Experiments reveal that HT may occur over distance up to 200Å and its efficiency strongly depends on the sequence of base pairs.[18] A key feature of the G-hopping mechanism is electron tunneling between nearest guanines. Each tunneling step is mediated by superexchange, the interaction with virtual states of intervening (AT) pairs. Short $(AT)_n$ bridges are usually not oxidized during the hole transfer. The mechanism is characterized by a rapid exponential decay of the CT rate with the distance between the donor and acceptor sites:

$$k_{da} = k_0 \exp\left[-\beta R_{da}\right] \tag{1}$$

The falloff parameter β determines the conductactivity of the bridge and it is typically in the range of 0.6-1.4 Å^{-1}. When $\beta = 0.7$ Å^{-1}, k_{da} decreases by an order of magnitude with every additional (AT) pair inserted between donor and acceptor and drops by two orders of magnitude when $\beta = 1.4$ Å^{-1}. The superexchange mechanism appears to be operative only at relatively short distances, $R \leq 15$Å. In contrast to single-step tunneling, the hopping mechanism allows long-range charge transfer. The distance between donor and acceptor is divided into several short tunneling steps (Figure 19-2), the probability of which is controlled by superexchange. The hopping rate is given by $k \propto N^{-\eta}$,[19] where N is the number of steps and the parameter η is between 1 and 2. The CT efficiency depends on the number and the length of the hopping steps as well as on the rate of hole trapping by water. When the trapping reaction is much slower than the hole transfer steps, the hole charge distribution over DNA will be controlled by relative energies of the radical cation states localized on different sites. The multi-step hopping mechanism has been successfully used to interpret a variety of experimental data.[6] However, when G's are separated by more than three AT pairs, hole hopping between adenine bases, A-hopping, may also become operative.[20] Experiments for sequences $G(A)_nG$ with $n > 3$, suggest a mechanism which involves endothermic oxidation of an adenine base by G^+, followed by hole hopping within the $(A)_n$ chain.[20] As we will see, fluctuations of DNA environment can induce hole transfer from G^+ to A.[21]

Relatively not much is known today about excess electron transport (EET) in DNA. EET implies the movement of radical-anion states through a stack. Of the four DNA bases, cytosine and thymine are most easily reduced, and therefore, anions of the pyrimidine bases should be charge carriers in excess electron transfer. Because thymine exhibits the strongest electron affinity, one can assume that an excess electron hops via thymine bases. The decisive step of the photorepair of thymine dimers by photolyase (Figure 19-1) is an electron transfer from photoactivated flavin to the dimer. Related systems have been used to study EET through DNA. Flavin (electron donor) can be incorporated into a duplex containing a thymine dimer (electron acceptor) separated from the donor by several AT pairs.[22] Several photochemical and spectroscopic studies on EET have been published recently.[23-25] Therefore, computational insight into the mechanistic and dynamic issues of excess electron transport in DNA is now of special interest.

While general theoretical aspects of charge transfer mediated by DNA appear to be understood,[26] many important mechanistic details on the process remain to explore. Computational modeling has been proven to be a powerful tool that increases our understanding of the mechanisms of charge transfer in DNA. Theoretical methods provide a variety of quantities that are difficult to obtain experimentally and allow a more detailed analysis of factors that influence the charge transfer process. A full account of the computational studies of charge migration through DNA is beyond the scope of this chapter. The paper is organized as follows. Section 2 outlines basics of the nonadiabatic electron transfer theory. Models and methods used in computational studies of CT in DNA are briefly described in Section 3. Section 4 describes estimation of microscopic parameters that control the rate of CT emphasizing the effects of structural fluctuations of DNA. In Section 5 we focus on excess charge delocalization over DNA. Section 6 considers several issues concerning mechanisms of charge transport in DNA. Finally, open questions on CT in DNA as well as perspectives of the computational study of this phenomenon are discussed.

2. BASICS OF ET THEORY

The quantum mechanical approach to electron transfer in chemical systems is considered in detail by Newton.[27] Usually the nonadiabatic ET is described in terms of diabatic states. The diabatic states are localized on the donor and acceptor sites and represent the initial and final states for ET. These states can be written as linear combinations of adiabatic electronic states. Usually they are expressed via two adiabatic states, and therefore, a two-state model is applied. In first order time-dependent perturbation theory, the CT rate is determined by the electronic coupling matrix element squared and the thermally weighted Frank-Condon factor

$$k_{CT} = \frac{2\pi}{\hbar} V_{da}^2 (FC).$$

In line with the Marcus theory,[28] the rate constant is expressed as

$$k_{CT} = \frac{2\pi}{\hbar\sqrt{4\pi\lambda k_B T}} V_{da}^2 \exp\left[-(\Delta G^\circ + \lambda)^2 / 4\lambda k_B T\right] \qquad (2)$$

In line with Equation (2), three parameters influence the reaction rate. V_{da} is the electronic coupling matrix element between donor and acceptor, the free energy ΔG° is the driving force of ET, and λ is the reorganization energy. All these parameters are depicted on Figure 19-3.

The donor-acceptor coupling V_{da} is the most important parameter that determines the dependence of k_{CT} on the distance between the donor and acceptor sites and their mutual orientation. When donor and acceptor are separated by a molecular bridge, the coupling is mediated by the super-exchange. Even in systems with single intervening base pair between d and a, the orbitals of the donor and acceptor do not overlap significantly and no direct coupling between these sites is provided across empty space.[29] In fact, virtual states of the bridge mediate the coupling and facilitate the CT process. Equation 2 is valid in the weak coupling limit ($V_{da} < kT \sim 0.026$ eV at root temperature). The electronic couplings between neighboring base pairs is found to be quite strong, ~ 0.1 eV.[30] However, the interaction between second neighbors is several times weaker. Consequently, ET in DNA related systems, where donor and acceptor separated by one or more intervening base pairs, should fall well within the nonadiabatic regime. Equation 2 is derived within the Condon approximation which in turn implies the electronic coupling to be essentially independent of the nuclear coordinates. This approximation seems to break down, however, when thermal fluctuations

Free Energy

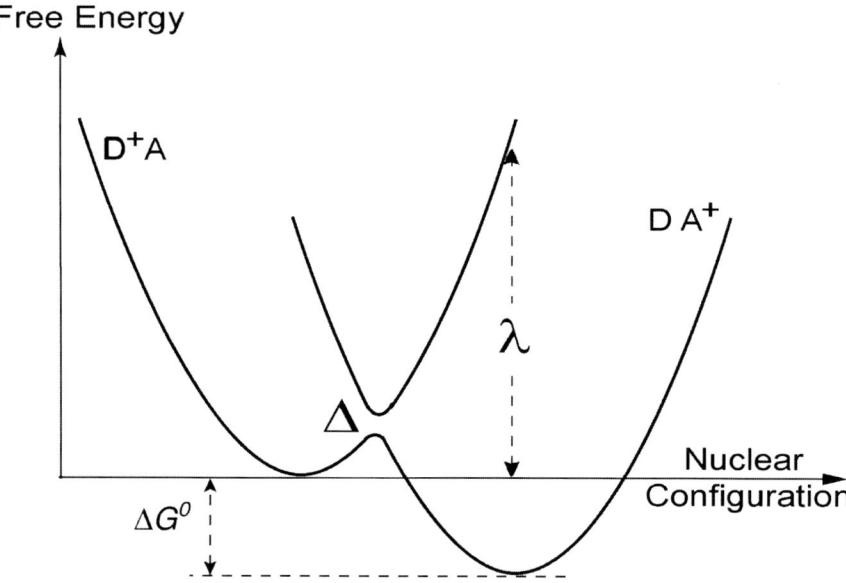

Figure 19-3. Free energy surfaces for nonadiabatic electron hole transfer. The splitting at the intersection of the curves is equal to twice the electronic coupling, $\Delta=2V_{da}$. The driving force ΔG° is the free energy for the ET reaction. The reorganization energy λ is equal to the change in the free energy to move the reactants to the nuclear configuration of the product.

of the DNA structure are taken into account. When donor and acceptor are equivalent, the diabatic states can be constructed from the plus and minus linear combinations of the lowest energy adiabatic states and the electronic coupling is equal to half the energy difference of these states. Because donor and acceptor in DNA are often nonequivalent, more elaborated methods must be employed. These methods and their application to estimate HT couplings have recently been reviewed.[31]

The driving force ΔG° is the difference of redox potentials of donor and acceptor. The redox potential of a nucleobase characterizes the localized hole state in DNA and linearly correlates with the vacuum ionization energy of the bases. ΔG° for charge shifting $D^{+}B A \rightarrow D B A^{+}$ is almost independent of the distance between donor and acceptor. However, for charge separation, $D B A \rightarrow D^{+} B A^{-}$, the Coulomb work for bringing D^{+} and A^{-} to the distance R_{da} has to be accounted for.

The reorganization energy λ is the change in the free energy to move the reactants to the product configuration without actually transferring the electron. The reorganization energy is usually divided into two parts, $\lambda = \lambda_{i} + \lambda_{s}$, the internal and solvent terms, respectively. The internal term λ_{i}

is associated with changes in the intra-molecular geometry of donor and acceptor caused by charge transfer when going from the equilibrium geometry of the initial state to the geometry of the final state. λ_i for the charge transfer reaction $d^+ + a \rightarrow d + a^+$ is a sum of the reorganization energies of donor and acceptor, $\lambda_i = \lambda_i(d) + \lambda_i(a)$:

$$\lambda_i(d) = E_+(d) - E_0(d) \; ; \; \lambda_i(a) = E_0(a^+) - E_+(a^+)$$

Energies $E_+(d)$ and $E_0(d)$ are calculated at geometries optimized for the donor oxidized state and the neutral state, respectively; $E_0(a^+)$ and $E_+(a^+)$ are calculated for a^+ using optimized geometries for a and a^+, respectively (see Figure 19-3). When the parabolas have a different curvature, λ_i is often estimated as the average of the corresponding values for the forward and back CT reactions.

The solvent term λ_s is associated with slow changes in the polarization of the surrounding medium. The magnitude of λ_s can be calculated from the classical dielectric continuum model developed by Marcus.[32] Within this model, the solvent medium is treated as a dielectric continuum with a polarization made up of two parts: a rapid electronic contribution and a slower orientational response. Because the time scale of the electron-transfer process is much faster than the solvent nuclear reorganization, only electronic polarization of the medium follows the changes in the charge distribution of solute. For the case of spherical donor and acceptor of radii r_d and r_a at the distance R_{da}, the solvent reorganization energy is given by

$$\lambda_s = \frac{q^2}{2} \left(\frac{1}{r_d} + \frac{1}{r_a} - \frac{2}{R_{da}} \right) \left(\frac{1}{\varepsilon_\infty} - \frac{1}{\varepsilon_0} \right),$$

where q is the charge transferred in the reaction, ε_∞ and ε_0 are the optical and static dielectric constants of the solvent.[32] Although this formula cannot provide quantitative estimation for the quasi one-dimensional DNA, it is useful for qualitative consideration. In particular, one can see that the reorganization energy depends upon the solvent polarity and should increases with the *d-a* distance. Delocalization of the transferring charge over neighboring bases may be described by the increasing r_d and r_a, and therefore, should reduce λ_s. The λ_s value may be estimated as the difference of solvation energies calculated with optical and static dielectric constants for the charge difference in final and initial states. Because in nonpolar solvents the parameter ε_∞ and ε_0 are very similar, λ_s should be small for ET reactions occurring in the nonpolar medium.

3. MODELS AND COMPUTATIONAL METHODS

3.1 DNA Models

In DNA, the ET channel consists of nearly parallel nucleobases. Each nucleobase is bound via sugar to a negatively charged phosphate group. So DNA duplexes are highly charged systems which are neutralized and stabilized by counter-ions Na^+ and water molecules. Although the sugar-phosphate backbone does not play a remarkable role in mediating the charge transfer, the position and dynamics of counter-ions and surrounding water may considerably affect ΔG^0 of the CT process.[21,33,34] The DNA structure is very flexible and characterized by essential conformational fluctuations. Thus, both internal and environmental degrees of freedom essentially modulate the ET rate over DNA stacks. The CT efficiency through DNA depends on the structure and dynamics of the duplex including the nature and the sequence of intervening base pairs, conformation and structural disorder along the helix, and medium effects.

Note that the photochemical charge separation in a DNA duplex seems to be unlikely without intercalated or bound chromophore. The excited singlet states of the nucleobases in DNA are extremely short-lived (<1 ps) and this plays a crucial role in protecting the molecule from damage by ultraviolet light. Hence, a charge injector should be inserted in the stack to initiate CT by photoexcitation. The efficiency of photoinduced charge transfer through the bridge depends on the redox characteristics of the charge injector and charge acceptor.[35,36] Note that intercalation of a chromophore into DNA may considerably change the structural parameters of the duplex and thereby affect the CT rate.

It has been recognized that charge migration in DNA can accompanied by proton transfer. Electron or hole transfer to a nucleobase will change its pKa-values and thereby can enforce rapid protonation/deprotonation of the base due to proton transfer between the nucleobase and surroundings. As a result, inter-strand proton transfer may interrupt the migration of electrons and holes in DNA by the conversion of radical ions to neutral radicals.[37] When the proton-transfer and electron-transfer events are separated, the first reaction step is energetically unfavorable; because of that a concerted process, proton-coupled electron transfer (PCET), often occurs. Thorp and co-workers observed a deuterium kinetic isotope effect by electron transfer in several DNA related systems which suggests that PCET can occur in DNA.[38] The coupling of electron transfer with the inter-base proton transfer may strongly affect CT efficiency.[39]

An accurate quantum mechanical description of systems which involve all these structural and dynamics features is hardly possible. So usually one considers remarkably simpler models focusing on one of the key factors or characteristics. Most theoretical studies of CT in DNA have been done for rigid systems or for models where only several degrees of freedom are coupled with electronic motion. It has been proven that simple but physically reasonable computational models may give very interesting and useful clues on the contribution of different effects in charge propagation in DNA.

3.2   Methods

There are two aspects in modeling ET in DNA: (1) proper description of electronic effects and (2) accounting for the dynamics of the π stack and its environment. In this context, Quantum Chemical Molecular Dynamics (QCMD) simulations appear to be the most promising approach. In this method, an electronic structure calculation has to be performed at each time step making the technique computationally very demanding. In their seminal paper, Car and Parrinello proposed an approach based on density functional theory and molecular dynamics, CPMD, in which the electronic terms are included in the motion equation and the coefficients of the basis functions are treated as dynamical variables.[40] CPMD makes it possible to solve the electronic structure problem for large systems accounting for coupling of electronic and nuclear coordinates. Application of the method to biological systems has already been reviewed.[41] Recently, Parrinello and co-workers have applied CPMD to study CT in DNA.[42] They studied a model consisting of 12 GC base pairs surrounded by solvated counter-ions (12 water molecules and two Na^+ ions per each base pair were included). The CPMD treatment of the system, which consists of 654 heavy atoms and 540 hydrogen atoms and possesses 3960 valence electrons, demonstrate the power of modern tools that become available for computational modeling. However, such calculations require enormous amount of computational resources and so far theoretical efforts have mostly been limited to simpler models.

A plausible computational approach to electron transfer in DNA is a combined QM/MD technique.[21,43] Conformational fluctuations of DNA and dynamics of solvated counter-ions are reasonably well reproduced by classical MD methods (see the contributions of Case and Cheatham to this book). Then several thousand snapshots extracted from the MD trajectory should be treated within a quantum chemical approach.

From a theoretical standpoint, ab initio methods provide the most accurate avenue for describing CT in DNA. However, these calculations are often computationally very demanding for extended systems. The high-level calculation (CASSCF, CASPT2) can be actually applied only for systems containing two base pairs whereas realistic models of DNA must include at least five base pairs. Therefore, less exact but more efficient methods have been employed. The Hartree-Fock and DFT methods appear to be sufficient for estimating CT parameters in stacks consisting of several base pairs.[30,44,45] Semiempirical methods may be very helpful for modeling CT in DNA.[36,46] In many studies, one employs model Hamiltonians based the tight binding approximation, in which only interactions between nearest sites (nucleobases or base pairs) are accounted for.[43,47-49]

4. CALCULATION OF CHARGE TRANSFER PARAMETERS

4.1 The Driving Force

Since electron holes are trapped at sites of minimum oxidation potentials, calculations of ionization energies of nucleobase sequences are useful for predicting the reactivity of different sites in DNA toward one-electron oxidation. In particular, the energy for hole transfer between two bases B and B' can be estimated as the difference of ionization energies of these bases embedded in the duplex. Saito and co-workers reported ionization potentials (IP) for XGY triplets calculated at the HF/6-31G* level.[44] The oxidation potential of G is found to be strongly influenced by adjacent 3'- and 5'-base pairs. Quantum chemical calculations of all possible triplets 5'-XBY-3' demonstrated that the oxidation potential of B in 5'-XBY-3' is considerably affected by the nature of 3'-Y and becomes smaller in the order C ~ T > A > G while the effect of the *preceding* base 5'-X is rather small.[46] Similar estimates have been derived recently from DFT calculations of triplets XGY.[45] The calculations suggest that the 5'-G in GG and both G on the 5'-side in GGG have the lowest oxidation potential in line with experimental findings.[44-46] Chemical modification of nucleobases changes their redox potentials and thereby can modulate CT properties of DNA.[48]

Also, the energetics of excess electron transfer in DNA has been estimated via differences of electron affinities (EA) of nucleobases B in trimers of Watson-Crick pairs.[50] It was shown that base stacking modulates the reduction potentials and that pyrimidines flanked by other pyrimidines, like in 5'-TTT-3' or 5'-CTC-3' sequences, are most easily reduced. Incorporation of a purine base close to the pyrimidine, however, makes it more difficult to reduce. The triads 5'-XCY-3' and 5'-XTY-3' where X, Y are C and T exhibit very similar EA values, and therefore, the corresponding radical-anion states should be approximately in resonance, favoring efficient transport of excess electrons in DNA.

4.2   Effect of DNA Environment on the Free Energy of Charge Transfer

Thus far, redox potentials of nucleobases have not been measured in the interior of a DNA double helix. Obviously, estimates derived for individual nucleosides disregard the influence of a structured environment. Schuster and co-authors showed the necessity of including the sugar-phosphate backbone, solvating water molecules, and counter-ions in models of charge transport in DNA.[33,34] They revealed that the ionization potential and the localization of the radical cation are strongly modulated by the location of counter-ions and the rate of CT is controlled by the probability of forming certain counter-ion configurations. Thus, hole migration can be gated by the configurational dynamics of the counter-ions. Recent QM/MD simulations have shown that the relative energies of the hole states in DNA are considerably affected by the local distribution of water molecules and counter-ions.[21]

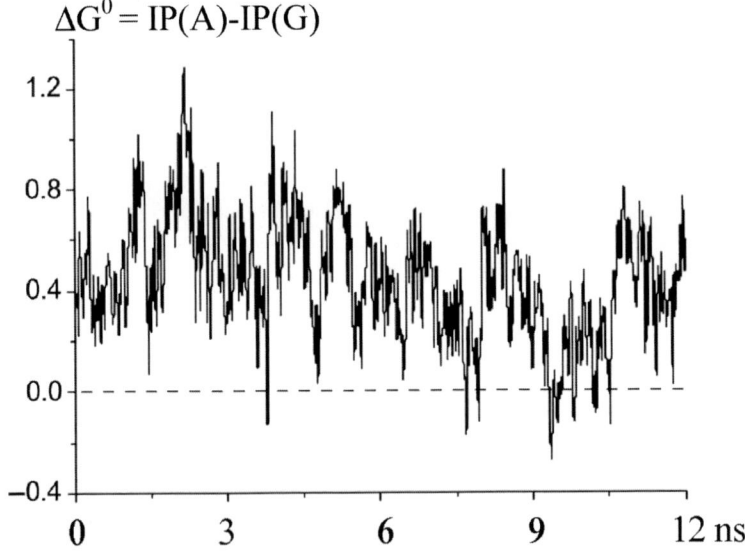

Figure 19-4. Fluctuation of the driving force for hole transfer between G and A bases[21].

The Figure 19-4 shows the energy of hole transfer from G_3 to A_6 in the duplex 5'-TTG$_3$T$_4$T$_5$T$_6$T$_7$T$_8$T$_9$T$_{10}$T$_{11}$G$_{12}$TT-3'. The average ΔG° value, ~ 0.4 eV, indicates that guanine is a stronger hole acceptor than adenine. Considerable fluctuations of IP were found along the MD trajectory. As seen, configurations of the system exist where hole transfer from G^+ to A is energetically feasible. The thermal motion of sodium ions and water molecules around DNA induces fluctuations of the redox potentials with a characteristic time of 0.3–0.4 ns. The fluctuations of ΔG° are large enough to make hole transfer from G^+ to A energetically feasible, and thus, allow a change-over from the accepted G-hopping mechanism to A-hopping. Note, that conformational changes of DNA π stacks do not cause significant fluctuations of ΔG°. To estimate the role of water surrounding DNA a modified duplex was studied in which all negatively charged phosphates were replaced by neutral methylphosphonate groups and all sodium ions were removed. The total fractions of time, when bridge sites are energetically preferred, are found to be similar for normal and modified duplexes (7–8%). These results suggest that the dynamics of water molecules strongly dominates the ΔG° fluctuations.

4.3 Electronic Coupling

The electronic coupling V_{da} between the diabatic electronic wave functions of the initial and final states can be estimated by using the generalized Mulliken-Hush formalism (GMH) developed by Cave and Newton.[51] In the two state model

$$V_{da} = \frac{(E_2 - E_1)|\mu_{12}|}{\sqrt{(\mu_2 - \mu_1)^2 + 4\mu_{12}^2}} \quad , \qquad (3)$$

where E_1 and E_2 is the energy of the ground and excited states, respectively, μ_{12} is the transition dipole moment between the states, (μ_2-μ_1) is the change of the dipole moment by excitation. The two-state model may be directly applied for system consisting of two base pairs. The HT electronic couplings between nucleobases and base pairs were calculated at the Hartree-Fock level.[29,30] These results permitted to analyze gross features of bridge specificity and directional asymmetry of V_{da} between nucleobases in DNA stacks. Detailed analysis showed that the electronic coupling between purine bases is notably affected through electrostatic and exchange interactions with pyrimidine bases.[52]

The nature of bridging bases B between donor and acceptor affects the charge migration through the π stack.[48] Two key characteristics of B can be varied (i) the energy of the virtual states of the bridge unit and (ii) the electronic coupling of B to adjacent bases. It was considered how one can control CT efficiency by modulating the intervening nucleobases.

4.3.1   Multi-State Effects

Usually, one employs a two-state model to calculate electronic couplings. An important advantage of the GMH method is that it is able to deal with multi-state situations where more than two adiabatic states enter into the description of the diabatic states of interests.[51] When electronic levels of the bridge or neighboring sites are energetically close to those of the donor and acceptor, the diabatic states may represent a combination of more than two adiabatic states. In such a situation, the multi-state model has be employed. Significant multi-state effects can arise even in rather short systems consisting of three base pairs.[53] For example, the V_{da} value in GAG predicted by the two-state model is five times larger than the coupling calculated with the multi-state method. The two-state scheme also cannot be applied to systems with the GG or GGG acceptor site. Although for some stacks, the two-state method provides accurate estimates of V_{da}, in general, this model fails to reproduce the electronic couplings in DNA calculated with the multi-state approach. In particular, the two-state model may lead to invalid estimates of V_{da} for systems with the tunneling gap less than 0.3 eV, e.g. for stacks with intra-strand purine bases.[53]

4.3.2   Excess Electron Transfer Coupling

The ability of DNA to mediate an excess electron is associated with the formation of radical-anion states of nucleobases. The QM treatment of systems carrying a negative charge, is more complicated than neutral or cationic species. In such systems the one-electron approximation may be insufficient and considerable basis set effects are expected. Recently, an efficient strategy has been discussed for calculating the EET matrix elements.[54] In a stack, the diabatic donor and acceptor states for EET can be approximated by radical-anion states of the separated base pairs. In one-electron approximation, a state of the excess electron is described by HOMO of the radical anion. It has been shown that the radical-anion states are well approximated by LUMOs of the neutral systems calculated without diffuse functions, and consequently, these molecular orbitals may be employed to estimate EET coupling matrix elements. The values based on LUMO's of neutral π-stacks are found to be in good agreement with the couplings obtained for the corresponding radical anion states. However, an extension of the basis set by including diffuse functions may lead to incorrect estimates of

EET couplings derived from data calculated for neutral systems. Thus, V_{da} for EET in DNA can be reasonably estimated in a similar manner as used to calculate hole transfer matrix elements. The EET matrix elements for models containing intra-strand T and C bases are essentially larger than the couplings of inter-strand pyrimidine bases. In general, ET couplings were found to be considerably smaller than the corresponding values for HT.[54]

4.3.3   Fluctuations of Electronic Couplings

Because the DNA structure is very flexible and characterized by essential conformational fluctuations, the question arises *How sensitive is the electronic coupling to the conformational changes.* It has been demonstrated that V_{da} in DNA stacks are very responsive to conformational changes and vary considerably with time.[55,56] Large changes in the electronic coupling may be found along the MD trajectory of a DNA stack. For a period of time of 10 ps the matrix element squared changes by 2-3 orders of magnitude depending on the number of intervening pairs between donor and acceptor. Thus, the conformational fluctuations of DNA considerably affect the efficiency of CT. The variation of V_{da} as a function of geometry is a clear breakdown of the Condon approximation. Consequently, instead of using Equation (2), one has to numerically integrate the CT rate along the MD trajectory for time period of ~1 ns. The calculations predict an increase in charge transfer efficiency due to DNA fluctuations as compared with the rigid duplex of ideal structure. It is worth noting that the effect of water molecules and counter-ions on the coupling is found to be rather small.

4.4   Reorganization Energy

The internal reorganization energy can be calculated with quantum chemical methods as described in Section 2. According to B3LYP/6-31G* calculations, hole transfer between two GC base pairs is associated with λ_i = 16.6 kcal/mol; while λ_i = 10.2 kcal/mol for CT between two AT pairs.[57]

  Several computational studies on the solvent reorganization energy for hole transfer in DNA have been published.[35,57-59] Electrostatic interactions in DNA significantly affect the outer-sphere contribution because of the highly charged backbones and the very polar core formed by nucleobases. As already noted, the Marcus spherical model cannot be applied to obtain quantitative estimates of λ_s in DNA. Within the classical approach, λ_s is derived from solution of the Poisson-Boltzmann equation. A multi-zone scheme is often employed for such calculations. Because of uncertainties concerning the construction of heterogeneous dielectric zones surrounding the donor and acceptor sites, the reorganization energy essentially depend on details of the calculation. The estimated values vary considerably among the different schemes. In systems with donor and acceptor separated by two intervening base pairs, λ_s ranges from ~31 kcal/mol[35] to 45.6 kcal/mol[58] and to ~69 kcal/mol.[57] Delocalization of the charge over two or more bases leads to essential decrease of reorganization energy. However, conformational fluctuations of a DNA stack have a rather small effect on the solvent reorganization energy.[57] The computational studies showed that the dependence of λ_s on the donor –acceptor distance is notable and has to be taken into account when estimating the CT rate in DNA.[35,57-59] The distance dependence of λ_s increases the falloff parameter β. For systems with a relatively small *d-a* separation, R_{da} ~6.8 Å, the corresponding contribution λ_s into β is quite large (~1.0 Å^{-1}), but it is nearly an order of magnitude smaller, 0.15 Å^{-1} for systems with R_{da} ~20 Å (five base pairs between D and A sites).[57]

LeBard et al. calculated reorganization energies for the processes of photoinduced hole injection and for the hole hopping between adjacent G sites connected by AT bridges of different length.[58] In addition to dielectric calculations, an explicit-solvent model has been also employed. The results received for hole injection and for hole hoping in hairpins are qualitatively similar. The magnitudes of the solvent reorganization energies obtained with explicit-solvent approach are smaller than those in standard dielectric continuum calculations. However, the slope of the dependence of λ_s on R_{da} does not depend significantly on the model. It has been demonstrated that the calculated solvent reorganization energy to be inconsistent with experimental estimates derived using Equation 2. However, a reasonably well agreement can be achieved when the Q-model[60] is employed for analysis of experimental data. The Q-model, in which the Hamiltonian of a two-state solute is linearly coupled to a harmonic solvent mode with different force constants in the initial and final states, predicts a much lower sensitivity of the overall rate constant to changes in the solvent reorganization energy than the Marcus theory. Thus, possible limitations of Equation 2 have been revealed.[59] Long-time CPMD calculations will provide results required to justify the model of charge transfer in DNA.

4.5   Quantum Chemical Study of CT Rate in DNA

4.5.1   Short-range Hole Transfer

The first quantum chemical study of hole transfer in DNA was carried out by Beratan and co-workers.[61] The electronic coupling was determined from semiempirical CNDO/2 calculations. It was shown that the π stacking interactions of nucleobases play a dominant role in providing the electronic coupling while the contribution of the sugar-phosphate backbone in V_{da} is small. Most important was the conclusion that the experimentally found long-range charge migration in DNA cannot be described with the superexchange mechanism. Recently, a comprehensive study of HT rates in DNA has been performed.[35] Several systems consisting of AT stacks of different length between donor and acceptor were computed. Guanines and 7-deaza-guanines (Z) were considered as the hole donor and acceptor sites. Because Z is easier to oxidize than G by 0.5 eV, duplexes containing Z bases have substantially larger tunneling gaps (the energy difference of hole states localized on the donor and the bridge) than the systems with guanines. The dependence of the electronic coupling on the donor-bridge energy gap has been analyzed. The values of β_{el} corresponding to decay of V_{da} were computed to be 0.95 and 1.46 Å^{-1}, for G- and Z-containing species, respectively. Because the electronic couplings of adjacent pairs Z-A and G-A are very similar, the difference in the distance dependence was attributed to the tunneling energy effect. Thus, the HT rates in DNA stacks with identical donor and acceptor (the driving force ΔG° in Equation 2 is equal to zero) sites depend critically on their oxidation potential because of the changes in the tunneling gap leading to variations in V_{da}. Also it has been shown that the tunneling gap has to be corrected by half the value of the reorganization energy.[35] While HT rates in DNA across one and two base pairs can be well described as single-step tunneling, for longer bridges the calculated rates are much smaller than observed. It was concluded that when three or more intervening AT base pairs are inserted between the donor and acceptor sites, instead of superexchange tunneling one should consider an alternative mechanism that implies thermal population of bridge states.

4.5.2   Photoinduced Charge Separation in DNA Hairpins

Most of data on the CT efficiency in DNA have been obtained in photo-chemical experiments. By photoexcitation of an intercalated chromophore the positive or negative charge is injected into DNA π stack. Consequently, to describe the kinetics of charge transfer the effects of the injector properties have to be explored. Beljonne et al.[36] addressed the distance dependence of the electronic coupling for charge injection in the model hairpins on the nature of intercalated chromophore. Photoinduced hole transfer from the electronically excited hole donor to the guanine acceptor was calculated using the semiempirical INDO/S method. The electronic characteristics of the initial and final states were employed within the GMH scheme, Equation 3, to estimate the electronic coupling. It was shown that the nature of the hole donor has a significant influence on the distance dependence of the photoinduced HT rate. In particular, the efficiency for hole injection is essentially affected by the energy gap between the donor and bridge electronic levels. Thus, both the injection energy and the chromophore-bridge coupling should be accounted for when estimating the relative CT rates through DNA stacks.

5.   EXCESS CHARGE DELOCALIZATION

An important issue that has not yet been resolved is delocalization of excess charge over neighboring base pairs in DNA. The description of charge transfer in π stacks depends on whether the excess charge is delocalized or not. The more extended the wave function, the better the conductance of the stack. Several mechanisms proposed for HT in DNA imply the hole delocalization over several nucleobases.[17,62] Different theoretical methods must be employed to describe propagation of localized and delocalized states. Delocalization of the charge over adjacent pairs is controlled by the interplay between the electronic coupling of base pairs, the internal reorganization energy and the interactions with polar environment. Recent computational studies give conflicting results. Basko and Conwell accounted for the solvation effects using a cylindrical cavity model within the tight-binding Hamiltonian.[63] The electron-hole wave function of a DNA π stacks in polar environment was found to be similar to that calculated for the isolated system where the hole charge is spread over five or more adjacent base pairs. By contrast, Beratan and co-workers showed that the interaction with surroundings considerably affects the charge distribution in DNA resulting in more localized hole state.[64]

Distribution of the hole charge in GG and GGG stacks has been recently considered using a quite simple but physically reliable model that takes into account all relevant interactions: the effect of flanking base pairs on the oxidation potential of guanine bases, the electronic coupling of GC pairs, the internal reorganization term and the solvation of the hole states.[65] In a system consisting of two GC pairs 1 and 2, the difference in charges on the guanine sites, $\Delta q = q_2 - q_1$, is determined by the difference of oxidation potentials $\Delta \varepsilon^0$, by the electronic coupling V_{12} as well as by the positive parameter σ that describes both the internal reorganization and solvation effects, $\sigma = -(\rho + \zeta)$. Δq may be found by solving the following equation

$$\Delta q = \frac{\Delta \varepsilon^0 - \sigma \Delta q}{\sqrt{\left(\Delta \varepsilon^0 - \sigma \Delta q\right)^2 + 4V_{12}^2}} \ .$$

The reorganization and solvation contributions, $\rho\Delta q$ and $\xi\Delta q$ respectively, depend on the charge distribution Δq. The reorganization parameter ρ is the difference of adiabatic and vertical ionization potentials of guanine, $\rho = \left(I_j^{ad} - I_j^{vert} \right)$. B3LYP/6-31G* calculations of a GC Watson-Crick pair predict ρ to be -0.36 eV. The parameter $\zeta = -0.80$ eV can be derived from solvation energies of oligomers calculated for states with the localized and delocalized hole charges. Thus, the reference value of the parameter σ is 1.16 eV. The magnitude of electronic coupling V_{12} depends on the stack conformation. In the dimer of ideal structure V_{12} is found to be about 0.080 eV which is very close to the average value of the coupling, 0.077 eV. So $V_{12} = 0.08$ eV is used.

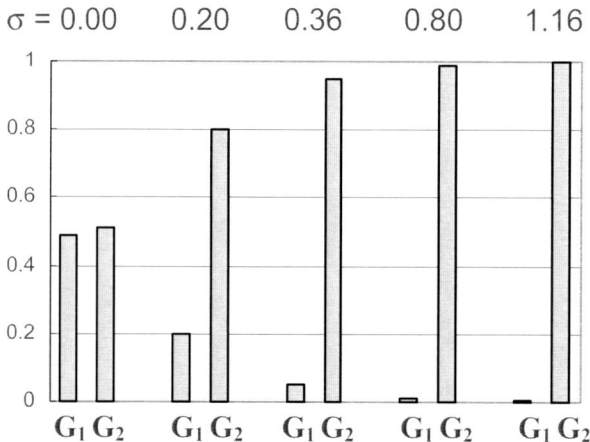

Figure 19-5. The effect of solvation and internal reorganization energy, parameter σ, on the charge delocalization in 5'-A-G$_1$G$_2$-G-3'.

Figure 19-5 compares the charge distributions in 5'-A-G$_1$G$_2$-G-3' calculated at different values of σ ranging from 0 to 1.16 eV. When $\sigma = 0$, i.e. the solvation and internal reorganization terms are neglected, the hole charge is almost equally delocalized over G$_1$ and G$_2$. At $\sigma = 0.20$ eV the charge ratio q_2 / q_1 is about 4:1. When only the reorganization term is taken into account, $\sigma = 0.36$ eV, the hole charge is distributed as $q_1 = 0.05$ and $q_2 = 0.95$. Therefore, even in isolated duplexes when all interactions with environment are switched off, the hole states should be essentially localized. The solvation term is twice as large as the reorganization energy, $\sigma = 0.80$ eV, and consequently, more significant localization of the positive charge is predicted, $q_1 = 0.010$ and $q_2 = 0.990$. When both effects are accounted for, $\sigma = 1.16$ eV, $q_1 = 0.005$ and $q_2 = 0.955$. Thus, the solvation and internal reorganization effects suppress essentially the charge delocalization in DNA leading to radical cation states confined to a single guanine sites. The result remains unchanged when considerable deviations of the electronic coupling from its average value are accounted for.[65,66]

Recently, a combined QM/MD study has been performed on a poly(dA)-poly(dT) duplex to examine the influence of DNA dynamics on the hole delocalization.[67] MD predicts that individual bases exhibit significant freedom of movement leading to large fluctuations in electronic couplings. The off-diagonal dynamical disorder results in charge localization of HOMO and LUMO of the stacks. Moreover, a large degree of localization for the wave functions is found for the states that are energetically close to the HOMO and LUMO. The off-diagonal disorder should reinforce the localization of an excess charge caused by the polar environment.

The hole charge localization in DNA has recently been investigated using the CPMD method.[42] The simulation was performed on a hydrated double stranded G-C DNA dodecamer. Six water molecules per nucleotide were considered while about 15 water molecules per each base are usually included in classical MD simulation. Thus, in the CPMD model, the solvent effects appear to be underestimated. The CPMD study revealed that the strongest localization occurs on those G-bases in which the angle between the vector normal to the plane of the base and the z-axis of DNA is large. It has been demonstrated that the hole can be localized either by proton shift from guanine to cytosine coupled to electron transfer or by a change in the solvation shell of the counter-ions. Thermal fluctuations appear to be essential for the charge localization which increases with the temperature.

6.   CHARGE TRANSFER MECHANISMS

As already noted, short-range CT in DNA is well described by the superexchange mechanism. However, even in this relatively simple model several important questions are still to answer. The main question concerns the limitation of Equation 2 when applied to ET in DNA. First, the Condon approximation (weak dependence of V_{da} on geometry) does not hold for DNA stacks.[26,55,56] Second, the assumption on linear coupling of donor and acceptor states to a harmonic solvent appears to be quite limited and should be replaced by the bilinear approximation (Q-model).[59,60] Third, the driving force for CT is quite sensitive to position of surrounding water and sodium ions around DNA, and therefore, dynamics of the environment essentially influences the charge transfer in DNA.[21] Thus, calculation of the parameters V_{da}, ΔG°, and λ and using them within Equation 2, probably will not give quantitative estimates of the CT rate constant. Moreover, these parameters cannot be determined independently. For instance, the electronic coupling mediated by the bridge is considerably influenced by the energy gap and the reorganization energy.[26] However, our current understanding of CT in DNA is mainly based on the Marcus theory and it would be completely incorrect to underestimate its role in model calculations and analysis of experimental data.

Long-distance charge transfer. In the G-hopping mechanism, hole hopes between neighboring guanine sites are mediated by superexchange. Many experimental data on HT have been interpreted within this model. However, for long $(AT)_n$ bridges ($n \geq 4$), this mechanism fails to rationalize the slight decrease of the CT efficiency with the lengths of the bridge. Accordingly, the G-hopping mechanism has been extended to A-hopping which involves intermediate radical-cation states localized on adenine bases.[9,20] The hole transfer from G^+ to A by means of thermal energy has been referred to as thermally induced hopping (TIH). Note that TIH appears to be unlikely for the energy gap estimated using the redox potentials of nucleobases measured in water. According to this data, the G^+ states should be more stable than the A^+ states by 0.4 eV or ~15kT.

Usually the CT efficiency is expected to increase when an AT base between donor and acceptor is replaced by a GC pair. Theoretical results obtained by Berlin et al. for long bridges suggest an inverted sequence dependence, namely, due to the thermally induced mechanism the exchange of AT by GC may reduce the hole transfer rate.[68,69]

Mechanisms implying charge delocalization. The hopping mechanism of a localized charge has been extended to delocalized radical cation states. Delocalization of the positive charge over several adjacent nucleobases has been postulated in several mechanisms. In the polaron model,[17,70] the radical cation hopping is thought to occur on the timescale of structural dynamics of DNA and its polar surroundings. The introduction of an excess charge into DNA will be accompanied by consequent structural changes lowering the energy of the system. These changes lead to the formation of a polaron, a distortion of the local environment around the radical cation, which has a shallow energy minimum. The hole becomes self-trapped. To overcome the energy barrier of the hole self-trapping the thermal energy is required.[71] Thus, initiated by thermal fluctuations, the radical cation migrates adiabatically from one polaron site to another. The size of the hopping step is controlled by the sequence of bases.

In the ion-gated model, a radical cation hops from one nucleobase to another depending on dynamics of surrounding sodium ions and water molecules. This mechanism has been combined with the polaron model in which delocalized-hole movement through the duplex is controlled by the configuration of polar environment.[71] Formation of a hole polaron on a long bridge of AT pairs has been considered recently.[72] The polaron consisting of four adenine bases is assumed to be stabilized by surrounding water.

The domain model[62] assumes hole delocalization over a transiently extended π-orbital without distorting the π stacks structure and without becoming trapped. A domain is thought to consist of 4-5 base pairs. CT-active conformations which allow the charge to hop among domains are generated due to thermal fluctuations of the stack conformation during the lifetime of the excited state of the intercalated chromophore.[62]

Thus, hole delocalization over several base pairs is assumed, at least implicitly, in several mechanisms of charge migration in DNA. However, there has been no experimental evidence of the charge delocalization. Recent computational and theoretical studies suggest that polar environment should essentially suppress charge delocalization in DNA.[35,65] Because of that, these mechanisms have to be justified.

7.   CONCLUDING REMARKS

Computational modeling of charge transfer through DNA is not only significant from the standpoint of predicting the CT efficiency in various systems but it is also important for understanding how and why the CT process occurs. It provides microscopic insights into system characteristics which are difficult or even hardly possible to analyze by experimental techniques. The simulations are also essential for verification of proposed mechanisms and analysis of its consistency and limitations. The mechanistic picture derived from theoretical models has significant implications for ongoing experimental studies of CT in DNA.

From the discussion presented above it has become clear that the charge transfer mediated by DNA cannot be represented as a simple function of the bridge sequence but is closely related to the dynamics of the π stack and its surroundings. Even more complicated picture should be considered in advanced systems including M-DNA and metallized DNA.[73] M-DNA, complexes formed between DNA and metal ions are of interest for the development of nano-electronic devices. In particular, molecular logic functions have been demonstrated in M-DNA. The conductivity of DNA can be considerably increased by metallization of DNA which implies coating of the double helix with 4d- and 5d-transition metals. These systems turn out to be very promising for DNA based nanoelectronics.

Despite considerable theoretical and computational efforts several important issues remain to be explored. This requires estimating a number of electronic, structural and dynamics characteristics of physically reason-able models. Although modern computational tools have been proven to be very useful for modeling of charge propagation in DNA, elaboration of efficient computational schemes specially designed to treat charge transfer in DNA related systems seems to be crucial for ongoing theoretical studies. In such an approach the quantum chemical description of the electronic structure should be combined with dynamics of the DNA π stack and its surroundings including explicit treatment of solvent molecules. Because extended models consisting of hundreds of atoms should be considered, the coupling of quantum and classical methods appears to be unavoidable. The following interactions are of special interest and should be treated accurately. The *stacking interaction* in π stacks is still a challenge for quantum-chemical methods[74] while the dispersion energy can reasonably well reproduced within modern force-field schemes. As has been already mentioned *proton transfer coupled to ET* may considerably affect charge movement in DNA. Therefore, the correct treatment of strong and weak hydrogen bonds as well as activation barriers for proton transfer is an essential prerequisite. The QM treatment of aperiodic systems containing many heavy transition metals is a critical problem. However, such calculations become very important because a proper description is required for contacts of DNA with metallic electrodes as well as for modeling metallized DNA and M-DNA. Semiempirical methods that are specially parametrized for such systems, may provide quite reliable description of the models.[75]

Electron transfer between donor and acceptor may be also viewed as a molecular conduction process,[76] in which the ET rate is replaced by electrical conductivity. Theoretical approaches to electron transmission and conduction in molecular junctions including DNA wires have recently been reviewed[76] and the relationship between the intramolecular electron transfer efficiency and the electrical conduction of molecular species has been discussed.[76-78] A combined approach based on quantum chemical modeling of intramolecular charge transfer and theoretical description of electron conduction processes will be essential for computational design of DNA based nano-electronic devices.

ACKNOWLEDGEMENTS

The work was supported by the Spanish *Ministerio de Educación y Ciencia*, Project No. CTQ2005-04563 from the *Dirección General Investigación.*

REFERENCES

1. Watson, J. D. & Crick, F. H. C. (1953). Molecular Structure of Nucleic Acids - a Structure for Deoxyribose Nucleic Acid. *Nature* **171**, 737-738.
2. Eley, D. D. & Spivey, D. I. (1962). Semiconductivity of Organic Substances .9. Nucleic Acid in Dry State. *Trans. Farad. Soc.* **58**, 411-415.
3. Murphy, C. J., Arkin, M. R., Jenkins, Y., Ghatlia, N. D., Bossmann, S. H., Turro, N. J. & Barton, J. K. (1993). Long-Range Photoinduced Electron-Transfer through a DNA Helix. *Science* **262**, 1025-1029.
4. Arkin, M. R., Stemp, E. D. A., Holmlin, R. E., Barton, J. K., Hormann, A., Olson, E. J. C. & Barbara, P. F. (1996). Rates of DNA-mediated electron transfer between metallo-intercalators. *Science* **273**, 475-480.
5. Hall, D. B., Holmlin, R. E. & Barton, J. K. (1996). Oxidative DNA damage through long-range electron transfer. *Nature* **382**, 731-735.
6. Schuster, G. B. Editor, (2004). *Long-Range Charge Transfer in DNA. Topics in Current Chemistry*, **236** and **237**.
7. Voityuk, A. A., Michel-Beyerle, M. E. & Rösch, N. (1996). A quantum chemical study of photoinduced DNA repair: On the splitting of pyrimidine model dimers initiated by electron transfer. *J. Am. Chem. Soc.* **118**, 9750-9758.
8. Rak, J.,Voityuk, A. A., Michel-Beyerle, M. E. & Rösch, N. (1999). Effect of Proton Transfer on the Anionic and Cationic Pathways of Pyrimidine Photo-Dimers Cleavage. A Computational Study. *J. Phys. Chem. A* **103**, 3569-3574.
9. Giese, B. (2002). Long-distance electron transfer through DNA. *Annu. Rev. Biochem.* **71**, 51-70.
10. Park, S. J., Lazarides, A. A., Mirkin, C. A., Brazis, P. W., Kannewurf, C. R. & Letsinger, R. L. (2000). The electrical properties of gold nanoparticle assemblies linked by DNA. *Angew. Chem. Int. Ed.* **39**, 3845-3852.
11. Park, S. J., Taton, T. A. & Mirkin, C. A. (2002). Array-based electrical detection of DNA with nanoparticle probes. *Science* **295**, 1503-1506.
12. Dekker, C. & Ratner, M. A. (2001). Electronic properties of DNA. *Physics World* **14**, 29-33.
13. Kasumov, A. Y., Kociak, M., Gueron, S., Reulet, B., Volkov, V. T., Klinov, D. V. & Bouchiat, H. (2001). Proximity-induced superconductivity in DNA. *Science* **291**, 280-282.
14. Porath, D., Bezryadin, A., de Vries, S. & Dekker, C. (2000). Direct measurement of electrical transport through DNA molecules. *Nature* **403**, 635-638.
15. Giese, B. (2000). Long distance charge transport in DNA: The hopping mechanism. *Acc. Chem. Res.* **33**, 631-636.
16. Lewis, F. D., Letsinger, R. L. & Wasielewski, M. R. (2001). Dynamics of photoinduced charge transfer and hole transport in synthetic DNA hairpins. *Acc. Chem. Res.* **34**, 159-170.
17. Schuster, G. B. (2000). Long-range charge transfer in DNA: Transient structural distortions control the distance dependence. *Acc. Chem. Res.* **33**, 253-260.
18. Boon, E. M. & Barton, J. K. (2002). Charge transport in DNA. *Curr. Opin. Struct. Biol.* **12**, 320-329.
19. Jortner, J., Bixon, M., Langenbacher, T. & Michel-Beyerle, M. E. (1998). Charge transfer and transport in DNA. *Proc. Natl. Acad. Sci. USA* **95**, 12759-12765.

20. Giese, B., Amaudrut, J., Kohler, A. K., Spormann, M. & Wessely, S. (2001). Direct observation of hole transfer through DNA by hopping between adenine bases and by tunneling. *Nature* **412**, 318-320.

21. Voityuk, A. A., Siriwong, K. & Rösch, N. (2004). Environmental fluctuations facilitate electron-hole transfer from guanine to adenine in DNA pi stacks. *Angew. Chem. Int. Ed.* **43**, 624-627.

22. Behrens, C., Cichon, M. K., Grolle, F., Hennecke, U. & Carell, T. (2004). Excess electron transfer in defined donor-nucleobase and donor-DNA-acceptor systems. In *Topics in Current Chemistry* **236**, 187-204.

23. Lewis, F. D., Liu, X. Y., Miller, S. E., Hayes, R. T. & Wasielewski, M. R. (2002). Dynamics of electron injection in DNA hairpins. *J. Am. Chem. Soc.* **124**, 11280-11281.

24. Giese, B., Carl, B., Carl, T., Carell, T., Behrens, C., Hennecke, U., Schiemann, O. & Feresin, E. (2004). Excess electron transport through DNA: A single electron repairs more than one UV-induced lesion. *Angew. Chem. Int. Ed.* **43**, 1848-1851.

25. Ito, T. & Rokita, S. E. (2004). Criteria for efficient transport of excess electrons in DNA. *Angew. Chem. Int. Ed.* **43**, 1839-1842.

26. Berlin, Y. A., Kurnikov, I. V., Beratan, D., Ratner, M. A. & Burin, A. L. (2004). DNA electron transfer processes: Some theoretical notions. In *Topics in Current Chemistry* **237**, 1-36.

27. Newton, M. D. (1991). Quantum Chemical Probes of Electron-Transfer Kinetics - the Nature of Donor-Acceptor Interactions. *Chem. Rev.* **91**, 767-792.

28. Marcus, R. A. & Sutin, N. (1985). Electron Transfers in Chemistry and Biology. *Biochim. Biophys. Acta* **811**, 265-322.

29. Voityuk, A. A., Rösch, N., Bixon, M. & Jortner, J. (2000). Electronic coupling for charge transfer and transport in DNA. *J. Phys. Chem. B* **104**, 9740-9745.

30. Voityuk, A. A., Jortner, J., Bixon, M. & Rösch, N. (2001). Electronic coupling between Watson-Crick pairs for hole transfer and transport in desoxyribonucleic acid. *J. Chem. Phys.* **114**, 5614-5620.

31. Rösch, N. & Voityuk, A. A. (2004). Quantum chemical calculation of donor-acceptor coupling for charge transfer in DNA. In *Topics in Current Chemistry*, **237**, 37-72.

32. Marcus, R. A. (1964). Chemical and Electrochemical Electron-Transfer Theory. *Annu. Rev. Phys. Chem.* **15**, 155-166.

33. Barnett, R. N., Cleveland, C. L., Joy, A., Landman, U. & Schuster, G. B. (2001). Charge migration in DNA: Ion-gated transport. *Science* **294**, 567-571.

34. Barnett, R. N., Cleveland, C. L., Landman, U., Boone, E., Kanvah, S. & Schuster, G. B. (2003). Effect of base sequence and hydration on the electronic and hole transport properties of duplex DNA: Theory and experiment. *J. Phys. Chem. A* **107**, 3525-3537.

35. Tong, G. S. M., Kurnikov, I. V. & Beratan, D. N. (2002). Tunneling energy effects on GC oxidation in DNA. *J. Phys. Chem. B* **106**, 2381-2392.

36. Beljonne, D., Pourtois, G., Ratner, M. A. & Bredas, J. L. (2003). Pathways for photoinduced charge separation in DNA hairpins. *J. Am. Chem. Soc.* **125**, 14510-14517.

37. Giese, B. & Wessely, S. (2001). The significance of proton migration during hole hopping through DNA. *Chem. Comm.*, 2108-2109.

38. Weatherly, S. C., Yang, I. V. & Thorp, H. H. (2001). Proton-coupled electron transfer in duplex DNA: Driving force dependence and isotope effects on electrocatalytic oxidation of guanine. *J. Am. Chem. Soc.* **123**, 1236-1237.

39. Carra, C., Iordanova, N. & Hammes-Schiffer, S. (2002). Proton-coupled electron transfer in DNA-acrylamide complexes. *J. Phys. Chem. B* **106**, 8415-8421.

40. Car, R. & Parrinello, M. (1985). Unified Approach for Molecular-Dynamics and Density-Functional Theory. *Phys. Rev. Lett.* **55**, 2471-2474.

41. Carloni, P., Rothlisberger, U. & Parrinello, M. (2002). The role and perspective of ab initio molecular dynamics in the study of biological systems. *Acc. Chem. Res.* **35**, 455-464.

42. Gervasio, F. L., Laio, A., Parrinello, M. & Boero, M. (2005). Charge localization in DNA fibers. *Phys. Rev. Lett.* **94**, N 158103.

43. Lewis, J. P., Cheatham, T. E., Starikov, E. B., Wang, H. & Sankey, O. F. (2003). Dynamically amorphous character of electronic states in poly(dA)-poly(dT) DNA. *J. Phys. Chem. B* **107**, 2581-2587.

44. Saito, I., Nakamura, T., Nakatani, K., Yoshioka, Y., Yamaguchi, K. & Sugiyama, H. (1998). Mapping of the hot spots for DNA damage by one-electron oxidation: Efficacy of GG doublets and GGG triplets as a trap in long-range hole migration. *J. Am. Chem. Soc.* **120**, 12686-12687.

45. Senthilkumar, K., Grozema, F. C., Guerra, C. F., Bickelhaupt, F. M. & Siebbeles, L. D. A. (2003). Mapping the sites for selective oxidation of guanines in DNA. *J. Am. Chem. Soc.* **125**, 13658-13659.

46. Voityuk, A. A., Jortner, J., Bixon, M. & Rösch, N. (2000). Energetics of hole transfer in DNA. *Chem. Phys. Lett.* **324**, 430-434.

47. Grozema, F. C., Siebbeles, L. D. A., Berlin, Y. A. & Ratner, M. A. (2002). Hole mobility in DNA: Effects of static and dynamic structural fluctuations. *ChemPhysChem* **3**, 536-547.

48. Voityuk, A. A. & Rösch, N. (2002). Quantum chemical modeling of electron hole transfer through pi stacks of normal and modified pairs of nucleobases. *J. Phys. Chem. B* **106**, 3013-3018.

49. Roche, S. (2003). Sequence dependent DNA-mediated conduction. *Phys. Rev. Lett.* **91**, N 108101.

50. Voityuk, A. A., Michel-Beyerle, M. E. & Roosch, N. (2001). Energetics of excess electron transfer in DNA. *Chem. Phys. Lett.* **342**, 231-238.

51. Cave, R. J. & Newton, M. D. (1997). Calculation of electronic coupling matrix elements for ground and excited state electron transfer reactions: Comparison of the generalized Mulliken-Hush and block diagonalization methods. *J. Chem. Phys.* **106**, 9213-9226.

52. Rak, J., Voityuk, A. A., Marquez, A. & Rösch, N. (2002). The effect of pyrimidine bases on the hole-transfer coupling in DNA. *J. Phys. Chem. B* **106**, 7919-7926.

53. Voityuk, A. A. (2005). Electronic couplings in DNA π-stacks: multi-state effects. *J. Phys. Chem. B* **109**, 17917-17921.

54. Voityuk, A. A. (2005). Estimates of electronic coupling for excess electron transfer in DNA. *J. Chem. Phys.* **123**, N 034903.

55. Voityuk, A. A., Siriwong, K. & Rösch, N. (2001). Charge transfer in DNA. Sensitivity of electronic couplings to conformational changes. *PCCP* **3**, 5421-5425.

56. Troisi, A. & Orlandi, G. (2002). Hole migration in DNA: a theoretical analysis of the role of structural fluctuations. *J. Phys. Chem. B* **106**, 2093-2101.

57. Tavernier, H. L. & Fayer, M. D. (2000). Distance dependence of electron transfer in DNA: The role of the reorganization energy and free energy. *J. Phys. Chem. B* **104**, 11541-11550.

58. Siriwong, K., Voityuk, A. A., Newton, M. D. & Rösch, N. (2003). Estimate of the reorganization energy for charge transfer in DNA. *J. Phys. Chem. B* **107**, 2595-2601.

59. LeBard, D. N., Lilichenko, M., Matyushov, D. V., Berlin, Y. A. & Ratner, M. A. (2003). Solvent reorganization energy of charge transfer in DNA hairpins. *J. Phys. Chem. B* **107**, 14509-14520.

60. Matyushov, D. V. & Voth, G. A. (2000). Modeling the free energy surfaces of electron transfer in condensed phases. *J. Chem. Phys.* **113**, 5413-5424.

61. Priyadarshy, S., Risser, S. M. & Beratan, D. N. (1996). DNA is not a molecular wire: Protein-like electron-transfer predicted for an extended pi-electron system. *J. Phys. Chem.* **100**, 17678-17682.

62. O'Neill, M. A. & Barton, J. K. (2004). DNA charge transport: Conformationally gated hopping through stacked domains. *J. Am. Chem. Soc.* **126**, 11471-11483.

63. Basko, D. M. & Conwell, E. M. (2002). Effect of solvation on hole motion in DNA. *Phys. Rev. Lett.* **88**, N 098102.

64. Kurnikov, I. V., Tong, G. S. M., Madrid, M. & Beratan, D. N. (2002). Hole size and energetics in double helical DNA: Competition between quantum delocalization and solvation localization. *J. Phys. Chem. B* **106**, 7-10.

65. Voityuk, A. A. (2005). Are radical cation states delocalized over GG and GGG hole traps in DNA? *J. Phys. Chem. B* **109**, 10793-10796.

66. Voityuk, A. A. (2005). Charge Transfer in DNA. Hole Charge is Confined to a Single Base Pair due to Solvation Effects. *J. Chem. Phys.* **122**, N 204904.

67. Lewis, J. P., Pikus, J., Cheatham, T. E., Starikov, E. B., Wang, H., Tomfohr, J. & Sankey, O. F. (2002). A comparison of electronic states in periodic and aperiodic poly(dA)-poly(dT) DNA. *Phys. Stat. Solid. B* **233**, 90-100.

68. Berlin, Y. A., Burin, A. L. & Ratner, M. A. (2002). Elementary steps for charge transport in DNA: thermal activation vs. tunneling. *Chem. Phys.* **275**, 61-74.

69. Grozema, F. C., van Duijnen, P. T., Berlin, Y. A., Ratner, M. A. & Siebbeles, L. D. A. (2002). Intramolecular charge transport along isolated chains of conjugated polymers: Effect of torsional disorder and polymerization defects. *J. Phys. Chem. B* **106**, 7791-7795.

70. Conwell, E. M. & Rakhmanova, S. V. (2000). Polarons in DNA. *Proc. Natl. Acad. Sci. USA* **97**, 4556-4560.

71. Schuster, G. B. & Landman, U. (2004). The mechanism of long-distance radical cation transport in duplex DNA: Ion-gated hopping of polaron-like distortions. In *Topics in Current Chemistry*, **236**, 139-161.

72. Conwell, E. M., Park, J. H. & Choi, H. Y. (2005). Polarons in DNA: Transition from guanine to adenine transport. *J. Phys. Chem. B* **109**, 9760-9763.

73. Richter, J. (2003). Metallization of DNA. *Physica E* **16**, 157-173.

74. Šponer, J. & Hobza, P. (2003). Molecular interactions of nucleic acid bases. A review of quantum-chemical studies. *Coll. Czech. Chem. Comm.* **68**, 2231-2282.

75. Thiel, W. (1996). Perspectives on semiempirical molecular orbital theory. *Adv. Chem. Phys.* **93**, 703-757.

76. Nitzan, A. (2001). Electron transmission through molecules and molecular interfaces. *Annu. Rev. Phys. Chem.* **52**, 681-750.

77. Nitzan, A. & Ratner, M. A. (2003). Electron transport in molecular wire junctions. *Science* **300**, 1384-1389.

78. Nitzan, A. (2001). A relationship between electron-transfer rates and molecular conduction. *J. Phys. Chem. A* **105**, 2677-2679.

Chapter 20

QUANTUM CHEMICAL CALCULATIONS OF NMR PARAMETERS

Wolfgang Schöfberger[1], Vladimír Sychrovský[2] and Lukáš Trantírek[3]

[1]Institute of Organic Chemistry, Johannes Kepler University Linz, Altenbergerstraße 69, A- 4040 Linz, Austria; [2]Institute of Organic Chemistry and Biochemistry, Academy of Sciences of the Czech Republic, Flemingovo n. 2., 166 10 Praha 6, Czech Republic; [3]Faculty of Biology, University of South Bohemia & Institute of Parasitology, Academy of the Sciences of the Czech Republic, Branišovská 31, CZ-37005 České Budějovice, Czech Republic

Abstract: The main goal of this chapter is to provide a basis for understanding basic connections between NMR experiment and *in silico* computer simulation techniques. We outline the capability of modern theoretical approaches to assist with interpretation of experimental NMR data in terms of molecular structure.

Key words: NMR; quantum chemical calculations; chemical shielding; spin-spin coupling

1.   INTRODUCTION

Modern high resolution nuclear magnetic resonance spectroscopy (NMR) provides detailed information on the structure and dynamics of biomolecules both in solution and solid state[1,2,3]. In general, the appearance of a NMR spectrum can be characterized by: i) a number of distinct signals ii) signal positions, ii) a signal area, intensity and line-width, iv) the so called hyperfine structure of individual signals. While the quantity of signals in the NMR spectra corresponds to a number of the magnetically non-equivalent nuclei, the remaining parameters are related to the electronic structure of the

J. Šponer and F. Lankaš (eds.), Computational Studies of RNA and DNA, 513–536.
© 2006 Springer.

molecule, and thus can be used as a source of the structural information. However, to extract structural information from the acquired NMR spectrum, the relationship between NMR parameters and molecular structure has to be known. There are two ways how such a relationship can be established. In the first approach, called empirical parameterization, the NMR parameters for model compounds with known structures are experimentally determined. The variation of some structural parameter across a set of model compounds is correlated with the variation of one or more NMR parameters. In the second approach the calculated NMR parameters can be directly correlated with the molecular structure. Here, we will provide a brief introduction to quantum chemical calculations of the NMR parameters. Main goal of this chapter is to provide the reader basis for understanding basic connections between NMR experiment and *in silico* computer simulation techniques.

2.   PARAMETERS OF THE NMR SPECTRUM

2.1   Signal Position – Chemical Shift

Electrons of a molecule placed into the magnetic field (B_0) respond by creating their own magnetic field, which opposes to the B_0. The external field B_0 is thus shielded by the magnetic field from electrons that is to a magnitude smaller than B_0. The effective magnetic field at the nucleus (B_{eff}) therefore generally differs from the applied field by the fraction "σ" called the chemical shielding, which is defined as $B_{eff} = B_0(1-\sigma)$. The same nuclei in the molecule can in principle experience a different effective magnetic field as a result of a diverse chemical and structural environment. The chemical shielding reflects a local symmetry of electronic distribution at the nuclear site. At this point, it is necessary to outline the connection between chemical shielding and the experimentally acquired NMR spectrum. In general, the NMR spectrum is a record of the intensity of the emission of electromagnetic radiation by the nucleus at the given frequency of radiation. The signal position is therefore defined by resonance frequency ω (called the Larmor frequency), which can be related to the effective magnetic field as:

$$\omega = -\gamma \cdot B_{eff} = -\gamma\left(B_0(1-\sigma)\right)$$

(1)

where γ is the magnetogyric ratio (nucleus specific constant). Equation (1) gives a physical basis for understanding relationships between electronic structure and the NMR spectrum. However, from an experimental point of view, the frequency axis in NMR spectroscopy is a substantial problem. As can be seen from Eq. (1), the resonance frequency depends on the magnitude of the magnetic field. Assuming that a spectrum was measured at the field of 11.7 Tesla for example, one could not directly compare the resonance frequencies with those obtained on a spectrometer operating at 20.0 Tesla. In this case, it would be necessary to perform extensive calculations in order to understand if two resonance frequencies measured at different field strengths actually belong to the same nucleus. The variations of resonance frequencies due to the differential chemical shielding of protons are pitifully small when compared to the overall resonance frequencies. Not only that the resonance frequencies acquired at different B_0 have to be scaled, but the absolute frequency scale ($\sim 10^6$ Hz) is not appropriate for reporting of the NMR spectra. To solve these problems at once, the relative scale was introduced using a NMR parameter called a chemical shift (δ). The chemical shift can be calculated for any resonance line according to the following equation:

$$\delta = \frac{(\nu_0 - \nu)}{\nu_0} \qquad (2)$$

In this equation, ν ($\nu = 2\pi\omega$) is the resonance frequency of any peak for which the chemical shift δ is to be calculated and ν_0 is the resonance frequency of a given nucleus in standard substance. It is noteworthy that by this definition the chemical shift δ is dimensionless. Thus δ values can now be compared on different spectrometers assuming that the same standard was used for calibration. The absolute frequency scale given in 10^6 Hz is, however, not readily available. For this reason, chemical shift values are normally multiplied by a factor 10^6. To avoid confusion with the original values calculated according to the Eq. (2) are appended with 'ppm' (parts per million) to indicate this scaling.

2.2 The Hyperfine-Structure of the NMR Signal

As a hyperfine structure of a NMR signal, we denote the fine splitting of resonance line(s) as a result of inter-nuclear spin-spin interactions. There are two mechanisms – indirect and direct spin-spin interaction.

2.2.1   Indirect Spin-Spin Interaction (Scalar Coupling or J-Coupling)

Indirect spin-spin coupling, which is often called scalar coupling or J-coupling, is the effect of mutual interaction between two nuclear spins mediated by electrons polarized by the nuclear magnetic moment(s). Although, scalar coupling is anisotropic due to the asymmetric environment surrounding the nuclear spin, the interaction is averaged to an isotropic value in solution by a rapid molecular tumbling motion. It should be emphasized that the coupling constant is a consequence of the nuclear magnetic moment and therefore independent of the external B_0 field.

Scalar couplings are exhibited in NMR spectra as the signal splitting of resonance lines of the coupled nuclei (Figure 20-1). The splitting of a resonance line is quantitatively described by the so called scalar coupling constant given in Hertz (Hz). The scalar coupling constant of two nuclei that are separated by N bonds is denoted $^N J$. For example, $^1 J_{CH}$ is the coupling of a ^{1}H directly joined to a ^{13}C-atom. Although scalar interaction propagates along the chemical bonds, the magnitude of scalar coupling is dramatically reduced as the number of bonds separating the nuclei increases. A typical value of $^1 J_{CH}$ is in the range of 120 to 250 Hz. $^2 J_{CH}$ is between 0 and 10 Hz. Both ^{3}J and ^{4}J give detectable effects in a spectrum, but couplings over larger numbers of bonds are often hardly detectable. The value of scalar coupling can be either positive or negative.

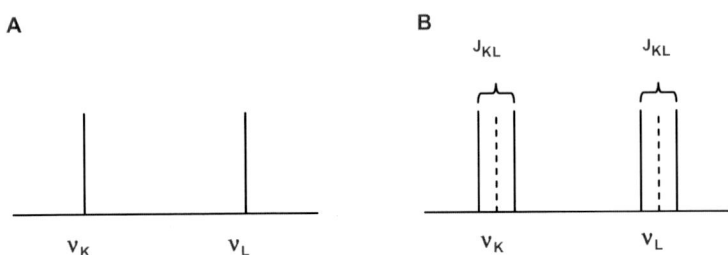

Figure 20-1. (A). Line splitting due to the zero (A) and non-zero (B) spin-spin coupling.

The indirect spin-spin couplings depend on the local distribution of electrons in a vicinity of the coupled nuclei. A specific change of the electronic environment is usually reflected in a modulation of J coupling magnitude. The scalar couplings are commonly used as a source of structural information (see Section 3.4.1 and 3.4.2).

2.2.2   Direct Spin-Spin (Dipolar) Interaction

Direct spin-spin (dipolar) interaction is caused by the direct interaction between coupled nuclei. (Figure 20-2). The dipolar field of nucleus K will change the resonance frequency of nucleus L by an amount that depends on the internuclear distance and the orientation of the internuclear vector relative to B_0. For a fixed orientation of the vector, say parallel to B_0, nuclear spin K can increase or decrease the total magnetic field at nucleus L, depending on whether K is parallel or antiparallel to B_0. In an ensemble of molecules, half of the K nuclei will be parallel to B_0, the other half antiparallel, and L will show two resonances, a doublet, separated in frequency by

$$D^{KL} = -\mu_0 \left(\frac{h}{2\pi} \right) \frac{\gamma_K \gamma_L}{4\pi^2 r_{KL}^3} \left\langle \left(3\cos^2 \theta - 1 \right) / 2 \right\rangle, \tag{3}$$

where θ is the angle between the inter-nuclear vector and B_0, the $<>$ brackets denote time or ensemble averaging, μ_0 is the magnetic permitivity of vacuum; h is Planck's constant; $\gamma_{K/L}$ is magnetogyric ratio of nucleus K and L and r_{KL} is the distance between nuclei K and L.

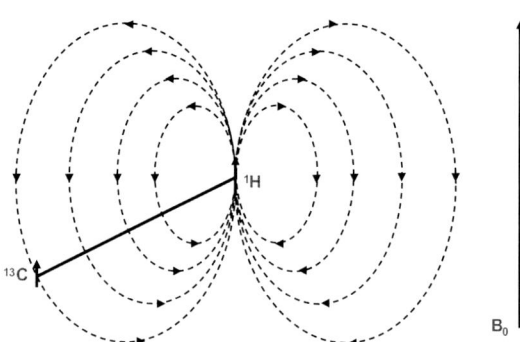

Figure 20-2. Magnetic dipole-dipole coupling, illustrated for a ^{13}C-^{1}H spin pair. The total magnetic field in the B_0 direction at the ^{13}C position can increase or decrease relative to B_0, depending on the orientation of the ^{13}C-^{1}H vector (see Eq. (3)) and the spin state of the proton (parallel or antiparallel to B_0).

In isotropic solution, rotational Brownian diffusion rapidly averages the inter-nuclear dipolar interaction described by the Eq. (3) to zero. In other words, in isotropic solution these interactions do not affect the hyperfine structure of NMR signals. However, in a liquid NMR experiment, it is possible to reach the intermediate case, where the compound under study is dissolved in a slightly anisotropic aqueous medium, in which not all orientations of the molecule are equally likely to occur. In such a case, the dipolar couplings no longer average to zero and are exhibited in NMR spectra such as adding to signal splitting caused by indirect spin-spin interactions. The dipolar couplings, which originate from interactions of proximate spins in partially ordered molecular system, are then called "residual dipolar couplings" (RDC). As can be seen from the Eq. (3), the RDC can serve as a valuable source of structural information as they depend on inter-atomic distance between interacting nuclei, as well as on the relative orientation of inter-atomic vector with respect to the external magnetic field. The structural evaluation of the RDCs essentially relies on the evaluation of simple geometric terms relating local molecule structure to the direction of the external magnetic field. However, the structural applications of RDCs are beyond of the scope of this text. Because of that, we will refer the interested reader to specialized literature on RDCs[3].

2.3 Signal Line-Width and Intensity – Relaxation of Nuclear Spin

Each signal in a spectrum can be characterized, besides its position and hyperfine structure, by its intensity, half-width and area under the peak. The area under a peak in an NMR spectrum corresponds to the number of magnetically equivalent nuclei. Note that this is in contrast to the peak height which varies with the line width of the peak. Basically, the NMR signal is a consequence of relaxation of excited nuclear spin states. As such, both signal intensity and half-width are related to the relaxation phenomenon. There are two pathways for the relaxation of nuclear spins: a) spin-lattice and b) spin-spin relaxations.

2.3.1 Spin-Lattice Relaxation

The spin-lattice relaxation (characterized by T_1 relaxation time) describes the recovery of z-magnetization to its thermal equilibrium at which populations of energy states reach the Boltzmann distribution. During T_1 relaxation, the exchange of energy between spin and the environment ("lattice") occurs due to various intra- and intermolecular interactions, including dipole-dipole relaxation (T_1^{DD}, chemical shift anisotropy (T_1^{CSA}), spin-rotation relaxation (T_1^{SR}), scalar coupling (T_1^{SC}), electric quadrupolar relaxation (T_1^{EQ}), interactions with unpaired electrons in paramagnetic compounds, etc. Spin-lattice relaxation can be summarized as:

$$\frac{1}{T_1} = \frac{1}{T_1^{DD}} + \frac{1}{T_1^{CSA}} + \frac{1}{T_1^{SR}} + \frac{1}{T_1^{SC}} + \frac{1}{T_1^{EQ}} + \frac{1}{T_1^{UE}} \ldots \ldots \tag{4}$$

The DD interaction is a dominant contribution to T_1 relaxation of protons in molecules in solution. Nuclear spins at the excited state can transit to the ground state via energy exchange with the surroundings or between nuclei. The energy exchanged to the lattice may be transformed into motions of translation, rotations and vibrations. The process of energy exchange is caused by the timed-dependent fluctuation of magnetic (or electric) fields at or near the Larmor frequency (ω). The fluctuating fields may be produced by vibrational, rotational, or translational motions of other surrounding nuclei, changes in chemical shielding, or the energy state of unpaired electrons. These time-dependent fluctuating fields have significant effects on nuclear relaxation. Consequently, molecular rotation and diffusion are the most efficient sources of nuclear relaxation in solution. For an aqueous solution of normal viscosity, T_1 relaxation is inversely proportional to correlation time τ_c $\sim 1/T_1$ (τ_c is defined as the time required for reorientation of a molecule about 1 radian). This parameter indicates that slower random motions are responsible for a shorter T_1 relaxation time. Because random molecular motions are influenced by the size of molecules, the magnitude of the correlation time depends significantly on molecular weight. For small molecules with molecular weights up to 100, τ_c falls to the interval of 10^{-12}–10^{-13}s, whereas macromolecules may have a τ_c as large as 10^{-8}s. Calculation of correlation time is a very complicated procedure in which many factors such as the shape of the molecule and different kinds of molecular motions must be taken into account. In general, τ_c is best estimated experimentally[4]. Spin-lattice relaxation affects the lifetime of the excited state. As such T_1 can influence the signal line-width. However, spin-lattice relaxation is normally too slow to contribute to the line width (see below). Knowing T_1 is, however, useful in the design of nuclear magnetization transfer experiments. The larger the T_1, the longer the duration and bigger the number of delays and pulses that can be introduced into the corresponding pulse sequences.

2.3.2 Spin-Spin Relaxation

T_2 relaxation (also known as spin-spin or transverse magnetization) describes the decay of transverse magnetization characterized by the Bloch equation.

$$\frac{dM_x}{dt} = -\frac{M_x}{T_2} \quad \text{or} \quad \frac{dM_y}{dt} = -\frac{M_y}{T_2} \tag{5}$$

In which the time constant T_2 is called the spin-spin or transverse relaxation time, which describes how fast transverse magnetization M_x or M_y decays to zero. Because transverse magnetizations M_x and M_y are detected parameters in a NMR experiment, T_2 relaxation time directly determines the signal line width in frequency domain – $\Delta v_{1/2} = 1/\pi T_2$, where $\Delta v_{1/2}$ is defined as the line width at half height of the signal amplitude. The T_2 relaxation process is adiabatic, which means that T_2 is not connected with population changes in the energy states. During spin-spin relaxation, transition of one nucleus from a high energy state to a lower one causes another nucleus to move simultaneously from the lower state to a higher one. There is no energy exchange with the environment and hence no gain or loss in energy of the molecular system. In practice, magnetic field inhomogeneity is the dominant contribution to the transverse relaxation. Each nucleus across a sample volume experiences a slightly different B_0 field. Consequently, some of the chemically equivalent nuclei rotate around the direction of the external magnetic field with different Larmor frequencies, resulting in the fanning-out of individual magnetization vectors. The net effect is a decrease of transverse magnetization. By taking into account the effect of B_0 inhomogeneity, transverse relaxation is described by the effective transverse relaxation time T_2^*.

$$M_y = M_0 \exp(-t/T_2^*) \tag{6}$$

and

$$\Delta v1/2 = 1/\pi T_2^* + \gamma \Delta B_0 \tag{7}$$

Equation (6) tells us that T_2^* is the time when the amplitude of a detected transverse magnetization has decayed by a factor of $1/e$. The first term in Eq. (7) represents the natural line width caused by spin-spin relaxation whereas the second term is the contribution of field inhomogeneity to the spectral line width. When molecules in non-viscous liquids are moving very rapidly T_1 equals T_2 for nuclei with a spin $1/2$, which results in very narrow signals. On the other hand, the signals are broad for macromolecules or solid-state samples with $T_1 > T_2$.

Relaxation (both spin-spin and spin-lattice) of nuclear magnetic excited states is principally due to the presence of a magnetic field which fluctuates at the transition frequencies of the spins. The principal mechanisms that generate fluctuations in the magnetic field at the nucleus are dipole-dipole interactions between spins and chemical shielding anisotropy. Often, more than one mechanism may operate at the same time. For proton NMR studies of large molecules, the dominate mechanism is dipole-dipole coupling between protons. Chemical shielding anisotropy is a dominate mechanism for carbon, fluorine and phosphorus at high magnetic field strength.

The oscillation or fluctuation of the local magnetic field is due to the presence of molecular motion. Thus, measurements of relaxation (T_1, T_2) can provide valuable and unique information on molecular motion. Relaxation rates for most of the above processes can be calculated with high accuracy from first principles. Therefore, it is possible to obtain detailed information on molecular motion from relaxation studies. The time scale of these motions is generally on the nano- to piko-second range and can determined using the following equations:

$$R_1 = \frac{1}{10} k_{dd} \left[J(\omega_C - \omega_H) + 3J(\omega_C) + 6J(\omega_C + \omega_H) \right] + \frac{2}{15} \Delta\sigma^2 \omega_C^2 J(\omega_C) \quad (8)$$

$$R_2 = \frac{1}{20} k_{dd} \left[J(\omega_C - \omega_H) + 3J(\omega_C) + 6J(\omega_C + \omega_H) + J(0) + 6J(\omega_H) \right]$$
$$+ \frac{1}{45} \Delta\sigma^2 \omega_C^2 \left[4J(0) + 3J(\omega_C^2) \right] + R_{ex} \quad (9)$$

R_1 and R_2 are relaxation rates $1/T_1$ and $1/T_2$, respectively. $k_{dd} = \gamma_H \gamma_C \hbar^2 / r^6{}_{CH}$; γ_C and γ_H are magnetogyric ratios for ^{13}C and ^1H nuclei, $\hbar$ is Plank's constant divided by 2π, ω is the Larmor frequency, $\Delta\sigma$ is the anisotropy of chemical shielding, and r_{CH} is the length of the C–H bond. R_1 and R_2 are described by five spectral densities, $J(0)$, $J(\omega_C)$, $J(\omega_H)$, $J(\omega_H-\omega_C)$, and $J(\omega_C+\omega_H)$ (for ^{13}C relaxation), as well as by a term to account for chemical exchange, R_{ex}. Eqs. (8)-(9) indirectly relate molecular structure with relaxation parameters such as signal line-width. As can be seen, the R_2 is modulated by both chemical shielding as well as interactions among proximate spins. At the same time, the relaxation parameters refer to molecular shape and internal and overall motion. We can access this information via analysis of the so-called "spectral-density-function".

2.3.3 Spectral Density Function

As already mentioned it is possible to separate the full process of nuclear relaxation (the return of magnetization to equilibrium) into two different processes: one by which the component M_z reaches its equilibrium value M_{z0}

and one in which phase coherence M_{xy} vanishes. The relaxation rates for these two processes for a spin-K relaxing exclusively through a dipolar coupling to another spin L (such as ^{13}C-^{1}H spin pair) are (as compared to Eqs. (8)-(9)):

$$R_1 = (1/T_1') = 1/10 \ (k_{DD})^2 \ [J(\omega_K\text{-}\omega_L) + 3J(\omega_K) + 6J(\omega_K + \omega_L)] \qquad (10)$$

$$R_2 = (1/T_2') = 1/20 \ (k_{DD})^2 \ [4J(0) + J(\omega_K\text{-}\omega_L) + 3J(\omega_K) + 6J(\omega_L) \\ + 6J(\omega_K + \omega_L)] \qquad (11)$$

The ω parameters are the corresponding resonance frequencies of nuclei K and L and the so-called spectral density functions $J(\omega)$s are given by $J(\omega) = \tau_c/(1 + \omega^2\tau_c^2)$ in most simple cases. The correlation time τ_c can be approximated by $\tau_c = MV/\eta RT$, where M and V are the molecular mass and volume, respectively. η is the viscosity and T the temperature. The distribution of the frequencies is usually expressed with spectral density function[5].

$$J^{AB}(\omega) = 2\int_0^\infty \cos(\omega \cdot t)\langle A(0)B(t)\rangle dt = 2\int_0^\infty \cos(\varpi \cdot t)G^{AB}(t)dt \qquad (12)$$

The symbols A(t) and B(t) stand for Hamiltonians relevant for fluctuating dipole-dipole interaction. The brackets represent the average over all molecules in the sample. The ensemble average of the product function, $\langle A(0)B(t)\rangle$, is also called the time correlation function $G^{AB}(t)$, which for the dipole-dipole relaxation is given by $G_0(t) = \frac{1}{5}\langle P_2\{\cos[\theta(0) - \theta(t)]\}\rangle$. Here, θ is the angle of the dipole-dipole vector relative to the external magnetic field. The correlation function $G^{AB}(t)$ decays exponentially for most simple case of isotropic rotational motion with the time constant τ_c as $G(t) = 1/5 \exp(-t/\tau_c)$. The corresponding spectral density function is given by:

$$J(\omega) = \frac{2}{5}\frac{\tau_c}{1 + (\omega\tau_c)^2} \qquad (13)$$

Overall motion consists of two motions the isotropic overall motion and the internal motion. Lipari and Szabo[5] assume that these two motions are independent. The time correlation function can then be factored into two contributions for overall isotropic tumbling, namely one from the overall motion $G_0(t)$, corresponding to $G(t)$, discussed above (Eq. 13) and one contribution representing internal motion $G_1(t)$, such as $G(t)=G_0(t).G_1(t)$. Internal motion can be described by a sum of exponential decaying terms $G_1(t) = \sum_i a_i e^{-t/\tau_1}$. The simplest description of $G_1(t)$ for internal motion is:

$$\tau_e = \frac{\int_0^\infty (G_1(t) - S^2) dt}{1 - S^2} \qquad (14)$$

From the definition of $G_1(t)$ it follows that $G_1(0)=0$, as presumed by Lipari and Szabo[5], $S=G_1(\infty)$ and the τ_e as the surface under the time correlation function. Finally this leads to a spectral density function for the isotropic overall motion of the form:

$$J(\omega) = \frac{2}{5} \left[\frac{S^2 \tau_c}{1 + (\omega \tau_c)^2} + \frac{(1 - S^2) \tau}{1 + (\omega \tau_c)^2} \right] , \qquad (15)$$

with $\tau^{-1} = \tau_c^{-1} + \tau_e^{-1}$ and $G_1(t) = S^2 + (1-S) \exp(-t/\tau_e)$ the overall correlation function. The spectral density function now contains two fitting parameters for each vector since the overall correlation time τ_c is expected to be the same for all nuclei in the molecule. This approach invokes no particular model to describe internal motion. S is a model-independent measure of the spatial restriction of the internal motion. In summary, this description of internal motion in terms of two parameters, τ_e and S^2, represents the total information content of NMR relaxation measurements. It is impossible to give complete and correct description of relaxation phenomena on just few pages. We refer the interested reader to specialized literature[6,7].

3. CALCULATION OF NMR PARAMETERS

There are two NMR parameters that can be calculated by quantum chemical methods and simultaneously can be correlated with the signal position and hyperfine structure of a NMR signal (see Section 2): i) chemical shielding, and ii) the indirect spin-spin coupling constant. Both NMR parameters are so called second order molecular properties, which mean that during their calculation, distinct terms arising due to the specific magnetic interactions are calculated as a mixed second order partial derivative of the total molecular energy. In other words, the magnetic interactions described by the magnetic Hamiltonian[8,9] are considered as perturbations of the wave function describing the molecular electronic system. The total energy of the molecular system E(B,M), inclusive of the homogeneous magnetic field B and the nuclear magnetic moment(s) M, and the ground state energy E_0 do not coincide. Corrections of the ground stated energy E_0 due to individual magnetic interactions can be expanded in a series of terms that are linear and bilinear in B and/or M, followed by terms of a higher order.

$$E(B,M) = E_0 + .. + \sum_K B \cdot \sigma(K) \cdot M_K + \sum_{K \neq L} M_K \cdot J(K,L) \cdot M_L + .. \quad (16)$$

The term σ(K) of the series, called the NMR shielding constant of nucleus K is bilinear in the magnetic induction B and the respective nuclear magnetic moment M_K. Similarly, the indirect NMR spin-spin coupling constant J(K,L) corresponds to the interaction of two nuclear magnetic moments M_K and M_L. The expansion term bilinear in B which is not explicitly given in the series of Eq. 16), corresponds to the molecular magnetizability. The terms that are only linear in either B or M do not contribute neither to σ nor to J. Also the expansion terms of higher order, which are not explicitly given in Eq. (16), can be safely neglected providing the strength of the external magnetic field corresponds to the strength in usual NMR experiment.

Both NMR constants σ and J are tensors represented by the 3 x 3 matrices reflecting the corrections to the total energy E(B,M) depend on the mutual orientation of the coupled vectors B(x,y,z) and M(x,y,z). The isotropic average of an NMR parameter, which reflects fast and isotropic tumbling of a molecule during NMR signal acquisition in a liquid phase experiment, is called the isotropic NMR constants.

$$P_{iso} = \frac{1}{3}\left(P_{xx} + P_{yy} + P_{zz}\right) \quad P \approx \sigma, J \quad (17)$$

Isotropic average calculated from the trace of the tensor of NMR parameter is usually implemented automatically in most of the calculation codes.

3.1   Chemical Shift (σ)

When considering the series in Eq. (16), the chemical shielding constant σ(K) can be calculated as the second order partial derivative of the total energy E(B,M) with respect to the nuclear magnetic moment M_K and the external magnetic field B.

$$\sigma(K) = \frac{\partial^2 E(B,M)}{\partial B \partial M_K} \quad | \, B, M_1, M_2, \ldots = 0 \tag{18}$$

By claiming the zero B and M after the second partial derivative is accomplished, the higher order terms of Eq. (16) are automatically neglected. The chemical shielding constant reflects the effective local magnetic field at the site of the nucleus as was discussed in Section 2.1. This local magnetic field consists of the external homogeneous magnetic field, the field of the nuclear magnetic moments and the magnetic field created by a movement of electrons. Each of these magnetic fields can be represented by its vector potential entering the magnetic Hamiltonian in a specific way[9,10]. Homogeneous magnetic field B is represented by the vector potential A(r).

$$B(r) = \left(\frac{\partial}{\partial x}, \frac{\partial}{\partial y}, \frac{\partial}{\partial z} \right) \times A(r), \text{ where } A(r) = \frac{1}{2} B_0 \times (r - R_0) \tag{19}$$

The homogeneous magnetic field B(r) with the magnitude B_0 created by the vector potential A(r) is clearly independent of the origin R_0. Furthermore, it is identical with the magnetic field created by the transformed vector potential,

$$A'(r) = A(r) + \left(\frac{\partial}{\partial x}, \frac{\partial}{\partial y}, \frac{\partial}{\partial z} \right) \times f(r) \tag{20}$$

where $f(r)$ is arbitrary scalar function of argument r. We would expect the calculated chemical shielding to be invariant with respect to the displacement of R_0 as well as with respect to the choice of $f(r)$. This is true only in the limit case of an exact wave function. For the approximate solution of the electronic problem the calculated chemical shielding would be dependent of the transformation used for $A(r)$. The problem concerning this choice is called the Gauge origin problem and can be surmounted by making the wave function gauge dependent with the efficiently placed origin R_0 of A(r). Two approaches besides the other (reviewed in ref.[11,-3]) are frequently used resulting in gauge independent NMR properties. The GIAO (Gauge Independent Atomic Orbitals) approach transforms the set of atomic orbitals by attaching the special phase factor[14]. A similar transformation is performed within the IGLO approach (Individual Gauge of Localized Orbitals) for the set of localized molecular orbitals[15,16].

The chemical shielding constant consists of two contributions usually called diamagnetic and paramagnetic shielding.

$$\sigma(K) = \langle \psi_0 \mid A_K \cdot A_0 \mid \psi_0 \rangle + 2 \cdot \mathrm{Re} \langle \psi_0 \mid A_K \cdot p \mid \psi_B \rangle \qquad (21)$$

The first one is calculated as a product of two vector potentials integrated over the ground state wave function Ψ_0. The vector potentials correspond to the magnetic field created by the nuclear magnetic moment M_K and the external magnetic field. The product of A_K and the momentum operator p appears as an integrand of the paramagnetic term. The first-order wave function Ψ_B, represented by the set of perturbed molecular orbitals with respect to the external field B, are needed in calculation of the paramagnetic shielding.

3.2   The Indirect Spin-Spin Coupling Constant (J)

Two natural coupling mechanisms may occur in a pair of nuclear spins: the spin-spin coupling mediated by electrons are called the indirect, scalar or J coupling and the through-space coupling called direct or dipolar (see Section 2.2). While the isotropic average of the indirect spin-spin coupling constant corresponds to the constant measured in a liquid state NMR experiment, the direct spin-spin coupling at the isotropic limit averages to zero.

The indirect NMR spin-spin coupling constant J(K,L) is calculated as the mixed second order partial derivative of the total energy by Eq. (16) with respect to the nuclear magnetic moments M_K and M_L of the coupled nuclei separated by n bonds.

$$^{n}J(K,L) = h \frac{\gamma_K}{2\pi} \frac{\gamma_L}{2\pi} \frac{\partial^2 E(B,M)}{\partial M_K \partial M_L} \quad \mid B, M_{1,2,\ldots K,\ldots L,\ldots} = 0 \qquad (22)$$

For the comparative reason it may be more convenient to use the reduced coupling constant Q, which is independent of the magnetogyric ratios γ_K and γ_L.

$$^nQ(K,L) = \frac{1}{h}\frac{2\pi}{\gamma_K}\frac{2\pi}{\gamma_L}{}^nJ(K,L) \tag{23}$$

The magnetic Hamiltonian describing the interactions of indirect spin-spin coupling consists of the following terms[9,10]:

$$\hat{H}^{DSO} = \sum_{K<L} M_K \left(\alpha^4 \sum_j \frac{r_{j,K}^T \cdot r_{j,L} \cdot I - r_{j,K} \cdot r_{j,L}^T}{r_{j,K}^3 \cdot r_{j,L}^3} \right) M_L$$

$$\hat{H}^{PSO} = \sum_K M_K \left(-i\alpha^4 \sum_j \frac{r_{j,K} \times \left(\frac{\partial}{\partial x_j}, \frac{\partial}{\partial y_j}, \frac{\partial}{\partial z_j} \right)}{r_{j,K}^3} \right) \tag{24}$$

$$\hat{H}^{FC} = \sum_K M_K \left(\frac{8\pi\alpha^2}{3} \sum_j \delta(r_{j,K}) \cdot s_j \right)$$

$$\hat{H}^{SD} = \sum_K M_K \left(\alpha^2 \sum_j \frac{3r_{j,K}^T s_j r_{j,K} - r_{j,K}^2 s_j}{r_{j,K}^5} \right)$$

where $\alpha \approx 1/137$ is the fine structure constant, $r_{j,K}$ is the coordinate of the *j-th* electron relative to the position of nucleus K, I is the unit matrix, i is the imaginary unit, $\delta(r_{i,A})$ is the Dirac delta function and s_j is the electronic spin operator of the *j*-th electron. Each of the magnetic Hamiltonians in Eq. (24) give rise to an individual contribution to the total indirect spin-spin coupling constant. The contributions are called accordingly: the Diamagnetic Spin-Orbit (DSO), the Paramagnetic Spin-Orbit (PSO), the Fermi-Contact (FC) and the Spin-Dipolar (SD).

$$J(K,L) = J^{DSO}(K,L) + J^{PSO}(K,L) + J^{FC}(K,L) + J^{SD}(K,L) \tag{25}$$

The DSO and PSO coupling terms originate from the interaction of the nuclear magnetic moment with the orbital momentum and the FC and SD

terms arise due to the interaction of the electronic spin magnetic moment
with the nuclear magnetic field of the coupled nucleus. While the DSO
Hamiltonian is bilinear in the nuclear magnetic moment M, the remaining
Hamiltonians in Eq. (24) are only linear. As a consequence the DSO
coupling contribution is calculated with the zero order ground state wave
function only, while the rest of the contributions require individual
calculation of the first order perturbed wave function.

$$J(K,L) = \left\langle \psi_0 \mid \hat{H}^{DSO}_{K,L} \mid \psi_0 \right\rangle + 2 \sum_{Pert \in PSO,FC,SD} \left\langle \psi_0 \mid \hat{H}^{Pert}_{K} \mid \psi^{Pert}_{L} \right\rangle \qquad (26)$$

The second order partial derivative present in Eq. (22) projects only those
terms that depend on the coupled nuclei K and L. In case of the DSO term,
the simultaneous coupling of nuclear magnetic moments M_K and M_L is
effective only. For the PSO, FC, and SD terms, the partial derivative is taken
with respect to nuclear magnetic moment M_K while the first order wave
function is calculated with respect to the perturbation M_L.

For the DFT implementations, the first order wave function $\Psi_L = \partial\Psi_0/\partial M_L$ can be calculated analytically using the Coupled Perturbed scheme
(CP DFT)[16,17] or using the finite perturbation theory[19-21]. Thus, the calculated
value of J(K,L) coupling constant need not be exactly the same with respect
to the interchange of nuclei K and L ($J(K,L){\neq}J(L,K)$). In other words, the
J-coupling value depends on the choice of the nucleus perturbing the
wave function in principle. For a sufficiently large atomic basis and well
converged $\Psi_{(L)}$ within the CP DFT approach, the calculated couplings J(K,L)
and J(L,K) are practically the same.

As we have emphasized that the DSO coupling term is calculated using
the ground state wave function. Evaluation of the PSO, FC, and SD terms
thus represents a majority of the effort needed for the calculation of
complete J-coupling. The elements of trace in J tensor account for the
calculation of nine separate perturbations for each coupling constant: one for
FC, three for PSO and six for SD. However, once the first order wave
function with respect to the chosen nucleus is obtained, all J-couplings in a
molecule between this perturbing nucleus and the rest of the nuclei in a
molecule can be calculated at the same computational cost by only replacing
the respective Hamiltonian for the remaining nuclei K in the definition Eq.
(26). Hence it is highly advisable to distribute the "perturbing" nuclei in a
molecule efficiently in order to reduce the computational effort as possible.
For example: the $^1J_{OH}$ and $^2J_{H,H}$ coupling constants in a water molecule
can be calculated by choosing as a perturbing nucleus only one of the
symmetrically placed hydrogen atoms.

The indirect spin-spin couplings depend on the local distribution of electrons in a vicinity of the coupled nuclei. Specific change of the electronic environment reflected by a change in J coupling can be caused by the rearrangement of the molecular structure or by different bonding situation occurring for the same couple of nuclei, for example. The effect of a different electronic environment along the spin-spin coupling pathway can be analyzed and visualized in terms of the electronic wave function or electron density modulation, which can help in the interpretation of calculated spin-spin couplings[22-25].

3.3   Comments on Calculations of NMR Properties

The calculated σ and J constant consists of two and four contributions, respectively, which originate from distinct magnetic perturbations. The mathematical form of these perturbations differs from case to case, which results in a more complex requirements put on the atomic basis as well as on the quantum chemical calculation method compared to other molecular properties[26]. The calculation of NMR spin-spin coupling constant is very demanding. Computational requirements needed for calculation of the NMR properties necessarily exceed those of the standard energy calculation.

Calculation of σ and J NMR constants has been implemented in a framework of many quantum-chemical calculation methods ranging from the relatively computationally inexpensive semi-empirical approaches (reviewed in [11,12]) till the high level *ab initio* calculation methods including the electron correlation (reviewed in [9,10]). Accuracy and reliability of the calculated NMR parameters is inproved accordingly. The computational requirements unfortunately exhibit correspondingly steep growth. The *ab initio* calculation methods beyond the HF approximation such as the MCSCF or CC approaches are therefore used mostly for accurate bench mark calculations in relatively small molecules. On the other hand, NMR calculations in the framework of the DFT approach are frequently done for relatively large molecular complexes with acceptable accuracy.

For the reliable calculation of chemical shifts and indirect spin-spin coupling constants, the effect of a solvent on calculated NMR parameters may be significant. In some cases it may provide the leading correction to a calculated NMR parameter [27]. In the case of a chemical shift calculation, one may rely on compensation of the solvent effect provided the nucleus in the target molecule and reference compound it is the same. However, this is not possible in the case of J-couplings when the absolute values are calculated. Two approaches based on different methodologies of the solvent model are frequently used. The natural way is to construct a complex of the target molecule with the explicit molecules of solvent. Unfortunately, the positions of solvent molecules as well as their dynamic behavior are usually not clearly defined, and the dependence of some NMR parameters on their orientation are not negligible[9,28]. When it is possible, the solvent pattern used in such a NMR calculation can be confirmed using a X-ray crystallography pattern. The dynamic motion of explicit solvent molecules can be also obtained by averaging their positions using the molecular dynamic simulation[29]. The effect of molecular motion itself can also affect the calculated NMR parameters[30-32]. The second solvent model relies on the effective mean polarization by dielectric cavity constructed around the target molecule[33,34]. This approach, called the implicit solvent model, is less computationally demanding than the explicit one and thus more straightforward in its application. However, since the construction of the cavity for the implicit solvent is not uniquely defined, it is highly advisable to test the performance of the implicit solvent against the explicit one.

3.4   Practical Examples

In order to extract structural information from NMR spectrum, the relationships between the spectral parameters and molecular structure has to be studied. As mentioned in Section 1, there are two ways how these relationships can be established: a) empirical parameterizations and b) application of quantum chemical calculations.

The empirical parameterizations/correlations are established based on the experimentally acquired NMR data from a set of the small model molecules of known structure. However, this approach suffers from severe drawbacks, such as difficulty to prepare suitable model compounds for calibration. In such a case, the application of quantum chemical calculations provides an alternative route. In this section, we show several recent examples of the successful applications of the quantum chemical calculations for the structural interpretation of NMR parameters in nucleic acids (NAs).

3.4.1   Dependence of Three-Bond Couplings on Torsion Angle

The classic application of J coupling constants to conformation analysis uses three-bond coupling constants to infer information about the magnitude of the torsion angle that connects two nuclei whose coupling is being monitored (the so called Karplus relationship). These applications are based on the assumption that the periodic behavior of vicinal proton coupling can be modeled as:

$$^{3}J_{HH'} = C_0 + C_1 \cos\phi + C_2 \cos 2\phi \qquad (28)$$

where ϕ is the corresponding torsion angle and C stand for empirical parameters. The Karplus relationships can be used for determination of the glycosidic torsion angle χ, which defines the orientation of the aromatic base with respect to the sugar unit in NAs. For this purpose, the interpretation of the measured three-bond carbon-proton scalar couplings $^3J_{C2/4-H1'}$ and $^3J_{C6/8-H1'}$ across the glycosidic bond using adequately parameterized Karplus equation is most often exploited. Previous studies of Karplus parameterizations[35] for the $^3J_{C2/4-H1'}$ and $^3J_{C6/8-H1'}$ have been restricted by a lack of experimental data for *syn* oriented pyrimidines, by available data covering only relatively narrow region of χ, and by uncertainties in the magnitudes of χ. These limitations enforced several approximations, namely a single Karplus parameterization for all types of nucleotides, the inclusion of the data for modified nucleosides and restriction of the curves maxima to $\chi \approx 60°$ and $\chi \approx 240°$ [35].

Very recently Munzarova and Sklenar[36,37], using the DFT calculation, demonstrated the existence of separate Karplus-like relationships for purine and pyrimidine nucleosides (Figure 20-3).

3.4.2 Dependence of Scalar Couplings on Hydrogen Bond Geometry

In biomolecular NMR, the indirect spin-spin coupling across the hydrogen bond is frequently used to detect various binding patterns of nuclei taking part in hydrogen bond (reviewed in[38]). A nonzero scalar coupling across the hydrogen bond (HB couplings) ideally detects the presence of a HB bond, which is a important stabilizing interaction in biomolecular complexes. Furthermore, the coupling value obtained for particular HB pattern responds sensitively to the bonding situation, i.e. to the inter-nuclear distance and namely to the bond/torsion angle variation among atoms involved in the HB. Detail analysis of the dependencies between the local electronic environment and HB couplings offers the potential to extract important structural information. An impressive example of such a structural NMR study for the P-O···H-N pattern in a complex of guanosine diphosphate with protein was recently presented by Mishima et al.[40] The theoretical calculation of the $^3J(P,N)$ and $^2J(P,H)$ coupling constants by Czernek and Bruschweiler[41] was further used for the calibration of the structural dependence for both NMR parameters. Another example of a similar detection by NMR was presented by Sychrovsky et al.[28] In their study they performed comparison of NMR parameters for the direct and water mediated binding pattern of metal cation to guanine base. Not only do the indirect couplings across the HB change due to the geometry variation but also other intermolecular NMR parameters as was shown by Sychrovsky et. al.[42] for the ^{1}J(N,C) coupling in guanine, which selectively responds to the type of base pairing within a DNA hairpin molecule.

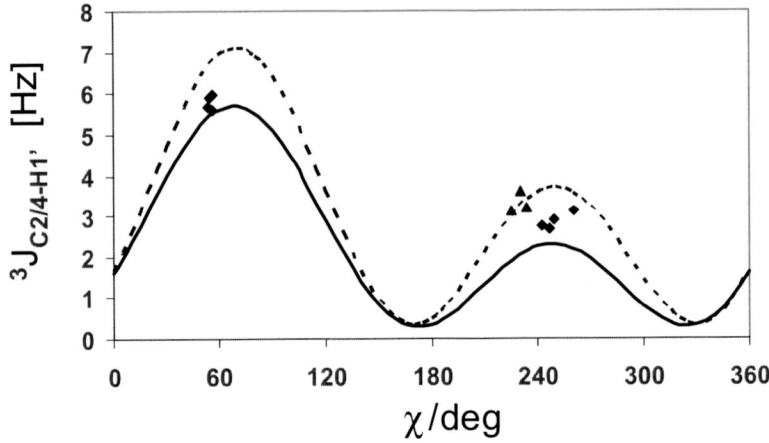

Figure 20-3. Parameterization for ^{3}J$_{C2/4-H1'}$ (χ). This parameterization is based on a theoretically calculated ^{3}J$_{C2/4-H1'}$ for deoxythymidine (solid line) and deoxyguanosine (dashed line). Theoretical results refer to data from [36,37]. Solid diamonds and triangles are indicative of experimental data for deoxyguanosine and deoxythymidine from[39].

3.4.3   Cross-Correlated Relaxation Rates vs. Torsion Angle

The cross-correlated relaxation rates between C1'-H1' dipole-dipole and N1/9 chemical shielding anisotropy (CSA) ($\Gamma_{N1/9,C1'H1'}$) can be used as a source of information on structure and dynamics around the glycosidic linkage in ribonucleic acids (RNAs)[43]. The modulation in $\Gamma_{N1/9,C1'H1'}$ is expressed as a simple geometric term relating the orientation between the principal axis of the N1/9 chemical shielding tensor and the C1'-H1' vector to the glycosidic torsion angle χ. In the case of a molecule undergoing isotropic rotational diffusion, and neglecting internal motions, the expression for the cross-correlated relaxation rates $\Gamma_{N1/9,C1'H1'}$ between the N1/9 chemical shielding tensor and C1'-H1' dipole-dipole vector is given by:

$$\Gamma_{N1/9,C1'H1'} = \frac{4}{15}\left(\frac{\mu_0\hbar}{4\pi}\right)\frac{\gamma_H\gamma_C}{r_{C1'H1'}^3}\gamma_N B_0 \sum_{i=1}^{3}\sigma_{ii}^N\left(\frac{3\cos^2\theta_{ii}-1}{2}\right)\tau_c \qquad (29)$$

where, γ_C, γ_H, γ_N denote the magnetogyric ratio of the nuclei ^{1}H, ^{13}C, and ^{15}N, respectively, $r_{C1'H1'}$ is the C1'-H1' inter-nuclear distance, B_0 is the strength of the magnetic field, σ_{ii} is the *ii-th* component of the nitrogen CS-tensor, θ_{ii} is the projection angle between C1'-H1' dipole-dipole vector and *ii-th* component of the nitrogen CS-tensor, and τ_c is the correlation time for isotropic rotation. In this methods, the interpretation of $\Gamma_{N1/9,C1'H1'}$ relies on the following assumptions: a) the magnitude and orientation of the N1/9 CS-tensor is known and b) N1/9 CS-tensor is independent of the local structure of the nucleoside.

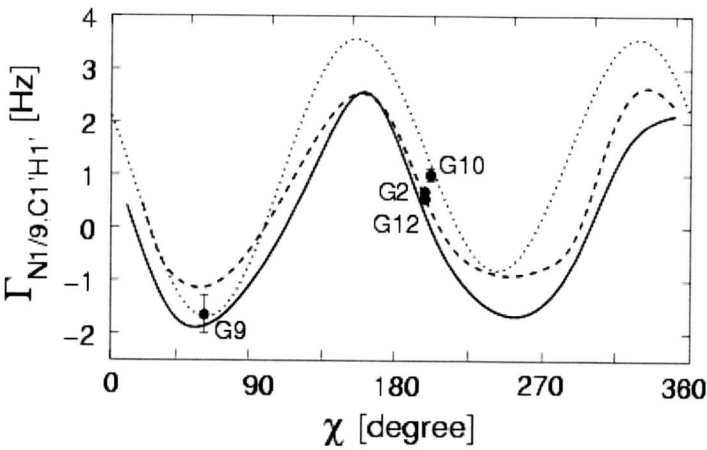

Figure 20-4. The $\Gamma_{N1/9,C1'H1'}(\chi)$ curve for dGua. Dashed and solid lines stand for $\Gamma_{N1/9,C1'H1'}(\chi)$ curves for C2'-endo and C3'-endo sugar pucker of the nucleotide, respectively. The lines were calculated using the conformationally dependent N1/9 CS tensors. Solid circles represent the experimental cross-correlated relaxation rates acquired from r(GGCACUUCGGUGCC)^{43}. Dotted line shows the dependence $\Gamma_{N1/9,C1'H1'}(\chi)$ calculated for the former parameterization assuming independency of both magnitude and orientation of N1/9 CS tensor on local DNA structure.

Very recently, Sychrovsky et al.[44] have demonstrated that latter assumption is not generally valid using the DFT calculation of N1/9 chemical shielding tensors as a function of the conformation of the glycosidic torsion angle and sugar ring. The N1/9 CS tensor magnitude and orientation notably changes with sugar conformation. These results indicated that accounting for the conformational dependent variability in a N1/9 chemical shielding tensor can be important for proper interpretation of experimental cross-correlated relaxation rates between theN1/9 CS tensor and C1'-H1' dipole-dipole in both DNA and RNA (Figure 20-4).

ACKNOWLEDGEMENT

The work has been supported by Grant Agency of the Czech Republic (203/05/0388 and P191) and Ministry of Education of the Czech Republic (1K04011).

REFERENCES

1. Cavanagh J., Fairbrother W. J., Palmer A. G. & Skelton N. J. (1996). *Protein NMR Spectroscopy: Principles and Practice*, Academic Press (San Diego) pp. 1-587.
2. Claridge T. D. W., *High-Resolution NMR Techniques in Organic Chemistry*, Pergamon Press, Oxford 1999.
3. Prestegard, J. H., Bougault, C. M. & Kishore, A. I. (2004). Residual dipolar couplings in structure determination of biomolecules. *Chem. Rev.* **104**, 3519-3540.
4. Bakhmutov V. I. (2004). *Practical NMR Relaxation for Chemists*, Wiley-VCH-Verlag.
5. Lipari G. & Szabo A. (1982). Model-free approach to the interpretation of nuclear magnetic resonance relaxation in macromolecules. 1. Theory and range of validity. *J. Am. Chem. Soc.* **104**, 4546-4559.
6. Palmer, A.G., (1997). Probing molecular motion by NMR *Curr. Op. Struct. Biol.* **7**, 732-737.
7. Korzhnev, D.M., Billeter, M., Arseniev, A.S., & Orekhov, V.Y. (2001). NMR studies of Brownian tumbling and internal motions in proteins. *Prog. Nucl. Man. Reson.* **38**, 197-266.
8. Fukui, H. (1999). Theory and calculation of nuclear spin-spin coupling constants *Prog. In NMR Spect.* **35**, 267-294.
9. Kaupp, M., Buhl, M. & Malkin, V.G. (2004). *Calculation of NMR and EPR parameters.* *Wiley-VCH Verlag.*
10. Helgaker, T., Jaszunski, M. & Ruud, K. (1999). Ab initio methods for the calculation of NMR shielding and indirect spin-spin coupling constants. *Chem. Rev.* **99**, 293-352.
11. Kowalewski J. (1982). Calculations of Nuclear Spin-Spin Couplings., *Ann. Rep. NMR Spect.* **12**, 81-176.
12. Contreras R.H., & Facelli J.C. (1993). Advances in Theoretical and Physical Aspects of Spin-Spin Coupling Constants *Ann. Rep. NMR Spec.* **27**, 255-355.
13. de Dios A. C. (1996). Ab initio calculations of the NMR chemical shift *J. Prog. NMR Spec.* **29**, 229-278.
14. Wolinski, K., Hinton, J. F. & Pulay, P. (1990). Efficient Implementation of the Gauge-Independent Atomic Orbital Method for Nmr Chemical-Shift Calculations. *J. Am. Chem. Soc.* **112**, 8251-8260.

15. Schindler, M., & Kutzelnigg, W. (1982). Theory of magnetic susceptibilities and NMR chemical shifts in terms of localized quantities. II. Application to some simple molecules *J. Chem. Phys.* **76**, 1919-1933.

16. Kutzelnigg, W., Fleischer, U., & Schindler, M. (1990). *In NMR Basisc Principles and Progress*; Springer: Berlin Vol. 23.

17. Sychrovsky, V., Grafenstein, J. & Cremer, D. (2000). Nuclear magnetic resonance spin-spin coupling constants from coupled perturbed density functional theory. *J. Chem. Phys.* **113**, 3530-3547.

18. Helgaker, T., Watson, M. & Handy, N. C. (2000). Analytical calculation of nuclear magnetic resonance indirect spin-spin coupling constants at the generalized gradient approximation and hybrid levels of density-functional theory. *J. Chem. Phys.* **113**, 9402-9409.

19. Malkin, V. G., Malkina, O. L., Casida, M. E. & Salahub, D. R. (1994). Nuclear-Magnetic-Resonance Shielding Tensors Calculated with a Sum-over-States Density-Functional Perturbation-Theory. *J. Am. Chem. Soc.* **116**, 5898-5908.

20. Malkin, V. G., Malkina, O. L. & Salahub, D. R. (1994). Calculation of Spin-Spin Coupling-Constants Using Density-Functional Theory. *Chem. Phys. Lett.* 221, 91-99.

22. Malkina, O. L. & Malkin, V. G., (2003). Visualization of nuclear spin-spin coupling pathways by real-space functions. *Ang. Chem.-Int. Ed.* **42**, 4335-4338.

21. Malkina, O. L., Salahub, D. R. & Malkin, V. G. (1996). Nuclear magnetic resonance spin-spin coupling constants from density functional theory: Problems and results. *J. Chem. Phys.* **105**, 8793-8800.

23. Tuttle, T., Grafenstein, J., Wu, A., Kraka, E. & Cremer, D. (2004). Analysis of the NMR spin-spin coupling mechanism across a H-bond: Nature of the H-bond in proteins. *J. Phys. Chem. B* **108**, 1115-1129.

24. Grafenstein, J. & Cremer, D. (2004). Systematic strategy for decoding the NMR spin-spin coupling mechanism: the J-OC-PSP method. *Magn. Reson. Chem.* **42**, S138-S157.

25. Grafenstein, J. & Cremer, D. (2004). One-electron versus electron-electron interaction contributions to the spin-spin coupling mechanism in nuclear magnetic resonance spectroscopy: analysis of basic electronic effects. *J. Chem. Phys.* **121**, 12217-32.

26. Helgaker, T., Jaszunski, M., Ruud, K. & Gorska, A. (1998). Basis-set dependence of nuclear spin-spin coupling constants. *Theor. Chem. Acc.* **99**, 175-182.

27. Autschbach, J. & Le Guennic, B. (2003). A theoretical study of the NMR spin-spin coupling constants of the complexes [(NC)(5)Pt-Tl(CN)(n)](n-) (n=0-3) and [(NC)(5)Pt-Tl-Pt(CN)(5)](3-). *J. Am. Chem. Soc.* **125**, 13585-13593.

28. Sychrovsky, V., Schneider, B., Hobza, P., Zidek, L. & Sklenar, V. (2003). The effect of water on NMR spin-spin couplings in DNA: Improvement of calculated values by application of two solvent models. *Phys. Chem. Chem. Phys.* **5**, 734-739.

29. Buhl, M. (2002). Structure, dynamics, and magnetic shieldings of permanganate ion in aqueous solution. A density functional study. *J. Chem. Phys. A* **106**, 10505-10509.

30. Ruud, K., Astrand, P. O. & Taylor, P. R. (2001). Zero-point vibrational effects on proton shieldings: Functional-group contributions from ab initio calculations. *J. Am. Chem. Soc.* **123**, 4826-4833.

31. Astrand, P. O. & Ruud, K. (2003). Zero-point vibrational contributions to fluorine shieldings in organic molecules. *Phys. Chem. Chem. Phys.* **5**, 5015-5020.

32. Ruden, T. A., Helgaker, T. & Jaszunski, M. (2004). The NMR indirect nuclear spin-spin coupling constants for some small rigid hydrocarbons: molecular equilibrium values and vibrational corrections. *Chem. Phys.* **296**, 53-62.

33. Mikkelsen, K. V., Ruud, K. & Helgaker, T. (1999). Solvent effects on the NMR parameters of H2S and HCN. *J. Comp. Chem.* **20**, 1281-1291.

34. Ruud, K., Frediani, L., Cammi, R. & Mennucci, B. (2003). Solvent effects on the indirect spin-spin coupling constants of benzene: The DFT-PCM approach. *Int. J. Mol. Sci.* **4**, 119-134.

35. Ippel, J. H., Wijmenga, S. S., deJong, R., Heus, H. A., Hilbers, C. W., deVroom, E., vanderMarel, G. A. & vanBoom, J. H. (1996). Heteronuclear scalar couplings in the bases and sugar rings of nucleic acids: Their determination and application in assignment and conformational analysis. *Magn. Reson. Chem.* **34**, S156-S176.

36. Munzarova, M. L. & Sklenar, V. (2002). Three-bond sugar-base couplings in purine versus pyrimidine nucleosides: a DFT study of Karplus relationships for (3)J(C2/4-H1') and (3)J(C6/8-H1') in DNA. *J. Am. Chem. Soc.* **124**, 10666-10667.

37. Munzarova, M. L. & Sklenar, V. (2003). DFT analysis of NMR scalar interactions across the glycosidic bond in DNA. *J. Am. Chem. Soc.* **125**, 3649-3658.

38. Grzesiek, S., Cordier, F., Jaravine, V. & Barfield, M. (2004). Insights into biomolecular hydrogen bonds from hydrogen bond scalar couplings. *Prog. Nucl. Magn. Reson. Spec.* **45**, 275-300.

39. Trantirek, L., Stefl, R., Masse, J. E., Feigon, J. & Sklenar, V. (2002). Determination of the glycosidic torsion angles in uniformly 13C-labeled nucleic acids from vicinal coupling constants 3J(C2)/4-H1' and 3J(C6)/8-H1'. *J Biomol NMR* **23**, 1-12.

40. Mishima, M., Hatanaka, M., Yokoyama, S., Ikegami, T., Walchli, M., Ito, Y. & Shirakawa, M. (2000). Intermolecular P-31-N-15 and P-31-H-1 scalar couplings across hydrogen bonds formed between a protein and a nucleotide. *J. Am. Chem. Soc.* **122**, 5883-5884.

41. Czernek, J. & Bruschweiler, R. (2001). Geometric dependence of (3h)J(P-31-N-15) and (2h)J(P-31-H-1) scalar couplings in protein-nucleotide complexes. *J. Am. Chem. Soc.* **123**, 11079-11080.

42. Sychrovsky, V., Vacek, J., Hobza, P., Zidek, L., Sklenar, V. & Cremer, D. (2002). Exploring the structure of a DNA hairpin with the help of NMR spin-spin coupling constants: An experimental and quantum chemical investigation. *J. Phys. Chem. B* **106**, 10242-10250.

43. Duchardt, E., Richter, C., Ohlenschlager, O., Gorlach, M., Wohnert, J. & Schwalbe, H. (2004). Determination of the glycosidic bond angle chi in RNA from cross-correlated relaxation of CH dipolar coupling and N chemical shift anisotropy. *J. Am. Chem. Soc.* **126**, 1962-1970.

44. Sychrovsky, V., Muller, N., Schneider, B., Smrecki, V., Šponer, J., Spirko, V. & Trantirek, L. (2005). Sugar Pucker Modulates the Cross-Correlated Relaxation Rates Across the Glycosidic Bond in DNA. *J. Am. Chem. Soc.* **127**, 14663-14667.

Chapter 21

THE IMPORTANCE OF ENTROPIC FACTORS IN DNA BEHAVIOUR: INSIGHTS FROM SIMULATIONS

Sarah A. Harris[a] and Charles A. Laughton[b]

[a]*Department of Physics and Astronomy, University of Leeds, Leeds LS2 9JT, UK;* [b]*School of Pharmacy and Centre for Biomolecular Sciences, University of Nottingham, University Park, Nottingham NG7 4RD, UK*

Abstract: When using simulation methods to analyse problems of molecular recognition (e.g. DNA-ligand interactions), or molecular behaviour (e.g. response of a DNA helix to mechanical stress), it is always easiest to focus on issues that relate directly to the forcefield-based description of the system; thus we analyse the strengths and numbers of non-bonded interactions, or measure how the internal energies of the molecules change with alterations to their conformation. This is therefore essentially an 'enthalpic view' of DNA behaviour, but in any spontaneous change that a system makes in response to a change in conditions, it is the free energy changes involved which need to be understood. In this chapter we focus on how simulation methods can enable us to access the entropic factors involved in DNA behaviour, factors which are frequently 'hidden' from a casual observation of molecular structures. We illustrate this approach with two examples taken from our own recent work: an analysis of the origins of cooperativity in ligand-DNA recognition, and an analysis of the behaviour of duplex DNA when stretched in atomic force microscopy experiments.

Key words: Molecular Dynamics, Entropy, Hoechst 33258, Atomic Force Microscopy

1.   INTRODUCTION

Molecular modelling studies of nucleic acids (NAs) and their complexes can provide valuable information unavailable by other experimental techniques. For example, they can provide predictions for the structures of molecules

J. Šponer and F. Lankaš (eds.), Computational Studies of RNA and DNA, 537–558.

that, for one reason or another, have not been able to be determined by techniques such as X-ray crystallography or NMR spectroscopy. In other cases, they can provide, through techniques such as molecular dynamics (MD), an understanding of the dynamics and flexibility of NAs that are difficult, if not impossible, to observe directly by experimental methods. Finally, they can allow us to provide explanations for structural and dynamic features of their behaviour based on the analysis of the energetics of the molecular system. It is this last application of molecular modelling that we focus on in this chapter, and in particular how new analytical methods can quantify less obvious aspects of the thermodynamics of NA structure and recognition that in the past have often been overlooked. We begin with a general overview of the energetic analysis of NA simulation data, and then give two examples from our own recent work where this analysis provided explanations for the behaviour of NA systems that until that point had been perplexing or controversial.

Figure 21-1 shows the structure of the complex between the 'Dickerson-Drew dodecamer' d(CGCGAATTCGCG)$_2$ and the minor groove binding drug Hoechst 33258, as determined by X-ray crystallography (PDB code 1D43[1]).

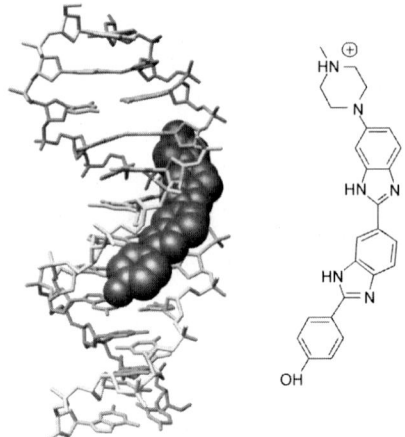

Figure 21-1. Left: structure of the complex between the DNA dodecamer d(CGCGAATTCGCG)$_2$ and Hoechst 33258[1]. Right: chemical structure of Hoechst 33258.

From a visual inspection of this structure, one can begin to try to understand the factors that promote the interaction between these two molecules. We see how the ligand is flat and crescent-shaped, so as to fit snugly into the helical minor groove of the DNA. As a result, we can predict that favourable van der Waals interactions between the two molecules are maximised. Secondly we note that the positively-charged piperidino group at the terminus of the ligand lies close to negatively-charged phosphate groups of the DNA, which we would expect to provide a strong electrostatic interaction. Thirdly, the ligand's inward-facing NH groups are also well-placed to make hydrogen-bonding interactions with H-bond acceptor groups on the DNA bases. This process of decomposing, or parsing, the total interaction between the two molecules into a number of components with different physical origins is conceptually very useful for trying to understand what drives and controls affinity and selectivity in the recognition process, but we must realise that in general these individual terms are not directly measurable experimentally – only their sum total.

The most straightforward application of molecular modelling to a system like this allows this process of energy parsing to be quantitated, so we can actually assign values, in kcal/mol or whatever units we choose, to electro-static, van der Waals, and hydrogen bonding interaction terms. We can also further subdivide these down to, if desired, individual atom-atom interaction terms. This is possible because this is the very way in which the atomistic, forcefield-based, approach to molecular modelling operates, where we calculate the total energy of the system as a sum of individual interactions. However, for this process to be accurate, we must be sure that the model 'system' that we analyse in this way does indeed reflect the experimental system. For example, it may seem obvious that electrostatic interactions between the, typically, positively charged minor-groove binding ligands such as furamidine and the negatively charged DNA are going to be very important, but, for example, Haq et al. in their detailed study[2] of the interaction of Hoechst 33258 with d(CGCAAATTTGCG)$_2$ showed that these are not very important at all, because the favourable electrostatic inter-actions between the two components that result from binding only replace approximately equally favourable electrostatic interactions between the individual components and the solvent (water and ions) that exist prior to binding. The lesson is straightforward – modelling studies must consider the full system, including solvent and ions, and must consider the full process, i.e. the equilibrium between the bound and unbound state.

Since it is an equilibrium that is being analysed, we require that our modelling process enables us to calculate free energy differences, ΔG, between the two states (eq. 1).

$$\Delta G = \Delta H - T\Delta S \qquad (1)$$

If we remember to include the solvent in our modelled system, and analyse both the bound and unbound states, then the type of energetic analysis discussed above can be used to calculate the molecular mechanics energies of the two states, E_{bound} and E_{free}. If we perform a well-equilibrated molecular dynamics simulation on each system of sufficient length, then the structures in the resulting trajectories can be regarded as Boltzmann-weighted ensembles and then we can estimate the ΔH term:

$$\Delta H = <E_{bound}> - <E_{free}> \tag{2}$$

However, in practice this is not so simple, mainly due to the problem in getting a well-sampled solvent environment. Using an explicit water model, the solute-solvent interaction energy terms can fluctuate greatly between snapshots, so that small differences between $<E_{bound}>$ and $<E_{free}>$ are difficult to measure reliably in the noise. The solution has been to ignore the explicit waters (and ions) in the snapshots and use a continuum model to calculate solute-solvent interaction terms. Two popular choices are the Poisson-Boltzmann (PB) and Generalised Born (GB) models. The GB model is gaining in popularity, as it is computationally less expensive and, if properly parameterised, apparently as accurate as the PB model, at least in straight-forward low-salt conditions [3,4].

So why not perform the initial simulations with a GB solvent model, since the expensive explicit solvent will be 'thrown away' in the analysis? We have examined this possibility in the context of drug-DNA interaction analysis and on the whole it does not seem to be worthwhile, for two, related, reasons [5]. Firstly, GB simulations on such systems can show reduced stability, probably due to the lack of a frictional damping effect that explicit water provides. Secondly, in order to provide reliable simulations the parameters of the GB model (in particular the size of the non-bonded cutoff) have to be set such that the computational cost of the simulations is only slightly less than performing them with explicit solvent. Newer implementations of the GB model that include a frictional term may however change this situation.

The next problem is that detailed calorimetric studies of drug-DNA interaction typically find that entropic factors can be very important in determining affinity and selectivity, indeed, in the case of minor groove binders they appear frequently to be dominant. How can we measure such terms from simulation data? The largest entropic term in most cases relates to the reorganisation of solvent that accompanies the binding process. The expulsion of ordered water molecules from the minor groove as the drug binds, and from the hydrophobic surface of the drug molecule too, are entropically highly favourable processes. Fortunately, the GB solvent model, with a surface area correction term (GB-SA model) includes this term: in other words, the MM/GB-SA analysis method provides a 'hybrid' energy term that includes all enthalpic components, plus solvent-related free energy components[6].

Two entropic terms remain unevaluated. Firstly there is the loss of rotational and translational degrees of freedom that accompany binding, ΔS_{r+t}, secondly there are changes in the configurational entropy of the molecules (internal degrees of freedom), ΔS_{conf}. For example, the binding of a ligand in the minor groove of DNA can stiffen the helix, reducing its configurational entropy. ΔS_{r+t} may be estimated from classical statistical mechanics, though alternative approaches (discussed below) may be more appropriate as they can take into account that these global degrees of freedom may not be completely lost, but appear as new vibrational degrees of freedom in the complex. The calculation of ΔS_{conf} has been seen as a difficult problem. The simplest option is to ignore it, assuming that it is a small term in comparison to the other entropic terms, particularly the hydrophobic term. But while this is true, it is very often the case that we are interested in understanding the origins of *differences* in binding affinities – for example between one ligand and another, or between one ligand and two alternative DNA sequences. In such cases, it can be much more dangerous to ignore this term, as the case studies we present later clearly illustrate.

The earliest approach to the calculation of configurational entropies was based on normal mode analysis and the statistical mechanics of a classical harmonic oscillator. The problem with this approach is that is calculates the configurational entropy from a highly optimised geometry that may bear little similarity to the configurations explored in an MD simulation. More recently two approaches have been developed that allow S_{conf} to be estimated directly from the MD data, due to Schlitter[7] (eqn. 3) and Andricioaei and Karplus[8] (eqn. 4). Both require the calculation of the eigenvalues, ω, of the mass-weighted coordinate covariance matrix from the simulation:

$$S \approx 0.5k \sum_i \ln\left(1 + \frac{e^2}{\alpha_i{}^2}\right) \qquad (3)$$

$$S = k \sum_i \frac{\alpha_i}{e^{\alpha_i} - 1} - \ln(1 - e^{-\alpha_i}) \qquad (4)$$

where $\alpha_i = \hbar\omega_i / kT$, and the sum is over all non-trivial vibrations. Although there are slight differences in the derivation of these two methods, in practice they give very similar results, at least in our hands. We now discuss some of the practicalities involved in determining configurational entropies from MD simulations using this approach.

Firstly, it is important that the snapshots from the MD simulation are all least-squares fitted to a common reference structure, so that all global rotational and translational motion is removed, otherwise S_{conf} will be 'contaminated' with S_{r+t}. (For a discussion of the application of this type of approach to the calculation of S_{r+t}, see ref. 9). Secondly, it is important that all eigenvalues are calculated. The eigenvalues resulting from the diagonalisation of a typical MD trajectory decay rapidly; though there are theoretically (3N-6) of them (i.e. maybe 2500 for a drug-DNA simulation), typically the top 50 or so will capture 90% of the variance in the simulation. But because of the logging operation involved in the entropy calculation, even small eigenvalues contribute significantly to the total. In order for the diagonalisation procedure to generate (3N-6) 'true' eigenvalues, it theoretically requires that the covariance matrix be constructed from the analysis of at least (3N-6) independent snapshots, so it is important that these are obtained. Thirdly, it is generally observed that the longer the trajectory that is analysed, the larger is the calculated entropy. This is not unexpected – longer simulations are more likely to access previously unsampled areas of configurational space. Fortunately, we find that for 'well behaved' drug-DNA simulations at least, there is a relationship between simulation length, t, and calculated entropy, $S(t)$:

$$S(t) = S_{inf} - at^{-n} \tag{5}$$

Where S_{inf} is the entropy for a simulation of infinite length, and a and n are fitting parameters. Justificaton for a relationship of this type, originally arrived at empirically, comes from the analysis of a quite general time series model with a stationary temporal covariance structure (IL Dryden et al., personal communication). Particular cases of the model include short-range dependence such as autoregressive models for the principal component projections (where the exponent n is 1), and long-range dependence models ($0 < n < 1$). It should be noted that for extrapolations using this approach to be reliable, simulations of the order of 5 ns are required for a typical drug-DNA system.

In summary then, if we wish to have a full understanding of what drives biological processes such as molecular recognition, then it is important that we have modelling methods that can probe both the enthalpic and entropic components of the system. The relationship between structure and enthalpy is one that it is straightforward to grasp – almost 'by eye' – but the relationship between entropy and dynamics can be less obvious and requires high-quality simulation data for quantitation. The way in which the methods of statistical physics have provided insights into molecular recognition problems unobtainable by other means, will now be illustrated through two case studies from our recent work.

2. DNA DYNAMICS AND INFORMATION TRANSFER

2.1 A Cooperative Ligand-DNA Interaction

One of the simplest examples of sequence selective recognition is provided by compounds which bind in the narrow minor groove of AT rich DNA. The ability of these molecules to discriminate between different DNA sequences has meant that they are of interest as potential therapeutic agents due to their ability to interfere with gene function. Consequently, they have been extensively studied both experimentally and theoretically. The small size of these systems relative to many protein or protein-DNA complexes means that they can be investigated in full atomic detail by X-ray crystallography, high resolution NMR and computationally by MD simulation. It has therefore been possible to relate changes in structure and dynamics at an atomic level to the free energy changes that drive sequence selective recognition. The selectivity of minor groove binding ligands therefore provides a useful model system for understanding molecular specificity in general.

As discussed above, the ligand Hoescht 33258 binds in the minor groove of DNA and recognises a target sequence of four AT base pairs. The dodecamer d(CTTTTGCAAAAG)$_2$ therefore contains two identical target sites separated by an intervening GC region which might be expected to isolate the two bound drugs. Unexpectedly, when the drug is gradually titrated into a solution of the DNA only the 2:1 complex and the unbound DNA are ever detected experimentally by NMR and the intermediate 1:1 drug-DNA complex is not observed[10]. This implies that the interaction of Hoescht 33258 with this sequence is highly cooperative; the association of the first ligand dramatically increases the probability that a second one will bind. A quantitative estimate of this cooperative effect indicates that binding a drug at the first site ensures that a second ligand is at least 2000 times more likely to bind at the other site. Consequently, there must be a difference in binding free energy $\Delta\Delta G$ of at least -4 kcal/mol between binding the first and second drug molecule. Figure 21-2 provides a schematic explanation of cooperativity in terms of the important thermodynamic changes driving drug-DNA complexation.

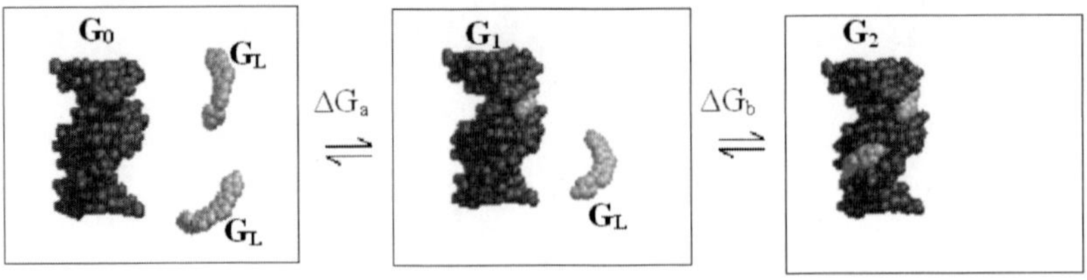

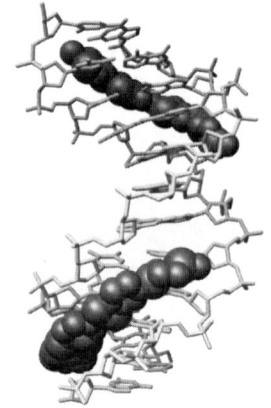

$$\Delta G_a = (G_1 + G_L) - (G_0 + 2G_L) \quad \Delta G_b = G_2 - (G_1 + G_L)$$

$$\Delta\Delta G = \Delta G_b - \Delta G_a = G_0 + G_2 - 2G_1$$

Figure 21 -2. Thermodynamic changes involved in the formation of the 2:1 complex.

2.2 Experimental Studies of the Ligand-DNA Complex

The structures of both the free DNA and the 2:1 drug-DNA complex have been obtained by high resolution NMR in order to gain some insight into the origins of this cooperativity. The structure of the 2:1 complex is shown in Figure 21-3.

Figure 21-3. NMR structure of the 2:1 complex between Hoechst33258 and d(CGCGAATTCGCG)$_2$.

When bound, the two drugs are separated by ~15Å with the intervening cavity filled with water molecules. Therefore, a direct interaction mechanism cannot explain cooperativity as there are no close molecular contacts between the two drugs that could contribute to $\Delta\Delta G$. Rather, this is an example of a purely allosteric interaction in which information concerning the occupancy of the first site is transferred by the DNA helix to the second binding site. Allostery is commonly thought to be due to long range conformational changes transmitted across a biomolecule in response to ligand binding. In this instance, cooperativity could be explained if binding the first drug molecule changed the shape of the DNA in such a way that the second could be more readily accommodated. However, comparing the structures of the DNA and the drug-DNA complex shows that the 2:1 complex is in fact very similar in shape to the unbound duplex, so that there are no conformational changes in the DNA that are large enough to explain the cooperativity observed.

2.3 Computer Simulations of the Ligand-DNA Complex

MD simulations of the unbound DNA, the 2:1 complex and the theoretical 1:1 complex were then used[11] to calculate key thermodynamic quantities related to the two binding events. The 1:1 complex can be generated *in silico* simply by removing one of the drugs from the 2:1 complex. To explain cooperativity, it is necessary to identify the origin of the *difference* in the free energy change on binding the first and second drug ($\Delta\Delta G$). This is made up of separate contributions from the changes in entropy and enthalpy, which can both be calculated from the computer simulations as previously described. A negative value of $\Delta\Delta G$ indicates cooperativity. The results are summarised in Table 21-1.

Table 21-1. Thermodynamic quantities calculated from the MD simulations. All values are in kcal/mol, for T=300K.

System	$E^{int}+G^{solv}$	$\Delta E^{int}+G^{solv}$	$\Delta\Delta E^{int}+G^{solv}$	TS_∞	ΔTS_∞	$\Delta\Delta TS_\infty$	$\Delta\Delta G$
Free DNA	-4375.6			830.6			
1:1 complex	-4404.6	-29.0	4.3	854.9	24.3	10.5	-6.2
2:1 complex	-4430.3	-25.7		889.7	38.4		

The established mechanism of allosteric interactions uses long range conformational changes in the biomolecule to transfer information between two distant sites. If the first drug "preorganises" the helix so that the interaction with the second is more favourable, then the $\Delta\Delta H$ term will be large and negative. However, the small positive value obtained for $\Delta\Delta H$ shows that the interaction energy between the drugs and the DNA is actually slightly *less* favourable in the 2:1 complex. This is consistent with the unexpected similarity between the structures of the unbound and 2:1 complexes. The simulations show that the DNA undergoes a slight structural rearrangement to accommodate each drug which is associated with a small energetic penalty. The DNA is placed under more strain in the 2:1 complex simply because it has to accommodate both drugs. Consequently, the interaction is enthalpically *anti-cooperative* for this system.

Cooperativity must therefore arise from differences in dynamic changes on binding the first and second drug, as is confirmed by the MD trajectories. The value obtained for T$\Delta\Delta$S is positive and consequently makes a negative contribution to $\Delta\Delta$G (which is correct for cooperativity). It is also large enough to overcome the anti-cooperative enthalpy term, providing an estimate of the cooperativity index which is consistent with experiment. Communication between distant sites is therefore predominantly an entropic effect. Accommodating each drug reduces the configurational entropy of the DNA due to steric clashes between the drugs and the minor groove. The interaction is cooperative because most of the unfavourable restriction in the flexibility of the whole helix is caused by the binding of the first drug. However, there is clearly a delicate interplay between the enthalpic and entropy components of this recognition process, and the cooperative effect would be expected to be extremely sensitive to changes in DNA sequence or the drug. This "balance" between opposing thermodynamic forces is very important biologically as it enables molecular recognition to be highly sensitive to minor changes in chemical composition or the environment.

2.4   Mechanism of Information Transfer

As the binding sites of the two molecules are physically separated, cooperativity implies that the DNA is able to pass information regarding occupancy from one binding site to the other. The structural parameter of most relevance to drug association is minor groove width. It is therefore possible to understand drug-induced changes in the mechanical properties of the helix simply by looking at changes in the flexibility of the minor groove. Figure 21-4 shows the changes in minor groove width associated with the three most important modes of flexibility of the system (which account for over 50% of the dynamics) for the unbound DNA, the 1:1 and 2:1 complexes.

The dynamic behaviour of the DNA is surprisingly simple, and clearly reveals the mechanism for information transfer between the two sites. The first three modes of oscillation into the minor groove resemble the modes of vibration of a simple string. Interference with the motion of the groove at one binding site will be transferred by these modes to the other site, just as pressing on a violin string alters the musical note that is played. The restriction in conformational freedom of the top binding site therefore also inhibits the flexibility of the bottom site as the two ends of the helix are dynamically coupled. In terms of free energy changes, the drug-DNA interaction is cooperative because in addition to paying an entropy penalty for binding to its own site, the first drug also does some of the unfavourable thermodynamic "work" required to stiffen the second binding site.

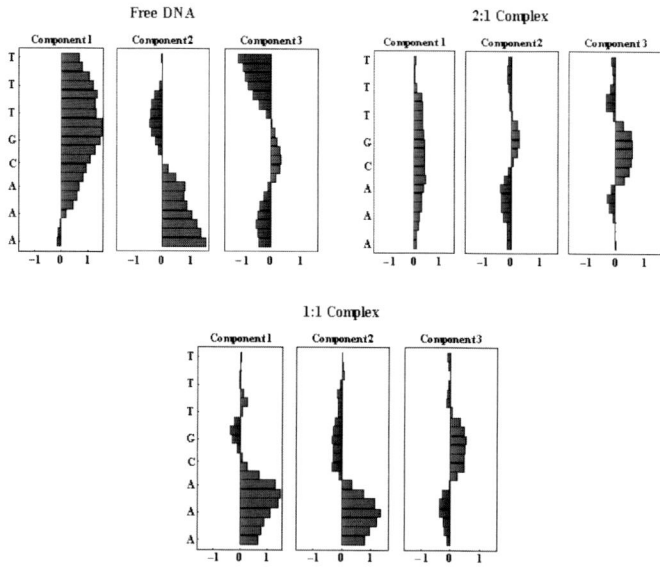

Figure 21-4. Groove width variations associated with the first three principal components of the dynamics of the DNA in the free, 1:1 and 2:1 systems.

The origin of cooperativity in this system differs from conventional allosteric interactions in that binding the first ligand predisposes the *dynamics* rather than structure of the molecule to accommodating a second. In this case the entropic portion of the free energy plays a dominant role in the selectivity of the ligand. This is an example of "allosteric communication without conformational change" originally suggested by Cooper and Dryden[12], who showed theoretically that considerable differences in binding free energy can be obtained from changes in conformational flexibility alone, without the need for changes in the average structure of the macromolecule. As is clear from cooperativity in this system, their analysis showed that the "information" contained within a biomolecule is carried not only by its unique molecular structure, but also by its flexibility and dynamic behaviour. The molecular origins of cooperativity in the absence of conformational change cannot be understood unless the dynamic properties

of the system are taken into account. Molecular discrimination which is
based on shape alone seems intuitively easier to understand simply because
recognition through dynamics has no analogue in the macroscopic world.
Rather, it is a unique consequence of the increased importance of thermal
fluctuations in biological systems which operate at the nanoscale.

3.   ENTROPY AND THE BIOMECHANICS OF DNA

3.1   The Biological Importance of DNA Mechanics

Duplex DNA is not generally found in its classical B-form conformation *in
vivo*. DNA must be very tightly packaged to fit within the nucleus, and is
therefore usually found either in a highly twisted *supercoiled* state, or
wrapped tightly around histone proteins into a densely compacted structure
(chromatin). DNA packing also affects its capacity to interact with other
biological molecules. The cell possesses an impressive range of machinery
for stretching, bending, twisting, unwinding, cutting and resealing DNA
strands to organise and copy the genetic material and to control the
production of proteins. Many of these DNA manipulation enzymes are
molecular motors capable of transforming chemical energy into a
mechanical force which deforms the duplex. Furthermore, reading the
genetic code involves separating the two strands of the duplex to expose the
bases for recognition by complementary RNA or DNA bases. Therefore, the
mechanical properties and deformability of DNA can be as important to
cellular function as the chemical composition of the bases themselves.

3.2   The Disagreement between Nanomanipulation
Experiments and MD Simulation

Advances in nanomanipulation techniques which make it possible to
measure the mechanical properties of single DNA molecules in the laboratory have provided considerable insight into the forces exerted by DNA
processing enzymes when they distort the duplex. Examples of experiments
to measure various mechanical properties of DNA are shown in Figure 21-5.
In particular, DNA stretching experiments have investigated the deformability of long polymeric DNA molecules (of order 50,000 base pairs) down
to oligomers of only 10 base pairs in length[13-16]. These measurements
characterise the biomechanical properties of the duplex by measuring the
force required to produce a given molecular extension. However, a
theoretical interpretation of these experiments has proven difficult mainly

because of a lack of understanding in the role of entropy and thermal fluctuations at the nanoscale.

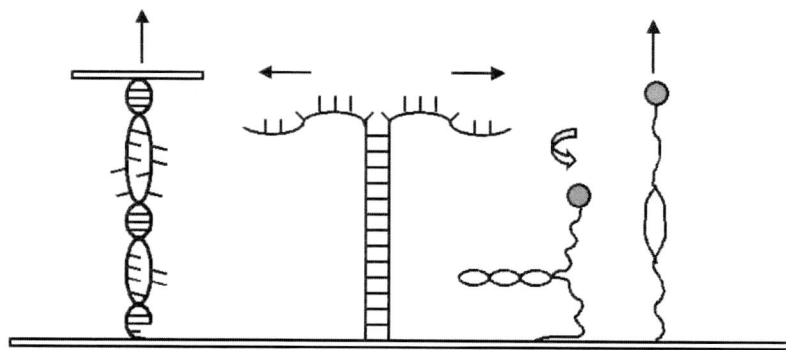

Figure 21-5. Measuring the mechanical properties of duplex DNA. Stretching the DNA molecule parallel to the helix axis induces melted domains in the molecule (far left). The DNA can also be unzipped by pulling perpendicular to the helix axis (centre left). Twisting the DNA includes supercoils (centre right), the molecule can then be denatured by stretching (far right).

The force required to stretch a thermodynamic object depends not only upon the way that the stress increases its internal energy, but also upon the way that its entropy is altered by the process. At very small forces (less than 10pN), the resistance of polymeric DNA to force is entirely due to the reduction of entropy on stretching. Long DNA sequences in solution behave like ideal, structureless polymers since the electrostatic repulsion between the two negatively charged sugar-phospate backbones prevents the DNA sticking to itself, therefore the molecules are curved and tangled into a random coil configuration which unravels when a force is applied. Beyond the entropic stretching regime, the straightened duplex behaves like a perfect elastic rod and deforms linearly with the applied force. All of these processes are completely reversible, and are well described by simple polymer theories. However, at forces of 65pN the mechanical stresses in the duplex are large enough to induce structural changes that are irreversible, indicating that the structure of the double helix is starting to fail under the applied tension. The plateau at 65pN evident in the force-extension curve shows that the stressed B-form helix suddenly extends to around 1.7 times its original length as the force is increased by just a few piconewtons. This behaviour cannot be described using simple polymer theories which neglect the internal structure of the molecule. Consequently, computer simulations in which the DNA is stretched in an analogous manner have been used to provide insight into the structural changes occuring at the atomic level. Pioneering simulation studies performed by Konrad and Bolonick[17] and Lebrun and Lavery[18] on a 12-mer DNA sequence obtained force-extension curves quailtatively similar to those obtained experimentally for much longer DNA sequences. These calculations suggested that the force plateau results from a structural transition from B-form DNA to a novel extended configuration known as S-ladder DNA. The structure of S-DNA is shown in Figure 21-6 (see inset).

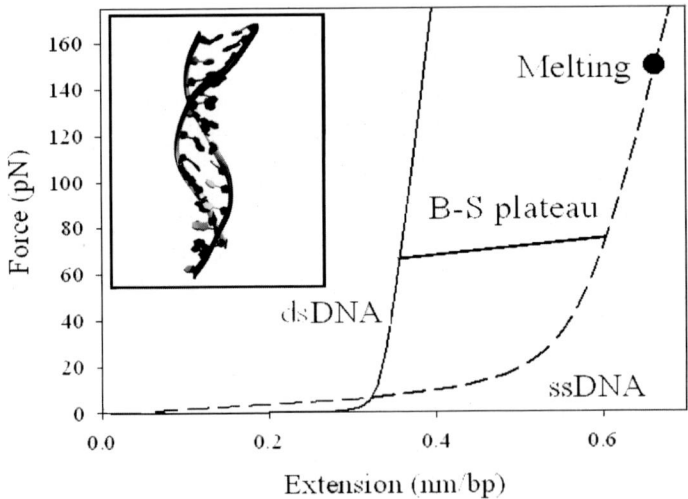

Figure 21-6. Experimental force-extension curve for long double stranded DNA (bold line) and single stranded DNA sequences (dashed line). The proposed structure for S-ladder DNA is also shown (see inset).

The B-helix to S-ladder transition can be explained by the changes in base stacking interactions within the helix. In the unstretched duplex, there are strong van der Waals interactions between stacked bases on the same polynucleotide strand. When the helix is switched into the S-ladder conformation, these stacking interactions are between bases on *opposite* DNA strands (as this becomes more energetically favourable when the DNA is extended). Unfortunately, these and similar calculations disagree with experimental stretching measurements for the short DNA sequences actually modelled in the simulations. As might be expected, short DNA sequences unbind at considerably lower forces than polymeric DNA (~30pN for a dodecamer) and the force plateau is not observed experimentally if the DNA contains less than 30 base pairs. This discrepancy is due to the six orders of magnitude gap in timescale accessible to the experiments (which take place over a microsecond timescale) and the simulations (which are limited to nanoseconds). Similar problems apply to calculations which mimic the mechanical unfolding of proteins by an applied force, which is currently a particularly intensive area of research for both experimentalists and molecular modellers. The serious limitations imposed by finite simulation timescales has lead to the common belief that it is impossible to use computer simulations to accurately represent single molecule manipulation experiments at the atomic level.

A thermodynamic analysis of the force-extension curves suggests an alternative explanation for the force plateau at 65pN. When the DNA is stretched, it should become thermodynamically unstable and unbind as soon as the reversible work performed by the external force becomes equal to the free energy holding the two strands together. Comparing the binding free energy of duplex DNA (known from conventional thermal melting studies) with the work done by the applied force (obtained by integrating under the force-distance curves) indicates that the duplex should melt into two separate strands at forces of around 65pN, precisely where the force-plateau is located. This implies that the plateau signifies a *force induced melting transition*, rather than the simple conformational change suggested by modelling studies. However, this does not explain why the duplex is able to resist relatively high forces (up to 150pN) before completely unbinding. To understand this observation, Williams *et al* proposed a model of DNA melting in which the kinetics of strand separation are particularly slow[19]. As the force plateau is reached and the DNA starts to go through the melting transition, regions appear in the duplex where the hydrogen bonding has been locally disrupted. However, these melted bubble-like regions are still separated by intact helical segments which hold the duplex together after the B-S transition and still provide resistance to the applied force. The kinetics associated with the growth and motion of the unpaired regions in partially melted DNA is expected to be very slow, a lot slower than the structural rearrangements that take place before B-S transition. Therefore, although single stranded DNA might be the most stable structure when the force plateau is reached, the new phase forms slowly on the timescale of the unbinding experiment. Consequently, larger forces are required to unbind the molecule than would be expected from equilibrium thermodynamics (which assumes the experiment is performed infinitely slowly). The effect is more pronounced at faster pulling rates so larger forces are required, as has been observed experimentally. This is because more irreversible work must be done against molecular friction if the experiment takes place further away from the equilibrium regime.

3.3   Quasi-static Simulations of DNA Stretching

Experiments which stretch short DNA sequences probe a completely different physical regime to those which use similar forces (< 50pN) to extend polymeric DNA. The free energy required to unbind an oligomer is not much bigger than the thermal energy scale. Therefore, when even a small force is applied, there is a reasonable probability that a short duplex will unbind as a result of spontaneous thermal fluctuations over the timescale of the experiment. Oligomer unbinding under force is therefore *thermally activated.* Consequently, the force required to unbind a short DNA strand is strongly dependent on the pulling rate. If the experiment takes place very slowly, there is a greater chance that the DNA will spontaneously adopt conformations that lead to the separation of the two strands due to its inherent thermal motion. Conversely, fast dissociation requires a larger force. AFM experiments take place over timescales of around a microsecond whereas computer simulations of forced unbinding will be limited to nanoseconds at best. Therefore, rather than attempting to reproduce the experiments literally by applying a steadily increasing force *in silico* (which would be impossible due to the gap in accessible timescales), we have measured the change in free energy as the DNA is extended so that the unbinding force can be calculated indirectly using thermodynamic arguments to identify where mechanical instability occurs. This new methodology provides results which are essentially time-independent or *quasi-static*, and which do not suffer from the artefacts associated with short simulation times[20].

To stretch the DNA, an initial "steered" MD simulation was performed in which the DNA sequence d(CGAAAAAAAACG)$_2$ was extended by 1Å per pull, followed by 10ps energetic relaxation. This simulation provided a measurement of the increase in enthalpy of the dodecamer as it is stretched by an external restraint (such as an AFM tip). The force-extension curve could then be calculated by differentiating the energy change with respect to the extension, as in equation 6.

$$F = \frac{dU}{dx} \tag{6}$$

The force-extension curve obtained in this way agrees with previous simulation studies, rather than with the experiments. The duplex does not become mechanically unstable until extensions of > 50Å (far larger than the 15Å measured experimentally) and considerably higher forces are required to unbind the two DNA strands (> 600pN compared to 30pN). Furthermore, the theoretical force-extension curve contains a distinctive plateau also obtained by previous simulation studies but which is only observed in experiments which stretch very long DNA sequences. However, the calculation so far has not considered any of the dynamic changes which might occur when DNA is extended as only the enthalpic portion of the free energy change has been obtained.

To calculate the entropic portion of the free energy change as the DNA is stretched, configurations of the DNA sampled from various points along the steered dynamics trajectory were used as the starting points for additional 5ns MD simulations of the DNA restrained at various extensions, as shown in Figure 21-7.

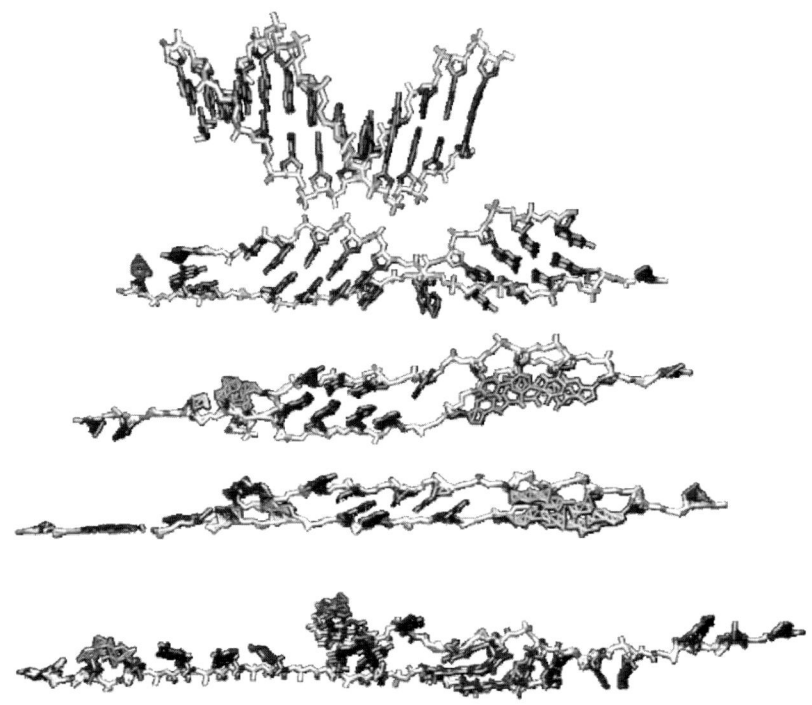

Figure 21-7. Examples of stretched DNA structures used to calculate the entropic portion of the free energy change.

The stretched DNA molecules were pinned by harmonic restraints to prevent the molecule relaxing back to its unstretched configuration, and the entropy of the duplex at each extension was calculated from the trajectories as previously described. These calculations show that the configurational entropy of the DNA initially decreases, indicating that the DNA stiffens as it is placed under tension as might be expected. However, as the DNA is stretched towards the force plateau, the duplex reaches an extension where it suddenly becomes so much more flexible that the entropy change outweighs the enthalpic term. This indicates that thermal fluctuations in the molecule become so large that the weakened energetic term is no longer strong enough to hold the two strands of the duplex together, and the molecule will simply shake itself apart under the applied tension if given sufficient time. These results therefore show that the stretched duplex becomes thermodynamically unstable and should unbind spontaneously before the force plateau is reached, as is observed experimentally. Our calculations do not measure forces directly, however, we can deduce the force required to unbind the DNA at a given loading rate if we know the extension where mechanical instability occurs. When the entropic contribution is included in the calculation, the simulations provide quantitative results for the force that are in excellent agreement with experimental measurements for a dodecamer. Clearly, atomistic computer simulations studies can be used to model single molecule stretching experiments, so long as the entropic contribution can be taken into account.

3.4   The Entropic Instability of S-ladder DNA

These numerical results can be easily understood by visual inspection of the MD trajectories. Figure 21-8 shows DNA configurations at the beginning and end of two of the "pinned" simulations just beyond the structural transition from B to S-form DNA. As the simulation proceeds, the S-DNA structure clearly develops a hole or *denaturation bubble* in the centre of the helix where one or more of the hydrogen bonds break spontaneously. These partially melted S-ladder structures that appear over nanosecond timescales are reminiscent of the early stages of DNA melting in the model proposed by Williams *et al*[19]. Our calculations indicate that all of the extended DNA duplexes in Figure 21-8 are entropically unstable and should unbind spontaneously. Similarly, the MD trajectories themselves imply that these structures are transient since the duplex disintegrates significantly even over the relatively short timescale of the simulations. Presumably, if it were possible to run simulations of S-ladder DNA over very long (i.e. microsecond) timescales, the unbound regions within the molecule would grow until the helix eventually unbinds entropically into two single strands. This

S-ladder DNA structure will certainly not resist force over the relatively long timescales of a nanomanipulation experiment. The force-plateau observed in DNA stretching experiments therefore signifies a melting transition, rather than a simple conformational change from B to S-ladder DNA. Unfortunately, the term B-S transition is still commonly used in the literature, even though the existence of the S-ladder DNA structure has been disputed.

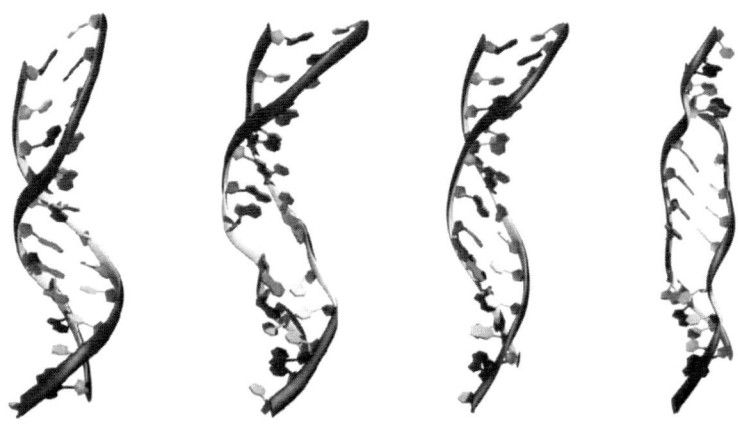

Figure 21-8. The entropic instability of S-ladder DNA. The initially partial S-ladder helix (left) restrainded at an extension of 23 Å develops a denaturation bubble after ~4ns (centre left). Two holes form in the S-DNA structure (centre right) restrained at an extension of 32 Å in under 1 ns (right).

These calculations show that thermal fluctuations and entropy play a significant role in governing the biomechanical behaviour of DNA. The slowly disintegrating structures shown in Figure 21-8 appear mechanically stable if only the enthalpic term is considered, but are clearly thermo-dynamically unstable when the entropy is also taken into account. This observation also explains why neglecting the entropy is equivalent to measuring the force directly using a very fast MD simulation. When the DNA is unbound rapidly transient structures appear to withstand large forces as insufficient time is allowed for them to unbind naturally; the entropy change is also constrained to be close to zero as the molecule can only explore a small region of conformational space over these limited timescales. This corresponds to the system being taken far away from equilibrium. A similar argument has been used to explain why long DNA sequences are stabilised by non-equilibrium effects so that unbinding occurs at higher forces (150pN) than would be expected from equilibrium thermodynamics

(65pN). Computer simulation studies which do not account for entropic effects operate considerably further from equilibrium than any nano-manipulation experiment, so the shape of the force-extension curves obtained using these methods is actually very reasonable.

3.5   Implications for the Action of Molecular Motors

This study of the biomechanics of DNA shows the close connection between force, thermodynamics and kinetics in biology. Duplex DNA must remain stable when not in use to protect the genetic message. Simultaneously, it must be easily taken apart by DNA processing motor proteins so that the genes can be read and copied. Our calculations show that the energy holding the two strands together can be overcome by relatively low forces as long as the process takes place sufficiently slowly for unbinding to be assisted by thermal fluctuations and entropic effects. Similarly, the molecular engine will operate most efficiently if it applies force over timescales which are slow compared to thermal fluctuations so that no energy is dissipated through molecular friction. The timescale of thermal fluctuations is set by the ambient temperature of biological systems. To produce a DNA processing motor, evolution has engineered a delicate thermodynamic balance between enthalpy and entropy, force and time. One aim of nanotechnology is to build similar machines which also operate at an atomic level, but which are designed to perform new technological functions. The design of such engines will be fundamentally different to an equivalent macroscopic machine where thermal fluctuations less significant.

4.   CONCLUSIONS

The power of molecular simulation lies in its ability to provide insights into biological and chemical processes that are impossible, or very difficult, to obtain by any other means. The analysis of free energy changes in terms of enthalpic and entropic components, and of these in terms of contributions from subsections of the system, is one example of this. The entropic terms have always been regarded as some of the most challenging to reliably compute, but developments in the methods of statistical physics, combined with the increasing availability of high-performance computing resources that make the extended molecular dynamics simulations required ever easier to achieve, mean that these challenges can now be met for increasingly complex and large systems.

REFERENCES

1. Quintana, J. R., Lipanov, A. A. & Dickerson, R. E. (1991) Low-temperature crystallographic analyses of the binding of Hoechst 33258 to the double-helical DNA dodecamer C-G-C-G-A-A-T-T-C-G-C-G. *Biochemistry* **30**, 10294-10306.
2. Haq, I., Ladbury, J. E., Chowdhry, B. Z., Jenkins, T. C. & Chaires, J. B. (1997) Specific binding of Hoechst 33258 to the d(CGCAAATTTGCG)(2) duplex: Calorimetric and spectroscopic studies. *Journal of Molecular Biology* **271**, 255-257.
3. Orozco, M. & Luque, F. J. (2000) Theoretical methods for the description of the solvent effect in biomolecular systems *Chem.Rev.* **100**, 4187-4225.
4. Still, W. C., Tempcyck, A., Hawley, R. C. & Hendrickson, T. (1990) Semianalytical treatment of solvation for molecular mechanics and dyamics *J. Am. Chem. Soc.* **112**, 6127-6129.
5. Sands, Z. A. & Laughton, C. A. (2004) Molecular Dynamics Simulations of DNA Using the Generalized Born Solvation Model: Quantitative Comparisons with Explicit Solvation Results. *J. Phys. Chem. B.* **108**, 10113-10119.
6. Kollman, P. A., Massova, I., Reyes, C., Kuhn, B., Huo, S., Chong, L., Lee, M., Lee, T., Duan, Y., Wang, W., Donini, O., Cieplak, P., Srinivasan, J., Case, D. A. & Cheatham III, T. E. (2000) Calculating Structures and Free Energies of Complex Molecules: Combining Molecular Mechanics and Continuum Models. *Accts. Chem. Res.* **33**, 889-897.
7. Schlitter, J. (1993) Estimation of absolute and relative entropies of macromolecules using the covariance matrix *Chem. Phys. Let.* **215**, 617-621.
8. Andricioaei, I. & Karplus, M. (2001) On the calculation of entropy from covariance matrices of the atomic fluctuations *J. Chem.Phys.* **115**, 6289-6292.
9. Carlsson, J. & Aqvist, J. (2005) Absolute and relative entropies from computer simulation with applications to ligand binding *J. Phys. Chem. B* **109**, 6448-6456.
10. Gavathiotis, E., Sharman, G. J. & Searle, M. S. (2000) Sequence-dependent variation in DNA minor groove width dictates orientational preference of Hoechst 33258 in A-tract recognition: solution NMR structure of the 2:1 complex with d(CTTTTGCAAAAG)₂. *Nucleic Acids Res.* **28** 728-735.
11. Harris, S. A., Gavathiotis, E., Searle, M. S., Orozco, M. & Laughton, C. A. (2001) Cooperativity in drug-DNA recognition: a molecular dynamics study. *J. Am. Chem. Soc.* **123**, 12658-12663.
12. Cooper, A. & Dryden, D. T. F. (1984) Allostery without conformational change: A plausible model. **11**, 103-109.
13. Clausen-Schaumann, H., Rief, M., Tolksdork, C. & Gaub, H. E. (2000) Mechanical stability of single DNA molecules. *Biophys. J.* **78**, 1997-2007.
14. Rief, M., Clausen-Schaumann, H. & Gaub, H. E. (1999) Sequence dependent mechanics of single DNA molecules. *Nat. Struct. Biol.* **6**, 346-349.
15. Strunz, T., Oroszlan, K., Schäfer, R. & Güntherodt, H. J. (1999) Dynamic force spectroscopy of single DNA molecules. *Proc. Natl. Acad. Sci. USA*, **96**, 11277-11282.
16. Pope, L. H., Davies, M. C., Laughton, C. A., Roberts, C. J., Tendler S. J. B. & Williams, P. M. (2001) Force-induced melting of a short DNA double helix. *Eur. Biophys. J.* **30**, 53-62.
17. Konrad, M. W. & Bolonick, J. I. (1996) Molecular dynamics simulation of DNA stretching is consistent with the tension observed for extension and strand separation and predicts a novel ladder structure. *J. Am. Chem. Soc.* **118**, 10989-10994.
18. Lavery, R. & Lebrun, A. (1999) Modelling DNA stretching for physics and biology. *Genetica*, **106**, 75-84.

19. Williams, M. C., Rouzina, I. & Bloomfield, V. A. (2002) Thermodynamics of DNA interactions from single molecule stretching experiments. *Acc. Chem. Res.* **35**, 159-166.
20. Harris, S. A., Sands, Z. A. & Laughton, C. A. (2005) Molecular dynamics simulations of duplex stretching reveal the importance of entropy in determining the biomechanical properties of DNA. *Biophys. J.* **88**, 1684-1691.

Chapter 22

SEQUENCE-DEPENDENT HARMONIC DEFORMABILITY OF NUCLEIC ACIDS INFERRED FROM ATOMISTIC MOLECULAR DYNAMICS

Filip Lankaš

Institute for Mathematics B, EPFL (Swiss Federal Institute of Technology), Lausanne, Switzerland

Abstract: This chapter is concerned with a method of deducing coarse-grained conformation and harmonic stiffness of nucleic acids from unrestrained atomistic molecular dynamics. It contains a brief introduction and a review of recent computational studies. The length scales of individual base pairs, basepair steps, and longer oligonucleotides are discussed. The results are compared with analogous data based on experimental investigations. Finally, limitations and challenges of the approach are exposed.

Key words: DNA and RNA mechanics, flexibility, molecular dynamics, conformational fluctuations

1. HIGH-DIMENSIONAL AND REDUCED MODELS

Many objects from everyday life, as well as many systems of biological interest, contain very large number of individual particles, such as atoms and molecules. The behaviour and properties of such macroscopic systems are governed by statistical laws, resulting from the very presence of a large number of particles forming the system. The study of these laws is the subject of thermodynamics and statistical physics[1]. A *macroscopic state (macrostate)* of such a system is typically described by a small number of parameters, such as energy and volume. A *closed system* is one that does not interact with any other systems; its energy and volume are then constant, and there is no exchange of matter with the surroundings. Consider a part of the system which is very small compared to the whole system but still contains a

559

J. Šponer and F. Lankaš (eds.), Computational Studies of RNA and DNA, 559–577.
© 2006 *Springer.*

very large number of particles – it can thus be also considered macroscopic, and its macroscopic state can again be described by a small number of parameters. Such a *subsystem* is not closed, but interacts in various ways with other parts of the system.

Let us take an example: a weight on a spring, hanging from a support and placed on a table somewhere in a school laboratory, may be considered a macroscopic system which, for many purposes, can be characterized by just two thermodynamic parameters, the spring length x and temperature T. It forms a small part of a large system comprising also the surrounding medium which as a whole can be assumed closed.

A particular macroscopic state can typically be realized by a huge number of different *microscopic states (microstates)* which represent the quantum states of the system or, in the classical approximation, small intervals of positions and momenta of all the particles. A closed system eventually reaches a state of *thermodynamic equilibrium*. The entropy S_0 of a closed system in equilibrium is related to the number of microstates W, which realize its macroscopic state. This relationship is expressed by the Boltzmann formula, $S_0 = k_B \ln W$, where the Boltzmann constant k_B determines the units of entropy. We can also write $W = \exp(S_0 / k_B)$. Assuming that every microstate occurs with the same probability, the probability of the equilibrium state, among all possible states of a system, will be $w = const.\exp(S_0 / k_B)$.

A remark is in order here. In reality, no system is, or ever can be, truly closed. Its interactions with the surroundings, however small, make its energy slightly imprecise. Furthermore we cannot measure the energy of any system with absolute precision. Therefore, the number of microstates W in the definition of entropy in fact corresponds to an ensemble of macrostates whose energies lie within a certain interval ΔE. It can be shown that, if the system has a large number of degrees of freedom, the value of entropy is essentially independent of the precise value of ΔE.

The physical quantities which describe a macroscopic system in equilibrium are almost always very close to their mean values; nevertheless due to the stochastic nature of the system's behaviour, deviations from the mean values (*fluctuations*) do occur and one may be interested to know their probability distribution.

Let us consider a closed system, and let x be a physical quantity describing the system or a part of it (in the former case x should not be a quantity which is constant for a closed system, such as energy). The quantity x is assumed to be a thermodynamic parameter, in the sense that a state of a given value of x (strictly speakinkg, of x within some small interval Δx) can still be realized by enormous number of microstates. One can consider the entropy, $S_t(x)$, associated with such a state *of the whole closed system,* which would again have the Boltzmannian interpretation given above: the quantity W in the definition of entropy will now be the number of micro-states which realize macrostates of energy within ΔE and x within Δx. Again, for a high-dimensional system the value of $S_t(x)$ will be essentially independent of ΔE and Δx. The number of microstates per unit interval of x is then $W/\Delta x = \exp(S_t(x)/k_B)/\Delta x$ which implies the probability density of the form $w = const.\exp(S_t(x)/k_B)/\Delta x$. If we absorb Δx into the constant, we finally obtain

$$w(x) = const.\exp(S_t(x)/k_B) \tag{1}$$

This formula was first proposed by A. Einstein (1910) in a work devoted to the problem of opalescence in fluids and fluid mixtures near the critical point[2].

The probability density (1) is thus proportional to $\exp(S_t)$, where S_t is the total entropy of the closed system; it can be equally well expressed in terms of the deviation ΔS_t of S_t with respect to its equilibrium value S_0: $\Delta S_t = S_t - S_0$. We then have $w = const.\exp(\Delta S_t/k_B)$.

The quantity ΔS_t allows for the following thermodynamic interpretation. Since the whole system is closed, its *equilibrium* entropy can be regarded as a function of energy alone: $S_0 = S_0(E_0)$. A fluctuation then brings the system out of equilibrium, so that its entropy S_t now differs from its equilibrium value S_0 for the same value of the total energy E_0 by $\Delta S_t = S_t - S_0 < 0$ and its state thus corresponds to some point (E_0, S_t) in the *E-S* plane. The same state, however, can be attained by initially considering the system in another equilibrium state, namely the one for which the *equilibrium* entropy is S_t – say, state (E_1, S_t) – and performing a reversible work on the system (thus S_t remains constant during the process) until it reaches the state (E_0, S_t). The amount of work necessary would be $R_{rev} = E_0 - E_1$. Since all the changes are assumed small, we can write $\Delta S_t = -(dS_0(E)/dE)R_{rev}$. But $dS_0(E)/dE = 1/T_0$, where T_0 is the equilibrium temperature of the system, which shows that $\Delta S_t = -R_{rev}/T_0$. We thus see that the change of entropy of the closed system related to a change in some thermodynamic quantity due to fluctuation is proportional to the work necessary to carry out reversibly the same change.

Consider now a special case, when we are interested in some small subsystem (body) within a closed system. The parts of the closed system outside the body can be considered as medium of such a large size that its temperature T_0 and pressure P_0 may be taken constant, equal to the equilibrium temperature and pressure of the whole closed system. Assume for a moment that, apart from interacting with the medium, the body is connected with an external device, which does work R on it. During the process, the body may also exchange heat and work with the medium. The work done by the medium is $P_0\Delta V_0$ and the heat transferred to it is $-T_0\Delta S_0$. The energy change of the body is thus $\Delta E = R + P_0\Delta V_0 - T_0\Delta S_0$ (the quantities without subscript refer to the body). Since $\Delta V_0 = -\Delta V$, $\Delta S + \Delta S_0 \geq 0$, we find $R \geq \Delta E - T_0\Delta S + P_0\Delta V$, where the equality holds for a reversible process: $R_{rev} = \Delta E - T_0\Delta S + P_0\Delta V$.

Taken together, we obtain the change of entropy of the whole closed system (body+medium) due to a fluctuation in the body as $\Delta S_t = -R_{rev}/T_0 = -(\Delta E - T_0\Delta S + P_0\Delta V)/T_0$ where ΔE, ΔS and ΔV are changes in the energy, entropy and volume of the body in the fluctuation. In the special case where the temperature and pressure of the body remain constant and equal to those of the medium, we have $R_{rev} = \Delta(E - TS + PV) = \Delta G$, where G is the Gibbs potential of the body. In this case the probability distribution of the fluctuations (1) adopts the form

$$w(x) = const.\exp(-\Delta G(x)/k_B T) \tag{2}$$

If the fluctuations are small, one can expand $G(x)$ in a Taylor series around x_0, the equilibrium value of x, which is given by $x_0 = \langle x \rangle$. Since $x = x_0$ corresponds to the thermodynamic equilibrium, G has minimum there and $(dG/dx)_{x=x_0} = 0$, so that the first nonzero term depending on x is quadratic and we have

$$w(x) = const.\exp(-(1/2)a(x - x_0)^2/k_B T) \tag{3}$$

where a is a positive constant. Thus $w(x)$ is a Gaussian distribution with

$$\langle x \rangle = x_0 \tag{4}$$

To obtain a, we calculate the second moment of x:

$$\left\langle (x - x_0)^2 \right\rangle = \int_{-\infty}^{+\infty} (x - x_0)^2 \, w(x)dx \Big/ \int_{-\infty}^{+\infty} w(x)dx \tag{5}$$

which gives

$$\left\langle (x - x_0)^2 \right\rangle = k_B T/a \tag{6}$$

This relation provides a if the standard deviation of x is known. The latter can be e.g. measured in experiment, or observed in simulations.

Now we may ask how much work we have to do on the system to deform it from $x = x_0$ to a particular value x. This work, if done reversibly, is just the free energy difference between the two states, so that we have $R_{rev} = (1/2)a(x - x_0)^2$.

Let's now return to our example, the weight suspended on a spring. Taking x as the length of the spring, we see that the constant a is nothing else than the spring's stiffness constant. Thus, we have established an alternative method to measure a spring constant: besides the conventional pulling of the spring by an external force and observing the elongation x, we can leave the whole system as is, not perturbing it at all, and just passively observe the fluctuations of x. Of course, this approach is probably impractical in the case of truly macroscopic springs from everyday life, but in case of "molecular springs" where fluctuations are relatively large compared to the size of the system, the approach can be quite useful: it allows for determining the spring length and stiffness without perturbing the system.

We may approach the problem from yet another perspective. Namely, we can try to represent our weight on a spring in a simplified way, as a point of mass m in a quadratic potential with the minimum at x_0 and the force constant a. While x was a thermodynamic quantity in the previous, high-dimensional model, in this reduced model it is the only degree of freedom. The Hamiltonian of the system is

$$H(x, p) = (1/2m)p^2 + (1/2)a(x - x_0)^2 \qquad (7)$$

where $p = dx/dt$ is the linear momentum of the point mass.

If the system is kept at constant temperature, the probability distribution of x is given by the (classical) canonical distribution, that is

$$w(x, p) = const.\exp(-H/k_B T) \qquad (8)$$

Luckily enough, the Hamiltonian is a sum of two terms the first of which, the kinetic energy, depends only on p, and the second, the potential energy, only on x. Thus if we are interested in the probability distribution of x alone, we can easily integrate out the term depending on p:

$$w(x) = \int_{-\infty}^{+\infty} w(x, p)dp = const.\exp(-(1/2)a(x - x_0)^2 / k_B T) \qquad (9)$$

where the integral over p, being just a number, was absorbed into the constant. Similarly, the distribution of p alone is

$$w(p) = const.\exp(-(1/2m)p^2 / k_B T) \qquad (10)$$

This enables us to easily extract m, x_0 and a, the parameters of the model, from observed fluctuations of x and p. We have

$$\langle p \rangle = \int_{-\infty}^{+\infty} pw(p)dp \Big/ \int_{-\infty}^{+\infty} w(p)dp = 0 \qquad (11)$$

$$\langle p^2 \rangle = \int_{-\infty}^{+\infty} p^2 w(p)dp \Big/ \int_{-\infty}^{+\infty} w(p)dp = mk_B T \qquad (12)$$

and similarly, using the probability distribution $w(x)$, we obtain

$$\langle x \rangle = x_0 \qquad (13)$$

$$\left\langle \left(x - \langle x \rangle \right)^2 \right\rangle = k_B T / a \qquad (14)$$

The external work necessary to deform the system from x_0 to a particular value x is then $(1/2)a(x - x_0)^2$.

Comparing these results with those for the high-dimensional model above, we note that the relations for extracting the equilibrium shape (length) x_0 and the stiffness constant a from fluctuations are the same for both models (of course, in the high-dimensional model there is no analogy to the point mass). The two models, although conceptually different, give the same prediction for the shape and stiffness of the system. As we'll see below, this is not true for analogous models for the shape and stiffness of DNA.

2.   INFERRING DNA SHAPE AND DEFORMABILITY FROM STRUCTURAL FLUCTUATIONS

A usual setup for atomistic molecular dynamics simulations of DNA oligomers consists of the oligomer itself, typically 10-30 basepairs long, surrounded by water molecules and cations to neutralize DNA's negative charge (extra pairs of cations and anions are sometimes added to mimic the presence of "added salt"). The system is let to evolve according to classical (Newtonian) dynamics while temperature and pressure are kept constant by some suitable representation of thermal and pressure reservoirs. The result of the simulation is a series of "snapshots" taken at discrete moments of time, each comprising the positions of all atoms of the system.

While the atomic positions fully define the structure of DNA, for many purposes it is convenient to describe the DNA conformation with a smaller set of parameters. Perhaps the most widely used are the helical (or basepair step) parameters, describing the mutual position and orientation of two adjacent base pairs. They bear conventional names: shift, slide, rise for translations, tilt, roll, twist for rotations. For each basepair step of the oligomer we thus have six parameters defining the conformation of that step. The helical parameters are calculated by first fitting to a given base a standard base with precisely defined, "canonical" geometry, using a least-squares procedure. Attached to the standard base is a coordinate frame[3]. One thus has a frame attached to each base in a basepair. Then, a procedure is employed to construct from these base frames one single frame for the whole base pair. The helical parameters are then calculated from the frames of adjacent base pairs in the oligomer, expressing in some way the relative translation and rotation of a basepair frame with respect to the preceding one. There are various programs available to perform this whole procedure, and the precise definition of the helical parameters differs from program to program[4,5].

The time series of snapshots can thus be transformed into time series of helical parameters, and one can analyze their fluctuations to obtain DNA equilibrium helical structure and deformability. The situation is analogous to that of the weight on a spring discussed above. Let us first consider the helical parameters to be thermodynamic quantities associated with a high-dimensional model. The free energy of the system will then depend on the values of the helical parameters of all steps of the oligomer. We now adopt a simplifying assumption that the total free energy is a sum of contributions, each depending only on the parameters of one step:

$$G = \sum_{a=1}^{N} G_a \qquad (15)$$

where N is the number of basepair steps in the oligomer.

We will denote the rotational parameters tilt, roll, twist by u_1, u_2, u_3 respectively, and the translational parameters shift, slide, rise by v_1, v_2, v_3. We refer to u_1, u_2, u_3 of step a collectively as $u^{(a)}$, and similarly $v^{(a)}$ comprises v_1, v_2, v_3 of step a. All six numbers associated with step a form a

6-tuple which we denote by $w^{(a)}$. According to Eq. (2), the probability density of the helical parameters of the whole oligomer is

$$p(w^{(1)},\ldots,w^{(N)}) = const.\exp(-G/k_BT) = const.\prod_{a=1}^{N}\exp(-G_a/k_BT)$$

and the distribution of a particular $w^{(a)}$ is obtained by integrating over the parameters of all the other steps,

$$p(w^{(a)}) = const.\exp(-G_a/k_BT) \tag{16}$$

So far we have not made any assumption about the functional form of G_a. In principle, G_a can be a complicated function with multiple minima corresponding to various metastable states ("substates") of the step. From now on we assume that there is just one single minimum for each G_a and that the step conformation exhibits small fluctuations around this minimum. We denote by $w_0^{(a)}$ the equilibrium value around which $w^{(a)}$ fluctuates:

$$w_0^{(a)} = \langle w^{(a)} \rangle \tag{17}$$

Expanding G_a in powers of $w^{(a)} - w_0^{(a)}$ around the equilibrium state, we again find that the linear term is zero because G_a has minimum at $w_0^{(a)}$ and the expansion truncated at the quadratic term gives

$$p(w^{(a)}) = const.\exp\left(-\left(w^{(a)} - w_0^{(a)}\right)^T K_a\left(w^{(a)} - w_0^{(a)}\right)/2k_BT\right) \tag{18}$$

The 6×6 matrix K_a with elements $(K_a)_{ij}$ is symmetric, positive definite (since G_a has minimum at $w_0^{(a)}$) and will be called the stiffness matrix of step a. Calculating the second moments of $w^{(a)}$ gives

$$\langle (w_i^{(a)} - w_{0i}^{(a)})(w_j^{(a)} - w_{0j}^{(a)}) \rangle = k_BT\left(K_a^{-1}\right)_{ij} \tag{19}$$

where K_a^{-1} is the inverse of K_a. This formula provides a way to determine K_a: we calculate the left-hand side of Eq. (19) as an average over the time series obtained from simulations – the values on the left-hand side taken together form a 6×6 matrix, the covariance matrix C_a with elements $(C_a)_{ij} = \langle (w_i^{(a)} - w_{0i}^{(a)})(w_j^{(a)} - w_{0j}^{(a)}) \rangle$ – and we use Eq. (19) to obtain K_a. Eq. (19) can be rewritten as $C_a = k_BTK_a^{-1}$ or

$$K_a = k_BTC_a^{-1} \tag{20}$$

so that the stiffness matrix is calculated as the inverse of the covariance matrix, multiplied by $k_B T$.

Let us now explore the other approach, i.e. we will assume a reduced model of DNA. A model of this type was proposed by Gonzalez and Maddocks[6]. In their approach the DNA oligomer is modeled by a chain of interacting rigid bodies, each corresponding to one base pair. The helical parameters are the only degrees of freedom of the model. The kinetic energy of the chain is a sum of kinetic energies of the rigid bodies (base pairs), and the potential energy is assumed to be a sum of contributions from individual basepair steps, which are assumed quadratic in $w^{(a)}$:

$$U = \sum_{a=1}^{N} U_a \tag{21}$$

$$U_a = \frac{1}{2}\left(w^{(a)} - w_0^{(a)}\right)^T K_a \left(w^{(a)} - w_0^{(a)}\right) \tag{22}$$

where $w_0^{(a)}$ is the configuration corresponding to the potential energy minimum. The system is considered to be kept at constant temperature, so that the canonical distribution (see Eq. (8)) can be assumed. There is, however, a very important point coming into play. In the canonical distribution, the Hamiltonian H is a function of the generalized coordinates ($w^{(a)}$ in this case) and their associated canonical momenta. However, the momenta conjugated to $w^{(a)}$ (that is, to $u^{(a)}, v^{(a)}$) are not the usual linear and angular momenta of rigid body motion. As a result, the chain kinetic energy depends on the momenta conjugated to $u^{(a)}, v^{(a)}$ as well as on $u^{(a)}, v^{(a)}$ themselves. This configurational dependence makes it impossible to factorize $\exp(-H/k_B T)$ as a product of a term depending only on coordinates and a second one depending only on momenta, a feature critical for our purposes. The solution is to change variables from $u^{(a)}, v^{(a)}$ and their conjugated momenta to $u^{(a)}, v^{(a)}$ and the usual linear and angular momenta, $p^{(a)}$ and $\pi^{(a)}$, so that now U depends only on $u^{(a)}, v^{(a)}$ and the kinetic energy only on $p^{(a)}$ and $\pi^{(a)}$. This change of variables, however, introduces a Jacobian factor into the probability distribution which now becomes:

$$p(w^{(a)}, p^{(a)}, \pi^{(a)}) = const. \exp(-H(u^{(a)}, v^{(a)}, p^{(a)}, \pi^{(a)})/k_B T)J_a \tag{23}$$

It can be shown that the Jacobian J_a for step a depends only on the angular variables $u^{(a)}$. Here the Hamiltonian is a sum of the potential energy depending only on the coordinates, and kinetic energy depending only on the momenta:

$$H = U(u^{(a)}, v^{(a)}) + \Phi(p^{(a)}, \pi^{(a)}) \tag{24}$$

thus the exponential in (23) can be factorized. Integrating (23) over the momenta gives the probability distribution of $w^{(a)}$ alone in the form

$$p(w^{(a)}) = const. \exp\left(-\left(w^{(a)} - w_0^{(a)}\right)^T K_a \left(w^{(a)} - w_0^{(a)}\right)/2k_B T\right) J_a \tag{25}$$

This formula then leads to the following relations between the fluctuations of $w^{(a)}$ and the model parameters:

$$\frac{\left\langle w^{(a)}/J_a \right\rangle}{\left\langle 1/J_a \right\rangle} = w_0^{(a)} \tag{26}$$

$$\frac{\left\langle \left(w_i^{(a)} - w_{0i}^{(a)}\right)\left(w_j^{(a)} - w_{0j}^{(a)}\right)/J_a \right\rangle}{\left\langle 1/J_a \right\rangle} = k_B T \left(K_a^{-1}\right)_{ij} \tag{27}$$

which are different from the analogous relations (17) and (19) for the high-dimensional model. Note, for instance, that the average $< w^{(a)} >$, which defines the equilibrium configuration in the high-dimensional model, is not the same as the minimum energy configuration $w_0^{(a)}$ of the reduced model. The stiffness matrices K_a are also different, so that the models predict different amount of work necessary to deform the system from one particular configuration to another. But if the fluctuations are small, then J_a will be nearly constant, the two models therefore giving close results.

The approach outlined above works for a broad class of definitions of the variables $u^{(a)}$ and $v^{(a)}$: $u^{(a)}$ are supposed to be independent coordinates for the relative rotation matrix connecting the two successive basepair frames, and $v^{(a)}$ are components of the translation vector connecting the origins of the two frames expressed in a certain "middle" frame, defined so that $v^{(a)}$ are invariant under rigid displacements of the chain ($u^{(a)}$ automatically satisfy this property as a result of their definition). Gonzalez and Maddocks have introduced a particular definition of $u^{(a)}$ and $v^{(a)}$ which they call *discretized rod variables*. In this particular case, the Jacobian factor takes the form

$$J_a = \frac{1}{1 + \frac{1}{4}\left\| u^{(a)} \right\|^2} \tag{28}$$

which is indeed close to a constant if the fluctuations of $u^{(a)}$ are small. However, the exact numerical effect of the Jacobian can only be inferred from analyzing real data.

3.   APPLICATIONS

The general high-dimensional model outlined above has been applied to estimate sequence-dependent DNA mechanical properties at different length scales. In particular, a sequence-dependent dinucleotide force field has been constructed[7]. There are ten unique dinucleotide steps, each of which is characterized by the minimum free energy conformation (defined by specifying values of the 6 helical parameters) and by a 6×6 symmetric, positive definite stiffness matrix. These parameters were inferred from atomic-resolution MD simulations of two 18-bp DNA oligomers, one of which contained a strong nucleosome positioning motif. Cyclization experiments on constructs involving the same sequences indicate that they have very different global mechanical properties, thus choosing them for our study, we could hope to cover a wide range of basepair step behaviour. The diagonal elements of the stiffness matrices (which correspond to stiffness constants associated with a change of only one helical parameter while the others are kept at their equilibrium values) are shown in Figure 22-1, in comparison with the same quantities inferred from a pool of crystal structures using an analogous approach[8]. The MD data exhibit some simple trends: in the case of roll, one can distinguish flexible YR, intermediate RR, and stiff RY steps. The YR steps are also the most flexible in tilt and partially in twist. A consistent increase of stiffness form YR through RR to RY is observed for rise, while shift and slide lack simple trends. By contrast, dependences in the x-ray data seem more complicated.

There is one subtle point to make: the x-ray stiffnesses are defined to within an unknown multiplicative factor, the "effective temperature" of the statistical ensemble. This factor cannot be inferred from the data but can be obtained by fitting the global bending stiffness of a DNA oligomer to the value known experimentally[9]. It turns out that the effective temperature of the x-ray ensemble is 295K[7].

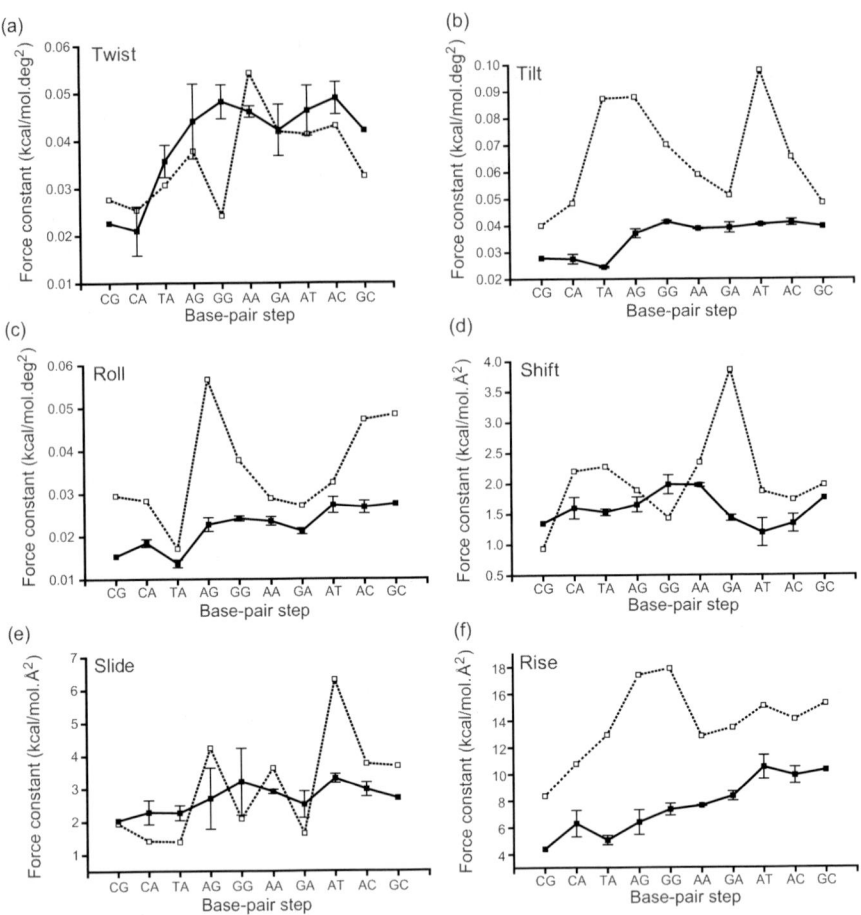

Figure 22-1. Diagonal force constants in basepair step harmonic deformation potentials obtained from molecular dynamics simulations[7] (*solid line*) as compared with those from the x-ray database study[8] (*dotted line*). Error bars in the simulation data indicate the range of values observed. From ref. 7, reproduced with permission.

Table 1. Average values of helical parameters in the basepair step model of DNA deformability inferred from MD simulations[7]. Angular parameters (tilt, roll, twist) are in degrees, translational parameters (shift, slide, rise) are in Ångströms.

Step	Tilt	Roll	Twist	Shift	Slide	Rise
CG	-0.59	8.75	27.93	-0.05	-0.76	3.26
CA	-0.01	8.36	25.69	0.18	-0.57	3.13
TA	0.46	10.33	29.74	-0.04	-1.02	3.34
AG	0.99	2.75	28.84	0.05	-1.33	3.38
GG	0.84	5.28	29.93	0.07	-1.64	3.61
AA	-1.52	2.31	31.04	-0.14	-1.04	3.31
GA	-0.28	2.27	33.32	-0.01	-0.92	3.37
AT	0.25	0.21	29.02	-0.05	-1.08	3.20
AC	-0.36	0.63	31.46	-0.12	-1.16	3.38
GC	0.25	1.23	33.13	0.22	-1.09	3.37

The same methodology was later used to obtain DNA mechanical properties at the base-pair level[10] – here the structural variables are shear, stretch, stagger, buckle, propeller, opening. Using the same simulations as above, one can find the average values and stiffness matrices for the AT and GC pair.

The general approach was also used to infer global mechanical properties of particular DNA oligomers (polyA, polyG, and alternating AT and GC)[11]. Here the structural variables are the total length of the oligomer, total twist, and two bending angles, to the grooves and to the backbone in the centre of the oligomer. A clear sequence-dependent signal was found in the data. Moreover, the values were within the range expected from experiments on sequence-averaged DNA. Some controversy has been associated with the experimentally determined torsional stiffness, different experiments giving values within a factor of three or more. Interestingly, the higher torsional stiffness (~100 nm) predicted in the simulations has been found in the most recent measurements[12]. Moreover, the twist-stretch coupling was calculated, its values being also consistent with experiments, and the values of twist-bend coupling, whose existence had been predicted theoretically[13], was calculated for the first time.

The method has since then been applied to other problems like a study of the flexibility of nucleosomal DNA[14], comparison of the mechanical properties of DNA and RNA[15, 16], and changes in flexibility upon DNA oxidative damage[17].

4.   COMPARISON TO EXPERIMENTALLY BASED FORCE FIELDS

Several groups have developed local (basepair step) force fields for DNA deformation using different methods. These force fields describe DNA deformability at various levels of detail and thus are not all directly comparable: the data of Olson and co-workers[8] (presented in more detail in the previous paragraph – see Figure 22-1) and the MD data of Lankaš et al.[7] contain the full stiffness matrix for the six conformational parameters, while other force fields usually report isotropic bending stiffness. Moreover, most force fields are relative and must be calibrated against a quantity known from elsewhere.

But some approximative comparison can be made. Here, we compare the force field of Robinson and co-workers[18] based on an Electron paramagnetic resonance technique, the one by De Santis and co-workers[19] inferred from thermal stability data, together with the deformability parameters by Olson and co-workers[8] (from an ensemble of crystal structures) and our MD data[7]. The common parameter to compare is the relative isotropic bending rigidity. This quantity was constructed as follows. First, an isotropic bending rigidity for the data of Olson et al. and for the MD data was calculated as harmonic

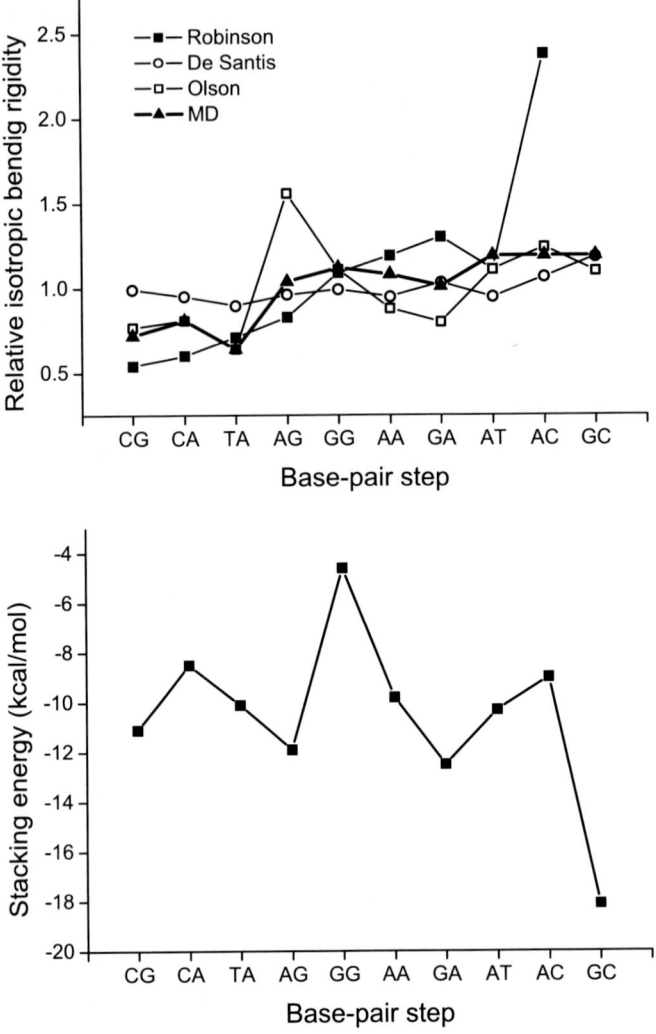

Figure 22-2. (a) Relative isotropic bending rigidity as a function of the dinucleotide sequence for basepair step force fields inferred from an EPR experiment[17], thermal stability data[18], ensemble of crystal structures[8] and atomistic molecular dynamics simulations[7]. Comparison with basepair step stacking energies (b) from modern *ab initio* quantum chemical calculations[27,28] indicates that none of the force fields is correlated with stacking energy values.

average of tilt and roll stiffness. The other two force fields already contain this value. Then, a "generic" value was computed as harmonic average of the values of the ten unique steps. The relative stiffness is then the stiffness for a particular step divided by the generic value.

The comparison is shown in Figure 22-2a. Note that the first three steps on the x-axis are of YR type, the last three are RY, and the four in the middle are RR. The Robinson data show a general tendency of increasing stiffness from YR through RR to RY. The value for GC is not available, since the reciprocal value of the force constant (the quantity inferred from the measurements) was zero within the sensitivity of the method; the most flexible and the stiffest steps thus differ by a factor of four or more. By contrast, the De Santis data show much smaller variations. The Olson parameters behave in a rather complicated way, changing between 0.8 and 1.6. The MD data exhibit a clear trend: they predict flexible YR steps, intermediate RR and stiff RY, spanning a range between 0.7 and 1.2. And so, the force fields differ both in the range of values and the trends they exhibit. Note, however, that the Olson and MD force fields are quite close to each other for YR and RY steps; for RR steps, the Olson data show substantial variations while the MD values change much less.

It should be stressed that this comparison regards only relative isotropic bending stiffness and says nothing about the flexibility in twist or in translational parameters, since those are not available for all the four force fields. As mentioned above, the trend in the MD data presented here in fact reflects the trend in roll stiffness, while tilt stiffness is low for YR and higher (and similar) for all the other steps.

It is interesting to compare these data with other related works. In their MD study, McConnel and Beveridge[20] assigned certain flexibility to each of the ten dinucleotide steps based on the area of their roll-tilt plots enclosing 98% of the MD data points. They found that YR steps are on average significantly more flexible than the others; the most flexible step was TA and the least flexible AT. Later, Thayer and Beveridge[21] divided the roll-tilt plots into quadrants and 10 concentric rings which enabled them to extract more detailed information about the distribution of MD data for use in their Hidden Markov model for DNA-protein interactions.

Packer and Hunter have calculated the stiffness with respect to slide in dinucleotides[22] and slide stiffness of the central step in tetranucleotides[23]. All the other conformational parameters are optimized. Their ratio of force constants between the stiffest and the most flexible step is more than 1:10. Among their findings one may notice that CG and GC exhibit similar flexibility at the dinucleotide level and that tetranucleotides with central AT step rank among the most flexible as well as the stiffest ones, depending on the base pairs flanking the central step.

An interesting question arises about a possible relationship between stiffness and stacking energy. The hypothesis that stacking energy may have a determining influence on DNA deformability was proposed by Hagerman[24]. De Santis and co-workers[19,25] report a very close correlation (0.96) between their stiffness scale and the stacking energies based on an early theoretical work[26]. However, the order of stability of the ten basepair steps predicted by these old semi-empirical calculations significantly differs from that predicted by the recent quantum chemical computations[27,28].

Modern electron correlation calculations represent almost converged estimates of stacking energies and provide reliable relative order of intrinsic stacking energy[29]. Comparing the data by De Santis and co-workers[19] with contemporary evaluation of base stacking in B-DNA geometries[27,28] leads to a correlation coefficient of only 0.3. Taking into account the quality of contemporary *ab initio* methods, we have to conclude that the thermal stability scale used by De Santis and co-workers is not correlated with intrinsic base stacking energies.

On the other hand, modern quantum chemical data are excellently reproduced by modern empirical force fields[27,28]. One might thus expect that the MD flexibility scale (calculated using the Cornell et al. force field[30]) will show better correlations with the stacking energy. But as can be seen by comparing Figures 22-2a and 22-2b, it is by no means the case. In fact, the stacking energies shown in Figure 22-2b are not correlated with any of the force fields in Figure 22-2a. Thus, we suggest that the DNA stiffness is not determined simply by the magnitude of base stacking but by a complex interplay of various factors such as potential energy profiles of the steps and balance between stacking and hydration effects.

5.   CONCLUSIONS AND PERSPECTIVES

Computational studies on DNA sequence-dependent mechanical properties have so far been focused mostly on two separate length scales: either they deal with DNA properties at the level of individual base pairs or basepair steps, or they investigate "global" properties of a multi-basepair fragment. In the former case the properties are supposed local, in the sense that the free energy needed to deform a longer fragment is assumed to be just a sum of the free energies needed to deform individual pairs and steps. Even though the mechanical properties of pairs and steps are mostly taken as independent of their sequence context, the context effects can in principle be taken into account while still maintaining the assumption of locality. This assumption, however, is almost certainly an oversimplification, since it has been observed that movements in neighbouring steps are correlated[7]. Hence, there is a need of data sets and theories that would allow for consistent passing from the base pair scale up to the scale of tens and hundreds of base pairs, taking the correlation effects (nonlocality) into account.

The first step obviously means creating a comprehensive database of sequence-dependent DNA properties at a scale longer than dinucleotides. The nearest higher level is the one of trinucleotides, but context effects and nonlocality can be more naturally studied at the level of tetranucleotides. Packer and Hunter[23] examined this level using a simplified molecular-mechanical model. Recently, an international consortium called ABC have completed a set of atomistic MD simulations of oligonucleotides comprising all the 136 possible tetranucleotide sequences[31,32]. While substantial analysis of the data has been already performed, the corresponding theoretical models which would allow for consistent transition among length scales and account for the nonlocality of DNA mechanical properties are still to be proposed. Such models, when supplied with the data, should ultimately form a basis for bioinformatics analysis tools capable for searching and comparing sequences according to their mechanical properties, thus exploring directly this "second code" embedded in a DNA sequence.

ACKNOWLEDGEMENTS

This work was supported by the Swiss National Science Foundation.

REFERENCES

1. Landau, L. D. & Lifshitz, E. M. (1980). *Statistical Physics, Part 1*. 3rd edit, Butterworth-Heinemann.
2. Einstein, A. (1910). Theorie der Opaleszenz von homogenen Flüssigkeiten und Flüssigkeitsgemischen in der Nähe des kritischen Zustandes. *Ann. Phys.* **33**, 1275-1298.
3. Olson, W. K., Bansal, M., Burley, S. K., Dickerson, R. E., Gerstein, M., Harvey, S. C., Heinemann, U., Lu, X.-J., Neidle, S., Shakked, Z., Sklenar, H., Suzuki, M., Tung, C.-S., Westhof, E., Wolberger, C. & Berman, H. M. (2001). A Standard Reference Frame for the Description of Nucleic Acid Base-pair Geometry. *J. Mol. Biol.* **313**, 229-237.
4. Lavery, R. & Sklenar, H. (1989). Defining the Structure of Irregular Nucleic Acids: Conventions and Principles. *J. Biomol. Struct. Dyn.* **6**, 655-667.
5. Lu, X. J. & Olson, W. K. (2003). 3DNA: a software package for the analysis, rebuilding and visualization of three-dimensional nucleic acid structures. *Nucleic Acids Res.* **31**, 5108-5121.
6. Gonzalez, O. & Maddocks, J. H. (2001). Extracting parameters for base-pair level models of DNA from molecular dynamics simulations. *Theor. Chem. Acc.* **106**, 76-82.
7. Lankaš, F., Šponer, J., Langowski, J. & T.E. Cheatham, I. (2003). DNA Basepair Step Deformability Inferred from Molecular Dynamics Simulations. *Biophys. J.* **85**, 2872-2883.
8. Olson, W. K., Gorin, A. A., Lu, X.-J., Hock, L. M. & Zhurkin, V. B. (1998). DNA sequence-dependent deformability deduced from protein-DNA crystal complexes. *Proc. Natl. Acad. Sci. USA* **95**, 11163-11168.
9. Matsumoto, A. & Olson, W. K. (2002). Sequence-dependent Motions of DNA: A Normal Mode Analysis at the Base-pair Level. *Biophys. J.* **83**, 22-41.
10. Lankaš, F., Šponer, J., Langowski, J. & T.E. Cheatham, I. (2004). DNA Deformability at the Base Pair Level. *J. Am. Chem. Soc.* **126**, 4124-4125.
11. Lankaš, F., Šponer, J., Hobza, P. & Langowski, J. (2000). Sequence-dependent Elastic Properties of DNA. *J. Mol. Biol.* **299**, 695-709.

12. Bryant, Z., Stone, M. D., Gore, J., Smith, S. B., Cozzarelli, N. R. & Bustamante, C. (2003). Structural transitions and elasticity from torque measurements on DNA. *Nature* **424**, 338-341.

13. Marko, J. F. & Siggia, E. D. (1994). Bending and Twisting Elasticity of DNA. *Macromolecules* **27**, 981-998.

14. Bishop, T. C. (2005). Molecular dynamics simulations of a nucleosome and free DNA. *J. Biomol. Struct. Dyn.* **22**, 673-685.

15. Noy, A., Perez, A., Lankaš, F., Luque, F. J. & Orozco, M. (2004). Relative Flexibility of DNA and RNA: a Molecular Dynamics Study. *J. Mol. Biol.* **343**, 627-638.

16. Perez, A., Noy, A., Lankaš, F., Luque, F. J. & Orozco, M. (2004). The relative flexibility of B-DNA and A-RNA duplexes: database analysis. *Nucleic Acids Res.* **32**, 6144-6151.

17. Barone, F., Lankaš, F., Spackova, N., Šponer, J., Karran, P., Bignami, M. & Mazzei, F. (2005). Structural and dynamic effects of single 7-hydro-8-oxoguanine bases located in a frameshift target DNA sequence. *Biophys. Chem.* **118**, 31-41.

18. Okonogi, T. M., Alley, S. C., Reese, A. W., Hopkins, P. B. & Robinson, B. H. (2002). Sequence-dependent dynamics of duplex DNA: The applicability of a dinucleotide model. *Biophys. J.* **83**, 3446-3459.

19. Scipioni, A., Anselmi, C., Zuccheri, G., Samori, B. & De Santis, P. (2002). Sequence-dependent DNA curvature and flexibility from scanning force microscopy images. *Biophys. J.* **83**, 2408-2418.

20. McConnell, K. J. & Beveridge, D. L. (2001). Molecular dynamics simulations of B,-DNA: Sequence effects on A-tract-induced bending and flexibility. *J. Mol. Biol.* **314**, 23-40.

21. Thayer, K. M. & Beveridge, D. L. (2002). Hidden Markov models from molecular dynamics simulations on DNA. *Proc. Natl. Acad. Sci. USA* **99**, 8642-8647.

22. Packer, M. J., Dauncey, M. P. & Hunter, C. A. (2000). Sequence-dependent DNA Structure: Dinucleotide Conformational Maps. *J. Mol. Biol.* **295**, 71-83.

23. Packer, M. J., Dauncey, M. P. & Hunter, C. A. (2000). Sequence-dependent DNA Structure: Tetranucleotide Conformational Maps. *J. Mol. Biol.* **295**, 85-103.

24. Hagerman, P. J. (1988). Flexibility of DNA. *Ann. Rev. Biophys. Biophys. Chem.* **17**, 265-86.

25. Anselmi, C., Santis, P. D., Paparcone, R., Savino, M. & Scipioni, A. (2002). From the sequence to the superstructural properties of DNAs. *Biophys. Chem.* **95**, 23-47.

26. Ornstein, R. L., Rein, R., Breen, D. L. & Macleroy, R. D. (1978). An optimized potential function for the calculation of nucleic acid interaction energies. I. Base stacking. *Biopolymers* **17**, 2341-2360.

27. Šponer, J., Florian, J., Ng, H.-L., Šponer, J. E. & Spackova, N. (2000). Local conformational variations observed in B-DNA crystals do not improve base stacking. Computational analysis of base stacking in d(CATGGGCCCATG)₂ B<->A intermediate crystal structure. *Nucleic Acids Res.* **28**, 4893-4902.

28. Šponer, J., Gabb, H. A., Leszczynski, J. & Hobza, P. (1997). Base-base and deoxyribose-base stacking interactions in B-DNA and Z-DNA. A quantum-chemical study. *Biophys. J.* **73**, 76-87.

29. Hobza, P. & Šponer, J. (2002). Toward true DNA base-stacking energies: MP2, CCSD(T), and complete basis set calculations. *J. Am. Chem. Soc.* **124**, 11802-11808.

30. Cornell, W. D., Cieplak, P., Bayly, C. I., Gould, I. R., Merz, K. M., Jr., Ferguson, D. M., Spellmeyer, D. C., Fox, T., Caldwell, J. W. & Kollman, P. A. (1995). A Second Generation Force Field for the Simulation of Proteins, Nucleic Acids, and Organic Molecules. *J. Am. Chem. Soc.* **117**, 5179-5197.

31. Beveridge, D. L., Barreiro, G., Byun, K. S., Case, D. A., T.E. Cheatham, I., Dixit, S. B., Giudice, E., Lankaš, F., Lavery, R., Maddocks, J. H., Osman, R., Seibert, E., Sklenar, H.,

Stoll, G., Thayer, K. M., Varnai, P. & Young, M. A. (2004). Molecular Dynamics Simulations of the 136 Unique Tetranucleotide Sequences of DNA Oligonucleotides. I. Research Design and Results on d(CpG) Steps. *Biophys. J.* **87**, 3799-3813.

32. Dixit, S. B., Beveridge, D. L., Case, D. A., T.E. Cheatham, I., Giudice, E., Lankaš, F., Lavery, R., Maddocks, J. H., Osman, R., Sklenar, H., Thayer, K. M. & Varnai, P. (2005). Molecular Dynamics Simulations of the 136 Unique Tetranucleotide Sequences of DNA Oligonucleotides. II: Sequence Context Effects on the Dynamical Structures of the 10 Unique Dinucleotide Steps. *Biophys. J.* **89**, 3721-3740.

Chapter 23

SIMULATION OF EQUILIBRIUM AND DYNAMIC PROPERTIES OF LARGE DNA MOLECULES

Alexander Vologodskii

Department of Chemistry, New York University, 31 Washington Place, New York, NY 10003

Abstract: The review considers computer simulation of DNA large-scale equilibrium and dynamic properties. Much attention is paid to DNA models used in the simulations and the choice of the model parameters. The Monte Carlo procedure for simulation of DNA equilibrium properties and Brownian dynamic simulation of DNA movement in solution are described for both linear and circular DNA molecules. Accuracy of the simulations is addressed by comparing measured and computed properties of the molecules.

Key words: DNA conformation, Monte Carlo, Brownian dynamics, wormlike chain, DNA supercoiling, DNA topology.

1.   INTRODUCTION

The typical DNA molecule, extracted from a cell, is very large. Although in the scale of a few base pairs the molecule is very rigid, small thermal fluctuations of the angles between adjacent base pairs result in a great variety of DNA conformations for molecules whose length exceeds a few hundreds base pairs (Fig. 1). Thus, conformational properties of such molecules have to be described in probability terms. We can talk, for example, about probability that two selected sites of a molecule will be juxtaposed or about the probability that the distance between the ends of the molecule has a particular value. We can also talk about average values of a particular property over the equilibrium ensemble of the molecule conformations, like the average value of the end-to-end distance. Of course,

J. Šponer and F. Lankaš (eds.), Computational Studies of RNA and DNA, 579–604.
© 2006 *Springer.*

the time required for the molecule to reach a conformation with a particular property is also a random variable with a specific distribution. Computer simulation of this kind of equilibrium and dynamic properties of the large DNA molecules is the subject of the current chapter. We will also present convincing evidence that the simulations have the capability to reliably predict these conformational properties and, therefore, can serve as an instrument in the studies of different biologically important problems involving DNA molecules.

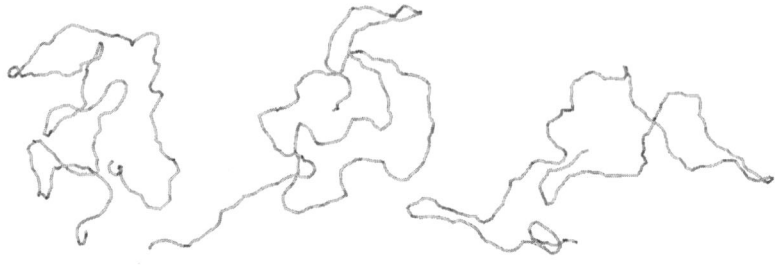

Figure 23-1. The projections of typical conformations of double-stranded DNA molecules 7,000 base pairs in length.

A large DNA molecule contains millions of atoms. Therefore, if we want to describe large-scale properties of such molecules, it is more practical to talk about conformation of the double helix axis rather than specify coordinates of all the atoms. For this level of description we can model DNA as a chain of straight segments with certain physical properties. Such phenomenological models of the double helix have only a few parameters which can be unambiguously found from experimental data. By using these simple models one can perform accurate estimations for nearly all of the statistical properties. It is interesting to consider the relationship between these models and all-atom models which are widely used for simulations of short oligonucleotides and proteins. We will analyze the issue at the beginning of the next section.

2. MONTE CARLO SIMULATION OF DNA EQUILIBRIUM PROPERTIES

2.1 From all-atom model to models of polymer statistical physics

Let us consider the all-atom model of a short DNA duplex. We assume that the duplex does not have an intrinsic curvature, so the minimum energy conformation of the duplex axis is straight (Although this assumption is not important, it simplifies the analysis). The relative coordinates of the duplex atoms are subject to thermal fluctuations, and as a result the two vectors specifying directions of the duplex ends, V_1 and V_2, are not parallel to each other in general (Fig. 2). Let us assume that V_1 is directed along the z axis of the coordinate system. We want to consider the projection of the duplex axis on the (x,z)-plane and analyze the angle θ between the projections of V_1 and V_2. We can also specify the vectors which are perpendicular to each base pair, v_i, and consider the angles between their projections, ϑ_i. The value of θ is the sum of the corresponding angles ϑ_i over adjacent base pairs of the duplex.

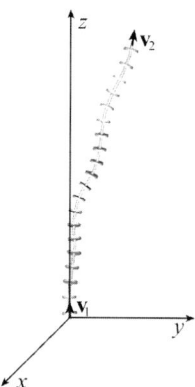

Figure 23-2. Diagram of a typical conformation of 20 bp DNA duplex. The angle θ, analyzed in the text, is formed by the projections of v_1 and v_2 on (x,z)-plane.

We do not know the distribution for ϑ_i, but we do know that the fluctuations of these angles around zero value are independent one from another (or nearly independent). Therefore, we can apply the Central Limit Theorem of the probability theory[1] to make an important conclusion on the distribution of θ, $P(\theta)$. According to the theorem, in a very good approximation, $P(\theta)$ is a Gaussian distribution:

$$P(\theta) = \sqrt{\frac{g}{2\pi}} \exp(-g\theta^2 / 2), \tag{1}$$

where g is a constant. On the other hand, this distribution of θ can be obtained in a very simple model. We can replace each half of the duplex by a straight segment assuming that the energy, E_b, of the resulting chain of two segments is specified by equation

$$E_b = \frac{g}{2} k_B T \theta^2, \tag{2}$$

where $k_B T$ is the Boltzmann temperature factor. Since the angle distribution has to be the Boltzmann distribution, where the probability of state i with energy E_i is proportional to $\exp(-E_i / k_B T)$, we will obtain the same $P(\theta)$ both for the all-atom model and for the model of two jointed straight segments. Thus, if we are interested only in the distribution $P(\theta)$, we can replace the all-atom model by the chain of two segments with harmonic potential. The benefits from this replacement are really enormous. First of all, we greatly simplified the model making it much more practical for any computations. Second, instead of hundreds of parameters specifying different types of atom-atom interactions, we have only one parameter, g. In principal, this parameter can be obtained in the simulations performed for the all-atom model of a short DNA duplex (which also should include water molecules and small ions). To do so, we need to calculate the distribution $P(\theta)$, so that we could obtain g by solving Eq. (1). The fluctuations of θ for short DNA fragments occur in a nanosecond time scale[2] which is accessible in today molecule dynamics simulations. Clearly, the calculation of g is a great challenge for such simulations and could serve as an important test of the simulation accuracy, since the value of g has been determined with great precision (see below). Two attempts of calculating g by the all-atom molecular dynamic simulation were made in the recent years.[3,4]

Modeling polymers as chains of straight segments has been used in polymer physics for more than seventy years. Such models are usually used to obtain general properties of polymer chains. In the case of DNA, however, due to its regularity and large bending rigidity, the models of this type are able to predict conformational properties of the double helix with great accuracy. Let us now consider the model, outlined above, in detail.[5]

A DNA molecule of N base pairs in length is modeled as a chain consisting of m rigid segments that are cylinders of equal length l (Fig. 3).

Figure 23-3. DNA model that is used to simulate large-scale conformational properties of the double helix. The model chain consists of the straight cylinders. The diagram illustrates a correspondence between the model cylinders (shown by black lines) and DNA molecule (shown by gray).

The value of l does not affect the simulation results if it is sufficiently small. For the majority of DNA conformational properties this condition means that one straight segment should correspond to less than 30 bp. The energy of the chain, E_b, is specified by the angles between the directions of adjacent segments i and $i+1$, θ_i (compare with Eq. 2):

$$E_b = \frac{1}{2} g k_B T \sum_{i=1}^{m} \theta_i^2 . \qquad (3)$$

The cylinders have a certain diameter, d, and are impenetrable one for another. The value of d accounts both for the geometrical diameter of the double helix and for electrostatic repulsion between negatively charged DNA segments. Thus, the value of d is larger than the geometric diameter. The concept of effective diameter is the simplest but sufficiently accurate way to account for the electrostatic interaction.[6,7] In this model g and d are the only two parameters.

Using this model and Monte Carlo (MC) methods we can construct a large set of chain conformations. The procedure for this conformational sampling must satisfy one condition: the probability that conformation i will appear in the set is proportional to $\exp(-E_i / k_B T)$. The Metropolis procedure described below is the most universal way of such statistical sampling. once the conformational set is constructed, we can perform its statistical analysis to estimate conformational properties of interest.

The model described above does not account for torsional orientations of DNA base pairs. These orientations do not affect conformational properties of linear or nicked circular DNA molecules, if the DNA is intrinsically straight. In some cases, however, one has to account for base pair torsional orientations explicitly. The corresponding model, described in Section 3.2, allows accounting for any bending and torsional orientation of base pairs in the minimum energy DNA conformation (sequence-dependent intrinsic curvature of the double helix). It also allows introduction of a sequence-dependent DNA bending rigidity and even local anisotropy of the bending rigidity. Of course, the sampling requires much more extensive computations in these cases. Thus, if sequence-dependent effects on DNA conformational properties is not a goal in a particular investigation, it is reasonable to use the simplest model with the minimal number of parameters representing the average values over DNA sequence.

2.2   Choice of parameters

The DNA bending rigidity, g, is the most important parameter of the model. Its value is directly related with DNA persistence length, a:

$$g = a / l \tag{4}$$

Eq. (4) assumes that $l \ll a$; in general case the relation between l and a is described in ref. (8). Since the concept of the persistence length is so important for the subject of this chapter, we repeat here its definition. Let us consider the angle, θ, between two chain segments separated by the contour length x. The cosine of this angle averaged over the equilibrium ensemble of the chain conformations, $\langle \cos \theta \rangle$, specifies the correlation between the chain local directions at two chosen points (Here the brackets $\langle \ \rangle$ mean that the averaging should be taken over the Boltzmann distribution of conformations). It is clear that $\langle \cos \theta \rangle$ decreases when x increases. It can be shown that for our model of a polymer chain, which is called *the discrete wormlike chain*, $\langle \cos \theta \rangle$ decreases exponentially (strictly speaking this is true only for an infinitely thin chain). The value of a specifies the rate of $\langle \cos \theta \rangle$ decrease:

$$\langle \cos \theta \rangle = \exp(-x / a). \tag{5}$$

So, the persistence length is a measure of the length over which the chain keeps its direction. One can find more about persistence length in ref. (9), for example. The value of a is an internal property of a polymer chain. It can be determined by comparison of measured and calculated values of chain properties. Of course, for such a determination it is better to use a property which strongly depends on a and does not depend on other parameters of the model chain (d in our case). For the double-stranded DNA the value of a is close to 50 nm or 150 bp.[10] With good accuracy the value of a does not depend on the salt concentration if $[\text{Na}^+] > 0.01$ M or $[\text{Mg}^{2+}] > 0.001$ M.[10]

The quantitative definition of the DNA effective diameter, d, is based on the concept of the second virial coefficient.[6] Since the electrostatic interaction is screened by small contraions, the value of d strongly depends on the type and concentration of salts in the DNA solution. There are conformational properties which depend strongly on *d*. the comparison of corresponding measured and computed characteristics allowed accurate determination of *d* for different ionic condition.[11-13] For solutions of sodium ions these empirical determinations of *d* are in a very good agreement with theoretical calculations based on the polyelectrolyte theory.[6] The dependence of *d* on [NaCl] and [MgCl$_2$] is shown on Fig. 4.

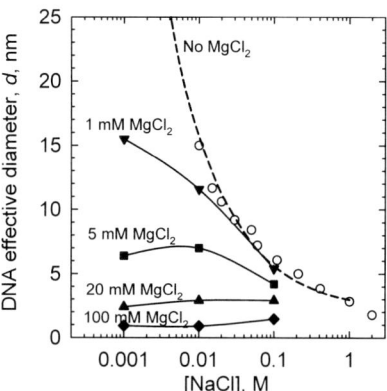

Figure 23-4. The DNA effective diameter, *d*, as a function of NaCl and MgCl$_2$ concentrations. The filled symbols correspond to mixtures of sodium and magnesium ions,[13] and the opened circles correspond to pure NaCl solutions.[12] The theoretical dependence for pure NaCl solutions is shown by the dashed line.[6]

In some cases we also need to account for the torsional rigidity of the double helix, *C*. Experimental data addressing the value of this parameter give rather broad range of values, from $2 \cdot 10^{-19}$ erg·cm till $4 \cdot 10^{-19}$ erg·cm.[5,14-19] Probably, taking the value in the middle of this range is a reasonable compromise at the moment.

2.3   Simulation procedure

The Metropolis-Monte Carlo procedure[20] is usually used for the statistical sampling of chain conformations. The procedure consists of consecutive displacements of the chain parts. At each step of the procedure a new trial conformation can be accepted or rejected. If the trial conformation is rejected, the current conformation must be added again to the constructed conformational set. The starting conformation is chosen arbitrarily.

At least two types of the displacements should be used in the simulations of linear molecules. In the first type, a subchain is rotated by a randomly chosen angle, ϕ^1, around the straight line connecting two randomly chosen vertices of the chain (Fig. 5A). In the second type of displacement a subchain includes one end of the chain, so the end-to-end distance of the chain can be changed. This randomly chosen subchain is rotated by a random angle, ϕ^2, around a randomly oriented axis passing through the internal end of the subchain (Fig. 5B). The values of ϕ^i are usually uniformly distributed over intervals $(-\phi_0^i, \phi_0^i)$, chosen so that about half of the trial conformations of each type are accepted. In general, one can introduced any type of displacement to increase the rate of sampling as long as it does not interrupt the principle of detailed balance.[21]

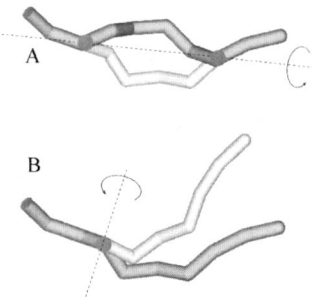

Figure 23-5. Trial motions of the model chain during MC simulations. Two current conformations of the chain are more shaded than the trial conformation. (A) In the crankshaft move a subchain is rotated by a random angle about an axis connecting two randomly chosen vertices. (B) In the second type of move a randomly chosen subchain includes one end of the chain. The subchain is rotated by a random angle around a randomly oriented axis passing through the internal end of the subchain.

Whether the new conformation is accepted is determined by applying the rules of the procedure:

a)   If the energy of the new conformation, E_{new}, is lower than the energy of the previous conformation, E_{old}, the new conformation is accepted;

b)   If the energy of the new conformation is greater than the energy of the previous conformation, then the new conformation is accepted with the probability $p = \exp[(E_{old} - E_{new}) / k_B T]$.

Forbidden conformations, which may appear during the displacements, such as conformations with intersecting segments, should be considered as having infinite energy. According to the acceptance rules such conformations are rejected.

Of course, there is a strong correlation between successive conformations constructed by the Metropolis procedure. Clearly, the constructed conformational set should be sufficiently large to overcome this correlation. Fortunately, modern computers allow constructing the sets of up to a few billions conformations and obtaining very accurate estimations of statistical properties.

2.4   Simulation of circular DNA molecules

If both strands of a DNA molecule are closed, the complementary strands form a link with a certain value of *the linking number, Lk* (see ref. (22), for example). *Lk* is a topological property whose value must be retained through all conformational changes. The value of *Lk* can be expressed as a sum of two other conformational characteristics, writhe, *Wr*, and twist, *Tw*:[23]

$$Lk = Wr + Tw. \tag{6}$$

The value of *Wr* depends only on the conformation of the double helix axis, and *Tw* is specified by the helix rotation around its local axis. Eq. (6) means that bending deformation of the DNA axis and its torsional deformation are connected to each other in closed circular DNA. We can specify the values of twist for each DNA segment explicitly and register their changes during the simulation. In this case Eq. (6) will be satisfied automatically. Although in some cases such a straightforward approach is unavoidable, when considering intrinsically straight DNA molecules we can account for the torsional deformation implicitly by using Eq. (6) to express *Tw* as $Lk - Wr$.[24,25] *Wr* is a function of closed curve geometry and can be calculated for any particular conformation of the model chain, since the geometry of the chain axis is always specified. One way to do it is to calculate the double integral:

$$Wr = \frac{1}{4\pi} \oint_C \oint_C \frac{(d\mathbf{r}_1 \times d\mathbf{r}_2) \cdot \mathbf{r}_{12}}{r_{12}^3} \tag{7}$$

where $\mathbf{r}_1$ and $\mathbf{r}_2$ are vectors, starting at an arbitrary chosen origin, whose ends run, upon integration, over the chain contour C, $\mathbf{r}_{12} = \mathbf{r}_2 - \mathbf{r}_1$. Different methods of *Wr* computation are analyzed in ref. (26). If we know the value of *Tw*, we can calculate the energy of torsional deformation E_t, associated with a particular conformation of DNA axis. The value of E_t should depend on the deviation of DNA *Tw* from its value in torsionally unstressed DNA, Tw_0, or Lk_0. The value of Lk_0 equals N/γ, where γ is the average number of base pairs per the double helix turn in the relaxed form of the considered

DNA. The difference $Lk - Lk_0 = \Delta Lk$ is called *the linking number difference*. It specifies the level of supercoiling in closed circular DNA. Subtracting Tw_0 from both sides of Eq. (6) we obtain:

$$\Delta Lk = Wr + Tw - Tw_0. \tag{8}$$

Thus, in calculating the energy of a particular conformation of closed circular DNA we have to account for the bending energy, E_b, calculated according to Eq. (3), and the torsional energy, E_t:

$$E_t = (2\pi C / L)(Tw - Tw_0)^2 = (2\pi C / L)(\Delta Lk - Wr)^2, \tag{9}$$

where L is the DNA length (in cm). The value of ΔLk should be considered as an initial parameter of the simulation and Wr is the writhe of the conformation.

Lk of the complementary strands is not the only topological characteristic of circular DNA. All circular DNA molecules, both nicked and closed, have topological characteristics specified by the topology of the DNA axis. An isolated circular molecule can have an unknotted conformation or form a knot of a particular type. Two or more circular DNAs can form different links (catenanes) in solution or can be unlinked. Since the Metropolis procedure used for conformational sampling does not exclude strand-passing during the chain deformation, we have to test the chain topological state if we are performing the simulation for a particular state. In such a simulation a trial conformation is rejected if its topology is different from that of the current conformation. If we want to calculate the equilibrium distribution of topological states, we construct the conformational ensemble regardless of topology, but have to determine topology of each conformation during the analysis of the constructed set.

To determine the topology of a particular conformation one has to calculate a topological invariant, a characteristic which depends on topology only. There are many topological invariants developed by mathematicians (see ref. (27), for example). The most convenient and computationally efficient topological invariant is the Alexander polynomial. The Alexander polynomials for the simplest knots and links are shown in Fig. 6. Algorithms have been developed to calculate this topological invariant.[28-30]

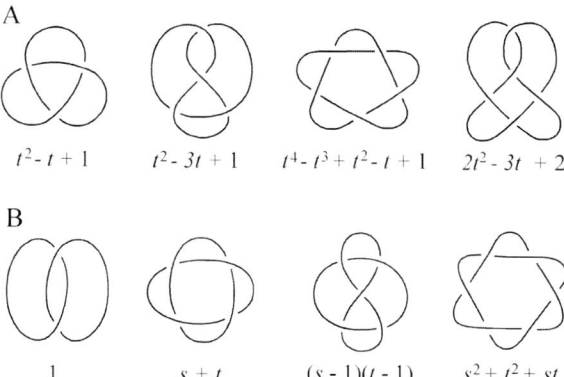

A

$t^2 - t + 1$ $t^2 - 3t + 1$ $t^4 - t^3 + t^2 - t + 1$ $2t^2 - 3t + 2$

B

1     $s + t$ $(s - 1)(t - 1)$ $s^2 + t^2 + st$

Figure 23-6. The simplest knots (A), links (B), and their Alexander polynomials. For unknotted contour the polynomial equals 1, and for a pair of unlinked contours it equals 0.

It should be noted that although the probability of knot appearance is relatively small for the molecules of a few kb in length,[12,31,32] the simulation procedure for supercoiled DNA generates many knots if the chain topology is not checked during each simulation step.[33]

2.5 Comparison between simulated and measured properties

How well can this simple model describe properties of actual DNA molecules? The only way to evaluate this is to compare computed and experimentally measured DNA properties. Over the last two decades many such comparisons were performed and showed very good agreement between the computational and experimental results. Simulations reproduce the experimental data on hydrodynamic properties of DNA molecules,[34-36], DNA cyclization,[37-41] equilibrium distributions of topological states,[5,12,31,32,42,43] elasticity of single DNA molecules,[44,45] and light and neutron scattering data on supercoiled DNA.[46-49] A detailed description of one comparison follows below.

In the mid 1990s it was known that the model works well for equilibrium conformations of linear and nicked circular DNA. However, it was not clear until that time how well the model describes the conformational properties of supercoiled DNA, a form that is characterized by frequent close approaches between distal segments of the molecule. Because these close approaches are rare in linear and open circular DNA molecules, the accurate description of intersegment electrostatic interaction is not so crucial for their properties. In supercoiled molecules, however, the electrostatic interaction between DNA segments is very important. Thus, it was critical to test how well the model can describe conformational properties of supercoiled molecules. This goal was achieved by comparing the computed and measured equilibrium linkage between supercoiled DNA and cyclizing linear molecules as illustrated in Fig. 7.[43]

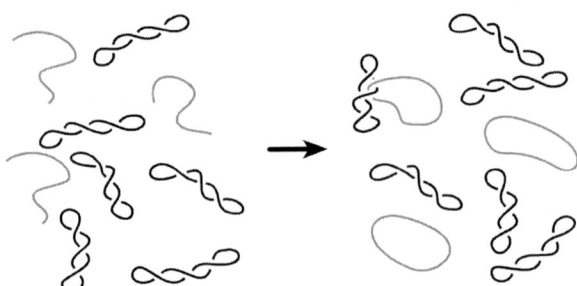

Figure 23-7. Probing conformational properties of supercoiled DNA by formation of topological links.[43] The diagram shows formation of links between supercoiled and cyclizing linear molecules. The cyclization occurred via long cohesive ends and resulted in nicked circular molecules. The simplest links shown here comprised at least 90 % of all links formed.

The probability P that a given open circular DNA will be linked with supercoiled molecules of concentration c can be expressed as

$$P = \int_0^\infty p(R)4\pi R^2 c \frac{N_A}{M} dR,$$ (10)

where $p(R)$ is the probability of linking of these two molecules if their centers of mass are separated by distance R, N_A is Avogadro's number, and M is the molecular weight of DNA. The equation has a simple interpretation because the term $4\pi R^2 dR \cdot cN_A/M$ is the probability of finding a supercoiled molecule in the volume element $4\pi R^2 dR$. Eq. (10) assumes that the concentration c is small enough so that we can ignore the formation of three or more linked molecules. We used Eq. (10) to calculate the values of P. This equation reduces the problem to calculating the equilibrium probability, $p(R)$, of the linking of two circular chains whose centers of masses are separated by the distance R. To obtain $p(R)$, we, first of all, constructed the equilibrium sets of conformations of closed chains. One set corresponded to supercoiled DNA 7 kb in length; the other set corresponded to open circles 10 kb in length. Then we placed all possible pairs of chain conformations from the two sets at a distance R between their centers of mass. During this placement, the segments of one chain could freely cross the segments of the other chain. Any conformation in which the chain segments overlapped was considered as forbidden. If the conformation had no overlapped segments, we determined the topological state of the chains by calculating the value of the Alexander polynomial for links, $\Delta(s,t)$. For each pair of chains conformations, we tested their topological states for 144 mutual orientations obtained by all possible 90° rotations of each molecule around the X, Y, and Z directions. $p(R)$ was calculated as the number of linked conformations divided by the total number of tested conformations of the two chains including both allowed and forbidden conformations. Fig. 8 shows $p(R)$ computed for two circular DNA molecules.

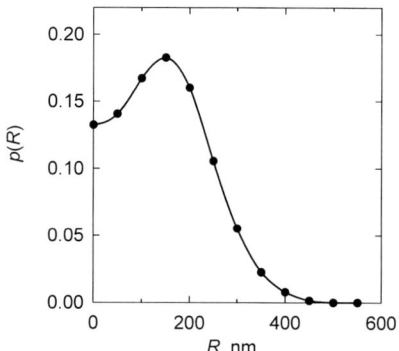

Figure 23-8. Probability of linkage between two relaxed DNA molecules 7,000 and 10,000 base pairs in length as a function of the distance between the chains centers of mass, R. The function strongly depends on solution ionic conditions. $p(R)$ shown in the figure corresponds to near physiological ionic conditions. For each value of R a fraction of conformations of two chains is forbidden because of overlap of one chain with the other. Since this fraction increases sharply as R diminishes, there is a decrease of $p(R)$ at small values of R.

For comparison with experimental data, it is convenient to introduce the constant B:

$$B = \frac{N_A}{M} \int_0^\infty p(R)4\pi R^2 dR .$$ (11)

This allows us to express P as

$$P = Bc .$$ (12)

The value of B does not depend on DNA concentration and reflects only the properties of particular circular DNAs and solution conditions. It depends on the lengths of both DNAs, the superhelix density, and on the ionic conditions, which change the conformational properties of supercoiled DNA and the probability of linking. For each particular conditions B was computed by numerical integration of $p(R)$.

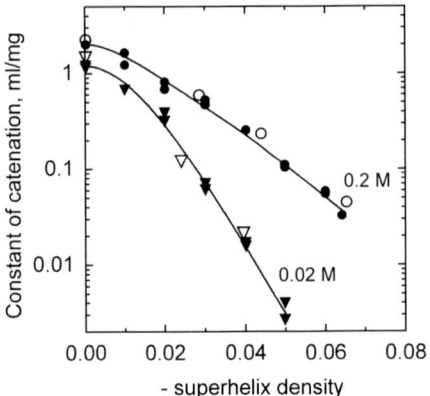

Figure 23-9. Measured and simulated probabilities of catenation as a function of supercoiling.[43] The experimental values of B (open symbols) are shown together with calculated results (filled symbols) for NaCl concentrations of 0.02 M ($\circ$, $\bullet$), 0.2 M ($\triangledown$, $\blacktriangledown$). The large changes of conformational properties of supercoiled DNA with ionic conditions result, in good approximation, from the change of intersegment electrostatic interactions, specified by the variation of DNA effective diameter.

Fig. 9 presents the measured and simulated values of B for solutions of two different concentrations of NaCl (0.02 M, 0.2 M) as a function of DNA superhelix density, $\Delta Lk/Lk_0$. Note that the simulated and measured values of B agree extremely well over the whole range of $\Delta Lk/Lk_0$ and NaCl concentrations studied, even though the range of B values exceeds two orders of magnitude. The data make it clear that conformations of supercoiled DNA vary greatly over this range of sodium ion concentrations. This work convincingly proved that the simulation predicts conformational properties of both relaxed and supercoiled DNA molecules with very good accuracy.

2.6 Special approaches addressing properties of rare conformations

Although the modern computers allow one to simulate up to 10^{10} conformations for a model chain corresponding to a DNA molecule of a few kb in length, it may be insufficient to estimate conformational properties or the probability of appearance of some rare conformations. In such cases different kinds of biased sampling can be used.

The method of "umbrella sampling" addresses calculation of conformational distributions under a specific conformational conditions.[50,51] For example, we may want to construct a conformational set under condition that chosen sites of the chain are juxtaposed in a proper orientation. The sampling is based in this case on introducing an artificial potential, $U(\mathbf{x})$, where $\mathbf{x}$ refers to the coordinates that define the mutual geometry of the specific sites. Although $U(\mathbf{x})$ can greatly increase the probability of the site juxtaposition, it does not disturb the conditional distribution since $U(\mathbf{x})$ has the same value for all conformations where the sites are juxtaposed ($\mathbf{x} = \mathbf{x}_0$). Indeed, the statistical weights of all conformations with the juxtaposed sites will be multiplied by the same factor, $\exp(-U(\mathbf{x}_0)/kT)$. The efficiency of this approach depends on proper choice of $U(\mathbf{x})$. The method was recently applied for the simulation of the juxtaposition of specific sites in supercoiled DNA.[52] The method can be also used to estimate probabilities of appearance of very rare conformations.[51]

Another approach was developed specifically for calculation of the cyclization probability of very short DNA fragments, when this probability is very small and direct MC sampling is inefficient.[53,54] Suppose we want to estimate $P(r_0)$, the probability of the conformations with end-to-end distance, r, less than a small value r_0. We choose a sequence of distances $r_0 < r_1 < ... < r_n$, where r_n is larger or equal to the chain contour length. Let $P(r_i)$ be the probability of conformations with $r < r_i$. We can also define the conditional probabilities, $P(r_i | r_{i+1})$, of conformations with $r < r_i$ in the subset of conformations with $r < r_{i+1}$. Since $P(r_i) = P(r_i | r_{i+1})P(r_{i+1})$ and $P(r_n) = 1$, the value of $P(r_0)$ can be calculated as

$$P(r_0) = \prod_{i=0}^{n-1} P(r_i | r_{i+1}) \qquad (13)$$

The sequence of distances $r_0 < r_1 < ... < r_n$ can be chosen so that all $P(r_i | r_{i+1})$ values are relatively large. This can always be achieved since $P(r_i | r_{i+1})$ approaches 1 when r_{i+1} approaches r_i. The large values of $P(r_i | r_{i+1})$ can be efficiently and accurately calculated by the Metropolis procedure. Each $P(r_i | r_{i+1})$ is calculated as the fraction of the conformations with $r < r_i$ in the subset of equilibrium conformations with $r < r_{i+1}$. These subsets are generated in the MC procedure by rejecting any trial conformation with $r > r_{i+1}$. The values $P(r_i | r_{i+1})$ are calculated sequentially from $P(r_0 | r_1)$ to $P(r_{n-1} | r_n)$). The starting conformation for each subset is the last conformation from the previous subset. The calculation of $P(r_0 | r_1)$ is started from a conformation with $r = 0$. The estimation shows that the best efficiency in estimating $P(r_0)$ is achieved when the values of $P(r_i | r_{i+1})$ are close to 0.2. Using this approach, one can speed up the computations by a few orders of magnitude as compared to the direct MC procedure. The generalized version of this procedure allows introducing different bending and torsional rigidity for each chain segment.[53] The value of $P(r_0)$ can be converted into experimentally measured value of the *j*-factor.[53,55]

3.   SIMULATION OF DNA DYNAMIC PROPERTIES

Although the Metropolis procedure described in the first part of the chapter resembles a dynamic simulation, it is a very artificial dynamics. The method was designed only to sample the equilibrium conformational ensemble. There is a method, however, to simulate the real dynamics of DNA molecules in this scale. This method, the Brownian dynamics (BD), is based on the Langevin description of molecular dynamics.[56] The traditional molecular dynamics simulation of large molecules in solution accounts for thermal motion by placing solvent molecules explicitly in the simulated system. Instead, the Langevin equations include the randomly fluctuating force exerted on the molecule subunits by the surrounding solvent. The equations of the BD are essentially simpler for numerical integration than the Langevin equations since the inertia term is skipped there and therefore they are the first order differential equations.[56] The fact that the inertia term can be skipped in our scale of molecular description is very important and deserves a separate analysis.

3.1   Movement without inertia

Let us consider movement of a bead in a viscous medium. The movement is described by the Newton equation

$$m\frac{dv}{dt} = F, \tag{14}$$

where m and v are the bead mass and velocity and F is the force acting on the bead. Let us assume that at moment $t = 0$ the velocity of the bead is v_o and the frictional force, $F = -fv$, where f is the frictional coefficient, is the only one acting on the bead. Under these conditions the equation solution is

$$v = v_o \exp(-t/\tau), \tag{15}$$

where $\tau = m/f$. Thus, the velocity exponentially diminishes with time. The value of f is specified by the Stokes formula:

$$f = 6\pi\eta r, \tag{16}$$

where r is the bead radius and η is the viscosity of the medium. If we also express m through the bead radius and density, ρ, we obtain that

$$\tau = \frac{2}{9}r^2\rho/\eta. \tag{17}$$

Eq. (17) shows that τ is proportional to r^2, so it changes dramatically when we go from the scale of our macroscopic world to the molecular scale. For $r = 1$ nm, $\eta = 0.01$ poise (the viscosity of water) and $\rho = 2$ g/cm, τ equals $4.4 \cdot 10^{-13}$ s $\cong 0.5$ ps. For this time interval the bead moves out of its position, on average, by ≈ 0.02 nm only, assuming that v_o corresponds to the average velocity of the bead at room temperature. This is a short distance even for the molecular scale. Thus, in a good approximation, we can say that the bead "forgets" nearly immediately about direction of its movement in water, as if it simply does not have any inertia.

Solving dynamic equations numerically one has to choose the integration time step, Δt. The value of Δt should be small enough so that the forces acting on the molecular subunits do not change much by the subunit displacements over time interval Δt. We can design a model for large-scale dynamic simulation which allows Δt to be set to 500 ps. In this case numerical solutions of the dynamic equations will not be affected by the subunits inertia since the momentum relaxation would occur for about 1/1000 of the time step. In traditional molecular dynamics, however, the forces acting between atoms change strongly over very short atom displacements, so one has to use Δt of $0.001 \div 0.1$ ps and the inertia term has to be taken into account.

A simple model, a large time step, and the possibility of skipping the inertia term allow extending BD simulations for time intervals up to 100 ms. These time intervals are sufficient to estimate characteristic rates for the majority of conformational rearrangements in DNA molecules of a few kb in length.

3.2   DNA model for the Brownian dynamics simulation

The model for the large-scale dynamic simulations is similar to the model used to simulate equilibrium properties of the double helix, although it has some specific features. It was originally developed by Allison et al.[2,57-59] and later extended for the case of supercoiled DNA.[60-64] The major difference between the two models is that infinitely rigid stretching and electrostatic potentials in the model for equilibrium simulation are replaced now by the corresponding smooth potentials, since smaller derivatives of intersubunit forces allow for a larger value of the integration timestep. The dynamic model also specifies hydrodynamic parameters of the subunits.

A DNA molecule of N base pairs in length is modeled as a chain consisting of m segments of equilibrium length l_0. The segments have no thickness. To specify the torsional deformation of the chain, the body-fixed coordinate (bfc) frame, defined by the unit vectors $\mathbf{f}_i, \mathbf{v}_i, \mathbf{u}_i$, is attached to each vertex of the chain so that $\mathbf{u}_i$ is directed along the segment i. The chain energy consists of the following five terms.

1. The bending energy, E_b, which is specified by Eq. (3).
2. The stretching energy is computed as

$$E_s = \frac{h}{2}\sum_{i=1}^{m}(l_i - l_0)^2 , \tag{18}$$

where l_i is the actual length of segment i, and h is the stretching rigidity constant. The stretching energy E_s is a computational device rather than an attempt to account for the actual stretching elasticity of the double helix. Smaller values of h allow larger timesteps in the BD simulations, but also imply larger departures from l_0. It is safe to choose $h = 100 k_B T / l_0^2$, so that the variance of l_i is close to $l_0^2 / 100$. It was shown that a further increase of h does not change, within the simulation accuracy, considered dynamic properties of the model chain.[65]

3. The energy of electrostatic intersegment interaction, E_e, is specified by the Debye-Hückel potential as a sum over all pairs of point charges located on the chain segments. The number of point charges placed on each segment, λ, is chosen to closely approximate continuous charges with the same linear density. The value of λ increases as the Debye length, $1/\kappa$, decreases.[62,64] The energy E_e is specified as

$$E_e = \frac{v^2 l_o^2}{\lambda^2 D}\sum_{i<j}^{N}\frac{\exp(-\kappa r_{ij})}{r_{ij}} , \tag{19}$$

where v is the effective linear charge density of the double helix, D is the dielectric constant of water, $N = kn\lambda$ is the total number of point charges, and r_{ij} is the distance between point charges i and j. The value of v corresponds here to the solution of the Poisson-Boltzmann equation for DNA modeled as a charged cylinder. This solution can be approximated well by the Debye-Hückel potential with an effective value of v chosen to match the electrostatic potential in the overlap region far from the cylindrical surface.[6]

4. The energy of torsional deformations is specified explicitly in the model.[60] Namely, Euler angles $\alpha_{i,i+1}, \beta_{i,i+1}, \gamma_{i,i+1}$ are used to describe the transformation from the bfc frame i to $i + 1$. In this representation, $\beta_{i,i+1}$ coincides with the bending angle, θ_i of Eq. (3). The torsional angle between frames i and $i + 1$ is defined as $\phi_{i,i+1} = \alpha_{i,i+1} + \gamma_{i,i+1}$. Thus the torsional energy is expressed as

$$E_t = \frac{C}{2l_0} \sum_{i=1}^{m} (\alpha_{i,i+1} + \gamma_{i,i+1} - \phi^0)^2, \tag{20}$$

where C is the DNA torsional rigidity constant, and ϕ^0 is the equilibrium twist of one model segment. The initial values of $\alpha_{i,i+1}$ and $\gamma_{i,i+1}$ can be chosen by many ways but they have to provide a particular value of torsional deformation of the model chain, namely

$$\sum_{i=1}^{n} (\phi_{i,i+1} - \phi^0) = \Delta Tw. \tag{21}$$

Together with the writhe, Wr, of the initial conformation, ΔTw specifies the linking number difference, $\Delta Lk = Wr + \Delta Tw$, for a particular simulation run.

5. The energy of the short-range repulsion between DNA segments, E_v, is added to the energy function to prevent passing one segment through another, since the electrostatic repulsion, specified above, does not exclude such passing. E_v can be introduced as

$$E_v = -\sum_{i,j}^{N} \mu r_{ij} \qquad\qquad \text{if } r_{ij} < 2 \text{ nm}$$

$$E_v = 0 \qquad\qquad \text{if } r_{ij} > 2 \text{ nm,}$$

where summation is performed over the same pairs of points as in Eq. (19). It was found that to eliminate strand passing it is sufficient to set $\mu = 35$ pN.

To account for hydrodynamic interactions of the DNA with solution and between different parts of the model chain we assumed that beads of radius a are located at each vertex of the chain. It is important to emphasize that the only interaction involving the beads is hydrodynamic. Thus, the presence of the beads does not affect equilibrium properties of the model chain. To account for the hydrodynamic interaction one can use the Rotne-Prager diffusion tensor, $\mathbf{D}_{ij}^n$.[66] The value of a is chosen to provide the experimentally measured values of the translational diffusion (sedimentation) coefficients. For $l_0 = 10$ nm the value of a equals 2.24 nm.[36,47,67]

Clearly, this model can be used for MC simulation of DNA equilibrium properties, although the corresponding computations are more time-consuming. On the other hand, all equilibrium properties can be obtained from Brownian dynamic simulations, and we will illustrate below that the properties obtained by the two methods are really identical. Still, estimating DNA equilibrium properties by BD simulation requires about 100 times more computer time than the corresponding MC simulation.

3.3   The simulation procedure

An algorithm for the BD simulation is specified by equations

$$\mathbf{r}_i^{n+1} = \mathbf{r}_i^n + \frac{\Delta t}{k_B T} \sum_j \mathbf{D}_{ij}^n \mathbf{F}_j^n + \mathbf{R}_i^n, \qquad (22)$$

where $\mathbf{r}_i^n$ and $\mathbf{r}_i^{n+1}$ are the current and the next positions of bead i, and $\mathbf{F}_j^n$ is the direct force acting on bead j. The sum over all j here is related with the hydrodynamic interaction between the chain subunits: the movement of each bead of the chain results in a certain movement of the surrounding solution and by such a way affects the displacement of bead i. The last term in Eqs. (22), $\mathbf{R}_i^n$, is a vector of Gaussian random numbers of mean zero and covariance matrix

$$\left\langle \mathbf{R}_i^n \mathbf{R}_j^n \right\rangle = 2\mathbf{D}_{ij}^n \Delta t . \qquad (23)$$

A procedure of generating $\mathbf{R}_i^n$ is described in ref. (56). This term accounts for the displacement of bead i by thermal motion. A large timestep used in the simulation allows the expression of the latter displacement over $\mathbf{D}_{ij}^n$.

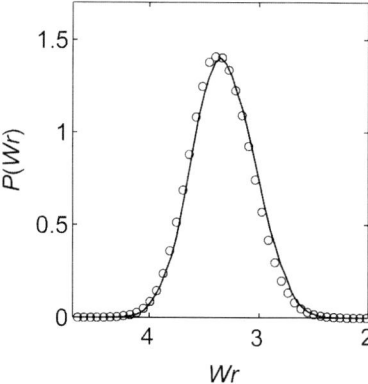

Figure 23-10. Comparison of the equilibrium distributions of *Wr* simulated by BD and MC procedures. The computations were performed for supercoiled DNA of 1500 bp in length with σ = -0.06, [NaCl] = 0.01 M. Both MC results (continuous line) and the BD data (circles) were obtained for the same DNA model.[62]

The results of the BD simulation must show the same DNA conformational properties as the MC procedure. Fig. 10 shows comparison of the simulation results for the distribution of *Wr* in supercoiled DNA molecules obtained by MC and BD methods.

3.4   Comparison Brownian dynamics simulations with experimental data

Until recently, there were no quantitative experimental data on the internal large-scale dynamics of DNA molecules, and therefore it was not possible to test how well the method describes this kind of motion. The fluorescence depolarization data[68,69] represented the only exception, although this process mainly depends on the DNA torsional motion. The translational and rotational diffusion coefficients, for which experimental values were known, are not sensitive to the rate of the internal dynamics, and can be equally well calculated by averaging over the equilibrium conformational ensemble.[34,36] These diffusion coefficients were also used to adjust the value of the hydrodynamic radius of the model beads and, therefore, cannot be used for critical testing of the method. This situation changed when Quake and co-authors performed a study of knot diffusion along stretched DNA.[70] They were able to tie different knots on a single DNA molecule by using optical tweezers and placing DNA into a medium with high viscosity. Stretching the knotted molecules resulted in highly localized knots (Fig. 11).

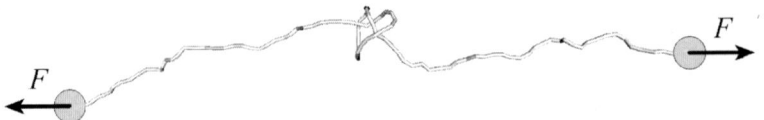

Figure 23-11. Diffusion of a tight knot along stretched DNA molecule. The diagram shows the experimental design of the system. The simulated conformation of DNA molecule 3000 bp in length was used for the illustration. Many times longer DNA molecules were used in the experiments.[70]

Monitoring position of the knots versus time, the researches measured the diffusion coefficients of different knots.[70] Thus, the study provided experimental data needed to test the accuracy of BD simulation of DNA internal motion. We simulated the diffusion of knots along stretched DNA molecules, calculated the knot diffusion coefficients and compared the results with the experimental data.[71] The computed and measured diffusion coefficients for three different knots are shown in Fig. 12.

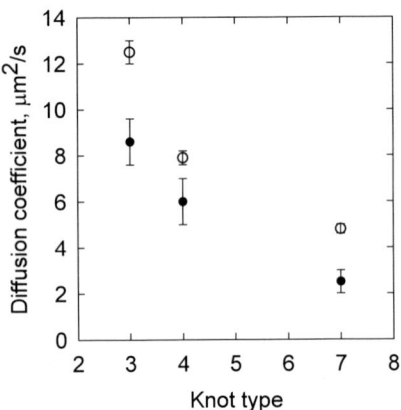

Figure 23-12. Diffusion coefficients of knots along stretched DNA molecules. The experimental data by Bao et al.[70] (○) and the results of the BD simulations[71] (●) were obtained for knots 3_1, 4_1, and 7_1.

Although there is a clear difference between computed and experimental diffusion coefficients for the studied knots, one should consider this as a reasonable agreement. Indeed, characteristic times of different dynamic processes in large DNA molecules vary along a very broad range, from nanoseconds for the torsional relaxation of the double helix[68] to milliseconds for juxtaposition of specific sites.[62,64] Thus, if the BD simulation allows one to estimate the rates of all these different processes within a factor of 2, we have a very valuable computational tool to address many kinetic problems of molecular biology. More experimental data on DNA dynamics are needed, however, for further investigation of the method accuracy.

ACKNOWLEDGEMENT

The work was supported by National Institutes of Health Grant GM 54215.

REFERENCES

1. Feller, W. (1968). *An introduction to probability theory and its applications*, 1. 2 vols, John Wiley & Sons, New York.
2. Allison, S. A. & McCammon, J. A. (1984). Multistep Brownian dynamics: application to short wormlike chains. *Biopolymers* **23**, 363-375.
3. Bruant, N., Flatters, D., Lavery, R. & Genest, D. (1999). From atomic to mesoscopic descriptions of the internal dynamics of DNA. *Biophys. J.* **77**, 2366-2376.
4. Lankas, F., Sponer, J., Hobza, P. & Langowski, J. (2000). Sequence-dependent elastic properties of DNA. *J. Mol. Biol.* **299**, 695-709.
5. Klenin, K. V., Vologodskii, A. V., Anshelevich, V. V., Klisko, V. Y., Dykhne, A. M. & Frank-Kamenetskii, M. D. (1989). Variance of writhe for wormlike DNA rings with excluded volume. *J. Biomol. Struct. Dyn.* **6**, 707-714.
6. Stigter, D. (1977). Interactions of highly charged colloidal cylinders with applications to double-stranded DNA. *Biopolymers* **16**, 1435-1448.
7. Vologodskii, A. V. & Cozzarelli, N. R. (1995). Modeling of long-range electrostatic interactions in DNA. *Biopolymers* **35**, 289-296.
8. Vologodskii, A. V. & Frank-Kamenetskii, M. D. (1992). Modeling supercoiled DNA. *Methods in Enzymology* **211**, 467-80.
9. Cantor, C. R. & Schimmel, P. R. (1980). *Biophysical chemistry*, 3. 3 vols, W. H. Freeman and Company, New York.
10. Hagerman, P. J. (1988). Flexibility of DNA. *Ann. Rev. Biophys. Biophys. Chem.* **17**, 265-286.
11. Brian, A. A., Frisch, H. L. & Lerman, L. S. (1981). Thermodynamics and equilibrium sedimentation analysis of the close approach of DNA molecules and a molecular ordering transition. *Biopolymers* **20**, 1305-1328.
12. Rybenkov, V. V., Cozzarelli, N. R. & Vologodskii, A. V. (1993). Probability of DNA knotting and the effective diameter of the DNA double helix. *Proc. Natl. Acad. Sci. USA* **90**, 5307-5311.
13. Rybenkov, V. V., Vologodskii, A. V. & Cozzarelli, N. R. (1997). The effect of ionic conditions on DNA helical repeat, effective diameter, and free energy of supercoiling. *Nucl. Acids Res.* **25**, 1412-1418.
14. Vologodskii, A. V., Anshelevich, V. V., Lukashin, A. V. & Frank-Kamenetskii, M. D. (1979). Statistical mechanics of supercoils and the torsional stiffness of the DNA. *Nature* **280**, 294-298.

15. Shore, D. & Baldwin, R. L. (1983). Energetics of DNA twisting. II. Topoisomer analysis. *J. Mol. Biol.* **170**, 983-1007.
16. Shore, D. & Baldwin, R. L. (1983). Energetics of DNA twisting. I. Relation between twist and cyclization probability. *J. Mol. Biol.* **170**, 957-981.
17. Horowitz, D. S. & Wang, J. C. (1984). Torsional rigidity of DNA and length dependence of the free energy of DNA supercoiling. *J. Mol. Biol.* **173**, 75-91.
18. Wu, P., Fujimoto, B. S. & Schurr, J. M. (1987). Time-resolved fluorescence polarization anisotropy of short restriction fragments: the friction factor for rotation of DNA about its symmetry axis. *Biopolymers* **26**, 1463-1488.
19. Bryant, Z., Stone, M. D., Gore, J., Smith, S. B., Cozzarelli, N. R. & Bustamante, C. (2003). Structural transitions and elasticity from torque measurements on DNA. *Nature* **424**, 338-41.
20. Metropolis, N., Rosenbluth, A. W., Rosenbluth, M. N., Teller, A. H. & Teller, E. (1953). Equation of state calculations by fast computing machines. *J. Chem. Phys.* **21**, 1087-1092.
21. Binder, K. & Heermann, D. W. (1997). *Monte Carlo simulations in statistical physics*, Springer, Berlin.
22. Vologodskii, A. V. (1999). Circular DNA. In *On-Line Biophysics Textbook* (Bloomfield, V., ed.), Vol. http://biosci.cbs.umn.edu/biophys/OLTB/supramol.html.
23. Fuller, F. B. (1971). The writhing number of a space curve. *Proc. Natl. Acad. Sci. USA* **68**, 815-819.
24. Hao, M.-H. & Olson, W. K. (1989). Global equilibrium configurations of supercoiled DNA. *Macromolecules* **22**, 3292-3303.
25. Vologodskii, A. V., Levene, S. D., Klenin, K. V., Frank-Kamenetskii, M. D. & Cozzarelli, N. R. (1992). Conformational and thermodynamic properties of supercoiled DNA. *J. Mol. Biol.* **227**, 1224-1243.
26. Klenin, K. & Langowski, J. (2000). Computation of writhe in modeling of supercoiled DNA. *Biopolymers* **54**, 307-317.
27. Murasugi, K. (1996). *Knot theory and its applications*, Birkhauser, Boston.
28. Vologodskii, A. V., Lukashin, A. V., Frank-Kamenetskii, M. D. & Anshelevich, V. V. (1974). Problem of knots in statistical mechanics of polymer chains. *Sov. Phys. JETP* **39**, 1059-1063.
29. Vologodskii, A. V., Lukashin, A. V. & Frank-Kamenetskii, M. D. (1975). Topological interaction between polymer chains. *Sov. Phys. JETP* **40**, 932-936.
30. Frank-Kamenetskii, M. D., Lukashin, A. V. & Vologodskii, M. D. (1975). Statistical mechanics and topology of polymer chains. *Nature* **258**, 398-402.
31. Klenin, K. V., Vologodskii, A. V., Anshelevich, V. V., Dykhne, A. M. & Frank-Kamenetskii, M. D. (1988). Effect of excluded volume on topological properties of circular DNA. *J. Biomol. Struct. Dyn.* **5**, 1173-1185.
32. Shaw, S. Y. & Wang, J. C. (1993). Knotting of a DNA chain during ring closure. *Science* **260**, 533-536.
33. Podtelezhnikov, A. A., Cozzarelli, N. R. & Vologodskii, A. V. (1999). Equilibrium Distributions of topological states in circular DNA: interplay of supercoiling and knotting. *Proc. Natl. Acad. Sci. USA* **96**, 12974-12979.
34. Hagerman, P. J. & Zimm, B. H. (1981). Monte Carlo approach to the analysis of the rotational diffusion of wormlike chains. *Biopolymers* **20**, 1481-502.
35. Hagerman, P. J. (1981). Investigation of the flexibility of DNA using transient electric birefringence. *Biopolymers* **20**, 1503-1535.

36. Rybenkov, V. V., Vologoskii, A. V. & Cozzarelli, N. R. (1997). The effect of ionic conditions on the conformations of supercoiled DNA. I. Sedimentation analysis. *J. Mol. Biol.* **267**, 299-311.

37. Levene, S. D. & Crothers, D. M. (1986). Topological distributions and the torsional rigidity of DNA. A Monte Carlo study of DNA circles. *J. Mol. Biol.* **189**, 73-83.

38. Taylor, W. H. & Hagerman, P. J. (1990). Application of the method of phage T4 DNA ligase-catalyzed ring-closure to the study of DNA structure. II. NaCl-dependence of DNA flexibility and helical repeat. *J. Mol. Biol.* **212**, 363-376.

39. Hagerman, P. J. (1990). Sequence-directed curvature of DNA. *Annu. Rev. Biochem.* **59**, 755-81.

40. Vologodskaia, M. & Vologodskii, A. (2002). Contribution of the intrinsic curvature to measured DNA persistence length. *J. Mol. Biol.* **317**, 205-213.

41. Du, Q., Smith, C., Shiffeldrim, N., Vologodskaia, M. & Vologodskii, A. (2005). Cyclization of short DNA fragments and bending fluctuations of the double helix. *Proc. Natl. Acad. Sci. USA* **102**, 5397-5402.

42. Vologodskii, A. V. & Cozzarelli, N. R. (1993). Monte Carlo analysis of the conformation of DNA catenanes. *J. Mol. Biol.* **232**, 1130-1140.

43. Rybenkov, V. V., Vologodskii, A. V. & Cozzarelli, N. R. (1997). The effect of ionic conditions on the conformations of supercoiled DNA. II. Equilibrium catenation. *J. Mol. Biol.* **267**, 312-323.

44. Vologodskii, A. V. (1994). DNA extension under the action of an external force. *Macromolecules* **27**, 5623-5625.

45. Vologodskii, A. V. & Marko, J. F. (1997). Extension of torsionally stressed DNA by external force. *Biophys. J.* **73**, 123-132.

46. Gebe, J. A., Delrow, J. J., Heath, P. J., Fujimoto, B. S., Stewart, D. W. & Schurr, J. M. (1996). Effects of Na+ and Mg2+ on the structures of supercoiled DNAs: comparison of simulations with experiments. *J. Mol. Biol.* **262**, 105-128.

47. Hammermann, M., Stainmaier, C., Merlitz, H., Kapp, U., Waldeck, W., Chirico, G. & Langowski, J. (1997). Salt effects on the structure and internal dynamics of superhelical DNAs studied by light scattering and Brownian dynamic. *Biophys. J* **73**, 2674-2687.

48. Klenin, K., Hammermann, M. & Langowski, J. (2000). Modeling Dynamic Light Scattering of Supercoiled DNA. *Macromolecules* **33**, 1459-1466.

49. Hammermann, M., Brun, N., Klenin, K. V., May, R., Toth, K. & Langowski, J. (1998). Salt-dependent DNA superhelix diameter studied by small angle neutron scattering measurements and Monte Carlo simulations. *Biophysical Journal* **75**, 3057-63.

50. McCammon, J. A. & Harvey, S. C. (1987). *Dynamics of proteins and nucleic acids*, Cambridge University Press, Cambridge, UK.

51. Klenin, K. V., Vologodskii, A. V., Anshelevich, V. V., Dykhne, A. M. & Frank-Kamenetskii, M. D. (1991). Computer simulation of DNA supercoiling. *J. Mol. Biol.* **217**, 413-419.

52. Grainge, I., Pathania, S., Vologodskii, A., Harshey, R. & Jayaram, M. (2002). Symmetric DNA sites are functionally asymmetric within Flp and Cre site-specific DNA recombination synapses. *J. Mol. Biol.* **320**, 515-527.

53. Podtelezhnikov, A. A., Mao, C., Seeman, N. C. & Vologodskii, A. V. (2000). Multimerization-Cyclization of DNA Fragments as a Method of Conformational Analysis. *Biophys. J* **79**, 2692-2704.

54. Podtelezhnikov, A. A. & Vologodskii, A. V. (2000). Dynamics of small loops in DNA molecules. *Macromolecules* **33**, 2767-2771.

55. Crothers, D. M., Drak, J., Kahn, J. D. & Levene, S. D. (1992). DNA bending, flexibility, and helical repeat by cyclization kinetics. *Meth. Enzymol.* **212**, 3-29.

56. Ermak, D. L. & McCammon, J. A. (1978). Brownian dynamics with hydrodynamic interactions. *J. Chem. Phys.* **69**, 1352-1360.

57. Allison, S. A. (1986). Brownian dynamics simulation of wormlike chains. Fluorescence depolarization and depolarized light scattering. *Macromolecules* **19**, 118-124.

58. Allison, S., Austin, R. & Hogan, M. (1989). Bending and twisting dynamics of short DNAs. Analysis of the triplet anisotropy decay of a 209 base pair fragment by Brownian simulation. *J. Chem. Phys.* **90**, 3843-3854.

59. Allison, S. A., Sorlie, S. S. & Pecora, R. (1990). Brownian dynamics simulations of wormlike chains - dynamic light scattering from a 2311 base pair DNA fragment. *Macromolecules* **23**, 1110-1118.

60. Chirico, G. & Langowski, J. (1994). Kinetics of DNA supercoiling studied by Brownian dynamics simulation. *Biopolymers* **34**, 415-433.

61. Chirico, G. & Langowski, J. (1996). Brownian dynamic simulations of supercoiled DNA with bent sequences. *Biophys. J.* **71**, 955-971.

62. Jian, H., Schlick, T. & Vologodskii, A. (1998). Internal motion of supercoiled DNA: Brownian dynamics simulations of site juxtaposition. *J. Mol. Biol.* **284**, 287-296.

63. Klenin, K., Merlitz, H. & Langowski, J. (1998). A Brownian dynamics program for the simulation of linear and circular DNA and other wormlike chain polyelectrolytes. *Biophysical Journal* **74**, 780-8.

64. Huang, J., Schlick, T. & Vologodskii, T. (2001). Dynamics of site juxtaposition in supercoiled DNA. *Proc. Natl. Acad. Sci. USA* **98**, 968-973.

65. Jian, H., Vologodskii, A. & Schlick, T. (1997). Combined wormlike-chain and bead model for dynamic simulations of long linear DNA. *J. Comp. Phys.* **73**, 123-132.

66. Rotne, J. & Prager, S. (1969). Variational treatment of hydrodynamic interaction in polymers. *J. Chem. Phys.* **50**, 4831-4837.

67. Langowski, J. & Giesen, U. (1989). Configurational and dynamic properties of different length superhelical DNAs measured by dynamic light scattering. *Biophys. Chem.* **34**, 9-18.

68. Thomas, J. C., Allison, S. A., Appelof, C. J. & Schurr, J. M. (1980). Torsional dynamics and depolarization of fluorescence of linear macromolecules. II. Flourescence polarization anisotropy measurements on a clean viral f29 DNA. *Biophys. Chem.* **12**, 177-188.

69. Song, L., Kim, U. S., Wilcoxon, J. & Schurr, J. M. (1991). Dynamic light scattering from weakly bending rods - estimation of the dynamic bending rigidity of the M13 virus. *Biopolymers* **31**, 547-567.

70. Bao, X. R., Lee, H. J. & Quake, S. R. (2003). Behaviour of complex knots in single DNA molecules. *Phys. Rev. Lett.* **91**, 265506.

71. Vologodskii, A. (2006). Brownian dynamic simulation of knot diffusion along stretched DNA molecule. *Biophys. J.*, **90**, 1594-1597.

Chapter 24

CHROMATIN SIMULATIONS
From DNA to chromatin fibers

J. Langowski and H. Schiessel

[1]*Division Biophysics of Macromolecules, German Cancer Research Center (DKFZ), Im Neuenheimer Feld 580, D-69120 Heidelberg, Germany*
[2]*Instituut-Lorentz, Universiteit Leiden, P.O. Box 9506, 2300 RA Leiden, The Netherlands*

Abstract: We give an overview over current coarse-grained models of DNA and the chromatin fiber. A short review of the major structural elements, interaction potentials and mechanical parameters relevant for chromatin structure is given. We then discuss the role of histone tails in nucleosome-nucleosome interaction and finally report some new results on the simulation of chromatin stretching by analytical and numerical models.

Key words: wormlike polymer chain, persistence length, Monte-Carlo models, Brownian dynamics, elasticity theory, single molecule stretching, nucleosome unrolling

1.   INTRODUCTION

The genomic DNA and the histone proteins compacting it into chromatin comprise most of the contents of the nucleus. In every human cell, for instance, $6\cdot10^9$ base pairs of DNA – that is, a total length of about 2 meters – must be packed into a more or less spheroid nuclear volume about 10–20 μm in diameter. This compaction must occur in such a way that the DNA molecule is still easily accessible to enzymes acting on it, such as replication, transcription and repair machineries, or regulatory factors.

Figure 24-1 gives an overview of the many length and time scales that have to be considered to describe DNA compaction. Describing such a complex system with molecular dynamics methods that have been successful in modeling medium-size proteins and protein-DNA complexes would be impossible with present computational means and not even desirable (in fact,

605

J. Šponer and F. Lankaš (eds.), Computational Studies of RNA and DNA, 605–634.
© 2006 *Springer.*

such an undertaking would be almost comparable to predicting the weather by solving the equations of motion of all the water molecules in the atmosphere). Thus, such a system must be described using some adequate approximation.

The principle of any such approximation will consist in defining suitable

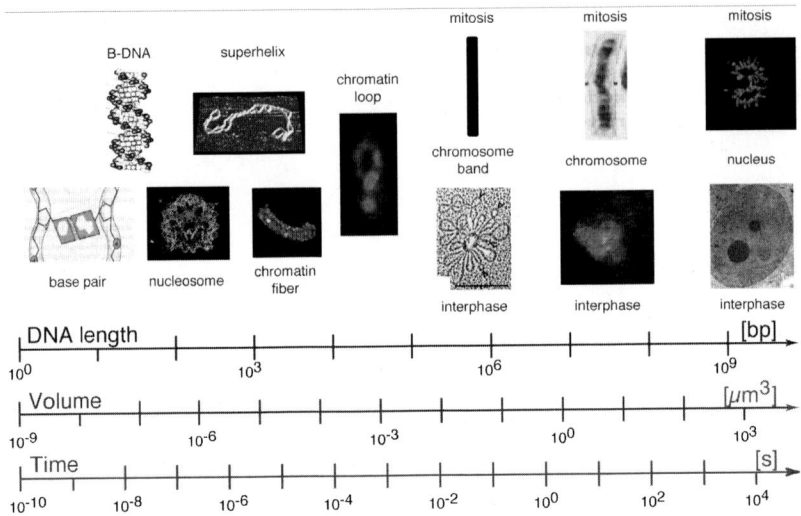

Figure 24-1: Time and length scales relevant in genome organization.

subunits of the molecule that behave like rigid object on the size and time scale considered. These objects interact through potentials that may in principle be derived from the interatomic force fields; however, in practice one mostly uses potentials that have been determined experimentally. To derive general principles of DNA and chromatin structure and dynamics, some models can be described analytically by closed equations; in many other cases, however, the potentials and the mechanics involved are more complicated and numerical simulations must be applied. Examples of both approaches will be given here.

The lowest level of DNA compaction in eukaryotic cells is the chromatin fiber: 147 bp stretches of DNA, which are wrapped in 1.65 left-handed turns about histone octamers formed by two copies each of the histones H2A, H2B, H3 and H4, alternate with free linker DNAs of 20-80 bp length. This repeating unit of the chromatin fiber is called the nucleosome; the histone octamer, together with the bound DNA, is the nucleosome core particle (NCP). The structure of the NCP has been determined by X-ray crystallography to atomic resolution[1-3]. At low ionic strengths, the polynucleosome chain forms a zig-zag, 'bead-on-a-string' structure (10 nm fiber), clearly seen on micrographs obtained by cryoelectron microscopy. Under physiological ionic conditions, the chromatin is more condensed (30 nm fiber). Its detailed structure in this state is yet unknown, however, recently the first high resolution crystallographic structure of a tetranucleosomes was published[4].

Two classes of models were proposed for the arrangement of NCPs inside the 30 nm fiber: the solenoid models[5-7] and the zig-zag models[8-13]. According to the solenoid model, the NCPs are packed one by one along a solenoid helix in the same order as they follow in the chain. The linker DNA is bent in order to provide a relatively small distance between the neighboring NCPs. On the contrary, in the zig-zag model, straight linkers connect the NCPs located on opposite sides of the fiber. The NCPs are also arranged in a helical order, but the neighbors in space are the second neighbors along the chain. Most recent experimental data on chromatin fibers[9,11,12,14-16], as well as the tetranucleosome crystallographic structure[4], are rather in favor of the zig-zag model, which is also energetically more favorable because the linker DNA does not need to bend.

The regular fiber geometry of the zig-zag model can be quantitatively described in terms of two parameters: the entry-exit angle α of the linker DNAs at each NCP and the twist angle β between successive NCPs on the chain. Therefore, in theoretical considerations, this model is often referred to as the two-angle model[13]. The angle α is strongly influenced by the linker histone (H1 or H5) located near the entry-exit region of the NCP. Through the DNA helix pitch, the angle β is coupled with the linker DNA length.

In reality, the 30 nm fiber structure is not quite so regular. It suffers transient fluctuations due to thermal motion. Statistical and dynamical properties of such a complex system can be understood only by means of numerical simulations. Because of the size of the system, standard all-atom molecular dynamics techniques that are successful for describing small and intermediate size biomolecules are not adequate. In order to deal with a system as large as the nucleosome or beyond, it must be appropriately 'coarse-grained', that is, groups of atoms must be comprised into larger units that are interacting via effective potentials. In recent years, several works dedicated to computer simulations of the 30 nm fiber were published[13,17-23]. In those works, the NCP was modeled by a sphere, an oblate ellipsoid, or a disk and the linker DNA was considered as a chain of straight segments with elastic joints. The "coarse-grained" energy defined in these systems was essentially based on the two-angle model, with α and β being adjustable parameters, on the known flexibility parameters of DNA and on reasonable assumptions for the internucleosome interaction.

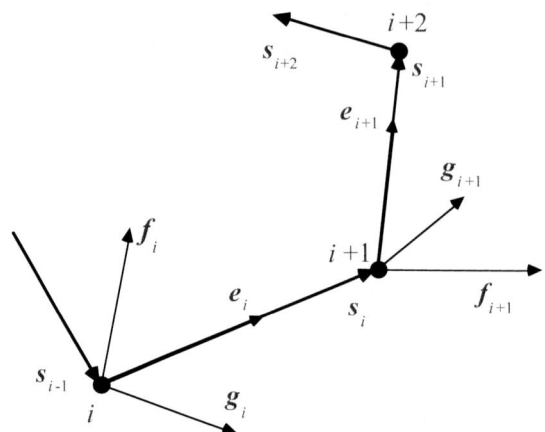

Figure 24-2. Section of a segmented polymer chain as used in the DNA and chromatin models described here.

2.   COARSE-GRAINED MODELS OF DNA

For setting up a model of the chromatin fiber, we must first be able to describe the mechanics of a long DNA chain. A 'coarse-grained' description of DNA can be achieved using a linear segmented chain. We see in Figure 24-1 that the motif of a 'linear elastic filament' is repeated through all size scales: DNA, as well as the chromatin fiber and to some extent its higher order structures may be approximated by a flexible wormlike chain. An approximate description for such a molecule is constructed by defining suitable segments, which behave like rigid cylinders on the time and length scale considered. Figure 24-2 schematizes the segmented chain geometry. The vector s_i defines the direction and length of segment i, f_i is a unit vector normal to the segment and g_i is an auxiliary vector that is used to take into account permanent bending of the DNA. The details of this chain geometry are given in[24].

2.1   DNA flexible Wormlike Chain – Nanomechanical Parameters

Neglecting all structural detail, four parameters are sufficient to describe the energetics of the segmented chain: elastic constants for bending, torsion and stretching and the interaction potential between chain segments (the latter also relating to the thickness of the molecule).

2.1.1 Bending Rigidity

The length of the segments must be chosen well below the *persistence length* L_p, which is a measure of the bending flexibility of the chain molecule, It is defined as the correlation length of the direction of the chain measured along its contour:

$$\langle \vec{u}(s)\vec{u}(s+s') \rangle = e^{-s'/L_p} \tag{1}$$

Here $\vec{u}(s)$ is a unit vector in the direction of the chain (e_i in Figure 24-2) and s resp. $s+s'$ are the positions along the chain contour, the angular brackets indicating the thermal average over all positions and chain conformations. Molecules shorter than L_p behave approximately like a rigid rod, while longer chains show significant internal flexibility. The bending elasticity A – the energy required to bend a polymer segment of unit length over an angle of 1 radian – is related to the persistence length by $L_p = A/k_BT$, k_B being Boltzmann's constant and T the absolute temperature. The energy required to bend two segments of the chain of length l by an angle θ with respect to one another is:

$$E_b = \frac{k_B T}{2} \frac{L_p}{l} \theta^2 \tag{2}$$

For DNA, L_p has been determined in a number of experiments (for a compilation, see[25]). While some uncertainties remain as regards the value at very high or low salt concentrations, the existing data agree on a consensus value of L_p = 45–50 nm (132–147 bp) at intermediate ionic strengths (10–100 mM NaCl and/or 0.1–10 µM Mg^{2+}).

2.1.2 Torsional Rigidity

The torsional rigidity C, defined as the energy required to twist a polymer segment of unit length through an angle of 1 radian, may be related in an analogous way to a *torsional persistence length* L_T through the directional correlation of a vector normal to the chain axis and with fixed orientation relative to the molecular structure of the polymer chain (e.g. for DNA one would use a vector pointing along the dyad axis for the first base pair and into a direction of $(n-1) \times 360°/10.5$ to the left of the dyad axis for the *n*-th base pair, assuming a helix period of 10.5 base pairs per turn). Again, C is related to L_T by $L_T = C/k_B T$ and the torsion energy between two segments of length l twisted by an angle ϕ:

$$E_T = \frac{k_B T}{2} \frac{L_T}{l} \phi^2 \qquad (3)$$

The torsional rigidity C has been measured by various techniques, including fluorescence polarization anisotropy decay[26-28] and DNA cyclization[29-31], and the published values converge on a torsional persistence length of 65 nm (191 bp).

2.1.3   Stretching Rigidity

The stretching elasticity of DNA has been measured by single molecule experiments[32,33] and also calculated by molecular dynamics simulations [34,35]. The stretching modulus σ of DNA is about 1500 pN, where $\sigma = F \cdot L_0 / \Delta L$ (ΔL being the extension of a chain of length L_0 by the force F). The stretching energy of a segment of length l that is stretched by Δl is:

$$E_{str} = \frac{1}{2} \frac{\sigma}{l} \Delta l^2 \qquad (4)$$

DNA stretching probably does not play a significant role in chromatin structural transitions, since much smaller forces are already causing large distortions of the 30 nm fiber (see below).

2.1.4   Intrachain Interactions

The average DNA helix diameter used in modeling applications such as the ones described here includes the diameter of the atomic-scale B-DNA structure and – approximately – the thickness of the hydration shell and ion layer closest to the double helix[36]. Both for the calculation of the electrostatic potential and the hydrodynamic properties of DNA (i.e. the friction coefficient of the helix for viscous drag) a helix diameter of 2.4 nm describes the chain best[24,37-39]. The choice of this parameter is supported by the results of chain knotting[40] or catenation[41], as well as light scattering[42] and neutron scatterin[39] experiments.

As pointed out in[24,43] DNA intrachain electrostatic repulsion can be adequately described by a Debye-Hückel electrostatic potential between two uniformly charged non-adjacent segments (i, j) in a 1-1 salt solution:

$$E_{ij}^{(e)} = \frac{v^2}{D} \iint d\lambda_i d\lambda_j \frac{e^{-\kappa r_{ij}}}{r_{ij}} \tag{5}$$

Here, the integration is done along the two segments, λ_i and λ_j are the distances from the segment beginnings, r_{ij} is the distance between the current positions at the segments to which the integration parameters λ_i and λ_j correspond; κ is the inverse of the Debye length, so that $\kappa^2 = 8\pi e^2 I/k_B TD$, I is the ionic strength, e the proton charge, D the dielectric constant of water, v the linear charge density which for DNA is equal to $v_{DNA} = -2e/\Delta$ where $\Delta = 0.34$ nm is the distance between base pairs. More details as to the normalization of the linear charge density etc. have been given in our earlier paper[24].

3. SIMULATION PROCEDURES

3.1 Monte-Carlo Simulations

The total energy of a segmented polymer chain is given by the sum of bending, twisting, stretching and intrachain interaction energies (eqs. 2-5):

$$E_{total} = E_b + E_T + E_{str} + E^{(e)} \tag{6}$$

To find a realistic chain conformation, one could now search for the conformation of minimum energy. Caution must be taken, however, that a simple search for the minimum of elastic energy will not be enough, because the system is highly degenerate: at equilibrium there exist a large number of possible conformations whose energy differs by much less than the thermal energy $k_B T$[44].

An ensemble of configurations at thermodynamic equilibrium, i.e. at minimum *free* energy, can be produced by the Metropolis Monte-Carlo algorithm [45]: Starting from a configuration i with Energy E_i, we generate configuration $i+1$ by a small statistical variation, e.g., by translating or rotating a group of atoms in the molecule. The new energy E_{i+1} is then calculated; the new configuration is counted into the average and taken as the new starting configuration if either $E_{i+1} < E_i$ or $X < e^{-(E_{i+1}-E_i)/k_B T}$, where $X \in [0,1]$ is a uniformly distributed random number. Otherwise the configuration i is counted again into the average. It can be shown that this procedure generates an ensemble of configurations of the molecule at thermodynamic equilibrium at temperature T.

For the DNA chain, a typical statistical variation (or 'Monte Carlo move') might be for instance a rotation of part of the molecule around a randomly taken axis. The characteristics of Monte Carlo procedures are described in more detail in the original papers such as [37,46].

3.2 Brownian Dynamics Simulations

For calculating the dynamics of DNA, equations of motion for the segmented DNA chain have to be set up using the intramolecular interaction potentials described above and including the thermal motion through a random force. This is the Brownian Dynamics (BD) method[47] which several groups applied to DNA in interpreting experimental data from fluorescence depolarization[48], dynamic light scattering[49,50], or triplet anisotropy decay[51]. Superhelical DNA has also been modeled by a BD approach[52,53]. The model allows to predict the kinetics of supercoiling and the internal motions of superhelical DNA over a time range of tens of milliseconds. This model has then been used in extensive studies of intramolecular reactions in superhelical DNA[54-58].

In the model, the equations of motion of a polymer chain of N segments in a viscous fluid are iterated numerically with a time step δt. The discrete equations of motion in the solvent for positions $\mathbf{r}_i$ and torsions ϕ_t of the i-th segment are then[59]:

$$\delta \mathbf{r}_i(t) = \delta t \frac{1}{k_B T} \sum_{j=1}^{N} \mathbf{D}_{ij} \mathbf{F}_j + \mathbf{R}_i$$

$$\delta \phi_i(t) = \delta t \frac{1}{k_B T} D_r T_i + S_i$$

(7)

where $\mathbf{D}_{ij}$ is a *hydrodynamic* interaction *matrix* (see below), D_r is the rotational diffusion coefficient (same for all segments), $\mathbf{F}_j$ and T_i are the forces and torques acting on segment j resp. i, and the random translations $\mathbf{R}_i$ and rotations S_i are sampled from Gaussian distributions with the following properties:

$$\langle \mathbf{R}_i \rangle = 0 \quad ; \quad \langle \mathbf{R}_i : \mathbf{R}_j \rangle = 2 \delta t \mathbf{D}_{ij}$$

$$\langle S_i \rangle = 0 \quad ; \quad \langle S_i S_j \rangle = 2 \delta t D_r$$

(8)

The (3N x 3N) $\underline{D}_{ij}$ matrix in the first line of Eqs.7,8 is the Rotne-Prager generalization of the Oseen tensor[60], which characterizes the hydrodynamic interaction between two spherical beads (eq. 5 and 6 in Chirico and Langowski[50]) For calculation of this matrix, the cylindrical DNA segments are approximated by beads with radius $r_b = 2.53$ nm. The model has been described in detail in several papers[24,50,52,61], and its code is available on request from the author.

4. NUCLEOSOMES

4.1 Nucleosome Structure

As mentioned above, the structure of the nucleosome core particle is known in detail from X-ray crystallography[1,2]. The histone octamer defines the wrapping path of the DNA, a left-handed helical ramp of 1 and 3/4 turns, 147 bp length and a ~28 Å pitch. This aggregate has a two-fold axis of symmetry (the dyad axis) that is perpendicular to the DNA superhelix axis. A schematic view of the NCP is given in Figure 24-3.

The regions where the wrapped DNA contacts the octamer surface are located where the minor grooves of the right-handed DNA double helix face inwards towards the surface of the octamer. There are 14 'sticking points' on the octamer surface, and while structural details for them are known, a reliable quantitative estimate of the free energy of binding per sticking point is still missing.

From studies of competitive protein binding to nucleosomal DNA[62,63] the adsorption energy per sticking point is estimated at $\sim 1.5 - 2 k_B T$. This number has to be taken with a grain of salt; first, it does not represent the *pure* adsorption energy but instead the *net* gain in energy that is left after the DNA has been bent around the octamer to make contact to the sticking point. A rough estimate of the deformation energy can be obtained from the DNA persistence length L_P of ~50 nm. Then the elastic energy required to bend the 127 bp of DNA around the octamer (10 bp at each terminus are essentially straight[1]) is given by

$$\frac{E_{elastic}}{k_B T} = \frac{L_P l}{2 R_0^2} \tag{9}$$

Here l is the bent part of the wrapped DNA, $\sim 127 \times 3.4$ Å $= 432$ Å and R_0 is the radius of curvature of the centerline of the wrapped DNA that is roughly 43 Å[1]. This leads to a bending energy of order $58 k_B T$, a number, however, that has again to be taken with caution since it is not clear whether the assumption of a homogeneous elastic filament holds up to such strong curvatures. Using these numbers nevertheless one can estimate the bending energy per ten basepairs, i.e., per sticking site, to be of order $60k_BT/14 \approx 4k_BT$ [64].

Together with the observation that the net gain per sticking point is $\sim 2k_BT$ this means that the pure adsorption energy is on average $\sim 6k_BT$ per binding site. Note that the huge pure adsorption energy of $\sim 6k_BT \times 14 \approx 85k_BT$ per nucleosome is cancelled to a large extend by the $\sim 58k_BT$ from the DNA bending, a fact that has important consequences for nucleosomal dynamics.

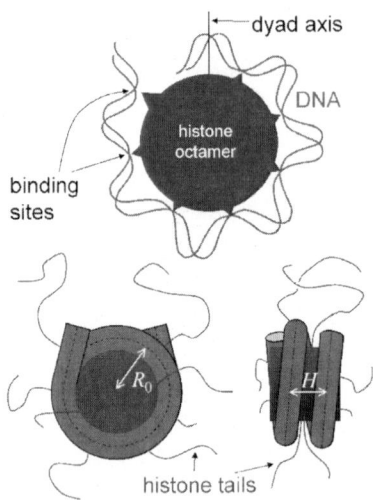

Figure 24-3. The top picture displays only the upper half of the wrapped DNA with its binding points to the histone octamer (located at the positions where the minor groove faces the octamer). At the bottom the full NCP is shown from the top and from the side. Also indicated are the 8 histone tails.

4.2   Nucleosome Unwrapping – Analytical Model

As has been shown experimentally in several cases, for large enough external stretching forces the DNA unwraps from the octamer and the nucleosome falls apart. It seems to be straightforward to estimate the critical force necessary to induce such an unwrapping from the net adsorption energy of the 50 nm DNA wrapped in the nucleosome, about $30k_BT$:

$$F_{crit} \approx \frac{30k_BT}{50nm} = 2.5pN \qquad (10)$$

The same critical force should be expected if there are several nucleosomes associated with the DNA fragment; all of them should unwrap at the same critical force. However, a recent experiment[65] on reconstituted chromatin fibers resulted in unwrapping forces very different from what Eq. (10) predicts. The experiment was performed on tandemly repeated nucleosome positioning sequences with up to 17 nucleosomes complexed at well-defined positions. When small forces ($F < 10 pN$) were applied for short times ($\sim 1 - 10 s$) the nucleosome unwrapped only partially by releasing the outer 60-70 bp of wrapped DNA in a gradual and equilibrium fashion. For higher forces ($F > 20 pN$) nucleosomes showed a pronounced sudden non-equilibrium release behavior of the remaining 80 bp – the latter force being much larger than expected from the above given equilibrium argument. To explain this peculiar finding Brower-Toland et al.[65] conjectured that there must be a barrier of $\sim 38 k_B T$ in the adsorption energy located after the first 70-80 bp and smeared out over not more than 10 bp which reflects some biochemical specificity of the nucleosome structure at that position. However, there is no experimental indication of such a huge specific barrier – neither from the crystal structure[2] nor from the equilibrium accessibility to nucleosomal DNA[62]. Kulic and Schiessel[66] argued that the barrier is caused by the underlying geometry and physics of the DNA spool rather than by a specific biochemistry of the nucleosome.

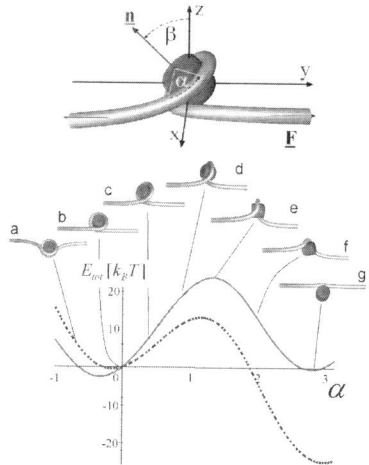

Figure 24-4. A nucleosome under tension. The top picture defines the two angles involved in the unwrapping process: the desorption angle α and the tilting angle β. The bottom shows the nucleosome unwrapping that involves a 180°-rotation of the octamer and the associated energy, Eq.(13), as a function of α for an applied tension of 6.5 pN. Unwrapping is only possible as an activated process going across a substantial barrier.

Summarizing the arguments in that work, we show in Figure 24-4 the model of a DNA spool under tension. The elastic energy of a WLC of length L is

$$E_{bend} = \frac{A}{2} \int_0^L ds\, \kappa^2(s) \qquad (11)$$

with $\kappa(s)$ being the curvature of the chain at point s along its contour (this is the continuum version of Eq. (2)). The DNA is assumed to be adsorbed on the protein spool surface along the predefined helical path with radius R_0 and pitch height H, with a pure adsorption energy density per wrapped length, k^a, given by the pure attraction of the binding sites (not including the bending contribution).

The degree of DNA adsorption is described by the desorption angle α which is defined to be zero for one full turn wrapped (cf. top of Figure 24-4). During unwrapping the spool needs to rotate transiently out of the plane while performing a full turn – as already pointed out by Cui and Bustamante[67]. Therefore a second angle, β, is introduced to describe the out-of-plane tilting of the spool, cf. Figure 24-4. When a tension F (along the Y–axis) acts on the two outgoing DNA arms the nucleosome will simultaneously respond with DNA deformation, spool tilting and DNA desorption from the spool.

The total energy of the system as a function of α and β has three contributions:

$$E_{tot}(\alpha,\beta) = E_{bend} + 2R_0 k^a \alpha - 2F\Delta y \qquad (12)$$

The first term in Eq. (12) is the deformation energy of the DNA chain, Eq. (11), the second describes the desorption cost and the third term represents the potential energy gained by pulling out the DNA ends, each by a distance Δy.

It is possible to work out the total energy on purely analytical grounds by calculating the shape and energy of the DNA arms accounting for the right boundary conditions at the points where the DNA enters and leaves the spool and at the DNA termini (that are assumed to be far from the spool). Instead of giving the full analytical expression of E_{tot} provided in Kulic and Schiessel[66], we merely present here a reasonable approximation in the limit for a flat spool with $R_0 \gg H$ and setting $\alpha = \beta$. In this case

$$E_{tot}(\alpha) \approx 2R\left[k^a - \frac{A}{2R_0^2} - F\right]\alpha + 2FR\cos\alpha\sin\alpha$$
$$+ 8\sqrt{AF}\left[1 - \sqrt{(1 + \cos^2\alpha)/2}\right]$$
(13)

In Figure 24-4 we plot the resulting energy landscape for a force of $F = 6.5$ pN. The dashed curve corresponds to the value $k^a = 2k_B T/nm$ as inferred from competitive protein binding data (see section 4.1); for the thick curve we assume a larger value, $k^a = 3k_B T/nm$ (see below).

4.3 Brownian Dynamics Simulation of Nucleosome Unwrapping

A similar unwrapping barrier as predicted above has been found in a numerical simulation of the unrolling of DNA from the histone core, which was performed in a recent application of BD (Klenin and Langowski, manuscript in preparation). A 200 bp DNA was represented as a wormlike chain as before, and attached to the surface of a cylinder 5 nm thick and 6 nm in diameter, assuming a binding energy of 2 kT per base pair, which was distributed uniformly along the contour of the DNA. This surface binding energy, for technical reasons even larger than that estimated in the analytical theory, is largely sufficient to overcome the elastic energy estimated from the persistence length (58 $k_B T$).

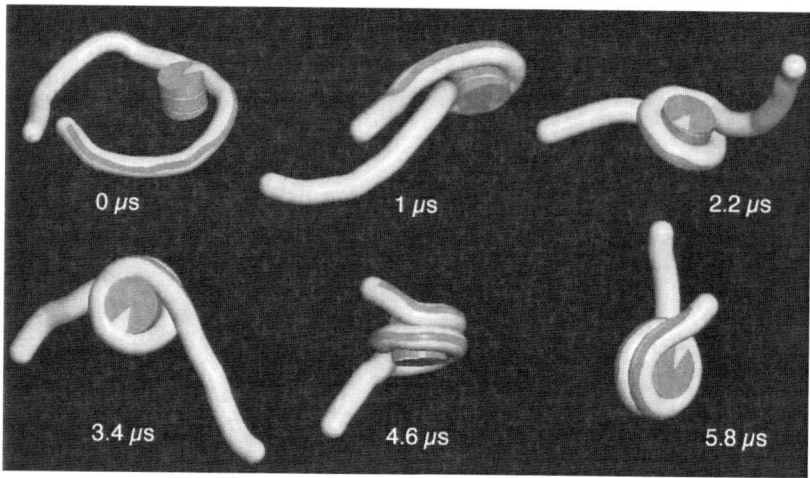

Figure 24-5a. BD simulation of wrapping of a 200 bp DNA on the histone core.

Figure 24-5a shows a trajectory from a BD simulation of the wrapping of the DNA on the histone core. Initially, the DNA is bound to one point on the histone octamer surface and then allowed to equilibrate. It is seen that under these conditions the DNA forms a superhelix on the cylinder surface within a few microseconds, the free linker DNA arms diverging slightly from the nucleosome surface (as earlier shown experimentally by Tóth et al.[68]).

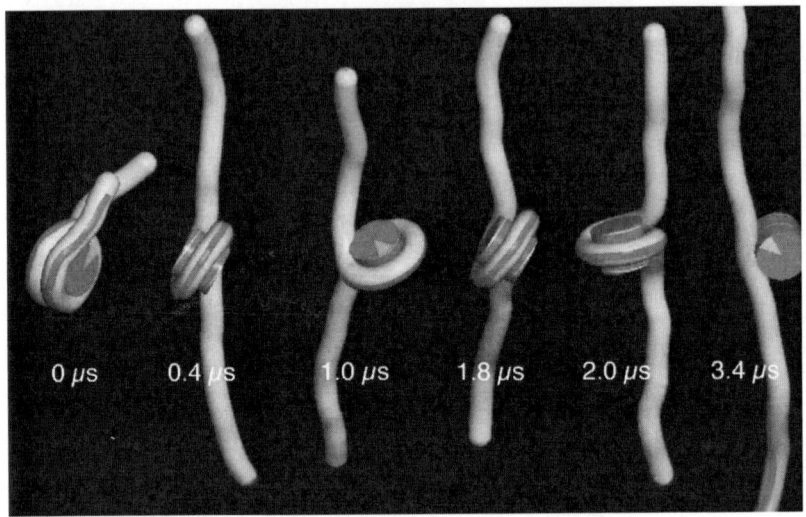

Figure 24-5b. BD simulation of unrolling of a 200 bp DNA from the histone core, using a stretching force of 16 pN.

Applying a stretching force of 16 pN to the DNA ends leads to unwrapping of the DNA (Figure 24-5b). It is remarkable that the unwinding of the first turn is quite fast (0.4 µs), after which a barrier has to be overcome by twisting the nucleosome cylinder (1.8 to 2 µs). Only after this twisting transition has been made, the second turn can unwind, which again is analogous to the arrest of nucleosome unrolling as observed in the single molecule stretching experiments by Brower-Toland et al.[65] The simulation also shows that the first turn is unrolled significantly faster than the second.

4.4   A Possible Mechanism for Nucleosome Unwrapping

The results of the analytical and numerical models agree well. The unwrapping experiment to which we compare them[65] utilized dynamical force spectroscopy (DFS) and exposed the nucleosomal array to a force F increasing at constant rate r_F, $F = r_F t$, and determined the most probable rupture force F^* as a function of loading rate. The rate of unwrapping is expected to be proportional to the Kramers' rate $\exp \Delta U - \pi R(F_{crit} - F)$ from which it can be shown that $F^* \propto \ln(r_F) + const.$

A detailed analysis shows that the rates over the barrier are much too fast in the model as compared to the rates at which nucleosomes unwrap in the experiment[66]. This forced us to critically reconsider the assumptions on which the model was based, especially the – at first sight – straightforward assumption that the adsorption energy per length is constant along the wrapping path. But this neglects an important feature of the nucleosome, namely that the two DNA turns interact. Clearly the turns are close enough to feel a considerable electrostatic repulsion, the exact amount of which is hard to be determined, e.g. due to the fact that the DNA is adsorbed on the low-dielectric protein core (image effects). Moreover, the presence of histone tails complicate things. It is known (see Section 5.2) that the tails adsorb on the nucleosomal DNA. If the nucleosome is fully wrapped the two turns have to share the cationic tails but if there is only one turn left, all these tails can in principle adsorb on this remaining turn. All these effects go in one direction: A remaining DNA turn on the wrapped nucleosome is much stronger adsorbed then a turn in the presence of the second turn wrapped. Indeed, very recent data by the same experimental group show that the force peeks of the discontinuous unwrapping events shift to substantially smaller values when the tail are partly removed or their charges partially neutralized[69].

The crucial point is now that the adsorption energy k^a was estimated from spontaneous unwrapping events of the second turn in the presence of the other turn [62,63] and thus k^a might have been strongly underestimated since the $k^a = 2k_B T/nm$ include the unfavorable repulsion from the other turn. To account for this we assumed that there is a different effective value of k^a for $\alpha > 0$ (less than one DNA turn) and for $\alpha < 0$ (more than one turn) [66]. Since the discontinuous unwrapping events observed in the experiment clearly correspond to the case where the last term is unwrapped (i.e. to the case $\alpha > 0$) we tuned the parameter k^a such that we can reproduce the DFS data in a satisfying way. From this we found that a value of $k^a = 3.0 - 3.5 k_B T/nm$ leads to a good agreement with the experimental data, a value that is *considerably* higher than the effective adsorption energy $k^a = 2k_B T/nm$ felt when a turn is unpeeled in the presence of the other turn.

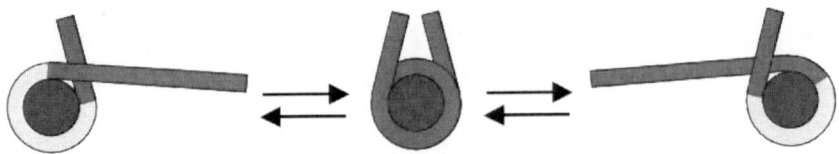

Figure 24-6. The site exposure mechanism allows access to DNA via spontaneous unwrapping[62,63]. The remaining turn (shown in yellow) has a stronger grip on the octamer and further unpeeling becomes too costly (first-second round difference[66]; see text for details).

This result might explain how the nucleosome can be transparent to DNA binding proteins and at the same time stable. When the nucleosome is fully wrapped each of two turns can easily unwrap spontaneously due to thermal fluctuations and therefore all DNA is transiently accessible for DNA binding proteins, cf. Figure 24-6. This has been proven experimentally via competitive protein binding by Widom and coworkers and has been termed the site exposure mechanism[62,63]; recently single molecule fluorescence resonance energy transfer measurements provide additional and more direct evidence for such conformational fluctuations[70,71]. What is, however, puzzling in this set of experiments is why the DNA stops to unpeel further once it encounters the dyad and why it does not fall apart. Our interpretation of the unwrapping data suggests that the reason for this is the first-second round difference as supported by the model calculations. Further simulations using the BD model will be used in the future to test this hypothesis and quantify its consequences.

5.   COARSE-GRAINED MODELS OF CHROMATIN

The next higher order structure into which DNA is packed in the eukaryotic nucleus, the chromatin fiber, can again be approximated by a flexible polymer chain. At physiological salt concentrations, this structure has a diameter of 30 nm and a linear mass density of about 5–6 nucleosomes per 10 nm fiber length, corresponding to approx. 100–120 base pairs / nm. Thus, the DNA is compacted by about a factor of 30–40 compared to the 3 base pairs / nm for the canonical B-DNA structure. As mentioned in the Introduction the internal geometry of this structure is still under discussion with more and more experimental evidence for the zig-zag geometry.

The interaction between nucleosomes plays an important role for the stability of the 30 nm fiber; recent experiments on liquid crystals of mononucleosomes[72-75] and also less concentrated mononucleosome solutions [76,77] show an attractive interaction that can be parameterized by an anisotropic

Lennard-Jones type potential[20]. Also, an electrostatic interaction potential has been computed using the crystallographic structure of the nucleosome[78]. A recent study[79] investigates the influence of tail bridging on internucleosome interaction (see below).

5.1 Persistence Length of Chromatin

For the chromatin fiber, estimates of L_p are controversial. Experimental, theoretical and simulation data support values over a wide range starting from $L_p = 30–50$ nm from scanning force microscopy (SFM) analysis of end-to-end distances of chromatin fibers on mica surface[80]. However, persistence lengths measured by SFM strongly depend on the binding conditions of the fiber to the mica[81]. Stretching chromatin fibers at low salt concentrations with optical tweezers suggests $L_p = 30$ nm[67], however, chromatin is known to form a very open structure at low salt. Small persistence lengths of 30–50 nm were also postulated from recombination frequencies in human cells[82] and formaldehyde cross-linking probabilities in yeast [83]. These data, however, are strongly influenced by the constraining of the chromatin chain inside a finite nuclear volume; also, the persistence length estimated from looping probabilities depends on the packing density, so these two parameters cannot be determined independently.

Other groups report stiffer fibers with L_p in the range of 100–200 nm, based on distance distributions for genetic marker pairs in human fibroblast nuclei[84-86] or recent experiments in budding yeast using in situ hybridization and live imaging techniques[87]. Stiffer fibers in the range of 200–250 nm are also supported by computer simulations by Mergell et al.[19]

Bystricky et al.[87] determined the persistence length and packing density of yeast chromatin independently by measuring the spatial distance between genetic markers both in fixed cells and *in vivo*. In the equation for the end-to-end distance of a wormlike chain with persistence length L_p,

$$\langle r^2 \rangle = 2L_p^2 \left(L_c/L_p - 1 + e^{-L_c/L_p} \right) \qquad (14)$$

the contour length L_c (in nm) is the ratio of the genomic distance d (in kb) divided by the linear mass density of the chromatin chain c (in bp/nm) or L_c = d/c. A fit of eq. 14 to the values of d and r from the distance measurements yields values for the persistence length $L_p = 183 \pm 76$ nm, and mass density c $= 142 \pm 21$ bp/nm.

Recent experiments [65,67,88-90] investigated the mechanical properties of the chromatin fiber by single molecule stretching techniques. For forces below 10-20 pN, the extension of the chromatin chain is defined by its elasticity and no structural transition occurs, whereas forces above 10-20 pN lead to the disintegration of nucleosomes. Nevertheless quantities like the stretching modulus of a chromatin fiber are still unclear. Stretching a nucleosome-assembled lambda-phage DNA extract with an optical tweezers, Bennink et al. [88] derived a stretching modulus of 150 pN for a salt concentration of 150 mM NaCl.

5.2   Tail Bridging for Nucleosome Attraction

Recent experiments point towards histone tail bridging as a simple mechanism for nucleosomal attraction[76,77,91]. The cationic histone tails extend considerably outside the globular part of the nucleosome as sketched schematically in Figure 24-3. Mangenot et al.[76] studied dilute solutions of NCPs by small angle X-ray scattering and suggested that the tails are the main elements responsible for the attraction (which is supported by the fact that the attraction disappears once the tails are removed[91]).

Strong theoretical support that tails are important in the interaction of nucleosomes within a chromatin fiber comes from a very recent computer simulation[17] where the NCP crystal structure has been mimicked by a cylinder with 277 charge patches (accounting for charged groups on the surface of the NCP) with all the tails anchored to it. By switching on and off the charges on the tails it was found that the tails play a crucial role in the electrostatic nucleosome-nucleosome and nucleosome-linker DNA interaction within that chromatin fiber model – especially leading to a stabilization of the fiber at physiological salt conditions. Even though this study shows the importance of tails for nucleosomal interaction, it does not reveal what is really the underlying physical mechanism.

In a recent study[79] we introduced a minimal model for an NCP with tails to test whether such a model shows similar features as the ones found for NCPs. Our NCP model, termed the eight-tail colloid, consists of a sphere with eight attached polymer chains (cf. inset of Figure 24-7). The sphere is a very coarse-grained representation of the NCP without the tails, i.e., the globular protein core with the DNA wrapped around. The sphere carries a central charge Z that represents the net charge of the DNA-octamer complex; since the DNA overcharges the cationic protein core, one has $Z < 0$[64]. The eight histone tails are modelled by flexible, positively charged chains grafted onto the sphere. All parameters in the model have been chosen to match closely the values of the NCP. All charged monomers and the central sphere experience an electrostatic interaction via the standard Debye-Hückel interaction (cf. also Eq. 5).

We demonstrated via BD simulation of a single eight-tail colloid[79] that a single colloid shows indeed similar features as the NCP, especially for small κ the tails are condensed onto the sphere and by increasing the screening the chains desorb. We then determined the interaction between two such complexes and found an attractive pair potential with a minimum of a few $k_B T$. The depth of the potential showed a non-monotonic dependence on the salt concentration which in turn was reflected in a non-monotonic dependence of the second virial coefficient A_2 with a minimum around conditions where the tails unfold. Again, all these observations are qualitatively similar to the experimental ones[76].

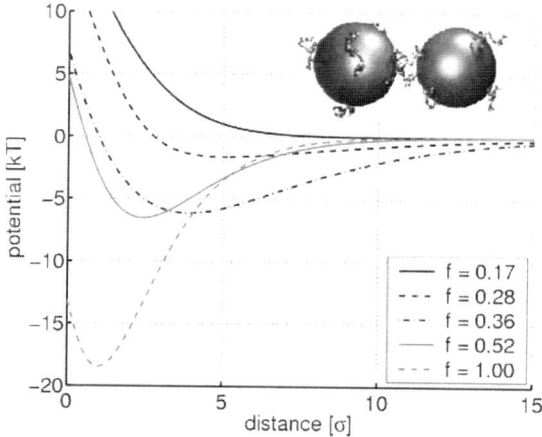

Figure 24-7. NCP-interaction potential with the NCPs represented by eight-tail colloids (inset); σ = 3.5 Å. Note the strong dependence of the potential depth on the charge fraction of the tail monomers (see[79] for details).

Most importantly, the interaction potential shows a strong dependence on the fraction of charged monomers in the tails, cf. Figure 24-7. Starting from a fraction f of 0.36 we find a depth of the potential of about 6 $k_B T$ that nearly disappears when f is reduced to 0.28. This suggests that the tail bridging can be used by the cellular machinery to control DNA compaction and genetic activity. It is in fact known that the cellular machinery is capable of controlling the charge state of the histone tails via the acetylation (the "discharging") and deacetylation (the "charging") of its lysine groups. Active, acetylated regions in chromatin are more open, inactive, deacetylated regions tend to condense locally and on larger scales as well[92]. For instance, chromatin fibers tend to form hairpin configurations once a sufficiently strong internucleosomal attraction has been reached[19,93]. This suggests a biochemical means by which the degree of chromatin compaction and genetic activity can be controlled via a physical mechanism, the tail-bridging effect.

5.3   Simulation of Chromatin Fiber Stretching

In the low-force regime, nucleosome unrolling does not play a significant role in the dynamics of the chromatin fiber, thus the nucleosome can be simply modeled as a rigid cylinder. In a recent study (Aumann et al. (2005) manuscript submitted) we simulated the stretching of 100-nucleosome chromatin fibers using our earlier Monte-Carlo model[20] and extracted the nanomechanical parameters of the 30 nm fiber from these simulations. The geometry used in these simulations is essentially the 'two-angle' model as described above. The chromatin fiber is approximated as a flexible polymer chain consisting of rigid ellipsoidal disks, 11 nm in diameter and 5.5 nm in height. Internucleosomal interactions are approximated by a more simple potential function than that discussed in the previous section: their attraction is modeled by an anisotropic Lennard-Jones type potential whose depth and minimum position depend on the relative orientation of the two nucleosoms (Gay-Berne-Potential). The disks are connected by linker DNA, which is represented by two cylindrical segments. Incoming and outgoing linker DNA are set 3.1 nm apart vertically on the NCP surface. The length of the linker DNA depends on the presence of linker histones and on the repeat length, which varies from organism to organism[94]. To explore the influence of the linker histone[12], simulations were performed with and without a stem motif added to each nucleosome. To simulate the stretching of the fiber, we added a pulling energy term E_{pull} to the total energy of the conformations during the MC steps, which is proportional to the x-component of the distance between the first and the last nucleosome of the fiber, $E_{pull} = -M \cdot |\vec{r}_{i,x} - \vec{r}_{N,x}|$, where M is the stretching modulus and $\vec{r}_{i,x}$ is the x-component of the position vector of nucleosome i.

Chains of 100 nucleosomes were simulated using 'pivot' Monte-Carlo moves where parts of the chain were rotated with respect to each other. As in our previous simulations[20], the linker DNA entry-exit angle α was taken as 26° for the initial conformation. This value converges to an effective angle α_{sim} in the range of experimental values between 35° and 45°[12] due to the electrostatic forces and thermal fluctuations. Simulations were done with either a condensed fiber as a starting conformation or an initial conformation where all segments are ordered in a straight line.

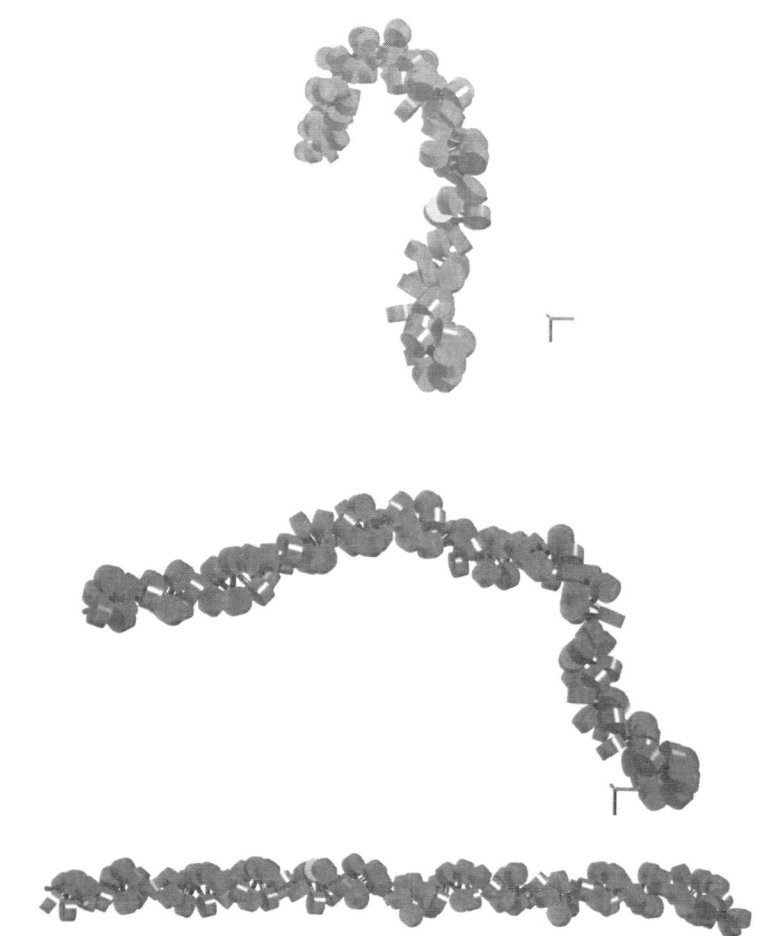

Figure 24-8. Stretching simulation of a fiber consisting of 100 nucleosomes (red), linker segments (blue) of repeat length l = 205 nm, an opening angle α_{init}=26° and a twisting angle β=90°. Top is a typical example of an equilibrated structure. Then the fiber is extended by applying an external pulling force F_{pull} = 5 pN and further equilibration initiated (center and bottom).

To check the statistics of the simulations, we calculated the autocorrelation function $G(\Delta N) = \langle X(N)X(N+\Delta N)\rangle / \langle X(N)^2 \rangle$, where N is the Monte Carlo step number and X either the energy, end-to-end distance or mass density of the fiber conformation at step N. ΔN is the number of steps separating two conformations, and the angular brackets denote the average over the complete trajectory. During the MC procedure, G(ΔN) decreases exponentially with a typical 'correlation length' N_{corr}. We considered two conformations statistically independent if they were separated by at least

N_{corr} steps on the trajectory. For the relaxation of the total energy of both systems, we found a maximum of $N_{corr} \approx 3600$ MC steps, for the end to end distance $N_{corr} \approx 3200$ and for the mass density $N_{corr} \approx 2600$. Thus, we performed $5 \cdot 10^5$ MC steps for the initial relaxation of the chain corresponding to more than 100 statistically independent conformations. After the equilibration the stretching potential was switched on and at least $3 \cdot 10^6$ MC steps were performed. For the final analysis, every 1000th conformation was used. Typical fiber conformations during the simulation are shown in Figure 24-8.

The bending and stretching rigidities of the modeled chromatin fiber are computed from the trajectories from the fluctuations in the bending angle or the fluctuation in the overall fiber length, respectively. The results show that the bending and the stretching stiffness of the chromatin fiber strongly depend on the local geometry of the nucleosome. Both the persistence length L_p, characterizing the bending stiffness of the fiber, and the stretching modulus ε, which describes the stretching stiffness of the fiber, decrease if either the linker lengths or the opening angle are increased, or the twisting

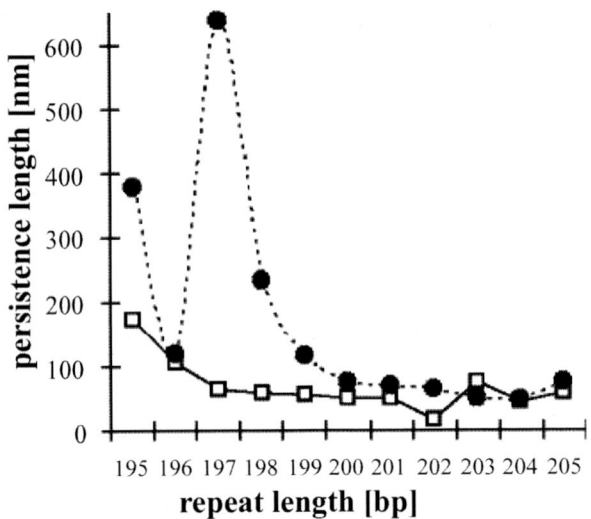

Figure 24-9. Persistence length of modeled 30 nm chromatin fibers with different nucleosomal repeats in the presence and absence of linker histone H1. The twisting angle between adjacent nucleosomes is adjusted to the canonical value of 360° per 10.5 bp. The persistence lengths of fibers with linker histone (closed symbols, dashed lines) are higher than for fibers without linker histone (open symbols, solid lines). This effect is stronger for short repeats and weakens with increasing repeat length. The peaks show that the twisting angle strongly influences the stiffness of the fiber, leading to a non-monotonous variation of L_p with nucleosome repeat.

angle is reduced. This behavior is independent of the presence of the linker histone H1. The latter decreases the opening angle α between the entry and exit of the linker DNA and as a result leads to a more condensed fiber structure for high salt concentrations[95]. This is in agreement with our simulations since the presence of the linker histone-induced stem motif yields higher persistence lengths thus stiffer fibers (Figure 24-9).

The other major result of the simulation comes from comparing the persistence length of the modeled fibers to that of a hypothetical rod from a isotropic elastic material having the same stretching rigidity as the chromatin fiber. Such a rod would have a bending rigidity 4-10 times higher than that actually measured, or simulated here. Thus, the chromatin fiber is less resistant to bending than to stretching. This property of the chromatin fiber is important for its ability to condense and decondense, for example to prevent or allow transcriptional access. Chromatin fibers thus seem to be packed more easily via dense loops than by a linear compression. The formation of hairpin structures has been observed in cryo-EM pictures under the presence of MENT, a heterochromatin protein that mediates higher order chromatin folding [93]. Some hairpin conformations could also be seen in simulations especially for higher internucleosomal attractions[96].

Double-stranded DNA is different from chromatin as far as the ratio of stretching and bending elasticity is concerned[97]: The stretching modulus ε of dsDNA for physiological salt conditions is estimated to ~1100 pN[33,98]. Assuming a homogenous elastic rod with a radius of 1 nm for DNA yields a bending persistence length of $L_p = 70$ nm, which is only a factor of 1.4 higher then the persistence length of 50 nm for dsDNA; thus, dsDNA is almost equally resistant to stretching and to bending.

As mentioned before, the exact value for the persistence length of the chromatin fiber is still under discussion, with estimates ranging from 30 nm to 260 nm. Some of the small values in this range were obtained at low salt concentrations, where a smaller persistence length compared to our results for high salt can be expected, since low salt is known to open the fiber. Other experiments resulting in small persistence lengths were done in constrained volumes by cross-linking procedures[82,83]. Under these circumstances, the condition of an unconstrained self-crossing walk is only fulfilled over short distances. Thus, for a given chain flexibility, the measured apparent persistence length will depend on the genomic separation and folding topology for which it is calculated[87]. The effect of spatial confinement on the apparent persistence length of a chromatin chain is further elaborated in a recent publication[99]. Furthermore, a persistence length in the range of the fiber diameter of 30 nm would lead to extremely irregular structures, which are hard to be reconciled with the concept of a "fiber".

Analysis of the distance distribution for genetic marker pairs in human fibroblast nuclei [84-86] provide higher values of $L_p = 100–140$ nm based on a wormlike chain model. Recent experiments in budding yeast using optimized *in situ* hybridization and live imaging techniques[87] report stiff interphase chromatin fibers estimating a persistence length of 120–200 nm. The persistence lengths that we obtain from the MC simulations for linker lengths of 10 and 15 bp are in the range 50-280 nm, decreasing with longer linkers.

Bennink et. al. derived 150 pN as stretching modulus for a salt concentration of 150 mM NaCl[88], using optical tweezers for the stretching of a nucleosome-assembled lambda-phage DNA extract of a *Xenopus laevis* egg with no linker histones attached and nucleosome repeat length of 200 bp. Our simulations yield a lower value of 40 pN already for a repeat length of 192 (no linker histone). One reason for this discrepancy may be the difference of 50 mM in the salt concentrations, since our simulation parameters have been calibrated for 100 mM NaCl. A lower salt concentration leads to a lower compaction thus to a lower stretching modulus. Furthermore the solution used in the tweezers experiment contains proteins known to act close to the entry-exit points similar to the linker histones. This is supported by the Monte-Carlo simulations, which for a repeat length of 200 bp (with stem) yields a stretching modulus in the range of 90-160 pN. Nevertheless the Gay-Berne potential used in the MC model is only an approximation of the nucleosome-nucleosome interaction, which is strongly dependent on salt conditions. To improve the quantitative predictions of our model, more detailed interaction potentials are need, as for instance the tail-bridging interaction outlined here. Moreover, for the modeling of nucleosome unwrapping and interpreting corresponding experiments[65,88], the DNA – nucleosome interaction has a decisive role. Inclusion of such potentials into the chromatin fiber model will provide a deeper insight in the architecture and behavior of the chromatin fiber as it undergoes biologically important modifications, as well as into its role in transcription and gene regulation.

REFERENCES

1. Luger, K., Mäder, A. W., Richmond, R. K., Sargent, D. F. & Richmond, T. J. (1997). Crystal structure of the nucleosome core particle at 2.8 Å resolution. *Nature* **389**, 251-260.
2. Davey, C. A., Sargent, D. F., Luger, K., Maeder, A. W. & Richmond, T. J. (2002). Solvent mediated interactions in the structure of the nucleosome core particle at 1.9 a resolution. *J Mol Biol* **319**, 1097-1113.
3. Richmond, T. J. & Davey, C. A. (2003). The structure of DNA in the nucleosome core. *Nature* **423**, 145-150.
4. Schalch, T., Duda, S., Sargent, D. F. & Richmond, T. J. (2005). X-ray structure of a tetranucleosome and its implications for the chromatin fibre. *Nature* **436**, 138-141.
5. Finch, J. T. & Klug, A. (1976). Solenoidal model for superstructure in chromatin. *Proc Natl Acad Sci USA* **73**, 1897-1901.

6. Thoma, F., Koller, T. & Klug, A. (1979). Involvement of histone H1 in the organization of the nucleosome and of the salt-dependent superstructures of chromatin. *J Cell Biol* **83**, 403-427.

7. Widom, J. & Klug, A. (1985). Structure of the 300A chromatin filament: X-ray diffraction from oriented samples. *Cell* **43**, 207-213.

8. Woodcock, C. L., Grigoryev, S. A., Horowitz, R. A. & Whitaker, N. (1993). A chromatin folding model that incorporates linker variability generates fibers resembling the native structures. *Proc Natl Acad Sci USA* **90**, 9021-9025.

9. Horowitz, R. A., Agard, D. A., Sedat, J. W. & Woodcock, C. L. (1994). The three-dimensional architecture of chromatin in situ: electron tomography reveals fibers composed of a continuously variable zig-zag nucleosomal ribbon. *J Cell Biol* 125, 1-10.

10. Woodcock, C. L. & Dimitrov, S. (2001). Higher-order structure of chromatin and chromosomes. *Curr Opin Genet Dev* 11, 130-135.

11. Leuba, S. H., Yang, G., Robert, C., Samori, B., van Holde, K., Zlatanova, J. & Bustamante, C. (1994). Three-dimensional structure of extended chromatin fibers as revealed by tapping-mode scanning force microscopy. *Proc Natl Acad Sci USA* **91**, 11621-11625.

12. Bednar, J., Horowitz, R. A., Grigoryev, S. A., Carruthers, L. M., Hansen, J. C., Koster, A. J. & Woodcock, C. L. (1998). Nucleosomes, linker DNA, and linker histone form a unique structural motif that directs the higher-order folding and compaction of chromatin. *Proc Natl Acad Sci USA* **95**, 14173-14178.

13. Schiessel, H., Gelbart, W. M. & Bruinsma, R. (2001). DNA folding: Structural and mechanical properties of the two- angle model for chromatin. *Biophys J* **80**, 1940-1956.

14. Friedland, W., Jacob, P., Paretzke, H. G. & Stork, T. (1998). Monte Carlo simulation of the production of short DNA fragments by low-linear energy transfer radiation using higher-order DNA models. *Radiation Research* **150**, 170-182.

15. Rydberg, B., Holley, W. R., Mian, I. S. & Chatterjee, A. (1998). Chromatin Conformation in Living Cells: Support for a Zig-Zag Model of the 30 nm Chromatin Fiber. *J Mol Biol* **284**, 71-84.

16. Dorigo, B., Schalch, T., Kulangara, A., Duda, S., Schroeder, R. R. & Richmond, T. J. (2004). Nucleosome arrays reveal the two-start organization of the chromatin fiber. *Science* 306, 1571-1573.

17. Sun, J., Zhang, Q. & Schlick, T. (2005). Electrostatic mechanism of nucleosomal array folding revealed by computer simulation. *Proc Natl Acad Sci U S A* **102**, 8180-8185.

18. Barbi, M., Mozziconacci, J. & Victor, J. M. (2005). How the chromatin fiber deals with topological constraints. *Phys Rev E Stat Nonlin Soft Matter Phys* **71**, 031910.

19. Mergell, B., Everaers, R. & Schiessel, H. (2004). Nucleosome interactions in chromatin: fiber stiffening and hairpin formation. *Phys Rev E Stat Nonlin Soft Matter Phys* **70**, 011915.

20. Wedemann, G. & Langowski, J. (2002). Computer simulation of the 30-nanometer chromatin fiber. *Biophys J* **82**, 2847-2859.

21. Beard, D. A. & Schlick, T. (2001). Computational Modeling Predicts the Structure and Dynamics of Chromatin Fiber. *Structure* 9, 105-114.

22. Katritch, V., Bustamante, C. & Olson, V. K. (2000). Pulling chromatin fibers: Computer simulations of direct physical micromanipulations. *J Mol Biol* **295**, 29-40.

23. Ehrlich, L., Munkel, C., Chirico, G. & Langowski, J. (1997). A Brownian dynamics model for the chromatin fiber. *Comput Appl Biosci* **13**, 271-279.

24. Klenin, K., Merlitz, H. & Langowski, J. (1998). A Brownian dynamics program for the simulation of linear and circular DNA and other wormlike chain polyelectrolytes. *Biophys J* **74**, 780-788.

25. Lu, Y., Weers, B. & Stellwagen, N. C. (2001). DNA persistence length revisited. *Biopolymers* **61**, 261-275.

26. Barkley, M. D. & Zimm, B. H. (1979). Theory of twisting and bending of chain macromolecules: analysis of the fluorescence depolarization of DNA. *J Chem Phys* **70**, 2991-3007.

27. Fujimoto, B. S. & Schurr, J. M. (1990). Dependence of the torsional rigidity of DNA on base composition. *Nature* **344**, 175-177.

28. Schurr, J. M., Fujimoto, B. S., Wu, P. & Song, L. (1992). Fluorescence studies of nucleic acids: dynamics, rigidities and structures. In *Topics in Fluorescence Spectroscopy* (Lakowicz, J. R., ed.), Vol. 3, pp. 137-229. Plenum Press, New York.

29. Shore, D. & Baldwin, R. L. (1983). Energetics of DNA twisting. I. Relation between twist and cyclization probability. *J Mol Biol* **179**, 957-981.

30. Horowitz, D. S. & Wang, J. C. (1984). Torsonal Rigidity of DNA and Length Dependence of the Free Energy of DNA Supercoiling. *J. Mol. Biol.* **173**, 75-91.

31. Taylor, W. H. & Hagerman, P. J. (1990). Application of the method of phage T4 DNA ligase-catalyzed ring-closure to the study of DNA structure I.NaCl-dependence of DNA flexibility and helical repeat. *J. Mol. Biol.* **212**, 363-376.

32. Cluzel, P., Lebrun, A., Heller, C., Lavery, R., Viovy, J. L., Chatenay, D. & Caron, F. (1996). DNA – An Extensible Molecule. *Science* **271**, 792-794.

33. Smith, S. B., Cui, Y. & Bustamante, C. (1996). Overstretching B-DNA: the elastic response of individual double-stranded and single-stranded DNA molecules. *Science* **271**, 795-799.

34. Lankaš, F., Šponer, J., Hobza, P. & Langowski, J. (2000). Sequence-dependent Elastic Properties of DNA. *J Mol Biol* **299**, 695-709.

35. Lankaš, F., Cheatham, I. T., Spackova, N., Hobza, P., Langowski, J. & Šponer, J. (2002). Critical effect of the n2 amino group on structure, dynamics, and elasticity of DNA polypurine tracts. *Biophys J* **82**, 2592-2609.

36. Schellman, J. A. & Stigter, D. (1977). Electrical double layer, zeta potential, and electrophoretic charge of double-stranded DNA. *Biopolymers* **16**, 1415-1434.

37. Klenin, K. V., Frank-Kamenetskii, M. D. & Langowski, J. (1995). Modulation of intramolecular interactions in superhelical DNA by curved sequences: a Monte Carlo simulation study. *Biophys J* **68**, 81-88.

38. Delrow, J. J., Gebe, J. A. & Schurr, J. M. (1997). Comparison of hard-cylinder and screened Coulomb interactions in the modeling of supercoiled DNAs. *Biopolymers* **42**, 455-470.

39. Hammermann, M., Brun, N., Klenin, K. V., May, R., Toth, K. & Langowski, J. (1998). Salt-dependent DNA superhelix diameter studied by small angle neutron scattering measurements and Monte Carlo simulations. *Biophys J* **75**, 3057-3063.

40. Rybenkov, V. V., Cozzarelli, N. R. & Vologodskii, A. V. (1993). Probability of DNA knotting and the effective diameter of the DNA double helix. *Proc Natl Acad Sci USA* **90**, 5307-5311.

41. Rybenkov, V. V., Vologodskii, A. V. & Cozzarelli, N. R. (1997). The effect of ionic conditions on DNA helical repeat, effective diameter and free energy of supercoiling. *Nucl Acids Res* **25**, 1412-1418.

42. Hammermann, M., Steinmaier, C., Merlitz, H., Kapp, U., Waldeck, W., Chirico, G. & Langowski, J. (1997). Salt effects on the structure and internal dynamics of superhelical DNAs studied by light scattering and Brownian dynamics. *Biophys J* **73**, 2674-2687.

43. Merlitz, H., Rippe, K., Klenin, K. V. & Langowski, J. (1998). Looping dynamics of linear DNA molecules and the effect of DNA curvature: a study by Brownian dynamics simulation. *Biophys J* **74**, 773-779.

44. Langowski, J., Olson, W. K., Pedersen, S. C., Tobias, I., Westcott, T. P. & Yang, Y. (1996). DNA supercoiling, localized bending and thermal fluctuations. *Trends Biochem Sci* **21**, 50.

45. Metropolis, N., Rosenbluth, A. W., Rosenbluth, M. N., Teller, A. H. & Teller, E. (1953). Equation of state calculations by fast computing machines. *The Journal of Chemical Physics* **21**, 1087-1092.

46. Gebe, J. A. & Schurr, J. M. (1994). Monte-Carlo Simulation of Circular DNAs – 1st Transition in Writhe and Twist Energy Parameters. *Biophys J* **66**, A156-A156.

47. Ermak, D. L. & McCammon, J. A. (1978). Brownian dynamics with hydrodynamic interactions. *J Chem Phys* **69**, 1352-1359.

48. Allison, S. A. (1986). Brownian dynamics simulation of wormlike chains. Fluorescence depolarization and depolarized light scattering. *Macromolecules* **19**, 118-124.

49. Allison, S. A., Sorlie, S. S. & Pecora, R. (1990). Brownian dynamics simulations of wormlike chains: Dynamic light scattering from a 2311 base pair DNA fragment. *Macromolecules* **23**, 1110-1118.

50. Chirico, G. & Langowski, J. (1992). Calculating hydrodynamic properties of DNA through a second-order Brownian dynamics algorithm. *Macromolecules* **25**, 769-775.

51. Allison, S. A., Austin, R. & Hogan, M. (1989). Bending and Twisting Dynamics of Short Linear DNAs - Analysis of the Triplet Anisotropy Decay of a 209-Base Pair Fragment by Brownian Simulation. *J Chem Phys* **90**, 3843-3854.

52. Chirico, G. & Langowski, J. (1994). Kinetics of DNA supercoiling studied by Brownian dynamics simulation. *Biopolymers* **34**, 415-433.

53. Wedemann, G., Munkel, C., Schoppe, G. & Langowski, J. (1998). Kinetics of structural changes in superhelical DNA. *Phys Rev E* **58**, 3537-3546.

54. Bussiek, M., Klenin, K. & Langowski, J. (2002). Kinetics of site-site interactions in supercoiled DNA with bent sequences. *J Mol Biol* **322**, 707-718.

55. Klenin, K. V. & Langowski, J. (2004). Modeling of intramolecular reactions of polymers: an efficient method based on Brownian dynamics simulations. *J Chem Phys* **121**, 4951-4960.

56. Klenin, K. V. & Langowski, J. (2001). Diffusion-Controlled Intrachain Reactions of Supercoiled DNA: Brownian Dynamics Simulations. *Biophys J* **80**, 69-74.

57. Klenin, K. V. & Langowski, J. (2001). Kinetics of intrachain reactions of supercoiled DNA: Theory and numerical modeling. *J Chem Phys* **114**, 5049-5060.

58. Klenin, K. V. & Langowski, J. (2001). Intrachain Reactions of Supercoiled DNA Simulated by Brownian Dynamics. *Biophys J* **81**, 1924-1929.

59. Dickinson, E., Allison, S. A. & McCammon, J. A. (1985). *J Chem Soc Faraday Trans 2* **81**, 591-601.

60. Rotne, J. & Prager, S. (1969). Variational treatment of hydrodynamic interaction in polymers. *J Chem Phys* **50**, 4831-4837.

61. Chirico, G. & Langowski, J. (1996). Brownian dynamics simulations of supercoiled DNA with bent sequences. *Biophys J* **71**, 955-971.

62. Polach, K. J. & Widom, J. (1996). A model for the cooperative binding of eukaryotic regulatory proteins to nucleosomal target sites. *J Mol Biol* **258**, 800-812.

63. Anderson, J. D. & Widom, J. (2000). Sequence and position-dependence of the equilibrium accessibility of nucleosomal DNA target sites. *J Mol Biol* **296**, 979-987.

64. Schiessel, H. (2003). The physics of chromatin. *J Phys Cond Mat* **15**, R699-R774.

65. Brower-Toland, B. D., Smith, C. L., Yeh, R. C., Lis, J. T., Peterson, C. L. & Wang, M. D. (2002). Mechanical disruption of individual nucleosomes reveals a reversible multistage release of DNA. *Proc Natl Acad Sci U S A* **99**, 1960-1965.

66. Kulic, I. M. & Schiessel, H. (2004). DNA Spools under Tension. *Phys Rev Lett* **92**, 228101-228104.

67. Cui, Y. & Bustamante, C. (2000). Pulling a single chromatin fiber reveals the forces that maintain its higher-order structure. *Proc Natl Acad Sci U S A* **97**, 127-132.

68. Tóth, K., Brun, N. & Langowski, J. (2001). Trajectory of nucleosomal linker DNA studied by fluorescence resonance energy transfer. *Biochemistry* **40**, 6921-6928.

69. Brower-Toland, B., Wacker, D. A., Fulbright, R. M., Lis, J. T., Kraus, W. L. & Wang, M. D. (2005). Specific contributions of histone tails and their acetylation to the mechanical stability of nucleosomes. *J Mol Biol* **346**, 135-146.

70. Li, G., Levitus, M., Bustamante, C. & Widom, J. (2005). Rapid spontaneous accessibility of nucleosomal DNA. *Nat Struct Mol Biol* **12**, 46-53.

71. Tomschik, M., Zheng, H., van Holde, K., Zlatanova, J. & Leuba, S. H. (2005). Fast, long-range, reversible conformational fluctuations in nucleosomes revealed by single-pair fluorescence resonance energy transfer. *Proc Natl Acad Sci U S A* **102**, 3278-3283.

72. Leforestier, A. & Livolant, F. (1997). Liquid crystalline ordering of nucleosome core particles under macromolecular crowding conditions: evidence for a discotic columnar hexagonal phase. *Biophys J* **73**, 1771-1776.

73. Livolant, F. & Leforestier, A. (2000). Chiral discotic columnar germs of nucleosome core particles. *Biophys J* **78**, 2716-2729.

74. Leforestier, A., Dubochet, J. & Livolant, F. (2001). Bilayers of nucleosome core particles. *Biophys J* **81**, 2414-2421.

75. Mangenot, S., Leforestier, A., Durand, D. & Livolant, F. (2003). X-ray diffraction characterization of the dense phases formed by nucleosome core particles. *Biophys J* **84**, 2570-2584.

76. Mangenot, S., Raspaud, E., Tribet, C., Belloni, L. & Livolant, F. (2002). Interactions between isolated nucleosome core particles: A tail-bridging effect? *Eur Phys J E* **7**, 221-231.

77. Mangenot, S., Leforestier, A., Vachette, P., Durand, D. & Livolant, F. (2002). Salt-induced conformation and interaction changes of nucleosome core particles. *Biophys J* **82**, 345-356.

78. Beard, D. A. & Schlick, T. (2001). Modeling salt-mediated electrostatics of macromolecules: the discrete surface charge optimization algorithm and its application to the nucleosome. *Biopolymers* **58**, 106-115.

79. Muehlbacher, F., Holm, C. & Schiessel, H. (2005). Controlled DNA compaction within chromatin: the tail-bridging effect. *Europhysics Letters, in press.*

80. Castro, C. (1994). Measurement of the elasticity of single chromatin fibers: the effect of histone H1. PhD Thesis, University of Oregon, Eugene.

81. Bussiek, M., Mucke, N. & Langowski, J. (2003). Polylysine-coated mica can be used to observe systematic changes in the supercoiled DNA conformation by scanning force microscopy in solution. *Nucleic Acids Res* **31**, e137.

82. Ringrose, L., Chabanis, S., Angrand, P. O., Woodroofe, C. & Stewart, A. F. (1999). Quantitative comparison of DNA looping *in vitro* and *in vivo*: chromatin increases effective DNA flexibility at short distances. *EMBO J* **18**, 6630-6641.

83. Dekker, J., Rippe, K., Dekker, M. & Kleckner, N. (2002). Capturing Chromosome Conformation. *Science* **295**, 1306-1311.

84. van den Engh, G., Sachs, R. & Trask, B. J. (1992). Estimating genomic distance from DNA sequence location in cell nuclei by a random walk model. *Science* **257**, 1410-1412.

85. Trask, B. J., Allen, S., Massa, H., Fertitta, A., Sachs, R., van den Engh, G. & Wu, M. (1993). Studies of metaphase and interphase chromosomes using fluorescence in situ hybridization. *Cold Spring Harb Symp Quant Biol* **58**, 767-775.

86. Yokota, H., van den Engh, G., Hearst, J., Sachs, R. K. & Trask, B. J. (1995). Evidence for the organization of chromatin in megabase pair-sized loops arranged along a random walk path in the human G0/G1 interphase nucleus. *J Cell Biol* **130**, 1239-1249.

87. Bystricky, K., Heun, P., Gehlen, L., Langowski, J. & Gasser, S. M. (2004). Long-range compaction and flexibility of interphase chromatin in budding yeast analyzed by high-resolution imaging techniques. *Proc Natl Acad Sci U S A* **101**, 16495-16500.

88. Bennink, M. L., Leuba, S. H., Leno, G. H., Zlatanova, J., de Grooth, B. G. & Greve, J. (2001). Unfolding individual nucleosomes by stretching single chromatin fibers with optical tweezers. *Nat Struct Biol* **8**, 606-610.

89. Leuba, S. H., Bennink, M. L. & Zlatanova, J. (2004). Single-molecule analysis of chromatin. *Methods Enzymol* **376**, 73-105.

90. Claudet, C., Angelov, D., Bouvet, P., Dimitrov, S. & Bednar, J. (2005). Histone octamer instability under single molecule experiment conditions. *J Biol Chem* **280**, 19958-19965.

91. Bertin, A., Leforestier, A., Durand, D. & Livolant, F. (2004). Role of histone tails in the conformation and interactions of nucleosome core particles. *Biochemistry* **43**, 4773-4780.

92. Tse, C., Sera, T., Wolffe, A. P. & Hansen, J. C. (1998). Disruption of higher-order folding by core histone acetylation dramatically enhances transcription of nucleosomal arrays by RNA polymerase III. *Mol Cell Biol* **18**, 4629-4638.

93. Grigoryev, S. A., Bednar, J. & Woodcock, C. L. (1999). MENT, a heterochromatin protein that mediates higher order chromatin folding, is a new serpin family member. *J Biol Chem* **274**, 5626-5636.

94. van Holde, K. E. (1989). *Chromatin*, Springer, Heidelberg.

95. van Holde, K. & Zlatanova, J. (1996). What determines the folding of the chromatin fiber. *Proceedings of the National Academy of Sciences of the USA* **93**, 10548-10555.

96. Mergell, B., Everaers, R. & Schiessel, H. (2004). Nucleosome interactions in chromatin: fiber stiffening and hairpin formation. *Phys Rev E* **70**, 011915.

97. Bustamante, C., Smith, S. B., Liphardt, J. & Smith, D. (2000). Single-molecule studies of DNA mechanics. *Curr Opin Struct Biol* **10**, 279-285.

98. Baumann, C. G., Smith, S. B., Bloomfield, V. A. & Bustamante, C. (1997). Ionic effects on the elasticity of single DNA molecules. *Proc Natl Acad Sci U S A* **94**, 6185-6190.

99. Gehlen, L. R., Rosa, A., Klenin, K., Langowski, J., Gasser, S. & Bystricky, K. (2005). Spatially confined polymer chains: implication of chromatin fiber flexibility and peripheral anchoring on telomere-telomere interaction. *J Phys Cond Matter,* in press.

INDEX

CHALLENGES AND ADVANCES IN
COMPUTATIONAL CHEMISTRY AND PHYSICS